6-8-06

D1198959

6-8-06

Electrical
Wiring
Residential

**Based on the 2005
NATIONAL ELECTRICAL CODE®**

Electrical
Wiring
Residential

FIFTEENTH EDITION

Ray C. Mullin

THOMSON

DELMAR LEARNING

Australia Canada Mexico Singapore Spain United Kingdom United States

THOMSON

DELMAR LEARNING

Electrical Wiring Residential, 15th Edition
by Ray C. Mullin

Vice President, Technology and Trades, SBU:
Alar Elken

Editorial Director:
Sandy Clark

Senior Acquisitions Editor:
Stephen Helba

Development Editor:
Jennifer A. Thompson

Marketing Director:
Dave Garza

Channel Manager:
Dennis Williams

Marketing Coordinator:
Casey Bruno

Production Director:
Mary Ellen Black

Production Coordinator:
Toni Hansen

Project Editor:
Ruth Fisher

Art/Design Specialist:
Mary Beth Vought

Senior Editorial Assistant:
Dawn Daugherty

Library of Congress Cataloging-in-Publication Data:

Mullin, Ray C.
  Electrical wiring residential /
Ray C. Mullin.—15th ed.
    p. cm.
  "Based on the 2005 National Electrical Code."
  Includes index.
  ISBN 1-4018-5019-7 (Softcover)
  ISBN 1-4018-5020-0 (Hardcover)
  1. Electric wiring, Interior. I. Title.
TK3285.M84 2004
621.319'24—dc22
                    2004025547

NOTICE TO THE READER

Publisher does not warrant or guarantee any of the products described herein or perform any independent analysis in connection with any of the product information contained herein. Publisher does not assume, and expressly disclaims, any obligation to obtain and include information other than that provided to it by the manufacturer.

The reader is expressly warned to consider and adopt all safety precautions that might be indicated by the activities herein and to avoid all potential hazards. By following the instructions contained herein, the reader willingly assumes all risks in connection with such instructions.

The publisher makes no representation or warranties of any kind, including but not limited to, the warranties of fitness for particular purpose or merchantability, nor are any such representations implied with respect to the material set forth herein, and the publisher takes no responsibility with respect to such material. The publisher shall not be liable for any special, consequential, or exemplary damages resulting, in whole or part, from the readers' use of, or reliance upon, this material.

Contents

Plans for Single-Family Dwelling

Foreword

THE IMPORTANCE OF PROPER TRAINING

Accurate materials and training are the two sides of the electrical safety coin. This coin is spent every day by various persons intrinsically involved in the electrical construction industry. Unfortunately, some spend it less wisely than others. Usually, the unwise spenders are those who rush to career, having neglected to acquire accurate materials and to focus on training themselves to a high level of proficiency.

Ray Mullin, author of this book, *Electrical Wiring—Residential*, has often stated, "The cost of education is small when compared to the price paid for ignorance." All too often, we, the citizens, pay the price for others' ignorance—ignorance of the codes, ignorance of proper wiring methods, ignorance of proper installation procedures, ignorance of design requirements, ignorance of product evaluations. This price becomes dear when our friends and family lose health or life or when our homes are destroyed.

Yet accurate materials are so easy to obtain. Ray Mullin is one technical writer who has paid his dues in the electrical industry; he put in his years as an apprentice, a journeyman, and then as a master electrician before beginning to write about his trade. *Electrical Wiring—Residential* contains accurate, up-to-date information about all aspects of residential wiring.

When installers and inspectors don't keep abreast of installation procedures and code requirements, things like cables across scuttle access to attics, improper spacing of receptacle outlets, improper short-circuit and ground-fault protection, and improper grounding of electrical systems, phone system, and CATV systems can lead to hazardous situations causing electrical shocks and fire. Not just anybody can install or inspect safe electrical systems. Trained professionals can, but even they must be constantly improving their knowledge and skills.

Because Ray Mullin cares about the electrical safety coin, he has striven to provide the most accurate information possible. It is up to each of us, however, to focus on the training. Some training can be acquired simply by reading the best books in our trade; some training can come through the online programs available; and other training, through participation in classes and seminars. In each instance, though, motivation and desire come from within—to know everything involved in our trade, to be totally proficient, to focus continually on improvement. As we seek both accurate information and training, we learn to spend the coin of safety to benefit others as well as ourselves. I commend you for acquiring *Electrical Wiring—Residential*; now I challenge you to make it part of yourself. I challenge you to spend the electrical safety coin wisely.

James W. Carpenter
CEO and Executive Director,
International Association of
Electrical Inspectors

INTENDED USE AND LEVEL

Electrical Wiring—Residential is intended for use in residential wiring courses at high schools, 2-year and 4-year colleges, as well as apprenticeship training programs. This comprehensive book guides readers, room by room, through the wiring of a typical residence and builds a foundation of knowledge by starting with the basic requirements of the *National Electrical Code,*® (*NEC*®)* then continuing on to the more advanced wiring methods. Each *Code* rule is presented through text, illustrations, examples, and wiring diagrams. In addition, an accompanying set of plans at the back of the book guides the reader through the wiring process by applying concepts learned in each unit to an actual residential building in order to understand and meet the requirements set forth by the *NEC*®.

An Important Note About Safety

In the educational field, it is pretty much a given that "Society will pay for education . . . one way or another." Proper training of a skilled trade is much better than hit or miss learning. Having to do the job over, having a house burn down, or having someone get electrocuted because of improper wiring is costly!

It really doesn't take any longer to do it right the first time than to have to do it over. You probably have heard the phrase "Measure twice . . . cut once. Measure once . . . cut twice." How true!

Electrical wiring is a skilled trade. Wiring should not be done by anyone not familiar with the hazards involved. It is a highly technical skill that requires much training. This material provides all of the electrical codes and standards information needed to approach house wiring in a safe manner. The *NEC*® defines a **qualified person** as "*One who has skills and knowledge related to the construction and operation of the electrical equipment and installation, and has received safety training on the hazards involved.*"**

Electrical Wiring—Residential will provide you with the know-how so you can wire houses that "Meet Code."

Electrical Wiring—Residential has become an integral part of approved (accredited) training programs by an increasing number of states that require residential electricians to have a residential license if they are going to wire homes and small apartments.

The *NEC*® has one thing in mind—safety! There is too much at stake to do less than what the *NEC*® requires. Anything less is unacceptable! The *NEC*® in *90.1(A)* makes it pretty clear. It states that "*The purpose of this Code is the practical safeguarding of persons and property from hazards arising from the use of electricity.*"**

Do not work on live circuits! Always de-energize the system before working on it! There is no compromise when it comes to safety! Many injuries and deaths have occurred when individuals worked on live equipment. The question is always: "Would the injury or death have occurred had the power been shut off?" The answer is "No!"

All mandatory safety-related work practices are found in the Federal Regulation Occupational Safety and Health Administration (OSHA), Title 29, Subpart S—Electrical, Sections 1910.331 through 1910.360.

**National Electrical Code*® and *NEC*® are registered trademarks of the National Fire Protection Association, Inc., Quincy, MA 02269.
**Reprinted with permission from NFPA 70-2005.

SUBJECT AND APPROACH

The 15th edition of *Electrical Wiring—Residential* is based on the 2005 *National Electrical Code* (*NEC*).® The *NEC*® is used as the basic standard for the layout and construction of residential electrical systems. In this text, thorough explanations are provided of *Code* requirements as they relate to residential wiring. To gain the greatest benefit from this edition, the student must use the *NEC*® on a continuing basis.

It is extremely difficult to learn the *NEC*® by merely reading it. This text brings together the rules of the *NEC*® and the wiring of an actual house. You will study the rules from the *NEC*,® and apply those rules to a true-to-life house wiring installation.

Take a moment to look at the Table of Contents. It is immediately apparent that you will not learn such things as how to drill a hole, tape a splice, fish a cable through a wall, use tools, or repair broken plaster around a box. These things you already know, or are learning on the job. The emphasis of this text is to teach you how to wire a house that "Meets Code." Doing it right the first time is far better than having to do it over because the electrical inspector turned down your job.

The first seven units in this book concentrate on basic electrical code requirements that apply to house wiring. This includes safety when working with electricity; construction symbols, plans, and specifications; wiring methods; conductor sizing; circuit layout; wiring diagrams; numerous ways to connect switches and receptacles; how to wire recessed luminaires (lighting fixtures); ground-fault circuit-interrupters (GFCIs); arc-fault circuit-interrupters (AFCIs); and surge suppressors.

The remaining units are devoted to the wiring of an actual house—room by room—circuit by circuit. All of these circuits come together for calculating the size of the main service. Since proper grounding is a key safety issue, the subject is covered in detail.

You will also learn about security systems, fire and smoke alarms, low-voltage remote-control wiring, swimming pools, standby generators, and you will be introduced to structured wiring for home automation.

You will find this text unique in that you will use the text, an actual set of plans and specifications, and the *NEC*®—all at the same time. It is perfect to learn house wiring and makes an excellent reference source for looking up specific topics relating to house wiring. The blueprint serves as the basis for the wiring schematics, cable layouts, and discussions provided in the text. Each unit dealing with a specific type of wiring is referenced to the appropriate plan sheet. All wiring systems are described in detail—lighting, appliance, heating, service entrance, and so on.

The house selected for this edition is scaled for current construction practices and costs. Note, however, that the wiring, luminaires (lighting fixtures), appliances, number of outlets, number of circuits, and track lighting are not all commonly found in a home of this size. The wiring may incorporate more features than are absolutely necessary. This was done to present as many features and *Code* issues as possible to give the student more experience in wiring a residence. Also included are many recommendations that are above and beyond the basic *NEC*® requirements.

Note: The *NEC*® (NFPA 70) becomes mandatory only after it has been adopted by a city, county, state, or other governing body. Until officially adopted, the *NEC*® is merely advisory in nature. State and local electrical codes may contain modifications of the *NEC*® to meet local requirements. In some cases, local codes will adopt certain more stringent regulations than those found in the *NEC*.® For example, the *NEC*® recognizes nonmetallic-sheathed cable as an acceptable wiring method for house wiring. Yet, the city of Chicago and surrounding counties do not permit nonmetallic-sheathed cable for house wiring. In these areas, all house wiring is done with electrical metallic tubing (thinwall).

There are also instances where a governing body has legislated action that waives specific *NEC®* requirements, feeling that the *NEC®* was too restrictive on that particular issue. Such instances are very rare. The instructor is encouraged to furnish students with any local variations from the *NEC®* that would affect this residential installation in a specific locality.

THE ELECTRICAL TRADE—TRAINING PROGRAMS

As you study *Electrical Wiring—Residential*, study with a purpose—to become the best residential wireman possible.

There will always be a need for skilled electricians! Qualified electricians almost always have work. It takes many hours of classroom and on-the-job training to become a skilled electrician. The best way to learn the electrical trade is through a training program approved by the U.S. Department of Labor (www.dol.gov). Many times an apprenticeship program is called "Earn While You Learn." These programs offer the related classroom training and the advantage of working on the job with skilled journeymen electricians. Completion of a Registered Apprenticeship program generally leads to higher pay, job security, higher quality of life, recognition across the country, and the opportunity for college credit and future degrees.

As a rule, these training programs require 144 to 180 hours of classroom-related technical training and 2000 hours of on-the-job training per year. Some programs have day classes and some have night classes. An electrical apprenticeship training program might run four to five years. The end-result—becoming a full-fledged licensed journeyman electrician capable of doing residential, commercial, and industrial electrical work. A residential electrician training program might run two to three years, with the training limited to the wiring of single- and multifamily dwellings. The end-result—receive a license limited to residential wiring.

To get into an apprenticeship program, the individual usually must have a high school education with at least one year of high school algebra, be at least 18 years old, must be physically in shape to perform the work electricians are called upon to do (e.g., climbing, lifting, work in inclement weather), and, most importantly, be drug free. There generally is a qualifying aptitude test to make sure the applicant has the ability to take on the responsibility of a rigid apprenticeship training program. In some areas, passing the high school equivalency General Education Development (GED) test is acceptable in place of a high school diploma.

Following completion of an apprenticeship program, continuing education courses are available to keep the journeyman up-to-date on Codes and other related topics and skills.

Journeymen electricians who have an interest in teaching apprentices will usually have to take instructor training courses. In certain programs, satisfactory completion of the required courses can lead to an associate degree. Others will go on to become foremen, supervisors, and contractors.

There are some areas where a "pre-apprenticeship" program is offered. To learn more about the careers possible in the electrical field, chat with your instructor, your local high school's guidance counselor, your vocational-technical-and adult education schools, an electrician, and an electrical contractor. Go "online" and search for electrical apprenticeship programs.

Your future is in your hands.

Some very important two-letter words that you should remember are:
IF IT IS TO BE, IT IS UP TO ME!

Job Titles

Most building codes and standards contain definitions for the various levels of competency of workers in the electrical industry. Here are some examples of typical definitions.

Apprentice shall mean a person who is required to be registered, who is in compliance with the provisions of this article, and who is working at the trade in the employment of a registered electrical contractor and is under the direct supervision of a licensed master electrician, journeyman electrician, or residential wireman.

Residential Wireman shall mean a person having the necessary qualifications, training, experience, and technical knowledge to wire for and install electrical apparatus and equipment for wiring one-, two-, three-, and four-family dwellings. A residential wireman is sometimes referred to as a *Class B Electrician.*

Journeyman Electrician shall mean a person having the necessary qualifications, training, experience, and technical knowledge to wire for, install, and repair electrical apparatus and equipment for light, heat, power, and other purposes, in accordance with standard rules and regulations governing such work.

Master Electrician means a person having the necessary qualifications, training, experience, and technical knowledge to properly plan, lay out, and supervise the installation and repair of wiring apparatus and equipment for electric light, heat, power, and other purposes, in accordance with standard codes and regulations governing such work, such as the *NEC.*®

Electrical Contractor means any person, firm, copartnership, corporation, association, or combination thereof who undertakes or offers to undertake for another the planning, laying out, supervising and installing, or the making of additions, alterations, and repairs in the installation of wiring apparatus and equipment for electrical light, heat, and power.

Most electrical inspectors across the country are members of the International Association of Electrical Inspectors. This organization publishes one of the finest technical bimonthly magazines devoted entirely to the *NEC*® and related topics. This organization is open to individuals that are not electrical inspectors. Electrical instructors, vo-tech students, apprentices, electricians, consulting engineers, contractors, and distributors are encouraged to join the IAEI so they can stay up to date on all *NEC*® issues, changes, and interpretations. An application form that explains the benefits of membership in the IAEI can be found in the Appendix of this text.

NEW TO THIS EDITION

Continuing in the tradition of previous editions, this edition thoroughly explains how *Code* changes affect house wiring installations. New and revised full-color illustrations supplement the explanations to ensure that electricians understand the new *Code* requirements. New photos reflect the latest wiring materials and components available on the market. Revised review questions test student understanding of the new content. New tables that summarize *Code* requirements offer a quick reference tool for students. Other reference aids are the tables reprinted directly from the 2005 edition of the *NEC.*® The extensive revisions for the fifteenth edition make *Electrical Wiring—Residential* the most up-to-date and well-organized guide to house wiring. Coverage of the *NEC*® has been expanded to well over 1000 *Code* references.

This text focuses on the technical skills required to perform electrical installations. It covers such topics as calculating conductor sizes, calculating voltage drop, determining appliance circuit requirements, sizing service, connecting electric appliances, grounding service and equipment, installing recessed luminaires (fixtures), and much more. These are critical skills that can make the difference between an installation that "Meets Code" and one that does not. The electrician must understand the reasons for following *Code* regulations to achieve an installation that is essentially free from hazard to life and property.

Note: Symbols have been added to indicate changes in the 2005 *National Electrical Code*® vs. the 2002 *National Electrical Code.*® ▶ ◀

This text might be called "Work in Progress." The author stays in touch with the latest residential wiring trends and the *National Electrical Code.*® Because the *NEC*® is revised every three years, this text follows the same cycle. *Electrical Wiring—Residential* has been carefully reviewed to editorially simplify, streamline, and improve its readability. Many diagrams have been simplified. Some units were reorganized so the *Code* requirements for the various applications are more uniform.

Much rewriting was done. The 2005 *NEC*® contains many editorial changes as well as renumbering and relocation of numerous *Code* references. All of these have been addressed in this edition of *Electrical Wiring—Residential*. Some text has been condensed and reformatted for ease in reading. Many diagrams have been simplified for clarity.

The Objectives have been fine-tuned for easier readability.

- To make room for new important *Code* issues and wiring diagrams, some not so critical text has been deleted. These changes were based on suggestions from instructors, inspectors, contractors, electricians, and other users from around the country. The author is always interested in input from these folks. Please send any suggestions and comments you might have to Thomson Delmar Learning. They will forward them to me.

- Basic requirements from the Americans with Disabilities Act (ADA) from the U.S. Department of Justice and the Fair Housing Act from the U.S. Department of Housing and Urban Development (HUD) are included in this new edition.

- Greatly expanded coverage of safety related issues, listing requirements, and information available from the Occupational Safety and Health Administration (OSHA) and the National Institute for Occupational Safety and Health (NIOSH). This section also tells what to do if you are involved with a possible electrocution on the job. This expanded material was added at the request of training directors and instructors using EWR in their courses. They wanted to learn more about on-the-job safety, and where they could find more educational material on the subject.

- The *NEC*® made the decision to no longer use the term "load center." Although still found in manufacturers' catalogs, the proper term to use is "panelboard." In the trade, they are still called *panels*, *panelboards*, and *load centers*.

- Many new wiring diagrams and illustrations.

- More illustrations of nonmetallic boxes.

- Additional text about masonry boxes.

- Changed the term "computed" to "calculated" throughout text per *NEC*® change.

- Complete rewrite of fluorescent lamps and ballasts for clarity.

- New coverage about airtight energy-efficient recessed lighting fixtures.

- Complete rewrite of the section on how many outlets are permitted on one circuit.

- New requirements for installation of overhead door operators.

- A complete rewrite of Unit 26—the requirements for smoke, heat, and carbon monoxide alarms. New location requirements. This was necessary because of the sweeping changes in the *National Fire Alarm Code*.

- More photos and *Code* requirements for nonmetallic boxes.

- More wiring diagrams for connecting switches.

- New text on color coding of conductors and nonmetallic-sheathed cables.

- Complete rewrite of how to design all of the essentials of a branch-circuit for motors.

- Expanded text on derating and correction factors for conductor sizing.

- Cautions about the hazards of jumpering (connecting) conductors between terminals of receptacles.

- Advantages and disadvantages of three-wire branch-circuits.

- More illustrations of how to connect equipment grounding conductors.
- New illustration explaining damp and wet locations.
- More information and illustrations about the different types of SE cable, their construction, and permitted use.
- New illustrations for running cable through EMT for physical protection and neater appearance.
- Reduction in the amount of coverage for low-voltage wiring, making it much simpler to understand.
- Expanded coverage and new illustration to clarify the *Code* requirements for bonding gas piping.
- More information on AFCI/GFCI testers.
- Latest information about the use of HACR circuit breakers.
- More information about compact fluorescent lamps.
- The latest *Code* requirements for installing ceiling fans.
- Added Key Terms to the Glossary in the back of the text.

SUPPLEMENT PACKAGE

An *Instructor's Guide* contains answers to all review questions included in the book and a blank service-entrance calculation form; also available in electronic format on the accompanying *E.resource*.
(Order #: 1-4018-5021-9)

An *E.resource* provides instructors with valuable classroom materials on CD-ROM:

- *PowerPoint Presentations* outline the important concepts covered in each unit. Extensively illustrated with photos, tables, and diagrams from the book, the presentation enhances classroom instruction. The PowerPoint also allows instructors to tailor the course to meet the needs of their individual class.

- A *Testbank* in the latest *ExamView* format provides instructors with approximately 1500 test questions to evaluate students as they work through each unit in the book. Answers and book page references are also provided. Completely editable, instructors may also wish to delete or add questions to meet the needs of their individual class.

- An *Image Library* contains nearly all the images in the book and can be used to enhance the PowerPoint presentation, or to create transparency masters and handouts.

- *Electronic Instructor's Guide* in Microsoft Word enables instructors to view and print answers to review questions contained in the book.

- *Electronic Blueprints* provide an online version of the drawings that are included at the back of the book, allowing instructors to project and reference in classroom presentations.

- *Video Clips* from the accompanying video series visually highlight important concepts presented in the book

(Order #: 1-4018-5024-3)

A *Lab Manual* provides over twenty exercises to aid students in learning both basic and complex wiring circuits. Each lab consists of a hands-on wiring exercise as well as *NEC®* drill problems. Students are also required to draw an electrical layout of wiring booths to familiarize themselves with electrical symbols. Allowing for instructor verification and student self-assessment, this manual is essential to applying important wiring concepts.
(Order #: 1-4018-5022-7)

Electrical Wiring Residential Video Series and CD Courseware: Available on VHS tapes or CD-ROM, *Electrical Wiring Residential Video Series* correlates directly with the *Electrical Wiring—Residential* book. Shot on various construction sites and enhanced with quality animations, each video is devoted to a specific, specialized topic. In addition to high-quality animations, questions have also been incorporated into each tape at strategic points to promote discussion aimed at enhancing viewers' understanding and preparing them for successful attainment of all learning objectives. Each tape is approximately 20 minutes in length, making it easy to incorporate into a comprehensive electrician education and training program.

The complete set includes: Video #1: Introduction to Electrical Installation, Video #2: Planning for Circuit Installation, Video #3: Ground Fault Circuit Interrupters, Video #4: Lighting by the Room, Video #5: Special Circuits, Video #6: Special Purpose Outlets, Video #7: Miscellaneous and Custom Installations, Video #8: Service Entrances.
(Video Set Order #: 0-7668-3188-4) (Courseware Order #: 0-7668-2435-7)

Also Available from Thomson Delmar Learning

Videos. The *House Wiring Video Series* is an integrated part of the *Residential Construction Academy House Wiring* package. The series contains a set of eight, 20-minute videos that provide step-by-step instruction for wiring a house. All the essential information is covered in this series, beginning with the important process of reviewing the plans and following through to the final phase of testing and troubleshooting. Need to know *NEC*® articles are highlighted, and Electrician's Tips and Safety Tips offer practical advice from the experts.

The complete set includes the following: Video #1: Safety and Safe Practices, Video #2: Hardware, Video #3: Tools, Video #4: Initial Review of Plans, Video #5: Rough-In, Video #6: Service Entrance, Video #7: Trim-Out, Video #8: Testing & Troubleshooting.

CD Courseware. This package also includes computer-based training that used video, animation, and testing to introduce, teach, or remediate the concepts covered in the videos. Students will be pre-tested on the material and then if needed, provided with the suitable remediation to ensure understanding of the concepts. Post-tests can be administered to ensure that students have gained mastery of all material.

For more information, visit www.residentialconstructionacademy.com.

Visit us at www.delmarelectric.com, now LIVE for the 2005 Code cycle!
This newly designed Web site provides information on other learning materials offered by Delmar, as well as industry links, career profiles, job opportunities, and more!

ABOUT THE AUTHOR

This text was written by Ray C. Mullin, former electrical instructor for the Wisconsin Schools of Vocational, Technical, and Adult Education. He is a former member of the International Brotherhood of Electrical Workers. He is a member of the International Association of

Electrical Inspectors, the Institute of Electrical and Electronic Engineers, and the National Fire Protection Association, Electrical Section, and has served on Code-Making Panel 4 of the *National Electrical Code.*®

Mr. Mullin completed his apprenticeship training and worked as a journeyman and supervisor for residential, commercial, and industrial installations. He has taught both day and night electrical apprentice and journeyman courses, has conducted engineering seminars, and has conducted many technical *Code* workshops and seminars at

International Association of Electrical Inspectors Chapter and Section meetings, and has served on their code panels.

He has written many technical articles that have appeared in electrical trade publications. He has served as a consultant to electrical equipment manufacturers regarding conformance of their products to industry standards, and on legal issues relative to personal injury lawsuits resulting from the misuse of electricity and electrical equipment. He has served as an expert witness.

Mr. Mullin presents his knowledge and experience in this text in a clear-cut manner that is easy to understand. This presentation will help students to fully understand the essentials required to pass the residential licensing examinations and to perform residential wiring that "meets Code."

Mr. Mullin is the author of *House Wiring with the NEC®*—a text that focuses entirely on the *National Electrical Code®* requirements for house wiring. He is co-author of *Electrical Wiring —Commercial, Illustrated Electrical Calculations*, and *The Smart House*. He has contributed technical material for Delmar's *Electrical Grounding* and to the International Association of Electrical Inspectors' texts *Soares Book On Grounding* and *Ferm's Fast Finder*.

He served on the Executive Board of the Western Section of the International Association of Electrical Inspectors, on their Code Clearing Committee, and, in the past, served as Secretary/ Treasurer of the Indiana Chapter of the IAEI.

Mr. Mullin is past Director, Technical Liaison for a major electrical manufacturer. In this position, he was deeply involved in electrical codes and standards as well as contributing and developing technical training material for use by this company's field engineering personnel.

Mr. Mullin attended the University of Wisconsin, Colorado State University, and the Milwaukee School of Engineering.

IMPORTANT NOTE

This edition of *Electrical Wiring—Residential* was completed after all normal steps of revising the *National Electrical Code® NFPA 70* were taken, and before the actual issuance and publication of the 2005 edition of the *NEC®*.

These steps include: NFPA solicits proposals for *2005 NEC®*, interested parties submit proposals to NFPA, proposals sent to Code-Making Panels (CMPs), proposals reviewed by CMPs and Technical Correlating Committee, Report on Proposals document published, interested parties submit comments on the proposals to NFPA, review of Comments by CMPs and Technical Correlating Committee, Report on Comments document published, review of all Proposals and Comments at NFPA Annual Meeting, new motions permitted to be made at the NFPA Annual Meeting. Finally, the Standard Council meets to review actions made at the NFPA Annual Meeting, and to authorize publication of the *NEC®*.

Every effort has been made to be technically correct, but there is always the possibility of typographical errors, or appeals made to the NFPA Board of Directors after the normal review process that could result in reversal of previous decisions by the CMPs.

If changes in the *NEC®* do occur after the printing of this text, these changes will be incorporated in the next printing.

The National Fire Protection Association has a standard procedure to introduce changes between *NEC® Code* cycles after the actual *NEC®* is printed. These are called "Tentative Interim Amendments," or TIAs. TIAs and typographical errors can be downloaded from the NFPA website, www.nfpa.org, to make your copy of the *Code* current.

Acknowledgments

I again wish to thank my wife, Helen, for her understanding and support while I devoted unlimited time attending meetings and working many hours on revising this edition of *Electrical Wiring—Residential*. Major revisions such as this take somewhere between 1000 and 1500 hours! Patience is a virtue!

As always, the team at Thomson Delmar Learning has done an outstanding job in bringing this edition to press. Their drive, dedication, and attention to minute details ensure that this text, without question, is the country's leading text on house wiring. They sure know how to keep the pressure on!

Special thanks to my good friend Jimmy Carpenter, Executive Director of the International Association of Electrical Inspectors for his inspiring FOREWORD to this text regarding the "Importance of Proper Training."

I would also like to thank many individuals in the electrical industry across the country. They are my friends and are outstanding *Code* experts. Madeline Borthick, Paul Dobrowsky, Joe Ellwanger, David Hittinger, Tim May, Rick Melcher, Bill Neitzel, Ron O'Riley, Mike O'Quinn, Cliff Rediger, Phil Simmons, Clarence Tibbs, Charlie Trout, Gordon Stewart, Ray Weber, the electrical staff at NFPA headquarters who responded timely to my questions on particular *Code* issues, the many Code-Making Panel members who clarified specific *Code* proposals for the 2005 *NEC*® and the other reviewers for their excellent suggestions that helped make this edition the most comprehensive and *Code*-compliant ever. Special thanks again to Robert Boiko for his technical input on water heaters and their safety related controls.

I wish I could name all of my friends in the electrical industry, but there are so many, it would take many pages to include all of their names. Thanks to all of you for your input. I apologize if I missed anyone.

The author and publisher would like to thank the following reviewers for their contributions:

DeWain Belote
Pinellas Tech Educational Center
St. Petersburg, FL

Larry Catron
Scott County Vo-Tech School
Gate City, VA

Martin Clark-Stone
SUNY Canton
Canton, NY

Warren DeJardin
Northeast Wisconsin Technical College
Oneida, WI

Charles Eldridge
Indianapolis Power & Light
Indianapolis, IN

David Gehlauf
Tri-County Vocational School
Glouster, OH

Thomas B. Lockett
Vatterott College
Quincy, IL

Richard Moore
Wake Tech CC
Raleigh, NC

Gary Reiman
Dunwoody Institute
Minneapolis, MN

Les Brinkley
Ashtabula County JVC
Ashtabula, OH

Robert Coy
KY Tech Elizabethtown
Elizabethtown, KY

Edward Dwyer
Triangle Tech
Pittsburgh, PA

Greg Fletcher
Kennebec Valley Tech College
Fairfield, ME

Michael Ross
San Jancintocc
Pasadena, TX

Fred Johnson
Champlain Valley Tech
Plattsburg, NY

Max Schroeder
Barton County Community College
Hoisington, KS

Cliff Redigar
Independent Electrical Contractors
Denver, CO

Melvin Titus
Cecil Co. School of Technology
Northeast, MD

Special thanks to Charlie Eldridge for his thorough review of *Code* content to ensure compliancy with the 2005 Edition of the *NEC.*®

Applicable tables and section references are reprinted with permission from NFPA 70-2005, *National Electrical Code,*® copyright © 2004, National Fire Protection Association, Quincy, MA 02269. This reprinted material is not the complete and official position of the NFPA on the referenced subject, which is represented only by the standard in its entirety.

General Information for Electrical Installations

OBJECTIVES

After studying this unit, the student should be able to

▶ understand the basic safety rules for working on electrical systems.

▶ access the Internet to obtain a virtual unlimited source of safety and technical related information.

▶ become familiar with important electrical codes, safety codes, and building codes such as NFPA 70, 70A, 70B, 70E, 73, OSHA, NIOSH, ADA, NRTL, and the ICC.

▶ learn about licensing, permits, plans, specifications, symbols, and notations.

▶ understand the role of the electrical inspector and the International Association of Electrical Inspectors.

▶ understand the metric system of measurement.

SAFETY IN THE WORKPLACE

Electricity is great when it is doing what it is intended to do, and that is to stay in its intended path. But electricity out of its intended path is dangerous, often resulting in fire, serious injury, or death.

Before getting into residential wiring and the *National Electrical Code®* (*NEC®*), we need to discuss on-the-job safety. Safety is not a joke! Electricians working on new construction, remodel work, maintenance, and repair work find that electricity is part of the work environment. Electricity is all around us, just waiting for the opportunity to get out of control. Repeat these words: **Safety First** . . . **Safety Last** . . . **Safety Always**!

Working on switches, receptacles, luminaires (fixtures), or appliances with the power turned on is dangerous. Turn off the power! In addition, check with a voltmeter to be sure the power is off.

The voltage level in a home is 120 volts between one "hot" conductor and the "neutral" conductor or grounded surface. Between the two "hot" conductors (line-to-line), the voltage is 240 volts.

An electrical shock is received when electrical current passes through the body. From basic electrical theory, you learned that line voltage appears across an open in a series circuit. Getting caught "in series" with a 120-volt circuit will give you a 120-volt shock. For example, open-circuit voltage between the two terminals of a single-pole switch on a lighting circuit is 120 volts when the switch is in the "OFF" position and the lamp(s) are in place. See Figure 1-1. Likewise, getting caught "in series" with a 240-volt circuit will give you a 240-volt shock.

Working on equipment with the power turned on can result in death or serious injury, either as a direct result of electricity (electrocution or burns) or from an indirect secondary reaction, such as falling off a

SWITCH
OFF

120-VOLT
SUPPLY

120

Figure 1-1   The voltage across the two terminals of the single-pole switch is 120 volts.

ladder or jerking away from the "hot" conductor into moving parts of equipment such as the turning blades of a fan. For example: A workman was seriously injured while working a "live" circuit that supplied a piece of equipment. He accidentally came into contact with a "hot" terminal, and reflex action caused him to pull his hand back into a turning pulley. The pulley cut deeply into his wrist, resulting in a tremendous loss of blood.

Dropping a metal tool onto live parts, allowing metal shavings from a drilling operation to fall onto live parts of electrical equipment, cutting into a "live" conductor and a "neutral" conductor at the same time, or touching the "live" wire and the "neutral' wire or a grounded surface at the same time can cause injury directly or indirectly.

A short circuit or ground fault can result in a devastating arc flash that can cause serious injury or death. The heat of an electrical arc has been determined to be hotter than the sun. Tiny hot "balls" of copper can fly into your eye or onto your skin. Figure 1-2 shows a disconnect switch that has been locked and tagged.

Lockout/tagout (sometimes called LOTO) is the physical restraint of all hazardous energy sources that supply power to a piece of equipment. It simply means putting a padlock on the switch and applying a warning tag on the switch.

Dirt, debris, and moisture can also set the stage for equipment failure and personal injury. Neatness and cleanliness in the workplace are a must.

What about Low-Voltage Systems?

Although circuits of less than 50 volts generally are considered harmless, don't get too smug when working on so-called low voltage. Low-voltage circuits are not necessarily low hazard. A slight "tingle" might cause a reflex. A capacitor that is discharging can give you quite a "jolt," causing you to jump or pull back.

In commercial work, such as telephone systems with large battery banks, there is extreme danger even though the voltage is "low." Think of a 12-volt car battery. If you drop a wrench across the battery terminals, you will immediately see a tremendous and dangerous arc flash.

It is the *current* that is the harmful component of an electrical circuit. *Voltage* pushes the current through the circuit. If you're not careful, you might become part of the circuit.

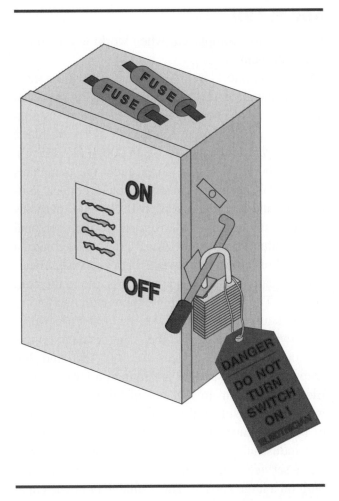

Figure 1-2 A typical disconnect switch with a lock and a tag attached to it. In the OSHA, ANSI, and NFPA standards, this is referred to as the *lockout/tagout* procedure.

Higher voltages can push greater currents through the body. Higher voltages like 240, 480, and 600 volts can cause severe skin burns and possibly out-of-sight injuries such as internal bleeding and/or destruction of tissues, nerves, and muscles.

It's the Law!

Not only is it a good idea to use proper safety measures as you work on and around electrical equipment, it is **required** by law. Electricians and electrical contractors need to be aware of these regulations. Practicing safety is a habit—like putting on your seat belt as soon as you get into your car.

The *NEC®* is full of requirements that are safety related. For example, *430.102(B)* requires that a dis-

connecting means be located in sight from the motor location and the driven machinery location. This section also has "in-sight" and "lock-off" requirements, discussed in detail in Unit 19.

The *NEC®* defines a **qualified person** as "*one who has skills and knowledge related to the construction and operation of the equipment and has specific safety training on the hazards involved.*" Merely telling someone or being told to "be careful" does not meet the definition of "proper training" and does not make the person qualified. An individual "qualified" in one skill might very well be "unqualified" in other skills.

According to *NFPA 70E, Electrical Safety in the Workplace*, circuits and conductors are not considered to be in an electrically safe condition until all sources of energy are removed, the disconnecting means is under lockout/tagout, and the absence of voltage is verified by an approved voltage tester.

The U.S. Department of Labor **Occupational Safety and Health Administration (OSHA)** regulations (Standards – 29 CFR) is the law! This entire standard relates to safety in the workplace for general industry. Specifically, Part 1910, Subpart S involves electrical safety requirements. The letters *CFR* stand for *Code of Federal Regulations*.

Key topics in the standard are: electric utilization systems, wiring design and protection, wiring methods, components and equipment for general use, specific purpose equipment and installations, hazardous (classified) locations, special systems, training, selection and use of work practices, use of equipment, selection and use of work practices, use of equipment, safeguards for personnel protection, and definitions (a mirror image of definitions found in the *NEC®*).

A direct quote from 1910.333 states that "Live parts to which an employee may be exposed shall be deenergized before the employee works on or near them, unless the employer can demonstrate that de-energizing introduces additional or increased hazards or is infeasible due to equipment design or operational limitations. Live parts that operate at less than 50 volts to ground need not be deenergized if there will be no increased exposure to electrical burns or to explosion due to electric arcs."

OSHA 1910.333(c)(2) states that "Only qualified persons may work on electric circuit parts or equipment that have not been deenergized under the procedures of paragraph (b) of this section. Such persons shall be capable of working safely on energized circuits and shall be familiar with the proper use of special precautionary techniques, personal protective equipment, insulating and shielding materials, and insulated tools."

OSHA 1910.399 defines a *qualified person* as "One familiar with the construction and operation of the equipment and the hazards involved," almost the same definition as that of the *NEC*.®

For the most part, turning the power off then locking and tagging the disconnecting means is the safest. As the OSHA regulations state, "A lock and a tag shall be placed on each disconnecting means used to de-energize circuits and equipment."

Part 1926 in the OSHA regulation (Standards – 29 CFR) deals with *Safety and Health Regulations for Construction*. Here we find the rules for anyone involved in the construction industry, not just electrical. A few of the topics are: medical services and first aid, safety training and education, recording and reporting injuries, housekeeping, personal protective equipment, means of egress, head protection, hearing protection, eye and face protection, ladders, scaffolds, rigging, hand and power tools, electrical requirements (a repeat of Part 1910, Subpart S), fall protection, and required signs and tags.

PERSONAL PROTECTIVE EQUIPMENT

Safety courses refer to personal protective equipment (PPE). These include such items as rubber gloves, insulating shoes and boots (footware suitable for electrical work are marked with the letters "EH"), face shields, safety glasses, hard hats, ear protectors, Nomex™, and similar products. OSHA 1910.132(f)(1) requires that "The employer shall provide training to each employee who is required by this section to use PPE."

Working on electrical equipment while wearing rings and other jewelry is not acceptable. OSHA states that "Conductive articles of jewelry and clothing (such as watch bands, bracelets, rings, key chains, necklaces, metalized aprons, cloth with conductive thread, or metal headgear) may not be worn if they might contact exposed energized parts. However, such articles may be worn if they are rendered nonconductive by covering, wrapping, or other insulating means."

ARC FLASH

Don't get too complacent when working on electrical equipment.

A major short circuit or ground fault at the main service panel or at the meter cabinet or base can deliver a lot of energy. On large electrical installations, an arc flash (also referred to as an arc blast) can generate temperatures of 35,000°F (19,427°C). This is hotter than the surface on the sun. This amount of heat will instantly melt copper, aluminum, and steel. The blast will blow hot particles of metal and hot gases all over, resulting in personal injury, fatality, and/or fire. An arc flash also creates a tremendous air pressure wave that can cause serious hearing damage and/or memory loss due to the concussion. The blast might blow the victim away from the arc source.

Don't be fooled by the size of the service. Typical residential services are 100, 150, and 200 amperes. Larger services are found on large homes. Electricians seem to feel out of harm's way when working on residential electrical systems and seem to be more cautious when working on commercial and industrial electrical systems. A fault at a small main service panel can be just as dangerous as a fault on a large service. The available fault current at the main service disconnect for all practical purposes is determined by the kVA rating and impedance of the transformer. Other major limiting factors for fault current are the size, type, and length of the service-entrance conductors. Available fault current can easily reach 22,000 amperes, as is evident by panels that have a 22,000/10,000-ampere series rating.

Short-circuit calculations are discussed in Unit 28.

Don't be fooled into thinking that if you cause a fault on the load side of the main disconnect that that main breaker will trip off and protect you from an arc flash. An arc flash will release the energy that the system is capable of delivering for as long as it takes the main circuit breaker to open. How much current (energy) the main breaker will "let through" is dependent upon the available fault current and the breaker's opening time.

▶ Although not required for house wiring, *NEC*® 110.16 requires that switchboards, panelboards, industrial control panels, meter socket enclosures, and motor control centers in commercial and industrial installations that are likely to be worked on while energized be field marked to warn qualified personnel of potential arc-flash hazard. ◀ More information on this subject is found in *NFPA 70E* and in the ANSI Standard Z535.4, *Product Safety Signs and Labels.*

TIP: When turning a standard disconnect switch "ON," don't stand in front of the switch. Instead, stand to one side. For example, if the handle of the switch is on the right, then stand to the right of the switch, using your left hand to operate the handle of the switch, and turn your head away from the switch. That way, if an arc flash occurs when you turn the disconnect switch "ON," you will not be standing in front of the switch. You will not have the switch's door fly into your face, and the molten metal particles resulting from the arc flash will fly past you.

Classifying Electrical Injuries

OSHA recognizes the four main types of electrical injuries as:

- Electrical shock (touching "live" line-to-line or line-to ground conductors) (ground-fault circuit interrupters are discussed in Unit 6)

- Electrocution (death due to severe electrical shock)

- Burns (from an arc flash)

- Falls (an electrical shock might cause you to lose your balance, pull back, jump, or fall off a ladder)

What to Do If You Are Involved with a Possible Electrocution

The following is taken in part from the OSHA, NIOSH, NSC regulations, and the American Heart Association recommendations. These are steps that should be taken in the event of a possible electrocution (cardiac arrest). You need to refer to the actual cardiopulmonary resuscitation (CPR) instructions for complete and detailed requirements, and to take CPR training.

- First of all, you must recognize that an emergency exists. Timing is everything. The time between the accident and arrival of paramedics is crucial. Call 911 immediately. Don't delay.

- Don't touch the person if he or she is still in contact with the live circuit.

- Shut off the power.

- Stay with the person while someone else contacts the paramedics, who have training in the basics of life support. In most localities, telephoning 911 will get you to the paramedics.

- Have the caller verify that the call was made and that help is on the way.

- Don't move the person.

- Check for bleeding; stop the bleeding if it occurs.

- If the person is unconscious, check for breathing.

- The *ABCs* of CPR are: *a*irway must be clear; *b*reathing is a must, either by the victim or the rescuer; and *c*irculation (check pulse).

- Perform CPR if the victim is not breathing—within 4 minutes is critical. If the brain is deprived of oxygen for more than 4 minutes, brain damage will occur. If it is deprived of oxygen for more than 10 minutes, the survival rate is 1 in 100. CPR keeps oxygenated blood flowing to the brain and heart.

- Defibrillation may be necessary to reestablish a normal heartbeat. Ventricular fibrillation is common with electric shock, which causes the heartbeat to be uneven and unable to properly pump blood.

- By now, the trained paramedics should have arrived to apply advanced care.

- When it comes to an electrical shock, *timing is everything!*

Electrical Equipment

For safety, it is very important that electrical equipment be "listed" by a nationally recognized testing laboratory (NRTL). This is a requirement of *NEC® 90.7.*

NEC® 110.3(B) Installation and Use states that "*Listed or labeled equipment shall be installed and used in accordance with any instructions included in the listing or labeling.*"

OSHA rules state that "All electrical products installed in the work place shall be listed, labeled, or otherwise determined to be safe by a Nationally Recognized Testing Laboratory (NRTL)."

Who Is Responsible for Safety?

You are!

The electrical inspector inspects electrical installations for compliance to the *NEC.* He or she is not really involved with on-the-job safety.

For on-the-job safety, OSHA puts the burden of responsibility on the employer. OSHA can impose large fines for noncompliance with its safety rules. But since it's your own safety that we are discussing, you share the responsibility by applying safe work practices, using the proper tools and PPE equipment the contractor furnishes, and installing "Listed" electrical equipment. Be alert to what's going on around you! Do a good job of housekeeping!

Tools

Using the proper tools for a job is vital to on-the-job safety.

OSHA Standard 1926.302 specifically covers the requirements for hand and power tools. The American National Standards Institute (ANSI) also has standards relating to tools.

If you want to learn more about tools, visit the Web site of the Hand Tools Institute at www.hti.org. The institute has a number of excellent safety education materials available. Of particular interest is its 90-plus-page publication *Guide to Hand Tools* that includes topics for selecting, proper use, maintaining, hazards involved, as well as special emphasis on eye protection using all types of hand tools.

Digital Multimeters

Some statistics show that more injuries occur from using electrical meters than from electric shock.

For safety, electricians should use quality digital multimeters that are *category rated*. The International Electrotechnical Commission (IEC) Standard 1010 for *Low Voltage Test, Measurement, and Control Equipment* rates the ability of a meter to withstand voltage transients (surges or spikes). This standard is very similar to UL Standard 3111. When lightning strikes a high line, or when utilities are performing switching operations, or when a capacitor is discharging, a circuit can "see" voltage transients that greatly exceed the withstand rating of the digital multimeter. The meter could explode, causing an arc flash (a fireball) that in all probability will result in

personal injury. A properly selected category-rated digital multimeter is able to withstand the spike without creating an arc blast. The leads of the meter are also able to handle high transient voltages.

Digital multimeters also are category rated based on the location of the equipment to be tested, because the closer the equipment is to the power source, the greater the danger from transient voltages.

Cat IV multimeters are used where the available fault current is high, such as a service entrance, a service main panel, service drops, and the house meter.

Cat III multimeters are used for permanently installed loads such as in switchgear, distribution panels, motors, bus bars, feeders, short branch-circuits, and appliance outlets where branch-circuit conductors are large and the distance is short.

Cat II multimeters are used on residential branch-circuits for testing loads that are plugged into receptacles.

Cat I multimeters are used where the current levels are very low, such as electronic equipment. Note that the lower the category rating, the lower is the meter's ability to withstand voltage transients. If you will be using the multimeter in all of the above situations, select the higher category rating.

Category-rated digital multimeters also contain fuses that protect against faults that happen when the meter is accidentally used to check voltage while it is inadvertently set in the current reading position.

To learn more about meters, visit the Web site of Fluke Corporation, www.fluke.com, for a wealth of technical information about the use of meters and other electrical and electronic measuring instruments.

Ladders

To learn more about ladders, visit the Web site of Werner Ladder Company. Download the *Climbing Pro Training Program*, a very comprehensive manual in which you will learn all about the selection and use of ladders. This manual covers safety rules such as: never work on a step ladder in which the spreaders are not fully locked into position; the 4 to 1 ratio, which means that the base of an extension ladder should be set back (S) one-fourth the length (L) of where the upper part of the ladder is supported (S = ¼ L); duty ratings such as for step ladders, do not stand higher than the second step from the top, and for extension ladders, do not stand higher than the fourth rung from top; plus many more safety tips. You can also download their four-page pamphlet entitled *Ladder Safety Tips*.

Ladders are labeled with their duty rating. Medium-duty commercial (Type II – 225#), heavy-duty industrial (Type I – 250#), and extra-heavy-duty (Type IA – 300#) ladders bear an OSHA compliance label. Light-duty household (Type III – 200#) ladders do not bear an OSHA logo.

TRAINING

If you want to learn more, visit manufacturers' Web sites. For example, Bussmann's Web site is www.Bussmann.com. It is easy to use and has a computer program for making arc-flash and fault-current calculations. This Web site also has a technical publication *Selecting Protective Devices*, or bulletin SPD, a 190-plus-page publication about overcurrent protection selection, application, *NEC,®* and safety. You can order a hard copy of the publication from them.

The OSHA Training Institute offers outreach training programs of interest to electricians, contractors, and instructors. The basic safety courses for general construction safety and health are the OSHA 10-hour and OSHA 30-hour courses. Instructors interested in becoming an outreach trainer for the 10- and 30-hour courses must complete the OSHA 500 course entitled "Trainer Course in Occupational Safety and Health Standards for the Construction Industry." To become an outreach trainer, a test must be passed. Before the end of four years, outreach trainers must take the OSHA 502 update course for the construction industry or the OSHA 502 update course "Update for Construction Industry Outreach Trainers." Completion cards are issued upon completion of these courses.

Other courses, publications, "free loan" videos, schedules of upcoming safety training seminars, and other important information relating to safety on the job are available from OSHA for electricians, contractors, and trainers.

Visit the OSHA Web site at www.OSHA.gov for everything there is to know about OSHA safety requirements in the workplace.

Another valuable source of safety information is the National Institute for Occupational Safety and Health, a division of the Department of Health and Human Services Center for Disease Control and Prevention (CDC). Check out their Web site

www.cdc.gov/niosh. They offer an excellent downloadable 80-plus-page manual on *Electrical Safety*.

The National Safety Council has a vast amount of information relative to all aspects of safety. Check out their Web site at www.nsc.org.

The Consumer Product Safety Commission offers many safety publications for downloading. Visit their Web site at www.cpsc.gov, click on Library, click on CPSC Publications, click on By General Category, and then click on Electrical Safety. Here you will see a list of CPSC publications about GFCIs, AFCIs, metal ladder hazards, home wiring hazards, repairing aluminum wiring, and others.

The *National Fire Protection Association* offers many publications, videos, and a training course relating to safety. Browse their Web site at www.nfpa.org.

NFPA 70E Standard for Electrical Safety in the Workplace and *NFPA 70B Recommended Practices for Electrical Equipment Maintenance* present much of the same text regarding electrical safety as does the OSHA regulation.

Accredited apprenticeship training programs incorporate safety training as an integral part of their curriculum.

SAFETY CANNOT BE COMPROMISED!

It is impossible to put a dollar value on a life.

Don't take chances! Use the right tools! Turn off the power. Follow a lockout/tagout procedure. Mark the tag with a description of exactly what that particular disconnect controls.

How many times have we heard "The person would not have been injured (or electrocuted) had he turned the power OFF"? How many more times can we say it? **Turn OFF the power before working on the circuit!**

Visit the Web sites of the various organizations mentioned earlier. The Web site list can also be found in the back of this text. These organization have a wealth of information about on-the-job safety educational material and safety training courses.

LICENSING AND PERMITS

Most communities, counties, and/or states require electricians and electrical contractors to be licensed. This usually means they have taken and have passed a test. To maintain a valid license, many states require electricians and electrical contractors to attend and satisfactorily complete approved continuing education courses consisting of a specified number of classroom hours over a given period of time. Quite often, a community will have a "Residential Only" license for electricians and contractors that limits their activity to house wiring.

Permits are a means for a community to permanently record electrical work to be done, who is doing the work, and to schedule inspections during and after the "rough-in" stage and in the "final"

stages of construction. Usually, permits must be issued prior to starting an electrical project. In most cases, homeowners are allowed to do electrical work in their own home where they live but not in other properties they might own.

Figure 1-3 is a simple application for an electrical permit form. Some permit application forms are much more detailed.

If you are not familiar with licensing and permit requirements in your area, it makes sense to check

---

APPLICATION FOR ELECTRICAL PERMIT
VILLAGE OF ANYWHERE, USA   1-234-567-8900, EXT. 1234

Date _____   Permit No. _____

Owner _____   Job Address _____

Telephone No. _____   Job Start Date _____

CONTRACTOR INFORMATION AND SIGNATURE

Electrical Contractor _____ Tel. No. _____

Address _____ City _____ State _____ Zip _____

Registration No. _____ City of Registration _____

Supervising Electrician (Please Print Name) _____

Supervising Electrician's Signature _____

Insurance Bond _____ Village Business License _____

SERVICE INSPECTIONS OR REVISIONS

Existing Service Size: Amps _____ Volts _____ No. of Circuits _____ No. Added _____

New Service Size: Amps _____ Volts _____ No. of Circuits _____

Type: Overhead _____ Underground _____

Service Installation Fees: 100-200 amps $50 _____ over 200 amps $75 _____

MOTORS AND AIR CONDITIONING EQUIPMENT

No. of Motors: up to 1 HP @ $50 _____   Over 1-10 HP @ $50 _____

                11-25 HP @ $25 _____   Over 25 HP @ $25 _____

Air Conditioner/Heat Pump:

No. of Tons _____ ($20 for first ton, $5 for each additional ton)   _____

Furnace (electric):   kW _____ Amps _____ ($25)   _____

Dryer (electric):   kW _____ Amps _____ ($10)   _____

Range, oven, cooktop (electric): Total kW _____ Amps _____ ($10 each)   _____

Water Heater (electric): kW _____ Amps _____ ($10)   _____

TYPE OF MISCELLANEOUS ELECTRIC WORK

_____

_____

Minimum Inspection Fee: $40.00   _____

Escrow Deposit, if applicable   _____

TOTAL DUE   _____

Figure 1-3   A typical Application for Electrical Permit.

this out with your local electrical inspector or building department before starting an electrical project. Not to do so could prove to be very costly. Many questions can be answered: You will find out what tests, if any, must be taken; which permits are needed; which electrical code is enforced in your community; minimum size electrical service; and so on. Generally, the electrical permit is taken out by an electrical contractor who is licensed and registered as an electrical contractor in the jurisdictional area.

For new construction or for a main electrical service change, you will also need to contact the electric utility.

Temporary Wiring

There is an ever-present electrical shock hazard on construction sites. The *NEC®* addresses this in *Article 590*. This is covered in Unit 6.

PLANS

An architect or electrical engineer prepares a set of drawings that shows the necessary instructions and details needed by the skilled workers who are to build the structure. These are referred to as **plans**, **prints**, **blueprints**, **drawings**, **construction drawings**, or **working drawings**. The sizes, quantities, and locations of the materials required and the construction features of the structural members are shown at a glance. These details of construction must be studied and interpreted by each skilled construction craftsperson—masons, carpenters, electricians, and others—before the actual work is started.

The electrician must be able to: (1) convert the two-dimensional plans into an actual electrical installation, and (2) visualize the many different views of the plans and coordinate them into a three-dimensional picture, as shown in Figure 1-4.

Figure 1-4 Three-dimensional view of house wiring.

The ability to visualize an accurate three-dimensional picture requires a thorough knowledge of blueprint reading. Because all of the skilled trades use a common set of plans, the electrician must be able to interpret the lines and symbols that refer to the electrical installation and also those used by the other construction trades. The electrician must know the structural makeup of the building and the construction materials to be used

Plans might be **black line prints**, which are simply photocopies, or they might be computer-aided drawings (CAD). Although pretty much a thing of the past, we still have blueprints. Blueprints are created by running yellowish, light-sensitive paper and transparent Mylar® with black images on it through a blueprint machine, where the two sheets are exposed to a bright light for a short period of time. The light-sensitive paper turns white in all places except where the black images are. Where the black images are, the light-sensitive paper turns blue. That is why it is called a blueprint.

A tear-out/fold-out set of full-size plans for the residence referred to throughout this text is included in the back of the book.

SPECIFICATIONS

The **specifications** (specs, for short) for the electrical work indicated on the plans for the residence discussed throughout this text are found in the back of the book, before the Appendix.

Working drawings are usually complex because of the amount of information that must be included. To prevent confusing detail, it is standard practice to include with each set of plans a set of detailed written specifications prepared by the architect.

These specifications provide general information to be used by all trades involved in the construction. In addition, specialized information is given for the individual trades. The specifications include information on the sizes, the type, and the desired quality of the standard parts to be used in the structure.

Typical specifications include a section on "General Clauses and Conditions," which is applicable to all trades involved in the construction. This section is followed by detailed requirements for the various trades—excavating, masonry, carpentry, plumbing, heating, electrical work, painting, and others.

In the electrical specifications, the listing of standard electrical parts and supplies frequently includes the manufacturers' names and the catalog numbers of the specified items. Such information ensures that these items will be of the correct size, type, and electrical rating, and that the quality will meet a certain standard. To allow for the possibility that the contractor will not always be able to obtain the specified item, the phrase "or equivalent" is usually added after the manufacturer's name and catalog number.

The specifications are also useful to the electrical contractor in that all of the items needed for a specific job are grouped together and the type or size of each item is indicated. This information allows the contractor to prepare an accurate cost estimate without having to find all of the data on the plans.

If there is a difference between the plans and specifications, the specifications will take preference. The electrical contractor should discuss the matter with the homeowner, architect, and engineer. The cost of the installation might vary considerably because of the difference(s), so obtain any changes to the plans and/or specifications in writing.

SYMBOLS AND NOTATIONS

The architect uses **symbols** and **notations** to simplify the drawing and presentation of information concerning electrical devices, appliances, and equipment. For example, an electric range outlet symbol is shown in Figure 1-5.

Most symbols have a standard interpretation throughout the country as adopted by the American National Standards Institute (ANSI). Symbols are described in detail in Unit 2.

A notation will generally be found on the plans (blueprints) next to a specific symbol calling attention to a variation, type, size, quantity, or other necessary information. In reality, a symbol might be considered to be a notation because symbols represent words, phrases, numbers, quantities, and so on.

Figure 1-5 An electric range outlet symbol.

Another method of using notations to avoid cluttering up a blueprint is to provide a system of symbols that refer to a specific table. For example, the written sentences on plans could be included in a table referred to by a notation. Figure 2-10 is an example of how this could be done. The special symbols that refer to the table would have been shown on the actual plan.

NATIONAL ELECTRICAL CODE® (NEC®)

The *NEC*® is the recognized electrical *Code* standard by everyone in the electrical industry.

The first sentence in the *NEC*® is found in *90.1*. This sentence lays the foundation for all electrical installations. It states: *"The purpose of this Code is the practical safeguarding of persons and property from hazards arising from the use of electricity."* It goes on to state that the *NEC*® *"contains provisions that are considered necessary for safety"* and that *"compliance therewith and proper maintenance will result in an installation that is essentially free from hazard but not necessarily efficient, convenient, or adequate for good service or future expansion of electrical use."**

As you study this text, you will learn how to "Meet Code" for wiring a typical house, not "Beat Code."

As *NEC*® requirements are discussed throughout this text, the sheer number of *Code* references can become mind boggling. To simplify using this text as a reference, in addition to the conventional "subject" index at the back of the text, there is a Cross Index, making it easy for you to pinpoint specific *Code* sections and articles found in this text.

The *NEC*® is published by the National Fire Protection Association, and is referred to as *NFPA 70*. The *NEC*® was first published in 1897. It is revised every three years so as to be as up-to-date as possible. The *NEC*® does not become law until adopted by official action of the legislative body of a city, municipality, county, or state. Because of the ever-present danger of fire or shock hazard through some failure of the electrical system, the electrician and the electrical contractor must use listed materials and must perform all work in accordance with recognized standards.

---

*Reprinted with permission from NFPA 70-2005.

One- and Two-Family Dwelling Electrical Code

NFPA 70A is the *National Electrical Code*® *Requirements for One- and Two-Family Dwellings*. It is a condensed version of the *NEC*® that contains excerpts directly from the *NEC*®-*NFPA 70*. It includes requirements and tables that relate only to house wiring. It does not include *Code* requirements for other types of construction.

Other Electrical Codes

In addition to the *NEC*,® you must also consider local and state electrical codes.

Code Arrangement

It is important to understand the arrangement of the *NEC*.®

- Introduction, *Article 90* — An introduction to the *National Electrical Code*.® Explains what is and what is not covered in the *NEC*,® the arrangement of the *NEC*,® who has the authority to enforce the *Code*, what mandatory rules are, what permissive rules are, and the basics of metric vs. inch-pound measurements found throughout the *NEC*.®

- Chapter 1 — General
  Article 100: Definitions
  Article 110: Requirements for Electrical Installations
  Applies to all electrical installations.

- Chapter 2 — Wiring and Protection
  Articles 200–299
  Applies to all electrical installations.

- Chapter 3 — Wiring Methods
  Articles 300–398
  Applies to all electrical installations.

- Chapter 4 — Equipment for General Use
  Articles 400–499
  Applies to all electrical installations.

- Chapter 5   Special Occupancies
                Articles 500–599

- Chapter 6   Special Equipment
                Articles 600–699

- Chapter 7   Special Conditions
                Articles 700–799

- Chapter 8   Communications Systems
                Articles 800–899
                The articles in Chapter 8 are
                not subject to the requirements
                of Articles 1 through 7 unless
                specifically stated.

- Chapter 9   Tables showing dimensional
                data for raceways and conduc-
                tors, resistance and reactance
                values of conductors, Class 2
                and Class 3 circuit power
                limitations.

- Annex A   Provides a comprehensive list
                of product safety standards from
                Underwriters Laboratories
                (UL).

- Annex B   Provides data for determining
                conductor ampacity where there
                is engineering supervision.

- Annex C   Has tables showing the
                maximum number of conductors
                permitted in various types of
                raceways.

- Annex D   Examples of load calculations.

- Annex E   Types of building construction.

- Annex F   Cross referencing of wiring
                method articles between 2005,
                2002, and 1999 *NEC®* editions.

- Annex G   Administration and Enforcement
                A comprehensive suggested
                typical electrical ordinance that
                could be adopted on a local
                level.

- Index   The alphabetical index for the
                NEC.®

Note: Annexes and Fine Print Notes (FPN) are informational only. They are not mandatory.

Easy to Find Things!

▶Finding things in the current edition of the *NEC®* is much easier than in past editions. All articles in Chapter 3 of the *NEC®* cover wiring methods. In each article, the same section number has been assigned for a particular requirement. Here are a few examples:

- *Scope* is found in *Section 999.1*

- *Permitted Uses* are found in *Section 999.10.*

- *Uses Not Permitted* are found in *Section 999.12.*

- *Securing and Supporting* is found in *Section 999.30.*

This is referred to as "the parallel numbering system."◀

What Are the Lines in the Margin of the *NEC®*?

Vertical lines found in the margin on the pages of the *NEC®* indicate that there has been a change in that specific part of the *NEC®* from the previous edition.

Who Writes the *Code*?

For each *Code* cycle, the NFPA solicits proposals from anyone interested in electrical safety. Anyone may submit a proposal to change the *NEC®* using the Proposal Form found in the back of the *NEC.®* Proposals received are then given to a specific Code-Making Panel (CMP) to accept, reject, accept in part, accept in principle, or accept in principle in part. These actions are published in the Report on Proposals (ROP). Individuals may send in their comments on these actions using the Comment Form found in the ROP. The CMPs meet again to review and take action on the comments received. These actions are published in the Report on Comments (ROC). Final action (voting) on proposals and comments is taken at the NFPA annual meeting.

Individuals who serve on Code-Making Panels are electrical inspectors, electrical contractors, electrical engineers, individuals from utilities, manufacturers, testing laboratories, the Consumer Product Safety Commission, insurance companies, and similar organizations. CMP members are appointed by the NFPA. The CMPs have 10 to 20 principal members, plus a similar number of alternate members. All of the CMPs have a good balance of representation.

Which Edition of the *NEC®* to Use?

This text is based on the 2005 edition of the *NEC®*. Some municipalities, cities, counties, and states have not yet adopted this edition and may still be using older editions. Check with your local electrical inspector to find out which edition of the *Code* is in force.

Copies of the latest edition of the *NEC®-NFPA 70* may be ordered from the following:

Thomson Delmar Learning
P.O. Box 8007
Clifton Park, NY 12065
Phone: 800-998-7498
www.delmarlearning.com

National Fire Protection Association
1 Batterymarch Park
Quincy, MA 02269-9101
Phone: 617-770-3000
www.nfpa.org

International Association of Electrical Inspectors
901 Waterfall Way, Suite 602
Richardson, TX 75080-7702
Phone: 800-786-4234
www.iaei.org

Also available is *NFPA 70A*, which is the *National Electrical Code® Requirements for One- and Two-Family Dwellings*. This *Code* contains excerpts from the *National Electrical Code®, NFPA 70* for wiring one- and two-family houses. You might refer to it as a condensed version of the *NEC®*. It does not include *Code* requirements for other types of construction.

Electrical Inspection Code for Existing Dwellings

This code is published by the National Fire Protection Association and is referred to as *NFPA 73*. It is a brief nine-page code that provides require-ments for evaluating installed electrical systems within and associated with existing dwellings to identify safety, fire, and shock hazards—such as improper installations, overheating, physical deterioration, and abuse. This code lists most of the electrical things to inspect in an existing dwelling that could result in a fire or shock hazard if not corrected. It only points out things to look for that are visible. It does not get into examining concealed wiring that would require removal of permanent parts of the structure. It also does not get into calculations, location requirements, and complex topics as does the *NEC®-NFPA 70*.

This code can be an extremely useful guide for electricians doing remodel work, and for electrical inspectors wanting to bring an existing dwelling to a reasonably safe condition. Many localities require that when a home changes ownership, the wiring must be brought up to some minimum standard, but not necessarily as extensively as would be the case for new construction.

Homes for the Physically Challenged

The *NEC®* does not address the wiring of homes for the physically or mentally challenged, or disabilities associated with the elderly.

Each installation for the physically challenged must be based upon the specific need(s) of the individual(s) who will occupy the home. Some physically challenged people are bedridden, require the mobility of a wheelchair, have trouble reaching, have trouble bending, etc. There are no hard and fast rules that *must* be followed—only many suggestions to consider.

Some of these are:

- Install more ceiling luminaires (lighting fixtures) instead of switching receptacles. Cords and lamps are obstacles.

- Install luminaires (lighting fixtures) having more than one bulb.

- Go "overboard" in the amount of lighting for all rooms, entrances, stairways, stairwells, closets, pantries, bathrooms, and so on.

- Use higher wattage bulbs—not to exceed the wattage permitted in the specific fixture.

- Consider installing luminaires (lighting fixtures) in certain areas (such as bathrooms and hallways) to be controlled by motion detectors.

- Consider installing exhaust fans in certain areas (such as laundries, showers, and bathrooms) that turn on automatically when the humidity reaches a predetermined value.

- Consider the height of switches and thermostats, usually 42 in. (1.0 m) or lower, instead of the standard 46–52 in. (1.15 to 1.3 m).

- Consider rocker-type switches instead of toggle type.

- Install pilot light switches.

- Consider "jumbo" switches.

- Be sure stairways and stairwells are well lit.

- Consider stair tread lighting.

- Position lighting switches so as not to be over stairways or ramps.

- Locate switches and receptacles to be readily and easily accessible—not behind doors or other hard-to-reach places.

- Consider installing wall receptacle outlets higher 24–27 in. (600 mm to 675 mm) than normal 12 in. (300 mm) height.

- Install lighted doorbell buttons.

- Chimes: consider adding a strategically located "dedicated" lamp(s) that will turn on when door bell buttons are pushed. The wiring diagram for this is found in Figure 25-25.

- Telephones: consider adding visible light(s) strategically located that will flash at the same time the telephone is ringing.

- Consider installing receptacle outlets and switches on the face of kitchen cabinets. Wall outlets and switches can be impossible for the physically challenged person to reach.

- Consider fire, smoke, and security systems directly connected to a central office for fast response to emergencies that do not depend upon the disabled to initiate the call.

- Consider installing the fuse box or breaker panel on the first floor instead of in the basement.

- Consider the advanced home systems concept of remote control of lighting, receptacles, appliances, television, telephones, and so on. The control features can make life much easier for a disabled person.

When involved with multifamily dwellings, check with your local building authority. They will have copies of the Americans with Disabilities Act (ADA) from the U.S. Department of Justice and the Fair Housing Act from the U.S. Department of Housing and Urban Development (HUD) that basically require that all units must have accessible light switches, electrical outlets, thermostats, and other environmental controls.

The American National Standards Institute (ANSI) publication *ANSI A117.1–2003, Standard for Accessible and Useable Buildings and Facilities,* contains many suggestions and considerations for buildings and facilities for the physically challenged. This and other standards can be ordered from:

American National Standards Institute
11 West 42nd Street
New York, NY 10036
Phone: 212-642-4900
www.ansi.com

The same standard is available from the International Code Council (ICC) under the publication number *ICC/ANSI A117.1-2003.*

Another virtually unlimited source of standards is:

Global Engineering Documents
15 Inverness Way East
Englewood, CO 80112
www.global.ihs.com

BUILDING CODES

The majority of the building departments across the country have for the most part adopted the *NEC®* rather than attempting to develop their own electrical codes. As you study this text, you will note numerous references to the *NEC,®* the electrical inspector, or the authority having jurisdiction.

An authority's level of knowledge varies. The electrical inspector may be full-time or part-time and may also have responsibility for other trades, such as plumbing or heating. The heads of the building departments in many communities are typically called the Building Commissioners or Directors of Development. Regardless of title, they are responsible for ensuring that the building codes in their communities are followed.

International Code Council (ICC)

Rather than writing their own codes, most communities for the most part, adopt the codes of the International Code Council (ICC). Over the years, the ICC has developed comprehensive and coordinated model construction codes. These include building, mechanical, plumbing, fire, energy conservation, existing building, fuel gas, sewage disposal, property maintenance, zoning, residential, and electrical codes. The ICC provides technical, educational, and informational products. Their address is:

ICC Headquarters
5203 Leesburg Pike, Suite 600
Falls Church, VA 22041
Phone: 703-931-4533
www.iccsafe.org

The *ICC International Residential Code* in Chapters 33 through 42 contains electrical provisions that were written and produced under the guidance of the *National Fire Protection Association*. The material in these chapters is copyrighted by the *National Fire Protection Association*. These provisions are a mirror image of the *National Electrical Code*® other than the layout and numbering system.

The *ICC International Electrical Code* is a rather brief 30-page administrative code. It contains only two pages of actual electrical provisions.

Electrical Wiring—Residential is based on the *National Electrical Code*.®

Because local electrical codes may differ from the *NEC*,® you should check with the local inspection authority to determine which edition of the *NEC*® is enforced, and what, if any, local requirements or amendments take precedence over the *NEC*.®

American National Standards Institute

The **American National Standards Institute (ANSI)** is an organization that coordinates the efforts and results of the various standards-developing organizations, such as those mentioned in pre-

vious paragraphs. Through this process, ANSI approves standards that then become recognized as American national standards. One will find much similarity between the technical information found in ANSI standards, the Underwriters Laboratories standards, the International Electronic and Electrical Engineers standards, and the *NEC*.®

International Association of Electrical Inspectors

The International Association of Electrical Inspectors (IAEI) is a nonprofit organization. The IAEI membership consists of electrical inspectors, building officials, electricians, engineers, contractors, and manufacturers throughout the United States and Canada. The major goal of the IAEI is to improve the understanding of the *NEC*.® Representatives of this organization serve as members of the various Code-Making Panels of the *NEC*,® and share equally with other members in the task of reviewing and revising the *NEC*.®

The IAEI publishes a bimonthly magazine— *The IAEI News*. It is devoted entirely to electrical code topics. Anyone in the electrical industry is welcome to join the IAEI. An application form is found after the Appendix of this text. Their address is:

International Association of Electrical
Inspectors
901 Waterfall Way, Suite 602
Richardson, TX 75080-7702
Phone: 800-786-IAEI
www.iaei.org

Code Definitions

The electrical industry uses many words (terms) that are unique to the electrical trade. These terms need clear definitions to enable the electrician to understand completely the meaning intended by the *Code*.

Article 100 of the *NEC*® is a "dictionary" of these terms. *Article 90* also provides further clarification of terms used in the *NEC*.® Here are a few examples:

Ampacity: The current, in amperes, that a conductor can carry continuously under the conditions of use without exceeding its temperature rating.

Approved: Acceptable to the authority having jurisdiction.

Authority Having Jurisdiction: The organization, office, or individual responsible for approving equipment, materials, an installation, or a procedure. See *90.4*.

▶**Dwelling Unit:** One or more rooms for the use of one or more persons as a housekeeping unit with space for eating, living, cooking, and sleeping, and permanent provisions for sanitation.◀ (**Note:** The terms *dwelling, residence, home,* and *house* are used interchangeably throughout this text.)

Fine Print Notes (FPN): Fine Print Notes (FPNs) are found throughout the *Code*. FPNs are "explanatory" in nature, in that they refer to other sections of the *Code*. FPNs also define things where further description is necessary. See *90.5*.

Listed: Equipment, materials, or services included in a list published by an organization that is acceptable to the authority having jurisdiction and concerned with evaluation of products or services, that maintains periodic inspection of production of listed equipment or materials or periodic evaluation of services, and whose listing states that either the equipment, material, or services meets appropriate designated standards or has been tested and found suitable for a specified purpose.

Shall: Indicates a mandatory rule, *90.5*. As you study the *NEC*, think of the word "shall" as meaning "must." Some examples found in the *NEC* where the word "shall" is used in combination with other words are: *shall be, shall have, shall not, shall be permitted, shall not be permitted,* and *shall not be required*.

Read *110.3(B)* Carefully!

One of the most far-reaching *NEC* rules is *110.3(B)*. This section states that "*Listed or labeled equipment shall be installed and used in accordance with any instructions included in the listing or labeling.*" This means that an entire electrical system and all of the system's electrical equipment must be *installed* and *used* in accordance with the *NEC* and the numerous standards to which the electrical equipment has been tested.

METRICS (SI) AND THE *NEC*

The United States is the last major country in the world not using the metric system of weights and measures as the primary system. For most of our lifetime, we have used the English system of weights and measures, also referred to as inch-pound and U.S. Customary. But this is changing!

Manufacturers are now showing both inch-pound and metric units in their catalogs. By law, plans and specifications for new governmental construction and renovation projects have used the metric system since January 1, 1994.

You may not feel comfortable with metric measurements, but metric measurements are here to stay. You might just as well become familiar with the metric system.

All measurements in the *NEC* are shown in both inch-pound and metric values.

For more information about metrics, refer to the "Metric System of Measurement" section found in the Appendix of this text.

Trade Sizes

A unique situation exists. Strange as it may seem, what electricians have been referring to for years has not been correct!

Raceway sizes have always been an approximation. For example, there has never been a ½-in. raceway! Measurements taken from the *NEC* for a few types of raceways are shown in Table 1-1.

You can readily see that the cross-sectional areas, critical when determining conductor fill, are different. It makes sense to refer to conduit, raceway, and tubing sizes as *trade sizes*. The *NEC* in *90.9(C)(1)* states that "where the actual measured

TRADE SIZE	INSIDE DIAMETER (I.D.)
½ Electrical Metallic Tubing	0.622 in.
½ Electrical Nonmetallic Tubing	0.560 in.
½ Flexible Metal Conduit	0.635 in.
½ Rigid Metal Conduit	0.632 in.
½ Intermediate Metal Conduit	0.660 in.

Table 1-1 Comparison of trade size vs. actual inside diameters.

size of a product is not the same as the nominal size, trade size designators shall be used rather than dimensions." This edition of *Electrical Wiring— Residential* uses the term *trade size* when referring to conduits, raceways, and tubing. For example, a ½-in. EMT is referred to as trade size ½ EMT.

The *NEC®* also uses the term *metric designator*. A trade size ½ EMT is shown as *metric designator 16 (½)*. A trade size 1 EMT is shown as *metric designator 27 (1)*. The numbers 16 and 27 are the metric designator values; the (½) and (1) are the trade sizes. The metric designator is the raceway's inside diameter—in rounded-off millimeters (mm). Table 1-2 shows some of the more common sizes of conduits, raceways, and tubing. A complete listing is found in *NEC® Table 300.1(C)*.

For ease in understanding, this text uses only the term *trade size* when referring to conduit and raceway sizes.

Conduit knockouts in boxes do not measure up to what we call them. Table 1-3 shows some examples.

Outlet boxes and device boxes use their nominal measurement as their *trade size*. For example, a 4 in. × 4 in. × 1½ in. does not have an internal cubic-inch volume of 4 in. × 4 in. × 1½ in. = 24 in.³ *Table 314.16(A)* shows this size box as having 21 cubic-in. volume. This table shows *trade sizes* in two columns—millimeters and inches.

In this text, a square outlet box is referred to as trade size 4 × 4 × 1½. Similarly, a single-gang device box would be referred to as a trade size 3 × 2 × 3 box.

METRIC DESIGNATOR AND TRADE SIZE	
Metric Designator	**Trade Size**
12	⅜
16	½
21	¾
27	1
35	1¼
41	1½
53	2
63	2½
78	3

Table 1-2 Trade sizes of raceways and their metric designator identification.

TRADE SIZE KNOCKOUT	ACTUAL MEASUREMENT
½	⅞ in.
¾	1³⁄₃₂ in.
1	1⅜ in.

Table 1-3 Comparison of knockout trade size vs. actual measurement.

Trade sizes for construction material will not change. A 2 × 4 is really a *name*, not an actual dimension. A 2 × 4 will keep its name forever. This is its *trade size*.

In this text, most measurements directly related to the *NEC®* are given in both inch-pound and metric units. In many instances, only the inch-pound units are shown. This is particularly true for the examples of raceway and box fill calculations, load calculations for square foot areas, and on the plans (drawings).

Because the *NEC®* rounded off most metric conversion values, a calculation using metrics results in a different answer when compared to the same calculation done using inch-pounds. For example, load calculations for a residence are based on 3 volt-amperes per square foot or 33 volt-amperes per square meter.

For a 40 ft × 50 ft dwelling: 3 VA × 40 ft × 50 ft = 6000 volt-amperes.

In metrics, using the rounded off values in the *NEC®*: 33 VA × 12 m × 15 m = 5940 volt-amperes.

The difference is small, but nevertheless, there is a difference.

To show calculations in both units throughout this text would be very difficult to understand and would take up too much space. Calculations in either metrics or inch-pounds are in compliance with the *NEC®* *90.9(D)*. In *90.9(C)(3)* we find that metric units are not required if the industry practice is to use inch-pound units.

It is interesting to note that the examples in *Chapter 9* of the *NEC®* use inch-pound units, not metrics.

Guide to Metric Usage

The metric system is a "base-10" or "decimal" system in that values can be easily multiplied or divided by 10 or "powers of 10." The metric system as we know it today is known as the International System of Units (SI), derived from the French term "le Système International d'Unités."

In the metric system, the units increase or decrease in multiples of 10, 100, 1000, and so on. For instance, one megawatt (1,000,000 watts) is 1000 times greater than one kilowatt (1000 watts).

By assigning a name to a measurement, such as a *watt*, the name becomes the unit. Adding a prefix to the unit, such as *kilo*, forms the new name *kilowatt*, meaning 1000 watts. Refer to Table 1-4 for prefixes used in the metric system.

The prefixes used most commonly are *centi*, *kilo*, and *milli*. Consider that the basic unit is a meter (one). Therefore, a centimeter is 0.01 meter, a kilometer is 1000 meters, and a millimeter is 0.001 meter.

Some common measurements of length and equivalents are shown in Table 1-5.

Electricians will find it useful to refer to the conversion factors and their abbreviations shown in Table 1-6.

Refer to the Appendix of this text for a comprehensive metric conversion table. The table includes information on how to "round off" numbers for practical use on the job.

PREFIX	SYMBOL	MULTIPLIER	SCIENTIFIC NOTATION (POWERS OF TEN)	VALUE
tera	T	1 000 000 000 000	10^{12}	one trillion (1 000 000 000 000/1)
giga	G	1 000 000 000	10^{9}	one billion (1 000 000 000/1)
mega	M	1 000 000	10^{6}	one million (1 000 000/1)
kilo	k	1 000	10^{3}	one thousand (1 000/1)
hecto	h	100	10^{2}	one hundred (100/1)
deka	da	10	10^{1}	ten (10/1)
unit		1	—	one (1)
deci	d	0.1	10^{-1}	one tenth (1/10)
centi	c	0.01	10^{-2}	one hundredth (1/100)
milli	m	0.001	10^{-3}	one thousandth (1/1 000)
micro	μ	0.000 001	10^{-6}	one millionth (1/1 000 000)
nano	n	0.000 000 001	10^{-9}	one billionth (1/1 000 000 000)
pico	p	0.000 000 000 001	10^{-12}	one trillionth (1/1 000 000 000 000)

Table 1-4   Metric prefixes, symbols, multipliers, powers, and values.

one inch	=	2.54	centimeters
	=	25.4	millimeters
	=	0.025 4	meter
one foot	=	12	inches
	=	0.304 8	meter
	=	30.48	centimeters
	=	304.8	millimeters
one yard	=	3	feet
	=	36	inches
	=	0.914 4	meter
	=	914.4	millimeters
one meter	=	100	centimeters
	=	1 000	millimeters
	=	1.093	yards
	=	3.281	feet
	=	39.370	inches

Table 1-5   Some common measurements of length and their equivalents.

inches (in) × 0.025 4	=	meters (m)
inches (in) × 0.254	=	decimeters (dm)
inches (in) × 2.54	=	centimeters (cm)
centimeters (cm) × 0.393 7	=	inches (in)
inches (in) × 25.4	=	millimeters (mm)
millimeters (mm) × 0.039 37	=	inches (in)
feet (ft) × 0.304 8	=	meters (m)
meters (m) × 3.280 8	=	feet (ft)
square inches (in²) × 6.452	=	square centimeters (cm²)
square centimeters (cm²) × 0.155	=	square inches (in²)
square feet (ft²) × 0.093	=	square meters (m²)
square meters (m²) × 10.764	=	square feet (ft²)
square yards (yd²) × 0.836 1	=	square meters (m²)
square meters (m²) × 1.196	=	square yards (yd²)
kilometers (km) × 1 000	=	meters (m)
kilometers (km) × 0.621	=	miles (mi)
miles (mi) × 1.609	=	kilometers (km)

Table 1-6   Useful conversions and their abbreviations.

NATIONALLY RECOGNIZED TESTING LABORATORIES (NTRL)

How does one know if a product is safe to use?

Manufacturers, consumers, regulatory authorities, and others recognize the importance of independent, "third-party" testing of products in an effort to reduce safety risks. Unless you have all of the necessary test equipment and knowledge of how to properly test a product for safety, the surest way is to accept the findings of a third-party testing agency.

Nationally recognized testing laboratories (NRTL) have the knowledge, wherewithal, and test equipment to test and evaluate products for safety.

An NRTL will perform tests on a product based on a specific nationally recognized safety standard. After the product has been tested and found to comply with the safety standard, the product is considered to be free from reasonably foreseeable risk of fire, electric shock, and related hazards. The product is then "listed" and will have a listing marking (label).

As you work with electrical products, make sure the product has a listing marked on it from a nationally recognized testing laboratory. If the product is too small to have a listing mark on the product itself, then look for the mark on the carton the product came in.

The following laboratories do a considerable amount of testing and listing of electrical equipment:

Underwriters Laboratories, Inc. (UL®)

Underwriters Laboratories, Inc. (UL), founded in 1894, is a highly qualified, nationally recognized testing laboratory. UL develops standards, and performs tests to these standards. Most reputable manufacturers of electrical equipment submit their products to UL where the equipment is subjected to numerous tests. These tests determine if the product can perform safely under normal and abnormal conditions to meet published standards. After UL tests and evaluates a product, and determines that the product complies with the specific standard, the manufacturer is then permitted to *label* its product with the UL Mark (see Figure 1-6). The products are then *listed* in a UL Directory.

UL *does not approve* a product. Rather, UL *lists* those products that conform to a specific safety standard. A UL Listing Mark on a product means that representative samples of the product have been tested and evaluated to nationally recognized safety standards with regard to fire, electric shock, and related safety hazards.

Do not confuse a UL marking in a circle with the markings found on *recognized components*. Recognized components that have passed certain tests are marked with the letters *RU* printed backwards (see Figure 1-7). By themselves, recognized components are not to be field-installed. They are intended for use in end-use products or systems that would ultimately be tested and listed, with the final assembly becoming a UL listed product. Some examples of recognized components are relays, ballasts, insulating materials, special switches, and so on.

Figure 1-6 The Underwriters Laboratories mark.

Figure 1-7 The *recognized components* mark.

Useful UL publications are:

Electrical Construction Equipment Directory (Green Book)

Electrical Appliance and Utilization Equipment Directory (Orange Book)

General Information for Electrical Equipment Directory (White Book)

It is extremely useful for an electrician, an electrical contractor, and/or an electrical inspector to refer to these directories when looking for specific requirements, permitted uses, limitations, and others for a certain product.

Many times the answer to a product-related question that cannot be found in the *NEC®* can be found in one of these directories.

The Green and Orange Directories provide technical information about a particular product, and they list the names and addresses of manufacturers and the manufacturers' identification numbers. The White Book provides technical information regarding a product but does not show manufacturers' names and addresses.

The best companion to the *NEC®* is probably the UL White Book.

These directories can be obtained by contacting:

Underwriters Laboratories, Inc.
333 Pfingsten Road
Northbrook, Illinois 60062-2096
Phone: 847-272-8800
www.ul.com

Underwriters Laboratories, Inc. and the Canadian Standards Association (CSA) have worked out an agreement whereby either agency can test, evaluate, and list equipment for the other agency. For example, UL might test and list an air-conditioner unit to the requirements of UL Standard 1995 (Heating and Cooling Equipment) because the Canadian Standard C22.2 No. 236-M90 is a mirror image of UL Standard 1995. One by one, the UL and CSA Standards are becoming similar.

When UL tests and lists products that comply to the requirements of a particular CSA standard, the UL logo shown in Figure 1-8 will appear with a "C" outside of and to the left of the circle.

Figure 1-8 The mark indicating that UL has tested and listed a product for compliance *only* with Canadian requirements.

This means the product has been tested and evaluated for compliance *only* with Canadian requirements.

This harmonization of Codes and Standards is going on in the world today as a result of the North American Free Trade Act (NAFTA). Discussions are also going on with Mexico. When all of this is finalized, electrical equipment standards may be the same in the United States, Canada, and Mexico.

CSA International

CSA International is the Canadian counterpart of Underwriters Laboratories, Inc. in the United States. CSA International is the source of the *Canadian Electrical Code (CEC)* and of the Canadian Standards for the testing, evaluation, and listing of electrical equipment in Canada. The *Canadian Electrical Code* is quite different from the *NEC.®* A Canadian version of *Electrical Wiring— Residential* is available in Canada.

The *Canadian Electrical Code* and CSA standards can be obtained by contacting:

CSA International
178 Rexdale Boulevard
Toronto, Ontario, CANADA
M9W 1R3
Phone: 416-747-4044
        800-463-6727
Fax: 416-747-2510
www.csa-international.org

Intertek Testing Services ITS

ITS is a nationally recognized testing laboratory. Its ETC SEMKO division provides testing, evaluation, labeling, listing, and follow-up service for the safety testing of electrical products. This is done in conformance to nationally recognized safety standards.

Information can be obtained by contacting:

Intertek Testing Services, NA, Inc.
ETL SEMKO
3933 US Route 11
Cortland, NY 13045
Phone: 607-753-6711
www.etlsemko.com

National Electrical Manufacturers Association NEMA

The National Electrical Manufacturers Association (NEMA) represents nearly 600 manufacturers of electrical products. NEMA develops electrical equipment standards, which, in many instances, are very similar to UL and other consensus standards. NEMA has representatives on the Code-Making Panels for the *NEC.* Additional information can be obtained by contacting:

National Electrical Manufacturers Association
1300 North 17th Street, Suite 1847
Rosslyn, VA 22209
Phone: 703-841-3200
www.nema.org

World Wide Web Sites

A comprehensive list of World Wide Web sites for manufacturers, organizations, and inspection and testing agencies is provided in the back of this text.

REVIEW

Note: Refer to the *NEC®* or the plans in the back of this text where necessary.

1. What is the purpose of specifications? _____

   _____

2. Refer to the specifications in the back of this text following Unit 32.

   a. What electrical codes must be conformed to?

      _____

      _____

      _____

   b. What is the wiring method to be used in the workshop?

      _____

   c. What size conductors are to be used for:

      The lighting branch-circuits: _____ AWG

      The small-appliance branch-circuits: _____ AWG

   d. What section of the specifications tells us the size of the service entrance to be installed for this residence?

      _____

3. What is done to prevent a plan from becoming confusing because of too much detail?

   _____

   _____

4. Name three requirements contained in the specifications regarding material.

   a. _____   c. _____

   b. _____

5. The specifications state that all work shall be done _____

   _____.

6. What phrase is used when a substitution is permitted for a specific item? _____

   _____

7. What is the purpose of an electrical symbol? _____

   _____

8. What is a notation? _____

   _____

9. Where are notations found? _____

10. List at least twelve electrical notations found on the plans for this residence. Refer to
    the plans at the back of the text. _____

    _____

    _____

    _____

    _____

    _____

11. What three parties must be satisfied with the completed electrical installation?

    a. _____   b. _____   c. _____

12. What code sets standards for electrical installation work? _____

    _____

13. What authority enforces the requirements set by the *NEC*®? _____

    _____

14. Does the *NEC*® provide minimum or maximum standards? _____

15. What do the letters *UL* signify? _____

    _____

16. What section of the *NEC*® states that all listed or labeled equipment shall be installed
    or used in accordance with any instructions included in the listing or labeling?

    _____

17. When the word "shall" appears in a *NEC*® reference, it means that it (must)(may) be
    done. (Underline the correct word.)

18. What is the purpose of the *NEC*®?

    _____

    _____

    _____

    _____

19. Does compliance with the *NEC®* always result in an electrical installation that is adequate, safe, and efficient? Why? _____

_____

_____

_____

20. Name two nationally recognized testing laboratories. _____

_____

_____

21. a. Do Underwriters Laboratories and the other recognized testing laboratories "approve" products? _____

    b. What do these testing laboratories do? _____

_____

_____

_____

22. a. Has the *NEC®* been officially adopted by the community in which you live? _____

    b. By the state in which you live? _____

    c. If your answer is YES to a. or b., are there amendments to the *NEC®*? _____

    d. If your answer is YES to c., list some of the more important amendments.

_____

_____

_____

_____

23. Does the *NEC®* make suggestions about how to wire a house that will be occupied by handicapped persons? _____

24. A junction box on a piece of European equipment is marked 200 cm³. Convert this to cubic inches. _____

_____

25. Convert 4,500 watts to Btu/hour. _____

_____

26. You will learn in Unit 3 that residential lighting loads are based on 3 volt-amperes per ft² (33 volt-amperes per m²). Determine the minimum lighting load required for an area of 186 m². Do calculations for both feet squared and meters squared so you can see the difference in answers. To convert meters squared to feet squared, refer to Table 1-6.

_____

_____

_____

_____

SAFETY RELATED QUESTIONS

27. What federal organization dictates requirements for work-related safety issues? Circle the correct answer.

    a. *NEC®*

    b. OSHA

    c. IAEI

28. What can you do to reduce or eliminate the possibility of receiving an electric shock when working on electrical circuits? Circle the correct answer.

    a. Turn off the power on the circuit you are working on.

    b. Turn off the power on the circuit you are working on, tag it, and lock the disconnect in the "OFF" position.

    c. Turn off the power on the circuit you are working on, then tell everyone else working with you not to turn the power back on.

29. What is the *NEC®* definition of a qualified person?

_____

30. Are you a *qualified person*? Explain your answer.

_____

_____

31. Are low-voltage systems totally safe? Explain.

_____

_____

32. What NFPA standard specifically covers safety in the workplace? _____

33. What do the letters PPE stand for?

_____

34. Explain briefly what an "arc flash" is.

_____

_____

35. Where might you obtain information about on-the-job safety and safety training?

_____

_____

_____

36. a. Is it a safe practice to use a step ladder in which the spreaders have not been fully locked into position? _____

    b. If the ladder collapses, who is at fault? _____

Electrical Symbols and Outlets

ELECTRICAL SYMBOLS

Electrical symbols used on an architectural plan show the location and type of electrical device required. A typical electrical installation as taken from a plan is shown in Figure 2-1.

The *NEC*® has many words unique to the electrical trade. For example:

▶A **device** is *"A unit of an electrical system that is intended to carry or control but not utilize electric energy."* ◀

An **outlet** is *"A point on the wiring system at which current is taken to supply utilization equipment."*

A **receptacle** is *"A contact device installed at the outlet for the connection of an attachment plug."*

Figure 2-1 Use of electrical symbols and notations on a floor plan.

ALLOW NOT LESS THAN 6 IN. (150 MM) OF CONDUCTOR MEASURED FROM WHERE THE CONDUCTORS EMERGE FROM CABLE OR RACEWAY TO THE END OF THE CONDUCTOR.

ALLOW NOT LESS THAN 3 IN. (75 MM) OF CONDUCTOR MEASURED FROM THE FRONT EDGE OF OPENING TO THE END OF THE CONDUCTOR WHERE THE OPENING IS LESS THAN 8 IN. (200 MM) IN ANY DIMENSION.

THESE DIMENSIONAL MINIMUMS DO NOT APPLY TO CONDUCTORS THAT ARE NOT SPLICED OR TERMINATED IN THE BOX.

RECEPTACLE OUTLET

Figure 2-2 **Where a receptacle is connected to the branch-circuit wires, the outlet is called a *receptacle outlet*. *NEC® 300.14* states the minimum length of conductors at an outlet, junction, or switch point for splices or the connection of luminaires (fixtures) or devices. Leaving the conductors too long can cause crowding of the wires in the box, leading to possible short circuits and/or ground faults.**

A **receptacle outlet** is *"An outlet where one or more receptacles are installed,"* Figure 2-2.

A **lighting outlet** is *"An outlet intended for the direct connection of a lampholder, a luminaire (lighting fixture), or a pendant cord terminating in a lampholder,"* Figure 2-3.

By definition, toggle switches and receptacles are devices because they carry current but do not consume power.

The term *opening* is widely used by electricians and electrical contractors when estimating the cost

of an installation. The term *opening* covers all lighting outlets, receptacle outlets, junction boxes, switches, etc. The electrician and/or electrical contractor will estimate a job at "X dollars per lighting outlet," "X dollars per switch," "X dollars per receptacle outlet," and so on. These estimates will include the time and material needed to complete the job. Each type of electrical *opening* is represented on the electrical plans as a symbol. The electrical openings in Figure 2-1 are shown by the symbols in Figure 2-4.

ANSI recently published a totally revised standard entitled *Symbols for Electrical Construction Drawings*. This was the first revision in over 20 years. Figure 2-5, Figure 2-6, Figure 2-7, Figure 2-8, Figure 2-9, and Figure 2-10 show the electrical symbols most commonly found on architectural and

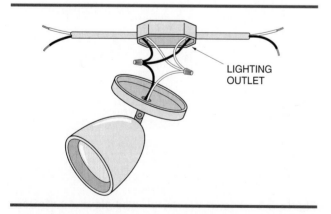

LIGHTING OUTLET

Figure 2-3 **When a luminaire (lighting fixture) is connected to the branch-circuit wires, the outlet is called a *lighting outlet*.**

RECEPTACLE OUTLET:

THREE-WAY SWITCH: S_3

CEILING LIGHTING OUTLET:

Figure 2-4 **Some common electrical symbols.**

OUTLETS	CEILING	WALL
SURFACE-MOUNTED INCANDESCENT		
LAMPHOLDER WITH PULL SWITCH		
RECESSED INCANDESCENT		
SURFACE-MOUNTED FLUORESCENT		
RECESSED FLUORESCENT		
SURFACE OR PENDANT CONTINUOUS ROW FLUORESCENT		
RECESSED CONTINUOUS ROW FLUORESCENT		
BARE LAMP FLUORESCENT STRIP		
SURFACE OR PENDANT EXIT		
RECESSED CEILING EXIT		
BLANKED OUTLET		
OUTLET CONTROLLED BY LOW-VOLTAGE SWITCHING WHEN RELAY IS INSTALLED IN OUTLET BOX		
JUNCTION BOX		

Figure 2-5 Lighting outlet symbols.

RECEPTACLE OUTLETS

SINGLE RECEPTACLE OUTLET	CLOTHES DRYER OUTLET
DUPLEX RECEPTACLE OUTLET	FAN OUTLET
TRIPLEX RECEPTACLE OUTLET	CLOCK OUTLET
DUPLEX RECEPTACLE OUTLET, SPLIT-CIRCUIT	FLOOR OUTLET
DOUBLE DUPLEX RECEPTACLE (QUADPLEX)	MULTIOUTLET ASSEMBLY; ARROW SHOWS LIMIT OF INSTALLATION. APPROPRIATE SYMBOL INDICATES TYPE OF OUTLET. SPACING OF OUTLETS INDICATED BY "X" INCHES.
WEATHERPROOF RECEPTACLE OUTLET	FLOOR SINGLE RECEPTACLE OUTLET
GROUND-FAULT CIRCUIT INTERRUPTER RECEPTACLE OUTLET	FLOOR DUPLEX RECEPTACLE OUTLET
RANGE OUTLET	FLOOR SPECIAL-PURPOSE OUTLET

SPECIAL-PURPOSE OUTLET (SUBSCRIPT LETTERS INDICATE SPECIAL VARIATIONS: DW = DISHWASHER. ALSO a, b, c, d, ETC. ARE LETTERS KEYED TO EXPLANATION ON DRAWINGS OR IN SPECIFICATIONS).

Figure 2-6 Receptacle outlet symbols.

SWITCH SYMBOLS	
S	SINGLE-POLE SWITCH
S_2	DOUBLE-POLE SWITCH
S_3	THREE-WAY SWITCH
S_4	FOUR-WAY SWITCH
S_D	DOOR SWITCH
S_{DS}	DIMMER SWITCH
S_G	GLOW SWITCH TOGGLE— GLOWS IN OFF POSITION
S_K	KEY-OPERATED SWITCH
S_{KP}	KEY SWITCH WITH PILOT LIGHT
S_{LV}	LOW-VOLTAGE SWITCH
S_{LM}	LOW-VOLTAGE MASTER SWITCH
S_{MC}	MOMENTARY-CONTACT SWITCH
M	OCCUPANCY SENSOR—WALL-MOUNTED WITH "OFF-AUTO" OVERRIDE SWITCH
M P	OCCUPANCY SENSOR—CEILING-MOUNTED "P" INDICATES MULTIPLE SWITCHES WIRE-IN PARALLEL
S_P	SWITCH WITH PILOT LIGHT ON WHEN SWITCH IS ON
S_T	TIMER SWITCH
S_R	VARIABLE-SPEED SWITCH
S_{WP}	WEATHERPROOF SWITCH

Figure 2-7   Switch symbols.

CIRCUITING

HOME RUN TO PANEL. NUMBER OF ARROWS INDICATES NUMBER OF CIRCUITS. LETTERS AND NUMBERS INDICATE CIRCUIT(S). FULL SLASHES INDICATE UNGROUNDED "HOT" (OR SWITCH LEG) CIRCUIT CONDUCTORS. HALF SLASHES INDICATE GROUNDED NEUTRAL CIRCUIT CONDUCTOR(S). NO SLASHES INDICATE ONE "HOT" AND ONE NEUTRAL.

BRANCH CIRCUIT. FULL SLASHES INDICATE UNGROUNDED "HOT" (OR SWITCH LEG) CIRCUIT CONDUCTORS. HALF SLASHES INDICATE GROUNDED NEUTRAL CIRCUIT CONDUCTOR(S). NO SLASHES INDICATE ONE "HOT" AND ONE NEUTRAL. DOTS INDICATE EQUIPMENT GROUNDING CONDUCTORS.

WIRING CONCEALED IN CONSTRUCTION IN FINISHED AREAS, EXPOSED IN UNFINISHED AREAS

CONDUIT CONCEALED IN OR UNDER FLOOR SLAB

CONDUIT TURNING UP

CONDUIT TURNING DOWN

CONDUIT ONLY (EMPTY)

SWITCH LEG; CONNECTS SWITCHED OUTLETS WITH CONTROL POINTS

Figure 2-8   Circuiting symbols.

electrical plans. Because some items may have more than one symbol, it is important to check the plans and specifications of any job you are working on for a symbol schedule to make sure you have interpreted the symbols correctly.

The dashed lines in Figure 2-1 run from the outlet to the switch or switches that control the outlet. These lines are usually curved so that they cannot be mistaken for invisible edge lines. Outlets shown on the plan without curved dashed lines are independent outlets and have no switch control.

A study of the plans for the single-family dwelling shows that many different electrical symbols are used to represent the electrical devices and equipment used in the building.

Be very careful when interpreting symbols because of the similarity between them, Figure 2-11.

When preparing electrical plans for commercial and industrial buildings, most architects and electrical engineers use symbols approved by ANSI wherever possible. For residential projects, usually only the basic symbols are used, showing the location of

Symbol	Description
⊣│││├	BATTERY
▢	BUZZER
⌒	CIRCUIT BREAKER
▽	DATA OUTLET
$\frac{xxAF}{yyAT}$ ▢	DISCONNECT SWITCH, FUSED; SIZE AS INDICATED ON DRAWINGS. "xxAF" INDICATES FUSE AMPERE RATING. "yyAT" INDICATES SWITCH AMPERE RATING.
xxA ▢	DISCONNECT SWITCH, UNFUSED; SIZE AS INDICATED ON DRAWINGS. "xxA" INDICATES SWITCH AMPERE RATING.
▣⊙	DOOR BELL
CH	DOOR CHIME
D	DOOR OPENER (ELECTRIC)
✕	FAN: CEILING-SUSPENDED (PADDLE)
✳	FAN: CEILING-SUSPENDED (PADDLE) FAN WITH LIGHT
⊠	FAN: WALL
⊣│││	GROUND
Ⓙ	JUNCTION BOX—CEILING

Symbol	Description
—Ⓙ	JUNCTION BOX—WALL
▨	LIGHTING OR POWER PANEL, RECESSED
▨	LIGHTING OR POWER PANEL, SURFACE
MD	MOTION DETECTOR
Ⓜ	MOTOR
②	MOTOR: "2" INDICATES HORSEPOWER
—∞—	OVERLOAD RELAY
▣	PUSHBUTTON
—▭—	SWITCH AND FUSE
▼	TELEPHONE OUTLET
▼W	TELEPHONE OUTLET— WALL-MOUNTED
▽	TELEPHONE/DATA OUTLET
Ⓣ L	THERMOSTAT—LINE-VOLTAGE
Ⓣ LV	THERMOSTAT—LOW-VOLTAGE
TS	TIME SWITCH
T	TRANSFORMER

Figure 2-9 Miscellaneous symbols.

SYMBOL	NOTATION
1	PLUGMOLD ENTIRE LENGTH OF WORKBENCH. OUTLETS 18 IN. (457 mm) O.C. INSTALL 48 IN. (1.2 m) TO CENTER FROM FLOOR. GFCI PROTECTED.
2	TRACK LIGHTING. PROVIDE 5 LAMPHOLDERS.
3	TWO 40-WATT RAPID START FLUORESCENT LAMPS IN VALANCE. CONTROL WITH DIMMER SWITCH.

Figure 2-10 Example of how certain notations might be added to a symbol when a symbol itself does not fully explain its meaning. The architect or engineer has a choice of explaining fully the meaning directly on the plan if there is sufficient room; if insufficient room, then a notation could be used.

SPLIT-CIRCUIT RECEPTACLE OUTLET

WEATHERPROOF RECEPTACLE OUTLET

Figure 2-11 Variations in significance of outlet symbols.

receptacles, switches, appliances, and luminaires (fixtures). Specific circuitry as shown in Figure 2-8 is generally left up to the electrician.

Later in this text, you will find suggested *cable layouts* for all of the branch-circuits in this residence. These *cable layouts* show the required number of conductors between outlets and switches.

LUMINAIRES (FIXTURES) AND OUTLETS

The term *luminaire* is found in the *NEC*.® In the United States, we have used the term *lighting fixture* for years. But since the *NEC*® is an international code, we must become familiar with some new

terms such as "luminaire." By definition, a luminaire is: "*A complete lighting unit consisting of a lamp or lamps together with the parts designed to distribute the light, to position and protect the lamps and ballast (where applicable), and to connect the lamps to the power supply.*"*

Simply stated for our purposes, a luminaire and a lighting fixture are one and the same.

The location of lighting outlets is determined by the amount and type of illumination required to provide the desired lighting effects. It is not the intent of this text to describe how proper and adequate lighting is determined. Rather, the text covers the proper methods of installing the circuits for such lighting.

Luminaire (fixture) manufacturers publish catalogs that provide a tremendous amount of information regarding recommendations for residential lighting. These publications are available at electrical distributors, lighting showrooms, and home centers.

Architects often include a specific amount of money in the specifications, referred to as a *fixture allowance*. Electrical contractors include this amount in their bid, and the choice of luminaires (fixtures) is then left to the homeowners. If the owner selects luminaires (fixtures) whose total cost exceeds the fixture allowance, the owner is expected to pay the difference between the actual cost and the fixture allowance. If the luminaires (fixtures) are not selected before the rough-in stage of wiring the house, an electrician will usually install outlet boxes such as those illustrated in Figure 2-20.

How luminaires (fixtures) are handled varies by job and geographical area you are working in. Sometimes the recessed luminaire (fixture) housings are included in the base bid, and the trims are included in the fixture allowance. Always read the specifications and study the plans carefully for important symbols and notations that might reference luminaires (fixtures).

Symbols Used to Represent Lighting Outlets and Switches

Typical lighting outlet symbols are shown in Figure 2-12. Typical switch symbols are shown in Figure 2-13.

---

*Reprinted with permission from NFPA 70-2005.

SYMBOL	TYPE OF LUMINAIRE (LIGHTING FIXTURE) OR OUTLET	BOXES THAT MAY BE USED
CEILING LIGHTING OUTLET	1. SURFACE-MOUNTED LUMINAIRE (LIGHTING FIXTURE)	(A) ADJUSTABLE BRACKET WITH OUTLET BOX. (B) OUTLET BOX WITH BRACKET FOR JOIST MOUNTING. (C) OUTLET BOX WITH CAPTIVE NAILS/BRACKET FOR JOIST MOUNTING. BOXES ARE "LISTED" FOR LUMINAIRE (LIGHTING FIXTURE) SUPPORT
RECESSED CEILING LIGHTING OUTLET	2. RECESSED LUMINAIRE (LIGHTING FIXTURE)	CABLE TERMINATES IN BOX THAT IS FURNISHED WITH LUMINAIRE (FIXTURE). SPECIAL CODE RULES GOVERNING THE INSTALLATION ARE FOUND IN *410.64* THROUGH *410.72* OF THE *NATIONAL ELECTRICAL CODE.*® SEE UNIT 7.
WALL LIGHTING OUTLET	3. WALL-MOUNTED LUMINAIRE (LIGHTING FIXTURE) ALSO CALLED A "SCONCE"	(A) OUTLET BOX WITH CAPTIVE NAILS/BRACKET FOR STUD MOUNTING. (B) ADAPTER RING FOR OUTLET BOX. BOX "LISTED" FOR LUMINAIRE (LIGHTING FIXTURE) SUPPORT.
DUPLEX RECEPTACLE OUTLET	4.	(A) SINGLE-GANG SWITCH (DEVICE) BOX. (B) 4 IN. SQUARE BOX WITH CAPTIVE NAILS FOR STUD MOUNTING. (C) 4 IN. SQUARE SINGLE-GANG RAISED ADAPTER RING.

Figure 2-12 Typical electrical symbols, items they represent, and boxes that could be used. *Photos courtesy* of Progress Lighting (1, 3, 4) and Halo/Cooper Lighting (2).

SYMBOL	FLUSH TOGGLE SWITCH	OPERATION	CONNECTIONS
S SINGLE-POLE		ON OFF	
S₂ DOUBLE-POLE		ON OFF	
S₃ THREE-WAY		POSITION 1 POSITION 2	
S₄ FOUR-WAY		1 2	
Sₚ SWITCH AND PILOT LIGHT		FOR CONTROLLING LIGHTS FROM ONE POINT WITH PILOT LIGHT INDICATION	ALSO AVAILABLE IN THREE-WAY TYPE OF CONTROLLING LIGHT FROM TWO POINTS WITH PILOT LIGHT INDICATION

Figure 2-13 Standard switches and symbols.

OUTLET, DEVICE, AND JUNCTION BOXES

Article 314 contains much detail about outlet boxes, device (switch) boxes, pull boxes, junction boxes, and conduit bodies (LB, LR, LL, C, etc.) all of which are used to provide the necessary space for making electrical connections, contain wiring, and support luminaires (fixtures), and so on. *Article 314* is quite lengthy. For the moment, we will focus on some of the basic *Code* requirements for outlet boxes, device boxes, and junction boxes. Additional *Code* requirements are covered on an "as needed" basis in this unit and in later units.

Screw sizes for typical electrical boxes are:

Device boxes and faceplates	No. 6-32
Outlet boxes	No. 8-32
Ceiling fan boxes (two sets of holes)	No. 10-32 and No. 8-32

Some Basic *Code* Requirements

- *NEC® 300.11* and *314.23:* Boxes must be securely mounted and fastened in place.

- *NEC® 300.15:* Where conduit, electrical-metallic tubing, nonmetallic-sheathed cable, Type AC cable, or other cables are installed, a box or conduit body must be installed at each conductor splice connection point, outlet, switch point, junction point, or pull point. This requirement is illustrated in Figure 2-14. Exceptions to this are found in *300.15(A)* through *(M)*. Typical exceptions are surface-mounted strip luminaires (fixtures), as shown in Figure 10-2. Most of these have a back plate that has a trade size ½ knockout intended for

the direct entry of a cable without the need of a box behind the luminaire (fixture).

- *NEC® 300.15(C):* Where cables enter or exit from conduit or tubing that is used to provide cable support or protection against physical damage, a fitting shall be provided on the end(s) of the conduit or tubing to protect the cable from abrasion. Figure 2-15 shows a transition from a cable to a raceway.

- *NEC® 314.16:* Boxes must be large enough for all of the enclosed conductors and wiring devices. Installing boxes too small for the number of conductors, wiring devices, luminaire (fixture) studs, and splices is a real problem! This topic is covered in detail later.

- *NEC® 314.17(C):* The exception allows multiple cables to be run through a single knockout opening in a nonmetallic box. This permission is not given for metallic boxes. See Figure 2-16.

- ▶*NEC® 314.20:* In walls or ceilings where the surface material is noncombustible, boxes, plaster rings, domed covers, extension rings, or listed extenders shall be mounted so that they will be set back not more than ¼ in. (6 mm). In walls or ceilings where the surface material is combustible, boxes, plaster rings, domed covers, extension rings, or listed extenders shall be set flush with the surface. ◀

- *NEC® 314.22:* A surface extension from a box of a concealed wiring system is made by mounting and mechanically securing an extension ring over the concealed box. An exception to this is that a surface extension is permitted to be made from the cover of a concealed box

Figure 2-14 A box (or fitting) must be installed wherever there are splices, outlets, switches, or other junction points as indicated by the "X." Unless listed for use without a box, such as a typical surface-mounted strip luminaire (fixture) that has a back plate with a trade size ½ knockout, point XX is a *Code* violation, *300.15.*

COMBINATION COUPLING TO JOIN EMT TO
AC CABLE OR FLEXIBLE METAL CONDUIT.

Figure 2-15 *NEC® 300.15(C)* permits a transition to be made from one wiring method to another wiring method. In this case, the armor of the Type AC cable is removed, allowing sufficient length of the conductors to be run through the conduit. A proper fitting must be used at the transition point, and the fitting must be accessible after installation. The conductors in Type AC cable are THHN.

Figure 2-16 Multiple cables may run through a single knockout opening in a nonmetallic box, *314.17(C), Exception.*

where the cover is designed so it is unlikely to fall off, or be removed if its securing means becomes loose. The wiring method must be flexible and arranged so that any required grounding continuity is independent of the connection between the box and cover. This is shown in Figure 2-17.

- *NEC® 314.25:* In completed installations, all boxes shall have a cover, faceplate, or luminaire (fixture) canopy. Do not leave any electrical boxes uncovered.

- *NEC® 314.27(A):* Boxes used at lighting outlets must be designed or installed so that a luminaire (lighting fixture) may be attached to it.

- *NEC® 314.27(A)* and *(B):* Listed outlet boxes are suitable for hanging luminaires (lighting fixtures) that weigh 50 lb. (23 kg) or less. Luminaires (lighting fixtures) that weigh more than 50 lb. (23 kg) must be supported independently of the outlet box, or the outlet box must be "listed" for the weight to be supported. Wall-mounted luminaires (lighting fixtures) of

A

VIOLATION. IT IS A VIOLATION OF THE *CODE* TO MAKE A RIGID EXTENSION FROM A COVER THAT IS ATTACHED TO A CONCEALED OUTLET BOX, *314.22.* IT WOULD BE DIFFICULT TO GAIN ACCESS TO THE CONNECTIONS INSIDE THE BOX.

B

THIS MEETS *CODE.* AN EXTENSION RING MUST BE MOUNTED OVER AND MECHANICALLY SECURED TO THE CONCEALED OUTLET BOX, *314.22.*

C

THIS MEETS *CODE.* IT IS PERMITTED TO MAKE A SURFACE EXTENSION FROM A COVER FASTENED TO A CONCEALED BOX WHERE THE EXTENSION WIRING METHOD IS FLEXIBLE AND WHERE A SEPARATE GROUNDING CONDUCTOR IS PROVIDED SO THAT THE GROUNDING PATH IS NOT DEPENDENT UPON THE SCREWS THAT ARE USED TO FASTEN THE COVER TO THE BOX. SEE *314.22,* EXCEPTION.

IT WOULD BE A CODE VIOLATION TO MAKE THE EXTENSION IF A FLEXIBLE WIRING METHOD WAS USED BUT HAD NO PROVISIONS FOR A SEPARATE EQUIPMENT GROUNDING CONDUCTOR.

Figure 2-17   How to make an extension from a flush-mounted box.

less than 6 lb. (3 kg) may be supported by a box, or by a plaster ring secured to a box, by not less than two No. 6-32 or larger screws, as in Figure 2-18.

- *NEC® 314.27(D):* Standard outlet boxes must not be used as the sole support for ceiling-suspended (paddle) fans unless they are specifically listed for this purpose. See Unit 9 for detailed discussion of this issue.

- *NEC® 314.29:* Conduit bodies, junction, pull, and outlet boxes must be installed so the wiring contained in them is accessible without removing any part of the building. Never install outlet boxes, junction boxes, pull boxes, or conduit bodies where they will be inaccessible behind or above permanent finished walls or ceilings. Should anything ever go wrong, it would be a nightmare, if not impossible, to troubleshoot and locate hidden splices. Wiring located above

"lay-in" ceilings is considered accessible because it is easy to remove a panel(s) to gain access to the space above the ceiling. The *NEC®* definition of *accessible* is "*Capable of being removed or exposed without damaging the building structure or finish, or not permanently closed in by the structure or finish of the building.*"*

The same accessibility requirement is true for underground wiring. Boxes must be accessible without excavating sidewalks, paving, earth, or other material used to establish the finished grade.

- *NEC® 314.71:* Pull and junction boxes must provide adequate space and dimensions for the installation of conductors, and they shall comply with the specific requirements of this section.

---

*Reprinted with permission of NFPA 70-2005.

METALLIC SWITCH
(DEVICE) BOX
OR DEVICE
PLASTER RING

ANY NONMETALLIC BOX

Figure 2-18   Wall-mounted luminaires (lighting fixtures) that do not weigh more than 6 lb. (3 kg) are permitted to be secured to these types of boxes and plaster rings by not less than two No. 6-32 or larger screws, *314.27(A), Exception.* For standard outlet boxes, the maximum weight is 50 lb. (23 kg), *314.27(B).*

- *NEC® 410.12:* This is another repeat of the requirement that boxes must be covered. The "cover" could be a blank cover, a luminaire (fixture) canopy, a lampholder, a switch or receptacle, or similar device with faceplate.
- ▶*NEC® 410.16(A):* Here we find permission to use outlet boxes and fittings to support luminaires (lighting fixtures) provided the requirements found in *314.27(A)* and *(B)* are met.◀

What to Do Where Recessed Luminaires (Lighting Fixtures) Will Be Installed. This topic is covered in detail in Unit 7.

NONMETALLIC OUTLET AND DEVICE BOXES

Today, most house wiring is done using nonmetallic boxes as permitted by *314.3*. Although not often done, *NEC® 314.3 (Exceptions)* allows metal raceways and metal-jacketed cables (BX) to be used with nonmetallic boxes provided all metal raceways or cables entering the box are bonded together to maintain the integrity of the grounding path to other equipment in the installation.

The house wiring system usually is formed by a number of specific circuits. Each circuit consists of runs of cable from outlet to outlet or from box to box. The residence plans show many branch-circuits for general lighting, appliances, electric heating, and other requirements. The specific *Code* rules for each of these circuits are covered in later units.

GANGED SWITCH (DEVICE) BOXES
Multiple-Gang Boxes

Where more than one wiring device is to be installed at one location, multiple-gang boxes are used. Nonmetallic boxes for wiring devices are available in two-gang, three-gang, and four-gang.

Metal device boxes can be ganged together by removing and discarding one side from both the first and third boxes, and both sides from the second (center) box.

Multiple gang-boxes are shown in Figure 2-19.

Box Fill

Figure 2-20 shows the dimensions of typical metal boxes and the number of conductors permitted in a given size box. This figure replicates *Table 314.16(A)* of the *NEC®.*

Nonmetallic boxes are marked with their cubic-inch volume. *Table 314.16(B)* shows the cubic-inch volume allowance required for different size conductors.

Figure 2-19   Examples of multiple-gang device boxes.

QUIK-CHEK BOX SELECTION GUIDE
FOR BOXES GENERALLY USED FOR RESIDENTIAL WIRING

DEVICE BOXES	WIRE SIZE	3x2x1½ (7.5 in.³)	3x2x2 (10 in.³)	3x2x2¼ (10.5 in.³)	3x2x2½ (12.5 in.³)	3x2x2¾ (14 in.³)	3x2x3 (16 in.³)	3x2x3½ (18 in.³)
	14 AWG	3	5	5	6	7	8	9
	12 AWG	3	4	4	5	6	7	8

THE NUMBER OF CONDUCTORS IS "PERGANG"

SQUARE BOXES	WIRE SIZE	4x4x1½ (21 in.³)	4x4x2⅛ (30.3 in.³)
	14 AWG	10	15
	12 AWG	9	13

OCTAGON BOXES	WIRE SIZE	4x1½ (15.5 in.³)	4x2⅛ (21.5 in.³)
	14 AWG	7	10
	12 AWG	6	9

HANDY BOXES	WIRE SIZE	4x2⅛x1½ (10.3 in.³)	4x2⅛x1⅞ (13 in.³)	4x2x2⅛ (14.5 in.³)
	14 AWG	5	6	7
	12 AWG	4	5	6

RAISED COVERS

WHERE RAISED COVERS ARE MARKED WITH THEIR VOLUME IN CUBIC INCHES, THAT VOLUME MAY BE ADDED TO THE BOX VOLUME TO DETERMINE MAXIMUM NUMBER OF CONDUCTORS IN THE COMBINED BOX AND RAISED COVER. NONMETALLIC RAISED COVERS ARE AVAILABLE.

MASONRY BOXES	WIRE SIZE	3¾x2x2½ (14 in.³)	3¾x2x3½ (21 in.³)
	14 AWG	7	10
	12 AWG	6	9
3-GANG BOX			

MASONRY BOXES HAVE LOTS OF ROOM. AVAILABLE UP TO 6 GANG. THE CONDUCTOR FILL IS FOR EACH GANG. REFER TO *TABLE 314.16(A)* AND *(B)*.

NOTE: BE SURE TO MAKE DEDUCTIONS FROM THE ABOVE MAXIMUM NUMBER OF CONDUCTORS PERMITTED FOR WIRING DEVICES, CABLE CLAMPS, FIXTURE STUDS, AND GROUNDING CONDUCTORS. THE CUBIC INCH (IN.³) VOLUME IS TAKEN DIRECTLY FROM *TABLE 314.16(A)* OF THE *NEC.*® NONMETALLIC BOXES ARE MARKED WITH THEIR CUBIC INCH CAPACITY.

Figure 2-20 Quik-chek box selection guide.

BOX MOUNTING

▶*NEC® 314.20* states that where noncombustible surface material (tile, gypsum, plaster, and like material) is used on walls or ceilings, boxes, plaster rings, domed covers, extension rings, or listed extenders must be mounted so that they will be set back not more than ¼ in. (6 mm) from the face of the surface material. Where combustible surface material (wood paneling) is used on walls or ceilings, boxes, plaster rings, domed covers, extension rings, or listed extenders must be set flush with the surface material, Figure 2-21.◀

CONCRETE, TILE, OR OTHER NONCOMBUSTIBLE MATERIAL

WOOD OR OTHER COMBUSTIBLE MATERIAL

A

BOX SET BACK NOT MORE THAN 1/4 IN. (6 mm)

BOX SET FLUSH WITH FINISHED SURFACE

REPAIR THESE GAPS SO GAPS OR OPENINGS ARE NO GREATER THAN 1/8 IN. (3 mm)

B

C

D

AN ASSORTMENT OF PLASTIC MOUNTING BLOCKS TO ATTAIN NEAT APPEARANCE WHEN MOUNTING LIGHTING FIXTURES AND RECEPTACLES WHERE OUTDOOR SIDING IS LAPPED.

Figure 2-21 Box position requirements in walls and ceilings constructed of various materials. If the box is not flush with the finished surface (as a result of improper installation) or if paneling, drywall, tile, or a mirror is added, then install a box extender, sometimes referred to as a "goof ring." One type is shown in (C). Box extenders that are ¼ in., ⅜ in., ½ in., or ¾ in. deep are also available for one- and two-gang boxes. Refer to *314.20* and *314.21*. For set-back requirements for panelboards, see *312.3* and *312.4*.

These requirements hold back, retard, and slow down the spread of fire caused by arcing at a loose connection, a short circuit, or a ground fault that might occur within the box. This significantly reduces the supply of oxygen needed to keep a fire burning. Closing the gaps around boxes further reduces the spread of fire.

Ganged sectional switch (device) boxes can be installed using a pair of metal mounting strips. These strips are also used to install a switch box between wall studs, Figure 2-22. When an outlet box is to be mounted at a specific location between joists, as for ceiling-mounted luminaires (fixtures), an adjustable bar hanger is used, as in Figure 2-12.

NEC® *314.23(B)(1)* requires that mounting nails or screws through a box shall not be more than ¼ in. (6 mm) from the back or ends of the box, as shown in Figure 2-23. This ensures that the nails or screws will not interfere with the wiring devices in the box. ►Furthermore, you are not allowed to use screws through a box unless the threads are protected by an approved means to avoid abrasion of conductor insulation. ◄

Why bother with these *Code* requirements that take up valuable wiring space and could lead to conductor damage? It is much better to use boxes that have external mounting means.

Most residences are constructed with wood framing. However, metal studs and joists are being used more and more. Figure 2-24 shows a few types of boxes for use with metal framing members.

The first one is a single-gang nonmetallic device box that can be riveted or screwed to the metal

MOUNTING STRIPS

Figure 2-22 Switch (device) boxes installed between studs using metal switch box supports, often referred to as "Kruse strips." If wood strips are used, they must have a cross-sectional dimension of not less than 1 in. × 2 in. (25 mm × 50 mm), *314.23(B)(2)*.

MAXIMUM 1/4 IN. (6 mm)

MAXIMUM 1/4 IN. (6 mm)

MORE THAN 1/4 IN. (6 mm)

MORE THAN 1/4 IN. (6 mm)

(A)   CORRECT

THE NAILS ARE KEPT WITHIN 1/4 IN. (6 mm) FROM THE BACK AND/OR ENDS OF THE BOX. THIS ALLOWS ROOM FOR THE WIRING DEVICE.

(B)   VIOLATION

THE NAILS HAVE BEEN INSTALLED MORE THAN 1/4 IN. (6 mm) FROM THE BACK AND/OR ENDS OF THE BOX. THESE NAILS WILL INTERFERE WITH THE WIRING DEVICE. IT MAY ALSO BE IMPOSSIBLE TO INSTALL A WIRING DEVICE DUE TO THE INTERFERENCE OF THESE NAILS.

A

B

Figure 2-23 Using nails or screws to install a sectional switch box, *314.23(B)(1)*.

Figure 2-24 Types of boxes that can be used with steel framing construction. Note the side clamps that "snap" directly onto the steel framing members.

stud. The second is a 4-in. square steel box with nonmetallic sheathed cable (Romex) clamps. The third is a 4-in., square steel box with armored cable (BX) clamps. The second and third boxes "snap" onto the metal stud, and do not require drilling the metal studs and fastening the box with nuts and bolts or screws. See Unit 4 for *Code* require-

ments about running cables through metal framing members.

Another type of device used for fastening electrical boxes to steel framing members is shown in Figure 2-25. The bracket is screwed to the steel framing members, then the box or boxes are attached to the bracket.

Figure 2-25 A bracket for use on steel framing members. The bracket spans the space between two studs. Note that more than one box can be mounted on this bracket. *Courtesy of* ERICO,® Inc.

Remodel (Old Work)

When installing switches, receptacles, and lighting outlets in remodel work where the walls and ceilings (paneling, drywall, sheetrock and plaster, or lath and plaster) are already in place, first make sure you will be able to run cables to where the switches, receptacles, and lighting outlets are to be installed. After making sure you can get the cables through the concealed spaces in the walls and ceilings, you can then proceed to cut openings the size of the specific type of box to be installed at each outlet location. Cables are then "fished" through the stud and joist spaces behind or above the finished walls and ceilings to where you have cut the openings. Then, boxes having plaster ears and snap-in brackets, Figure 2-26, can be inserted from the front, through the holes already cut at the locations where the wall or ceiling boxes are to be installed. After the boxes are snapped into place through the hole from the front, they become securely locked into place.

Another popular method of fastening wall and ceiling boxes in existing walls and ceilings is to use boxes that have plaster ears. Insert the box through the hole from the front. Then position a metal support into the hole on each side of the box. Be careful not to let the metal support fall into the hole. Next, bend the metal support over and into the box, (see Figure 2-27). Two metal supports are used. These metal switch box supports are known as Madison Hold-Its™.*

---

*Hold-It™ is a brand name trademark owned by Madison Equipment Co., Cleveland, OH.

Figure 2-27 Metal device box supports referred to as Madison Hold-Its™.

NOTE: AFTER INSERTING THE BOX INTO THE HOLE CUT IN THE DRYWALL, WHEN SCREWS ARE TURNED, "EARS" FLIP UPWARD BEHIND THE DRYWALL, TIGHTLY LOCKING THE BOX IN PLACE.

Figure 2-26 Nonmetallic boxes are used for remodel work. When boxes are inserted into the hole cut in the drywall, they securely lock into place. This type of box is not for use in new work.

Also available for old work are boxes that have a screw-type support on each side of the box, as in Figure 2-28. This type of box has plaster ears. It is inserted through the hole cut in the wall or ceiling, then the screws are tightened. This pulls up a metal support tightly behind the wall or ceiling, firmly holding the box in place. The action is very similar to that of Molly screw anchors.

Because these types of remodel boxes are supported by the drywall or plaster, as opposed to new work where the boxes are supported by the framing members, be careful not to hang heavy luminaires (lighting fixtures) from them. Be sure that there is proper support if a ceiling fan is to be installed. This is covered in Unit 9.

Spread of Fire

To protect lives and property, fire must be contained and not be allowed to spread.

In addition to the *NEC,*® you must also become familiar with building codes.

Building codes such as the International Code Council (ICC) *International Residential Code* generally do not require fire-resistance-rated walls and ceilings in one- and two-family buildings. Some communities have amendments to the basic building codes, such as requiring ⅝-in. (16-mm) gypsum wallboard [instead of the normal ½-in. (13-mm)] on the ceilings and on the walls between the garage and

Figure 2-28 These boxes are commonly used in remodel work. The box is installed from the front, then the screws are tightened. The box is adequately supported.

living areas and requiring *draft stopping* where ducts, cables, and piping run through bottom (sole) and top plates. Draft stopping is generally done by the insulation contractor. Townhouses, condominiums, and other multi-occupancy buildings do have fire-resistance rating requirements. To avoid costly mistakes, check with your local building officials to determine exactly what the requirements are!

In walls, partitions, and ceilings that are fire-resistance-rated, special consideration must be given when installing wall and ceiling electrical boxes "back-to-back" or in the same stud or joist space of common walls or ceilings, as might be found in multifamily buildings. Electrical wall boxes installed back-to-back or installed in the same stud or joist space defeat the fire-resistance rating of the wall or ceiling.

What Is Fire-Resistance Rating?

A building assembly's (walls, ceilings, partitions, gypsum, dry wall, plaster, etc.) fire-resistance rating refers to the period of time the assembly will serve as a barrier to the spread of fire. It is an indication of how long it can hold back a fully developed blaze before it spreads to adjacent areas of the building and how long the assembly can function structurally after being exposed to a fire.

While undergoing specific tests, passage of flames and/or structural collapse are the determining factors for establishing fire-resistance ratings.

To be truly fire-resistance-rated, the construction and materials must conform to very rigid requirements, as set forth in the UL Standards discussed below.

Ratings of fire-resistance materials are expressed in hours. Terms such a 1-hour fire resistance rating, 2-hour fire resistance rating, and so on, are used.

UL Standard 263 *Fire Tests of Building Construction and Materials* and NFPA 251 cover fire resistance ratings. In UL Standard 263, and in all building codes, you will find requirements such as: "the surface area of individual metallic outlet or switch boxes shall not exceed 16 square inches" (i.e., a 4-in. square box is 4 in. × 4 in. = 16 in.²), "the aggregate surface area of the boxes shall not exceed 100 in.² per 100 ft² of wall surface," "boxes located on opposite sides of walls or partitions shall be separated by a minimum horizontal distance of 24 in. (600 mm)," "the metallic outlet or switch boxes shall be securely fastened to the studs and the opening in the wallboard facing shall be cut so that the clearance between the box and the wallboard does not exceed ⅛ in. (3 mm)," and "the boxes shall be installed in compliance with the *NEC.*®"

The UL *Fire Resistance Directory* covers fire-resistant materials and assemblies, and contains a rather detailed listing of *Outlet Boxes and Fittings Classified for Fire Resistance (CEYY)*, showing all manufacturers' product part numbers for *nonmetallic boxes* to be installed in walls, partitions, and ceilings that meet the fire-resistance standard. This category also covers special-purpose boxes for installation in floors.

When using nonmetallic boxes in fire-resistance-rated walls, the restrictions are more stringent than for metallic boxes. For example, the UL *Fire Resistance Directory* shows that one manufacturer of nonmetallic boxes limits the product to be installed so that no opening exceeds 10 in.^{2} (64.5 cm^2). Another manufacturer limits the product to be installed so that no opening exceeds 25 in.^{2} (161 cm^2).

Molded fiberglass outlet and device boxes have become available, and are listed for use in fire-resistance-rated partitions without the use of putty pads, mineral wool batts, or fiberglass batts, provided a minimum horizontal distance of 3 in. (75 mm) is maintained between boxes and the boxes are not back-to-back. See Figure 2-31B.

If you are dealing with fire-resistance-rated construction, be sure to carefully read the box manufacturer's instructions and "Listing" details.

For both metallic and nonmetallic electric outlet boxes, the *minimum* distance between boxes on opposite sides of a fire-rated wall or partition is 24 in. (600 mm). There is an exception, which is discussed later.

For both metallic and nonmetallic electric outlet boxes, the maximum gap between the box and the wall material is ⅛ in. (3 mm). This holds true for fire-resistance-rated walls and non-fire-resistance-rated walls.

To maintain the hourly fire rating of a fire-rated wall that contains electrical outlet and switch boxes, the UL *Fire Resistance Directory* lists an *intumescent* (expands when heated) fire-resistant material that comes in "pads."

These "putty pads" are moldable, and can be used to wrap metal electrical boxes or inserted on the inside back wall of metal electrical boxes. When exposed to fire, the material expands and chars, sealing off the opening to prevent the spread of flames and limit the heat transfer from the fire-resistance-rated side of the wall to the non-fire-resistance-rated side of the wall. It is not to be used on nonmetallic boxes.

The ⅛ in. (3 mm) thickness provides a 1-hour fire-resistance rating. A 2-hour rating is obtained with ¼ in. (6 mm) thickness.

When this material is properly installed, the 24-in. (600-mm) separation between boxes is not required. However, the electrical outlet boxes *must not* be installed back-to-back. Using this material meets the requirements of *NEC*® 300.21 of the *NEC*® and other building codes.

The International Code Council (ICC) *International Residential Code* states that "fireblocking" be provided to cut off all concealed draft openings (both vertical and horizontal) and to form an effective fire barrier between stories, and between a top story and the roof space. This includes openings around vents, pipes, ducts, chimneys, and fireplaces at ceiling and floor level. Oxygen supports fire. Firestopping cuts off or significantly reduces the flow of oxygen. Most communities that have adopted the ICC code require firestopping around electrical conduits. In residential work, this is generally done by the insulation installer, before the drywall installer closes up the walls.

Be careful when roughing in wall boxes where wood paneling is to be installed. Some communities require that ½-in. (13-mm) gypsum board be installed on the studs behind finished wood paneling to establish a fire-resistance rating. ▶Remember that where the finish wall or ceiling surface is combustible material such as wood paneling, the front edge of the box, plaster ring, domed cover, extension ring, or listed extender shall be set flush with the wall surface. Refer to *NEC*® 314.20 and Figure 2-21.◀

The previous discussion makes it clear that the electrician must become familiar with certain building codes, in addition to the *NEC.*® Knowing the *NEC*® is not enough!

It is extremely important to check with the electrical inspector and/or building official for complete clarification on the matter of fire-resistance requirements before you find yourself cited with serious violations and lawsuits regarding noncompliance with fire codes. Changes can be costly, and will delay completion of the project. Improper wiring is hazardous. The *NEC®* addresses this issue in *300.21*.

Figure 2-29, Figure 2-30, Figure 2-31, and Figure 2-32 illustrate the basics for installing electrical outlet

and switch boxes in fire-resistance-rated partitions. This is an extremely important issue. If in doubt, check with your local electrical inspector and/or building official.

THESE ELECTRICAL WALL BOXES MUST NOT BE MOUNTED BACK TO BACK.

TO "MEET *CODE*," ELECTRICAL WALL BOXES MOUNTED ON OPPOSITE SIDES OF THE WALL MUST BE AT LEAST 24 IN. (600 MM) APART.

THIS SIDE OF WALL HAS A 1-HOUR FIRE-RESISTANCE RATING.

Figure 2-29 Building codes prohibit electrical wall boxes to be mounted back-to-back in fire-resistance-rated walls unless specific installation procedures, as illustrated in Figure 2-30, Figure 2-31, and Figure 2-32, are followed. One of these requirements is that electrical wall boxes must be kept at least 24 in. (600 mm) apart. This is to maintain the integrity of the fire-resistance rating of the wall.

THIS SIDE OF WALL IS NON-FIRE-RESISTANCE-RATED.

THIS SIDE OF WALL HAS A 1-HOUR FIRE-RESISTANCE RATING.

Figure 2-30 Building codes permit boxes to be mounted back-to-back or in the same wall cavity when the fire-resistance rating is maintained. In this detail, gypsum board has been installed in the wall cavity between the boxes, creating a new fire resistance barrier. Verify if this is acceptable to the building authority in your area.

THIS SIDE OF WALL IS NON-FIRE-RESISTANCE-RATED.

3 IN. (75 MM) MIN.

THIS SIDE OF WALL HAS A 1-HOUR FIRE-RESISTANCE RATING.

Figure 2-31 Building codes permit electrical outlet and device boxes to be mounted in the same stud space under certain conditions. (A) Here we see "Wall Opening Protective Material" packed around one outlet box. (B) Here we see molded fiberglass boxes in the same stud space with a minimum horizontal clearance of 3 in. (75 mm). These boxes are listed for this application. In both (A) and (B), the boxes are not permitted to be back-to-back.

Figure 2-32 This illustration shows an electrical outlet box that has been covered with fire-resistance "putty" to retain the wall's fire-resistance rating per building codes.

BOXES FOR CONDUIT WIRING

Some local electrical ordinances require conduit rather than cable wiring. Conduit wiring is discussed in Unit 18. Examples of conduit fill (how many wires are permitted in a given size conduit) are presented.

When conduit is installed in a residence, it is quite common to use 4-in. square boxes trimmed with suitable plaster covers, Figure 2-33(A). There are sufficient knockouts in the top, bottom, sides, and back of the box to permit a number of conduits to run to the box. Plenty of room is available for the conductors and wiring devices. Note how easily these 4-in. square outlet boxes can be mounted back-to-back by installing a small fitting between the boxes. This is illustrated in Figure 2-33(B) and Figure 2-33(C).

Four-inch square outlet boxes can be trimmed with one-gang or two-gang plaster rings where wiring devices will be installed. Where luminaires (lighting fixtures) will be installed, a plaster ring having a round opening should be installed.

▶Any and all unused openings in electrical equipment must be closed! This includes unused knockouts in boxes, spaces for circuit breakers in a panel, meter socket enclosures, and all other similar electrical equipment.◀ The closing material shall provide substantially the same protection as the wall of the equipment. See *110.12(A)*, *408.7*, and Figure 2-34.

Close the Gap Around Boxes!

NEC® *314.21* requires that openings around electrical boxes shall be repaired so that "there will be no gaps or open space greater than ⅛ in. (3 mm) at the edge of the box or fitting." This is to minimize the spread of fire. The UL *Fire Resistance Directory* mirrors the *NEC*® rule by stating that "the outlet or switch boxes shall be securely fastened to the studs and the opening in the wallboard facing shall be cut so that the clearance between the box and the wallboard does not exceed ⅛ in." See Figure 2-21(B).

This is easier said than done. The electrician installs the wiring (rough-in), followed by an inspection by the electrical inspector. Next comes the drywall/plaster/panel installer, who many times cuts out the box openings much larger than the outlet or switch box. In most cases, the walls and ceilings are painted or wallpapered before the electrician returns to install the receptacles, switches, and luminaires (lighting fixtures). For gaps or openings greater than ⅛ in. (3 mm) around the electrical boxes, should the electrician repair the gap with patching plaster, possibly damaging or marring the finished wall? Whose responsibility is it? Is it the electrician's? Is it the drywall/plaster/panel installer's?

The electrician should check with the drywall/plaster/panel installer to clarify who is to be responsible for seeing to it that gaps or openings around electrical outlet and switch boxes do not exceed ⅛ in. (3 mm). *314.21* puts the responsibility on the electrician.

YOKE

We use the term "yoke." The *NEC*® uses the term "yoke." Yet, the *NEC*® does not define a "yoke." What is a "yoke"? It is the metal or polycarbonate mounting strap on which a wiring device or devices are attached. The yoke in turn is attached to a device box or device cover with No. 6-32 screws. One yoke

Figure 2-33   (A) shows electrical metallic tubing run to a 4-in. square box. (B) and (C) show 4-in. square boxes attached together for "back-to-back" installations. When mounting boxes back-to-back, be sure to consider possible transfer of sound between the rooms. IMPORTANT: See text ("Spread of Fire") for details of fire-resistance ratings and the restrictions related to installing electrical boxes back-to-back.

KNOCKOUT
CLOSURES

Figure 2-34 Unused openings in boxes must be closed according to *110.12(A)*. This is done to contain electrical short-circuit problems inside the box or panel and to keep rodents out. Note that the conductors *do not* meet the 3-in. (75-mm) and 6-in. (150-mm) conductor length requirements of *300.14*. This is a good example of a *Code* violation. See Figure 2-2 for proper installation regarding conductor length at boxes.

may have one, two, or three wiring devices attached to it.

Visit an electrical distributor or home center. You will find many combination wiring devices on a single yoke or strap. Selecting one of these combination devices may eliminate the need to use combinations of interchangeable wiring devices.

SPECIAL-PURPOSE OUTLETS

Special-purpose outlets are usually indicated on the plans. These outlets are described by a notation and are usually detailed in the specifications. The plans indicate special-purpose outlets by a triangle inside a circle with subscript letters. In some cases, a subscript number is added to the letter.

When a special-purpose outlet is indicated on the plans or in the specifications, the electrician must check for special requirements. Such a requirement may be a separate circuit, a special 240-volt circuit, a special grounding or polarized receptacle, or other preparation.

A set of plans and specifications for this residence is found at the end of this text. The specifications include a *Schedule of Special Purpose Outlets*.

NUMBER OF CONDUCTORS IN BOX

NEC® 314.16 dictates that outlet boxes, switch boxes, and device boxes should be large enough to provide ample room for the wires in that box, without having to jam or crowd the wires into the box. Jamming the wires into the box can not only damage the insulation on the wires, but can result in heat buildup in the box that can further damage the insulation on the wires. The *Code* specifies the maximum number of conductors allowed in standard metal outlet, device, and junction boxes; see *Tables 314.16(A)* and *(B)*. Nonmetallic boxes are marked by the manufacturer with their cubic-inch capacity.

When conductors are the same size, the proper metal box size can be selected by referring to *Table 314.16(A)*. When conductors are of different sizes and nonmetallic boxes are installed, refer to *Table 314.16(B)* and use the cubic-inch volume for the particular size of wire being used.

Table 314.16(A) and *Table 314.16(B)* do not take into consideration the space taken by luminaire (fixture) studs, cable clamps, hickeys, switches, pilot lights, or receptacles that may be in the box. These require additional space. Table 2-1 shows the additional allowances that must be considered for different situations.

Table 314.16(A) Metal Boxes

Box Trade Size			Minimum Volume		Maximum Number of Conductors*						
mm	in.		cm³	in.³	18	16	14	12	10	8	6
100 × 32	(4 × 1¼)	round/octagonal	205	12.5	8	7	6	5	5	5	2
100 × 38	(4 × 1½)	round/octagonal	254	15.5	10	8	7	6	6	5	3
100 × 54	(4 × 2⅛)	round/octagonal	353	21.5	14	12	10	9	8	7	4
100 × 32	(4 × 1¼)	square	295	18.0	12	10	9	8	7	6	3
100 × 38	(4 × 1½)	square	344	21.0	14	12	10	9	8	7	4
100 × 54	(4 × 2⅛)	square	497	30.3	20	17	15	13	12	10	6
120 × 32	(4¹¹⁄₁₆ × 1¼)	square	418	25.5	17	14	12	11	10	8	5
120 × 38	(4¹¹⁄₁₆ × 1½)	square	484	29.5	19	16	14	13	11	9	5
120 × 54	(4¹¹⁄₁₆ × 2⅛)	square	689	42.0	28	24	21	18	16	14	8
75 × 50 × 38	(3 × 2 × 1½)	device	123	7.5	5	4	3	3	3	2	1
75 × 50 × 50	(3 × 2 × 2)	device	164	10.0	6	5	5	4	4	3	2
75 × 50 × 57	(3 × 2 × 2¼)	device	172	10.5	7	6	5	4	4	3	2
75 × 50 × 65	(3 × 2 × 2½)	device	205	12.5	8	7	6	5	5	4	2
75 × 50 × 70	(3 × 2 × 2¾)	device	230	14.0	9	8	7	6	5	4	2
75 × 50 × 90	(3 × 2 × 3½)	device	295	18.0	12	10	9	8	7	6	3
100 × 54 × 38	(4 × 2⅛ × 1½)	device	169	10.3	6	5	5	4	4	3	2
100 × 54 × 48	(4 × 2⅛ × 1⅞)	device	213	13.0	8	7	6	5	5	4	2
100 × 54 × 54	(4 × 2⅛ × 2⅛)	device	238	14.5	9	8	7	6	5	4	2
95 × 50 × 65	(3¾ × 2 × 2½)	masonry box/gang	230	14.0	9	8	7	6	5	4	2
95 × 50 × 90	(3¾ × 2 × 3½)	masonry box/gang	344	21.0	14	12	10	9	8	7	4
min. 44.5 depth	FS — single cover/gang (1¾)		221	13.5	9	7	6	6	5	4	2
min. 60.3 depth	FD — single cover/gang (2⅜)		295	18.0	12	10	9	8	7	6	3
min. 44.5 depth	FS — multiple cover/gang (1¾)		295	18.0	12	10	9	8	7	6	3
min. 60.3 depth	FD — multiple cover/gang (2⅜)		395	24.0	16	13	12	10	9	8	4

*Where no volume allowances are required by 314.16(B)(2) through (B)(5).

Table 314.16(B) Volume Allowance Required per Conductor

Size of Conductor (AWG)	Free Space Within Box for Each Conductor	
	cm³	in.³
18	24.6	1.50
16	28.7	1.75
14	32.8	2.00
12	36.9	2.25
10	41.0	2.50
8	49.2	3.00
6	81.9	5.00

Reprinted with permission from NFPA 70-2005.

SELECTING THE CORRECT SIZE BOX

Selecting the correct size box depends on the mix of conductors and wiring devices that will be contained in the particular box. Here are some examples.

When using a nonmetallic box where all conductors are the same size. A nonmetallic box is marked as having a volume of 22.8 in.³ The box contains no luminaire (fixture) stud or cable clamps. How many 14 AWG conductors are permitted in this box?

From *Table 314.16(B)*, the volume requirement for a 14 AWG conductor is 2.00 in.³ Therefore, the

• If box contains no fittings, devices, fixture studs, cable clamps, hickeys, switches, receptacles, or equipment grounding conductors:	• Refer directly to *Table 314.16(A)* or *Table 314.16(B)*.
• **Clamps.** If box contains one or more internal cable clamps:	• Add a single-volume based on the largest conductor in the box.
• **Support Fittings.** If box contains one or more luminaire (fixture) studs or hickeys:	• Add a single-volume for each type based on the largest conductor in the box.
• **Device or Equipment.** If box contains one or more wiring devices on a yoke:	• Add a double-volume for each yoke based on the largest conductor connected to a device on that yoke.
• **Equipment Grounding Conductors.** If a box contains one or more equipment grounding conductors:	• Add a single-volume based on the largest equipment grounding conductor in the box.
• **Isolated Equipment Grounding Conductor.** If a box contains one or more additional "isolated" (insulated) equipment grounding conductors as permitted by *250.146(D)* for "noise" reduction:	• Add a single-volume based on the largest equipment grounding conductor in the box.
• ▶For conductors less than 12 in. (300 mm) long between the raceway entries for the conductor that is looped through the box without being spliced: ◀	• ▶Add a single-volume for each conductor that is looped through the box. ◀
• ▶For conductors 12 in. (300 mm) or longer between the raceway entries for the conductor that is looped through the box without being spliced: ◀	• ▶Add a double-volume for each conductor that is looped through the box. ◀
• For conductors that originated outside of the box and terminate inside the box:	• Add a single-volume for each conductor that originates outside the box and terminates inside the box.
• If no part of the conductor leaves the box, as with a "jumper" wire used to connect three wiring devices on one yoke, or pigtails:	• Don't count this (these). No additional volume required.
• For small equipment grounding conductors or not more than four conductors smaller than 14 AWG that originate from a luminaire (fixture) canopy or similar canopy (like a fan) and terminate in the box:	• Don't count this (these). No additional volume required.
• For small fittings, such as locknuts, bushings, and wire connectors:	• Don't count this (these). No additional volume required.

Table 2-1 Quick checklist for possibilities to be considered when determining proper size boxes. See *314.16(B)(1)* through *314.16(B)(5)*.

maximum number of 14 AWG conductors permitted in this box is:

$$\frac{22.8}{2} = 11.4 \text{ (Round down to 11 conductors)}$$

When using a metal box where all conductors are the same size.   A box contains one luminaire (fixture) stud and two internal cable clamps. Four 12-AWG conductors enter the box.

four 12 AWG conductors	4
one luminaire (fixture) stud	1
two cable clamps (only count one)	1
Total	6

Referring to *Table 314.16(A)* and *Table 314.16(B)*, a 4 in. × 1½ in. octagon box is suitable for this example.

When using a metal or nonmetallic box where the conductors are different sizes.   What is the minimum cubic-inch volume required for a box that will contain one internal cable clamp, one switch, and one receptacle, all mounted on one yoke? Two 14 AWG and two 12 AWG conductors enter the box. The wiring method is armored cable.

two 14 AWG conductors @ 2 in.³ per wire	= 4.00 in.³
two 12 AWG conductors @ 2.25 in.³ per wire	= 4.50 in.³
one cable clamp @ 2.25 in.³	= 2.25 in.³
one switch and one receptacle on one yoke @ 2.25 in.³ × 2	= 4.50 in.³
Total	15.25 in.³

Select a box having a minimum volume of 15.25 in.³ of space. The cubic-inch volume may be marked on the box; otherwise refer to the second column of *Table 314.16(A)* entitled "Minimum Volume Capacity."

The *Code* in *314.16(A)(2)* requires that all boxes *other* than those listed in *Table 314.16(A)* be durably and legibly marked by the manufacturer with their cubic-inch capacity. Nonmetallic boxes have their cubic-inch volume marked on the box. When sectional boxes are ganged together, the calculated volume is the total cubic-inch volume of the assembled boxes.

When installing a box that will have a raised cover attached to it.   When marked with their cubic-inch volume, the *additional* space provided by plaster rings, domed covers, raised covers, and extension rings is permitted to be used when determining the overall volume. This is illustrated in Figure 2-35.

How many 12 AWG conductors are permitted in this box and raised plaster ring? Refer to Figure 2-36 and *Table 314.16(A)* and *Table 314.16(B)*.

$$\frac{25.5 \text{ in.}^3 \text{ of total space}}{2.25 \text{ in.}^3 \text{ per 12 AWG conductor}} = \text{Eleven 12 AWG conductors}$$

ADDITIONAL WIRING SPACE PROVIDED BY THE
RAISED COVER IS 2 IN. x 3 IN. x 3/4 IN. = 4 1/2 IN.^{3}

Figure 2-35 Raised cover. Raised covers are sometimes called plaster rings.

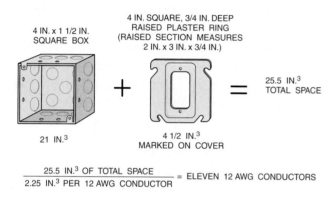

4 IN. x 1 1/2 IN.
SQUARE BOX

4 IN. SQUARE, 3/4 IN. DEEP
RAISED PLASTER RING
(RAISED SECTION MEASURES
2 IN. x 3 IN. x 3/4 IN.)

25.5 IN.^{3}
TOTAL SPACE

21 IN.^{3}

4 1/2 IN.^{3}
MARKED ON COVER

$$\frac{25.5 \text{ IN.}^3 \text{ OF TOTAL SPACE}}{2.25 \text{ IN.}^3 \text{ PER 12 AWG CONDUCTOR}} = \text{ELEVEN 12 AWG CONDUCTORS}$$

Figure 2-36 When marked with their cubic-inch volume, the volume of the cover may be added to the volume of the box to determine the total cubic-inch volume.

This calculation actually resulted in 11.33. The conductor fill numbers in *Table 314.16(A)* were created by dropping any excess above the whole number after the calculations were made. Following the same procedure, this conductor fill calculation dropped off the excess.

Electricians and electrical inspectors have become very aware of the fact that GFCI receptacles, dimmers, and timers take up a lot more space than regular wiring devices. Therefore, it is a good practice to install boxes that will provide plenty of room for the wires, instead of pushing, jamming, and crowding the wires into the box.

Figure 2-37 shows that transformer leads 18 AWG or larger must be counted when selecting a proper size box.

Figure 2-37 Transformer leads 18 AWG or larger must be counted when selecting the proper size box, *314.16(B)(1)*. In the example shown, the box contains four conductors.

The minimum required size of equipment grounding conductors is shown in *Table 250.122*. The equipment grounding conductors are the same size as the circuit conductors in cables having 14-AWG, 12-AWG, and 10-AWG circuit conductors. Thus, for normal house wiring, box sizes can be calculated using *Table 314.16(A)*. Refer to Figure 2-38.

Figure 2-20, a quik-chek box selection guide, shows some of the most popular types of boxes used in residential wiring. This guide also shows the box's cubic-inch volume.

FOR CODE PURPOSES,
THIS BOX CONTAINS
FOUR CONDUCTORS.

Figure 2-38 Four or fewer luminaire (fixture) conductors, smaller than 14 AWG, and/or small equipment grounding conductors that originate in the luminaire (fixture) and terminate in the box, need not be counted when calculating box fill, *314.16(B)(1), Exception.*

Figure 2-39 illustrates typical wiring using cable and conduit. Note in this figure how the conductor count is determined.

Using electrical metallic tubing (EMT) makes it possible to "loop" conductors through the box, which counts as one conductor only. Looping conductors through a box is not possible when using cable as the wiring method.

The basic rules for box fill are summarized in Table 2-1. Boxes that are intended to support ceiling fans are discussed in detail in Unit 9.

Means for Connecting Equipment Grounding Conductors in a Box

NEC® 314.40(D) requires that "*a means shall be provided in each metal box for the connection of **equipment grounding conductors**.*" Generally, these are tapped No. 10-32 holes in the box that are marked GR, GRN, GRND, or similar identification. The screws used for terminating the equipment grounding conductor must be hexagonal and must be green. Electrical distributors sell green, hexagonal No. 10-32 screws for this purpose.

METAL CONDUIT

A ▶A CONDUCTOR LOOPED THROUGH A BOX IS COUNTED AS:
• ONE CONDUCTOR (A SINGLE-VOLUME COUNT) IF LESS THASN 12-IN. (300-mm) LONG BETWEEN THE RACEWAYS ENTRIES.
• TWO CONDUCTOR (A DOUBLE-VOLUME COUNT) IF 12-IN. (300-mm) OR LONGER BETWEEN THE RACEWAY ENTRIES.◀

CABLE (EACH WITH TWO CIRCUIT CONDUCTORS PLUS ONE EQUIPMENT GROUNDING CONDUCTOR.)

B WHEN USING CABLE, THE CONDUCTOR COUNT IN THIS ILLUSTRATION IS FIVE. THE TWO EQUIPMENT GROUNDING CONDUCTORS ARE COUNTED AS ONE.

CABLE AND INTERNAL CLAMPS (EACH CABLE HAS TWO CIRCUIT CONDUCTORS PLUS ONE EQUIPMENT GROUNDING CONDUCTOR.)

C WHEN USING CABLE AND A BOX THAT HAS INTERNAL CABLE CLAMPS, THE CONDUCTOR COUNT IN THIS ILLUSTRATION IS SIX. THE TWO EQUIPMENT GROUNDING CONDUCTORS ARE COUNTED AS ONE. THE TWO CABLE CLAMPS ARE COUNTED AS ONE.

Figure 2-39   Example of conductor count for both metal conduit and cable installations.

HEIGHT OF RECEPTACLE OUTLETS

There are no hard-and-fast rules for locating most outlets. A number of conditions determine the proper height for a switch box. For example, the height of the kitchen counter backsplash determines where the switches and receptacle outlets are located between the kitchen countertop and the cabinets.

The residence featured in this text is electrically heated and is discussed in Unit 23. The type of electric heat could be:

- electric furnace (as in this text).

- electric resistance heating buried in ceiling plaster or "sandwiched" between two layers of drywall material.

- electric baseboard heaters.

- heat pump.

Let us consider the electric baseboard heaters. In most cases, the height of these electric baseboard units from the top of the unit to the finished floor seldom exceeds 6 in. (150 mm). The important issue here is that the manufacturer's receptacle accessories may have to be used to conform to the receptacle spacing requirements, as covered in *210.52*.

Electrical receptacle outlets are not permitted to be located above an electric baseboard heating unit. Refer to the section "Location of Electric Baseboard Heaters in Relation to Receptacle Outlets," in Unit 23.

It is common practice among electricians to consult the plans and specifications to determine the proper heights and clearances for the installation of electrical devices. The electrician then has these dimensions verified by the architect, electrical engineer, designer, or homeowner. This practice avoids unnecessary and costly changes in the locations of outlets and switches as the building progresses.

POSITIONING OF RECEPTACLES

Although no *NEC*® rules exist on positioning receptacles, there is always the possibility of a metal wall plate coming loose and falling downward onto the blades of an attachment plug cap that is loosely plugged into the receptacle. This can create sparks that could result in a fire, burns, and potential shock. Positioning the *equipment grounding* conductor slot or the *grounded* conductor slot on top will minimize these hazards.

The equipment grounding blade on a three-wire attachment plug cap is longer than the other blades. Thus, when inserted into a receptacle, in agreement with *250.124(A)*, the equipment grounding blade is first to make and last to break.

There is also the argument that with the equipment grounding slot to the bottom, if a plug cap comes loose and starts to fall out, the longer equipment grounding blade would be the last connection to be disconnected. See Figures 2-40, 2-41, 2-42, and 2-43.

Figure 2-40 **Grounding hole to the top. A loose metal faceplate could fall onto the grounding blade of the attachment plug cap, but no sparks would fly.**

Figure 2-41 **Grounding hole to the bottom. A loose metal faceplate could fall onto both the grounded neutral and "hot" blades. Sparks would fly.**

Figure 2-42 **Grounded neutral blades on top. A loose metal faceplate could fall onto these grounded neutral blades. No sparks would fly.**

NOT RECOMMENDED

Figure 2-43 "Hot" terminal on top. A loose metal faceplate could fall onto these live blades. If this were a split-circuit receptacle fed by a three-wire, 120/240-volt circuit, the short would be across the 240-volt line. Sparks would fly.

TYPICAL HEIGHTS FOR SWITCHES AND OUTLETS

There are no hard and fast *Code* rules regarding the heights for switches and receptacles; local preferences prevail. Typical heights for switches and outlets are shown in Table 2-2. These dimensions usually are satisfactory. However, the electrician must check the blueprints, specifications, and details for measurements that may affect the location of a particular outlet or switch. The cabinet spacing, available space between the countertop and the cabinet, and the tile height may influence the location of the outlet or switch. For example, if the top of the wall tile is exactly 4 ft (1.2 m) from the finished

SWITCHES	
Regular	46 in. (1.15 m). Some homeowners want wall switches mounted 30 in.–36 in. (750 mm–900 mm) so as to be easy for children to reach and easy for adults to reach when carrying something. The choice is yours!
Between counter and kitchen cabinets. Depends on backsplash.	44 in.–46 in. (1 m–1.15 m)
RECEPTACLE OUTLETS	
Regular	12 in. (300 mm)
Between counter and kitchen cabinets. Depends on backsplash.	44 in.–46 in. (1 m–1.15 m)
In garages	46 in. (1.15 m) Minimum 18 in. (450 mm)
In unfinished basements	46 in. (1.15 m)
In finished basements	12 in. (300 mm)
Outdoors (above grade or deck)	18 in. (450 mm)
WALL LUMINAIRE (FIXTURE) OUTLETS	
Outdoor entrance bracket luminaires (fixtures)	72 in. (1.8 m). If luminaire (fixture) is "upward," mount wall box lower. If luminaire is "downward," mount wall box higher. A good appearance for typical outdoor wall lanterns is when the center of wall box is approximately 10 in. (0.254 m) below the top of the door. The final choice is yours.
Inside wall brackets	5 ft (1.5 m)
Side of medicine cabinet or mirror Above medicine cabinet or mirror	You need to know the measurement of the medicine cabinet. Check the rough-in opening of medicine cabinet measurement of mirror. Mount electrical wall box approx. 6 in. (150 mm) to center above rough-in opening or mirror. Medicine cabinets that come complete with luminaires (fixtures) have a wiring compartment with a conduit knockout(s) in which the supply cable or conduit is secured using the appropriate fitting. Many strip luminaires (fixtures) have a backplate with a conduit knockout in which the supply cable or conduit is secured using the appropriate fitting. Where to bring in the cable or conduit takes careful planning.

Note: All dimensions are from finished floor to center of the electrical box. If possible, try to mount wall boxes for luminaires (lighting fixtures) based upon the type of luminaire (fixture) to be installed. Verify all dimensions before "roughing in." If wiring for physically handicapped, the above heights may need to be lowered in the case of switches, and raised in the case of receptacles.

Table 2-2 Typical heights for switches, receptacles, and wall luminaire (lighting fixture) outlets. Local height preferences prevail.

WHERE A WALL IS PARTIALLY TILED, A SWITCH OR CONVENIENCE OUTLET MUST BE LOCATED ENTIRELY OUT OF THE TILE AREA (A) OR ENTIRELY WITHIN THE TILE AREA (B), (C).

THE FACEPLATE IN (D) DOES NOT "HUG" THE WALL PROPERLY. THIS INSTALLATION IS CONSIDERED UNACCEPTABLE BY MOST ELECTRICIANS, AND IS A VIOLATION OF THE LAST SENTENCE OF *NEC® 404.9(A)*.

Figure 2-44   Locating an outlet on a tiled wall.

floor line, a wall switch should not be mounted 4 ft (1.2 m) to center. This is considered poor workmanship. The switch should be located entirely within the tile area or entirely out of the tile area, as in Figure 2-44. This situation requires the full cooperation of all craftspersons involved in the construction job.

Faceplates

Faceplates for switches and receptacles shall be installed so as to completely cover the wall opening and seat against the wall surface, *404.9(A)* and *406.5*.

Faceplates should be level! No matter how good the wiring inside the wall is, no matter how good the connections and splices are, no matter how proper the equipment grounding is, the only thing the homeowner will see is the wiring devices and the faceplates. That establishes the impression the homeowner has of the overall wiring installation. Because of the elongated slots in the yoke of a wiring device, it is rather easy to level a wiring device and faceplate on a single-gang box even if the box is not level. This is not true for multi-gang boxes. It is extremely important to make sure multi-gang boxes are level. If a multi-gang box is not level, it will be virtually impossible to make the wiring devices and faceplate level.

Another nice detail that can improve the appearance is to align the slots in all of the faceplate mounting screws in the same direction, as in Figure 2-45.

ALIGN SCREW—SLOTS LIKE THIS,

NOT LIKE THIS.

Figure 2-45   Aligning the slots of the faceplate. Mounting screws in the same direction makes the installation look neater. Make sure the faceplate is level.

REVIEW

Note: For assistance in finding the answers to these review questions, refer to the *NEC,*® the plans in the back of this text, and the specifications found in the Appendix.

PART 1—ELECTRICAL FEATURES

1. What does a plan show about electrical outlets? _____

_____

_____

_____

2. What is an outlet? _____

_____

_____

3. Match the following switch types with the proper symbol.
   a. single-pole   **S$_p$**
   b. three-way   **S$_4$**
   c. four-way   **S**
   d. single-pole with pilot light   **S$_3$**

4. The plans show dashed lines running between switches and various outlets. What do these dashed lines indicate? _____

_____

5. Why are dashed lines usually curved? _____

6. a. What are junction boxes used for? _____

   b. Are junction boxes normally used in wiring the first floor? Explain. _____

_____

_____

   c. Are junction boxes normally used to wire exposed portions of the basement? Explain. _____

_____

7. How are standard sectional switch (device) boxes mounted? _____

_____

_____

_____

8. a. What is an offset bar hanger? _____

_____

_____

   b. What types of boxes may be used with offset bar hangers? _____

_____

_____

9. What methods may be used to mount luminaires (lighting fixtures) to an outlet box fastened to an offset bar hanger? _____

_____

_____

10. What advantage does a 4-in. octagon box have over a 3¼-in. octagon box?

_____

_____

11. What is the size of the opening of a switch (device) box for a single device? _____

_____

_____

12. The space between a door casing and a window casing is 3½ in. (88.9 mm). Two switches are to be installed at this location. What type of switches will be used?

_____

_____

13. Three switches are mounted in a three-gang switch (device) box. The wall plate for this assembly is called a _____ plate.

14. The mounting holes in a device (switch) box are tapped for No. 6-32 screws. The mounting holes in an outlet box are tapped for No. 8-32 screws. The mounting holes in metal boxes for attaching equipment grounding conductors are tapped for _____ screws.

15. a. How high above the finished floor in the living room are switches located?

_____

    b. How high above the garage floor are switches located?

_____

16. a. How high above the finished floor in the living room are receptacles located?

_____

    b. How high above the garage floor are receptacles located?

_____

17. Outdoor receptacle outlets in this dwelling are located _____ _____ in. above grade.

18. In the spaces provided, draw the correct symbol for each of the descriptions listed in (a) through (r).

a. _____ Lighting panel

b. _____ Clock outlet

c. _____ Duplex outlet

d. _____ Outside telephone

e. _____ Single-pole switch

f. _____ Motor

g. _____ Duplex outlet, split-circuit

h. _____ Lampholder with pull switch

i. _____ Weatherproof outlet

j. _____ Special-purpose outlet

k. _____ Fan outlet

l. _____ Range outlet

m. _____ Power panel

n. _____ Three-way switch

o. _____ Push button

p. _____ Thermostat

q. _____ Electric door opener

r. _____ Multi-outlet assembly

19. The front edge of a box installed in a combustible wall must be _____ with the finished surface.

20. List the maximum number of 12 AWG conductors permitted in a

    a. 4 in. × 1½ in. octagon box. _____

    b. 4¹¹⁄₁₆ in. × 1½ in. square box. _____

    c. 3 in. × 2 in. × 3½ in. device box. _____

21. When a switch (device) box is nailed to a stud, and the nail runs through the box, the nails must not interfere with the wiring space. To accomplish this, keep the nail

    _____ (Select a, b, or c.)

    a. halfway between the front and rear of the box.

    b. a maximum of ¼ in. (6 mm) from the front edge of the box.

    c. a maximum of ¼ in. (6 mm) from the rear of the box.

22. Hanging a ceiling luminaire (fixture) directly from a plastic outlet box is permitted only if _____

    _____

    _____

    _____

23. It is necessary to count luminaire (fixture) wires when counting the permitted number of conductors in a box according to *314.16*.

    (True)   (False)   Underline or circle the correct answer.

24. *Table 314.16(A)* allows a maximum of ten wires in a certain box. However, the box will have two cable clamps and one fixture stud in it. What is the maximum number of wires allowed in this box? _____

    _____

25. When laying out a job, the electrician will usually make a layout of the circuit, taking into consideration the best way to run the cables and/or conduits and how to make up the electrical connections. Doing this ahead of time, the electrician determines exactly how many conductors will be fed into each box. With experience, the electrician will probably select two or three sizes and types of boxes that will provide adequate space to "meet Code." *Table 314.16(A)* of the *Code* shows the maximum number of conductors permitted in a given size box. In addition to counting the number of conductors that will be in the box, what is the additional volume that must be provided for the following items? Enter *single* or *double* volume allowance in the blank provided.

    a. one or more internal cable clamps: _____ volume allowance.

    b. for a fixture stud: _____ volume allowance.

    c. for one or more wiring devices on one yoke: _____ volume allowance.

    d. for one or more equipment grounding conductors: _____ volume allowance.

26. Is it permissible to install a receptacle outlet above an electric baseboard heater?

    _____

    _____

27. What is the maximum weight of a luminaire (lighting fixture) permitted to be hung directly from an outlet box in a ceiling? _____ lb.

28. Two 12/2 and two 14/2 nonmetallic-sheathed AWG cables enter a box. Each cable has an equipment grounding conductor. The 12 AWG conductors are connected to a receptacle. Two of the 14 AWG conductors are connected to a toggle switch. The other two 14 AWG conductors are spliced together since they serve as a switch loop. The box contains two cable clamps. Calculate the minimum cubic-in. volume required for the box.

29. Using the same number and size of conductors as in question 28, but using electrical metallic tubing, calculate the minimum cubic-in. volume required for the box. There will be no separate equipment grounding conductors, nor will there be any clamps in the box.

30. To allow for adequate conductor length at electrical outlet and device boxes to make up connections, *300.14* requires that not less than [3 in. (75 mm), 6 in. (150 mm), 9 in. (225 mm)] of conductor length be provided. This length is measured from where the conductor emerges from the cable or raceway to the end of the conductor. For box openings having any dimension less than 8 in. (200 mm), the minimum length of conductor measured from the box opening in the wall to the end of the conductor is [3 in. (75 mm), 6 in. (150 mm), 9 in. (225 mm)]. Circle the correct answers.

31. When wiring a residence, what must be considered when installing wall boxes on both sides of a common partition that separates the garage and a habitable room? _____

_____

_____

32. Does the *NEC*® allow metal raceways to be used with nonmetallic boxes?

Yes _____ No _____ , *NEC*® _____

PART 2—STRUCTURAL FEATURES

1. To what scale is the basement plan drawn? _____

2. What is the size of the footing for the steel Lally columns in the basement? _____
   _____

3. To what kind of material will the front porch lighting bracket luminaire (fixture) be attached? _____

4. Give the size, spacing, and direction of the ceiling joists in the workshop. _____
   _____

5. What is the size of the lot on which this residence is located? _____
   _____

6. The front of the house is facing which compass direction? _____

7. How far is the front garage wall from the curb? _____

8. How far is the side garage wall from the property lot line? _____

9. How many steel Lally columns are in the basement, and what size are they? _____
   _____

10. What is the purpose of the I-beams that rest on top of the steel Lally columns? _____
    _____

11. To make sure that switch boxes and outlet boxes are set properly in the garage walls and ceilings, we need to know the thickness of the gypsum wallboard used in these locations. In the garage, what is the thickness and type of the gypsum wallboard?

    a. on the "warm walls" of the garage? _____

    b. on the "cold walls" of the garage? _____

    c. on the ceiling of the garage? _____

12. Where is access to the attic provided? _____

13. Give the thickness of the outer basement walls. _____

14. What material is indicated for the foundation walls? _____

15. Where are the smoke detectors located in the basement? _____
    _____

16. What is the ceiling height in the basement workshop from bottom of the joists to the floor?
    _____

17. Give the size and type of the front door. _____
    _____

18. What is the stud size for the partitions between the bathrooms in the bedroom area where substantial plumbing is to be installed? _____

19. Who is to furnish the range hood? _____
    _____

20. Who is to install the range hood? _____

Determining the Required Number and Location of Lighting and Small-Appliance Circuits

OBJECTIVES

After studying this unit, the student should be able to

▶ understand the *NEC®* requirements for calculating branch-circuit sizing and loading.

▶ understand the term *volt-amperes per square foot*.

▶ calculate the occupied floor area of a residence.

▶ determine the minimum number of lighting and small-appliance branch-circuits.

▶ know where receptacle outlets and lighting outlets are required.

It is standard practice in the design and planning of dwelling units to permit the electrician to plan the circuits. Thus, the residence plans do not include layouts for the various branch-circuits. The electrician may follow the general guidelines established by the architect. However, any wiring systems designed and installed by the electrician must conform to the *NEC®*, as well as local and state code requirements.

This unit focuses on lighting branch-circuits and small-appliance branch-circuits. The circuits supplying the electric range, oven, clothes dryer, and other specific circuitry not considered to be a lighting branch-circuit or a small-appliance branch-circuit are covered later in this text. Refer to the index for the specific circuit being examined.

BASICS OF WIRE SIZING AND LOADING

The *NEC®* establishes some very important fundamentals that weave their way through the decision-making process for an electrical installation. They are presented here in brief form, and are covered in detail as required throughout this text.

The *NEC®* defines a **branch-circuit** as *"the circuit conductors between the final overcurrent device protecting the circuit and the outlet(s)."* See Figure 3-1. In the residence discussed in this text, the wiring to wall outlets, the dryer, the range, and so on, are all examples of a branch-circuit.

The *NEC®* defines a **feeder** as *"All circuit conductors between the service equipment, the source of a separately derived system, or other power supply source and the final branch-circuit overcurrent device."** In the residence discussed in this text, the wiring between Main Panel A and Subpanel B is a feeder.

The ampacity (current-carrying capacity) of a conductor must not be less than the rating of the

*Reprinted with permission from NFPA 70-2005.

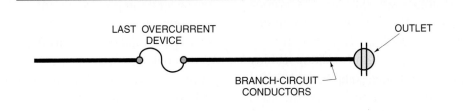

Figure 3-1 The branch-circuit is that part of the wiring that runs from the final overcurrent device to the outlet. The rating of the overcurrent device, not the conductor size, determines the rating of the branch-circuit.

overcurrent device protecting that conductor, *210.19* and *210.20*. A common exception to this is a motor branch-circuit, where it is quite common to have overcurrent devices (fuses or breakers) sized larger than the ampacity of the conductor. Motors and motor circuits are covered specifically in *Article 430*. The **ampere** rating of the branch-circuit overcurrent protective device (fuse or circuit breaker) determines the rating of the branch-circuit. For example, if a 20-ampere conductor is protected by a 15-ampere fuse, the circuit is considered to be a 15-ampere branch-circuit, *210.3*.

Standard branch-circuits that serve more than one receptacle outlet or more than one lighting outlet are rated 15, 20, 30, 40, and 50 amperes. A branch-circuit that supplies an individual load can be of any ampere rating, *210.3*.

Where the ampacity of the conductor does not match up with a standard rating of a fuse or breaker, the next higher standard size overcurrent device may be used, provided the overcurrent device does not exceed 800 amperes, *240.4(B)*. This deviation is not permitted, however, if the circuit supplies receptacles where "plug-connected" appliances, and so on, could be used, because too many "plug-in" loads could result in an overload condition, *240.4(B)(1)*. You may go to the next standard size overcurrent device only when the circuit supplies other than receptacles for cord- and plug-connected portable loads.

For instance, when a conductor having an allowable ampacity of 25 amperes (see *Table 310.16*) 12 AWG Type TW is derated to 70%:

$$25 \times 0.70 = 17.5 \text{ amperes}$$

It is permitted to use a 20-ampere overcurrent device if the circuit supplies *only* fixed lighting outlets or other fixed loads.

If the previous example were to supply receptacle outlets, then the rating of the overcurrent device would have to be dropped to 15 amperes, otherwise it is possible to overload the conductors by plugging in more load than the conductors can safely carry.

The allowable ampacity of conductors commonly used in residential occupancies is found in *Table 310.16*. The ampacities in *Table 310.16* are subject to **correction factors** that must be applied if high ambient temperatures are encountered—for example, in attics. See the bottom of *Table 310.16*.

Conductor ampacities are also subject to a **derating factor** if more than three current-carrying conductors are installed in a single raceway or cable. See *Table 310.15(B)(2)*.

▶Most general-use receptacle outlets in a residence are included in the general lighting load calculations, *220.14(J)*.◀

Receptacle outlets connected to the 20-ampere small-appliance branch-circuits in the kitchen, dining room, laundry, and workshop are not considered part of general lighting load. Additional load values must be added into the calculations for these receptacle outlets. This is discussed later in this text.

The minimum lighting load for dwellings is 3 **volt-amperes** per square foot. See *220.12(A)* and *Table 220.12*.

Continuous Loads

The *NEC®* defines **continuous load** as "*a load where the maximum current is expected to continue for three hours or more.*" *Continuous* loads shall not exceed 80% of the rating of the branch-circuit. General lighting outlets and receptacle outlets in residences are not considered to be continuous loads.

Certain loads in homes are considered to be continuous and must be treated accordingly. Examples are electric water heaters ▶(*422.13*)◀, central electric heating ▶[*424.3(B)*]◀, snow-melting cables ▶(*426.4*)◀, and air-conditioning equipment (*440.32*). For these loads, the branch-circuit rating, the conductors, and the overcurrent device shall not be less than 125% of the rating of the equipment. Mathematically, sizing the conductors and overcurrent device at 125% of the load is the same as loading the conductors and overcurrent device to 80%.

For example, an electric furnace with a nameplate rating of 40 amperes would require the supply conductors and overcurrent protection to be not less than:

$$40 \times 1.25 = 50 \text{ amperes}$$

VOLTAGE

▶All calculations throughout this text use voltage values of 120 volts and 240 volts. This complies with the requirements of *220.5*, which lists the various voltage values to use for different electrical systems.◀ This is repeated in the second paragraph of *Annex D* of the *NEC*.®

The word *nominal* means "in name only, not a fact." For example, in our calculations, we use 120 volts even though the actual voltage might be 110, 115, or 117 volts. This provides us with a uniformity that, without it, would lead to misleading and confusing results in our calculations.

Most public service commissions require that the voltage for residential services be held to ±5 percent of the nominal voltage as measured at the service point.

CALCULATING LOADS

When wiring a house, it is all but impossible to know which appliances, lighting, heating, and other loads will be turned on at the same time. Different families lead different lifestyles. There is tremendous diversity. There is a big difference between "connected load" and "actual load." Who knows what will be plugged into a wall receptacle, now or in the future? It's a guess at best. Over the years, the *NEC*® has developed procedures for calculating loads in typical one- and two-family homes.

The rules for doing the calculations are found in *Article 220*. For lighting and receptacles, the computations are based on volt-amperes per square foot.

For the small-appliance circuits in kitchens and dining rooms, the basis is 1500 volt-amperes per circuit. For large appliances such as dryers, electric ranges, ovens, cooktops, water heaters, air conditioners, heat pumps, and so on, which are not all used continuously or at the same time, there are demand factors to be used in the calculations. Following the requirements in the *NEC*,® the various calculations roll together in steps that result in the proper sizing of branch-circuits, feeders, and service equipment.

As you work your way through this text, examples are provided for just about every kind and type of load found in a typical residence.

Inch-pounds vs. Metrics When Calculating Loads

As discussed in Unit 1, converting inch-pound measurements to metric measurements and vice versa results in odd fractional results. Adding further to this problem are the values rounded off when the *NEC*® Code Making Panels did the metric conversions. When square feet is converted to square meters and the unit loads are calculated for each, the end results are different; close, but nevertheless different. To show both calculations would be confusing as well as space consuming. Many of the measurements in this text are shown in both inch-pound units and metric units. Load calculations throughout this text will use inch-pound values only, which is in agreement with the *Examples* given in *Annex D* of the 2005 *NEC*.®

CALCULATING FLOOR AREA

The general lighting load for a dwelling is based on the square footage of the dwelling. Here is how this is done.

First Floor Area

To estimate the total load for a dwelling, the occupied floor area of the dwelling must be calculated. Note in the residence plans that the first floor area has an irregular shape. In this case, the simplest method of calculating the occupied floor area is to determine the total floor area using the outside dimensions of the dwelling. Then, the areas of the following spaces are subtracted from the total area: open porches, garages, or other unfinished or unused spaces if they are not adaptable for future use, *220.12*.

Many open porches, terraces, patios, and similar areas are commonly used as recreation and entertainment areas. Therefore, adequate lighting and receptacle outlets should be provided.

For practicality, the author has chosen to round up dimensions for the determining of total square footage and to round down dimensions for those areas (garage, porch, and portions of the inset at the front of the house) not to be included in the computation of the general lighting load. This produces a slightly larger result as opposed to being on the conservative side.

Figure 3-2 shows the procedure for calculating the total square footage of this residence.

Basement Area

The basement area of a home is not generally

included in the calculating of general lighting loads because it could be classified as an *"unfinished space not adaptable for future use,"* 220.12. Yet in this residence, more than half of the total basement area is finished off as a recreation room, which certainly is considered living area. The workshop area also is intended to be used.

To simplify the calculation for this residence, we will consider the entire basement as usable space. The basement square footage area is the same as the area of the first floor.

The combined occupied area of the dwelling is found by adding the first floor and basement areas together:

First Floor	1616 ft² (149 m²)
Basement	1616 ft² (149 m²)
Total	3232 ft² (298 m²)

APPROX. 56 FEET (17 m)

56 ft × 36 ft = 2016 ft²
(17 m × 11 m = 185 m²)

FIRST FLOOR AREA =	2016 ft² (185 m²)
DEDUCT AREAS	− 195 ft² (18 m²)
NOT TO BE	
INCLUDED PER *220.12*	− 121 ft² (11 m²)
	− 84 ft² (8 m²)
TOTAL AREA	1616 ft² (148 m²)

APPROX. 36 FEET (11 m)

15 ft × 13 ft = 195 ft²
(4.6 m × 3.96 m = 18 m²)

11 ft × 11 ft
= 121 ft²
(3.4 m × 3.4 m = 11 m²)

12 ft × 7 ft = 84 ft²
(3.7 m × 2.1 m = 8.0 m²)

Figure 3-2 Determining the first floor square footage area. The basement area is the same as the first floor. See text for explanation of how the dimensions of the residence were rounded off to make the calculations much simpler and much more practical for the electrician and electrical inspector to calculate.

DETERMINING THE MINIMUM NUMBER OF LIGHTING BRANCH-CIRCUITS

Table 220.12 of the *NEC®* tells us that the minimum load requirement for dwelling units is 3 volt-amperes (VA) per square ft (0.093 m²) of occupied area.

To determine the minimum number of 15-ampere lighting branch-circuits required for a residence, here are the simple steps.

Step 1: $\dfrac{3 \text{ volt-amperes} \times \text{square ft}}{120 \text{ volts}} = \text{amperes}$

Step 2: $\dfrac{\text{amperes}}{15} = \dfrac{\text{minimum number of 15-ampere}}{\text{lighting branch-circuits}}$

This equates to a minimum of:

- one 15-ampere lighting branch-circuit for every 600 ft² (55.8 m²).

- one 20-ampere lighting branch-circuit for every 800 ft² (74.4 m²).

Note: 20-ampere lighting branch-circuits are rarely used in residential installations.

In this residence, let's calculate the load for the total occupied area of 3232 ft² (298 m²):

$$3232 \times 3 = 9696 \text{ volt-amperes}$$

$$\text{Amperes} = \frac{\text{volt-amperes}}{\text{volts}} = \frac{9696}{120} = 80.8 \text{ amperes}$$

The minimum number of 15-ampere lighting branch-circuits is:

$$\frac{80.0 \text{ amperes}}{15 \text{ amperes}} = \frac{5.39 \text{ branch-circuits}}{(\text{round up to } 6)}$$

Since we cannot have a fraction of a branch-circuit, we rounded up to a minimum of six 15-ampere lighting branch-circuits.

We get the same answer using 600 ft² (55.8 m²) for each 15-ampere lighting branch-circuit as follows:

$$\frac{3232 \text{ ft}^2}{600} = \frac{5.39 \text{ branch-circuits}}{(\text{round up to } 6)}$$

No matter which method we use, a minimum of six 15-ampere lighting branch-circuits is required.

In *Annex D* of the *NEC®* we find many examples of load calculations. Here, we also find that ▶"*Except where the calculations result in a major fraction of an ampere (0.5 or larger), such fractions*

are permitted to be dropped."◀ **CAUTION: This permission is for load calculations such as in Step 1. This permission to drop fractions is *not* to be used in Step 2 because it is impossible to have a fraction of a branch-circuit. That is why we rounded up from 5.39 to arrive at a minimum of six 15-ampere branch-circuits.**

Load calculations for small-appliance branch-circuits, and other major appliance branch-circuits are in addition to the general lighting loads. These calculations are discussed throughout this text as they appear.

▶The *NEC®* in *210.11(B)* states that the calculated load "*shall be evenly proportioned among multi-outlet branch circuits within the panelboard(s).*"◀

Table 3-1 is the schedule of the 15-ampere lighting and 20-ampere small-appliance branch-circuits in this residence.

SUMMARY OF 15-AMPERE LIGHTING AND 20-AMPERE RECEPTACLE CIRCUITS

15-Ampere Circuits		**20-Ampere Circuits**	
A14	Bathrooms, Hall Ltg.	A18	Workbench Receptacles
A15	Front Entry/Porch	A20	Workshop Receptacles, Window Wall
A16	Front Bedroom Ltg. & Outdoor Receptacle	A22	Master Bath Receptacle
A17	Workshop Ltg.	A23	Hall Bath Receptacle
A19	Master Bedroom Ltg. & Outdoor Receptacle		
A21	Study/Bedroom Ltg.		
B7	Kitchen Ltg.	B13	Kitchen North & East Wall Receptacles
B9	Wet Bar Ltg. & Receptacles	B15	Kitchen Receptacles, West Countertop
B10	Laundry, Rear Entry, Powder Room, Attic Ltg.	B16	Kitchen Receptacles, South Countertop, & Island
B11	Recreation Room Receptacles	B18	Washer Receptacle
B12	Recreation Room Ltg.	B20	Laundry Room Receptacles & Outdoor Receptacle
B14	Garage, Post Light	B21	Powder Room Receptacle
B17	Living Room & Outdoor Receptacle	B22	Refrigerator

Note: Panel "A" is the main panel located in the workshop. Panel "B" is the subpanel located in the corner of the recreation room.

Table 3-1

240.4(D) tells us that the maximum overcurrent protection is:

15 amperes for 14 AWG copper conductors

20 amperes for 12 AWG copper conductors

30 amperes for 10 AWG copper conductors

In conformance to *210.3*, a branch-circuit is rated according to the rating or setting of the over-current device. For example, a lighting branch-circuit is rated 15 amperes even if 12 AWG copper conductors are installed.

TRACK LIGHTING LOADS

Track lighting loads are considered to be part of the general lighting load for residential installations, based upon the 3 volt-amperes per square foot as previously discussed. There is no need to add more wattage (volt-amperes) to the load calculations for track lighting in homes.

As required by *410.101(B)*, the connected load on a lighting track shall not exceed the rating of the track. Also, be sure that a lighting track is supplied by a branch-circuit having a rating not more than that of the track. There is more on track lighting in Units 12 and 13.

DETERMINING THE NUMBER OF SMALL-APPLIANCE BRANCH-CIRCUITS

NEC® 220.52(A) and *(B)* state that an additional load of not less than 1500 volt-amperes shall be included for each two-wire small-appliance circuit and each laundry circuit.

The receptacle for the refrigerator in the kitchen is permitted to be supplied by an individual 15- or 20-ampere branch-circuit rather than connecting it to one of the two required 20-ampere small appliance circuits. This permission is found in *210.52(B)(1), Exception No. 2*. Because this separate branch-circuit diverts the refrigerator load from the receptacles that serve the countertop areas, it is not considered to be a small-appliance branch-circuit. An additional 1500 volt-amperes *does not* have to be included in the load calculations, *220.52(A), Exception*. There is nothing in the *Code* that prohibits adding another 1500 volt-amperes to the load calculations for this additional branch-

circuit, as this would result in more capacity in the electrical system.

NEC® 210.11(C)(1) requires that two or more 20-ampere small-appliance circuits must be provided for supplying receptacle outlets, as specified in *210.52(B)*.

NEC® 210.52(B)(1) states that the 20-ampere appliance circuits shall serve only the receptacle outlets in the kitchen, pantry, dining room, breakfast room, or similar rooms of dwelling units. *NEC® 210.52(B)(2)* also requires that these 20-ampere small-appliance circuits shall not supply other outlets.

Exceptions in *210.52(B)(2)* allow clock receptacles, and receptacles that are installed to plug in the cords of gas-fired ranges, ovens, or counter-mounted cooking units in order to power up their timers and ignition systems, Figure 3-3.

NEC® 210.52(B)(3) requires that not less than two 20-ampere small-appliance circuits must supply the receptacle outlets serving the countertop areas in kitchens.

RECEPTACLE BEHIND CLOCK SOLELY TO SUPPLY AND SUPPORT ELECTRIC CLOCK

20-AMPERE SMALL APPLIANCE BRANCH-CIRCUIT IN KITCHEN, DINING ROOM, PANTRY, AND BREAKFAST ROOM

RECEPTACLE INSTALLED TO PLUG IN GAS-FIRED APPLIANCE TO OPERATE IGNITION

Figure 3-3 A receptacle installed solely to supply and support an electric clock, or a receptacle that supplies power for clocks, clock timers, and electric ignition for gas ranges, ovens, and cooktops may be connected to a 20-ampere small-appliance circuit or to a general-purpose lighting circuit, *210.52(B)(2), Exceptions*.

Food waste disposers and/or dishwashers shall not be connected to these required 20-ampere small-appliance circuits. They must be supplied by other circuits, as is discussed later. Generally, 12 AWG conductors are used in small-appliance branch-circuits. The use of this larger conductor, rather than a 14 AWG conductor, helps improve appliance performance and lessens the danger of overloading the circuits.

A complete list of the branch-circuits for this residence is found in Unit 27. The circuit directories for Main Panel A and Subpanel B show clearly the number of lighting, appliance, and special-purpose circuits provided.

RECEPTACLE OUTLET BRANCH-CIRCUIT RATINGS

The receptacles in the kitchen, breakfast area, workshop, powder room, bathrooms, and laundry room are connected to 20-ampere branch-circuits. All other receptacles are connected to 15-ampere branch-circuits.

SUMMARY OF WHERE RECEPTACLE AND LIGHTING OUTLETS MUST BE INSTALLED IN RESIDENCES

The following is a recap of the *NEC®* requirements for receptacle and lighting outlets. We will look at these *Code* rules, room by room, later in this text.

Remember, the *NEC®* provides *minimum* requirements. There is nothing wrong with installing more receptacles than the *NEC®* minimum. You will never have enough receptacles! Think about it. A duplex receptacle can't handle computers, cable modems, small UPS devices, printers, stereos, VCRs, VDRs, CD players, CD burners, shredders, television sets, radios, fax machines, answering machines, typewriters, adding machines, pencil sharpeners, lamps, telephones, and others.

Consider spacing receptacles closer than the minimum. Consider quadplex receptacles or two duplex receptacles at each receptacle outlet location and consider surge suppression receptacles or surge protection multiple plug-in strips. See Unit 6 for a discussion of surge protection.

TIP: Some electricians install a receptacle below the wall switch in bedrooms, living rooms, family rooms, and similar rooms. As is usually the case, the required receptacles will be behind furniture, making it difficult to reach for plugging in a vacuum cleaner or other temporarily used appliance. A receptacle below the wall switch will be very easy to get to. A real plus! See Figure 3-4 for an explanation of *receptacle* and *receptacle outlet*.

Receptacle Outlets (125 Volts, Single-Phase, 15 and 20 Amperes)

Ground-Fault Circuit-Interrupter (GFCI): This protection is required in specific locations in a home. GFCIs are discussed in Unit 6.

Arc-Fault Circuit-Interrupter (AFCI): Required in bedrooms. AFCIs are discussed in Unit 6.

- Wall receptacles must be placed so that no point measured horizontally along the floor line in any wall space is more than 6 ft (1.8 m) from a receptacle outlet. Fixed room dividers and railings are considered to be walls for the purpose of this requirement. See *210.52(A)(1)*.

- Any wall space 24 in. (600 mm) or more in width must have a receptacle outlet, *210.52(A)(2)(1)*.

- Nonsliding fixed glass panels on exterior walls are considered to be wall space and are to be figured in when applying the rule that no point along the floor line is more than 6 ft (1.8 m) from a receptacle. Sliding panels on exterior walls are not considered to be a wall, *210.52(A)(2)*.

- Receptacle outlets located in the floor more than 18 in. (450 mm) from the wall are not to be counted as meeting the required number of wall receptacle outlets, *210.52(A)(3)*. An example of this would be a floor receptacle installed under a dining room table for plugging in warming trays, coffee pots, or similar appliances.

- Hallways 10 ft (3.0 m) or longer in homes must have at least one receptacle outlet, *210.52(H)*. See Figure 3-5.

- Circuits supplying wall receptacle outlets for lighting loads are generally 15-ampere circuits, but are permitted by the *Code* to be 20 amperes.

- A receptacle outlet shall be installed in an accessible location within 25 ft (7.5 m) of heat-

A RECEPTACLE OUTLET WHERE ONE OR MORE RECEPTACLES WILL BE INSTALLED.

A RECEPTACLE OUTLET WITH A SINGLE RECEPTACLE. (ONE CONTACT DEVICE)

A RECEPTACLE OUTLET WITH A MULTIPLE (DUPLEX) RECEPTACLE. THIS IS TWO RECEPTACLES. (TWO CONTACT DEVICES)

A RECEPTACLE OUTLET WITH TWO MULTIPLE (DUPLEX) RECEPTACLES. THIS IS FOUR RECEPTACLES. (FOUR CONTACT DEVICES)

HALLWAY

12 FT (3.7 m)

8 FT (2.5 m)

L-SHAPED HALLWAY

6 FT (1.8 m)

Figure 3-5   Determination of hallway length. Hallways 10 ft (3.0 m) or longer must have at least one receptacle outlet, *210.52(H)*.

Figure 3-4   The branch-circuit wiring and the box where one or more receptacles will be installed is defined by the *Code* as a *receptacle outlet*. The receptacle itself, whether a single or multiple device (one, two, or three receptacles) on one strap (yoke) is defined as a *receptacle*. For house wiring, receptacle loads are considered to be part of the general lighting load calculations, *220.14(J)*.

ing, air-conditioning, and refrigeration equipment. If the equipment is located on the side of the house, the receptacles required on the front

and rear of the house might not be within 25 ft (7.5 m) of the equipment, in which case an additional receptacle is required. Do *not* connect this receptacle outlet to the load side of the equipment disconnecting means. See *210.63*.

- A receptacle is *not* required on the rooftop of one- and two-family dwellings to serve heating, air-conditioning, and refrigeration equipment if the required outside receptacles are within 25 ft (7.5 m) of the equipment.

Basements

- At least one receptacle outlet must be installed in addition to the laundry outlet, *210.52(G)*.

- Connect to a 15- or 20-ampere circuit.

- GFCI protection is *not* required for receptacles that are not readily accessible, or for a single receptacle or duplex receptacle for two appliances that are not easily moved, are cord- and plug-connected, and are located in a dedicated space, *210.8(A)(5), Exceptions No. 1* and *No. 2*. Examples of this include the receptacle for laundry equipment, a sump pump, a water softener, a freezer, or a refrigerator.

- In finished basements, the required number, spacing, and location of receptacles, lighting outlets, and switches shall be in conformance to *210.52(A)* and *210.70*.

- In finished basements, GFCI protection for receptacles is *not* required unless the receptacles are located within 6 ft (1.8 m) from the outer edge of a wet bar sink and are intended to serve the countertop areas around the wet bar sink.

- Unfinished basements, or unfinished portions of basements that are not generally "lived in," such as work areas and storage areas, are not subject to the spacing requirements for receptacles as specified in *210.52(A)*.

- In each separate unfinished basement area that is not intended to be "lived in," such as storage areas or work areas, at least one 125-volt, single-phase, 15- or 20-ampere receptacle must be installed, *210.52(G)*. This receptacle(s) must be GFCI protected, *210.8(A)(5)*. This includes receptacles that are an integral part of porcelain or plastic pull chain or keyless lampholders.

- See the "Laundry Area" section in this chapter for receptacle outlet requirements in laundry areas.

- Refer to *210.8(A)*, *210.52(G)*, and *210.70*.

- ▶Any receptacle within 6 ft (1.8 m) of a laundry, utility, or wet bar sink must be GFCI protected. ◀

Small-Appliance Receptacle Outlets

- At least two 20-ampere small-appliance circuits must supply the receptacles that serve the countertop surfaces in the kitchen, *210.11(C)(1)* and *210.52(B)(3)*. Two small-appliance branch-circuits is the minimum. It would be advisable to run additional 20-ampere branch-circuits to these areas because of the heavy concentration of appliances.

- These two 20-ampere small-appliance circuits may also supply the receptacles in the dining area, pantry, breakfast room, and the refrigerator receptacle, *210.52(B)(1)*. *Exception No. 2* permits connecting the refrigerator receptacle to an individual 15- or 20-ampere circuit. **TIP: There are too many nuisance power outages when circuit breakers trip or fuses open because too many appliances are plugged in and used at the same time in the kitchen. Many large combination refrigerator/freezers today draw quite a bit of current. It is better to be safe than sorry. Connect the refrigerator receptacle to an individual 15- or 20- ampere branch-circuit. Being behind the refrigerator, the receptacle does not have to be GFCI protected. Only receptacles that serve the countertops are required to be GFCI protected, *210.8(A)(6)*.**

  See Unit 12 for more details about receptacles for refrigerators.

- Space countertop receptacles in kitchens according to *210.52(C)*.

- GFCI protection must be provided for all receptacles in kitchens that serve countertop areas, *210.8(A)(6)*. GFCI protection may be GFCI receptacles or GFCI circuit breakers, as discussed in detail in Unit 6.

- These two or more 20-ampere small-appliance circuits shall not supply other outlets, *210.52(B)(2)*. Exceptions to this are: (1) a clock receptacle and (2) a receptacle(s) for plugging in a gas-fired range, oven, or counter-mounted cooktop which are permitted to be supplied by one of the 20-ampere small-appliance circuits.

- Allow 1500 volt-amperes for each 20-ampere small-appliance branch-circuit when computing loads for determining the size of the service-entrance equipment, *220.52(A)*.

- Refer to *210.8(A)(6)*, *210.11(C)(1)*, *210.52(B)(1)*, *(2)*, and *(3)*, *210.52(C)*, and *220.52(A)*.

Countertops in Kitchens (See Figure 3-6)

- Receptacles that serve countertops in kitchens must be GFCI protected. This includes a receptacle located inside an "appliance garage."

- Receptacles are not permitted to be installed "face-up" in countertops.

Figure 3-6 Floor plan showing placement of receptacles that serve countertop surfaces in kitchens and wet bar areas in conformance to *210.52(C)*. All 120-volt receptacles that serve countertops must be GFCI protected, *210.8(A)(6)* and *(7)*. Receptacles that are located behind the refrigerator, under the sink for plugging in a food waste disposer, or inside the cabinets for plugging in a microwave oven do not have to be GFCI protected because they do not serve the countertop surfaces. Drawing is not to scale. See text for detailed explanation.

- Receptacles must be installed above all countertop spaces 12 in. (300 mm) or wider.

- Receptacles must be installed not more than 20 in. (500 mm) above a countertop.

- Receptacles above countertop spaces shall be positioned so that no point along the wall line is more than 24 in. (600 mm), measured horizontally, from a receptacle outlet in that space. Example: A section of a countertop measures 5 ft 3 in. (1.6 m) along the wall line. The minimum number of receptacles is 5 ft 3 in. ÷ 4 ft = 1+. Therefore, install two receptacles in this space.

- Receptacles may be installed below countertops only:
  - where the construction is for the physically impaired, or
  - on islands and peninsulas where there is

no way to mount a receptacle within 20 in. (500 mm) above the countertop.

For either of these two conditions:
- Receptacles shall be mounted not more than 12 in. (300 mm) below the countertop.
- No receptacles shall be installed below a countertop that extends more than 6 in. (150 mm) beyond its base cabinet. The logic to this rule is the popular use of stools where the countertop extends beyond the base cabinet, increasing the possibility of someone pulling coffee pots and similar appliances off of the countertop.

- Peninsulas and islands that have a long dimension of 24 in. (600 mm) or greater and a short dimension of 12 in. (300 mm) or greater, must have at least one receptacle. Measure the peninsula from the edge where the peninsula connects to the wall base cabinet.

- If a countertop, peninsula, or island has a range, refrigerator, or sink that creates separate countertop spaces on both sides of these or similar items, each space is considered a separate space. Receptacles shall be installed to comply with the required number of receptacles in these spaces.

- Receptacle outlets not readily accessible, such as behind a refrigerator, under the sink for plugging in a food waste disposer or dishwasher, or behind an appliance that is fastened in place, are not considered part of the receptacle outlet requirement for serving countertops. These receptacles are not required to be GFCI protected.

- Receptacles inside an "appliance garage" are *in addition* to the receptacles required to serve countertops. They must be GFCI protected. Although permitted to be connected to one of the two required 20-ampere small-appliance circuits, consider running a separate 20-ampere branch-circuit.

- A receptacle located inside an upper cabinet that serves a "hang below cabinet" microwave oven should be connected to an individual branch-circuit. This is discussed in Unit 20.

- ▶Receptacles are generally not required on the wall behind a range or sink. However, a receptacle is required:
  - if the range or sink is parallel to the wall, and the distance from the wall to the range or sink is 12 in. (300 mm) or greater.
  - if the range or sink is corner mounted on an angle, and the distance from the corner to the range or sink is 18 in. (450 mm) or greater.
  This is illustrated in *Figure 210.52* in the *NEC.*

- Wet bar sinks: Receptacles installed within 6 ft (1.8 m) from the outside edge of a wet bar sink must be GFCI protected.

- See Unit 6 for details on GFCI protection.

- See *210.8(A)(6) and (7), 210.11(C), 210.52(C), 220.52(A), and 406.4(E).*

Laundry Area

Automatic clothes washers draw a large amount of current during certain operating cycles. Here are the requirements for receptacles for the laundry area:

- At least one receptacle is required within 6 ft (1.8 m) of the intended location of the washer in the laundry area. Quite often, electricians install a duplex receptacle between the washer and gas clothes dryer to serve both appliances.

- ▶Any receptacle within 6 ft (1.8 m) of a laundry, utility, or wet bar sink must be GFCI protected. ◀

- This receptacle(s) must be supplied by a separate 20-ampere branch-circuit.

- This branch-circuit shall not supply any other outlets outside the laundry area.

- This branch-circuit and receptacle(s) are in addition to other required receptacles and branch-circuits.

- GFCI protection is generally required for receptacles in unfinished basements. Exempt from this rule are single or duplex receptacles that serve appliances that are not easily moved and that are located in dedicated spaces such as those for a washer and gas dryer, which are plugged into a duplex receptacle.

- If you install a single-receptacle on a 20-ampere branch-circuit, it must be rated 20 amperes.

- If you install a duplex receptacle on a 20-ampere branch-circuit, it may be rated 15 amperes or 20 amperes.

- Figure 1500 volt-amperes for this laundry branch-circuit when calculating the size of the service-entrance conductors.

- Refer to *210.8(A)(5), 210.8(A)(7), 210.11(C)(2), 210.21(B)(1), 210.21(B)(3), 210.50(C), 210.52(F) and (G), 220.14(J), and 220.52(B).*

- Allow 1500 volt-amperes for this laundry branch-circuit when computing loads for determining the size of the service-entrance equipment.

- Refer to *210.11(C)(2), 210.21(B)(1), 210.8(A)(5), 210.50(C), 210.52(F), and 220.16(B).*

Bathroom Receptacles

Here is a recap of the *NEC®* requirements for receptacles in bathrooms:

- At least one receptacle must be installed in a bathroom.

- Receptacle(s) in bathrooms must be GFCI protected.

- The receptacle(s) must be installed within 3 ft (900 mm) of the outside edge of each basin:
  - on the wall or partition adjacent to the basin.
  - ▶on the side or face of the basin cabinet not more than 12 in. (300 mm) below the countertop.◀

- The receptacle(s) must be supplied by at least one separate 20-ampere branch-circuit and must not supply other loads such as lighting outlets.

- The 20-ampere branch-circuit and receptacle is in addition to other required outlets and branch-circuits.

- The separate 20-ampere branch-circuit is permitted to supply more than one bathroom, although it is not recommended because of the use of high-wattage hair dryers. It is better to run a separate 20-ampere branch-circuit to each bathroom.

- If the separate 20-ampere branch-circuit supplies only one bathroom, other outlets in that bathroom are permitted to be connected to it. The other permitted loads are 80% if they are cord- and plug-connected, and 50% if they are fixed or fastened-in-place equipment.

- Instead of the "only one bathroom" exception, run a separate 20-ampere branch-circuit to each bathroom. That's what the plans for the residence call for.

- Don't be left in the dark. Do not connect bathroom lighting to this separate 20-ampere receptacle branch-circuit.

- Receptacles are not permitted face up on counters or other work surfaces.

- A receptacle on a medicine cabinet or luminaire (fixture) does not qualify as the required receptacle. It is virtually impossible to conform to the requirements that call for a separate circuit to be within 3 ft (900 mm) of the basin rim and be GFCI protected.

- No additional load need be added for this separate 20-ampere branch-circuit. The separate branch-circuit requirement merely redirects the receptacle load off the lighting branch-circuit.

- See *210.8(A)(1)*, *210.11(C)(3)*, *210.23(A)(1)* and *(2)*, *210.52(D)*, *220.14(J)*, and *406.4(E)*.

Outdoors

- ▶At least one receptacle must be installed in front and another in back of a one-family dwelling. These receptacles must be *accessible at grade level* and not more than 6½ ft (2.0 m) above grade. See Figure 3-7. The same is true for each unit of a multifamily dwelling where the entrance/egress is at grade level.

  The *NEC®* wording *accessible at grade level* means that the receptacles must be accessible from grade level, which could be the grass, sidewalk, driveway, deck, or patio.◀

  Most electrical inspectors will accept a receptacle on a porch or deck if the receptacle can be reached from grade level. A receptacle inside a screened-in porch or deck would not meet the intent of *210.52(E)*.

- Additional receptacles may be installed higher, such as receptacles used for plugging in ice- and snow-melting heating cables and/or decorative outdoor lighting.

AT LEAST ONE OUTDOOR RECEPTACLE OUTLET REQUIRED IN FRONT AND ANOTHER IN BACK OF HOUSE.

Figure 3-7   Outdoor receptacle requirements for residences, *210.52(E)*, require that for one-family dwellings, and each unit of a two-family dwelling that is at grade level, at least one receptacle outlet must be installed in the front and another in the back. These receptacles must not be mounted more than 6½ ft (2.0 m) above grade.

- For servicing HVAC equipment, *NEC®* 210.63 requires that a 125-volt, single-phase, 15- or 20-ampere-rated receptacle be installed:
  – at an accessible location.
  – on the same level as the HVAC equipment.
  – within 25 ft (7.5 m) of the equipment.

  This receptacle shall not be connected to the load side of the equipment disconnecting means.

  The outdoor receptacles required by *210.52(E)* (one in the front, one in the back) *might* or *might not* meet the requirements of *210.63.* If the HVAC equipment is located on a roof, in an attic, in a crawl space, or in a similar location, a receptacle must be installed in that location so as to be on the same level as the equipment.

- Connect to a 15- or 20-ampere circuit.

- *All* outdoor receptacles must be GFCI protected. An exception is that personnel GFCI protection is *not* required for an outdoor receptacle(s) installed for snow- and ice-melting equipment (i.e., heating cables) that is not **readily accessible**, and is supplied by a dedicated branch-circuit.

- Ground-fault protection of equipment (GFPE) must be provided for "fixed" outdoor deicing and snow-melting equipment. GFPE can be a circuit breaker, a receptacle, or be self-contained in the equipment. GFPE devices for *equipment protection* trip at approximately 30 milliamperes. When used for deicing and snow melting, mineral-insulated metal-sheathed cable embedded in noncombustible material such as concrete does not require gound-fault protection.

- See *210.8(A)(3)*, *210.52(E)*, *210.63*, and *426.28*.

Garages

- At least one receptacle outlet must be provided.

- Connect to a 15- or 20-ampere circuit.

- All receptacles must be GFCI protected. An exception to this is receptacles that are not readily accessible, such as a receptacle mounted on the ceiling for plugging in a garage door opener. Also exempt from the GFCI requirement are single receptacles or duplex receptacles intended for cord- and plug-connected appliances that are not easily moved and are located in dedicated spaces. Examples are a freezer, refrigerator, or central vacuum equipment.

- If the garage is detached, the *Code* does not require electric power, but if electric power is provided, then at least one GFCI receptacle outlet must be installed.

  Receptacles are not required in a storage shed (accessory building), but if installed, they must be GFCI protected if the floor is at or below grade level.

- See *210.8(A)(2)* and *210.52(G)*.

Receptacles in Other Locations. All of the required receptacles discussed above are *in addition* to any receptacles that are part of luminaires (lighting fixtures) or appliances, are located inside cabinets, or are located more than 5½ ft (1.7 m) above the floor, *210.52*.

GFCI Exemption Recap

- Under certain conditions, GFCI protection is *not* required for receptacle outlets in garages, in crawl spaces at or below ground level, or in unfinished basements if the receptacles are:
  – not readily accessible.
  – single-type receptacles or duplex-type receptacles located in dedicated spaces to be used for cord- and plug-connected appliances that are not easily moved.

  Examples of these would be such things as overhead door openers, refrigerators, freezers, water softeners, central vacuum equipment or sump pumps. These would be the receptacles that are dedicated to specific plug-in appliances. The homeowner would normally not unplug the appliance to plug in an extension cord. These receptacles are exempt from the GFCI requirement because they are not intended to be used for plugging in extension cords for portable electric tools or portable appliances. These GFCI-exempt receptacles are *in addition* to the receptacles required by *210.52(G)* for garages and basements. See *210.8(A)*.

Specific Loads. Appliances such as electric ranges, ovens, air conditioners, electric heat, heat pumps, water heaters, and similar loads are discussed later in this text. See the Index for their location.

Receptacle Definitions. The *NEC®* in *Article 100* defines the following:

- receptacle: A receptacle is a contact device installed at the outlet for the connection of an attachment plug.
- single receptacle: A single receptacle is a single contact device with no other contact device on the same yoke.
- multiple receptacle: A multiple receptacle is two or more contact devices on the same yoke.

See Figure 3-4 for an illustration of these definitions.

Receptacle and Branch-Circuit Ratings

Table 3-2 shows the requirements of *210.21(B)(3)* and *Table 210.21(B)(3)* where the branch-circuit supplies two or more receptacles or outlets in homes.

Where the branch-circuit supplies a single receptacle, the receptacle must have a rating not less than the branch-circuit ampere rating, *210.21(B)(1)*.

Arc-Fault Circuit Interrupters (AFCIs). See Unit 6 for complete information about AFCI requirements in bedrooms.

Replacing Existing Receptacles. See Unit 6 for detailed discussion and illustrations about replacing existing receptacles in older homes.

Lighting Outlets *(Section 210.70)*

The *NEC®* contains minimum requirements for providing lighting for dwellings.

Switched Lighting Outlets. Per *210.70*, lighting outlets in dwellings must be installed as shown in Figure 3-8. Here is a summary of these requirements.

Habitable Rooms. In every habitable room and in every bathroom, at least one wall-switch-controlled lighting outlet must be installed. Switched receptacles meet the intent of this requirement except in kitchens and bathrooms. However, if plug-in swag lighting as illustrated in Figure 3-9 is used in a dining room, a switched receptacle connected to a general lighting branch-circuit must be installed. This receptacle must be in addition to the receptacles supplied by the required 20-ampere small-appliance branch-circuits, *210.52(B)(1)*, *Exception No. 1*, Figure 3-10.

Occupancy motion sensors may be used if: (1) they are in addition to the required wall switch, or (2) they have a manual override and are located where the wall switch would normally be located.

Additional Locations. At least one wall switch-controlled lighting outlet must be installed in hallways, stairways, attached garages, and detached garages with electric power, *210.70(A)(2)(a)*.

The *NEC®* does *not* require that a detached garage have electric power, but if it does have electric power, then it must have at least one GFCI receptacle, and it must have wall switch-controlled lighting.

Some electrical inspectors consider a tool shed or other similar accessory building to be a storage or utility room that comes under the requirements of *210.70(A)(3)*, thus requiring a wall switch-controlled lighting outlet.

At least one wall switch-controlled lighting outlet must be installed to provide illumination on the exterior side of outdoor entrances or exits that have

BRANCH-CIRCUIT RATING	RECEPTACLE RATING
15-ampere lighting branch-circuit: circuit has two or more receptacles or outlets.	15-ampere maximum
20-ampere small-appliance branch-circuit: circuit has two or more receptacles.	15 or 20 amperes
Individual 20-ampere branch-circuit: circuit has only a single receptacle connected. An example of this is a dedicated branch-circuit for a microwave oven.	20 amperes

Table 3-2 This table shows the ampere rating for receptacles used on 15- and 20-ampere branch-circuits. See *Table 210.21(B)(3)*.

ATTIC

BEDROOM HALLWAY BATHROOM

OUTDOOR ENTRANCE

INSIDE ATTACHED GARAGE

OUTDOOR ENTRANCE

STAIRWAY

ANY HABITABLE ROOM

UTILITY ROOM

LIGHTING OUTLET NOT REQUIRED OUTSIDE OF VEHICLE DOOR

BASEMENT

HABITABLE RECREATION ROOM

STAIRWAY

UNDERFLOOR SPACE USED FOR STORAGE

○ LIGHTING OUTLET MUST BE WALL SWITCH-CONTROLLED.

● LIGHTING OUTLET(S) MUST BE SWITCH-CONTROLLED IF SPACE IS USED FOR STORAGE OR IF SPACE CONTAINS EQUIPMENT THAT WILL REQUIRE SERVICING, SUCH AS AIR CONDITIONING, FURNACES, AND SO ON. THE LIGHTING OUTLET(S) MUST BE LOCATED NEAR THE EQUIPMENT. THE SWITCH MUST BE LOCATED NEAR ENTRY TO THE SPACE.

⊗ IF PERMANENT STAIRWAY HAS SIX OR MORE RISERS, INSTALL SWITCHES ON BOTH LEVELS.

EXCEPTIONS:
• SWITCH-CONTROLLED RECEPTACLES MAY BE INSTALLED IN PLACE OF LIGHTING OUTLETS, EXCEPT IN KITCHENS AND BATHROOMS.
• REMOTE, CENTRAL, OR AUTOMATIC CONTROL OF LIGHTING IS ACCEPTABLE FOR HALLWAYS, STAIRWAYS, AND OUTDOOR ENTRANCES.

Figure 3-8 Lighting outlets required in a typical dwelling unit.

grade-level access to dwelling units, attached garages, and detached garages if they have electric power. A sliding glass door is considered to be an outdoor entrance. A vehicle door is not considered to be an outdoor entrance. See *210.70(A)(2)(b)*.

Although not necessarily good design practice,

• One switch could control the required outdoor lighting at more than one entrance.

• One switch could control both the required outdoor and indoor lighting of a garage.

• One outdoor luminaire (fixture), properly located, could meet the required outdoor lighting for both the house entrance and the garage entrance.

Stairways. Where there are six stair risers or more on an interior stairway, the *NEC®* requires that wall switch-control led stairway lighting must be provided on each floor level and any landing level that has an entryway. *210.70(A)(2)(c)*.

This switching requirement includes permanent stairways to basements and attics but does not

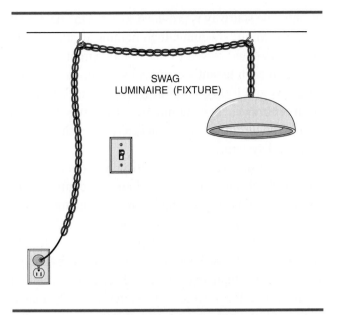

Figure 3-9 A receptacle controlled by a wall switch is acceptable as the required lighting outlet in rooms other than a kitchen or bathroom, _210.70(A)(1), Exception._

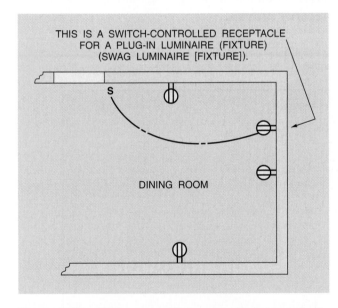

Figure 3-10 When a switch-controlled receptacle outlet is installed in rooms where the receptacle outlets normally would be 20-ampere small-appliance circuits (breakfast room, dining room, and so on.), it must be in addition to the required small-appliance receptacle outlets, _210.52(B)(1), Exception No. 1._

include fold-down storable ladders commonly installed for access to residential attics.

The _NEC_® does not tell us how to illuminate the stairway.

The _International Residential Code_ is more specific about stairway lighting. It requires indoor stairways to have lighting that will illuminate the landings and treads, and that the lighting shall be located in the immediate vicinity of each landing at the top and bottom of the stairs. This _Code_ also requires that the switching must be accessible at the top and bottom of the stairway without using any step to reach a switch.

**If You Don't Install Switches, What Else is Acceptable?** Remote control, central control, motion sensors, photocells, and other automatic devices may be installed instead of wall switches for the control of lighting in hallways, stairways, and at outdoor entrances. For example, an outdoor post light controlled by a photocell does not require a separate switch.

This is permitted by the _Exception_ to _210.70(A)(2)(a)_, _(b)_, and _(c)_.

**Closet Lighting.** The _NEC_® does not require lighting in closets, but some local codes do. Check this out in your locality. See Unit 8 for a detailed explanation of _Code_ requirements for closet lighting.

**Basements, Attics, Storage, and Other Equipment Spaces.** At least one switch-controlled lighting outlet must be installed in spaces that are used for storage or contain equipment that requires servicing, such as attics, underfloor spaces, utility rooms, and basements. Examples of equipment that may need servicing are furnaces, air conditioners, heat pumps, sump pumps, and so on. The control is permitted to be a wall switch or a lighting outlet that has an integral switch such as a pull chain switch. The lighting outlet must be at or near the equipment that may require servicing. The control must be at the usual point of entrance to the space.

The Number of Branch-Circuits Required, and How to Determine the Number of Lighting and Receptacle Outlets to Be Connected to One Branch-Circuit

This unit has explained the _NEC_® rules on where receptacle and lighting outlets are to be installed. The question of _how many_ receptacle outlets and lighting outlets should be connected to each branch-circuit is covered in detail in Unit 8, where we begin the actual layout of the wiring for the residence discussed throughout this text.

Common Areas in Two-Family or Multifamily Dwellings

NEC® 210.25 contains specific requirements for circuits that serve common areas of two-family or multifamily dwellings. Be careful when wiring a two-family or multifamily dwelling. For areas such as a common entrance, common basement, or common laundry area, the branch-circuits for lighting, central fire/smoke/security alarms, communication systems, or any other electrical requirements that serve the common area *shall not* be served by circuits from any individual dwelling unit's power service. Common areas will require separate circuits, associated panels, and metering equipment. The reasoning for this requirement is that power could be lost in one occupant's electrical system, resulting in no lights or alarms for the other tenant(s) using the common shared area.

For example, a two-family dwelling would probably require two separate stairways to the basement. If only one stairway is provided to serve the basement of a two-family dwelling, then lighting would have to be provided and served from a branch-circuit originating from each tenant's electrical system, so that each tenant would have control of the lighting. For a common entrance, two luminaires (lighting fixtures) would be installed and connected to each tenant's electrical system.

Likewise, fire/smoke/security systems in a two-family dwelling shall be installed and connected to branch-circuits originating from each tenant's electrical panel. This becomes an architectural design issue. This is not generally a problem in two-family dwellings, but can become a problem in dwellings having more than two tenants. Rather than having to install a third watt-hour meter and associated electrical equipment on a two-family dwelling, the design of the structure can be such that there are no common areas, Figure 3-11. A fourplex will probably require a fifth watt-hour meter, four watt-hour meters for the tenants and one watt-hour meter to serve common area lighting and the fire/smoke/security systems.

Figure 3-11 This outline of a two-family dwelling shows a common rear entry, a common stairway to the basement, and a common front entry. Note that each entry has one luminaire (lighting fixture). These common area luminaires (lighting fixtures) are not permitted to be connected to either of the tenant's electrical systems in the individual dwelling units, *210.25*. The choices are: (1) architect/designer must redo the layout of the building to eliminate the common entrances, (2) provide two luminaires (lighting fixtures) in each entry, with each luminaire (fixture) connected to the respective individual tenant's electrical circuit, or (3) install a third meter and panel to supply the electrical loads in the common areas.

REVIEW

Note: In the following questions the student may refer to the *NEC®* to obtain the answers. To help the student find these answers, references to *Code* sections are included after some of the questions. These references are not given in later reviews, where it is assumed that you are becoming familiar with the *Code*.

1. What is the meaning of calculated load? _____

   _____

2. How are branch-circuits rated? See *NEC® 210.3*. _____

   _____

3. How is the rating of the branch-circuit protective device affected when the conductors used are of a larger size than called for by the *Code*? See *NEC® 210.3*. _____

   _____

4. What dimensions are used when measuring the area of a building? See *NEC® 220.12*.

   _____

5. What spaces are not included in the floor area when computing the load in volt-amperes per square foot? See *NEC® 220.12*. _____

   _____

6. What is the unit load per square foot for dwelling units? See *Table 220.12*. _____

   _____

7. According to *NEC® 210.50(C)*, a laundry equipment outlet must be placed within _____ ft (_____ m) of the intended location of the laundry equipment.

8. How is the total load in volt-amperes for lighting purposes determined? See *NEC® 220.12*. _____

   _____

   _____

9. How is the total lighting load in amperes determined? _____

   _____

10. How is the required number of branch-circuits determined? _____

    _____

11. What is the minimum number of 15-ampere lighting circuits required if the dwelling has an occupied area of 4000 ft² (368 m²)? Show all calculations.

12. How many lighting branch-circuits are provided in this dwelling? _____

13. a. What is the minimum load allowance for small-appliance circuits for dwellings? See *NEC® 220.52(A)*. _____

    b. A separate 15-ampere branch-circuit is run to the receptacle outlet behind the refrigerator instead of connecting it to one of the two 20-ampere small-appliance branch-circuits that are required in kitchens. For this separate circuit, an additional 1500 volt-amperes (shall) (does not have to) be added to the load calculations for dwellings. Circle the correct answer.

14. What is the smallest size wire that can be used in a branch-circuit rated at 20 amperes?

    _____

15. How is the load determined for outlets supplying specific appliances? See *NEC® 220.14*. _____

    _____

16. What type of circuits must be provided for receptacle outlets in the kitchen, pantry, dining room, and breakfast room? See *NEC® 210.11(C)(1)*. _____

    _____

17. How is the minimum number of receptacle outlets determined for most occupied rooms? See *NEC® 210.52(A)*. _____

    _____

    _____

    _____

    _____

    _____

    _____

    _____

    _____

18. In a single-family dwelling, what types of overload protection for circuits are used? See *NEC® 240.6*. _____

    _____

    _____

    _____

    _____

19. The *NEC®* in *Article 100* defines a continuous load as "a load where the maximum current is expected to continue for three hours or more." According to *210.19(A)* and *210.20(A)*, the branch-circuit conductors and overcurrent protection for a continuous load shall be sized at not less than (100%) (125%) (150%) of the continuous load. Circle the correct answer.

20. The minimum number of outdoor receptacles for a residence is _____,

    NEC® _____. State the location. _____

21. The *Code* indicates the rooms in a dwelling that are required to have switched lighting outlets or switched receptacles. Write yes (switch required) or no (switch not required) for the following areas: See *NEC® 210.70(A)*.

    a. attic _____

    b. stairway _____

    c. crawl space (where used for storage) _____

    d. hallway _____

    e. bathroom _____

    f. closet _____

22. Is a receptacle required in a bedroom on a 3 ft (900 mm) wall space behind the door? The door is normally left open._____

23. The *Code* requires that at least one 20-ampere circuit feeding a receptacle outlet must be provided for the laundry. May this circuit supply other outlets?_____

24. In a basement, at least one receptacle outlet must be installed in addition to the receptacle outlet installed for the laundry equipment. This additional receptacle outlet and any other receptacle outlets in unfinished areas must be _____ protected.

25. *NEC® 210.8(A)(2)* requires that *all* receptacles installed in garages must have GFCI protection, but there are certain exceptions. What are these exceptions?

    _____

    _____

    _____

    _____

26. Define a branch-circuit. _____

    _____

27. Although the *Code* contains many exceptions to the basic overcurrent protection requirements for conductors, in general, the rating of the branch-circuit overcurrent device must (not be less than) (not be more than) the ampacity of a conductor. Circle the correct answer.

28. The rating of a branch-circuit is based upon (circle the correct answer)
    a. the rating of the overcurrent device
    b. the length of the circuit
    c. the branch-circuit wire size

29. a. A 25-ampere branch-circuit conductor is derated to 70%. It is important to provide proper overcurrent protection for these conductors. The derated conductor ampacity is _____

    _____

    b. If the connected load is a "fixed" nonmotor, noncontinuous load, the branch-circuit overcurrent device may be sized at (20) (25) amperes. Circle the correct answer.

c. If the above circuits supply receptacle outlets, the branch-circuit overcurrent device must be sized at not over (15) (20) (25) amperes. Circle the correct answer.

30. Small-appliance receptacle outlets are (included) (not included) in the 3 volt-ampere per square foot calculations. Circle the correct answer.

31. If a homeowner wishes to have a switched receptacle outlet for a swag lamp in a dining room, may this switched receptacle outlet be considered to be one of the receptacle outlets required for the 20-ampere small-appliance circuits? _____

32. How many receptacle outlets are required on a 13 ft (4.0 m) wall space between two doors? Refer to *NEC*® *210.52(H)*. Draw the outlets in on the diagram.

33. Is a receptacle required in a hallway in a home when the hallway is 8 ft (2.5 m) long?

_____

34. A split-level home has one stairway that has six risers between two levels of the home. Which of the following choices meets *Code*? Choose the correct answer.
    a. Two 3-way switches must be installed (one switch at the top of the stairs, one switch at the bottom of the stairs) to control the lighting for the stairway.
    b. One single-pole switch is permitted on either level to control the lighting for the stairway.
    c. No switches are necessary for the stairway lighting because the lighting on the upper level and the lower level provides enough illumination on the stairway.

35. A sliding glass door is installed in a family room that leads to an outdoor deck. The sliding door has one 4 ft (1.22 m) sliding section and one 4 ft (1.22 m) permanently mounted glass section. Inside the recreation room, what does the *Code* say about wall receptacle outlets relative to the receptacle's position near the sliding glass door?

_____

_____

_____

_____

_____

_____

_____

_____

36. The *Code* permits certain other receptacles to be connected to a 20-ampere small appliance circuit. What are they? Which section of the *NEC®* supports your answer?

_____

_____

_____

_____

_____

_____

37. What is the minimum number of 20-ampere small-appliance circuits required by the *Code* to supply the receptacle outlets that serve the countertop surfaces in the kitchen? In which section of the *Code* is this requirement found? _____

38. Does the *Code* allow a 120-volt receptacle located behind a gas range to be supplied by a small-appliance circuit? (Yes) (No) Circle the correct answer.

39. When determining the location and number of receptacles required, fixed room dividers and railings (shall be) (need not be) considered to be wall space. Circle the correct answer.

40. No point along the wall line shall be more than 6 ft (1.8 m) from a receptacle outlet. This requirement (does apply) (does not apply) to unfinished residential basements. Circle the correct answer.

41. Receptacle outlets that serve countertop surfaces in kitchens are required to be _____ protected, and shall be installed so that no point along the wall line above the base cabinets is more than [18 in. (450 mm)] [24 in. (600 mm)] [36 in. (900 mm)] from a receptacle. Circle the correct answer.

42. In kitchens, the *Code* requires that at least (one) (two) receptacle(s) be installed to serve the countertop surfaces on an island or peninsula. Circle the correct answer. Where in the *Code* is this requirement found? *NEC®* _____

43. In the past, it was common practice to connect the lighting and receptacle(s) in a bathroom to the same circuit. Because of overloads caused by high-wattage grooming appliances, the *Code* now requires that these receptacle(s) be supplied by a separate 20-ampere branch-circuit. Where in the *Code* is this requirement found? *NEC®* _____

44. If a residence has two bathrooms, the *Code* states:
   a. The receptacles in each bathroom must be connected to a separate 20-ampere branch-circuit. For two bathrooms, this would mean two circuits. (True) (False) Circle the correct answer.
   b. The receptacles in both bathrooms are permitted to be served by the same 20-ampere branch-circuit. (True) (False) Circle the correct answer.

45. In your own words, explain the GFCI exemptions for receptacle outlets installed in unfinished basements and garages. Give some examples.

_____

_____

_____

_____

46. *NEC®* _____ of the *Code* prohibits connecting lighting and/or fire/smoke/ security systems to an individual tenant's electrical source where common (shared) areas are present, in which case the lighting provided for both tenants is connected on one tenant's electrical source.

47. When installing weatherproof outdoor receptacles high up under the eaves of a house intended to be used for plugging in deicing and snow-melting cable for the rain gutters, must these receptacles be GFCI protected for personnel protection? (Yes) (No) Circle the correct answer.

Conductor Sizes and Types, Wiring Methods, Wire Connections, Voltage Drop, Neutral Sizing for Services

OBJECTIVES

After studying this unit, the student should be able to

▶ determine the current-carrying capacity (ampacity) of conductors.

▶ understand overcurrent protection for conductors and maximum loading of branch-circuits.

▶ understand aluminum conductors and the possible fire hazards if they are not properly installed.

▶ know the *NEC®* installation requirements for all types of cables and raceways.

▶ understand the special ampacity ratings of service-entrance conductors.

▶ make voltage-drop calculations.

▶ learn an alternate cost and time-saving method permitted to bring nonmetallic-sheathed cables into the top of a surface-mounted panel.

CONDUCTORS

Throughout this text, all references to conductors are for copper conductors, unless otherwise stated.

Wire Size

The copper wire used in electrical installations is graded for size according to the American Wire Gauge (AWG) Standard. The wire diameter in the AWG standard is expressed as a whole number. The higher the AWG number, the smaller the wire. AWG sizes vary from fine, hairlike wire used in coils and small transformers to very large diameter wire required in industrial wiring to handle heavy loads.

The wire may be a single strand (solid conductor), or it may consist of many strands. Each strand of wire acts as a separate conducting unit. The wire size used for a circuit depends on the maximum current to be carried. The *NEC®* in *Table 210.24* shows that the minimum conductor size for branch-circuit wiring is 14 AWG. Smaller size conductors are permitted for bell wiring, thermostat wiring, communications wiring, intercom wiring, luminaire (fixture) wires, and similar low-energy circuits.

Don't Get Confused!

In the past, conductor size was shown as, for example, "No. 12 AWG." This same conductor now appears in the *NEC®* as "12 AWG." The "12" is a size, not a quantity.

Table 4-1 shows typical applications for different size conductors.

Solid and Stranded Conductors

For residential wiring, sizes 14, 12, and 10 AWG conductors are generally solid when the wiring method is nonmetallic-sheathed cable or armored cable.

In a raceway, the preference is to use stranded 10 AWG conductors because of the flexibility and ease of handling and pulling in the conductors.

NEC® 310.3 requires that conductors 8 AWG and larger must be stranded where installed in a raceway.

Ampacity

Ampacity means "the current, in amperes, that a conductor can carry continuously under the conditions of use without exceeding its temperature rating." This value depends upon the conductor's cross-sectional area, whether the conductor is copper or aluminum, and the type of insulation around the conductor. Ampacity values, also referred to as current-carrying capacity, are found in *Article 310*. The most commonly used conductor ampacities are found in *Table 310.16*.

The allowable ampacity values in the tables are valid where there are no more than three current-carrying conductors in a raceway or cable, and where the temperature does not exceed 86°F (30°C).

When there are more than three current-carrying conductors in a raceway or cable, the allowable ampacity values are adjusted according to the factors in *Table 310.15(B)(2)(a)*.

When the ambient temperature exceeds 86°F (30°C), the allowable ampacity values are corrected according to the factors found below the tables.

If both conditions are present, then both penalties (adjustment and correction factors) must be applied. More on conductor ampacities, derating, adjusting, and correction factors is presented in Unit 18.

Conductors must have an ampacity not less than the maximum load that they are supplying, as shown in Figure 4-1. All conductors of a specific branch-circuit must have an ampacity of the branch-circuit's

CONDUCTOR APPLICATIONS CHART		
Conductor Size	**Overcurrent Protection**	**Typical Applications** (Check wattage and/or ampere rating of load to select the correct size conductors based on *Table 310.16*.)
20 AWG	Class 2 circuit transformers provide overcurrent protection.	Telephone wiring is usually 20 or 22 AWG.
18 AWG	7 amperes. See *Table 430.72(B)* and *725.23*. Class 2 circuit transformers provide overcurrent protection.	Low-voltage wiring for thermostats, chimes, security, remote control, home automation systems, etc. For these types of installations, 18 or 20 AWG conductors can be used depending on the connected load and length of circuit.
16 AWG	10 amperes. See *Table 430.72(B)* and *725-23*. Class 2 circuit transformers provide overcurrent protection.	Same applications as above. Good for long runs to minimize voltage drop.
14 AWG	15 amperes	Typical lighting branch-circuits.
12 AWG	20 amperes	Small-appliance branch-circuits for the receptacles in kitchens and dining rooms. Also laundry receptacles and workshop receptacles. Often used as the "home run" for lighting branch-circuits. Some water heaters.
10 AWG	30 amperes	Most clothes dryers, built-in ovens, cooktops, central air conditioners, some water heaters, heat pumps.
8 AWG	40 amperes	Ranges, ovens, heat pumps, some large clothes dryers, large central air conditioners, heat pumps.
6 AWG	50 amperes	Electric furnaces, heat pumps.
4 AWG	70 amperes	Electric furnaces, feeders to subpanels.
3 AWG and larger	100 amperes	Main service-entrance conductors, feeders to subpanels, electric furnaces.

Table 4-1 Conductor Applications Chart

Figure 4-1   Branch-circuit conductors shall have an ampacity not less than the maximum load to be served, *210.19(A)*. See *210.23* for permissible loads.

rating, as shown in Figure 4-2. There are exceptions to this rule, such as taps for electric ranges (see Unit 20.)

Ampacity of Flexible Cords

Table 4-2 shows the allowable ampacities for some sizes of flexible cords. Refer to *Table 400.5(A)* in the *NEC®* for other specific types and sizes of flexible cords.

▶Overcurrent protection requirements for extension cords are found in *240.5*.◀

Conductor Sizing

The diameter of wire is usually given in a unit called a "mil." A *mil* is defined as one-thousandth of an inch (0.001 inch). Mils squared are known as "circular mils."

Table 8 in *Chapter 9* of the *NEC®* clearly shows that conductors are expressed in AWG numbers from 18 (1620 circular mils) through 4/0 (211,600 circular mils). Wire sizes larger than 4/0 are expressed in circular mils.

CONDUCTORS MUST HAVE MINIMUM 20-AMPERE RATING

Figure 4-2   All conductors in this circuit supplying readily accessible receptacles shall have an ampacity of not less than the rating of the branch-circuit. In this example, 20-ampere conductors must be used, *210.19(A)* and *Table 210.24*.

Large conductors, such as 500,000 circular mils, are generally expressed as 500 kcmils. Since the letter *k* designates 1000, the term *kcmils* means "thousand circular mils." This is much easier to express in both written and verbal terms.

Older texts used the term *MCM*, which also means "thousand circular mils." The first letter *M* refers to the Roman numeral that represents 1000. Thus, 500 MCM means the same as 500 kcmils.

ALLOWABLE AMPACITIES FOR FLEXIBLE CORDS

Conductor Size AWG Copper	Cords in Which Three Conductors Carry Current Example: black, white, red, plus equipment grounding conductor	Cords in Which Two Conductors Carry Current Example: black, white, plus equipment grounding conductor
14	15	18
12	20	25
10	25	30
8	35	40
6	45	55
4	60	70

Table 4-2   Allowable ampacities for some of the more common flexible cords used for residential applications. These values were taken from *Table 400.5(A)*, *NEC®*. Refer to the *NEC®* table for other specific types and sizes.

Roman numerals are no longer used in the electrical industry for expressing conductor sizes.

Overcurrent Protection for Conductors

Conductors must be protected against overcurrent by fuses or circuit breakers rated not more than the ampacity of the conductors, *Section 240.4*.

NEC® 240.4(B) permits the use of the next higher standard overcurrent device rating as shown in *240.6(A)*. This permission is granted *only* when the overcurrent device is rated 800 amperes or less. For example, from *Table 310.16*, a 6 Type TW 140°F (60°C) conductor has an allowable ampacity of 55 amperes. In *240.6(A)*, we find that the next higher standard size overcurrent device is 60 amperes. Nonstandard ampere ratings of fuses and breakers are also permitted.

NEC® 240.4(D) spells out the maximum overcurrent protection for small branch-circuit conductors. Table 4-3 illustrates the requirements of *240.4(D)*.

There are exceptions for motor branch-circuits and HVAC equipment. Motor circuits are discussed in Unit 19 and Unit 22. HVAC circuits are discussed in Unit 23.

Branch-Circuit Rating

The rating of the overcurrent device (OCD) determines the rating of a branch-circuit, as shown in Figure 4-3.

PERMISSIBLE LOADS ON BRANCH-CIRCUITS *(210.23)*

The *NEC®* is very specific about the loads permitted on branch-circuits. Here is a recap of these requirements.

- The load shall not exceed the branch-circuit rating.

CONDUCTOR SIZE	MAXIMUM OVERCURRENT PROTECTION
14 AWG copper	15 amperes
12 AWG copper	20 amperes
10 AWG copper	30 amperes

Table 4-3 Maximum overcurrent protection for 14, 12, and 10 AWG copper conductors.

Figure 4-3 All three of the above branch-circuits are rated as 15-ampere circuits, even though larger conductors were used for some other reason, such as solving a voltage-drop problem. The rating of the overcurrent device (OCD) determines the rating of a branch-circuit, *210.3*.

- The branch-circuit must be rated 15, 20, 30, 40, or 50 amperes when serving two or more outlets.
- An individual branch-circuit can supply any size load.
- 15- and 20-ampere branch-circuits:
    a. May supply lighting, other equipment, or both types of loads.
    b. Cord- and plug-connected equipment shall not exceed 80 percent of the branch-circuit rating.
    c. Equipment fastened in place shall not exceed 50 percent of the branch-circuit rating if the branch-circuit also supplies lighting, other cord- and plug-connected equipment, or both types of loads.
- The 20-ampere small-appliance circuits in homes shall not supply other loads, *210.11(C)(1)*.
- 30-ampere branch-circuits may supply equipment such as dryers, cooktops, etc. Cord- and plug-connected equipment shall not exceed 80 percent of the branch-circuit rating.
- 40- and 50-ampere branch-circuits may supply cooking equipment that is fastened in place, such as an electric range.
- Over 50-ampere-rated branch-circuits are for electric furnaces, large heat pumps, air-conditioning equipment, large double-ovens, and similar large loads.

ALUMINUM CONDUCTORS

The conductivity of aluminum is not as great as that of copper for a given size. For example, checking *Table 310.16*, an 8 AWG Type THHN copper conductor has an allowable ampacity of 55 amperes. An 8 AWG Type THHN aluminum or copper-clad aluminum conductor has an ampacity of 45 amperes.

In *240.4(D)*, the maximum overcurrent protection for a 12 AWG copper conductor is 20 amperes but only 15 amperes for a 12 AWG aluminum or copper-clad aluminum conductor.

Aluminum conductors have a higher resistance compared to a copper conductor of the same size. When considering voltage drop, a conductor's resistance is a key ingredient. Voltage-drop calculations are discussed later on in this unit.

Common Connection Problems

Some common problems associated with aluminum conductors when not properly connected may be summarized as follows:

- A corrosive action is set up when dissimilar wires come in contact with one another when moisture is present.

- The surface of aluminum oxidizes as soon as it is exposed to air. If this oxidized surface is not penetrated, a poor connection results. When installing aluminum conductors, particularly in large sizes, an inhibitor (antioxidant) is brushed onto the aluminum conductor, then the conductor is scraped with a stiff brush where the connection is to be made. The process of scraping the conductor breaks through the oxidation, and the inhibitor keeps the air from coming into contact with the conductor. Thus, further oxidation is prevented. Aluminum connectors of the compression type usually have an inhibitor paste already factory-installed inside of the connector.

- Aluminum wire expands and contracts to a greater degree than does copper wire for an equal load. This is referred to as *creep* or *cold flow*. This factor is another possible cause of a poor connection. Crimp connectors for aluminum conductors are usually longer than those for comparable copper conductors, thus resulting in greater contact surface of the conductor in the connector.

- Older technology aluminum such as Alloy 1350 experienced the above problems. The newer technology aluminum such as AA-8000 has much better conductivity, creep resistance, and strength. Still, extreme care must be taken when terminating aluminum conductors. The manufacturer of the aluminum conductor will provide detailed installation instructions.

Proper Installation Procedures

Proper, trouble-free connections for aluminum conductors require terminals, lugs, and/or connectors that are suitable for the type of conductor being installed.

Terminals on receptacles and switches must be suitable for the conductors being attached. Table 4-4 shows how the electrician can identify these terminals. Listed connectors provide proper connection when properly installed.

See *110.14*, *404.14(C)*, and *406.2(C)*.

TYPE OF DEVICE	MARKING ON TERMINAL OR CONDUCTOR	CONNECTOR PERMITTED
15- or 20-ampere receptacles and switches	CO/ALR	aluminum, copper, copper-clad aluminum
15- and 20-ampere receptacles and switches	NONE	copper, copper-clad aluminum
30-ampere and greater receptacles and switches	AL/CU	aluminum, copper, copper-clad aluminum
30-ampere and greater receptacles and switches	NONE	copper only
Screwless pressure terminal connectors of the push-in type	NONE	copper or copper-clad aluminum
Wire connectors	AL	aluminum
Wire connectors	AL/CU or CU/AL	aluminum, copper, copper-clad aluminum
Wire connectors	CC	copper-clad aluminum only
Wire connectors	CC/CU or CU/CC	copper or copper-clad aluminum
Wire connectors	CU or CC/CU	copper only
Any of the above devices	COPPER OR CU ONLY	copper only

Table 4-4 Terminal identification markings and the acceptable types of conductors permitted to be connected to a specific type of terminal.

Electrical Fires Caused by Improper Connections of Aluminum Conductors

Many electrical fires have been caused by improper connections and terminations of aluminum conductors. The problem is not the aluminum conductors. The problem is with the connections and terminations.

Many of these problems occurred in the meter base and at the main panel where aluminum service-entrance conductors were installed years ago. In the course of time, the connections failed. Troubles also happened where aluminum conductors were terminated on receptacles and switches.

The Consumer Product Safety Commission (CPSC) (Washington, D.C. 20207) publishes an excellent free booklet (#516) entitled *Repairing Aluminum Wiring* that explains what you can do with connections and terminations for the small aluminum wires that were installed years ago for the branch-circuit wiring of receptacles and switches.

The CPSC states that older homes wired with aluminum conductors are 55 times more likely to have one or more connections that will reach "Fire Hazard Conditions" than are homes wired with copper.

Although the *NEC*, notably in *110.5, 110.14, 240.4(D), 310.2(B), 310.14, Table 310.16* and other sections, recognizes aluminum conductors in sizes 12 AWG and larger, there are some local electrical codes that do not allow aluminum conductors to be used. Check this out before using aluminum conductors.

Some of the signs of potential problems might be:

- Wall switches and receptacles that feel warm to the touch.

- Smell of burning plastic or rubber at switches, receptacle, main panel, or meter.

- Flickering lights.

- Lights getting bright or dim.

- Circuits that don't work; some circuits are on, others are off.

- TV shifts to another channel for no apparent reason.

- Appliances with electronic controls shut off or change settings for no apparent reason.

- Security system sounds off for no apparent reason.

Wire Connections

Wire connectors are known in the trade by such names as *screw terminal, pressure terminal connector, wire connector, Wing-Nut,® Wire-Nut,® Scotchlok,® split-bolt connector, pressure cable connector, solderless lug, soldering lug, solder lug,* and others.

All cartons and manufacturers' literature show the size, number, and types of conductors permitted.

Solder-type lugs and connectors are rarely, if ever, used today. In fact, connections that depend upon solder are *not* permitted for connecting service-entrance conductors to service equipment, *230.81.* Nor is solder permitted for grounding and bonding connections, *250.8.* The labor costs and time spent make the use of the solder-type connections prohibitive.

Solderless connectors designed to establish *Code*-compliant electrical connections are shown in Figure 4-4.

As with the terminals on wiring devices (switches and receptacles), wire connectors must be marked AL when they are to be used with aluminum conductors. This marking is found on the connector itself, or appears on or in the shipping carton.

Connectors marked **AL/CU** are suitable for use with aluminum, copper, or copper-clad aluminum conductors. This marking is found on the connector itself, or it appears on or in the shipping carton.

Connectors not marked *AL* or *AL/CU* are for use with copper conductors only.

Unless specially stated on or in the shipping carton, or on the connector itself, conductors made of copper, aluminum, or copper-clad aluminum *may not* be used in combination in the same connector.

When combinations are permitted, the connector will be identified for the purpose and for the conditions when and where they may be used. The conditions usually are limited to dry locations only. There are some "twist-on" wire connectors that have recently been listed for use with combinations of copper and aluminum conductors.

The preceding data are found in *110.14* and in the Underwriters Laboratories (UL) Standards. When terminating large conductors such as service-entrance conductors, be sure to read any instructions that might be included with the equipment where these conductors will be terminated. Terminations for all electrical connections to devices and equipment shall be torqued as required by the manufacturer of such electrical device or equipment, *110.12(D).*

CRIMP CONNECTORS USED TO SPLICE AND TERMINATE 20 AWG TO 500 KCMIL ALUMINUM-TO-ALUMINUM, ALUMINUM-TO-COPPER, OR COPPER-TO-COPPER CONDUCTORS.

PROPERLY CRIMP
THEN TAPE

CONNECTORS USED TO CONNECT WIRES TOGETHER ON COMBINATIONS OF 18 AWG THROUGH 6 AWG CONDUCTORS. THEY ARE TWIST-ON, SOLDERLESS, AND TAPELESS.

*WIRE-NUT® and WING-NUT® are registered trademarks of IDEAL INDUSTRIES, INC. Scotchlok® is a registered trademark of 3M.

WIRE CONNECTORS VARIOUSLY KNOWN AS
WIRE-NUT,® WING-NUT,® AND SCOTCHLOK.®

CONNECTORS USED TO CONNECT WIRES TOGETHER IN COMBINATIONS OF 16, 14, AND 12 AWG CONDUCTORS. THEY ARE CRIMPED ON WITH A SPECIAL TOOL, THEN COVERED WITH A SNAP-ON INSULATING CAP.

CRIMP-TYPE WIRE CONNECTOR
AND INSULATING CAP

SOLDERLESS CONNECTORS ARE AVAILABLE IN SIZES 14 AWG THROUGH 500 KCMIL CONDUCTORS. THEY ARE USED FOR ONE SOLID OR ONE STRANDED CONDUCTOR ONLY, UNLESS OTHERWISE NOTED ON THE CONNECTOR OR ON ITS SHIPPING CARTON. THE SCREW MAY BE OF THE STANDARD SCREWDRIVER SLOT TYPE, OR IT MAY BE FOR USE WITH AN ALLEN WRENCH OR SOCKET WRENCH.

SOLDERLESS CONNECTORS

COMPRESSION CONNECTORS ARE USED FOR 8 AWG THROUGH 1000 KCMIL CONDUCTORS. THE WIRE IS INSERTED INTO THE END OF THE CONNECTOR, THEN CRIMPED ON WITH A SPECIAL COMPRESSION TOOL.

COMPRESSION CONNECTOR

SPLIT-BOLT CONNECTORS ARE USED FOR CONNECTING TWO CONDUCTORS TOGETHER, OR FOR TAPPING ONE CONDUCTOR TO ANOTHER. THEY ARE AVAILABLE IN SIZES 10 AWG THROUGH 1000 KCMIL. THEY ARE USED FOR TWO SOLID AND/OR TWO STRANDED CONDUCTORS ONLY, UNLESS OTHERWISE NOTED ON THE CONNECTOR OR ON ITS SHIPPING CARTON.

SPLIT-BOLT CONNECTOR

Figure 4-4   Types of wire connectors.

Old Houses Wired with Aluminum Conductors

If you come across a house wired with aluminum conductors, it was probably wired back in the mid sixties or early seventies. You will very likely find problems where the conductors are terminated on switches and receptacles. Take special care when changing out switches and receptacles or making splices. Use wiring devices that bear the AL/CU marking. You can also use special "pigtails" designed to make the connection between aluminum and copper conductors and terminals. One such connector is COPALUM®—a compression (crimp) connector. Apply an antioxidant to the splice or connection as recommended by the manufacturer of the connector.

CONDUCTOR INSULATION

The *NEC*® requires that all conductors be insulated, *310.2(A)*. There are a few exceptions, such as the permission to use a bare neutral conductor for services, and bare equipment grounding conductors.

What Is Thermoplastic Insulation?

Thermoplastic insulation is the most common. It is like ice cream. It will soften and melt if heated above its rated temperature. It can be heated, melted, and reshaped. Thermoplastic insulation will stiffen at temperatures colder than 14°F (minus 10°C). Typical examples of thermoplastic insulation are Types THHN, THHW, THW, THWN, and TW.

What Is Thermoset Insulation?

Thermoset insulation can withstand higher and lower temperatures. It is like baking a cake. Once the ingredients have been mixed, heated, and formed, it can never be reheated and reshaped. If heated above its rated temperature, it will char and crack. Typical examples of thermoset insulation are Types RHH, RHW, XHH, and XHHW.

Table 310.13 lists the various conductor insulations and their applications.

The allowable ampacities of copper conductors are given in *Table 310.16* for various types of insulation. *Table 310.15(B)(6)* is a table of conductor ampacities that are greater than the ampacities found in *Table 310.16*. This table is permitted to be used *only* for dwelling unit 120/240-volt, three-wire, single-phase service-entrance conductors, service lateral conductors, and feeder conductors that serve as the main power feeder to a dwelling unit. This table is based upon the tremendous known load diversity found in homes. These special ampacities must not be used for feeder conductors that do not serve as the main power feeder to a dwelling. An example of this would be the feeder from the Main Panel A to Panel B in the residence discussed later on in this text.

Table 4-5 shows the special ampacities discussed above. This table is a mirror image of *Table 310.15(B)(6)* in the *NEC*.®

The insulation covering wires and cables used in house wiring is usually rated at 600 volts or less. Exceptions to this statement are low-voltage wiring and luminaire (fixture) wiring.

SPECIAL AMPACITY RATINGS FOR RESIDENTIAL 120/240-VOLT, 3-WIRE, SINGLE-PHASE SERVICE-ENTRANCE CONDUCTORS, SERVICE-LATERAL CONDUCTORS, AND FEEDER CONDUCTORS THAT SERVE AS THE MAIN POWER FEEDER TO A DWELLING UNIT. CONDUCTOR TYPES RHH, RHW, RHW-2, THHN, THHW, THW, THW-2, THWN, THWN-2, XHHW, XHHW-2, SE, USE, USE-2.

Copper Conductor AWG for Insulation of RH-RHH-RHW-THW-THWN-THHW-THHN-XHHW-USE	Aluminum or Copper-Clad Aluminum Conductors (AWG)	Allowable Special Ampacity
4	2	100
3	1	110
2	1/0	125
1	2/0	150
1/0	3/0	175
2/0	4/0	200
3/0	250 kcmil	225
4/0	300 kcmil	250
250 kcmil	350 kcmil	300
350 kcmil	500 kcmil	350
400 kcmil	600 kcmil	400

Note: Where the conductors will be exposed to temperatures in excess of 86°F (30°C), apply the correction factors found at the bottom of *Table 310.16*. **Note:** Conductors having a suffix "2," such as THWN-2, are permitted for continuous use at 194°F (90°C) in wet or dry locations.

Table 4-5

Older Types of Conductor Insulation

Although Type THHN/THWN is the most common building wire used today, older types of conductor insulation might still be found, such as Types A, RH, RHH, RHW, RU, RUH, RUW, T, TW, and THW. Some of these are still shown in the *NEC®* but may no longer be manufactured. Some are still manufactured but are difficult, if not impossible, to find at an electrical supply house. Others are available only as a special-order product.

WET, DAMP, DRY, AND SUNLIGHT LOCATIONS

Conductors are listed for specific locations. Be sure that the conductor you use is suitable for the location. Table 4-6 shows some of the more typical conductors used in house wiring. Here are some definitions taken directly from the *NEC®*.

- **Damp Location.** Locations protected from weather and not subject to saturation with water or other liquids but subject to moderate degrees of moisture. Examples of such locations include partially protected locations under canopies, marquees, roofed open porches, and like locations; and interior locations subject to moderate degrees of moisture, such as some basements, some barns, and some cold-storage warehouses. (See Figure 11-4.)

- **Dry Location.** A location not normally subject to dampness or wetness. A location classified as dry may be temporarily subject to dampness or wetness, as in the case of a building under construction.

TRADE NAME	TYPE LETTER	MAXIMUM OPERATING TEMP.	APPLICATION PROVISIONS	INSULATION	AWG	OUTER COVERING
Heat-resistant thermoplastic	THHN	194°F (90°C)	Dry and damp locations	Flame-retardant, heat-resistant thermoplastic	14 through 1000 kcmil	Nylon jacket or equivalent
Moisture- and heat-resistant thermoplastic	THHW	167°F (75°C) 194°F (90°C)	Wet location Dry location	Flame-retardant, moisture- and heat-resistant thermoplastic	14 through 1000 kcmil	None
Moisture- and heat-resistant thermoplastic	THWN Note: If marked THWN-2, OK for 194°F (90°C) in dry or wet locations	167°F (75°C)	Dry and wet locations	Flame-retardant, moisture- and heat-resistant thermoplastic	14 through 1000 kcmil	Nylon jacket or equivalent
Moisture- and heat-resistant thermoplastic	THW Note: If marked THW-2, OK for 194°F (90°C) in dry or wet locations	167°F (75°C)	Dry and wet locations	Flame-retardant, moisture- and heat-resistant thermoplastic	14 through 2000 kcmil	None
Underground feeder and branch-circuit single conductor or multiconductor cable. See *Article 339*	UF	140°F (60°C) 167°F (75°C)	See *Article 339* See *Article 339*	Moisture-resistant Moisture-resistant	14 through 4/0	Integral with insulation

Table 4-6   Typical conductors used for residential wiring.

- **Wet Location.** Installations underground or in concrete slabs or masonry in direct contact with the earth; locations subject to saturation with water or other liquids, such as vehicle washing areas; and locations exposed to weather and unprotected. (See Figure 11-4).

- **Locations Exposed to Direct Sunlight.** ▶Insulated conductors and cables used where exposed to direct rays of the sun shall be listed, or listed and marked as being "sunlight resistant."◀

See *NEC® 310.8, 310.10, 310.13,* and *Table 310.13* for additional information.

Conductor Temperature Ratings

Conductors are also rated as to the temperature they can withstand. For example:

	Celsius	Fahrenheit
Type TW	60°	140°
Type THWN	75°	167°
Type THHN	90°	194°

Without question, conductors with Type THHN insulation are the most popular and most commonly used, particularly in the smaller sizes, because of their small diameter (easy to handle; more conductors of a given size permitted in a given size raceway) and their suitability for installation where high temperatures are encountered, such as attics, buried in insulation, and supplying recessed fixtures.

The temperature rating of nonmetallic-sheathed cable is 194°F (90°C). However, *334.80* states that the allowable ampacity is that of the 140°F (60°C) column in *Table 310.16.*

Multiple-Type Designations

Some conductors are listed for more than one application. For example, a conductor marked THHN/THWN is rated 600 volts, 90°C (194°F) when used in dry locations, *and* 600 volts, 75°C (167°F) when used in wet locations. As always, read the surface marking on the conductor insulation, on the tag, on the reel, or on the carton. The *NEC®* covers *Conductors for General Wiring* in *Article 310.*

The Weakest Link of the Chain

A conductor has two ends!

Selecting a conductor ampacity based solely on the allowable ampacity values found in *Table 310.16* is a very common mistake that can prove costly.

In any given circuit, we have an assortment of electrical components that have different maximum temperature ratings. Our job is to find out which component is the "weakest link" in the system and do our circuit design based on the lowest rated component in the circuit. Maximum temperature ratings are found in the *NEC®* and in the UL Standards.

- **Circuit breakers:** UL 489. Generally marked "75°C Only" or "60°C/75°C" (140°F/167°F).

- **Conductors:** *Table 310.13, Table 310.16,* and UL 83.

- **Disconnect switches:** UL 98. Generally marked "75°C Only" or "60°C/75°C" (140°F/167°F).

- **Panelboards:** UL 67. Panelboards are generally marked 167°F (75°C). The temperature rating of a panelboard has been established as an assembly with circuit breakers in place. Do not use the circuit breakers' temperature marking by itself. It is the panelboard temperature marking that counts.

- **Receptacles and attachment-plug caps:** UL 498. Most 15- and 20-ampere branch-circuits have receptacles that have a maximum temperature rating of 140°F (60°C). Some 30-ampere receptacles are rated 140°F (60°C) and some are rated for 167°F (75°C). Larger 50-ampere receptacles are rated 167°F (75°C).

- **Snap switches:** UL 20. Most 15- and 20-ampere branch-circuits have snap switches with a maximum temperature rating of 140°F (60°C).

- **Wire splicing devices and connectors:** Wire-Nuts,® Wing-Nuts,® Twisters,® and Scotchloks® for copper conductors (UL 486) and for aluminum conductors (UL 486B). These types of connectors are rated 221°F (105°C).

Table 4-7 indicates the correct applications for conductors and terminations.

Another key *NEC®* requirement is found in *110.14(C),* "Termination Limitations," where we find the following:

For circuits rated 100 amperes or less, or marked for conductor sizes 14 through 1 AWG, unless otherwise identified, the terminals on wiring devices, switches, breakers, motor controllers, and other electrical equipment are based upon the ampacity of

TERMINATION RATING	CONDUCTOR INSULATION RATING		
	60°C	75°C	90°C
60°C	OK	OK (at 60°C ampacity)	OK (at 60°C ampacity)
75°C	NO	OK	OK (at 75°C ampacity)
60/75°C	OK	OK (at 60°C or 75°C ampacity)	OK (at 60°C or 75°C ampacity)
90°C	NO	NO	OK (Only if equipment has a 90°C rating.)

Table 4-7 Types of conductor temperature ratings permitted for various terminal temperature ratings in conformance to the *NEC*® and the UL Standards.

140°F (60°C) conductors. It is acceptable to install conductors having a higher temperature rating, such as 90°C (194°F) THHN, but you must use the 140°F (60°C) ampacity values.

For circuits rated over 100 amperes, or marked for conductors larger than 1 AWG, the conductors installed must have a minimum 167°F (75°C) rating. The ampacity of the conductors is based on the 167°F (75°C) values. It is acceptable to install conductors having a higher temperature rating, such as 194°F (90°C) THHN, but you must use the 167°F (75°C) ampacity values.

When using high-temperature insulated conductors where adjustment factors are to be applied, such as *derating* for more than three current-carrying conductors in a raceway or cable, or *correcting* where high temperatures are encountered, the final results of these adjustments (in amperes) must comply with the requirements of *110.14(C)*. Similar requirements are found in the UL Standards.

Examples of applying *derating factors* and *correcting factors* are found in Unit 18.

Temperature ratings of conductors and cables are discussed later in this unit.

In *240.4(D)*, there is a built-in code compliance limitation for applying small branch-circuit conductors. Here we find that the maximum overcurrent protection is 15 amperes for 14 AWG, 20 amperes for 12 AWG, and 30 amperes for 10 AWG.

From *Table 310.16* we find that:

* 14 AWG copper conductors' allowable ampacity is:

20 amperes in the 60°C (140°F) column
20 amperes in the 75°C (167°F) column
25 amperes in the 90°C (194°F) column

* 12 AWG copper conductors' allowable ampacity is:
25 amperes in the 60°C (140°F) column
25 amperes in the 75°C (167°F) column
30 amperes in the 90°C (194°F) column

* 10 AWG copper conductors' allowable ampacity is:
30 amperes in the 60°C (140°F) column
35 amperes in the 75°C (167°F) column
40 amperes in the 90°C (194°F) column

There are exceptions to this maximum overcurrent protection rule for motor, air-conditioner, and heat pump branch-circuits. These exceptions are covered elsewhere in this text.

In summary, we cannot arbitrarily use the ampacity values for conductors as found in *Table 310.16*. Conductor ampacity is not a "stand-alone" issue. We must also consider the temperature limitations of the equipment, such as panelboards, receptacles, snap switches, receptacles, connectors, and so on. The lowest maximum temperature-rated device in an electrical system is the weakest link, and it is the weakest link that we must base our ultimate conductor ampacity and conductor insulation decision on.

Knob-and-Tube Wiring

Older homes were wired with open, individual, insulated conductors supported by porcelain *knobs* and *tubes*. In its day, knob-and-tube wiring served its purpose well. This text does not cover knob-and-tube wiring because this method is no longer used in new construction. On occasion, it is necessary to make some modifications to knob-and-tube wiring when doing remodel work. The *NEC*® covers this wiring method in *Article 394*.

Watch out when working on old knob-and-tube wiring. You will possibly find the old rubber insulation on the conductors dried out and brittle from the many years of heat generated by luminaires (lighting fixtures) and from the extreme heat in attics, particularly when the conductors are completely buried in thermal insulation. The conductor insulation might fall apart when touched or moved. This is a potential fire hazard!

If you do find this situation, one fix is to slide insulation tubing (available from electrical distributors) over the conductors to reinsulate them.

Neutral Conductor Size

Figure 4-5 shows how *line-to-line* loads and *line-to-neutral* loads are connected. The grounded neutral (white) conductor Ⓝ for residential services and feeders is permitted to be smaller than the "hot" ungrounded phase (black, red) conductors L1 and L2 only when the neutral conductor is properly and adequately sized to carry the maximum unbalance loads calculated according to *215.2, 220.61,* and *230.42.*

NEC® 215.2 relates to feeders and refers us to Article 220 for calculation requirements.

NEC® 230.42 relates to services, and refers us to Article 220 for calculation requirements.

Focusing on the neutral conductor, *220.61* states that *"the feeder or service neutral load shall be the maximum unbalance of the load determined by this Article."*

NEC® 220.61 further states that *"the maximum unbalanced load shall be the maximum net calculated load between the neutral and any one ungrounded conductor."*

In a typical residence, we find a number of loads that carry little or no neutral current, such as an electric water heater, electric clothes dryer, electric oven and range, electric furnace, and air conditioner.

Checking Figure 4-5, you can readily see that load (C) is connected to a straight 240-volt, 2-wire circuit and has no neutral current. Note that load (E) is connected to a 120/240-volt 3-wire circuit, and in this hookup the 120-volt motor (M) will result in current flowing in the neutral conductor. Loads (B) are 120-volt loads that are connected line to neutral.

Thus we find logic in the *NEC®* which permits reducing the neutral size on services and feeders where the calculations prove that the neutral will be carrying less current than the "hot" phase conductors.

See Unit 20 for sizing neutral conductors for electric range branch-circuits and Unit 29 for sizing neutral service-entrance conductors.

VOLTAGE DROP

Low voltage can cause lights to dim, some television pictures to "shrink," motors to run hot, electric heaters to not produce their rated heat output, and appliances to not operate properly.

Low voltage in a home can be caused by:

• wire that is too small for the load being served

• a circuit that is too long

• poor connections at the terminals

• conductors operating at high temperatures having higher resistance than when operating at lower temperatures

Figure 4-5   Diagram showing line-to-line and line-to-neutral loads.

A simple formula for calculating voltage drop on single-phase systems considers only the dc resistance of the conductors and the temperature of the conductor. See *Table 8, Chapter 9* in the *NEC®* for the dc resistance values. The more accurate formulas consider ac resistance, reactance, temperature, spacing, in metal conduit and in nonmetallic conduit, *NEC® Table 9, Chapter 9*. Voltage drop is covered in great detail in *Electrical Wiring—Commercial* (Thomson Delmar Learning). The simple voltage-drop formula is more accurate with smaller conductors and gets less and less accurate as conductor size increases. It is sufficiently accurate for voltage-drop calculations necessary for residential wiring.

To Find Voltage Drop in a Single-Phase Circuit:

$$E_d = \frac{K \times I \times L \times 2}{CMA}$$

To Find Conductor Size for a Single-Phase Circuit:

$$CMA = \frac{K \times I \times L \times 2}{E_d}$$

In the above formulas:

E_d = allowable voltage drop in volts.

K = approximate resistance in ohms per circular-mil foot at 167°F (75°C).

- for uncoated copper wire, use 12 ohms.*
- for aluminum wire, use 20 ohms.*

I = current in amperes flowing through the conductors.

L = length in feet from beginning of circuit to the load.

CMA = cross-sectional area of the conductors in circular mils.**

We use the factor of 2 for single-phase circuits because there is voltage drop in both conductors, to and from the connected load.

To Find Voltage Drop and Conductor Size for a Three-Phase Circuit

Although residential wiring does not use three-phase systems, commercial and industrial systems do. To make voltage drop and conductor sizing calculations for a three-phase system, substitute the factor of 1.73 (the square root of 3) in place of the factor 2 in the voltage-drop formulas.

Code References to Voltage Drop

For branch-circuits, refer to *210.19(A), FPN No. 4.* The recommended maximum voltage drop is 3 percent, as shown in Figure 4-6.

*Derived values from *Tables 8* and *9, Chapter 9, NEC®.*
**From Table 4-8 in this text or from *Table 8, Chapter 9, NEC®.*

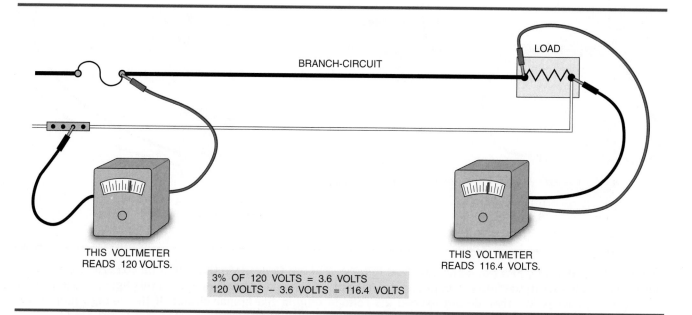

THIS VOLTMETER READS 120 VOLTS.

THIS VOLTMETER READS 116.4 VOLTS.

3% OF 120 VOLTS = 3.6 VOLTS
120 VOLTS − 3.6 VOLTS = 116.4 VOLTS

BRANCH-CIRCUIT

LOAD

Figure 4-6   Maximum recommended voltage drop on a branch-circuit is 3%, *210.19(A), FPN No. 4.*

For feeders, refer to *215.2(A)*, *FPN No. 2*. The recommended maximum voltage drop is 3 percent. See Figure 4-7.

When both branch-circuits and feeders are involved, the total voltage drop should not exceed 5 percent. This is shown in Figure 4-7.

According to *90.5(C)*, *FPNs* are explanatory only, and are not to be enforced as *Code* requirement.

There is nothing in the *NEC®* that dictates what the incoming voltage to a home must be. Incoming voltage is really determined by the electric utility and is affected by the type, size, and length of the service conductors, transformer(s), and primary lines. Electric utilities come under the jurisdiction of the *National Electrical Safety Code® (NESC)*. A local public service commission, usually at the state level, mandates maximum over- and undervoltages. This is generally within the 5 percent range of nominal voltage.

Examples of Voltage-Drop Calculations

EXAMPLE 1: What is the approximate voltage drop on a 120-volt, single-phase circuit consisting of 14 AWG copper conductors where the load is 11 amperes and the distance of the circuit from the panel to actual load is 85 ft (25.91 m)?

SOLUTION:

$$E_d = \frac{K \times I \times L \times 2}{CMA}$$

$$E_d = \frac{12 \times 11 \times 85 \times 2}{4110}$$

$$E_d = 5.46 \text{ volts drop}$$

Note: Refer to Table 4-8 for the CMA value.

This exceeds the voltage drop permitted by the *Code*, which is:

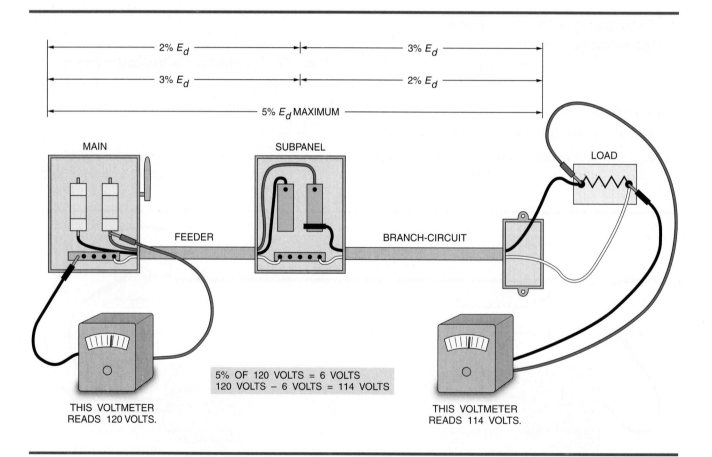

Figure 4-7 *210.19(A), FPN No. 4* and *215.2(A) FPN No. 2* of the *Code* state that the total voltage drop from the beginning of a feeder to the farthest outlet on a branch-circuit should not exceed 5%. In this figure, if the voltage drop in the feeder is 3%, then do not exceed 2% voltage drop in the branch-circuit. If the voltage drop in the feeder is 2%, then do not exceed 3% voltage drop in the branch-circuit.

CONDUCTOR SIZE, AWG	CROSS-SECTIONAL AREA IN CIRCULAR MILS
18	1620
16	2580
14	4110
12	6530
10	10,380
8	16,510
6	26,240
4	41,740
3	52,620
2	66,360
1	83,690
0	105,600
00	133,100
000	167,800
0000	211,600

Table 4-8 Circular mil cross-sectional area for many of the more common conductors.

$$3\% \text{ of } 120 \text{ volts } = 3.6 \text{ volts}$$

Let's try it again using 12 AWG conductors:

$$E_d = \frac{12 \times 11 \times 85 \times 2}{6530}$$

$$E_d = 3.44 \text{ volts drop}$$

This "meets Code."

EXAMPLE 2: Find the wire size needed to keep the voltage drop to no more than 3 percent on a single-phase, 240-volt, air-conditioner circuit. The nameplate reads: "Minimum Circuit Ampacity 40 Amperes." The circuit originates at the main panel located approximately 65 ft (19.81 m) from the air-conditioner unit. No neutral is required.

SOLUTION: Checking *Table 310.16*, the conductors could be 6 AWG Type TW copper or 8 AWG Type THHN copper.

The permitted voltage drop is:

$$E_d = 240 \times 0.03 = 7.2 \text{ volts}$$

Let's see what voltage drop we might experience if we installed the 8 AWG Type THHN conductors.

$$E_d = \frac{12 \times 40 \times 65 \times 2}{16,510}$$

$$= 3.78 \text{ volts drop}$$

This is well below the permissible 7.2 volts drop and would make an acceptable installation.

A Word of Caution When Using High-Temperature Conductors

Note in *Table 310.16* that conductors fall into three classes of temperature ratings—60°C, 75°C, and 90°C. For a given conductor size, we find that the allowable ampacity of a 90°C insulated conductor is greater than that of a 60°C insulated conductor. Therefore, be careful when selecting conductors based on their ability to withstand high temperatures. For example:

- 8 AWG THHN (90°C) copper has an ampacity of 55 amperes.
- 8 AWG THW (75°C) copper has an ampacity of 50 amperes.
- 8 AWG TW (60°C) copper has an ampacity of 40 amperes.

We learned earlier in this unit that according to *110.14(C)*, we use the 60°C column of *Table 310.16* for circuits rated 100 amperes or less, and the 75°C column for circuits rated over 100 amperes. However, when the equipment is marked as being suitable for 75°C, we can take advantage of the higher ampacity of 75°C conductors. This oftentimes results in a smaller size conductor and a correspondingly smaller raceway.

But installing smaller conductors might result in excessive voltage drop.

Therefore, after selecting the proper size conductor for a given load, it is always a good idea to run a voltage-drop calculation to make sure the voltage drop is not excessive.

What Is the Advantage of High-Temperature Conductors?

They can withstand the high temperatures found in attics and high-temperature climates.

Another big advantage of the higher ampacities of high-temperature insulation is that the *NEC*® permits the higher ampacity values to be used as the starting point when applying derating and correcting factors. Example calculations of this are found in Unit 18.

Effect of Voltage Variation

Unit 19 has more information on the effect voltage differences have on wattage output of appliances and on electric motors.

Table 310.16 Allowable Ampacities of Insulated Conductors Rated 0 Through 2000 Volts, 60°C Through 90°C (140°F Through 194°F), Not More Than Three Current-Carrying Conductors in Raceway, Cable, or Earth (Directly Buried), Based on Ambient Temperature of 30°C (86°F)

	Temperature Rating of Conductor (See Table 310.13.)						
	60°C (140°F)	75°C (167°F)	90°C (194°F)	60°C (140°F)	75°C (167°F)	90°C (194°F)	
Size AWG or kcmil	Types TW, UF	Types RHW, THHW, THW, THWN, XHHW, USE, ZW	Types TBS, SA, SIS, FEP, FEPB, MI, RHH, RHW-2, THHN, THHW, THW-2, THWN-2, USE-2, XHH, XHHW, XHHW-2, ZW-2	Types TW, UF	Types RHW, THHW, THW, THWN, XHHW, USE	Types TBS, SA, SIS, THHN, THHW, THW-2, THWN-2, RHH, RHW-2, USE-2, XHH, XHHW, XHHW-2, ZW-2	Size AWG or kcmil
	COPPER			ALUMINUM OR COPPER-CLAD ALUMINUM			
18	—	—	14	—	—	—	—
16	—	—	18	—	—	—	—
14*	20	20	25	—	—	—	—
12*	25	25	30	20	20	25	12*
10*	30	35	40	25	30	35	10*
8	40	50	55	30	40	45	8
6	55	65	75	40	50	60	6
4	70	85	95	55	65	75	4
3	85	100	110	65	75	85	3
2	95	115	130	75	90	100	2
1	110	130	150	85	100	115	1
1/0	125	150	170	100	120	135	1/0
2/0	145	175	195	115	135	150	2/0
3/0	165	200	225	130	155	175	3/0
4/0	195	230	260	150	180	205	4/0
250	215	255	290	170	205	230	250
300	240	285	320	190	230	255	300
350	260	310	350	210	250	280	350
400	280	335	380	225	270	305	400
500	320	380	430	260	310	350	500
600	355	420	475	285	340	385	600
700	385	460	520	310	375	420	700
750	400	475	535	320	385	435	750
800	410	490	555	330	395	450	800
900	435	520	585	355	425	480	900
1000	455	545	615	375	445	500	1000
1250	495	590	665	405	485	545	1250
1500	520	625	705	435	520	585	1500
1750	545	650	735	455	545	615	1750
2000	560	665	750	470	560	630	2000
	CORRECTION FACTORS						
Ambient Temp. (°C)	For ambient temperatures other than 30°C (86°F), multiply the allowable ampacities shown above by the appropriate factor shown below.						Ambient Temp. (°F)
21–25	1.08	1.05	1.04	1.08	1.05	1.04	70–77
26–30	1.00	1.00	1.00	1.00	1.00	1.00	78–86
31–35	0.91	0.94	0.96	0.91	0.94	0.96	87–95
36–40	0.82	0.88	0.91	0.82	0.88	0.91	96–104
41–45	0.71	0.82	0.87	0.71	0.82	0.87	105–113
46–50	0.58	0.75	0.82	0.58	0.75	0.82	114–122
51–55	0.41	0.67	0.76	0.41	0.67	0.76	123–131
56–60	—	0.58	0.71	—	0.58	0.71	132–140
61–70	—	0.33	0.58	—	0.33	0.58	141–158
71–80	—	—	0.41	—	—	0.41	159–176

* See 240.4(D).

Reprinted with permission from NFPA 70-2005.

APPROXIMATE CONDUCTOR SIZE RELATIONSHIP

There is a definite relationship between the circular mil area and the resistance of conductors. The following rules explain this relationship.

Rule One: For wire sizes up through 0000, every third size doubles or halves in circular mil area.

Thus, a 1 AWG conductor is 2 times larger than a 4 AWG conductor (83,690 versus 41,470). Thus, a "0" wire is one-half the size of "0000" wire (105,600 versus 211,600).

Rule Two: For wire sizes up through 0000, every consecutive wire size is approximately 1.26 times larger or smaller than the preceding wire size.

Thus, a 3 AWG conductor is approximately 1.26 times larger than a 4 AWG conductor (41,740 × 1.26 = 52,592). A 2 AWG conductor is approximately 1.26 times smaller than a 1 AWG conductor (83,690 ÷ 1.26 = 66,420).

Try to fix in your mind that a 10 AWG conductor has a cross-sectional area of 10,380 circular mils, and that it has a resistance of 1.2 ohms per 1000 ft (300 m). The resistance of aluminum wire is approximately 2 ohms per 1000 ft (300 m). By remembering these numbers, you will be able to perform voltage-drop calculations without having the wire tables readily available.

EXAMPLE: What is the approximate cross-sectional area, in circular mils, and resistance of a 6 AWG copper conductor?

WIRE SIZE (AWG)	CMA (In Circular Mils)	OHMS PER 1000 FT
10	10,380	1.2
9		
8		
7	20,760	0.6
6	26,158	0.0476

Table 4-9 "Every third size" and the 1.26 relationship between the circular-mil area (CMA) and the resistance in ohms per 1000 ft of copper conductors sizes 10 AWG through 6 AWG.

SOLUTION: Note in Table 4-9 that when the CMA of a wire is doubled, then its resistance is cut in half. Inversely, when a given wire size is reduced to one-half, its resistance doubles.

NONMETALLIC-SHEATHED CABLE
(ARTICLE 334)

The *NEC*® and UL Standard 719 describe the construction details of nonmetallic-sheathed cable. The following is a summary of these details.

Description

Most electricians still refer to nonmetallic-sheathed cable as *Romex*, a name chosen years ago by the Rome Wire and Cable Company. Romex® is now a registered trademark of Southwire Company.

Nonmetallic-sheathed cable is a factory assembly of two or more insulated conductors having an outer sheath of moisture-resistant, flame-retardant, nonmetallic material. This cable is available with two or three current-carrying conductors. The conductors range in size from 14 AWG through 2 AWG for copper conductors, and from 12 AWG through 2 AWG for aluminum or copper-clad aluminum conductors. Two-wire cables contain one black conductor, one white conductor, and one bare equipment grounding conductor. Three-wire cables contain one black, one white, one red, and one bare equipment grounding conductor. Equipment grounding conductors are permitted to have green insulation, but bare equipment grounding conductors are the most common.

▶*NEC*® *334.108* requires that nonmetallic-sheathed cable have an insulated or a bare equipment grounding conductor. ◄

Figure 4-8 clearly shows a bare equipment grounding conductor. Sometimes the bare equipment grounding conductor is wrapped with paper or fiberglass, which acts as a filler. The equipment grounding conductor is *not* permitted to be used as a current-carrying conductor.

Types of Nonmetallic-Sheathed Cable

UL lists three types of nonmetallic-sheathed cables:

- Type NM-B cable has a flame-retardant, moisture-resistant, nonmetallic outer jacket. The conductors are rated 194°F (90°C). The ampacity is based on 140°F (60°C).

- Type NMC-B cable has a flame-retardant, moisture-resistant, fungus-resistant, and corrosion-resistant nonmetallic outer jacket. The conductors are rated 194°F (90°C). The ampacity is based on 140°F (60°C).

- Type NMS-B cable contains the conventional insulated power conductors as well as telephone, coaxial, home entertainment, and signaling conductors all in one cable. This type of cable is used for home automation systems using the latest in digital technology. *Article*

Figure 4-8 A nonmetallic-sheathed Type NM-B cable showing (A) black "ungrounded" (hot) conductor, (B) bare equipment "grounding" conductor, and (C) white "grounded" conductor. *Courtesy of* Southwire Company.

780 recognizes this wiring method for closed loop and programmed power distribution. Type NMS-B cable has a moisture-resistant, flame-retardant, nonmetallic outer jacket.

Although somewhat confusing, the UL standards use the suffix B, whereas the *NEC®* refers to nonmetallic-sheathed cables as Type NM, NMC, and NMS. The conductors in these cables are rated 194°F (90°C). Types NM, NMC, and NMS cable identified by the markings NM-B, NMC-B, and NMS-B meet this 194°F (90°C) requirement. See *334.112*.

Why the Suffix *B*?

The suffix "B" means that the conductors have 194°F (90°C) insulation.

Older nonmetallic-sheathed cable contained 140°F (60°C) rated conductors. There were many problems with the insulation becoming brittle and breaking off due to the extremely high temperatures associated with recessed and surface-mounted luminaires (fixtures): fires resulted. In hot attics, after applying correction factors, the adjusted allowable ampacity of the conductors could very well be zero. To solve this dilemma, since December 17, 1984, UL has required that the conductors be rated 194°F (90°C), and that the cable be identified with the suffix *B*. This suffix makes it easy to differentiate the newer 194°F (90°C) from the old 140°F (60°C) cable.

The ampacity of the conductors in nonmetallic-sheathed cable is based on that of 140°F (60°C) conductors. The 194°F (90°C) ampacity is permitted to be used for derating purposes, such as might be necessary in attics. The final derated ampacity must not exceed the ampacity for 140°F (60°C) conductors. More specifics in *334.80*.

Table 310.16 shows the allowable ampacities for conductors.

Overcurrent Protection for Small Conductors

Overcurrent protection for small copper conductors, in conformance to *240.4(D)*, is:

- 14 AWG   15 amperes
- 12 AWG   20 amperes
- 10 AWG   30 amperes

A typical nonmetallic-sheathed cable is illustrated in Figure 4-8.

Color Coding of Type NM-B Jacket

To make installations easier for the electrician and the electrical inspector to identify the conductor size, color coded nonmetallic-sheathed cable is available.

Conductor Size	Color of Jacket
14 AWG	White
12 AWG	Yellow
10 AWG	Orange
8 and 6 AWG	Black

New Types of NM-B Cable

At least one manufacturer offers a four-conductor Type NM-B that can be used for two branch-circuits. This cable is used where sharing of the neutral is *not wanted*, as in a multiwire branch-circuit, or for branch-circuits supplying GFCIs and AFCIs, where sharing the neutral is *not permitted*.

This cable contains:

One bare equipment grounding conductor

Pair 1: one black and one white

Pair 2: one black and one white (with a red stripe)

Because all four conductors in this cable are current-carrying conductors, the ampacity of the conductors must be adjusted (derated) according to *310.15(B)(2)*.

Where Nonmetallic-Sheathed Cable May Be Used

Table 4-10 shows the uses permitted for Types NM-B, NMC-B and NMS-B.

Equipment Grounding Conductors

Equipment grounding conductors are selected based upon the rating or setting of the overcurrent device protecting a given branch-circuit or feeder. See *Table 250.122*. The equipment grounding conductor in a nonmetallic-sheathed cable is sized according to this in Table 4-11.

Installation

Nonmetallic-sheathed cable is probably the least expensive of the various wiring methods. It is relatively lightweight and easy to install. It is widely used for dwelling unit installations on circuits of 600 volts or less. Figure 4-9 shows an example of a stripper used for stripping nonmetallic-sheathed

	TYPE NM-B	TYPE NMC-B	TYPE NMS-B
May be used on circuits of 600 volts or less.	Yes	Yes	Yes
May be run exposed or concealed in dry locations.	Yes	Yes	Yes
May be installed exposed or concealed in damp and moist locations.	No	Yes	No
Has flame-retardant and moisture-resistant outer covering.	Yes	Yes	Yes
Has fungus-resistant and corrosion-resistant outer covering.	No	Yes	No
May be used to wire one- and two-family dwellings, or multifamily dwellings.	Yes	Yes	Yes
May be embedded in masonry, concrete, plaster, adobe, fill.	No	No	No
May be installed or fished in the hollow voids of masonry blocks or tile walls where not exposed to excessive moisture or dampness.	Yes	Yes	Yes
May be installed or fished in the hollow voids of masonry blocks or tile walls where exposed to excessive moisture.	No	Yes	No
In outside walls of masonry block or tile.	No	Yes	No
In inside wall of masonry block or tile.	Yes	Yes	Yes
May be used as service-entrance cable.	No	No	No
Must be protected against physical damage.	Yes	Yes	Yes
May be run in shallow chase of masonry, concrete, or adobe if protected by a steel plate at least 1/16 in. (1.6 mm) thick, then covered with plaster, adobe, or similar finish.	No	Yes	No

Table 4-10 Uses permitted for nonmetallic-sheathed cable.

Rating or Setting of Overcurrent Device	MINIMUM SIZE (AWG) EQUIPMENT GROUNDING CONDUCTOR	
	Copper	Aluminum or Copper-Clad Aluminum
15	14	12
20	12	10
30	10	8
40	10	8
60	10	8
100	8	6
200	6	4

Table 4-11 Minimum size equipment grounding conductors is based on the rating or setting of the overcurrent device protecting that particular branch-circuit. This table is based on *Table 250.122* of the *NEC*.

Figure 4-9 This stripper is used to remove the outer jacket from nonmetallic-sheathed cable. *Courtesy of Seatek Co., Inc.*

cable. A knife also works fine with care. The installation of nonmetallic-sheathed cable must conform to the requirements of *Article 334*. Refer to Figure 4-10, Figure 4-11, Figure 4-12, Figure 4-13, Figure 4-14, and Figure 4-15.

- The cable must be strapped or stapled not more than 12 in. (300 mm) from a box or fitting.

- Do not staple flat nonmetallic-sheathed cables on edge, Figure 4-14.

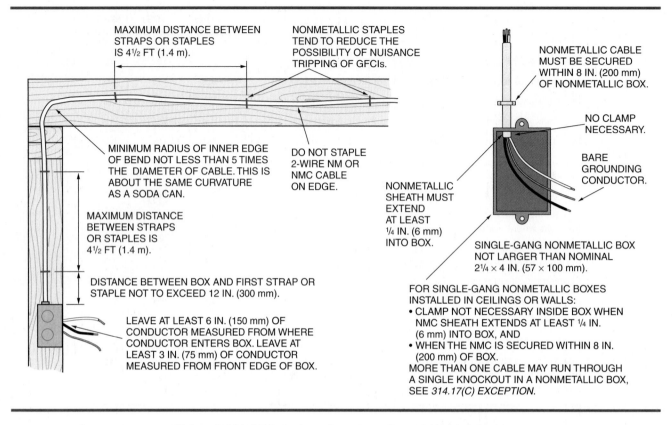

MAXIMUM DISTANCE BETWEEN STRAPS OR STAPLES IS 4½ FT (1.4 m).

NONMETALLIC STAPLES TEND TO REDUCE THE POSSIBILITY OF NUISANCE TRIPPING OF GFCIs.

NONMETALLIC CABLE MUST BE SECURED WITHIN 8 IN. (200 mm) OF NONMETALLIC BOX.

NO CLAMP NECESSARY.

MINIMUM RADIUS OF INNER EDGE OF BEND NOT LESS THAN 5 TIMES THE DIAMETER OF CABLE. THIS IS ABOUT THE SAME CURVATURE AS A SODA CAN.

DO NOT STAPLE 2-WIRE NM OR NMC CABLE ON EDGE.

BARE GROUNDING CONDUCTOR.

NONMETALLIC SHEATH MUST EXTEND AT LEAST ¼ IN. (6 mm) INTO BOX.

MAXIMUM DISTANCE BETWEEN STRAPS OR STAPLES IS 4½ FT (1.4 m).

SINGLE-GANG NONMETALLIC BOX NOT LARGER THAN NOMINAL 2¼ × 4 IN. (57 × 100 mm).

DISTANCE BETWEEN BOX AND FIRST STRAP OR STAPLE NOT TO EXCEED 12 IN. (300 mm).

FOR SINGLE-GANG NONMETALLIC BOXES INSTALLED IN CEILINGS OR WALLS:
• CLAMP NOT NECESSARY INSIDE BOX WHEN NMC SHEATH EXTENDS AT LEAST ¼ IN. (6 mm) INTO BOX, AND
• WHEN THE NMC IS SECURED WITHIN 8 IN. (200 mm) OF BOX.
MORE THAN ONE CABLE MAY RUN THROUGH A SINGLE KNOCKOUT IN A NONMETALLIC BOX, SEE *314.17(C) EXCEPTION.*

LEAVE AT LEAST 6 IN. (150 mm) OF CONDUCTOR MEASURED FROM WHERE CONDUCTOR ENTERS BOX. LEAVE AT LEAST 3 IN. (75 mm) OF CONDUCTOR MEASURED FROM FRONT EDGE OF BOX.

Figure 4-10   Installation of nonmetallic-sheathed cable.

NONMETALLIC SHEATHED CABLE

PROTECT CABLE FROM ABRASION ON BOTH ENDS, *300.15(C).*

PROTECT CABLE WITH RIGID METAL CONDUIT, INTERMEDIATE METAL CONDUIT, ELECTRICAL METALLIC TUBING, SCHEDULE 80 PVC RIGID NONMETALLIC CONDUIT, OR OTHER APPROVED MEANS. PROTECT AT LEAST 6 IN. (150 mm) ABOVE FLOOR.

Figure 4-11   Installation of exposed nonmetallic-sheathed cable where passing through a floor. See *334.15(B).*

NONMETALLIC-SHEATHED CABLES

FIRE OR DRAFT STOPPING WITH THERMAL INSULATION OR SEALING FOAM

Figure 4-12   (A) When nonmetallic-sheathed cables are "bundled" or "stacked" for distances of more than 24 in. (600 mm), their ampacities must be reduced (derated, adjusted), or ▶(B) when fire-stopping material is used where more than two nonmetallic-sheathed cables pass through wood framing members, their ampacities must be reduced according to *310.15(B)(2)(a).* See *334.80.*◀

Figure 4-13 (A) One-hole strap used for conduit, EMT, large nonmetallic-sheathed cable and armored cable. (B) Metal staple for armored cable and nonmetallic-sheathed cable. (C) Insulated staple for nonmetallic-sheathed cable. Certain types of insulated and noninsulated staples can be applied with a staple gun.

- The intervals between straps or staples must not exceed 4½ ft (1.4 m).

- Exposed runs of cables can be subjected to lots of abuse. A child may pull or stand on the cables and adults may use the cables to hang things on. *NEC® 334.15* requires that cables be protected against physical damage where necessary. For example, where exposed in a garage with open studs, run the cables on the sides of the studs. Where cables are run horizontally from/through stud to stud, protection is required. Some inspectors require protective boards over exposed horizontal cable runs, or sheet rock, or other wall finish up to the ceiling, or at least 7 ft (2.13 m) up. For protection of cables in attics, see Unit 15.

- The 4½ ft (1.4 m) securing requirement is not needed where nonmetallic-sheathed cable is run *horizontally* through holes in wood or metal framing members (studs, joists, rafters, etc.). The cable is considered to be adequately supported by the framing members, *334.30.*

- The inner edge of the bend shall have a minimum radius not less than five times the cable diameter, *334.24.*

- The cable must not be used in circuits of more than 600 volts.

- **CAUTION:** **Be very careful when installing staples. Driving them too hard can squeeze and damage the insulation on the conductors, causing short circuits and/or ground faults.**

- See Figure 4-10 for special conditions when using single-gang nonmetallic device boxes.

- Nonmetallic-sheathed cable must be protected where passing through a floor by at least 6 in. (150 mm) of rigid metal conduit, intermediate metal conduit, electrical metallic tubing, Schedule 80 PVC rigid nonmetallic conduit, or other approved means, *334.15(B).* A fitting (bushing or connector) must be used at both ends of the conduit to protect the cable from abrasion, Figure 4-11.

CODE VIOLATION
334.30

- CABLE ON EDGE

"MEETS *CODE*"
334.30

- CABLE LIES FLAT

Figure 4-14 It is a *Code* violation to staple flat nonmetallic-sheathed cables on edge, *334.30.*

IT IS OK TO FASTEN CABLES NOT SMALLER THAN TWO 6 AWG OR THREE 8 AWG CONDUCTORS DIRECTLY TO BOTTOM OF JOISTS, *334.15(C)*.

CABLES OF ANY SIZE MAY BE RUN THROUGH BORED HOLES IN JOISTS, RAFTERS, AND STUDS. THEY ARE CONSIDERED TO BE ADEQUATELY SUPPORTED, *334.30*.

CABLES OF ANY SIZE MAY BE RUN ON THE SIDES OF JOISTS. THEY MUST BE SECURED (STAPLED OR STRAPPED), *334.15(A)*, *334.30*.

*MINIMUM RADIUS OF INNER EDGE OF BEND NOT LESS THAN 5 TIMES THE DIAMETER OF CABLE. THIS IS ABOUT THE SAME CURVATURE AS A SODA CAN.

THIS IS A RUNNING BOARD.

CABLES OF ANY SIZE MAY BE RUN PARALLEL TO SIDES OR FACE OF JOISTS. KEEP CABLES AT LEAST 1¼ IN. (32 mm) FROM EDGE OF FRAMING MEMBER, *300.4(D)*.

CABLES SMALLER THAN TWO 6 OR THREE 8 AWG CONDUCTORS MUST BE RUN THROUGH BORED HOLES OR BE RUN ON THE SURFACE OF A RUNNING BOARD, *334.15(C)*.

Figure 4-15   The requirements for running nonmetallic-sheathed cable in unfinished basements.

- When nonmetallic-sheathed cables are "bundled" or "stacked" for distances of more than 24 in. (600 mm), their ampacities must be reduced (adjusted, derated) according to *Table 310.15(B)(2)(a)*. When cables are bunched together, the heat generated by the conductors cannot easily dissipate. See Figure 4-12A.

- ▶When more than two nonmetallic-sheathed cables pass through a wood framing member that is to be fire-stopped with thermal insulation or sealing foam, their ampacities must be derated according to *Table 310.15(B)(2)(a)*. See *334.80* and Figure 4-12B.◀

- When Type AC and Type MC cables are "bundled" or "stacked," derating is not necessary if the following is true:

  - the conductors are 12 AWG copper
  - if the bundling or stacking is not more than 24 in. (600 mm)
  - there are no more than three current-carrying conductors in each cable
  - there are no more than 20 current-carrying conductors in the bundle
  - When there are more than 20 current-carrying conductors in the bundle, a 60% adjustment factor shall be applied. See *310.15(B)(2), Exception No. 5.*

- The cable may be run in unsupported lengths of not more than 4½ ft (1.4 m) between the last point of support of the cable and a luminaire (fixture) or other equipment within an accessible ceiling. For this last 4½ ft (1.4 m), the 12-in.

(300-mm) and 4½-ft (1.4-m) securing requirements mentioned above are not necessary if the nonmetallic-sheathed cable is used as a luminaire (fixture) whip to connect luminaires (fixtures) of other equipment. See *334.30*. Luminaire (fixture) whips are discussed in Unit 7 and Unit 17.

Figure 4-15 shows the installation of nonmetallic-sheathed cable in unfinished basements, *334.15(C)*.

See Unit 15 and Figures 15-13 and Figure 15-14 for additional text and diagrams covering the installation of cables in attics.

Figure 4-10 shows the *Code* requirements for securing nonmetallic-sheathed cable when using nonmetallic boxes.

The *Code* in *314.40(D)* requires that all metal boxes have provisions for the attachment of an equipment grounding conductor. Figure 4-16 shows a gang-type switch (device) box that is tapped for a screw by which the grounding conductor may be connected underneath. Figure 4-17 shows an outlet box, also with provisions for attaching grounding conductors. Figure 4-18 illustrates the use of a small grounding clip.

NEC® 410.20 requires that there be a provision whereby the equipment grounding conductor can be attached to the exposed metal parts of luminaires (lighting fixtures).

Figure 4-17   Outlet box.

Figure 4-18   Method of attaching ground clip to a metal switch (device) box. See *250.148*.

Are Cables Permitted in Raceways?

Yes, but only if certain conditions are met.

NEC® 300.15(C) permits using a listed conduit or tubing without a box where the conduit or tubing is used to protect a nonmetallic-sheathed cable against physical damage. Figure 4-19 shows an example of this in a basement where the wiring method in the ceiling is NM cable, and a switch or receptacle is in a box mounted on the wall. An EMT is installed from the box to a point above the ceiling joists, where the EMT is secured to the side of the joists. The NM cable is then run inside the EMT, and a fitting is installed on the upper end of the tubing to protect the cable against abrasion. ▶The

Figure 4-16   Gang-type switch (device) box.

SLIDE
NM CABLE
INTO EMT

NM CABLE

STAPLES

▶NONMETALLIC BUSHING
OR ADAPTER, OR
COMBINATION EMT
TO NM COUPLING◀

STRAPS

▶CABLE INSIDE RACEWAY
DOES NOT REQUIRE
SECURING, *334.30.*◀

EMT

LISTED METAL CONDUIT OR
TUBING AND METAL BOX SHALL
BE GROUNDED

SLIDE CABLE ALL THE WAY INTO
EMT SO CABLE JACKET ENDS AT
LEAST ¼ IN. (6 mm) INTO BOX.

EMT CONNECTOR

GFCI RECEPTACLE

HANDY BOX

Figure 4-19 ▶**An example of EMT protecting a nonmetallic-sheathed cable from physical damage where the cable is run down the wall of an unfinished basement. This also makes a neat installation. The EMT and the metal box must be grounded. Physical protection and grounding are required by *334.15(C)*. NEC® *300.18(A), Exception* also recognizes this type of installation.** ◀

cable does not require securing inside the raceway, *334.30.* ◀ The finished job is neat.

The conductors inside a nonmetallic-sheathed cable meet all of the UL requirements for Type

THHN conductors but are not surface marked in any manner.

NEC® *334.15(B)* requires protection of nonmetallic-sheathed cable from physical damage where necessary. This section lists different types of raceways permitted to be used for protection of the cable.

The raceway fill for a cable in a conduit or tubing is based on the allowable percentage fill values specified in *Table 1, Chapter 9*.

Alternate Method of Installing NMC above Panel

Nonmetallic-sheathed cables are usually secured to a panel, box, or cabinet, with a cable clamp or connector. For surface-mounted enclosures only,

312.5(C), Exception provides an alternate method for running nonmetallic-sheathed cables as illustrated in Figure 4-20.

Identifying Nonmetallic-Sheathed Cables

One nice thing about nonmetallic-sheathed cable, particularly that with a white outer jacket, is that at panelboards and device and outlet boxes, you can use a permanent marking pen to mark the outer jacket with a few words indicating what the cable is for. Examples: *FEED TO ATTIC; FEED TO L.R.;*

A NONMETALLIC-SHEATHED CABLE SECURED WITHIN 12 IN. (300 mm) FROM WHERE IT EMERGES FROM RACEWAY.

B RACEWAY COMES OUT OF TOP OF PANEL AND MUST NOT PENETRATE STRUCTURAL CEILING.

C A FITTING IS NEEDED ON BOTH ENDS TO PREVENT ABRASION.

D RACEWAY MUST BE SEALED TO PREVENT DEBRIS FROM GETTING INTO PANEL.

E OUTER SHEATH OF CABLE MUST EXTEND INTO PANEL AT LEAST ¼ IN. (6 mm)

F SECURE THE RACEWAY.

G DO NOT EXCEED PERCENT FILL AS SHOWN IN *TABLE 1, CHAPTER 9, NEC.*® USE LARGEST DIMENSION OF CABLE AS ITS DIAMETER FOR CALCULATING THE CABLE'S CROSS-SECTIONAL AREA, *CHAPTER 9, TABLE 1, NOTE 9.*

H IF RACEWAY IS NOT OVER 24 IN. (600 mm): MAXIMUM CABLE FILL IS 60%. DERATING CONDUCTOR AMPACITY *IS NOT* NECESSARY. IF RACEWAY IS OVER 24 IN. (600 mm): MAXIMUM CABLE FILL IS 40%. DERATING CONDUCTOR AMPACITY *IS* NECESSARY. *NEC.*® *310.15(B)(a), EXCEPTION NO. 3* AND *CHAPTER 9, TABLE 1, NOTE 4.*

NONFLEXIBLE RACEWAY NOT LESS THAN 18 IN. (450 mm) NOR MORE THAN 10 FT (3.0 m).

SURFACE-MOUNTED PANEL

MAIN BONDING JUMPER

MAIN

Figure 4-20 An alternate method for running nonmetallic-sheathed cables into a surface-mounted panel. This is permitted for surface-mounted panels only, *312.5(C), Exception.*

FEED TO DINING ROOM RECEPTACLES; FEED TO PANEL; BC#6; TO OTHER 3-WAY SWITCH; SWITCH LEG TO CEILING LIGHT; TO OUTSIDE LIGHT; ETC.

By marking the cables when they are installed, confusion when making the connections is reduced considerably.

Another great idea is to make a sleeve by cutting off a short 2 in. length of NM cable, pull out the conductors, then mark an identifying statement on the sleeve. Slide the sleeve over the specific conductors inside the panel or box.

Multifamily Dwellings

Nonmetallic-sheathed cable is permitted to be installed in one- and two-family dwellings, and in multifamily dwellings where the building construction is Type III, Type IV, or Type V. See *334.12(A)*. Definitions of construction types are found in building codes, the glossary of this text, and in *Annex E* of the *NEC*.®

For years, the *NEC*® restricted nonmetallic-sheathed cable to a maximum of three floors above finished grade. This limitation was removed in the *2002 NEC*.®

ARMORED CABLE (TYPE AC) AND METAL-CLAD CABLE (TYPE MC)

Type AC and Type MC cables look very much the same (see Figure 4-21), yet there are significant differences. Table 4-12 compares some of these differences.

To remove the outer metal armor, a tool like that shown in Figure 4-23 may be used.

If a hacksaw is used, make an angle cut on one of the armor's convolutions as long as the conductors need to be, perhaps 10 to 12 in. Be careful not to cut too deep or you might cut into the conductor insulation. Bend the armor at the cut and snap it off. Then slide the 10–12 in. of armor off the end of the cable. See Figure 4-24.

Most electricians, when using armored cable, will make three cuts with their hacksaw. The middle cut is where the cable and conductors will be cut off completely. The first and third cuts are where the armor will be snapped off. This saves time because it actually prepares two ends of the cable.

Most electricians still refer to armored cable as BX, supposedly derived from an abbreviation of the Bronx in New York City, where armored cable was once manufactured. BX was a trademark owned by the General Electric Company, but over the years it has become a generic term.

Figure 4-21 Flexible armored cable. *Courtesy of* AFC Cable Systems, Inc.

TYPE AC (*ARTICLE 320*)	TYPE MC (*ARTICLE 330*)
Article 320: This article contains all of the installation requirements for the installation of Type AC armored cable.	**Article 330:** This article contains all of the installation requirements for the installation of Type MC metal-clad cable.
320-1. Definition: Type AC cable is a fabricated assembly of insulated conductors in a flexible metallic enclosure.	**330.1. Definition:** Type MC cable is a factory assembly of one or more insulated circuit conductors with or without optical fiber members and enclosed in an armor of interlocking metal tape or a smooth or corrugated metallic sheath.
Number of Conductors: 2 to 4 current-carrying conductors plus bonding wire. It may also have a separate equipment grounding conductor.	**Number of Conductors:** Any number of current-carrying conductors. At least one manufacturer has a "Home Run Cable" that has conductors for more than one branch circuit, plus the equipment grounding conductors.
Conductor Size and Type: 14 AWG thru 1 AWG for copper conductors. 12 AWG thru 1 AWG for aluminum conductors will be marked "AL" on a tag on the carton or reel. 12 AWG thru 1 AWG for copper-clad aluminum conductors will be marked "*AL (CU-CLAD)*" or "*Cu-Clad Al*" on a tag on the carton or reel.	**Conductor Size and Type:** 18 AWG thru 2000 kcmils for copper conductors. 12 AWG thru 2000 kcmils for aluminum or copper-clad aluminum conductors. 12 AWG thru 2000 kcmils for copper-clad aluminum conductors will be marked "*AL (CU-CLAD)*" or "*Cu-Clad Al*" on a tag on the carton or reel.
Color Coding: Two-conductor: one black, one white. Three-conductor: one black, one white, one red. Four-conductor: one black, one white, one red, one blue. These are in addition to any equipment grounding and bonding conductors.	**Color Coding:** Two-conductor: one black, one white. Three-conductor: one black, one white, one red. Four-conductor: one black, one white, one red, one blue. These are in addition to any equipment grounding and bonding conductors.
Bonding & Grounding: Has a 16 AWG bare bonding wire. Bonding wire is in continuous direct contact with the metal armor. The bonding wire and armor act together to serve as the acceptable equipment ground. The bonding wire does not have to be terminated, just folded back over the armor. It can even be cut off where it comes out of the armor. Figure 4-22 shows how the bonding strip is folded back over the armor. Never use the bonding strip as a grounded neutral conductor or as a separate equipment grounding conductor. It is an internal bonding strip—nothing more!	**Bonding & Grounding:** Has a separate green insulated equipment grounding conductor. May have two equipment grounding conductors for isolated ground requirements, such as for computer wiring. Unless specifically listed for grounding, the jacket of MC cable is not to be used as an equipment grounding conductor. The sheath of smooth or corrugated tube Type MC is listed as acceptable as the required equipment grounding conductor. Some MC cables have a full size bare aluminum conductor in continuous, direct contact with the metal armor. This bare conductor and the metal act together to serve as the acceptable equipment ground.
Insulation: Type ACTHH has 194°F (90°C) thermoplastic insulation, by far the most common. Type ACTH has 167°F (75°C) thermoplastic insulation. Type AC has thermoset insulation. Type ACHH has 194°F (90°C) thermoset insulation.	**Insulation:** Type MC has 194°F (90°C) thermoplastic insulation. Some Type MC has thermoset insulation. Type MC cable is available with a PVC outer jacket suitable for direct burial.
Conductors are individually wrapped with a flame-retardant fibrous cover: light brown Kraft paper. *Author's comment:* This is an easy way to distinguish Type AC from Type MC.	No individual wrap. Has a polyester (Mylar) tape over all conductors. Has no bare bond wire. *Author's comment:* These are easy ways to distinguish Type AC from Type MC.
Armor: Galvanized steel or aluminum. Armored cable with aluminum armor is marked *Aluminum Armor*.	**Armor:** Galvanized steel, aluminum, or bronze. Three types: *Interlocked* (*most common*). Requires a separate equipment grounding conductor. The armor by itself is not an acceptable equipment grounding conductor. *Smooth tube.* Does not require a separate equipment grounding conductor. *Corrugated tube.* Does not require a separate equipment grounding conductor.

Table 4-12 Comparisons of Type AC and Type MC cables.

(continues)

TYPE AC (*ARTICLE 320*)	TYPE MC (*ARTICLE 330*)
Covering over armor: None	**Covering over armor:** Available with PVC outer covering.
Locations: OK for dry locations. Not OK for damp or wet locations. Not OK for direct burial.	**Locations:** OK for wet locations if so listed. OK for direct burial is so listed. OK to be installed in a raceway.
Insulating bushings: Required to protect conductor insulation from damage. Referred to as "antishorts," these keep any sharp metal edges of the armor from cutting the conductor insulation. Usually supplied in a bag with cable. Figure 4-22(A) shows an antishort being inserted. Figure 4-22(B) shows the recommended way of bending the bonding strip over the antishort to hold the antishort in place.	**Insulating bushings:** Recommended but not required to protect conductor insulation from damage. Usually in a bag supplied with cable.
Minimum radius of bends: 5× diameter of cable. For small size cables, this is about the same curvature as a soda can.	**Minimum radius of bends:** Smooth Sheath: • 10× for sizes not over ¾ in. (19 mm) in diameter. • 12× for sizes over ¾ in. (19 mm) and not over 1½ in. (38 mm) in diameter. • 15× for sizes over 1½ in. (38 mm) in diameter. Corrugated or interlocked: • 7× diameter of cable.
Not available with "super neutral."	Available with "super neutral" when required for computer branch-circuit wiring. For example, three 12 AWG and one 10 AWG neutral.
Not available with fiber-optic cables.	Available with power conductors and fiber-optic cables in one cable.
Support: Not more than 4½ ft (1.4 m) apart. Not more than 12 in. (300 mm) from box or fitting. Support not needed where cable is fished through walls and ceilings. Cables run horizontally through holes in studs, joists, rafters, etc., are considered supported by the framing members. Support not needed when used as a fixture whip within an accessible ceiling for lengths not over 6 ft (1.8 m) from last point of support.	**Support:** Not more than 6 ft (1.8 m) apart. Not more than 12 in. (300 mm) from box or fitting for cables having four or fewer 10 AWG or smaller conductors. Support not needed where cable is fished through walls and ceilings. Cables run horizontally through holes in studs, joists, rafters, etc., are considered supported by the framing members. Support not needed when used as a fixture whip within an accessible ceiling for lengths not over 6 ft (1.8 m) from last point of support.
Voltage: 600 volts or less	**Voltage:** 600 volts or less. Some MC has a voltage rating not to exceed 2000 volts.
Ampacity: Use the 60°C column of *Table 310.16*.	**Ampacity:** Use the 60°C column of *Table 310.16*. For 1/0 and larger conductors, OK to use the 75°C column.
Thermal insulation: If buried in thermal insulation, the conductors must be rated 194°F (90°C).	**Thermal insulation:** No mention in *NEC*.® Type MC has 194°F (90°C) thermoplastic insulation.
Connectors: Set screw-type connectors not permitted with aluminum armor. Listed connectors are suitable for grounding purposes.	**Connectors:** Set screw-type connectors not permitted with aluminum armor. Listed connectors are suitable for grounding purposes.
UL Standard 4. Because UL standards have evolved in numerical order, it is apparent that the armored cable standard is one of the first standards to be developed.	**UL Standard 1569.**

Table 4-12 (*continued*)

Figure 4-22 Antishort bushing prevents cutting of the conductor insulation by the sharp metal armor. *Courtesy of* AFC Cable Systems, Inc., New Bedford, MA.

Figure 4-23 This is an example of a tool that precisely cuts the outer armor of armored cable, making it easy to remove the armor with a few turns of the handle. *Courtesy of* Seatek Co., Inc.

Figure 4-24 Using a hacksaw to cut through a raised convolution of the cable armor. Don't cut too deep into the armor or you will damage the conductors.

Figure 4-21 shows armored cable with the armor removed to show the conductors that are wrapped with paper and the bare bonding wire.

Table 4-13 shows the uses and installation requirements for Type AC and Type MC cables for residential wiring.

USES & INSTALLATION IN DWELLINGS	Type AC	Type MC
a) Branch-circuits & feeders	Yes	Yes
b) Services	No	Yes
c) Exposed & concealed work in:		
Dry locations	Yes	Yes
Wet locations	No	Yes (See 300.10(A)(12) for specifics)
d) Where not subject to physical damage	Yes	Yes
e) May be embedded in plaster finish on masonry walls or run through the hollow spaces of such walls if these locations are not considered damp or wet.	Yes	Yes
f) Run or fished in air voids of masonry block or tile walls where not exposed to excessive moisture or dampness. Masonry in direct contact with earth is considered to be a wet location.	Yes	Yes
g) In any raceway	No	Yes
h) Direct burial	No	Yes, if identified for such use.
i) In accessible attics or roof spaces, must be protected by guard strips at least as high as the cable when run across top of floor joists, or within 7 ft (2.1 m) of the floor or floor joists when the cable is run across the face of studs or rafters. If there is no permanent stairway or ladder, protection is needed only within 6 ft (1.8 m) from edge of scuttle hole or entrance to attic. No guard strips or running boards are necessary for cables run parallel to the sides of joists, rafters, or studs. See Unit 15 for additional discussion and diagrams regarding physical protection of cables in attics.	Yes	Yes

Table 4-13   Uses and installation in dwellings (Type AC and Type MC).

INSTALLING CABLES THROUGH WOOD AND METAL FRAMING MEMBERS (300.4)

Wood Construction

When wiring a house, nonmetallic-sheathed cables are run through holes drilled in studs, joists, and other framing members. Typically, holes are drilled in the approximate centers of wood framing members. To avoid possible damage to the cables, make sure that the edge of the hole is at least 1¼ in. (32 mm) from the nearest edge of the framing member, *300.4(A)(1)*. Refer to Figure 4-25.

You have probably noticed that a 2 × 4 measures 1½ in. × 3½ in., and that a 2 × 3 measures 1½ in. × 2½ in. This can present a real problem!

Steel Plates

If the 1¼ in. (32 mm) from the edge of hole to the edge of framing member cannot be maintained, or if the cable is laid in a notch, a steel plate at least ¹⁄₁₆ in. (1.6 mm) thick must be used to protect the cable from errant nails or drywall screws, *300.4(A)(1)*. ▶In all cases where the *NEC*® requires ¹⁄₁₆ in. (1.6 mm) steel plates, listed and marked steel plates

① CABLES RUN THROUGH FRAMING MEMBERS (JOISTS, RAFTERS, STUDS) MUST NOT BE LESS THAN 1 ¼ IN. (32 mm) FROM THE NEAREST EDGE OF THE FRAMING MEMBER. IF THIS CLEARANCE CANNOT BE MAINTAINED, THE CABLE MUST BE PROTECTED BY A STEEL PLATE AT LEAST ¹/₁₆ IN. (1.6 mm) THICK.

MORE THAN 1 ¼ IN. (32 mm)

STEEL PLATE

SOMETIMES IT IS POSSIBLE TO DRILL HOLE TO ONE SIDE OF CENTER SO AS TO NEED ONLY ONE ¹/₁₆ IN. (1.6 mm) THICK STEEL PLATE.

CABLES

1 ¼ IN. (32 mm)

2 x 4

NOTCHED STUDS

2 x 2

LESS THAN 1 ¼ IN. (32 mm)

③ AN ALTERNATIVE TO USING STEEL PLATES IS TO USE CABLE "STAND OFF" DEVICES. SEE FIGURE 4-27.

▶ NOTE: LISTED AND MARKED STEEL PLATES ARE PERMITTED TO BE LESS THAN ¹/₁₆ IN. (1.6 mm) THICK.◀

② METAL PLATE, ¹/₁₆ IN. (1.6 mm) THICK NEEDED TO PROTECT CABLES LAID IN NOTCH IN STUD.

Figure 4-25 Methods of protecting nonmetallic and armored cables so nails and screws that "miss" the studs will not damage the cables, *300.4(A)(1)* and *(2)*. Because of their strength, intermediate and rigid metal conduit, rigid nonmetallic conduit, and electrical metallic tubing are exempt from this rule per the *Exceptions* found in *300.4*. See Figure 4-27 for alternative methods of protecting cables.

are permitted to be less than ¹/₁₆ in. (1.6 mm) because they have passed tough mechanical tests. This permission is found in *300.4(A)(1) Exception 2, 300.4(A)(2) Exception 2, 300.4(B)(2) Exception, 300.4(D) Exception 3,* and *300.4(E) Exception 2.* ◀ See Figure 4-25.

Instead of using steel plates, Figure 4-26 illustrates devices that can be used to *stand off* the cables from a framing member.

Where cables and nonmetallic raceways are run along the sides (parallel) of framing members, be sure to keep them more than 1¼ in. (32 mm) from

Figure 4-26 Devices that can be used to meet the *Code* requirement to maintain a 1¼-in. (32 mm) clearance from framing members.

PIPE

STEEL PLATE
1/16 IN. (1.6 mm)
THICK

►LISTED AND
MARKED STEEL
PLATES ARE
PERMITTED TO BE
LESS THAN 1/16 IN.
(1.6 mm) THICK. ◄

1 1/4 IN. (32 mm)

Figure 4-27   Where it is necessary to cross pipes or other similar obstructions, making it impossible to maintain at least 1¼-in. (32-mm) clearance from the framing member to the cable, install a steel plate at least ¼₆ in. (1.6 mm) thick to protect the cable from possible damage from nails or screws. See *300.4(A)(1)* and *(A)(2)*.

the edge of the framing member. Otherwise, use steel plates. See Figure 4-27.

Metal Construction

Metal studs are being used more and more. They don't burn and termites don't eat them. They are stronger than wood, and can be set 20 in. (500 mm) on center instead of the usual 16-in. (400-mm) spacing. Metal studs are 10 times more conductive than wood, so treatment of insulation must be addressed.

Where nonmetallic-sheathed cables pass through holes or slots in metal framing members, the cable must be protected by a listed bushing or listed grommet securely fastened in place prior to installing the cable. The bushing or grommet shall:

- remain in place during the wall finishing process,
- cover the complete opening, and
- be listed for the purpose of cable protection.

This information is found in *300.4(B)(1)* and *334.17*. See Figure 4-28.

Bushings that snap into a hole sized for the specific size bushing (½ in., ¾ in., etc.), or a two-piece snap-in bushing that fits just about any size precut hole in the metal framing members are acceptable to use, as shown in Figure 4-28. These types of bushings (insulators) are not required in metal framing members when the wiring method is armored cable, EMT, FMC, or electrical nonmetallic tubing.

When running nonmetallic-sheathed cable or electrical nonmetallic tubing through metal framing members where there is a likelihood that nails or screws could be driven into the cables or tubing, a steel plate, steel sleeve, or steel clip at least ¼₆ in. (1.6 mm) thick must be installed to protect the cables and/or tubing. This is spelled out in *300.4(B)(2)*. See Figure 4-27 and Figure 4-28.

CABLES IN GROOVES

NEC® 300.4(E) addresses the wiring problems associated with cutting grooves into building material because there is no hollow space for the wiring. Nonmetallic-sheathed cable and some raceways must be protected when laid in a groove. Protection must be provided with a ¼₆-in. (1.6-mm) steel plate, sleeve, or equivalent, or the groove must be deep enough to allow not less than 1¼-in. (32-mm) free space for the entire length of the groove. This situation can be encountered where styrofoam insulation building blocks are grooved to receive the electrical cables, then covered with wallboard, wood paneling, or other finished wall material. Another typical example is where solid wood planking is installed on top of wood beams. The top side of the planking is covered with roofing material, and the bottom side is exposed and serves as the finished ceiling. The only way to install wiring for ceiling luminaires (fixtures) and fans is to groove the planking in some manner, as shown in Figure 4-29.

METAL
STUDS

SHARP
EDGES

Figure 4-28 *NEC® 300.4(B)* **and** *334.17* **require protection of nonmetallic-sheathed cable where it runs through holes in metal framing members. Shown are two types of listed bushings (grommets) that can be installed to protect nonmetallic-sheathed cable from abrasion. Electricians generally avoid using nonmetallic-sheathed cable through metal studs and joists. They use electrical metallic tubing, flexible metal conduit, armored cable (Type AC or MC), or electrical nonmetallic tubing. Where there is a likelihood that nails or screws might penetrate the nonmetallic-sheathed cable, ⅟₁₆-in. (1.6-mm) thick steel plates, sleeves, or clips must be installed to protect the cable or electrical nonmetallic tubing.**

This additional protection is not required when the raceway is intermediate metal conduit *(Article 342)*, rigid metal conduit *(Article 344)*, rigid non-metallic conduit *(Article 352)*, or electrical metallic tubing *(Article 358)* (see Figure 4-28).

NEC® 334.30(B)(1), and *300.4(D), Exception 2*

allow cables to be "fished" between outlet boxes or other access points without additional protection.

In the recreation room of the residence plans, precautions must be taken so that there is adequate protection for the wiring. The walls in the recreation room are to be paneled. Therefore, if the carpenter

►LISTED AND MARKED STEEL
PLATES ARE PERMITTED TO
BE LESS THAN ¹⁄₁₆ IN. (1.6 mm)
THICK.◄

Figure 4-29   Example of how a solid wood framing member or styrofoam insulating block might be grooved to receive the nonmetallic-sheathed cable. After the cable is laid in the groove, a steel plate not less than ¹⁄₁₆ in. (1.6 mm) thick must be installed over the groove to protect the cable from damage, such as from driven nails or screws. See *300.4(E)*.

uses 1 × 2 or 2 × 2 furring strips, as in Figure 4-30, nonmetallic-sheathed cable would require the additional ¹⁄₁₆-in. (1.6-mm) steel plate protection or the use of *stand off* devices similar to those illustrated in Figure 4-26.

Some building contractors attach 1-in. insulation to the walls, then construct a 2 × 4 wall in front of the basement structural foundation wall, leaving 1-in. spacing between the insulation and the back side of the 2 × 4 studs. This makes it easy to run cables or conduits behind the 2 × 4 studs, and eliminates the need for the additional mechanical protection, as shown in Figure 4-31.

Watch out for any wall partitions where 2 × 4 studs are installed "flat," as in Figure 4-32. In this case, the choice is to provide the required mechanical protection as required by *300.4*, or to install the wiring in EMT.

See Unit 15 concerning the installation of cables in attics.

Building Codes

When complying with the *National Electrical Code*, be sure to also comply with the building code. Building codes are concerned with the weakening of framing members when drilled or notched. Meeting the requirements of the *NEC®* can be in conflict with building codes and vice versa. Here is a

FOUNDATION WALL

1 x 2 OR 2 x 2
FURRING STRIPS

DRYWALL
OR PANELING

Figure 4-30   When 1 × 2 or 2 × 2 furring strips are installed on the surface of a basement wall, additional protection, as shown in Figure 4-25, Figure 4-26, Figure 4-28, and Figure 4-29, must be provided.

Figure 4-31 This sketch shows how to insulate a foundation wall and how to construct a wall that provides space to run cables and/or raceways that will not require the additional protection as shown in Figure 4-25, Figure 4-26, Figure 4-28, and Figure 4-29.

Figure 4-32 When 2 × 4 studs are installed "flat," such as might be found in nonbearing partitions, cables running parallel to or through the studs must be protected against the possibility of having nails or screws driven through the cables. This means protecting the cable with ¹⁄₁₆-in. (1.6-mm) steel plates or equivalent for the entire length of the cable, or installing the wiring using intermediate metal conduit, rigid metal conduit, rigid nonmetallic conduit, or EMT. See Figure 4-25, Figure 4-26, Figure 4-27, and Figure 4-28.

recap of some of the key building code requirements found in the ICC *International Residential Code*.

Studs:

- Cuts or notches in exterior or bearing partitions shall not exceed 25% of the stud's width.

- Cuts or notches in nonbearing partitions shall not exceed 40% of the stud's width.

- Bored or drilled holes in any stud:

  – shall not be greater than 40% of the stud's width, and

  – shall not be closer than ⁵⁄₁₆ in. (16 mm) to the edge of the stud, and

  – shall not be located in the same section as a cut or notch.

Joists:

- Notches on the top or bottom of joists shall not exceed one-sixth the depth of the joist, shall not be longer than one-third the width of the joists, and shall not be in the middle third of the joist.

- Holes bored in joists shall not be within 2 in. (50 mm) of the top or bottom of the joist or any other hole, and the diameter of any such hole shall not exceed one-third the depth of the joist. If the joist is also notched, the hole shall not be closer than 2 in. (50 mm) to the notch.

Engineered Wood Products

Today, we find many homes constructed with engineered wood I-joists and beams. The strength of

these factory-manufactured framing members conform to various building codes. It is critical that certain precautions be taken relative to drilling and cutting these products so as not to weaken them. In general,

- Do not cut, notch, or drill holes in trusses, laminated veneer lumber, glue-laminated members, or I-joists unless specifically permitted by the manufacturer.

- Do not cut, drill, or notch flanges. Flanges are the top and bottom pieces.

- Do not cut holes too close to supports or to other holes.

Figure 4-33   Nonmetallic-sheathed cable may pass through cold air return, joist, or stud spaces in dwellings, but only if it is run at right angles to the long dimension of the space, *300.22(C), Exception*.

- For multiple holes, the amount of web to be left between holes must be at least twice the diameter of the largest adjacent hole. The web is the piece between the flanges.
- Holes not over 1½ in. (38 mm) usually can be drilled anywhere in the web.
- Do not hammer on the web except to remove knockout prescored holes.
- Check and follow the manufacturer's instructions!

INSTALLATION OF CABLES THROUGH DUCTS

NEC® 300.22 is extremely strict as to what types of wiring methods are permitted for installation of cables through ducts or plenum chambers. These stringent rules are for fire safety.

The *Code* requirements are somewhat relaxed by *300.22(C), Exception*. In this exception, permission is given to run nonmetallic-sheathed and armored cables in joist and stud spaces (i.e., cold air returns), but only if they are run perpendicular to the long dimensions of such spaces, as illustrated in Figure 4-33.

CONNECTORS FOR INSTALLING NONMETALLIC-SHEATHED AND ARMORED CABLE

The connectors shown in Figure 4-34 are used to fasten nonmetallic-sheathed cable and armored cable to the boxes and panels in which they terminate. These connectors clamp the cable securely to each outlet box. Many boxes have built-in clamps and do not require separate connectors.

The question continues to be asked: May more than one cable be inserted in one connector? Unless the UL listing of a specific cable connector indicates that the connector has been tested for use with more than one cable, the rule is *one cable, one connector*. This will be marked on the carton.

INTERMEDIATE METAL CONDUIT (ARTICLE 342), RIGID METAL CONDUIT (ARTICLE 344), RIGID NONMETALLIC CONDUIT (ARTICLE 352), AND ELECTRICAL METALLIC TUBING (ARTICLE 358)

Some communities do not permit cable wiring. They require that the wiring method be a raceway—metallic or nonmetallic. You will have to check this out before starting a project.

The following comparison chart provides the highlights of the *NEC*® requirements for different types of raceways that might be used in a residence. For more detailed information, refer to the manufacturer's installation instructions, the specific *NEC*® Article, and the UL White Book.

Grounding: All listed fittings for listed metallic raceways and listed metal jacketed cables are tested to carry a specified amount of fault current for a specified length of time, making the fittings acceptable for grounding when properly installed.

In this residence, the wiring method for all of the exposed wiring in the workshop is to be electrical metallic tubing (EMT). Wiring with EMT is covered in detail in Unit 18.

USED FOR NONMETALLIC-SHEATHED CABLE

USED FOR ARMORED CABLE

Figure 4-34 Cable connectors.

CONDUIT AND TUBING				
	INTERMEDIATE METAL CONDUIT IMC *Article 342*	**RIGID METAL CONDUIT RMC** *Article 344*	**RIGID NONMETALLIC CONDUIT RNC** *Article 352*	**ELECTRICAL METALLIC TUBING EMT** *Article 358*
Bends	Not more than the equivalent of four quarter bends between pull points. 360° total. Manufactured bends available. Field bends made with benders.	Not more than the equivalent of four quarter bends between pull points. 360° total. Manufactured bends available. Field bends made with benders.	Not more than the equivalent of four quarter bends between pull points. 360° total. Manufactured bends available. Field bends made with benders that electrically heat the PVC to just the right temperature.	Not more than the equivalent of four quarter bends between pull points. 360° total. Manufactured bends available. Field bends made with benders.
Bend radius	▶See *Table 2, Chapter 9, NEC.®*◀	▶See *Table 2, Chapter 9, NEC.®*◀	▶See *Table 2, Chapter 9, NEC.®*◀	▶See *Table 2, Chapter 9, NEC.®*◀
Bushings	Where 4 AWG or larger conductors are installed, smoothly rounded insulating fittings or the equivalent must be used, *300.4(F)* and *312.6(C)*. See Figure 27-29 and Figure 27-30. Not required where the IMC terminates in a hub or boss. See Figure 27-16.	Where 4 AWG or larger conductors are installed, smoothly rounded insulating fittings or the equivalent must be used, *300.4(F)* and *312.6(C)*. See Figure 27-29 and Figure 27-30. Not required where the RMC terminates in a hub or boss. See Figure 27-16.	Where 4 AWG or larger conductors are installed, smoothly rounded insulating fittings or the equivalent must be used, *300.4(F)* and *312.6(C)*. See Figure 27-29 and Figure 27-30. Not required where the RNC terminates in a hub or boss. See Figure 27-16.	Where 4 AWG or larger conductors are installed, smoothly rounded insulating fittings or the equivalent must be used, *300.4(F)* and *312.6(C)*. See Figure 27-29 and Figure 27-30. Not required where the IMC terminates in a hub or boss. See Figure 27-16.
Cable permitted inside	▶Yes, if permitted by the specific cable *Code* article.◀	▶Yes, if permitted by the specific cable *Code* article.◀	▶Yes, if permitted by the specific cable *Code* article.◀	▶Yes, if permitted by the specific cable *Code* article.◀
Cinder fill	Yes, where subject to permanent moisture if protected on all sides by noncinder concrete not less than 2 in. (50 mm), or if at least 18 in. (450 mm) under the fill, or if it has corrosion protection.	Yes, where subject to permanent moisture if protected on all sides by noncinder concrete not less than 2 in. (50 mm), or if at least 18 in. (450 mm) under the fill, or if it has corrosion protection.	Yes.	Yes, where subject to permanent moisture if protected on all sides by noncinder concrete not less than 2 in. (50 mm), or if at least 18 in. (450 mm) under the fill, or if it has corrosion protection.
Concealed in walls, ceilings, and floors	Yes.	Yes.	Yes.	Yes.
Concrete above grade	Yes.	Steel: Yes. Aluminum: Yes.	Yes.	Steel: Yes. Aluminum: Yes, if it has supplemental corrosion protection.
Concrete on grade	Yes.	Yes.	Yes.	Steel: Yes Aluminum: Yes, if it has supplemental corrosion protection.
Concrete below grade	Yes.	Steel: Yes. Aluminum: Yes, but only if it has supplementary corrosion protection such as bitumastic paint, special tape wraps, or PVC coated.		Steel: Yes, if it has supplemental corrosion protection. Aluminum: Yes, if it has supplemental corrosion protection.

(continues)

(continued)

	INTERMEDIATE METAL CONDUIT IMC *Article 342*	RIGID METAL CONDUIT RMC *Article 344*	RIGID NONMETALLIC CONDUIT RNC *Article 352*	ELECTRICAL METALLIC TUBING EMT *Article 358*
Conductor fill	As determined by *Table 1, Chapter 9, NEC.®* Inside interior diameter greater than RMC of same trade size. More room for conductors. See Unit 18 for examples of conductor fill calculations.	As determined by *Table 1, Chapter 9, NEC.®* See Unit 18 for examples of conductor fill calculations.	As determined by *Table 1, Chapter 9, NEC.®* See Unit 18 for examples of conductor fill calculations.	As determined by *Table 1, Chapter 9, NEC.®* See Unit 18 for examples of conductor fill calculations.
Conductors, when to install?	After complete raceway system is installed, *300.18(A).*	After complete raceway system is installed, *300.18(A).*	After complete raceway system is installed, *300.18(A).*	After complete raceway system is installed, *300.18(A).*
Corrosive conditions	Yes, if protected by corrosion protection and judged suitable for the condition. Listing will indicate this.	Yes, if protected by corrosion protection and judged suitable for the condition. Listing will indicate this. Aluminum resists most corrosive atmospheres.	Yes. Like those found around swimming pool areas.	Yes, if protected by corrosion protection and judged suitable for the condition. Listing will indicate this.
Dry and damp locations	Yes.	Yes.	Yes.	Yes.
Expansion fittings for thermal expansion	Not required.	Not required.	Yes. Refer to *Table 352.44(A)* and *(B).* Without expansion fittings, RNC in direct sunlight will pull out of hubs of boxes, or will pull the boxes off the wall.	Not required.
Exposed	Yes.	Yes.	Yes.	Yes.
Fittings	Threaded and threadless.	Threaded and threadless compression. For aluminum conduit, aluminum threaded fittings recommended, but cadmium plated or galvanized fittings are satisfactory in most installations.	Attached with solvent-type cement.	Set screw and threadless compression.
Grounding	Is considered to be an equipment grounding conductor.	Is considered to be an equipment grounding conductor.	Where required, a separate equipment grounding conductor must be installed.	Is considered to be an equipment grounding conductor.
Joints	Use listed couplings, connectors, and fittings.	Use listed couplings, connectors, and fittings.	Use proper cement.	Use listed couplings, connectors, and fittings.
Length	10 ft	10 ft	10 ft	10 ft
Listed	Yes.	Yes.	Yes.	Yes.
Material	Steel—galvanized. Is as strong as same trade size RMC.	Steel—galvanized. Also aluminum.	Polyvinyl chloride (PVC). Schedule 40 (heavy wall thickness). Schedule 80 (extra-heavy wall thickness). O.D. same for both. I.D. of Schedule 40 is greater than that of Schedule 80.	Steel—galvanized. Also aluminum.

(continues)

continued

	INTERMEDIATE METAL CONDUIT IMC *Article 342*	RIGID METAL CONDUIT RMC *Article 344*	RIGID NONMETALLIC CONDUIT RNC *Article 352*	ELECTRICAL METALLIC TUBING EMT *Article 358*
Rough edges	Must be properly reamed to remove rough edges.	Must be properly reamed to remove rough edges.	Must be properly reamed to remove rough edges.	Must be properly reamed to remove rough edges.
Support of conduit bodies	Yes.	Yes.	Yes. Conduit bodies not to contain device or support luminaires (lighting fixtures) or other equipment.	Yes.
Securing and supporting	Maximum 3 ft (900 mm) from box or conduit body. See *Table 344.30(B)(2)* for distance between supports. See Figure 4-35.	Maximum 3 ft (900 mm) from box or conduit body. See *Table 344.30(B)(2)* for distance between supports. See Figure 4-35.	Maximum 3 ft (900 mm) from box or conduit body. See Table *352.30(B)* for distance between supports. See Table 4-14 and Figure 4-35.	Maximum 3 ft (900 mm) from box or conduit body. At least every 10 ft (3 m) between supports. See Figure 4-35.
Sizes	Trade sizes ½ thru 4.	Trade sizes ½ thru 6.	Trade sizes ½ thru 6.	Steel: Trade sizes ½ thru 4. Aluminum: Trade sizes 2 thru 4.
Threaded	Yes. Same threads as RMC. ▶Field cut threads shall be coated with a listed electrically conductive corrosion-resistant compound.◀	Yes. ▶Field cut threads shall be coated with a listed electrically conductive corrosion-resistant compound.◀	No.	No.
Underground in soil.	Yes.	Yes.	Yes.	Steel: Yes, if it has supplemental corrosion protection. Aluminum: Yes, if it has supplemental corrosion protection.
Weight	Medium. For example, trade size ½: 100 ft (30.5 m) approx. 62 lb. (28.1 kg).	Heavy. For example, trade size ½: 100 ft (30.5 m) approx. 82 lb. (37.2 kg). Aluminum is approx. ⅓ as heavy as steel. For example, trade size ½: 100 ft (30.5 m) approx. 28 lb. (12.7 kg).	Very light. For example, trade size ½ Schedule 40: 100 ft (30.5 m) approx. 17 lb. (7.7 kg); Schedule 80 approx. 21 lb. (9.53 kg).	Light. For example, trade size ½: 100 ft (30.5 m) approx. 30 lb. (13.5 kg).
Wet locations	Yes, if raceway, fittings, bolts, screws, straps, etc., are of corrosion-resistant material.	Yes. if raceway, fittings, bolts, screws, straps, etc., are of corrosion-resistant material.	Yes, if raceway, fittings, bolts, screws, straps, etc., are of corrosion-resistant material.	Yes, if raceway, fittings, bolts, screws, straps, etc., are of corrosion-resistant material.

RIGID NONMETALLIC CONDUIT TRADE SIZE	MAXIMUM SPACING BETWEEN SUPPORTS
½, ¾, 1	3 ft (900 mm)
1¼, 1½, 2	5 ft (1.5 m)

Table 4-14   Maximum distance between supports for rigid nonmetallic conduit.

NOT MORE THAN 3 FT (900 mm) BETWEEN
SUPPORT AND OUTLET BOX, JUNCTION BOX, CABINET,
CONDUIT BODY, OR CONDUIT TERMINATION

NOT MORE THAN 10 FT (3.0 m)
BETWEEN SUPPORTS

Figure 4-35 *Code* requirements for securing and supporting raceways are found in *342.30(A)* and *(B)* (intermediate metal conduit), *344.30* (rigid metal conduit), *352.30* (rigid nonmetallic conduit), and *358.30* (electrical metallic tubing). There are exceptions to the measurements indicated in the above illustration where raceways are "fished" through finished or concealed partitions or where structural members do not readily permit fastening according.

FLEXIBLE CONNECTIONS

The installation of some equipment requires flexible connections, both to simplify the installation and to stop the transfer of vibrations. In residential wiring, flexible connections are used to connect such equipment such as attic fans, food waste disposers, dishwashers, air conditioners, heat pumps, and recessed luminaires (fixtures).

The following chart shows the *NEC*® requirements for three types of commonly used flexible conduit.

FLEXIBLE CONNECTIONS			
	FLEXIBLE METAL CONDUIT FMC *Article 348*	**LIQUIDTIGHT FLEXIBLE METAL CONDUIT LFMC** *Article 350*	**LIQUIDTIGHT FLEXIBLE NONMETALLLIC CONDUIT LFNC** *Article 356*
	See Figure 4-37(A).	See Figure 4-37(B).	See Figure 4-37(C).
Bends	Not more than the equivalent of four quarter bends between pull points. 360° total.	Not more than the equivalent of four quarter bends between pull points. 360° total.	Not more than the equivalent of four quarter bends between pull points. 360° total.
Bend radius	▶See *Table 2, Chapter 9, NEC.*®◀	▶See *Table 2, Chapter 9, NEC.*®◀	▶See *Table 2, Chapter 9, NEC.*®◀
Bushings	Where 4 AWG or larger conductors are installed, smoothly rounded insulating fittings or the equivalent must be used, *300.4(F)* and *312.6(C)*.	Where 4 AWG or larger conductors are installed, smoothly rounded insulating fittings or the equivalent must be used, *300.4(F)* and *312.6(C)*.	Where 4 AWG or larger conductors are installed, smoothly rounded insulating fittings or the equivalent must be used, *300.4(F)* and *312.6(C)*.
Cable permitted inside	▶Yes, if permitted by the specific cable *Code* article.◀	▶Yes, if permitted by the specific cable *Code* article.◀	▶Yes, if permitted by the specific cable *Code* article.◀
Concealed in walls, ceilings, and floors	Yes.	Yes.	Yes.
Concrete above grade	No.	Yes, if marked "Direct Burial."	Yes, if marked "Direct Burial."
Concrete on grade	No.	Yes, if marked "Direct Burial."	Yes, if marked "Direct Burial."

(continues)

(continued)

	FLEXIBLE METAL CONDUIT FMC *Article 348*	LIQUIDTIGHT FLEXIBLE METAL CONDUIT LFMC *Article 350*	LIQUIDTIGHT FLEXIBLE NONMETALLLIC CONDUIT LFNC *Article 356*
Concrete below grade	No.	Yes, if marked "Direct Burial."	Yes, if marked "Direct Burial."
Conductor fill	As determined by *Tables 1* and *4, Chapter 9, NEC.*® For trade size ⅜, use *Table 348.22.*	As determined by *Table 1, Chapter 9, NEC.*® For trade size ⅜, use *Table 348.22.*	As determined by *Table 1, Chapter 9, NEC.*® For trade size ⅜, use *Table 348.22.*
Conductors, when to install?	After complete raceway system is installed, *300.18(A).*	After complete raceway system is installed, *300.18(A).*	After complete raceway system is installed, *300.18(A).*
Corrosive conditions	Yes, if protected by corrosion protection and judged suitable for the condition. Listing will indicate this.	Yes, if protected by corrosion protection and judged suitable for the condition. Listing will indicate this. Aluminum resists most corrosive atmospheres.	Yes. Like those found around swimming pool areas.
Dry and damp locations	Yes.	Yes.	Yes.
Exposed	Yes.	Yes.	Yes.
Fittings	Must be listed. Sizes larger than trade size ¾ will be marked GRND if acceptable for grounding. Do not conceal angle-type fittings. See Figure 4-36.	Must be listed. Sizes larger than trade size ¾ will be marked GRND if acceptable for grounding. Do not conceal angle-type fittings. See Figure 4-36.	Must be listed. Do not conceal angle-type fittings. See Figure 4-36.
Grounding	When terminated with listed fittings, it is considered to be an EGC: • not over 6 ft (1.8 m) and the overcurrent device does not exceed 20 amperes. See Figure 4-39 and Figure 4-40. Longer than 6 ft (1.8 m), not acceptable as a grounding means. Install a separate equipment grounding conductor sized per *Table 250.122.* ►Always install an equipment grounding conductor where FMC will be flexing after installation.◄ See *250.118(5).*	When terminated with listed fittings, it is considered to be an EGC: • in trade sizes ⅜ and ½ not over 6 ft (1.8 m) and the overcurrent device does not exceed 20 amperes. • in trade sizes ¾, 1, and 1¼ not over 6 ft (1.8 m) and the overcurrent device does not exceed 60 amperes. Trade size 1½ and larger, not acceptable for grounding. See Figure 4-39, Figure 4-40, Figure 4-41, and Figure 4-42. Longer than 6 ft (1.8 m), LFMC is not acceptable as a grounding means. Install a separate equipment grounding conductor sized per *Table 250.122.* ►Always install an equipment grounding conductor where LFMC will be flexing after installation.◄ See *250.118(6).*	If required such as when used for flexibility, a separate equipment grounding conductor must be installed. Size EGC per *250.122.* See Figure 4-41.
Length	No limit.	No limit.	Not longer than 6 ft (1.8 m) unless secured per *356.30.*
Listed	Must be listed. See UL 1.	Must be listed. See UL 360.	Must be listed. See UL 1660.

(continues)

(continued)

	FLEXIBLE METAL CONDUIT FMC *Article 348*	LIQUIDTIGHT FLEXIBLE METAL CONDUIT LFMC *Article 350*	LIQUIDTIGHT FLEXIBLE NONMETALLLIC CONDUIT LFNC *Article 356*
Material	Steel and aluminum. Must not be used where subject to physical damage.	Steel with nonmetallic outer jacket. Must not be used where subject to physical damage.	Nonmetallic. Three types. See *356.2* and UL White Book. Must not be used where subject to physical damage.
Rough edges	Must be properly trimmed to remove rough edges.	Must be properly trimmed to remove rough edges.	Must be properly trimmed to remove rough edges.
Securing and supporting	Maximum 12 in. (300 mm) from box, cabinet, or fitting.	Maximum 12 in. (300 mm) from box, cabinet, or fitting.	Maximum 12 in. (300 mm) from box, cabinet, or fitting.
	Maximum 4½ ft (1.4 m) between supports.	Maximum 4½ ft (1.4 m) between supports.	Maximum 4½ ft (1.4 m) between supports.
	Waived if fished through walls and ceilings—where flexibility is needed and the flex is not more than 3 ft (900 mm) from where it is terminated, or as a whip not over 6 ft (1.8 m). See Figure 4-38, Figure 7-2, Figure 7-6, and Figure 7-10.	Waived if fished through walls and ceilings—where flexibility is needed and the flex is not more than 3 ft (900 mm) from where it is terminated, or as a whip not over 6 ft (1.8 m). See Figure 4-38, Figure 7-2, Figure 7-6, and Figure 7-10.	Waived if fished through walls and ceilings—where flexibility is needed and the flex is not more than 3 ft (900 mm) from where it is terminated, or as a whip not over 6 ft (1.8 m). See Figure 4-38, Figure 7-2, Figure 7-6, and Figure 7-10.
	Also waived when supported horizontally through framing members not more than 4½ ft (1.4 m) apart and secured within 12 in. (300 mm) of box, cabinet, or fitting.	Also waived when supported horizontally through framing members not more than 4½ ft (1.4 m) apart and secured within 12 in. (300 mm) of box, cabinet, or fitting.	Also waived when supported horizontally through framing members not more than 3 ft (900 mm) apart and secured within 12 in. (300 mm) of box, cabinet, or fitting.

(continues)

Figure 4-36   Various fittings for (A) flexible metal conduit, (B) liquidtight flexible metal conduit, and (C) liquidtight flexible nonmetallic conduit.

Figure 4-37   (A) Flexible metal conduit. (B) Flexible liquidtight metal conduit. (C) Flexible liquidtight nonmetallic conduit. *Courtesy of* Electri-Flex Co.

(continued)

	FLEXIBLE METAL CONDUIT FMC *Article 348*	LIQUIDTIGHT FLEXIBLE METAL CONDUIT LFMC *Article 350*	LIQUIDTIGHT FLEXIBLE NONMETALLLIC CONDUIT LFNC *Article 356*
Sizes	Steel: Trade sizes ½ thru 4. Aluminum: Trade sizes ½ thru 3. Trade size ⅜ OK for enclosing leads to a motor or as a luminaire (fixture) whip not longer than 6 ft (1.8 m).	Trade sizes ½ thru 4. Trade size ⅜ OK for enclosing leads to a motor or as a luminaire (fixture) whip not longer than 6 ft (1.8 m).	Trade sizes ½ thru 4. Trade size ⅜ OK for enclosing leads to a motor or as a luminaire (fixture) whip not longer than 6 ft (1.8 m).
In sunlight	Yes.	Yes.	Yes, if listed and marked "Outdoor."
Temperature	Limited to the temperature rating of the conductors.	Limited to 60°C (140°F) unless marked otherwise. Also limited to temperature rating of the conductors.	Limited to 60°C (140°F) unless marked otherwise. Also limited to temperature rating of the conductors. Can become brittle in extreme cold.
Underground in soil.	No.	Yes, if marked "Direct Burial."	Yes, if marked "Direct Burial."
Uses	Where flexibility is required. Do not use where subject to physical damage.	Where flexibility is required. Do not use where subject to physical damage.	Where flexibility is required. Do not use where subject to physical damage.
Wet locations	Yes, but only if water will not enter any enclosures through the convolutions of the flex.	Yes.	Yes.

Figure 4-38   Some of the more common places where flexible connections could be used.

Figure 4-39   If a flexible metal conduit is serving as an equipment ground return path, the total length shall not exceed 6 ft (1.8 m). Refer to *250.118(5)c*, and *250.118(6)d*.

ELECTRICAL NONMETALLIC TUBING (ENT) *(ARTICLE 362)*

Electrical nonmetallic tubing (ENT) is covered in *Article 362* of the *NEC®.* It is a pliable, lightweight, corrugated raceway made of PVC. It requires

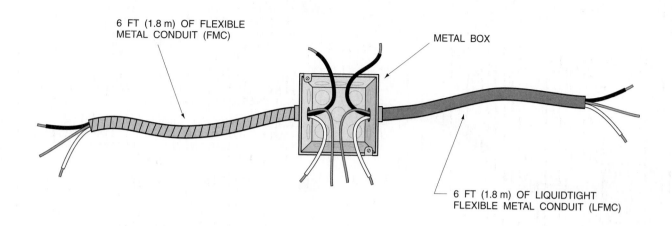

6 FT (1.8 m) OF FLEXIBLE
METAL CONDUIT (FMC)

METAL BOX

6 FT (1.8 m) OF LIQUIDTIGHT
FLEXIBLE METAL CONDUIT (LFMC)

Figure 4-40   The "combined" length of flexible metal conduit and liquidtight flexible metal conduit shall not exceed 6 ft (1.8 m) if it is to serve as the same ground return path, *250.118(5)c*, and *250.118(6)d*. In the above illustration, a separate equipment grounding conductor would have to be provided.

LIQUIDTIGHT FLEXIBLE
METALLIC CONDUIT
OR
LIQUIDTIGHT FLEXIBLE
NONMETALLIC CONDUIT

AIR CONDITIONER

Figure 4-41   Use of liquidtight flexible metal conduit or flexible liquidtight nonmetallic conduit.

fittings made specifically for the ENT. Figure 4-43 shows color-coded ENT and some fittings. Color-coded ENT can be used to differentiate power, communication, and fire protection signaling system.

Although not often used in typical house wiring, some electricians find ENT very practical for providing a raceway system for installing structured wiring conductors and cables. Structured wiring is discussed in Unit 31.

SERVICE-ENTRANCE CABLE
(ARTICLE 338)

Service-entrance cable is covered in *Article 338* of the *National Electrical Code.*®

Service-entrance cable is defined as "*a single conductor or multi-conductor assembly provided with or without an overall covering, primarily used for services.*"

Where permitted by local electrical codes, service-entrance cable is most often run from the utility's **service point** to the meter base, then from the meter base to the main panel.

It is also permitted as a branch-circuit to connect major electrical appliances and as a feeder to a subpanel. For these situations, the choice of whether to use Type SE cable or nonmetallic-sheathed cable becomes a matter of choice, the major issue being the cost of the cable. Some Type SE cable with aluminum conductors may cost less than that for nonmetallic-sheathed cable with copper conductors. The labor for the installation of either type is the same.

There are two styles of Type SE cable: SER and SEU. Table 4-15 shows the *NEC*® installation requirements and the key differences in construction of these cables. See Table 4-15 and Figure 4-44.

(A) PERMITTED AS A GROUNDING MEANS IF IT IS NOT OVER TRADE SIZE 1¼, IS NOT OVER 6 FT (1.8 m) LONG, AND IS CONNECTED BY FITTINGS LISTED FOR GROUNDING. THE 6 FT (1.8 m) LENGTH INCLUDES THE TOTAL LENGTH OF ANY AND ALL FLEXIBLE CONNECTIONS IN THE RUN. SEE FIGURE 4-40.

(B) MINIMUM TRADE SIZE ½. TRADE SIZE ⅜ PERMITTED AS A LUMINAIRE (FIXTURE) WHIP NOT OVER 6 FT (1.8 m) LONG.

(C) WHEN USED AS THE GROUNDING MEANS, THE MAXIMUM OVERCURRENT DEVICE IS 20 AMPERES FOR TRADE SIZES ⅜ AND ½, AND 60 AMPERES FOR TRADE SIZES ¾, 1, AND 1¼.

(D) NOT SUITABLE AS A GROUNDING MEANS UNDER ANY OF THE FOLLOWING CONDITIONS:
• TRADE SIZES 1½ AND LARGER
• TRADE SIZES ⅜ AND ½ WHEN OVERCURRENT DEVICE IS GREATER THAN 20 AMPERES
• TRADE SIZES ⅜ AND ½ WHEN LONGER THAN 6 FT (1.8 m)
• TRADE SIZES ¾, 1 AND 1¼ WHEN OVERCURRENT DEVICE IS GREATER THAN 60 AMPERES
• TRADE SIZES ¾, 1 AND 1¼ WHEN LONGER THAN 6 FT (1.8 m)
FOR THESE CONDITIONS, INSTALL A SEPARATE EQUIPMENT GROUNDING CONDUCTOR SIZED PER *TABLE 250.122.*

Figure 4-42   Grounding capabilities of liquidtight flexible metal conduit (LFMC). See *Article 350.*

Figure 4-43   Color-coded ENT, a coupling, a threaded connector, and a snap-in connector.

SERVICE-ENTRANCE CABLES TYPE SE, SER, SEU, AND USE COMPARISONS

TYPE	COVERING	INSULATION	USES PERMITTED	BENDS	AMPACITY
SE Service-Entrance Cable Conforms to UL Standard 854: Service-Entrance Cables. See Figure 4-44.	Flame-retardant, moisture- and sunlight-resistant covering.	Type RHW, XHHW, or THWN insulation. Sizes 12 AWG and larger for copper. Sizes 10 AWG and larger for aluminum or copper-clad aluminum. If aluminum conductors, jacket will be marked "AL." If copper-clad aluminum conductors, jacket will be marked "AL (CU-CLAD)" or "Cu-Clad AL". If letter "H" on insulation, temperature rating 167°F (75°C), wet or dry locations. If letter "HH" on insulation, temperature rating 167°F (75°C) wet/194°F (90°C) dry locations. If marked with a "-2", suitable for wet or dry locations as high as 194°F (90°C). May have one conductor uninsulated. May have reduced neutral as permitted by 220.61 and 310.15(B)(6).	600 volts or less. Above ground. Suitable for use where exposed to sun. Use "sill plate" where cable is bent to pass through a wall. See Figure 27-3. Permitted for interior wiring (feeder to a subpanel, ranges, dryers, cooktops, ovens) where all conductors are insulated. Must be installed according to the requirement in *Article 334* (Nonmetallic-sheathed cable). A bare, uninsulated conductor permitted only for use as an equipment grounding conductor. For existing installations only, may have a bare neutral, *250.140*. Not permitted for new work. Grounding frames of electric ranges and dryers to the neutral was permitted prior to the 1996 *NEC*® No longer permitted for new work.	Internal radius of bend not to exceed 5 times the diameter of the cable.	See *Table 310.16*. Refer to the 75°C and 90°C columns. For single-phase dwelling services, see *310.15(B)(6)*.
SER Conforms to UL Standard 854: Service-Entrance Cables. Subject to the same rules as Type SE.	Same as above.	Same as above. Available with two insulated phase conductors and one insulated neutral. Also available with two insulated phase conductors, one insulated neutral, and one bare equipment grounding conductor. Available with reduced neutral as permitted by 220.61 and 310.15(B)(6). Type XHHW-2 conductors. Rated 194°F (90°C), wet or dry.	Same as above.	Same as above.	Refer to the 75°C and 90°C columns. For single-phase dwelling services, see *310.15(B)(6)*.

Table 4-15

(continues)

SERVICE-ENTRANCE CABLES TYPE SE, SER, SEU, AND USE COMPARISONS (*continued*)

TYPE	COVERING	INSULATION	USES PERMITTED	BENDS	AMPACITY
SEU Conforms to UL Standard 854: Service-Entrance Cables. Subject to the same rules as Type SE.	Same as above.	Type XHHW-2 conductors. Rated 194°F (90°C), wet or dry. Type THHN/THWN Rated 90°C, dry; 75°C, wet. Easily identified by the bare wraparound neutral. Available with two and three insulated conductors plus the bare wraparound neutral. Available with reduced neutral as permitted by 220.61 and 310.15(B)(6).	Same as above.	Same as above.	Refer to the 75°C and 90°C columns. For single-phase dwelling services, see 310.15(B)(6).
USE Service-Entrance Conductor. Individual conductors may be bundled in an assembly. Conforms to UL Standard 854: Service-Entrance Cables. USE cable is shown in Figure 4-45.	Moisture-resistant covering. Not required to have flame-retardant covering.	Insulation equivalent to Type RHW or XHHW. The temperature rating of the cable is the same as the temperature rating of the individual conductors. If this is not marked on the outer jacket, then the cable is rated 167°F (75°C). If Type USE-2, insulation equivalent to Type RHW-2 or XHHW-2. Rated 194°F (90°C), wet or dry. Cabled single conductor may have bare copper conductor with the assembly. May have reduced neutral if multiple conductor assembly, as permitted by 220.61 and 310.15(B)(6).	Same as above except USE is also permitted for underground, direct burial. Permitted to emerge above ground in meter bases and similar enclosures. If run to a termination box inside of building after emerging from underground, shall not exceed 1.8 m (6 ft) in length inside building. Not permitted for inside premise wiring.	Same as above.	Refer to the 75°C and 90°C columns. For single-phase dwelling services, see 310.15(B)(6).

Table 4-15 (*continued*)

Figure 4-44   Type SEU service-entrance cable. The major characteristic is the Uninsulated wraparound neutral conductor. See Table 4-15 for *NEC®* requirements for the permitted uses of Type SE cables.

Figure 4-45   Type SER service-entrance cable. The major characteristics are the insulated neutral conductor and bare equipment grounding conductor. See Table 4-15 for *NEC®* requirements for the permitted uses of Type SE cables.

REVIEW

Note: Refer to the *Code* or the plans where necessary.

1. The largest size solid conductor permitted to be installed in a raceway is (10 AWG) (8 AWG) (6 AWG). Circle the correct answer.

2. What is the minimum branch-circuit wire size that may be installed in a dwelling?
   _____

3. What exceptions, if any, are there to the answer for question 2? _____
   _____
   _____
   _____
   _____
   _____

4. What determines the ampacity of a wire? _____
   _____
   _____
   _____
   _____
   _____

5. What unit of measurement is used for the diameter of wires? _____

6. What unit of measurement is used for the cross-sectional area of wires? _____
   _____

7. What is the voltage rating of the conductors in Type NMC cable? _____

8. Indicate the allowable ampacity of the following Type THHN (copper) conductors. Refer to *Table 310.16*.

   a. 14 AWG _____ amperes   d. 8 AWG _____ amperes

   b. 12 AWG _____ amperes   e. 6 AWG _____ amperes

   c. 10 AWG _____ amperes   f. 4 AWG _____ amperes

9. What is the maximum operating temperature of the following conductors? Give the answer in degrees Fahrenheit and in degrees Celsius.

   a. Type XHHW _____   c. Type THHN _____

   b. Type THWN _____   d. Type TW   _____

10. What are the colors of the conductors in nonmetallic-sheathed cable for

    a. two-wire cable? _____ , _____

    b. three-wire cable? _____ , _____ , _____

11. For nonmetallic-sheathed cable, may the uninsulated conductor be used for purposes other than grounding? _____

12. What size equipment grounding conductor is used with the following sizes of nonmetallic-sheathed cable?

    a. 14 AWG _____   c. 10 AWG _____

    b. 12 AWG _____   d. 8 AWG _____

13. Under what condition may nonmetallic-sheathed cable (Type NM) be fished in the hollow voids of masonry block walls? _____

    _____

    _____

14. a. What is the maximum distance permitted between straps on a cable installation?

       _____

    b. What is the maximum distance permitted between a box and the first strap in a cable installation? _____

    c. Does the *NEC*® permit two-wire Romex stapled on edge? _____

    d. When Type NMC cable is run through holes in studs and joists, must additional support be provided? _____

15. What is the difference between Type AC and Type ACT armored cables? _____

    _____

    _____

    _____

    _____

    _____

    _____

    _____

16. Type ACT armored cable may be bent to a radius of not less than _____ times the diameter of the cable.

17. When armored cable is used, what protection is provided at the cable ends?

_____

_____

_____

_____

18. What protection must be provided when installing a cable in a notched stud or joist, or when a cable is run through bored holes in a stud or joist where the distance is less than 1¼ in. (32 mm) from the edge of the framing member to the cable, or where the cable is run parallel to a stud or joist and the distance is less than 1¼ in. (32 mm) from the edge of the framing member to the cable? _____

_____

_____

_____

_____

19. For installing directly in a concrete slab, (armored cable) (nonmetallic-sheathed cable) (conduit) may be used. Circle the correct method of installation.

20. Circle the correct answer to the following statements:

   a. Type SE service-entrance cable is for aboveground use.   True   False

   b. Type USE service-entrance cable is suitable for direct burial
      in the ground.   True   False

   c. Type SE cable with an uninsulated neutral is permitted for
      hooking up an electric range.   True   False

   d. Type SE cable with an insulated neutral is permitted for
      hooking up an electric range.   True   False

   e. Are Types SE and USE service-entrance cables used in your area?   Yes   No

21. When running NMC through a bored hole in a stud, the nearest edge of the bored hole shall not be less than _____ in. from the face of the stud unless the cable is protected by a ¹⁄₁₆ in. (1.6 mm) metal steel plate.

22. Where is the main service-entrance panel located in this residence? _____

_____

23. a. Is nonmetallic-sheathed cable permitted in your area for residential wiring?

   _____

   b. From what source is this information obtained? _____

24. Is it permitted to use flexible metal conduit over 6 ft (1.8 m) in length as a grounding means? (Yes) (No)   Circle the correct answer.

25. Liquidtight flexible metal conduit may serve as a grounding means in sizes up to and including _____ in. where used with listed fittings.

26. The allowable current-carrying capacity (ampacity) of aluminum wire, or the maximum overcurrent protection in the case of 14 AWG, 12 AWG, and 10 AWG conductors, is less than that of copper wire for a given size, insulation, and temperature of 86°F (30°C). Refer to *Table 310.16* and *240.4(D)* and complete the following table. Important: where the ampacity of the conductor does not match the rating of a standard fuse or circuit breaker as listed in *240.6(A)*, *240.4(B)* permits the selection of the next "higher" standard rating of fuse or circuit breaker if the next higher standard rating does not exceed 800 amperes. Nonstandard ampere ratings are also permitted.

WIRE (AWG-KCMIL)	COPPER		ALUMINUM	
	Ampacity	Overcurrent Protection	Ampacity	Overcurrent Protection
12 AWG THHN				
10 AWG THHN				
3 AWG THW				
0000 THWN				
500 KCMIL THWN				

27. It is permissible for an electrician to connect aluminum, copper, or copper-clad aluminum conductors together in the same connector. (True) (False). Circle the correct answer.

28. Terminals of switches and receptacles marked CO/ALR are suitable for use with _____ , _____ , and _____ conductors.

29. Wire connectors marked AL/CU are suitable for use with _____ , _____ , and _____ conductors.

30. A wire connector bearing no marking or reference to AL, CU, or ALR is suitable for use with (copper) (aluminum) conductors only. Circle the correct answer.

31. When Type NM or NMC cable is run through a floor, it must be protected by at least _____ in. (_____ mm) of _____

_____

_____

32. When nonmetallic-sheathed cables are bunched or bundled together for distances longer than 24 in. (600 mm), what happens to their current-carrying ability?

_____

33. In diagrams A and B, nonmetallic-sheathed cable is run through the cold air return. Which diagram "meets Code"? A _____ B _____ Check one.

34. The marking on the outer jacket of a nonmetallic-sheathed cable indicates the letters NMC-B. What does the letter **B** signify?_____

_____

_____

35. A 120-volt branch-circuit supplies a resistive heating load of 10 amperes. The distance from the panel to the heater is approximately 140 ft. Calculate the voltage drop using (a) 14 AWG, (b) 12 AWG, (c) 10 AWG, (d) 8 AWG copper conductors. See *210.19(A)*, *FPN No. 4* and *215.2(A)(3)*, *FPN No. 2*.

36. In question 35, it is desired to keep the voltage drop to 3% maximum. What minimum size wire would be installed to accomplish this 3% maximum voltage drop? See *210.19(A)*, *FPN No. 4* and *215.2(A)(3)*, *FPN No. 2*.

_____

_____

_____

_____

_____

37. *NEC® 215.2*, *220.61*, *230.42*, and *310.15(B)(6)* state that the neutral conductor of a residential service or feeder can be smaller than the phase conductors, but only:

_____

_____

_____

38. The allowable ampacity of a 4 AWG THHN from *Table 310.16* is 95 amperes. What is this conductor's ampacity if connected to a terminal listed for use with 60°C wire?

_____

39. If, because of some obstruction in a wall space, it is impossible to keep an NMC cable at least 1¼ in. (32 mm) from the edge of the stud, then it shall be protected by a metal plate at least _____ in. thick.

40. The recessed fluorescent luminaires (fixtures) installed in the ceiling of the recreation room of this residence are connected with trade size ⅜ flexible metal conduit. These flexible connections are commonly referred to as fixture whips. Does the flexible metal conduit provide adequate grounding for the luminaires (fixtures), or must a separate equipment grounding conductor be installed? _____

_____

_____

_____

_____

41. Flexible liquidtight metal conduit will be used to connect the air-conditioner unit. It will be trade size ¾. Must a separate equipment grounding conductor be installed in this FLMC to ground the air conditioner properly?_____

_____

_____

_____

_____

42. What size overcurrent device protects the air-conditioning unit?_____

_____

43. May the 20-ampere small-appliance branch-circuits in the kitchen of a residence also supply the lighting above the kitchen sink and under-cabinet lighting? Check the correct answer. Yes _____ No _____

44. A 30-ampere branch-circuit is installed for an electric clothes dryer. What is the maximum ampere rating of a dryer permitted by the *NEC*® to be cord-and-plug connected to this circuit? Check the correct answer.

    a. _____ 30 amperes

    b. _____ 24 amperes

    c. _____ 15 amperes

45. In many areas, metal framing members are being used in residential construction. When using nonmetallic-sheathed cable, what must be used where the cable is run through the metal framing members?_____

46. Are set-screw-type connectors permitted to be used with armored cable that has an aluminum armor? _____

47. Most armored cable today has 90°C conductors. What is the correct designation for this type of armored cable?_____

48. If you saw two different types of SE cables, how would you distinguish them?

    _____

    _____

    _____

    _____

Switch Control of Lighting Circuits, Receptacle Bonding, and Induction Heating Resulting from Unusual Switch Connections

OBJECTIVES

After studying this unit, the student should be able to

▶ identify the grounded and ungrounded conductors in cable or conduit (color coding).

▶ identify the various types of toggle switches for lighting circuit control.

▶ select a switch with the proper rating for the specific installation conditions.

▶ describe the operation that each type of toggle switch performs in typical lighting circuit installations.

▶ demonstrate the correct wiring connections for each type of switch per *Code* requirements.

▶ understand the various ways to bond wiring devices to the outlet box.

▶ understand how to design circuits to avoid heating by induction.

WARNING! Improper wiring can result in fires, personal injuries, and electrocutions! All wiring must be done in conformance to the *NEC.* Anything less is unacceptable!

CONDUCTOR IDENTIFICATION
(*ARTICLES 200* AND *210*)

Before making any electrical connections, an electrician must understand conductor color coding.

Grounded Conductor: For alternating-current circuits, the *NEC®* requires that the grounded (identified) conductor have an outer finish that is either continuous white or gray, or have an outer finish (not green) that has three continuous white stripes along the conductor's entire length. In multiwire circuits, the grounded conductor is also called a *neutral* conductor.

Ungrounded Conductor: An ungrounded conductor must have an outer finish that is a color other than green, white, gray, or three continuous white stripes. This conductor is commonly called the "hot" conductor. You will get a shock if this "hot" conductor and a grounded conductor or grounded surface such as a metal water pipe are touched at the same time.

Popular colors are black, white, red, blue, green, yellow, brown, orange, and gray. Conductors 1 AWG and larger are generally available in black only.

Three Continuous White Stripes: In house wiring, you will probably never run across a conductor that has three continuous white stripes. This method of identifying conductors is done by the cable manufacturer, and would be found on larger size conductors and some service-entrance cables such as those used for a mobile home feeder (two black, one black that has three white stripes, one green).

Grounded Neutral Conductor

For residential wiring, the white grounded conductor is commonly referred to as the *neutral*. However, this conductor is not always a neutral conductor as you will learn.

For branch-circuits and feeders, the grounded neutral conductor must be insulated, *310.2(A)*.

For residential services, the grounded neutral conductor is permitted to be insulated or bare, *230.41*.

For residential wiring, the 120/240-volt electrical system is grounded by the electric utility at their transformer, and again by the electrician at the main service. This is in accordance with *250.20(B)(1)* requiring that the system be grounded so the maximum voltage to ground on the ungrounded conductors does not exceed 150 volts. *250.26(2)* requires that

the neutral conductor be grounded for single-phase, three-wire systems.

Electricians often use the term "neutral" whenever they refer to a white grounded circuit conductor. Proposals have been made to include a definition of a neutral in the *NEC®*—with no success. Even electrical engineering textbooks are evasive as to whether the definition of a neutral refers only to the actual source (transformer), or down into the system such as a two-wire branch-circuit.

A neutral conductor is:

- the conductor that carries only the unbalanced current from the other conductors, as in the case of multiwire circuits of three or more conductors, and

- the conductor where the voltage from every other conductor to it is equal under normal operating conditions.

By these definitions, the white grounded conductor in a two-wire circuit is not truly a neutral conductor. In a multiwire three-wire branch-circuit, the white grounded conductor is a neutral conductor. See Figure 5-1.

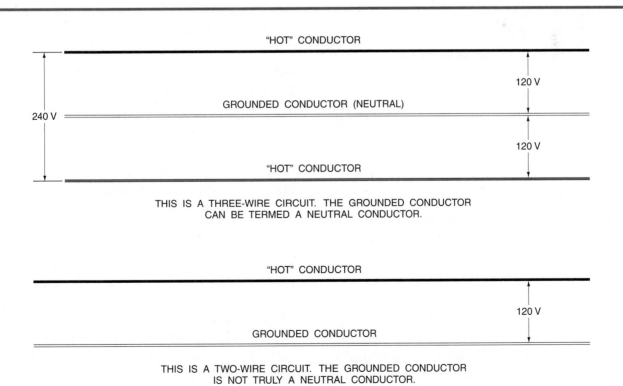

THIS IS A THREE-WIRE CIRCUIT. THE GROUNDED CONDUCTOR CAN BE TERMED A NEUTRAL CONDUCTOR.

THIS IS A TWO-WIRE CIRCUIT. THE GROUNDED CONDUCTOR IS NOT TRULY A NEUTRAL CONDUCTOR.

Figure 5-1   Definition of a true neutral conductor.

For residential wiring, if you are comfortable calling the white grounded conductor a neutral, continue to do so. Do not get hung up on technicalities.

For those of you who will broaden your training into commercial and industrial wiring, you will find that there are electrical systems such as a 3-phase Delta corner grounded system where the grounded conductor is a phase conductor and is not a neutral conductor. This is covered in detail in the *Electrical Wiring—Commercial* text.

A neutral conductor in residential wiring is always a grounded conductor—but a grounded conductor is not always a neutral conductor.

Multiwire Branch-Circuit

The *Code* defines a multiwire branch-circuit as "*A branch circuit consisting of two or more ungrounded conductors having a voltage between them, and a grounded conductor having equal voltage between it and each ungrounded conductor of the circuit and that is connected to the neutral or grounded conductor of the system.*"*

For additional information relative to the hazards involved with multiwire circuits, refer to Figure 17-5, Figure 17-6, Figure 17-7, Figure 17-8, and Figure 17-9 in Unit 17.

Color Coding (Cable Wiring)

The conductors in nonmetallic-sheathed cable (Romex) are color coded as follows.

Two-wire:	one black ("hot" phase conductor)
	one white (grounded "identified" conductor)
	one bare (equipment grounding conductor)
Three-wire:	one black ("hot" phase conductor)
	one white (grounded "identified" conductor)
	one red ("hot" phase conductor)
	one bare (equipment grounding conductor)
Four-wire:	one black ("hot" phase conductor)
	one white (grounded "identified" conductor)
	one red ("hot" phase conductor)

one blue ("hot" phase conductor)
one bare (equipment grounding conductor)

Four-wire nonmetallic-sheathed cable and armored cable are not commonly used for house wiring. In fact, it may be impossible to find this type of cable in stock at electrical supply houses, hardware stores, or home centers. It may need to be ordered special. The cost is very high when compared to two-wire and three-wire cables. It is generally easier and more economical to lay out the wiring using two-conductor and three-conductor cables.

Color Coding (Raceway Wiring)

When the wiring method is a raceway such as EMT, the choice of colors for the ungrounded "hot" conductors is virtually unlimited, with the following restrictions:

Green: reserved for use as an equipment grounding conductor only. See *250.119* and *310.12(B)*.

White or Gray: Reserved for use only as a grounded circuit conductor (neutral). See Table 5-1 for more information about identifying conductors.

What about Conductors with Gray Insulation?

The *NEC*® recognizes a conductor with gray insulation to be used as a grounded conductor. See *200.6(A)* and *(B)*.

The *NEC*® prohibits a conductor with gray insulation to be used as an ungrounded "hot" conductor. See *200.6(A)* and *(B)*.

For years, conductors with gray insulation were used in commercial and industrial installations as both grounded conductors and ungrounded conductors. Past editions of the *NEC*® used the term *natural gray*, but no one really knew what it was. Was it "dirty white"? Was it "light gray"? Was it "dark gray"? Confusion was the end result. To bring an end to the confusion, all references to the term *natural gray* have been deleted in the *NEC*.®

You might be saying to yourself "no cable that I use for house wiring has a gray conductor in it." Right! But in commercial and industrial work, there

---

*Reprinted with permission from NFPA 70-2005.

TO CHANGE FROM THIS	TO THIS	DO THIS
red, black, blue, etc.	an *equipment grounding* conductor	**Conductors 6 AWG or smaller:** No reidentification permitted for conductors 6 AWG and smaller. An equipment grounding conductor must be bare, covered, or insulated. If covered or insulated, the outer finish must be green or green with yellow stripes. ►The color must run the entire length of the conductor and must encircle the conductor.◄ **Conductors larger than 6 AWG:** At time of installation, at both ends and at every point where the conductor is accessible, do any of the following for the entire exposed length: • strip off the covering or insulation • paint the covering or insulation green • use green tape, adhesive labels, or green heat shrink tubing. ►Identification must encircle the conductor.◄ See *210.5(B)*, *250.119*, and *310.12(B)*.
red, black, blue, etc.	a *grounded* conductor	**Conductors 6 AWG or smaller:** For the conductor's entire length, the insulation must be white, gray, or have three continuous white stripes on other than green insulation. Reidentification in the field is not permitted when the wiring method is a raceway. **Conductors larger than 6 AWG:** For the conductor's entire length, the insulation must be white, gray, or have three continuous white stripes on other than green insulation. At terminations, reidentification may be white paint, white tape, or white heat shrink tubing. This marking must encircle the conductor or insulation. Do this marking at time of installation. ►Identification must encircle the conductor.◄ See *200.2*, *200.6(A)* and *(B)*, *200.7(A)*, *(B)*, and *(C)*, *210.5(A)*, and *310.12(A)*.
white, gray	an *ungrounded* (hot) conductor	**No change permitted:** Insulated conductors that are white, gray, or have threecontinuous white stripes are to be used as grounded conductors only. There is no provision in the *NEC*® for reidentifying white or gray conductors for use as ungrounded conductors in raceways. The only exception to use white or gray insulated conductors as ungrounded conductors is for cable wiring. See *200.7(A)*, *(B)*, and *(C)*, and *310.12(C)*.

Table 5-1 Changing colors of conductor insulation in a raceway or cable.

are situations where two systems are installed in the same raceway, such as a 120/208-volt and a 277/480-volt circuit. This is an example of where you might want to use a white insulated conductor for the 120/208-volt system neutral, and a gray insulated conductor for the 277/480-volt system neutral.

Changing Colors When Conductors Are in a Cable

The basic rule in *200.7* is that grounded conductors must be white, gray, or have three continuous white stripes on other than green insulation.

For cable wiring such as nonmetallic-sheathed cable or armored cable, *200.7(C)(1) and (2)* permits the white or gray conductor to be used for single-pole, three-way, or four-way switch loops. These sections require that when used for a switch "loop," the conductor that is white, gray, or marked with three continuous white stripes is to be used for the supply to the switch, and not as the return conductor from the switch to the switched outlet. Also, the conductor must be permanently reidentified to indicate its use by painting or other effective means, at its terminations, and at each location where the conductor is visible and accessible. See Figure 5-2.

Figure 5-2 **(A) Shows a white conductor reidentified with black plastic electrical tape. (B) Shows an assortment of easy to use, 6 in. long heat shrink tubing that is slid over the conductor to be reidentified, then shrunk tightly over the conductor by applying heat as specified in the instructions.**

How to Reidentify Conductor Insulation

Some of the insulation colors available are black, white, red, blue, green, yellow, brown, orange, and gray. Conductor 1 AWG and larger, unless special ordered, are available in black only.

Plastic electrical tape, heat shrink tubing, and fast-drying "touch-up" paint are generally considered acceptable ways to permanently reidentify conductor insulation. You might want to test the paint on a conductor to make sure that it does not have a harmful effect on the insulation, *NEC® 310.9.* Whichever method you choose to reidentify a conductor, be sure it encircles the insulation.

Heat shrink tubing (UL recognized) with a shrink ratio of 2:1 is available at electrical distributors in packages of assorted colors and sizes, for example, 6 in. (150 mm) long. See Figure 5-2B.

Some electrical inspectors have very strong feelings about what they will accept for reidentifying conductors. They might not consider plastic tape as being a "permanent" material. Check it out!

It is pretty obvious that the intent of the *Code* is to permanently reidentify a white conductor when it is used as a "hot" conductor. If not properly reidenti-fied, someone someday will get a surprise. An accident waiting to happen—such "surprises" can prove to be fatal!

The wiring diagrams that follow are good examples of where permanent reidentification of the white conductor is necessary.

Changing Colors When Conductors Are in a Raceway

Occasionally, it may be necessary to change the actual color of the conductor's insulation. Table 5-1 provides guidelines for doing this.

CONNECTING WIRING DEVICES

Terminating and splicing conductors is a skill. The connections must be tight so as not to overheat under load.

▶*NEC® 110.3(B)* requires us to follow the man-ufacturers' instructions regarding the listing and labeling of all electrical products. *NEC® 110.14* provides clear requirements for making electrical connections.◀

Table 4-4 shows in detail the terminal identifica-tion markings found on wiring devices (switches and receptacles) indicating the types of conductors permitted to be connected to them.

Most commonly used wiring devices have screw terminals.

Some wiring devices have back-wiring holes only. For these, the conductor insulation is stripped off for the desired length, inserted into the hole, and held in place by the device's internal spring pressure mechanism. These wiring devices are referred to as having *push-in* terminals.

For convenience, some wiring devices have both screw terminals and back-wiring holes, as illustrated in Figure 5-3 and Figure 5-4. These can be connected or hooked up in different ways. One way is if the conductor insulation is stripped off for the desired length and inserted into the back hole, then the screw terminal is properly tightened. In another way, the conductor insulation is stripped off for the desired length and terminated under the screw terminal, in which case the back-wiring holes are not used.

Using both the back-wiring terminals and the screw terminals, it would be possible to terminate as

Figure 5-4 Exploded view of a grounding-type receptacle, showing all internal parts. *200.10(B)* requires that terminals on receptacles, plugs, and connectors that are intended for the ground-circuit conductor be substantially white (usually silver colored), or marked with the word "white" or the letter "W" adjacent to the terminal. The terminal for the ungrounded "hot" conductor is usually brass colored.

many as four conductors to the white terminal, and four conductors to the brass terminal. This is a violation of *110.3(B)*. UL does not test receptacles with so many conductors connected to the device. A much better way is shown in Figure 5-5 where all of the necessary splicing is done independent of the receptacle. Short lengths of wire called *pigtails* are included in the splice, and these pigtails in turn connect to the receptacle. Removal of the receptacle would not have any effect on the splice.

Installing Switches and Receptacles

▶ Always make sure that the yokes of switches and receptacles are rigidly held in place so they do not twist. Not tightly securing wiring devices is

Figure 5-3 Grounding-type receptacle detailing various parts of the receptacle.

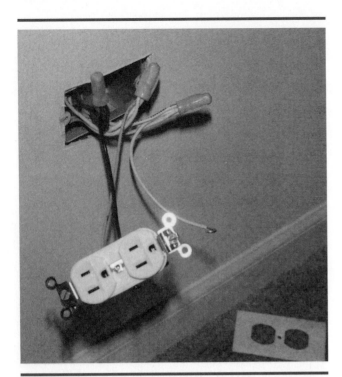

Figure 5-5 **A receptacle connected with pigtails. This particular installation has one outlet switched and one outlet on continuously. Six conductors enter the box because the wiring runs from receptacle to receptacle around the room. A 4-in. square outlet box trimmed with a single-gang plaster (dry-wall) cover is used. Removal of the receptacle does not affect the circuit to the other receptacles.**

Figure 5-6 **Slot configuration of receptacles.**

A 15-ampere, 125-volt, NEMA 5-15R receptacle.
Short slot: "hot" conductor.
Long slot: grounded conductor.
Horseshoe slot: equipment grounding conductor.

A 20-ampere, 125-volt, NEMA 5-20R receptacle.
Short slot: "hot" conductor.
Long "T" slot: grounded conductor.
Horseshoe slot: equipment grounding conductor.

A 15-ampere, 250-volt, NEMA 6-15R receptacle. Slot arrangement will not accept standard 125-volt attachment plug cap; could be used for a 240-volt window air conditioner.
Short slots: "hot" conductors.
Horseshoe slot: equipment grounding conductor.

A 20-ampere, 250-volt, NEMA 6-20R receptacle. Slot arrangement will not accept standard 125-volt attachment plug cap; could be used for a 240-volt window air conditioner.
Short slot: "hot" conductor.
Long "T" slot: grounded conductor.
Horseshoe slot: equipment grounding conductor.

unsafe as well as being an example of sloppy work-manship, *406.4(A)*. ◄

Receptacle Configuration

Figure 5-6 illustrates some of the common receptacle configurations used in homes in conformance to *Table 210.21(B)(3)*. Range and dryer receptacles are discussed in Unit 20.

PUSH-IN TERMINATIONS

Be careful when using screwless push-in terminals.

Screwless push-in terminals on receptacles are listed by Underwriters Laboratories (UL) for use *only* with solid 14 AWG copper conductors. They *are not* to be used with:

• aluminum or copper-clad aluminum conductors.

• stranded conductors.

• 12 AWG conductors. By design, the holes are large enough to take only a 14 AWG solid conductor.

Push-in terminals for 12 AWG solid copper conductors are still permitted on snap switches.

TOGGLE SWITCHES *(ARTICLE 404)*

UL refers to them as snap switches. Manufacturers and electricians refer to them as toggle switches. They are one and the same.

The most frequently used switch in lighting circuits is the toggle flush switch, Figure 5-7. When mounted in a flush switch box, the switch is concealed in the wall with only the insulated handle or toggle protruding.

Figure 5-8 shows how switches are weatherproofed.

Toggle Switch Ratings

UL lists toggle switches used for lighting circuits as *general-use snap switches*. The UL requirements are the same as *404.14(A)* and *(B)* of the *NEC*®.

Figure 5-7   Toggle flush switches. *Courtesy of* Hubbell, Incorporated.

Figure 5-8   Switch is protected by a weatherproof cover.

AC/DC General-Use Snap Switches
[404.14(B)]

The requirements for general-use snap switches include the following:

- alternating-current (ac) or direct-current (dc) circuits.

- resistive loads not to exceed the ampere rating of the switch at rated voltage.

- inductive loads not to exceed one-half the ampere rating of the switch at rated voltage unless otherwise marked.

- tungsten filament lamp loads not to exceed the ampere rating of the switch at 125 volts when marked with the letter **T**.

- for switches marked with a horsepower rating, a motor load shall not exceed the rating of the switch at rated voltage.

Why "T" Rating?

A tungsten filament lamp draws a very high momentary inrush current at the instant the circuit is energized. This is because the *cold resistance* of tungsten is very low. For instance, the cold resistance of a typical 100-watt lamp is approximately 9.5 ohms. This same lamp has a *hot resistance* of 144 ohms when operating at 100% of its rated voltage.

Normal operating current would be

$$I = \frac{E}{R} = \frac{120}{144} = 0.83 \text{ ampere}$$

The instantaneous inrush current could be as high as

$$I = \frac{E}{R} = \frac{170 \text{ (peak voltage)}}{9.5} = 17.9 \text{ amperes}$$

This instantaneous inrush current drops off to normal operating current in about 6 cycles (0.10 second). The contacts of T-rated switches are designed to handle these momentary high inrush currents. See Unit 13 for more information pertaining to inrush currents.

The ac/dc general-use snap switch normally is not marked ac/dc. However, it is always marked with the current and voltage rating, such as 10A-125V or 5A-250V-T.

AC General-Use Snap Switches
[404.14(A)]

This is the type most commonly used for house wiring projects. Alternating-current general-use snap switches are marked "ac only," in addition to identifying their current and voltage ratings. A typical switch marking is 15A, 120-277V ac. The 277-volt rating is required on 277/480-volt systems. Other requirements include:

- ac only.

- resistive and inductive loads not to exceed the ampere rating of the switch at the voltage involved.

- tungsten-filament lamp loads not to exceed the ampere rating of the switch at 120 volts.

- motor loads not to exceed 80% of the ampere rating of the switch at rated voltage. A UL requirement is that the load shall not exceed 2 horsepower.

What Are the Switching Conductors Called?

In *200.7(C)(2)*, we find a few unique words that need to be explained. There are also local and regional expressions for these words. It is always best to use the terms found in the *NEC*.

- *Switch loop:* The conductors that run between (to and from) a switched outlet and the switch controlling that outlet.

- *Supply:* The conductor that runs from the source to the switch. It might also be referred to as the *feed*, *hot*, *hot leg*, or *supply leg*.

- *Return:* The conductor that runs from the switch back to the switched outlet. It might also be referred to as the *switch leg* or *switched leg*.

Conductor Color Coding for Switch Connections

The insulated conductors in nonmetallic-sheathed cable or armored cable are black-white, black-white-red, or, in some cases, black-white-red-blue.

We usually think of the black insulated conductor as the supply to a switch and a red insulated conductor as the return. However, when wiring with nonmetallic-sheathed cable or armored cable, we are "stuck" with the colors found in the cables. We must not use a white conductor as the return conductor from a switch, *200.7(C)(2)*. A color-coding scheme needs to be established for the supplies and returns. In the case of three-way and four-way switches, we have the *travelers*. *Travelers* have also been known to be called *dummies*, *runners*, and *kickers*, depending on what part of the country you are in. The wiring diagrams in Figures 5-11 through 5-22 show the recommended color schemes when wiring with cable.

When the wiring method is a raceway or electrical metallic tubing (EMT), the choices of insulation colors are virtually unlimited.

Always connect a white wire to the white (silverish) terminal or to the white wire of a lampholder or receptacle, *200.9* and *200.10*. Always connect the black switch leg conductor (or red in some cases) to the black wire (or dark brassy terminal) of a lampholder or receptacle.

In cables, always reidentify white conductors when they are used as ungrounded (hot) conductors. This reidentification must be done wherever the conductors are visible and accessible. These requirements are found in *200.7(C)(2)*. See Table 5-1.

Never use a green-colored insulation for a grounded or ungrounded conductor. Green is reserved for equipment grounding conductors. See *210.5* and *250.119*. See Table 5-1.

Connecting Switches, Receptacles, and Lighting Outlets

A conductor carrying an alternating current produces a magnetic field (flux) around the conductor. This magnetic field extends outside of the conductor. The greater the current, the stronger the magnetic field. In a 60-Hz (pronounced 60-hertz) circuit, the current and magnetic field reverses 120 times each cycle. If this single conductor is run through a steel raceway, a steel jacketed cable (i.e., Type AC or MC), a steel locknut, a steel bushing, or through a knockout in a steel box, the alternating magnetic field "induces" heat in the steel. Heat in the steel, in turn, can harm the insulation on the conductor. When all conductors of the same circuit are run through the same raceway, the magnetic fields around the conductors are equal and opposite, thereby canceling one another out.

To prevent induction heating and to keep the impedance of a circuit as low as possible where conductors are run in metal raceways, metal jacketed cables, and in metal enclosures, the installation must follow certain *NEC®* requirements.

Induction Heating

NEC® 300.3(B) requires that *"all conductors of the same circuit and, where used, the grounded conductor, all equipment grounding conductors, and bonding conductors shall be contained within the same raceway, auxiliary gutter, cable tray, trench, cable, or cord, unless otherwise permitted."** There are four exceptions: for paralleled installations, for grounding conductors, for nonferrous wiring methods, and certain enclosures.

NEC® 300.20(A) requires that *"Where conductors carrying alternating current are installed in metal enclosures or metal raceways, they shall be so arranged as to avoid heating the surrounding metal by induction. To accomplish this, all phase conductors and, where used, the grounded conductor and all equipment grounding conductors shall be grouped together."**

NEC® 404.2(A) requires that *"Three-way and four-way switches shall be so wired that all switching is done only in the ungrounded circuit conductor. Where in metal raceways or metal-armored cables, wiring between switches and outlets shall be in accordance with 300.20(A)."** Switch loops do not require a grounded conductor.

Figure 5-9, and Figure 5-10 illustrate the intent of the *NEC®* to prevent induction heating.

*Reprinted with permission from NFPA 70-2005.

Figure 5-9   A few examples of arranging circuitry to avoid induction heating according to *300.20* when metal raceway or metal cables (BX) are used. Always ask yourself, "Is the same amount of current flowing in both directions in the metal raceway?" If the answer is "no," there will be induction heating of the metal that can damage the insulation on the conductors.

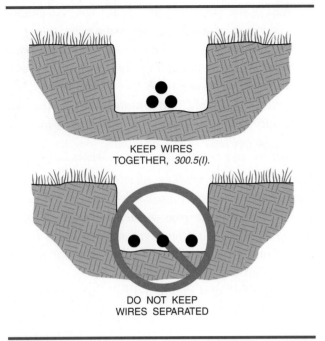

Figure 5-10   Arrangement of conductors in trenches. Refer to Unit 16 for complete information pertaining to underground wiring.

Don't Do It!

Unusual switch connections may or may not conform to the *NEC*, depending upon how the circuitry is run. Some electricians have found that they can save substantial lengths of three-wire cable when hooking up three-way and four-way switches by using two-wire cable.

Connections as shown in Figure 5-11 are in violation of *300.3(B)*, *300.20(A)*, and *404.2(A)* when metal boxes and/or steel jacketed Types AC and MC are used. If nonmetallic-sheathed cable and plastic boxes are used, this connection is not a *Code* violation because there would be no induction heating. If the framing members are steel, we have a *Code* violation because of the induction heating issue. The circuitry in Figure 5-11 results in greater impedance and greater voltage drop than if wired in a conventional manner. High impedance will cause the tripping time of a breaker or opening time of a fuse to increase. Another possible problem would be that of picking up "noise" from the magnetic field surround-

Figure 5-11   This connection is a violation of *300.3(B)*, *300.20(A)*, and *404.2(A)* when using steel jacketed cables, steel raceways, or steel outlet and device boxes. See text for a full explanation. This is not a recommended way of connecting three-way and four-way switches.

ing the conductor. This could affect the operation of a computer or telephone located in the vicinity, particularly if non-twisted telephone cables are used. The circuitry is very confusing and is certainly an odd way to do a simple job.

Because of possible adverse interference created by magnetic fields, and confusion for someone who at some later date may be involved with that circuit for troubleshooting, repair, replacement, or remodel, the circuitry in Figure 5-11 is strongly discouraged. **Don't do it.**

Is It a Light, a Lamp, a Fixture?

Let's not get hung up on words! We say things such as "Turn on the light."

Light is "The radiant energy that is capable of exciting the retina and producing a visual sensation."

A **lamp** is "A generic term for a man-made source of light."

A bulb is "A glass envelope—the glass component part used in a bulb assembly."

A **luminaire** is "A complete lighting unit consisting of a lamp or lamps together with the parts designed to distribute the light, to position and protect the lamps and ballast (where applicable), and to connect the lamps to the power supply." The *NEC®* uses the term *luminaire* as the internationally used term for *lighting fixture.*

Use whatever term that you feel comfortable with. Most of us will continue to say "Turn on the light."

Switch Types

Switches are available in single-pole, three-way, four-way, double-pole, dimmer, and speed control types. They are available in snap (toggle), push, rocker, slider, rotary, electronic, and with or without pilot lights. Faceplates are commonly available in white, brown, and ivory plastic, stainless steel, brass, chrome, and wood.

In all of the following wiring diagrams, the equipment grounding conductors are not shown as this is a separate issue covered elsewhere in this text.

Single-Pole Switches. A single-pole switch is marked with "ON" and "OFF" positions. Used when the load is to be controlled from one switching point, a single-pole switch is connected in series with the ungrounded (hot) conductor.

Single-Pole Switch, Feed at Switch. In Figure 5-12, the 120-volt source enters the switch box where the white grounded conductor is spliced directly through to the lamp. The black conductor from the supply circuit connects to one terminal on the switch. The black conductor in the cable coming from the outlet box at the lamp is connected to the other terminal on the switch. Because a single-pole switch connection is simply a series circuit, it makes no difference which terminal is the supply and which terminal is the load. This layout needs no reidentification of the white conductor.

Single-Pole Switch, Feed at Light. In Figure 5-13, the 120-volt source enters the outlet box at the

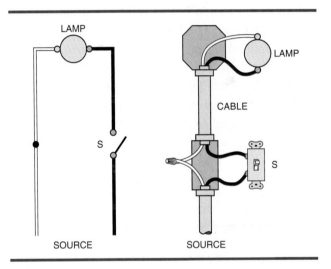

Figure 5-12   **Single-pole switch in circuit with feed at switch.**

Figure 5-13   **Single-pole switch in circuit with feed at light.**

lamp. The switch loop is a two-wire cable. The white wire of a cable is permitted in *200.7(C)(2)* to be used as the supply (feed) to the switch. When so used, *200.7(C)(2)* requires that the white conductor be reidentified in the ceiling outlet box and in the device box. The white conductor is not permitted to be used as the return (switch leg) from the switch to the switched outlet. Color plastic tape is most often used to reidentify a conductor. For this diagram, black tape would be suitable.

Single-Pole Switch, Feed at Switch, Receptacle "On" Continuously.

Figure 5-14 shows the 120-volt source entering the switch box where the switch is located. The switch controls the lamp. The receptacle remains "hot" at all times. The white conductor in the switch box is spliced straight through to the outlet box. The black conductor in the switch box is spliced through to the outlet box and also feeds the switch. The red conductor in the three-wire cable is the switch leg, connecting to the switch and to the lamp. A two-wire cable from the outlet box carries the circuit to the receptacle. The circuitry is such that no reidentification of the white conductor is necessary.

Three-Way Switches.

Three-way switches are used when a load is to be controlled from two locations. The term "three-way" is very misleading. Three-way switches have three terminals. Maybe they should have been named "three-terminal" switches. One terminal is called the "common" and is darker in color than the other two terminals. The other two terminals are called "traveler terminals."

Figure 5-15 shows the two positions of a three-way switch. Actually, a three-way switch is a single-pole, double-throw switch.

There is no "ON/OFF" marking on a three-way switch, as can be seen in Figure 5-16.

Three-Way Switch Control, Feed at Switch.

Figure 5-17 shows the 120-volt source entering at the first switch location. A three-wire cable is run between the two switches. A two-wire cable is run from the second switch to the lamp. The white conductors in both switch boxes are spliced directly through to the lamp. The red and black conductors in the three-wire cable are the "travelers." The black conductor in the three-wire cable is reidentified with red tape, consistent with keeping both "travelers" red in color. The black conductor from the source is

Figure 5-14   Ceiling outlet controlled by single-pole switch with live receptacle outlet and feed at switch.

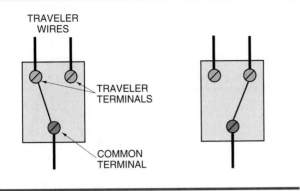

Figure 5-15 Two positions of a three-way switch.

Figure 5-16 Toggle switch: three-way snap switch.
Courtesy of **Pass & Seymour.**

Figure 5-17 Circuit with three-way switch control. The feed is at the first switch. The load is connected to the second switch. The black and red conductors are used for the travelers. The black conductor has been reidentified with red tape.

connected to the "common" terminal of the first switch. The black conductor in the cable coming from the outlet box at the lamp is connected to the "common" terminal of the second switch. The circuitry is such that no reidentification of the white conductor is necessary.

Three-Way Switch Control, Feed at Light.
Figure 5-18 shows the 120-volt source feeding into the outlet box at the lamp. A two-wire cable is run from the outlet box to the first switch location. A three-wire cable is run from the first switch to the second switch. In all locations where the white wire is to be used as an ungrounded "hot" conductor, it

must be reidentified, *200.7(C)(2)*. In this diagram, reidentification of the white conductor in the outlet box is with black tape.

The black conductor of the two-wire cable is connected to the "common" terminal of the first switch. At the first switch, the white conductor in the two-wire cable is reidentified with black tape. The white conductor of the three-wire cable is reidentified with red tape. The red conductor and the reidentified white conductor in the three-wire cable are the "travelers." The white conductor is reidentified with red tape, consistent with keeping both "travelers" red in color.

At the second switch location, the black conductor of the three-wire cable is connected to the "common" terminal of the switch. The white conductor of

Figure 5-18 Circuit with three-way switch control and feed at light.

the three-wire cable is reidentified with red tape. The red conductor and the reidentified white conductor in the three-wire cable are the "travelers." Here again, the white conductor is reidentified with red tape, so that the "travelers" are red in color.

Three-Way Switch Control, Feed at Light (Alternate Connection). Figure 5-19 shows the 120-volt source entering at the outlet box. The white conductor from the 120-volt source at the outlet box goes directly to the lampholder. In all locations where the white wire is to be used as an ungrounded "hot" conductor, it must be reidentified, *200.7(C)(2)*. From the outlet box to the switch on the left, the black conductor in the three-wire cable serves as the switch return to the lamp, and the red conductor and the reidentified white conductor are the "travelers." From the outlet box to the switch on the right, the black conductor serves as the feed to the three-way switch, and the red conductor and the reidentified white conductor are the "travelers." For uniformity, reidentification of the white conductors used for "travelers" through the outlet box and in the switch boxes is with red tape.

When wiring with cable, this connection is not the best way to run the cables. Always try to design your cable runs so the travelers run between switches, not through the outlet box at the luminaire (fixture).

Four-Way Switches. Four-way switches are used where switch control is needed from more than two locations. The term "four-way" is misleading.

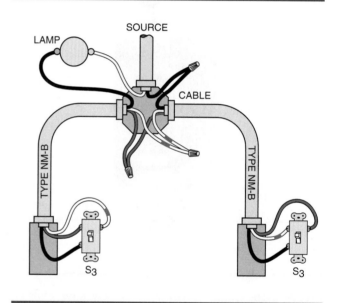

Figure 5-19 Circuit with a three-way switch control with feed at the lamp. This is not the best way to wire multiple switches when wiring with cable. It is best to run the travelers between switches, not through the outlet box. Less confusion—fewer connections.

Four-way switches have four terminals. Maybe four-way switches should have been called "four-terminal" switches. A four-way switch does not have "ON" and "OFF" markings. Four-way switches are connected to the "travelers" *between* two three-way switches. One three-way switch is fed by the source—the other three-way switch is connected to the load. Make sure that the "travelers" are connected to the proper "pairs" of terminals on a four-way switch; otherwise the switching will not work.

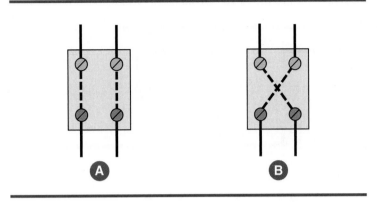

Figure 5-20   Two positions of a four-way switch.

Figure 5-20 shows the internal switching of a four-way switch.

Figure 5-21 shows a lamp controlled from three switching points. Because the white conductor from the source is connected straight through all of the switch boxes and on to the lamp, no reidentification of the white wire is necessary. The black conductor from the source is connected to the common terminal on the first three-way switch. The black conductor at the third switch is connected to the common terminal on the three-way switch. The black and red conductors in the three-wire cables are the "travelers"

and are connected to the proper "traveler" terminals of the three-way and four-way switches. The black conductors in the three-wire cables have been reidentified with red tape, consistent with keeping both "travelers" red in color. At the lamp, the white conductor of the two-wire cable connects to the white conductor (or silverish terminal) of the lampholder, and the black conductor connects to the black conductor (or dark brassy terminal) of the lampholder.

Figure 5-22 shows a lamp controlled from three switching points. The supply comes into the outlet box where the lamp is located. The white conductor

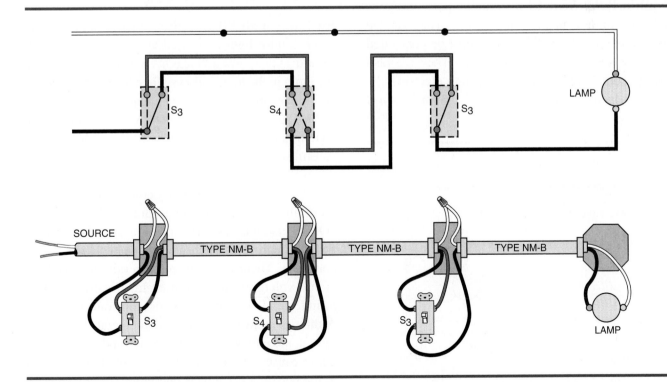

Figure 5-21   Circuit with switch control at three locations—feed at switch.

Figure 5-22 Circuit with switch control at three locations—feed at lamp.

is required to be reidentified at every point where it is used as an ungrounded "hot" conductor, *200.7(C)(2)*.

Double-Pole Switches. Double-pole switches are not common in residential work. They can be used when two separate circuits must be controlled with one switch, as in Figure 5-23. They are also used to control 240-volt loads, such as electric heat, motors, electric clothes dryers, and similar 240-volt loads, as in Figure 5-24. When line-voltage thermostats for electric heat are double-pole, they switch both ungrounded "hot" conductors, and are marked with "ON" and "OFF" positions.

Switches with Pilot Lights. There are instances when a pilot light is desired at the switch location. Figure 5-25 shows a "locator" type of switch where the lamp in the toggle glows when the switch is in the "OFF" position. *This is sometimes referred to as a "glow" or "lit-handle" switch.* A resistor inside the switch is connected in series with the pilot light. Be careful when working on this type of switch because when the switch is in the "OFF" position, the load and the pilot light are actually connected in "series." Even though these switches are marked with an "OFF" position, they are not truly off. The circuit's path through the switch and connected load is still complete. There is a voltage drop (almost full open circuit voltage) across the pilot light/resistor, and an infinitesimal voltage drop across the connected load. Although there is not enough voltage at the light to light it, there may be enough voltage and current present to cause a person to "flinch," which might result in the person jerking back and falling off a ladder.

Figure 5-26 shows a true pilot light that is "on" when the light is on.

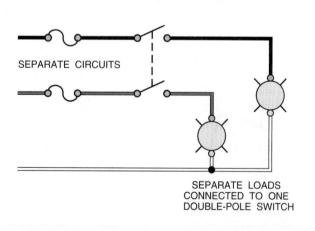

Figure 5-23 Application of a double-pole switch.

Figure 5-24   Double-pole (2-pole) disconnect switch.

Figure 5-25   In (A) the switch is in the "OFF" position. The neon lamp in the handle of the switch glows and the load is off. This is a series circuit. The neon lamp has extremely high resistance. In a series circuit voltage divides proportionately to the resistance. Therefore, the neon lamp "sees" line voltage (120 volts) for all practical purposes and the load "sees" zero voltage. The neon lamp glows.

In (B) the switch is in the "ON" position. The neon lamp is shunted out and therefore has zero voltage across it. Thus, the neon lamp does not glow. Full voltage is supplied to the load. This type of switch might be referred to as a *locator* because it glows when the load is off and does not glow when the load is on.

Figure 5-26   A true pilot light. The pilot lamp is an integral part of the switch. When the load is turned on, the pilot light is also on. When the load is turned off, the pilot light is also off. This switch has three terminals because it requires a grounded circuit conductor.

COMBINATION WIRING DEVICES

In the space of one standard wiring device, it is possible to install combination wiring devices. Figure 5-27A shows three types of combination wiring devices. Other combinations are available. Figure 5-27B shows an interchangeable line of wiring devices, assembled on the metal yoke by the electrician. Switches, receptacles, and neon and incandescent pilot lights are available. Combination wiring devices offer a neater appearance as opposed to having two-gang or three-gang faceplates at a given location. Another common use is where wall space is limited, such as between a door casing and a window casing where there is not enough space for a multigang faceplate. Be sure the wall box has plenty of room for the number of cable clamps, wiring devices, and conductors.

Miscellaneous Connections. Figure 5-28 and Figure 5-29 are diagrams that somewhat combine parts of the previous wiring diagrams. These connections become rather complicated using cable because of the limitation on the available insulation colors. These connections are more suited to raceway wiring.

Do You Need the Grounded Circuit Conductor at the Switch? Maybe. But in most instances, the answer is *NO*. See *404.2(A), Exception.*

However, some electronic dimmers, motion sensors, and lighting controls need both the grounded and ungrounded circuit conductors present. These devices need a small voltage drop across them to operate. Check the wiring diagram to determine if the grounded circuit conductor is required at the switch location. See Figure 5-30.

Illuminated toggles, a rocker, or a faceplate (load "OFF) do not require the grounded circuit conductor. These are merely "in series" with the load. See Figure 5-25.

True pilot lights (load "ON") also require the grounded circuit conductor. See Figure 5-26.

Many electricians anticipate using electronic switching devices and always run the full branch-circuit (grounded and ungrounded conductors) to the switch location, then up to the lighting outlet as shown in Figures 5-12 and 5-14. This eliminates the need to reidentify a white conductor when used as a "hot" switch leg. This is a matter of choice.

Proposals to change the *NEC®* have been made to require that a branch-circuit grounded conductor

Figure 5-27 Combination wiring devices. Many combinations are available on a single yoke: slider, rocker, toggle, three-way plus two single poles, four single-pole switches, switches and pilot lights, switches and receptacles. *Photos courtesy of* Leviton Manufacturing Co., Inc.

THE WIRING METHOD IN THIS WIRING DIAGRAM IS A RACEWAY SYSTEM. WHEN CABLE IS USED, THE CONDUCTOR COLORS ARE DIFFERENT THAN IN THIS DIAGRAM, AND SOME CONDUCTORS MUST BE REIDENTIFIED.

Figure 5-28 A lighting outlet controlled by two three-way switches. The receptacle is "live" at all times. The wiring method is a raceway. When cable is used, the conductor colors are different from those in this wiring diagram, and some conductors must be reidentified. A typical example of this might be between a house and a detached garage or other outbuilding.

THE WIRING METHOD IN THIS WIRING DIAGRAM IS A RACEWAY SYSTEM. WHEN CABLE IS USED, THE CONDUCTOR COLORS ARE DIFFERENT THAN IN THIS DIAGRAM, AND SOME CONDUCTORS MUST BE REIDENTIFIED.

Figure 5-29 Two lighting outlets controlled by two three-way switches. The receptacle is "live" at all times. An example of this hookup might be between a house and a detached garage or other outbuilding, where one lighting outlet is on the house and the other lighting outlet is on or in the garage or outbuilding.

Figure 5-30 An X-10 switch. To operate properly, it requires the presence of the grounded circuit conductor. *Never* connect the white wire on the switch to an equipment grounding conductor in the box. This is hazardous and is a violation of *250.6(A)* and *250.24(A)(5)*.

be present at every switch location. To date, these proposals have been rejected by the Code-Making Panel (CMP) as being unrealistic because most switching locations do not need the grounded circuit conductor; what to do with the conductor when it is not used; and the possible need for a larger wall box because of the extra conductor.

Never use the equipment grounding conductor as a substitute for the circuit's grounded conductor. This could prove to be deadly.

As always, read the manufacturers' instructions.

Does "OFF" Really Mean "OFF?"

The answer is YES. When a switching device has a marked "OFF" position, it must completely disconnect the ungrounded conductor(s) to the load it serves when turned to the "OFF" position. UL Standards require this. Also see *404.15(B)*.

Electronic controls, such as some motion (occupancy) sensors, some dimmers, some fan speed controls, and X-10 transmitting and receiving devices (see Unit 31), require a small voltage drop across the device to operate. When these devices are turned to what appears to be the "OFF" mode, there is still an ever-so-slight amount of voltage present at the load. The device is really not in a full "OFF" position. This can be hazardous to someone working on a luminaire (fixture), for example, thinking it is totally "OFF."

Refer to Unit 6 for additional information relating to the effects of an electrical shock.

Bonding and Grounding at Receptacles and Switches

A metal box is considered to be adequately grounded when the wiring method is armored cable, nonmetallic-sheathed cable with ground, or a metal raceway such as EMT, *250.118*. A separate equipment grounding conductor can also provide the required grounding, *250.134(B)*. Equipment grounding conductors may be bare or green.

Figure 5-31 illustrates a switch that has an equipment grounding conductor terminal. Using a switch of this type provides an easy means for providing the required grounding of a metal faceplate when using nonmetallic boxes.

Grounding and bonding of the equipment grounding conductor to a metal box, switch, or receptacle is important. Figure 5-32, Figure 5-33, and Figure 5-34 show how an equipment grounding conductor of a nonmetallic-sheathed cable can be attached to a metal box and to the equipment grounding screw of a receptacle. Most metal boxes have a No. 10-32 tapped hole for securing a green hexagon-shaped equipment grounding screw.

Other than for special "isolated ground" receptacles, the metal yoke of a receptacle is integrally bonded to the equipment grounding terminal of the receptacle.

Figure 5-3 and Figure 5-4 show a receptacle that has a special "self-grounding" feature for the No. 6-32 mounting screws that results in effective

Figure 5-31 Single-pole toggle switch. Note that this switch has a terminal for the connection of an equipment ground conductor. Thus, when a metal faceplate is installed, it will be grounded. See *404.9(A)* **and** *(B)*. *Courtesy of* **Pass & Seymour.**

grounding of the metal yoke and equipment grounding terminal without the use of a separate equipment grounding conductor.

To ensure the continuity of the equipment grounding conductor path, *250.148* requires that where more than one equipment grounding conductor enters a box, they shall be spliced with devices "suitable for the use." Splices shall not depend on solder. The splicing must be done so that if a receptacle or other wiring device is removed, the continuity of the equipment grounding path shall not be interrupted. This is clearly shown in Figure 5-33. Splicing must be done in accordance with *110.14(B)*.

Grounding Metal Faceplates and Switches

Assuming that an equipment grounding conductor has been carried to and properly connected at all metallic and nonmetallic boxes, we now need to be concerned with carrying that equipment grounding means to receptacles, switches, and metal faceplates.

Figure 5-32 Connections for grounding-type receptacles.

Figure 5-33   One method of connecting the grounding conductor using a special grounding-type wire connector. See *250.146* and *250.148*.

Figure 5-34   Method of attaching grounding clip to switch (device) box. See *250.146* and *250.148*.

Metal faceplates for receptacles shall be grounded, *406.5(B)*. This is accomplished through the No. 6-32 screw that secures the faceplate to the metal yoke on the receptacle.

A similar requirement for the grounding of metal faceplates is found in *314.25(A)*.

Snap switches and dimmer switches must provide a means to ground metal faceplates, *404.9(B)*.

This is a requirement whether metallic or nonmetallic face plates are installed. The actual grounding of the faceplate is through the No. 6-32 screws that secure the faceplate to the metal yoke on the switch.

In *UL Standard 20*, we find the requirement that flush-type switches intended for mounting in a flush-device box shall provide a grounding means. In most cases this is a green hexagon-shaped screw connected to the metal yoke of the switch. This can be seen on the switches illustrated in Figure 5-7, Figure 5-16, and Figure 5-31. The grounding means could also be a special self-grounding clip on the metal yoke, as illustrated in Figure 5-3 and Figure 5-4, a bare copper wire, a copper wire with green insulation, or a copper wire with green insulation having one or more yellow stripes.

Connecting a bare equipment grounding conductor to the equipment grounding terminal on a switch complies with *406.5(B)* for the grounding of a metal faceplate. Where there are multiple switches, there is no need to attach an equipment grounding conductor on all of the switches. One connection is enough.

If the equipment grounding path is accomplished through a self-grounding clip on the metal yoke of a switch or receptacle, you do not have to connect an equipment grounding conductor to the green hexagon-shaped grounding screw on the switch or receptacle.

Many wiring devices have small pieces of cardboard holding the No. 6-32 mounting screws from falling out. ▶Remove at least one of these small pieces of cardboard when direct metal-to-metal contact between the yoke of the wiring device and the box is needed, *250.146(A)*.◀ An example of this would be a receptacle secured to a handy box. If there is no metal-to-metal contact, then an equipment grounding conductor must be connected to the grounding terminal on the wiring device or the wiring device must have the special self-grounding clips discussed above. See Figure 5-32 and Figure 5-33.

For surface-mounted boxes of the type shown in Figures 18-8 and 18-9, where cover-mounted receptacles are used, the metal-to-metal requirement mentioned earlier is not acceptable. Some other "listed" means for providing satisfactory ground continuity must be used, *250.146(A)*. This section makes it clear that there must be direct metal-to-metal contact between the metal box and the device yoke to be considered an acceptable means of grounding the receptacle equipment grounding terminal and the metal cover. ▶At least one of the insulating washers must be removed to ensure direct metal-to-metal contact.◀

In general, where no effective grounding means exists in the switch box, it is acceptable to replace a snap switch with another snap switch that does not have a provision for grounding.

▶However, in existing installations where no effective grounding means is present, and where the snap switch is located within reach of the earth, a conductive floor (i.e., concrete, tile), or other conductive surfaces (i.e., metal pipes, metal ducts), use nonmetallic faceplates, or provide GFCI protection for the wiring to that particular snap switch. See *404.9(B), Exception*.◀

CAUTION: *Never* make connections (jumpering) between the different colored terminals on receptacles. Lives have been lost because of this!

- Jumpering between the brass and silver terminals is a violation of *200.10* and *200.11*.

- Jumpering between the silver and green hexagon-shaped terminals is a violation of *250.24(A)(5)* and *250.142(B)*.

- Jumpering between the brass and green hexagon-shaped terminals is a violation of *250.126*.

COMMON *CODE* VIOLATION— MAKING "TAPS"

As defined in *240.2*, a "tap" is a conductor that has overcurrent protection ahead of its point of supply that exceeds the ampacity of the conductor. In general, conductors shall be protected at their ampacities per *310.15*. Tap conductors are excluded from the basic rule, and this is found in *240.4(E)*. Some examples of "taps" are small luminaire (fixture) wires. In some case, taps can be cost-effective for hooking up electric ranges, counter-mounted cooktops, and wall-mounted ovens. This is discussed in Unit 20.

Table 5-2 shows the maximum overcurrent protection for small conductors as permitted by *240.4(D)*.

Figure 5-35 illustrates a *Code* violation. The switch leg (loop) conductors are part of the branch-circuit wiring and are not to be considered a "tap." Although the connected load may be well within the ampacity of the switch leg, the switch leg must also be capable of handling short circuits and ground faults. The switch leg shown in Figure 5-35 must be the same size as the branch-circuit conductors, which are 12 AWG 20-ampere conductors protected by a 20-ampere circuit breaker.

Connecting Receptacles to 20-Ampere Branch-Circuits

Another very common *Code* violation is using short 14 AWG pigtails to connect receptacles to 20-ampere small-appliance branch-circuit conductors in a kitchen or dining room. These short pigtails must be rated 20 amperes because they are an *extension* of the 20-ampere branch-circuit wiring . . . they are not "taps." Figure 5-5 illustrates this type of connection.

CONDUCTOR SIZE	MAXIMUM OVERCURRENT PROTECTION
14 AWG copper	15 amperes
12 AWG copper	20 amperes
10 AWG copper	30 amperes

Table 5-2

PANEL

LUMINAIRE (FIXTURE)

12 AWG CONDUCTORS

20 A

NEUTRAL

14 AWG CONDUCTORS

Figure 5-35 *Violation:* The electrician has run a 20-ampere branch-circuit consisting of 12 AWG conductors to the outlet box where a lumiaire (fixture) will be installed. 14 AWG conductors are run as the switch legs between the lighting outlet and the switch. The switch legs are part of the branch-circuit and must have an ampacity equal to that of the branch-circuit ampere rating, 20 amperes. Switch legs are not "taps." For the same reasons, it is a *Code* violation to use short "pigtails" of 14 AWG wire to connect receptacles to the 12 AWG conductors fed from a 20-ampere branch-circuit.

Consider Using Larger Size Conductors for Home Runs

To minimize voltage drop, sometimes 12 AWG conductors are used for long "home runs" (50 ft or longer) even though the branch-circuit overcurrent device is rated 15 amperes. A "home run" is the branch-circuit wiring from the panel to the first outlet or junction box in that circuit. This is permitted by *210.3*, which states, *"where conductors of higher ampacity are used for any reason, the ampere rating or setting of the specified overcurrent device shall determine the circuit rating."* In this case, 14 AWG conductors are permitted for the switch legs. But this leads to confusion later when someone removes the cover from the panel and wonders why 12 AWG branch-circuit conductors are connected to 15-ampere circuit breakers. For house wiring, it is best to use the same size conductor throughout a given branch-circuit.

Specifications for commercial and industrial installations oftentimes will call for a "home run" of 50 ft or more to be 10 AWG conductors, even though the branch-circuit rating is 20 amperes.

*Reprinted with permission from NFPA 70-2005.

Table 210.24 is a summary of branch-circuit requirements. This table lists branch-circuit ratings, branch-circuit conductor sizes, tap conductor sizes, overcurrent protection, outlet (lampholder and receptacle) ratings, maximum loads, and permissible loads.

TIMERS

Timers are unique in that they provide automatic control of electrical loads. Timers are also referred to as time clocks.

Timers are used where a load is to be controlled for specific "ON/OFF" times of the day or night. The capabilities of timers are endless.

Figure 5-36 shows two types of spring-loaded timers that are connected in series with the load—the same as a standard wall switch. A typical application is in a bathroom for the control of an exhaust fan or an electric heater. These timers install in a standard single-gang device box. Because they take up quite a bit of space, make sure the wall box is large enough to meet the box-fill requirements of *314.16*.

Figure 5-37 is a 24-hour time clock commonly used to control security lighting, decorative lighting, or energy management. This particular time clock

Figure 5-36   Spring-loaded timers. *Courtesy of* Invensys Climate Control.

has four adjustable and removable "pins" that control two "ON" and two "OFF" operations (additional "pins" can be added). There is an "override" switch for manual control.

Figure 5-38 is an electronic 24-hour, 7-day time clock. It can be programmed to skip certain days, and up to sixteen events can be programmed. There is an "override" switch for manual control.

Figure 5-37   A 24-hour time clock. *Courtesy of* Invensys Climate Control.

Figure 5-38   An electronic time clock. *Courtesy of* Invensys Climate Control.

REVIEW

Note: Refer to the *Code* or the plans where necessary.

1. The identified grounded circuit conductor must be _____ or _____ in color.

2. Explain how lighting switches are rated. _____

   _____

   _____

   _____

   _____

   _____

3. A T-rated switch may be used to its _____ current capacity when controlling an incandescent lighting load.

4. What switch type and rating is required to control five 300-watt tungsten filament lamps on a 120-volt circuit? Show calculations. _____

   _____

   _____

   _____

   _____

   _____

5. List four types of lighting switches.

   a. _____   c. _____

   b. _____   d. _____

6. To control a lighting load from one control point, what type of switch would be used?

   _____

7. Single-pole switches are always connected to the _____ wire.

8. Complete the connections in the following arrangement so that both ceiling light outlets are controlled from the single-pole switch. Assume the installation is in cable.

WHITE		WHITE		WHITE	
	LAMP		LAMP		SWITCH
120-VOLT SOURCE		RED			
BLACK		BLACK		BLACK	

9. Complete the connections for the diagram. Installation is cable.

10. A three-way switch may be compared to a _____ switch.

11. What type of switch is installed to control a luminaire (fixture) from two different control points? How many switches are needed and what type are they? _____

_____

12. Complete the connections in the following arrangement so that the lamp may be controlled from either three-way switch.

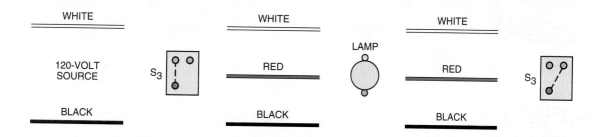

13. When connecting four-way switches, care must be taken to connect the "travelers" to the _____ terminals.

14. Show the connections for a ceiling outlet that is to be controlled from any one of three switch locations. The 120-volt feed is at the light. Use color pens or pencils. Assume the installation is in cable.

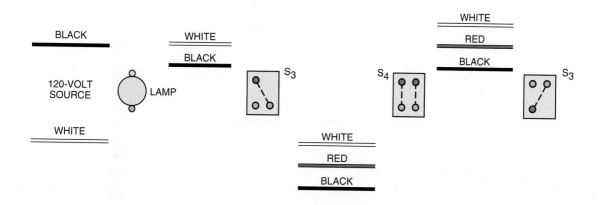

15. Match the following switch types with the correct number of terminals for each.

    Three-way switch   Two terminals

    Single-pole switch   Four terminals

    Four-way switch   Three terminals

16. When connecting single-pole, three-way, and four-way switches, they must be wired so that all switching is done in the _____ circuit conductor.

17. What section of the *Code* emphasizes the fact that all circuiting must be done so as to avoid the damaging effects of induction heating? _____

18. If you had to install an underground three-wire feeder to a remote building using three individual conductors, which of the following installations "meets Code"? Circle the correct installation.

A   B

19. Is it always necessary to attach the bare equipment grounding conductor of a nonmetallic-sheathed cable to the green hexagon-shaped grounding screw on a receptacle? Explain.

    _____

    _____

    _____

    _____

    _____

    _____

20. List the methods by which an equipment grounding conductor is connected to a device box.

    _____

    _____

    _____

    _____

    _____

    _____

21. When two nonmetallic-sheathed cables (Romex) enter a box, is it permitted to bring both of the bare equipment grounding conductors directly to the grounding terminal of a receptacle, using the terminal as a splice point? _____

22. This installation is in electrical metallic tubing. Receptacle outlet "B" is only a few feet from switch "C." Receptacle outlet "A" is only a few feet from switch "D." The electrician saved a considerable amount of wire by picking up the supply (feed) for switch "C" from receptacle "B," and the switch leg (return) from switch "D." What is wrong with this installation?

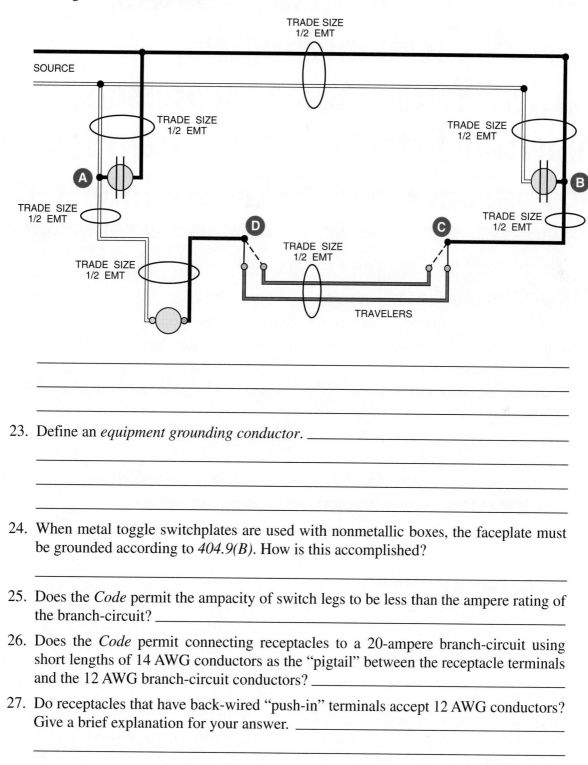

_____

_____

_____

23. Define an *equipment grounding conductor.* _____

_____

_____

_____

24. When metal toggle switchplates are used with nonmetallic boxes, the faceplate must be grounded according to *404.9(B)*. How is this accomplished?

_____

25. Does the *Code* permit the ampacity of switch legs to be less than the ampere rating of the branch-circuit? _____

26. Does the *Code* permit connecting receptacles to a 20-ampere branch-circuit using short lengths of 14 AWG conductors as the "pigtail" between the receptacle terminals and the 12 AWG branch-circuit conductors? _____

27. Do receptacles that have back-wired "push-in" terminals accept 12 AWG conductors? Give a brief explanation for your answer. _____

_____

_____

Ground-Fault Circuit Interrupters, Arc-Fault Circuit Interrupters, Transient Voltage Surge Suppressors, Immersion Detection Circuit Interrupters, and Appliance Leakage Current Interrupters

OBJECTIVES

After studying this unit, the student should be able to

▶ understand how GFCIs, AFCIs, IDCIs, and ALCIs operate.

▶ understand *NEC®* requirements of where and how GFCIs and AFCIs are to be installed and connected.

▶ understand why AFCIs and GFCIs should never be used on a shared neutral branch-circuit.

▶ understand the important *NEC®* requirements for replacing existing receptacles.

▶ know the rules for providing GFCI protection on construction sites.

▶ understand the basics of transient voltage surge suppressors.

ELECTRICAL SHOCK HAZARDS

There can be no compromise where human life is involved!

Many injuries have occurred and many lives have been lost because of electrical shock. Coming in contact with live wires or with an appliance or other equipment that is "hot" spells danger. Problems arise in equipment when there is a breakdown of insulation because of wear and tear, defective construction, or misuse of the equipment. Insulation failure can result when the "hot" ungrounded conductor in the appliance comes in contact with the metal frame of the appliance. If the equipment is not properly grounded, the potential for an electrical shock is present.

The shock hazard exists whenever the user can touch both the defective equipment and grounded surfaces such as water pipes, stainless steel sinks, metal faucets, grounded metal luminaires (fixtures), earth, concrete in contact with the earth, or water.

A severe shock can cause considerably more damage to the human body than is visible. A person may suffer internal hemorrhages, destruction of tissues, nerves, and muscles. Further injury can result from a fall, cuts, burns, or broken bones.

The effect of an electric current passing through

a human body varies, depending upon circuit characteristics (i.e., current, frequency [60 Hz is the worst], voltage, body contact resistance [open wounds are extremely hazardous as compared to a callused hand]), internal body resistance, the path of current, duration of contact, and environmental conditions (e.g., humidity).

Experts generally agree on values of 800 to 1000 ohms as typical body and contact resistance, but the resistance can vary from a few hundred ohms to many thousand ohms. Let's see how important the role of resistance is for a person coming in contact with a 120-volt circuit.

Take a dry hand, for example:

$$I = \frac{E}{R} = \frac{120 \text{ volts}}{100,000 \text{ ohms}} = \begin{array}{l} 0.0012 \text{ ampere} \\ (1.2 \text{ milliamperes}) \end{array}$$

This would probably be a little tingle.

Now take a wet hand and standing barefoot on the ground:

$$I = \frac{E}{R} = \frac{120 \text{ volts}}{1,000 \text{ ohms}} = \begin{array}{l} 0.12 \text{ ampere} \\ (120 \text{ milliamperes}) \end{array}$$

This would probably be fatal.

EFFECT OF ELECTRIC SHOCK		
	Current in milliamperes @ 60 hertz	
	Men	Women
• cannot be felt	0.4	0.3
• a little tingling—mild sensation	1.1	0.7
• shock—not painful—can still let go	1.8	1.2
• shock—painful—can still let go	9.0	6.0
• shock—painful—just about to point where you can't let go—called "threshold"—you may be thrown clear	16.0	10.5
• shock—painful—severe—can't let go—muscles immobilize—breathing stops	23.0	15.0
• ventricular fibrillation (usually fatal) length of time . . . 0.03 sec. length of time . . . 3.0 sec.	1000 100	1000 100

Table 6-1

Figure 6-1 shows a time/current curve that indicates the amount of current that a normal, healthy adult can stand for a certain time. Table 6-1 shows

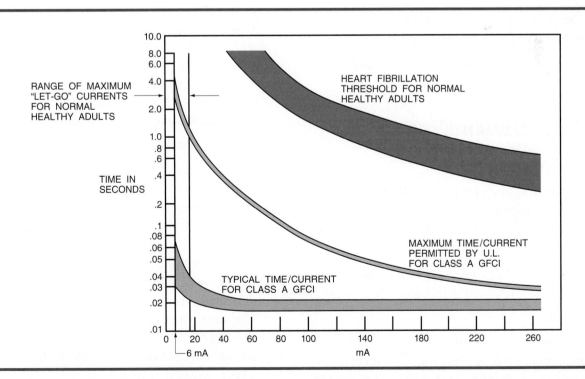

Figure 6-1   The time/current curve shows the tripping characteristics of a typical Class A GFCI. Note that if you follow the 6-mA line vertically to the crosshatched typical time/current curve, you will find that the GFCI will open in from approximately 0.035 second to just less than 0.1 second. One electrical cycle is ⅟₆₀ of a second (0.0167 second). An air bag in an automobile inflates in approximately ⅟₂₀ of a second (0.05 second).

expected sensations at various current levels in milli-amperes.

The path of current is very important. Where and how is contact to the live conductor made? What are you touching with the other hand? Where and in what are you standing? The current flow might be finger to finger, hand to hand, head to hand, head to foot, or any other combination such as a hand holding a "hot" wire, a hand holding a pair of pliers, a hand holding an electric drill, a hand grasping a grounded metal pipe, two hands grasping a grounded metal pipe, a hand in water, standing in water, hand to foot, foot to foot, and so on. Hand to foot can be fatal since the current is probably passing through the heart and lungs.

Generally, voltages of less than 50 volts to ground are considered safe. Voltages 50 volts and greater are considered lethal. All ac systems (with few exceptions) of 50 volts to 1000 volts that supply premise wiring are required to be grounded, *250.20(B)*. This was covered in Unit 1.

Some circuits of less than 50 volts must be grounded according to *250.20(A)*:

- if the circuit is supplied by a transformer where the primary exceeds 150 volts to ground

- if the circuit is supplied by a transformer where the primary supply is ungrounded

- if the circuit is overhead wiring outside of a building

CODE REQUIREMENTS FOR GROUND-FAULT CIRCUIT INTERRUPTERS *(210.8)*

To protect against electric shocks, *210.8(A)* requires that personnel ground-fault protection be provided for *all* 125-volt, 15- and 20-ampere receptacle outlets in dwellings as follows:

- in bathrooms. See Figure 10-7 for the definition of a bathroom.

- outdoors.

- in kitchens, for all receptacles that serve the countertop surfaces. This includes receptacles installed on islands and peninsulas, as in Figure 6-2.
  GFCI protection is *not* required for:
  – a receptacle installed solely for a clock.

  – a receptacle installed inside of an upper kitchen cabinet for plugging in a microwave oven that is fastened to the underside of the cabinet.
  – a receptacle installed below the kitchen sink for plugging in a food-waste disposer or dishwasher.

- ▶Any receptacle within 6 ft (1.8 m) of a laundry, utility, or wet bar sink must be GFCI protected, as in Figure 6-2.◀

- in attached or detached garages.
  GFCI protection is *not* required for:
  – receptacles that are not readily accessible, such as a receptacle installed on the ceiling for an overhead garage door opener.
  – a single receptacle, or a duplex receptacle installed for two cord- and plug-connected appliances in a dedicated space where the appliances are not easily moved from one location to another, such as a freezer, refrigerator, washer, or gas clothes dryer.

- in other buildings (sheds, workshops, or storage buildings).
  GFCI protection is *not* required for:
  – receptacles that are not readily accessible.
  – a single receptacle, or a duplex receptacle installed for two cord- and plug-connected appliances in a dedicated space where the appliances are not easily moved from one location to another, such as a freezer, refrigerator, washer, or gas clothes dryer.

- in crawl spaces that are at or below grade.

- in unfinished basements. Porcelain or plastic lampholders that have a plug on them also must be GFCI protected. See the exceptions listed below.

  Some individuals try to get around the requirement for GFCI personnel protection in basements by calling the basement "finished," even though it is clearly evident that it is not "finished." If the basement is intended to be "finished," then the *Code* requirements for the spacing of receptacles, required lighting outlets, and switch controls would have to be followed. This is covered in Unit 3.

- **WARNING: Never plug a sump pump into a GFCI-protected receptacle! The possibility of the GFCI nuisance tripping "OFF"—**

Figure 6-2 In kitchens, *all* 125-volt, single-phase, 15- and 20-ampere receptacles that serve countertop surfaces (walls, islands, or peninsulas) must be GFCI protected. The key words are "that serve countertop surfaces." GFCI personnel protection is *not* required for the refrigerator receptacle, a receptacle under the sink for plugging-in a food waste disposer, or a receptacle installed inside cabinets above the countertops for plugging in a microwave oven attached to the underside of the cabinets. ▶All 125-volt, single-phase, 15- and 20-ampere receptacles within 6 ft (1.8 m) of a wet bar, laundry, or utility room sink must be GFCI protected. See *210.8(A)(6)* and *210.8(A)(7)*.◀ Drawing is not to scale.

unknown to the homeowner—could result in a flooded basement. Since a single receptacle is exempt from the GFCI requirement, it is better to install a single receptacle for a sump pump. The same logic applies to plugging a refrigerator or freezer into a GFCI-protected receptacle.

- ▶GFCI protection is not required for a single receptacle that supplies only a permanently installed fire alarm or burglar alarm system.◀

- there is a special rule for residential underground branch-circuit wiring that can result in a labor cost savings when the underground branch-circuit wiring is GFCI protected. *Table 300.5* of the *NEC®* indicates that for residential underground 120-volt or less branch-circuits that are rated 20 amperes or less, the underground

wiring need be buried only 12 in. (300 mm) below grade when the circuit is GFCI protected. An example of this would be the Type UF cable supplying the post light in front of the residence. Type UF cable must be buried at least 24 in. (600 mm) below grade if the circuit is not GFCI protected. If the underground circuit is installed in rigid or intermediate metal conduit, a depth of not less than 6 in. (150 mm) is required. Refer to Unit 16 for more detailed information regarding residential underground wiring.

- for swimming pools, see Unit 30.

- for the branch-circuit supplying heated floors in bathrooms, and hydromassage bathtub, spa, and hot tub locations, *424.44(G)*. This GFCI requirement is regardless of the type of flooring or heating cables used.

GFCI Receptacles and Breakers

The *Code* requirements for ground-fault circuit protection can be met in many ways. Figure 6-3 illustrates a GFCI circuit breaker installed on a branch-circuit. A ground fault in the range of 4 to 6 milliamperes or more shuts off the entire circuit. If, for example, a GFCI circuit breaker were to be installed on Circuit B14, a ground fault at any point in the circuit would shut off the garage lighting, garage receptacles, rear garage door outside bracket luminaire (fixture), post light, and overhead door opener.

When a GFCI receptacle is installed, then only that receptacle is shut off when a ground fault greater than 6 milliamperes occurs, as in Figure 6-4.

GFCI receptacles break both the ungrounded (hot) and grounded conductors.

GFCI circuit breakers break only the ungrounded (hot) conductor.

Figure 6-5 shows the effect of a GFCI installed as part of a feeder supplying 15- and 20-ampere receptacle branch-circuits. This is rarely used in residential wiring.

Figure 6-6 is a pictorial view of how a ground-fault circuit interrupter operates. Figure 6-7 shows the actual internal wiring of a GFCI.

Nuisance Tripping

Occasionally, a GFCI will trip for seemingly no reason. This could be "leakage" in extremely long runs of cable from a GFCI circuit breaker in a panel to the protected branch-circuit wiring. One manufacturer's literature specifies a maximum one-way length of 250 ft for the branch-circuit.

The allowable "leakage" for listed appliances is so small that the appliances should not contribute to nuisance tripping of GFCIs. However, if nuisance tripping can be traced to an appliance, the appliance should be checked by a qualified appliance repairman.

Nuisance tripping occasionally can be traced to moisture somewhere in the circuit wiring, receptacles, or lighting outlets.

Some electricians are of the opinion that using nonmetallic staples (see Figure 4-13[C]) will reduce the possibility of nuisance tripping. No actual testing has been done to prove or disprove this opinion.

What a GFCI Does

A GFCI monitors the current balance between the ungrounded "hot" conductor and the grounded conductor. As soon as the current flowing through the "hot" conductor is in the range of 4 to 6 milliamperes more than the current flowing in the

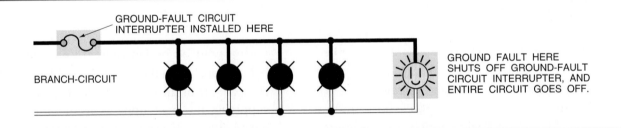

Figure 6-3   Ground-fault circuit interrupter as a part of the branch-circuit overcurrent device.

Figure 6-4   Ground-fault circuit interrupter as an integral part of a receptacle outlet.

Figure 6-5   Ground-fault circuit interrupter as part of the feeder supplying a panel that serves a number of 15- and 20-ampere branch-circuits. This is permitted by *215.9*, but is rarely—if ever—used in residential wiring.

Figure 6-6   Basic principle of how a ground-fault circuit interrupter operates.

"return" grounded conductor, the GFCI senses this unbalance and trips (opens) the circuit off. The unbalance indicates that part of the current flowing in the circuit is being diverted to some path other than the normal return path along the grounded return conductor. If the "other" path is through a human body, as illustrated in Figure 6-6, the outcome could be fatal.

UL Standard No. 943 covers ground-fault circuit interrupters.

- Class "A" GFCI devices are designed to trip when a ground fault in the range of 4 to 6 milli-

Figure 6-7 **Ground-fault circuit interrupter internal components and connections. Receptacle-type GFCIs switch both the phase (hot) and grounded conductors. Note that when the test button is pushed, the test current passes through the test button, the sensor, then back around (bypasses, outside of) the sensor, then back to the opposite circuit conductor. This is how the "unbalance" is created, then monitored by the solid-state circuitry to signal the GFCI's contacts to open. Note that because both normal "load" currents pass through the sensor, no unbalance is present.**

amperes ($^4/_{1000}$ to $^6/_{1000}$ of an ampere) or more occurs.

- Class "B" GFCI devices are pretty much obsolete. They were designed to trip on ground faults of 20 milliamperes ($^{20}/_{1000}$ of an ampere) or more. They were used only for underwater swimming pool lighting installed before the adoption of the 1965 *NEC.*® For this application, Class "A" devices were too sensitive and would nuisance trip.

What a GFCI Does *Not* Do

- It *does not* protect against electrical shock when a person touches both circuit conductors at the same time (two "hot" wires, or one "hot" wire and one grounded neutral wire) because the current flowing in both conductors is the same. Thus, there is no unbalance of current for the GFCI to sense and trip.

- It *does not* limit the magnitude of ground-fault current. It *does* limit the length of time that a ground fault will flow. In other words, you will still receive a severe shock during the time it takes the GFCI device to trip "off." See Figure 6-1.

- It *does not* sense solid short circuits between the "hot" conductor and the grounded "neutral" conductor. The branch-circuit fuse or circuit breaker provides this protection.

- It *does not* sense solid short circuits between two "hot" conductors. The branch-circuit fuse or circuit breaker provides this protection.

- It *does not* sense and protect against the damaging effects of arcing faults, such as would occur with frayed extension cords. This protection is provided by an arc-fault circuit interrupter (AFCI) discussed later in this unit.

- It *does not* provide overload protection for the branch-circuit wiring. It provides *ground-fault protection only*.

GROUND-FAULT CIRCUIT INTERRUPTERS IN RESIDENCE CIRCUITS

In this residence, GFCI personnel protection must be provided for receptacle outlets installed outdoors, in the bathrooms, in the garage, in the workshop, specific receptacles in the basement area, and in the kitchen for those receptacles that serve countertops. ▶All 125-volt, single-phase, 15- and 20-ampere receptacles within 6 ft (1.8 m) of a wet bar, laundry, or utility room sink must be GFCI protected. See *210.8(A)(6)* and *210.8(A)(7).*◀ The GFCI protection can be provided by GFCI circuit breakers or by GFCI receptacles of the type shown in Figure 6-8. This is a design issue that confronts the electrician.

GROUND-FAULT CIRCUIT INTERRUPTER AS AN INTEGRAL PART OF A DUPLEX GROUNDING-TYPE CONVENIENCE RECEPTACLE.

B

A SINGLE-POLE GROUND-FAULT CIRCUIT-INTERRUPTER CIRCUIT BREAKER. TWO-POLE GFCI CIRCUIT BREAKERS ARE ALSO AVAILABLE.

Figure 6-8 (A) A ground-fault circuit-interrupter receptacle, *courtesy of* Pass & Seymour, and (B) a ground-fault circuit-interrupter circuit breaker. The switching mechanism of a GFCI receptacle opens both the ungrounded "hot" conductor and the grounded conductor. The switching mechanism of a GFCI circuit breaker opens the ungrounded "hot" conductor only. *Courtesy of* Schneider Electric.

The electrician must decide how to provide the GFCI personnel protection required by *210.8*. These receptacle outlets are connected to the circuits listed in Table 6-2.

Swimming pools also have special requirements for GFCI protection. These requirements are covered in Unit 30.

GFCIs operate properly only on grounded electrical systems, as is the case in all residential, condo, apartment, commercial, and industrial wiring. The GFCI will operate on a two-wire circuit even though an equipment grounding conductor is not included with the circuit conductor. In this residence, equipment grounding conductors are in the cables. In the case of metal conduit, the equipment grounding conductor is the metal conduit.

Never ground a system neutral conductor except at the service equipment. A GFCI would be inoperative.

Never connect the neutral of one circuit to the neutral of another circuit.

When a GFCI feeds an isolation transformer (separate primary winding and separate secondary winding), as might be used for swimming pool underwater luminaires (fixtures), the GFCI *will not*

CIRCUITS	GFCI RECEPTACLE LOCATION
A15	Front porch receptacle
A16	Outdoor receptacle on rear of residence outside of Front Bedroom
A18	Workbench receptacles
A19	Outdoor receptacle on rear of residence outside of Master Bedroom
A20	Workshop receptacles on window wall
A22	Master Bath receptacle
A23	Bath (off hall) receptacle
B9	Wet-bar (Recreation Room) receptacles
B13	Kitchen receptacles
B14	Garage receptacles
B15	Kitchen receptacles
B16	Kitchen receptacles
B17	Outdoor receptacle outside of Living Room next to sliding door
B20	Outdoor receptacle outside of laundry
B21	Powder Room receptacle

Table 6-2 Circuit numbers and general location of GFCI-protected receptacles. GFCI protection can be provided by using GFCI receptacles or GFCI circuit breakers.

detect any ground faults on the secondary of the transformer.

Ground-fault circuit interrupters may be installed on other circuits, in other locations, and even when rewiring existing installations where the *Code* does not specifically call for GFCI protection.

Do not connect fire and smoke alarms to a circuit that is protected by a GFCI breaker. A nuisance tripping of the GFCI would render the fire alarm and smoke detectors inoperative. This requirement is found in *NFPA 72, The National Fire Alarm Code.*

FEEDTHROUGH GROUND-FAULT CIRCUIT INTERRUPTER

The decision to use more GFCIs rather than trying to protect many receptacles through one GFCI becomes one of economy and practicality. GFCI receptacles are more expensive than regular receptacles. Here is where knowledge of material and labor costs comes into play. The decision must be made separately for each installation, keeping in mind that GFCI protection is a safety issue recognized and clearly stated in the *Code.* However, the actual circuit layout is left up to the electrician.

Figure 6-9 illustrates how a feedthrough GFCI receptacle supplies many other receptacles. Should a ground fault occur anywhere on this circuit, all eleven receptacles lose power—not a good circuit layout. Attempting to locate the ground-fault problem, unless it's obvious, can be very time-consuming. The use of more GFCI receptacles is generally the more practical approach.

Figure 6-10 shows how a feedthrough receptacle is connected into a circuit. Figure 6-11 shows a feedthrough GFCI receptacle connected midway into a circuit. In both diagrams, the feedthrough GFCI receptacle and all downstream outlets are ground-fault protected.

In UL Standard 943, we find that a GFCI receptacle may have black-and-white line wire leads, red-and-gray load wire leads, and a green lead for the equipment grounding conductor as illustrated in Figure 6-10. GFCI receptacles may have ordinary screw terminals: brass color for the "hot" conductors, silver color for the grounded conductors, and a green hexagon-shaped screw for the equipment grounding conductor. The line and load connections are clearly marked.

Where the wiring method is a grounded method such as electrical metallic tubing, flexible metal conduit, or armored cable, the grounding terminal of the GFCI receptacle is considered adequately grounded by the metallic grounded wiring method,

FEEDTHROUGH GFCI

Figure 6-9 Illustration showing one GFCI feedthrough receptacle protecting ten other receptacles. Although this might be an economical way to protect many receptacle outlets with only one GFCI feedthrough receptacle, there is great likelihood that the GFCI will nuisance trip because of leakage currents. Use common sense and the manufacturer's recommendations as to how many devices should be protected through one GFCI.

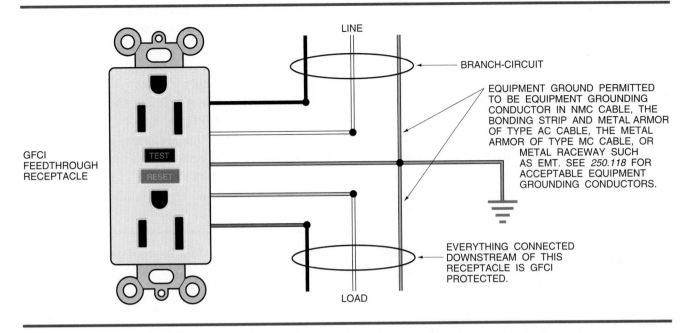

LINE

BRANCH-CIRCUIT

EQUIPMENT GROUND PERMITTED
TO BE EQUIPMENT GROUNDING
CONDUCTOR IN NMC CABLE, THE
BONDING STRIP AND METAL ARMOR
OF TYPE AC CABLE, THE METAL
ARMOR OF TYPE MC CABLE, OR
METAL RACEWAY SUCH
AS EMT. SEE *250.118* FOR
ACCEPTABLE EQUIPMENT
GROUNDING CONDUCTORS.

GFCI
FEEDTHROUGH
RECEPTACLE

TEST

RESET

EVERYTHING CONNECTED
DOWNSTREAM OF THIS
RECEPTACLE IS GFCI
PROTECTED.

LOAD

Figure 6-10 Connecting feedthrough ground-fault circuit interrupter into circuit.

GFCI FEEDTHROUGH RECEPTACLE

BRANCH-
CIRCUIT

LINE

TEST

RESET

LOAD

THE GFCI RECEPTACLE AND ALL
OUTLETS CONNECTED TO THE LOAD
SIDE OF THE GFCI RECEPTACLE ARE
GROUND-FAULT PROTECTED.

Figure 6-11 Feedthrough GFCI receptacle installed midway in a branch-circuit.

and a separate connection to the green lead or to the green hexagon-shaped screw is not necessary. This was discussed and illustrated in Unit 5.

A "safety yellow" adhesive label must cover the load terminals, or be wrapped around the load wire leads. The label must be marked with wording like this: *"Attention! The load terminals under this label are for feeding additional receptacles. Miswiring can leave this outlet without ground-fault protection. Read instructions prior to wiring."*

The most important factors to be considered are the continuity of electrical power and the economy of the installation. The decision must be made separately for each installation.

WARNING: Do not reverse the line and load connections on a feedthrough GFCI receptacle. This would result in the feedthrough GFCI receptacle itself still being "live" even though the GFCI mechanism has tripped "off" and has shut off the circuit downstream from the feedthrough GFCI receptacle. It is easy to tell if a GFCI receptacle has the line and load leads reversed. Push the test button. The GFCI will trip. If the GFCI receptacle is still "hot," it has been wired incorrectly.

▶The latest UL Standard requires that a GFCI receptacle must trip to the "OFF" position if the receptacle is miswired, such as a reversal of the line-load connections.◀

WARNING: When hooking up GFCIs or AFCIs, *never share a neutral* such as when using a multiwire branch-circuit. A GFCI OR AFCI device will not work! Be careful when making connections! Follow the manufacturer's instructions. GFCI and AFCI circuit breakers and receptacles have both the ungrounded conductor and the grounded conductor going in and coming out of them. This allows the sensor inside of the GFCI or AFCI to detect any unbalance current. Think of it this way: "what goes in must come out—we hope." In a multiwire circuit, the neutral carries the unbalanced current of the circuit. The return current on the neutral may not be the same as the current flowing in the ungrounded conductor. Figure 6-12(A) is a three-wire, multiwire circuit. Trace the current flow and note that the GFCI or AFCI will sense a current unbalance of 5 amperes and trip "OFF." In fact, as long as the unbalance exists, it will never be able to be turned "ON."

Figure 6-12(B) and Figure 6-12(C) show a two-

Figure 6-12 These diagrams show the wrong and right way to connect loads to GFCI or AFCI devices. Never use a GFCI or AFCI circuit breaker on a shared neutral branch-circuit.

wire circuit. Figure 6-12(B) is an improper connection of a feedthrough GFCI receptacle. Note how the GFCI detects a current unbalance of 1 ampere and will trip "OFF." As long as the unbalance exists, the GFCI receptacle will continue to trip "OFF" until the misconnection is corrected. Figure 6-12(C) illustrates the proper connection of a GFCI receptacle.

Other Features for GFCI Receptacles

Look for these features when purchasing GFCI receptacles.
GFCI receptacles

- have a lockout feature so the GFCI cannot be reset if the internal GFCI circuit is not functioning properly.

- have a built-in line/load reversal feature that prevents the GFCI from being reset if the line and load connections are reversed.

- have an indicator light that glows when the test button is pushed if the line/load connections are properly connected. This same indicator light glows when the GFCI has tripped.

TESTING AND RECORDING OF TEST DATA FOR GFCI RECEPTACLES

Refer to Figure 6-8 and note that the GFCI receptacle has "T" (Test) and "R" (Reset) buttons. Pushing the test button places a small ground fault on the circuit. If operating properly, the GFCI receptacle should trip to the "OFF" position. Pushing the reset button will restore power. The operation is similar for the GFCI circuit breakers.

Because of the fragile nature of the electronic circuitry in GFCIs, test them often, particularly in high lightning strike areas of the country. GFCIs and other delicate electronic circuitry are susceptible to transient voltage surges.

Underwriters Laboratories requires that detailed installation and testing instructions be included in the packaging for GFCI receptacle or circuit breaker. Instructions are not only for the electrician but also for homeowners, and must be left in a conspicuous place so homeowners can familiarize themselves with the receptacle, its operation, and its need for testing. Figure 6-13 shows a chart homeowners can use to record monthly GFCI testing.

OCCUPANT'S TEST RECORD

TO TEST, depress the "TEST" button, the "RESET" button should extend. Should the "RESET" button not extend, the GFCI will not protect against electrical shock. Call qualified electrician.

TO RESET, depress the "RESET" button firmly into the GFCI unit until an audible click is heard. If reset properly, the "RESET" button will be flush with the surface of the "TEST" button.

This label should be retained and placed in a conspicuous location to remind the occupants that for maximum protection against electrical shock, each GFCI should be tested monthly.

Year	Jan	Feb	Mar	Apr	May	Jun	Jul	Aug	Sep	Oct	Nov	Dec

Figure 6-13   Homeowner's testing chart for recording GFCI testing dates.

ARC-FAULT CIRCUIT INTERRUPTERS (AFCIs)

Arc-fault circuit interrupters (AFCIs) are "the new kid on the street."

To meet the requirements of *210.12*, Circuits A16, A19, and A21 are protected by AFCI circuit breakers installed in Panel A.

▶*NEC® 210.12* requires that listed AFCI protection be provided for **all** 120-volt, single-phase, 15- and 20-ampere branch-circuits that supply outlets in dwelling unit bedrooms.◀ Their purpose is to protect the branch-circuit wiring and all receptacle outlets, lighting/fan outlets, and 120-volt smoke alarm outlets in bedrooms. As you will learn in Chapter 26, smoke alarms are required to have battery backup, so the smoke alarm will continue to provide protection even if the AFCI trips.

Closet lighting is not specifically mentioned in *210.12*. However, in many instances the closet lighting is connected to the same branch-circuit as the bedroom lighting and receptacles and would be protected by an AFCI circuit breaker supplying that circuit. You might even have some hallway lighting or hallway receptacles on that circuit. This does not present a problem. It's a matter of thinking through the wiring layout for the most cost-effective arrangement.

After you study AFCIs, you may have to reconsider how you lay out a branch-circuit where AFCI protection is mandatory.

Why AFCI Protection?

Electrical arcing is an "early" event in the history of a typical electrical fire. Arcing is considered one of the leading causes of electrical fires in homes. Think about frayed cords, cords under carpets, cords pinched under legs of furniture, and nails being driven through plaster into cables in the wall. The Consumer Product Safety Commission recently cited 150,000 residential electrical fires annually in the United States, 850 deaths, 6000 injuries, and more than $1.5 billion in property loss. These statistics are pretty convincing of the need for greater electrical protection. The heat of an arc is extremely high, known to reach 10,000°F or more. Hot particles of metal expelled by the arc are enough to ignite most surrounding combustible materials.

We have all been guilty of blaming electrical fires on overloaded circuits, overloaded extension cords, and too many extension cords plugged into a receptacle. The heat of an overloaded cord by itself is generally not hot enough to ignite surrounding combustible materials. An overload condition can cause the breakdown of the insulation on the conductors, allowing the conductors to touch one another. This sets the stage for an arcing condition between the conductors, or between the "hot" conductor and ground. We now have the ingredients of an electrical fire.

For example, let's say that the cord on an electric iron becomes frayed and the conductor(s) break. The iron draws 10 amperes. The arcing of 10 amperes trying to bridge the gap in the broken conductor(s) develops the tremendous heat capable of igniting surrounding combustible material, yet 10 amperes flowing in the circuit will not trip a 15-ampere circuit breaker or open a 15-ampere fuse. This condition is not an overload and initially is not a short circuit. Conventional overcurrent devices "sees" this as normal. They are not designed to detect and open under current levels below their trip setting. What is needed is a device that can detect the unique characteristics of an arcing condition—an AFCI.

Types of Arcs

Arcs come in a few varieties—line-to-line, line-to-neutral, or line-to-ground. These are further classified as:

Series Arcs: When a conductor that is carrying current breaks or when a loose connection occurs, it is a "series" arcing fault because the current is trying to travel in its intended path.

Parallel Arcs: When arcing occurs between the black ("hot") and white grounded (neutral) circuit conductors, or between the black ("hot") conductor and ground, it is referred to as a "parallel" arc. These conditions might be caused by a nail being driven through a cable, a staple driven too tightly, or a clamp or connector squeezing the cable too tightly. The current is traveling outside of its intended path.

A broken or frayed cord could result in a "series" or "parallel" arc, or both.

Why Bedrooms?

Consumer Product Safety Commission studies indicate that a high percentage of residential electrical fires occur in bedrooms. The Code-Making Panels saw the benefit of arc-fault protection and introduced the AFCI requirement for residential bedrooms. As time passes, we can expect AFCI protection requirements to extend to other rooms of a home, in the same manner that ground-fault circuit interrupter (GFCI) protection spread to more locations over a number of *Code* editions.

Types of AFCI Devices

UL Standard 1699 specifies the sensing and tripping characteristics of an AFCI. AFCIs are available in the following types:

1. *Branch/feeder AFCI:* This is an AFCI circuit breaker installed in a panel. It provides arc-fault protection for the entire branch-circuit. It also provides somewhat limited protection for extension cords and other things plugged into a receptacle. This is the most common type of AFCI used today.

2. *Outlet AFCI:* This is an AFCI receptacle. It provides protection for cord sets and other things that are plugged into the receptacle. They are more sensitive than branch/feeder AFCIs, and offer greater arc-fault protection for cord-and-plug-connected equipment.

   Some AFCI receptacles have the technology to "look" upstream as well as downstream to sense arcing anywhere on the branch-circuit wiring.

   ▶A feedthrough AFCI receptacle installed at the first receptacle is permitted by *210.12(B), Exception No. 1.* The maximum length measured along the conductors from the branch-circuit breaker to the AFCI receptacle shall not exceed 6 ft (1.8 m). The conductors shall be in a metal raceway or metal sheathed cable, *210.12(B) Exception No. 2.*◀

   A "feedthrough" AFCI receptacle provides arc-fault protection for that receptacle and all downstream outlets connected to that receptacle.

3. *Combination AFCI:* This combines the features of a branch/feeder AFCI and an outlet AFCI.

4. *Portable AFCI:* A plug-in device with one or more receptacles. It provides AFCL protection for cords that are plugged into the device. It is similar to the portable GFCI devices.

5. *Cord AFCI:* A plug-in device that provides AFCI protection to the power cord connected to it. It has no additional outlets.

UL Standards for AFCIs and GFCIs

Type	Conforms to requirements of	If circuit breaker type, also conforms to requirements of
AFCI	UL 1699	UL489
GFCI	UL 943	UL 489
AFCI/GFCI combination	UL 1699 and 943	UL 489

How Does an AFCI Work?

An *arc-fault circuit interrupter (AFCI)* minimizes the hazards of arcing faults.

AFCI devices are designed to sense the rapid fluctuations of current flow typical of an arcing (sputtering) condition. It will recognize the unique characteristic that is sometimes referred to as the "signature" of a dangerous arc. It will de-energize the circuit when dangerous arcing is detected.

An AFCI can distinguish between dangerous arcing and harmless arcing. It will not nuisance trip when a switch is turned on or off, when something is plugged into or unplugged from a receptacle, when a light bulb burns out, when brushes arc in universal motors, or when similar harmless arcing occurs.

What Types of AFCI Devices Do Electricians Install?

The most common AFCIs that electricians install today are the circuit-breaker type. Circuit-breaker AFCIs are the same width as a standard circuit breaker so they will fit into a standard electrical panel. See Figure 6-14.

How to Connect AFCIs

When installing a 125-volt, single-pole AFCI circuit breaker, you first "snap" the breaker into the panel. Then the white lead from the AFCI circuit breaker is connected to the neutral bar in the panel.

Figure 6-14 An arc-fault circuit interrupter (AFCI) circuit breaker. *Courtesy of* Square D/Schneider Electric.

The black branch-circuit conductor is connected to the brass-colored load side terminal on the breaker. The white branch-circuit conductor is connected to the silver-colored load side terminal on the breaker. This is the same as for GFCI circuit breakers discussed earlier in this unit.

Never share the neutral of a multiwire branch-circuit when using GFCIs and AFCIs. See Figure 6-12. Two-pole AFCIs and GFCIs are available. Always follow the instructions furnished with these devices.

AFCIs in Existing Homes

The mandatory requirement in *210.12* for AFCI protection is for new work only. It is not retroactive. But it's not a bad idea to consider installing AFCIs on existing installations. It is easy to do. Just replace existing circuit breaker(s) that feed bedrooms with an AFCI circuit breaker(s). In older homes, the insulation on the original wiring is rated 140°F (60°C). Over the years, the insulation becomes brittle. For example, in and above a luminaire (fixture), the heat from the luminaire (fixture) literally bakes and chars the insulation on the conductors. AFCI protection could "catch" an arcing problem in its early stages. Today, conductors used in house wiring are rated 194°F (90°C), so the deterioration of conductor insulation because of heat has become a thing of the past.

Testing AFCIs and GFCIs

AFCI/GFCI combination testers are used to test AFCIs and GFCIs to make sure they are working properly.

What Does the Future Hold?

Over time, as more field experience is compiled, we will learn more about the effectiveness of AFCIs in preventing fires. As usage increases, prices will drop. Down the road, you can expect to see many changes in the requirements for AFCI devices. You will see AFCI/GFCI combination devices.

REPLACING EXISTING RECEPTACLES

The *NEC®* is very specific on the type of receptacle permitted to be used as a replacement for an existing receptacle. These *Code* rules are found in *406.3(D)*.

In *406.3(D)(2)* we find a retroactive ruling requiring that when an existing receptacle is replaced in *any* location where the present code requires GFCI protection (bathrooms, kitchens, outdoors, unfinished basements, garages, etc.), the replacement receptacle must be GFCI protected. This could be a GFCI-type receptacle, or the branch-circuit could be protected by a GFCI-type circuit breaker.

It is important to note that on existing wiring systems, such as "knob and tube," or older non-metallic-sheathed cable wiring that did not have an equipment grounding conductor, that a GFCI receptacle will still properly function. It does not need an equipment grounding conductor. Thus, protection against lethal shock is still provided. See Figure 6-7.

Replacing Existing Two-Wire Receptacles Where Grounding Means Does Exist

When replacing an existing receptacle (Figure 6-15[A]) where the wall box is properly grounded (Figure 6-15[E]) or where the branch-circuit wiring contains an equipment grounding conductor (Figure 6-15[D]), at least two easy choices are possible for the replacement receptacle:

1. The replacement receptacle must be of the grounding type (Figure 6-15[B]) unless . . .

2. The replacement receptacle is of the GFCI type (Figure 6-15[C]).

REPLACING A RECEPTACLE WHERE GROUNDING MEANS EXISTS.

A — OLD TWO-WIRE RECEPTACLE

B — GROUNDING-TYPE RECEPTACLE — GRND

C — GFCI RECEPTACLE — GRND

D — EQUIPMENT GROUNDING CONDUCTOR / GROUNDED CONDUCTOR / HOT CONDUCTOR

E — GROUNDED BOX

Figure 6-15   If a wall box (E) is properly grounded by any of the methods specified in *250.118*, or if the branch-circuit wiring contains an equipment grounding conductor (D), an existing receptacle of the type shown in (A) *must* be replaced with a grounding-type receptacle (B) or with a GFCI receptacle (C). See *406.3(D)(1)*.

If the enclosure is not grounded, *250.130(C)* permits an equipment grounding conductor to be run from the enclosure to be grounded, or from the green hexagon grounding screw of a grounding-type receptacle, to any accessible point on the grounding electrode system or to any accessible point on the grounding electrode conductor. Note however, that this permission is only for existing installations. See Figure 6-18.

Be sure that the equipment grounding conductor of the circuit is connected to the receptacle's green hexagon-shaped grounding terminal.

Do not connect the white grounded circuit conductor to the green hexagon-shaped grounding terminal of the receptacle.

Do not connect the white grounded circuit conductor to a metal box.

Replacing Existing Two-Wire Receptacles Where Grounding Means Does Not Exist

When replacing an existing two-wire non-grounding-type receptacle (Figure 6-16[A]), where the box is not grounded (Figure 6-16[E]), or where

an equipment grounding conductor has not been run with the circuit conductors (Figure 6-16[D]), four choices are possible for selecting the replacement receptacle:

1. The replacement receptacle may be a non-grounding type (Figure 6-16[A]).

2. The replacement receptacle may be a GFCI type (Figure 6-16[C]).

   • The GFCI replacement receptacle must be marked "*No Equipment Ground*," as in Figure 6-17.

   • The green hexagonal grounding terminal of the GFCI replacement receptacle does not have to be connected to any grounding means. It can be left "unconnected." The GFCI's trip mechanism will operate properly when ground faults occur anywhere on the load side of the GFCI replacement receptacle. Ground-fault protection is still there. Refer to Figure 6-7.

   • Do not connect an equipment grounding conductor from the green hexagonal grounding terminal of a replacement GFCI

REPLACING A RECEPTACLE WHERE GROUNDING MEANS DOES NOT EXIST.

A OLD TWO-WIRE RECEPTACLE

B GROUNDING-TYPE RECEPTACLE

C GFCI RECEPTACLE

D GROUNDED CONDUCTOR / NO EQUIPMENT GROUNDING CONDUCTOR RUN WITH CIRCUIT CONDUCTORS / HOT CONDUCTOR

E NONGROUNDED BOX

Figure 6-16 If a box (E) is *not* grounded, or if an equipment grounding conductor has *not* been run with the circuit conductors (D), an existing nongrounding-type receptacle (A) may be replaced with a nongrounding-type receptacle (A), a GFCI-type receptacle (C), a grounding-type receptacle (B) if supplied through a GFCI receptacle (C), or a grounding-type receptacle (B) if a separate equipment grounding conductor is run from the receptacle to any accessible point on the grounding electrode system, *250.130(C)*. See Figure 6-18. Most electricians would replace an existing nongrounding-type receptacle with a GFCI type rather than install an equipment grounding conductor, which could be very labor-intensive. See *406.3(D)(3)*.

REPLACEMENT GFCI RECEPTACLE WHERE NO GROUNDING MEANS EXISTS. MUST BE MARKED "NO EQUIPMENT GROUND."

IMPORTANT:
WHERE AN EXISTING NONGROUNDING-TYPE RECEPTACLE IS REPLACED WITH A GROUNDING-TYPE RECEPTACLE, AND IS SUPPLIED THROUGH ANOTHER GFCI RECEPTACLE, IT MUST BE MARKED "GFCI PROTECTED" AND "NO EQUIPMENT GROUND." WITHOUT THIS REQUIRED MARKING, IT WOULD BE DIFFICULT TO KNOW IF THAT RECEPTACLE IS GFCI PROTECTED, OR IF THE GROUNDING TERMINAL ON THAT RECEPTACLE IS TRULY GROUNDED. MANUFACTURERS OF RECEPTACLES PROVIDE PRESSURE-SENSITIVE LABELS FOR THIS PURPOSE.

DOWNSTREAM REPLACEMENT GROUNDING-TYPE RECEPTACLE FED THROUGH A REPLACEMENT GFCI RECEPTACLE. MUST BE MARKED "GFCI PROTECTED" AND "NO EQUIPMENT GROUND."

Figure 6-17 Where a grounding means *does not* exist, there are certain marking requirements as presented in this diagram. *Do not* connect an equipment grounding conductor between these receptacles. See *406.3(D)(3)*.

receptacle (Figure 6-17) to any other downstream receptacles that are fed through the replacement GFCI receptacle.

The reason this is not permitted is that if at a later date someone saw the conductor connected to the green hexagonal grounding terminal of the downstream receptacle, there would be an immediate assumption that the other end of that conductor had been properly connected to an acceptable grounding point of the electrical system. The fact is that the so-called equipment grounding conductor had been connected to the replacement GFCI receptacle's green hexagonal grounding terminal that was not grounded in the first place. We now have a false sense of security, a real shock hazard.

3. The replacement receptacle may be a grounding type (Figure 6-16[B]) if it is supplied through a GFCI-type receptacle (Figure 6-16[C]).

  • The replacement receptacle must be marked "*GFCI Protected*" and "*No Equipment Ground,*" as in Figure 6-17.

  • The green hexagonal equipment grounding terminal of the replacement grounding-type receptacle (Figure 6-16[B]) need not be connected to any grounding means. It

may be left "unconnected." The upstream feedthrough GFCI receptacle (Figure 6-16[C]) trip mechanism will work properly when ground faults occur anywhere on its load-side. Ground-fault protection is still there. Refer to Figure 6-7.

  • Do not connect an equipment grounding conductor from the green hexagonal grounding terminal of a replacement GFCI receptacle (Figure 6-17) to any other downstream receptacles that are fed through the replacement GFCI receptacle.

4. The replacement receptacle may be a grounding type (Figure 6-16[B]) if:

  • an equipment grounding conductor, sized per *Table 250.122*, is run from the replacement receptacle's green hexagonal grounding screw and properly connected to a listed ground clamp at some point on the first 5 ft (1.5 m) of the metal water piping as it enters the building. You are *not* permitted to make the equipment ground connection to the interior metal water piping anywhere beyond the first 5 ft (1.5 m). See *250.130(C)* and the *FPN*. Also read *250.50* for the explanation of a "grounding electrode system." This permission is only for replacing a receptacle in existing installations. See Figure 6-18.

REPLACEMENT RECEPTACLE

GROUNDING SCREW

14 AWG OR 12 AWG GREEN OR BARE EQUIPMENT GROUNDING CONDUCTOR.

GROUNDING ELECTRODE SYSTEM. SEE *250.50*

BECAUSE OF THE POSSIBLE HIGH COST OF INSTALLING AN EQUIPMENT GROUNDING CONDUCTOR PER THE *NEC* REQUIREMENTS, MOST ELECTRICIANS WOULD OPT TO MAKE THE REPLACEMENT WITH A GFCI-TYPE RECEPTACLE.

LISTED GROUNDING CLAMP

MAKE CONNECTION ON FIRST 5 FT (1.5 m) OF INTERIOR METAL WATER PIPE, OR TO ANY ACCESSIBLE POINT ON THE GROUNDING ELECTRODE CONDUCTOR.

Figure 6-18 In old installations where no equipment ground exists, one option is to run an equipment grounding conductor to an accessible point on the grounding electrode system per *250.130(C)*. This can be very expensive laborwise. The better choice is to make the replacement with a GFCI receptacle. Label it *"No Equipment Ground"* using the pressure-sensitive label furnished with the GFCI receptacle. See Figure 6-16 and Figure 6-17.

PERSONNEL GROUND-FAULT PROTECTION FOR ALL TEMPORARY WIRING

Because of the nature of construction sites, there is a continual presence of shock hazard that can lead to serious personal injury or death through electrocution. Workers are standing in water, standing on damp or wet ground, or in contact with steel framing members. Electric cords and cables are lying on the ground, subject to severe mechanical abuse. All of these conditions spell danger.

All 125-volt, single-phase, 15-, 20- and 30-ampere receptacle outlets that are not part of the permanent wiring of the building and that will be used by the workers on the construction site must be GFCI protected, *590.6(A)*. Receptacle outlets that are part of the actual permanent wiring of a building and are used by personnel for temporary power are also required to be GFCI protected.

For receptacles *other* than 125-volt, single-phase, 15-, 20-, and 30-ampere receptacles, there are two options for providing personnel protection. In *590.6(B)(1)*, GFCI protection is one of the choices. In *590.6(B)(2)*, a written *Assured Equipment Grounding Conductor Program* is the second choice. This program shall be continuously enforced on the construction site. A designated person must keep a written log, ensuring that all electrical equipment is properly installed and maintained according to the applicable requirements of *250.114, 250.138, 406.3(C)*, and *590.4(D)*. Because this option is difficult to enforce, many Authorities Having Jurisdiction (AHJs) require GFCI protection as stated in *590.6(B)(1)*.

Figure 6-19 is a listed manufactured power outlet for providing GFCI protected temporary power on construction sites. It meets the requirements of *590.6(A)*.

Most electricians and other construction site workers carry their own listed portable ground-fault circuit-interrupter cord sets to be sure they are protected against shock hazard. These are shown in Figure 6-20 and also meet the requirements of *590.6(A)*.

Portable GFCI devices are easy to carry around. They have "open neutral" protection should the neutral in the circuit supplying the GFCI open for whatever reason.

Figure 6-19 A temporary power box used on construction sites. This particular type is connected with a 50-ampere, 120/240-volt power cord. It has six 20-ampere, GFCI-protected, 125-volt receptacles plus additional 30-ampere and 50-ampere Twist-Lock® receptacles. *Courtesy of* **Hubbell, Incorporated, Wiring Device Division.**

Portable GFCI devices are available with manual reset, which is advantageous should a power outage occur or if the GFCI is unplugged, so that equipment (drills, saws, etc.) will not start up again when the power is restored. This could cause injury to anyone using the equipment when the power is restored.

Portable devices are also available that reset automatically. These should be used on lighting, engine heaters, water (sump) pumps and similar equipment where it would be advantageous to have the equipment start up as soon as power is restored.

Temporary Power for Construction Sites

In many parts of the country, it is common for the electrical contractor to furnish and install a panel for all trades to use at a construction site of a new home. Figure 6-21 shows a typical pedestal that might be secured to a pole or set in the ground. It contains a meter base, a main breaker, and GFCI branch breakers for 120-volt receptacles.

Figure 6-20   Plug- and cord-connected "portable" GFCI devices are easy to carry around. They can be used anywhere when working with electrical tools and extension cords. These devices operate independently and are not dependent upon whether or not the branch-circuit is GFCI protected. These portable GFCI in-line cords are relatively inexpensive and should always be used when using portable electric tools. *Courtesy of* Hubbell Incorporated, Wiring Device Division.

Figure 6-21   An example of a temporary construction site combination meter, breakers, and receptacles. The breakers and receptacles are under the lift up cover.

IMMERSION DETECTION CIRCUIT INTERRUPTERS (IDCIs) AND APPLIANCE LEAKAGE CURRENT INTERRUPTERS (ALCIs)

Major culprits for electrical shock in the home are personal grooming appliances. The worst offender was found to be hand-held hair-drying appliances, most commonly used near a wash basin, or even worse, next to a water-filled bathtub with a person in it. Accidentally dropping or pulling the appliance into the tub can be deadly.

Because older homes probably do not have GFCI receptacles in bathrooms, *immersion detection circuit interrupters (IDCIs)* and *appliance leakage current interrupters (ALCIs)* were introduced to provide people with shock protection.

IDCI- and ALCI-protected appliances are easily recognized by the large attachment plug cap on the cord. IDCI and ALCI devices are an integral part of the attachment plug cap on hand-held hair-drying appliances such as electric hair dryers, stylers, and heated air combs. These devices might be manually resettable or nonresettable.

The operation of an IDCI or ALCI device depends upon a "third" wire probe in the cord, which is connected to a sensor inside the appliance.

This probe detects leakage current in the range of 4 to 6 milliamperes when conductive liquid (e.g., water) enters the appliance and makes contact with any live part inside the appliance and the internal "sensor." IDCI and ALCI devices cut off power to the appliance regardless of whether the appliance switch is in the "ON" or "OFF" position.

UL requires that the plug cap be marked: *"Warning: To Reduce the Risk of Electric Shock, Do Not Remove, Modify, or Immerse This Plug."*

Never replace the special attachment plug cap on these appliances with a standard plug cap. That would make the appliance unsafe!

IDCIs and ALCIs are a requirement of *NEC®* *422.41* and UL Standard 859.

TRANSIENT VOLTAGE SURGE SUPPRESSOR (TVSS)

Transient voltage surge suppressor (TVSS) *Code* rules that are permanently installed are found in *Article 285* in the *NEC.®* Always install TVSS devices that are listed under the requirements of UL Standard 1449. Carefully read and follow the instructions furnished with the TVSS.

In today's homes, we find many electronic appliances (television, stereo, personal computer, fax equipment, word processor, VCR, VDR, CD player, digital stereo equipment, microwave oven), all of which contain many sensitive, delicate electronic components (printed circuit boards, chips, microprocessors, transistors, etc.).

Voltage transients, called *surges* or *spikes*, can stress, degrade, and/or destroy these components. They can cause loss of memory in the equipment or "lock up" the microprocessor.

The increased complexity of electronic integrated circuits makes this equipment an easy target for "dirty" power that can and will affect the performance of the equipment.

Voltage transients cause abnormal current to flow through the sensitive electronic components. This energy is measured in joules. A *joule* is the unit of energy when one ampere passes through a one-ohm resistance for one second (like wattage, only on a much smaller scale).

Line surges can be line-to-neutral, line-to-ground, and line-to-line.

Transients

Transients are generally grouped into two categories:

- **Ring wave:** These transients originate within the building and can be caused by copiers, computers, printers, HVAC cycling, spark igniters on furnaces, water heaters, dryers, gas ranges, ovens, motors, and other inductive loads.

- **Impulse:** These transients originate outside of the building and are caused by utility company switching, lightning, and so on.

To minimize the damaging results of these transient line surges, service equipment, panelboards, load centers, feeders, branch-circuits, and individual receptacle outlets can be protected with a TVSS, as shown in Figure 6-22.

A TVSS contains one or more metal-oxide varistors (MOV) that clamp the transients by absorbing the major portion of the energy (joules) created by the surge, allowing only a small, safe amount of energy to enter the actual connected load.

The MOV clamps the transient in times of less than 1 nanosecond, which is one billionth of a second, and keeps the voltage spike passed through to the connected load to a maximum range of 400 to 500 peak volts.

Typical TVSS devices for homes are available as an integral part of receptacle outlets that mount in the same wall boxes as regular receptacles. They look the same as a normal receptacle and may have an audible alarm that sounds when an MOV has failed. The TVSS device may also have a visual indication, such as a light-emitting diode (LED) that glows continuously until an MOV fails. Then the LED starts to flash on and off, as in Figure 6-23.

TVSS devices are available in plug-in strips (Figure 6-24) and as part of desktop computer hardware power direction stands commonly placed under a monitor for plugging in a monitor, printer, scanner, and other peripheral equipment. These minimize the accumulation of the multitude of cords under and behind the equipment.

A TVSS on a branch-circuit will provide surge suppression for all of the receptacles on the same circuit. These surges are classified as Category A low-level surges.

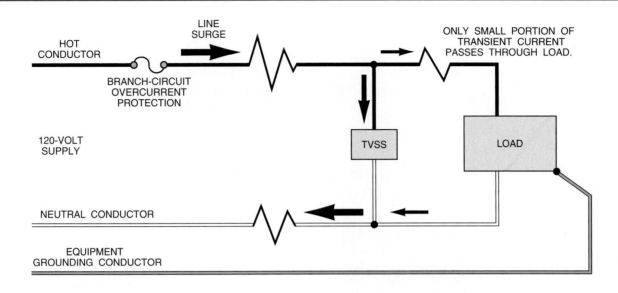

Figure 6-22   A transient voltage surge suppressor absorbs and bypasses (shunts) transient currents around the load. The MOV dissipates the surge in the form of heat.

Figure 6-23   Surge suppressor protector. *Courtesy of* Hubbell, Incorporated, Wiring Device Division.

Figure 6-24   Surge suppressors. *Courtesy of* Hubbell, Incorporated, Wiring Device Division.

"Whole-house" surge protectors are available that offer surge protection for the entire house. See Figure 6-25. These are generally available at electrical supply houses and are installed at the main electrical panel of a home.

When a whole-house surge protector is installed, it is still a good idea to install surge protectors at the plug-in locations to more closely protect against low level surges at the computer or other delicate electronic equipment location.

Noise

Electromagnetic "noise," not to be confused with audible noise, is external interference that can be generated in varying degrees by all types of electrical and electronic equipment. You might recognize "noise" as snow on a TV screen, or as humming/buzzing/static on a radio or telephone. Some typical

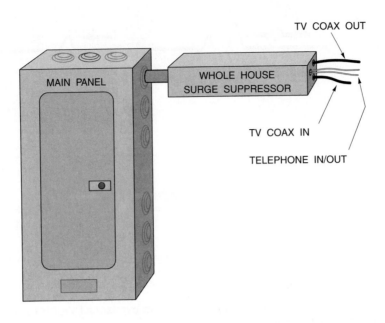

Figure 6-25 A whole-house transient voltage surge suppressor installed next to the main service panel. *Article 285* in the *NEC®* requires that these devices be listed, and have their short-circuit rating marked on the device. Be sure to follow the manufacturer's installation and connection instructions.

sources of noise are fluorescent lamps and ballasts, motors running or being switched "ON" and "OFF," loose electrical connections, or improper grounding. Even equipment running in the area such as X-ray equipment can be the culprit.

Although low-level noise is not physically damaging to electronic equipment, computers can lose memory, perform wrong calculations, have program malfunction, or lose Internet connection.

Noise comes from electromagnetic interference (EMI) and radio frequency interference (RFI).

EMI is usually caused by ground currents of very low values from such things as motors, utility switching loads, and lightning, and is transmitted through metal conduits.

RFI is noise, like the buzzing heard on a car radio when driving under a high-voltage transmission line. This interference "radiates" through the air from the source. Noise can be radiated or conducted from one residence to another. In commercial buildings, this radiated noise can be picked up by the grounding system of the building.

Consumer and commercial products today must meet Federal Communication Commission requirements for emission of electromagnetic and electronic noise so as not to interfere with communications equipment. Note your cell phone has very limited amount of transmitting power.

In commercial work, *isolated ground receptacles* are installed in computer rooms to minimize "noise."

In residential work, isolated ground receptacles are rarely if ever used.

Various types of electronic "noise' eliminators are available from electronic stores that filter out and virtually eliminate undesirable interference.

Here are some related UL Standards:

- UL 498 Attachment Plugs and Receptacles
- UL 943 Ground-Fault Circuit Interrupters
- UL 1283 Electromagnetic Interference Filters
- UL 1363 Relocatable Power Taps
- UL 1449 Transient Voltage Surge Suppressors
- UL 1644 Immersion Detection Circuit Interrupters
- UL 1699 Arc-Fault Circuit Interrupters

REVIEW

1. Explain the operation of a ground-fault circuit interrupter. Why are GFCI devices used? Where are GFCI receptacles required? _____

   _____

   _____

   _____

   _____

   _____

2. Residential GFCI devices are set to trip a ground-fault current above _____ milliamperes.

3. Where must GFCI receptacles be installed in residential garages? _____

   _____

   _____

   _____

   _____

   _____

4. The *Code* requires GFCI protection for certain receptacles in the kitchen. Explain where these are required. _____

   _____

   _____

5. Is it a *Code* requirement to install GFCI receptacles in a fully carpeted, finished recreation room in the basement? _____

6. A homeowner calls in an electrical contractor to install a separate circuit in the basement (unfinished) for a freezer. Is a GFCI receptacle required? _____

7. GFCI protection is available as (a) branch-circuit breaker GFCI, (b) feeder circuit breaker GFCI, (c) individual GFCI receptacles, (d) feedthrough GFCI receptacles. In your opinion, for residential use, what type would you install? _____

8. Extremely long circuit runs connected to a GFCI branch-circuit breaker might result in
   a. nuisance tripping of the GFCI
   b. loss of protection
   c. the need to reduce the load on the circuit
   Circle the correct answer.

9. If a person comes in contact with the "hot" and grounded conductors of a two-wire branch-circuit that is protected by a GFCI, will the GFCI trip "OFF"? Why?

   _____

   _____

   _____

   _____

10. What might happen if the line and load connections of a feedthrough GFCI receptacle were reversed? _____

_____

_____

_____

_____

11. May a GFCI receptacle be installed as a replacement in an old installation where the two-wire circuit has no equipment grounding conductor? _____

12. What two types of receptacles may be used to replace a defective receptacle in an older home that is wired with knob-and-tube wiring where no equipment grounding means exists in the box? _____

_____

13. You are asked to replace a receptacle. Upon checking the wiring, you find that the wiring method is conduit and that the wall box is properly grounded. The receptacle is of the older style two-wire type that does not have a grounding terminal. You remove the old receptacle and replace it with (Circle the letter of the correct answer.)
    a. the same type of receptacle as the type being removed.
    b. a receptacle that is of the grounding type.
    c. a GFCI receptacle.

14. What color are the terminals of a standard grounding-type receptacle? _____

_____

15. What special shape and color are the grounding terminals of receptacle outlets and other devices? _____

_____

16. Construction sites can be dangerous because of the manner in which extension cords, portable electrical tools, and other electrical equipment are used and abused. To reduce this hazard, *590.6(A)* of the *NEC®* requires that all 125-volt, single-phase, 15-, 20-, and 30-ampere receptacle outlets used for temporary wiring during construction must be _____ protected.

17. In your own words, explain why the *Code* does not require certain receptacle outlets in kitchens, garages, and basements to be GFCI protected. _____

_____

_____

_____

_____

_____

_____

_____

18. Circuit A1 supplies GFCI receptacles ① and ②. Circuit A2 supplies GFCI receptacles ③ and ④. Receptacles ① and ③ are feedthrough type. Using colored pencils or marking pens, complete all connections.

19. The term *TVSS* is becoming quite common. What do the letters stand for? _____
    _____
    _____

20. Transients (surges) on a line can cause spikes or surges of energy that can damage delicate electronic components. A TVSS device contains one or more _____ _____ _____ that bypass and absorb the energy of the transient.

21. Undesirable "noise" on a circuit can cause computers to lock up, lose their memory, and/or cause erratic performance of the computer. This "noise" does not damage the equipment. The two types of this "noise" are EMI and RFI. What do these letters mean?
    _____
    _____
    _____
    _____

22. Can TVSS receptacles be installed in standard device boxes? _____

23. Some line transients are not damaging to electronic equipment, but can cause the equipment to operate improperly. The effects of these transients can be minimized by installing _____ _____ _____ .

24. Briefly explain the operation of an immersion detection circuit interrupter (IDCI).

_____

_____

_____

_____

_____

_____

25. What range of leakage current must trip an IDCI? _____

_____

_____

26. What amount of current flowing through a male human being will cause muscle contractions that will keep him from letting go of the live wire? _____ milliamperes

27. Other than for a few exceptions, the *NEC*® requires that all systems having a voltage of over _____ volts shall be grounded.

28. In an old house, an existing nongrounding non-GFCI-type receptacle in a bathroom needs to be replaced. According to *406.3(D)*, this replacement receptacle shall be (circle the correct letter):

    a. a nongrounding-type receptacle.

    b. a GFCI-type receptacle or be GFCI protected in some other way, such as changing the breaker that supplies the receptacle to a GFCI type.

29. Do all receptacle outlets in kitchens have to be GFCI protected? Explain. _____

_____

_____

_____

30. What special device is required for all 120-volt outlets in bedrooms? _____

_____

31. When installing GFCI and AFCI circuit breakers in a panel, it is permissible to connect them to a multiwire branch circuit. (True) (False). Circle the correct answer.

Luminaires (Fixtures), Ballasts, and Lamps

OBJECTIVES

After studying this unit, the student should be able to

▶ **understand luminaire (fixture) terminology, such as Type IC and Type NON IC.**

▶ **understand the *NEC*® requirements for installing and connecting surface and recessed luminaires (fixtures).**

▶ **realize that thermal insulation may have to be kept away from recessed luminaires (fixtures).**

▶ **understand thermal protection requirements for recessed luminaires (fixtures).**

▶ **know how to use "fixture whips."**

▶ **understand energy-saving ballasts and lamps.**

▶ **understand what a Class P ballast is.**

WHAT IS A LUMINAIRE (FIXTURE)?

NEC® defines a *luminaire (fixture)* as "*a complete lighting unit consisting of a lamp or lamps together with the parts designed to distribute the light, to position and protect the lamps and ballast (where applicable), and to connect the lamps to a power supply.*"* "Luminaire" is the international term for "lighting fixture" and is used throughout the *NEC*® In the *NEC*® the term is followed by the word "fixture."

TYPES OF LUMINAIRES (FIXTURES)

There are literally thousands of different types of luminaires (fixtures) from which to choose to satisfy certain needs, wants, desires, space requirements, and last but not least, price considerations.

*Reprinted with permission from NFPA 70-2005.

Note in Table 7-1, that whether the luminaire (fixture) is incandescent or fluorescent, the basic categories are surface mounted, recessed mounted, and suspended ceiling mounted.

The *Code* Requirements

Article 410 sets forth the requirements for installing luminaires (fixtures). The electrician must "meet Code" with regard to mounting, supporting, grounding, live-parts exposure, insulation clearances,

FLUORESCENT	INCANDESCENT
• Surface	• Surface
• Recessed	• Recessed
• Suspended Ceiling	• Suspended Ceiling

Table 7-1

supply conductor types, maximum lamp wattages, and so forth.

Probably the two biggest contributing factors to fires caused by luminaires (fixtures) are installing lamp wattages that exceed that for which the luminaire (fixture) has been designed, and burying recessed luminaires (fixtures) under thermal insulation when the luminaire (fixture) has not been designed for such an installation.

UL tests, lists, and labels luminaires (fixtures) that are in conformance to their standards. Always install luminaires (fixtures) that bear the UL label.

In addition to the *NEC*, the UL *Electrical Construction Materials Directory* (Green Book) and the UL *General Information for Electrical Equipment* (White Book), and manufacturers' catalogs and literature are excellent sources of information about luminaires (fixtures).

Read the Label

NEC® 110.3(B) states that *"Listed or labeled equipment shall be installed and used in accordance with any instructions included in the listing or labeling."*

It is important to carefully read the label and any instructions furnished with a luminaire (fixture). Most *Code* requirements can be met by simply following this information.

Here are a few examples of label and instruction information:

- *Maximum lamp wattage*

- *Type of lamp*

- *For supply connections, use wire rated for at least _____°F (°C)*

- *Type-IC*

- *Type Non-IC*

- *Suitable for wet locations*

- *Thermally protected*

Installing and Connecting Luminaires (Fixtures)

The circuit conductors in a wall or ceiling box where luminaires (fixtures) are to be installed are usually:

white—the "identified" grounded conductor, and

black—the ungrounded "hot" conductor. A "hot" switch leg might also be red or other color but never white or green.

Most surface-mounted luminaires (fixtures) will have a black and a white conductor in the canopy, making it easy to match these conductors to the circuit conductors in the box—white to white, black to black.

Chain-suspended luminaires (fixtures) usually come with a flexible, flat parallel conductor cord that weaves through the links of the chain. In a cord, it's a little more difficult to make a distinction between the "hot" conductor and the "identified" conductor.

To ensure proper polarity when making up the cord connections, generally connect as follows:

- The conductor with round insulation connects to the black "hot" circuit conductor.

- The conductor with grooved or raised insulation is the "identified" conductor that connects to the white circuit conductor.

- In some instances, the "identified" conductor will be tinned so it will have a silver color.

- The bare equipment grounding conductor (EGC) from the luminaire (fixture) connects to the green hexagon-shaped screw in a metal electrical box or on the luminaire's (fixture) mounting bar. Nonmetallic boxes will have a bare EGC from the nonmetallic-sheathed cable(s) to which the luminaire's (fixture) bare EGC is connected.

Surface-Mounted Luminaires (Fixtures)

These are easy to install. It is simply a matter of following the manufacturers' instructions furnished with the luminaires (fixtures). The luminaire (fixture) is attached to a ceiling outlet box or wall outlet box using luminaire (fixture) studs, hickeys, bar straps, or luminaire (fixture) extensions. Do not exceed the maximum lamp wattage marked on the luminaire (fixture).

Special attention must be given when installing surface-mounted luminaires (fixtures) on low-density cellulose fiberboard, Figure 7-1. Because of the potential fire hazard, *410.76(B)* states that fluorescent luminaires (fixtures) that are surface mounted

Figure 7-2 Typical recessed luminaire (fixture).

Figure 7-1 If they are to be mounted on *low-density cellulose fiberboard*, surface-mounted fluorescent luminaires (fixtures) must be marked "Suitable for Surface Mounting on Low-Density Cellulose Fiberboard."

on this material must be spaced at least 1½ in. (38 mm) from the fiberboard surface unless the fixture is marked "Suitable for Surface Mounting on Low-Density Cellulose Fiberboard."

The *Fine Print Note* to *410.76(B)* explains combustible low-density cellulose fiberboard as sheets, panels, and tiles that have a density of 20 lb./ft³ (320 kg/m³) or less that are formed of bonded plant fiber material. It does not include fiberboard that has a density of over 20 lb./ft³ (320 kg/m³) or material that has been integrally treated with fire-retarding chemicals to meet specific standards. Solid or laminated wood does not come under the definition of "combustible low-density cellulose fiberboard."

To obtain fire-resistance ratings, most acoustical ceiling panels today are made from mineral wool fibers or fiberglass.

Recessed Luminaires (Fixtures)

In the trade, recessed luminaires (fixtures) are commonly referred to as "cans."

Installing recessed luminaires (fixtures) requires more attention than simple surface mounted luminaires (fixtures). The *Code* requirements for the installation of recessed luminaires (fixtures) are found in *410.64* through *410.72*. UL Standard 1571 covers recessed incandescent luminaires (fixtures).

Recessed luminaires (fixtures) as shown in Figure 7-2 have an inherent heat problem. They

must be suitable for the application and must be installed properly.

It is absolutely essential for the electrician to know early in the roughing-in stages of wiring a house what types of luminaires (fixtures) are to be installed. This is particularly true for recessed-type luminaires (fixtures). To ensure that such factors as location, proper and adequate framing, and possible obstructions have been taken into consideration, the electrician must work closely with the general building contractor, carpenter, plumber, heating contractor, the thermal insulation installer, and the other building trades.

Figure 7-3(A) shows a Type Non-IC (insulated ceiling) recessed luminaire (fixture) where a minimum 3-in. (75-mm) clearance between the sides of the luminaire (fixture) and thermal insulation is needed. Above the luminaire (fixture), thermal insulation must be installed so as not to entrap heat in the cavity. How far should the thermal insulation be above the luminaire? It is a judgment call: 3 in. (75 mm) should suffice. Check the manufacturer's instructions.

Figure 7-3(B) shows a Type IC recessed luminaire (fixture) that may be completely buried in thermal insulation.

Figure 7-3(C) is a sketch that summarizes the minimum clearances for a Type Non-IC recessed luminaire (fixture) that is not listed for direct burial in thermal insulation, *410.66(B)*.

Recessed luminaires (fixtures) are available for both new work and remodel work. Remodel work recessed luminaires (fixtures) can be installed from below an existing ceiling by cutting a hole in the

Figure 7-3   (A) is a Type Non-IC recessed luminaire (fixture) that requires clearance between the luminaire (fixture) and the thermal insulation. (B) is a Type IC recessed luminaire (fixture) that may be completely covered with thermal insulation. (C) shows the required clearances from thermal insulation for a Type Non-IC luminaire (fixture). See *410.66*. *Photos courtesy of* Halo/Cooper Lighting (A, B).

ceiling, bringing power to the luminaire (fixture), making the electrical connections in the integral junction box on the luminaire (fixture), then installing the housing through the hole.

Energy-Efficient Housings

Many states have residential energy-efficiency requirements, particularly for recessed luminaires (fixtures). The most widely followed regulations are the State of California Title 24 requirements, and the State of Washington Restricted Air Flow Requirements.

Basically, these recessed "cans" have gasketed airtight housings that prevent heated or cooled air to escape into attics and other unconditioned spaces in the house. They are listed for use in insulated ceilings and in direct contact with insulation.

Although energy-efficiency requirements generally are not applicable to new one- and two-family homes, they might have been adopted by your state or community. Check with your local electrical inspector and/or building official to see if such laws are applicable in your area for one- and two-family dwellings.

Thermal Protection

To protect against the hazards of overheating, UL and the *NEC®* in *410.65(C)* require that recessed luminaires (fixtures) be equipped with an integral thermal protector, as in Figure 7-4. These devices will cycle on and off repeatedly until the heat problem has been resolved. There are two exceptions to

Figure 7-4   One type of recessed luminaire (fixture) thermal cutout.

this requirement. Thermal protection is not required for recessed incandescent luminaires (fixtures) designed for installation directly in a concrete pour, or for recessed incandescent luminaires (fixtures) that are constructed so as to not exceed "temperature performance characteristics" equivalent to thermally protected luminaires (fixtures). These luminaires (fixtures) are so identified.

What the Junction Box (Wiring Compartment) on the Recessed Luminaire (Fixture) Housing Is For

Most recessed luminaires (fixtures) come equipped with a junction box (wiring compartment) that is an integral part of the luminaire (fixture) housing. This is clearly shown in Figure 7-2 and Figure 7-5. For most residential-type recessed luminaire (fixture) installations, nonmetallic-sheathed cable or armored cable can be run directly into this junction box. These cables have conductors rated 194°F (90°C). Check the marking on the luminaire (fixture) for supply conductor temperature limitations.

Figure 7-6 shows the clearances normally needed between recessed luminaires (fixtures) and combustible framing members. This figure also summarizes the *NEC*® requirements for connecting the supply conductors to a recessed luminaire (fixture). Note in one case that the branch-circuit wiring is brought directly into the integral junction box on

the luminaire (fixture). In the other case, a "fixture whip" is run from an adjacent junction box to the luminaire (fixture). In some instances, the luminaire (fixture) comes with a flexible metal "fixture whip" and in other situations, the "fixture whip" is supplied by the electrician. "Fixture whips" are discussed later.

In *410.11*, we find that "*Branch-circuit wiring, other than 2-wire or multiwire branch-circuits supplying power to luminaires (fixtures) connected together, shall not be passed through an outlet box that is an integral part of a luminaire (fixture) unless the luminaire (fixture) is identified for through-wiring.*"*

In *410.31*, we find that "*luminaires (fixtures) shall not be used as a raceway for circuit conductors unless listed and identified for use as a raceway.*"*

Some recessed luminaires (fixtures) are marked "identified for through-wiring," in which case branch-circuit conductors in addition to the conductors that supply the luminaire (fixture) are permitted to be run through the outlet box or wiring compartment on the luminaire (fixture). A typical label will look like Figure 7-7.

The manufacturer of the luminaire (fixture), following UL requirements, will determine the maximum number, size, and temperature rating for the

---

*Reprinted with permission of NFPA 70-2005.

Figure 7-5 Roughing-in box of a recessed luminaire (fixture) with mounting brackets and junction box (wiring compartment).

ALL INCANDESCENT RECESSED LUMINAIRES (FIXTURES) MUST HAVE THERMAL PROTECTION AND BE SO MARKED. THERMAL PROTECTION NOT REQUIRED IF IDENTIFIED AS BEING INHERENTLY PROTECTED BY DESIGN AND CONSTRUCTION, *410.65(C)*.

FOR "FIXTURE WHIPS," THE MAXIMUM DISTANCE TO LAST SUPPORT IS 6 FT (1.8 m) EXCEPT FOR NM AND NMC, WHERE THE MAXIMUM DISTANCE TO SUPPORT IS 4½ FT (1.32 m).

BRANCH-CIRCUIT CONDUCTORS WITH INSULATION SUITABLE FOR THE TEMPERATURE REQUIREMENTS MARKED ON THE "LISTED" PREWIRED RECESSED LUMINAIRE (FIXTURE) MAY BE RUN DIRECTLY TO JUNCTION BOX ON LUMINAIRE (FIXTURE), *410.67(B)*

JUNCTION BOX AT LEAST 12 IN. (300 mm) FROM LUMINAIRE (FIXTURE), *410.67(C)*

IF FLEX METAL CONDUIT, MAY BE TRADE SIZE 3/8, *348.20(A)*

"LISTED" PREWIRED LUMINAIRE (FIXTURE) WITH JUNCTION BOX

THESE BOXES MUST BE ACCESSIBLE, *314.29*

"LISTED" LUMINAIRE (FIXTURE) WITHOUT JUNCTION BOX

CLEARANCES:
• TYPE NON IC: AT LEAST 1/2 IN. (13 mm) FROM COMBUSTABLE MATERIAL, EXCEPT AT SUPPORT, *410.66(A)(1)*
• TYPE IC: NO MINIMUM CLEARANCE, *410.66(A)(2)*

KEEP INSULATION AT LEAST 3 IN. (75 mm) FROM LUMINAIRE (FIXTURE) UNLESS IT IS SUITABLE FOR DIRECT CONTACT WITH INSULATION, *410.66(B)*

ADJACENT COMBUSTIBLE MATERIAL TEMPERATURE NOT TO EXCEED 94°F (90°C), *410.65(A)* AND *410.68*

Figure 7-6 Requirements for installing recessed luminaires (fixtures).

Maximum of ___ ___ AWG through branch-circuit conductors suitable for at least ___°C (___°F) permitted in a junction box (___ in ___ out).

Figure 7-7 A typical label found on a recessed luminaire (fixture) specifying the maximum number and size of conductors permitted to be run through the wiring compartment on the luminaire (fixture).

conductors entering and leaving the outlet box or wiring compartment, and will so mark the label.

Only those luminaires (fixtures) that have been tested for the extra heat generated by the additional branch-circuit conductors will bear the label marking shown in Figure 7-7.

These fixtures have been tested for the added heat generated by the additional branch-circuit conductors.

Figure 7-8 shows three recessed luminaires (fixtures), each with an integral outlet box. If these luminaires (fixtures) are marked "Identified for through-wiring," then it would be permitted to run conductors through the outlet boxes *in addition* to the conductors that supply the luminaires (fixtures). If the "Identified for through-wiring" marking is not found on the luminaire (fixture), then it is a *Code* violation to run any conductors through the box other than the conductors that supply the luminaires (fixtures) that are "daisy chained."

Some electricians prefer to run the branch-circuit wiring to the junction box on the recessed luminaire (fixture), then drop the switch loop to the switch box location. Others prefer to run the branch-circuit to the switch location, then to the junction box on the luminaire (fixture). Either way is acceptable.

Recessed Luminaire (Fixture) Trims

Recessed incandescent luminaires (fixtures) are listed by Underwriters Laboratories with specific trims. Luminaire (fixture)/trim combinations are

Figure 7-8   The only conductors permitted to run into or through the junction boxes on the recessed luminaires (fixtures) are those that supply the luminaires (fixtures). Conductors other than those that supply the luminaires (fixtures) are permitted to run through the outlet boxes if the luminaire (fixture) is marked "Identified For Through-Wiring." See *410.11* and *410.31*.

marked on the label. Do not use trims that are not listed for use with the particular recessed luminaire (fixture). Installing a trim not listed for use with a specific recessed luminaire (fixture) can result in overheating and possible on-off cycling of the luminaire's (fixture's) thermal protective device. Mismatching luminaires (fixtures) and trims is a violation of *110.3(B)* of the *NEC*,® which states that "*Listed or labeled equipment shall be installed or used in accordance with any instructions included in the listing or labeling.*"

Sloped Ceilings

Be very careful when selecting recessed luminaires (fixtures). Conventional recessed luminaires (fixtures) are listed for installation in flat ceilings, not in sloped ceilings! Recessed luminaires (fixtures) intended for installation in a sloped ceiling are submitted to UL specifically for testing in a sloped ceiling. Look for the marking "Sloped Ceiling." Instructions furnished with recessed luminaires (fixtures) provide the necessary information, such as being suitable for sloped ceilings having a 2/12, 6/12, or 12/12 pitch. Check the manufacturer's catalog and the UL listing.

Suspended Ceiling Lay-In Luminaires (Fixtures)

Figure 7-9 illustrates luminaires (fixtures) installed in a typical suspended ceiling.

The recreation room of this residence has a "dropped" suspended acoustical paneled ceiling. Luminaires (fixtures) installed in a "dropped" ceil-

Figure 7-9   Suspended ceiling luminaires (fixtures).

ing must bear a label stating "Suspended Ceiling Luminaire (Fixture)." These types of luminaires (fixtures) are not classified as recessed luminaires (fixtures) because in most cases, there is a great deal of open space above and around them.

Support of Suspended Ceiling Luminaires

In *410.16(C)* we find that when framing members of a suspended ceiling grid are used to support luminaires (fixtures), all framing members must be securely fastened together and to the building structure. It is the responsibility of the installer of the ceiling grid to do this.

All lay-in suspended ceiling luminaires (fixtures) must be securely fastened to the ceiling grid members with bolts, screws, or rivets. Listed clips supplied by the manufacturer of the luminaire (fixture) are also permitted to be used. Clips that meet this criteria are shown in Figure 17-1.

The logic behind all of the "securely fastening" requirements is that in the event of a major problem such as an earthquake or fire, the luminaires (fixtures) will not fall down and injure someone.

Connecting Suspended Ceiling Luminaires (Fixtures)

The most common way to connect suspended ceiling lay-in luminaires (fixtures) is to complete all of the wiring above the ceiling using conventional wiring methods.

Then from accessible outlet boxes strategically placed above the ceiling near the intended location of the lay-in luminaires (fixtures), a flexible connection not more than 6 ft (1.8 m) long is made between the outlet boxes and the luminaires (fixtures). See Figure 7-10. Lay-in ceiling tiles provides access to these outlet boxes.

In the electrical trade, these flexible connections are referred to as "fixture whips."

The *NEC*® does not have a definition of a "fixture whip."

Don't confuse a "fixture whip" with a "tap." Most fixture whips in residential installations have conductors the same size as the branch-circuit conductors, so the "whip" is really an extension of the branch-circuit, and by the *NEC*® definition is not really a "tap."

A "tap" is defined in *240.2* as a conductor that has overcurrent protection ahead of its point of supply that exceeds the maximum overcurrent protection for that conductor. This is quite common in commercial installations where the branch-circuit overcurrent protection is rated 20 amperes, and the tap conductors are rated 15 amperes. Tap conductors must be in a suitable raceway or Type AC or MC cable of at least 18 in. (450 mm) but not more than 6 ft (1.8 m) in length.

A fixture whip might be:

- armored cable (AC), *320.30(D)(3)*.

- metal clad cable (MC), *330.30(D)(2)*.

- nonmetallic-sheathed cable (NM and NMC), *334.30(B)(2)*.

- flexible metal conduit (FMC): OK to use trade size ³⁄₈, *348.30(A), Exceptions No. 3* and *No. 4*.

FLEXIBLE METAL CONDUIT AND LIQUIDTIGHT FLEXIBLE METAL CONDUIT SUITABLE AS EQUIPMENT GROUNDING MEANS IF NOT OVER 6 FT (1.8 m) LONG BRANCH-CIRCUIT OVERCURRENT DEVICE IS NOT OVER 20 AMPERES, AND FITTINGS ARE LISTED FOR GROUNDING, *250.118(6)* AND *250.118(7)*.

TIE WIRES

THIS SPACE ACCESSIBLE FROM BELOW CEILING BY LIFTING OUT PANELS.

SUSPENDED CEILING LUMINAIRE (FIXTURE)

SUSPENDED CEILING

OUTLET BOX

FIXTURE WHIP. SEE TEXT FOR WIRING METHODS PERMITTED FOR FIXTURE WHIPS.

CONDUCTORS IN FLEXIBLE CONNECTION (FIXTURE WHIP) MUST BE SUITABLE FOR TEMPERATURE REQUIREMENTS AS SPECIFIED ON LUMINAIRE (FIXTURE) LABELING, *410.67*.

Figure 7-10 A suspended ceiling luminaire (fixture) connected with a flexible fixture whip between an outlet box and the luminaire (fixture).

If trade size ⅜, length is not to exceed 6 ft (1.8 m), *348.20(A)(2)*.

- liquidtight flexible metallic conduit (LFMC): OK to use trade size ⅜, *350.30(A), Exceptions No. 3* and *No. 4*. If trade size ⅜, length is not to exceed 6 ft (1.8 m), *350.20(A)*.

- liquidtight flexible nonmetallic conduit (LFNC): OK to use trade size ⅜, *356.30(2)*. If trade size ⅜, length is not to exceed 6 ft (1.8 m), *356.20(A)*.

- electrical nonmetallic-tubing (ENT), *362.30(A)*. Trade size smaller than ½ is not permitted, *362.20(A)*.

Note: For the above wiring methods, the maximum distance to the last point of support is 6 ft (1.8 m), except for NM and NMC where the maximum distance to the last point of support is 4½ feet (1.32 m).

Grounding Luminaires (Fixtures)

All luminaires (fixtures) must be grounded, *410.18(A)*. Assuming that the wiring is a metallic wiring method or nonmetallic-sheathed cable with ground, you have an acceptable equipment ground at the lighting outlet. Grounding of a surface-mounted luminaire (fixture) is easily accomplished by securing the luminaire (fixture) to the properly grounded metal outlet box with the hardware provided with the luminaire (fixture). Metal outlet boxes have a tapped No. 10-32 hole in which to insert a green hexagonal grounding screw. The bare copper equipment grounding conductor in the nonmetallic-sheathed cable is usually terminated under this screw. Grounding and bonding was discussed in Unit 5. If the outlet box is nonmetallic, the small bare equipment grounding conductor from the luminaire (fixture) is connected to the equipment grounding conductor in the outlet box. For suspended ceiling luminaires (fixtures), grounding of the luminaire (fixture) is accomplished by using metallic fixtures whips or nonmetallic-sheathed cable with ground between the outlet box and the luminaire (fixture). Always follow the manufacturers' instructions for making the electrical connections.

What about replacing a luminaire (fixture) in an older home? If there is no equipment grounding means in the outlet box, then a luminaire (fixture) that is made of insulating material and has no exposed conductive parts must be used, *410.18(B)*.

The exception to *410.18(B)* allows a metallic replacement luminaire (fixture) to be installed and connected to a system that does not have an equipment grounding conductor. In this case, a separate equipment grounding conductor must be installed in conformance with *250.130(C)*. This requirement was discussed in Unit 6 in the topic of replacing receptacles, so it will not be repeated here. Acceptable types of equipment grounding conductors are listed in *250.118*.

In Unit 17, Figure 17-1 and Figure 17-2 have additional information regarding recessed luminaire (fixture) installations such as those in the recreation room of this residence.

Luminaires (Fixtures) in Closets

There are special requirements for installing luminaires (fixtures) in closets. This is covered in Unit 8.

Luminaires (Fixtures) in Bathrooms

Because of the electrical shock hazards associated with electricity and water, there are special restrictions for luminaires (fixtures) installed in bathrooms. This is covered in Unit 10.

FLUORESCENT BALLASTS AND LAMPS, INCANDESCENT LAMPS

The following is a brief introduction to a complex subject. For much more information, check out lamp and ballast manufacturers' Web sites listed in the Appendix of this text.

From an electrician's point of view, lamps and light bulbs mean the same thing. Use any one of the words, and everyone will know what you mean.

Always read the label on a luminaire (fixture) for the type and maximum wattage of the lamp to be installed in that luminaire (fixture).

Fluorescent Ballasts

An incandescent lamp contains a filament that has a specific "hot" resistance value when operating at rated voltage. When energized, the lamp will provide the light output for which the lamp was designed.

A fluorescent lamp, on the other hand, cannot be connected directly to a circuit. It needs a ballast.

Fluorescent lamps do not have a filament running from end-to-end. Instead, they have a filament at each end. The filaments are connected to a ballast, as in Figure 7-11. The ballast is needed to (1) control power, (2) control voltage to heat the filaments, (3) control voltage across the lamp to start the arc within the tube, and (4) limit the current flowing through the lamp. The arc is needed to ionize the gas and vaporize the droplets of mercury inside the lamp.

Without a ballast, the lamp would probably not start. But if it did start, it would "run away" with itself. The current flowing through the gases within the lamp would rise rapidly, destroying the lamp in a very short time.

Ballast Types

Preheat. Preheat ballasts are connected in a simple series circuit, Figure 7-11. They are easily identified because they have a "starter." One type of starter is automatic and looks like a small roll of lifesavers with two "buttons" on one end.

Another type of starter is a manual "ON/OFF" switch that has a momentary "make" position just beyond the "ON" position. When you push the switch on and hold it there for a few seconds, the lamp filaments glow. When the switch is released, the start contacts open, an arc is initiated within the lamp, and the lamp lights up.

Preheat lamps have two pins on each end.

Preheat lamps and ballasts are not used for dimming applications.

Rapid Start. Probably the most common type used today. Rapid start ballasts/lamps do not require

a starter. The lamps start in less than 1 second. For reliable starting, ballast manufacturers recommend that there be a grounded metal surface within ½ in. (12.7 mm) of the lamp and running the full length of the lamp, that the ballast be grounded, and that the supply circuit originates from a grounded system. T5 rapid start lamps do not require a grounded surface for reliable starting.

Rapid start lamps have two pins on each end.

Rapid start lamps can be dimmed using a special dimming ballast. See Unit 13.

Instant Start. Instant start lamps do not require a starter. These ballasts provide a high-voltage "kick" to start the lamp instantly. They require special fluorescent lamps that do not require preheating of the lamp filaments. Because instant start fluorescent lamps are started by brute force, they have a shorter life (as much as 40% less) than rapid start lamps when older style magnetic ballasts are used. With electronic ballasts, satisfactory lamp life can be expected.

Instant start lamps have one pin on each end.

Instant start ballasts/lamps cannot be used for dimming applications.

Dimming Ballasts. Special dimming ballasts and dimmers are needed for controlling the light output of fluorescent lamps. Rapid start lamps are used. Incandescent lamp dimmers cannot be used to control fluorescent lamps. An exception to this is that dimmers marked "Incandescent Only" can be used to dim compact fluorescent lamps. Dimming ballasts are discussed in Unit 13.

Figure 7-11 Simple series circuit.

Mismatching of Fluorescent Lamps and Ballasts

Be sure to use the proper lamp for a given ballast. Mismatching a lamp and ballast may result in poor starting and poor performance, as well as shortened lamp and/or ballast life. The manufacturer's ballast and/or lamp warranty may be null and void.

T8 lamps are designed to be used interchangeably on magnetic or electronic rapid start ballasts or electronic instant start ballasts. Lamp life is reduced slightly when used with an instant start ballast.

Voltage

Operating a ballast at an over-voltage condition will cause the it to run hot and shorten its life.

Operating a ballast at an under-voltage situation can result in premature lamp failure and unreliable starting.

Most ballasts today will operate satisfactorily within a range of +5% to –7% of their rated voltage. The higher quality CBM certified ballasts will operate satisfactorily within a range of ±10%.

Sound Rating

Most ballasts will hum, some more than others.

Ballasts are sound rated and are marked with letters A through F. A is the quietest, and F is the noisiest. Look for an "A" or "B" sound rating for residential applications.

Magnetic ballasts (core and coil) hum when the metal laminations vibrate because of the alternating current reversals. This hum can be magnified by the luminaire (fixture) itself, and/or the surface the luminaire (fixture) is mounted on.

Electronic ballasts have little, if any, hum.

CAUTION: Do not insert spacers, washers, or shims between a ballast and the luminaire (fixture) to make the ballast more quiet. This will cause the ballast to run much hotter, and could result in shortened ballast life and possible fire hazard. Instead, replace the noisy ballast with a quiet, sound-rated one. Sometimes checking and tightening the many nuts, bolts, and screws of the luminaire (fixture) will solve the problem.

Energy-Saving Ballasts

The market for magnetic (core and coil) ballasts is shrinking!

The National Appliance Energy Conservation Amendment of 1988, Public Law 100-357 prohibited manufacturers from producing ballasts having a power factor of less than 90%. Ballasts that meet or exceed the federal standards for energy savings are marked with a letter "E" in a circle. Dimming ballasts and ballasts designed specifically for residential use were exempted.

Today's electronic ballasts are much more energy efficient than the older style magnetic ballasts (core and coil). Energy-saving ballasts might cost more initially, but the payback is in the energy consumption saving over time.

Old-style fluorescent ballasts gets very warm and might consume 14 to 16 watts, whereas an electronic ballast might consume 8 to 10 watts. Combined with energy-saving fluorescent lamps that use 32 or 34 watts instead of 40 watts, energy savings are considerable. You are buying light, not heat.

When installing fluorescent luminaires (fixtures), check the label on the ballast that shows the actual volt-amperes that the ballast and lamp will draw in combination. *Do not* attempt to use lamp wattage only when making load calculations, because this could lead to an overloaded branch-circuit. For example, a high-efficiency ballast might draw a total of 42 volt-amperes, whereas an old-style magnetic ballast might draw 102 volt-amperes. Refer to Table 7-2 for other comparisons.

The higher the power factor rating of a ballast, the more energy efficient. Look for a power factor rating in the mid to high 90s.

Energy-Saving Lamps

The National Energy Policy Act of 1992 enacted restrictions on lamps. In October 1995, the common 4-ft, 40-watt T12 linear medium bi-pin fluorescent lamp was eliminated. This was replaced by an energy-efficient 34-watt T12 lamp, a direct replacement for the discontinued lamp.

Some lamps may be designated F40T12/ES, but the lamp draws 34 watts instead of 40 watts. The "ES" stands for "energy saving." ES is a generic designation. Manufacturers may use other designations such as "SS" for SuperSaver, "EW" for Econ-o-Watt, "WM" for Watt-Miser, and others.

The older, high wattage incandescent R30, R40, and PAR38 lamps were also discontinued and

BALLAST	LINE CURRENT	LINE VOLTAGE	LINE VOLT/AMPERES	LAMP WATTAGE	LINE POWER FACTOR
No. 1	0.35	120	42	40	0.95 (95%)
No. 2	0.45	120	54	40	0.74 (74%)
No. 3	0.55	120	66	40	0.61 (61%)
No. 4	0.85	120	102	40	0.39 (39%)

Table 7-2 Various line currents, volt-amperes, wattages, and overall power factor for various single-lamp fluorescent ballasts. These values were taken from actual ballast manufacturers' data. For two-lamp ballasts, double the values in this table.

replaced with lower wattage lamps. See the sections on "Lamp Shapes" and "Lamp Diameters" in this unit.

T12 lamps are still found in 4-ft shop lights and square luminaires (fixtures) that use U-tube lamps. Most newer square luminaires (fixtures) have U-tube T8 lamps. In new commercial installations, the T8 lamp has taken over from the T12 lamp.

Energy saving fluorescent lamps use up to 80% less energy than incandescent lamps of similar brightness. Fluorescent lamps can last 13 times or more longer than incandescent lamps.

It has been estimated that the total electric bill savings across the country will exceed $250 billion over the next 15 years. Table 7-2 compares the power factor of various types of ballasts. Note the difference in line current for the different types of ballasts.

Cold Temperature and Fluorescent Lamps

Conventional fluorescent lamps will operate satisfactorily in ambient temperatures of 60°F (16°C) or more.

In cold temperatures, standard fluorescent lamps will start poorly or possibly not even start. Special ballasts and cold-weather fluorescent lamps are needed. Refer to the manufacturer's literature for instructions relating to cold temperature.

Lamp Designations for Rapid Start and Preheat Fluorescent Lamps

Example: F40T12/WWX/RS

"F"—fluorescent

"40"—wattage

"T"—tubular shape

"12"—diameter of lamp in eighths of an inch

"WWX"—color (warm white deluxe)

"RS"—rapid start

"FC"—shape (circular)

"FB" or "FU"—U-shaped, bent

Lamp Shapes

"A"—standard shape, general use

"P" or "PS"—pear shaped (150 watts and larger)

"C"—cone shape, like a night light or Christmas tree bulb.

"F"—flame shape, decorative

"G"—globular

"PAR"—parabolic shape like a bowl, concentrates light

"R"—reflector type; might have reflective material near the base or at the bottom of lamp, concentrates light.

Burn Base Down. Some lamps will be designated as "Burn Base Down." Be sure to follow these words of warning.

Lamp Diameters

There is a simple way to determine the diameter of a lamp at its widest measurement. The industry uses an "eighths of an inch rule." For example, the diameter of a T5 fluorescent lamp is $5 \div 8 = 0.625$ in.

The diameter of a T8 fluorescent lamp is $8 \div 8 = 1$ in. The diameter of a T12 lamp is $12 \div 8 = 1\frac{1}{2}$ in.

Incandescent lamps follow the same system. The diameter of an A21 lamp is $21 \div 8 = 2.625$ in. The diameter of an R30 lamp is $30 \div 8 = 3\frac{3}{4}$ in.

Watts versus Volt-Amperes

It is very easy to overload a branch-circuit that supplies fluorescent lighting. The most common mistake of making load calculations for fluorescent lighting loads is to use lamp wattage instead of the volt-amperes and total current draw as marked on the ballast's label.

The culprits are old-style, low-power factor, low-efficiency ballasts!

Let's make a comparison of possibilities for the recreation room lighting where there are six recessed fluorescent luminaires (fixtures), each containing two 2-lamp ballasts. The nameplate on each ballast indicates a line current rating of 0.70 ampere at 120 volts, which is 84 volt-amperes. The lamps are marked 40-watts.

The total current in amperes is

$$12 \times 0.70 = 8.4 \text{ amperes}$$

The total power in volt-amperes is

$$8.4 \times 120 = 1008 \text{ volt-amperes}$$

The total lamp wattage is

$$40 \times 24 = 960 \text{ watts}$$

If we had calculated the load for these luminaires (fixtures) based upon wattage, the result would have been

$$\frac{W}{E} = \frac{960}{120} = 8 \text{ amperes}$$

This result is not much different than the 8.4 amperes using data from the ballast nameplate.

Had we used low-cost, low-efficiency ballasts like No. 4 shown in Table 7-2, the result would be entirely different. The ballasts in Table 7-2 are single-lamp ballasts. We double the current values for a good approximation of a similar 2-lamp ballast. Using the low-power factor ballast, we find that the total current in amperes is:

$$12 \times 0.85 \times 2 = 20.4 \text{ amperes}$$

The total power in volt-amperes is

$$20.4 \times 120 = 2,448 \text{ volt-amperes}$$

or

$$12 \times 102 \times 2 = 2,448 \text{ volt-amperes}$$

The total lamp wattage is

$$24 \times 40 = 960 \text{ watts}$$

As previously shown, if we use lamp wattage to calculate the current draw:

$$I = \frac{W}{E} = \frac{960}{120} = 8 \text{ amperes}$$

But we really have a current draw of 20.4 amperes, which would overload the 15-ampere branch-circuit B12. In fact, a load of 20.4 amperes is too much for a 20-ampere branch-circuit. Two 15-ampere branch-circuits would have been needed to hook up the recreation room recessed fluorescent luminaires (fixtures). Remember, *do not* load any circuit to more than 80% of the branch-circuit's rating. That is 16 amperes for a 20-ampere branch-circuit and 12 amperes for a 15-ampere branch-circuit. And that is the maximum.

It is apparent that the *installed* cost using cheap luminaires (fixtures) is greater and more complicated than if high-quality luminaires (fixtures) using energy-efficient ballasts and lamps had been used. However, after the initial installation, the energy savings add up significantly.

Low power factor ballasts means higher light bills! High-efficiency, high-power factor ballasts should always be used. More on the wiring of the lighting in the recreation room is covered in Unit 17.

Other Considerations

- Because heat trapped by insulation around and on top of a luminaire (fixture) can shorten the life of a ballast, always follow the manufacturer's installation requirements and the requirements found in *Article 410*. There is a "nonscientific" rule of thumb referred to as the "half life rule." It means that for every 21°F (10°C) above electrical equipments (motors, conductors, transformers, etc.) recommended maximum operating temperature, the expected life of that equipment is cut roughly in half.

Raising the temperature another 21°F (10°C) will again cut the expected life in half. A 21°F (10°C) temperature rise (**Note:** This is a temperature difference, not an absolute temperature measurement), above the ballast's rated temperature of 194°F (90°C), can reduce the ballast's life by one-half. This "half-life rule" is also true for conductors, motors, transformers, and other electrical equipment.

- Fluorescent lamps that are intensely blackened on both ends should be replaced. Operating a two-lamp ballast with only one lamp working will cause the ballast to run hot, and will shorten the life of the ballast. Severe blackening of one end of the lamp can also ruin the ballast. A flickering lamp should be replaced.

- Poor starting of a fluorescent lamp can be caused by poor contact in the lampholder, poor grounding, excessive moisture on the outside of the tube, or cold temperature (approximately 50°F [10°C]), as well as dirt, dust, and grime on the lamp.

- Most ballasts will operate satisfactorily within a range of +5% to −7% of their rated voltage. The higher quality CBM-certified ballasts will operate satisfactorily within a range of ±10%.

- Most T12 fluorescent lamps sold today are of the energy-efficient type. Generally, energy-efficient fluorescent lamps should not be used with old-style magnetic ballasts. To do so could result in poor starting, reduced lamp life, flickering, or spiraling. Home centers carry the most common T12 fluorescent lamps. Electrical supply houses usually carry different types of fluorescent lamps that can be matched with different types of ballasts. Check the markings on the ballast and the lamp carton, as well as the instructional literature available at the point-of-sale of the ballast or lamp, and/or in the manufacturer's catalogs.

- Lamp life generally is rated in "X" number of hours of operation based on 3 hours per start. Frequent switching results in shortened expected lamp life. Inversely, leaving the lamps on for long periods of time extends the expected lamp life. Vibration, rough handling, cleaning, and so on, shortens lamp life.

- Be careful of the type of conductors you use to connect fluorescent luminaires (fixtures). Branch-circuit conductors within 3 in. (75 mm) of a ballast must have an insulation temperature rating of not less than 194°F (90°C), *410.33*. Type THHN conductors, the conductors in non-metallic-sheathed cable and in Type ACTHH armored cable are rated 194°F (90°C).

Voltage Limitations

The maximum voltage allowed for residential lighting is 120 volts between conductors, per *210.6(A)*.

In or on a home, lighting equipment that operates with an open-circuit voltage over 1000 volts is not permitted, *410.80(B)*. This pretty much eliminates most neon lighting systems for decorative lighting purposes, as shown in Figure 7-12.

Incandescent Lamp Life at Different Voltages

Operating an incandescent lamp at other than rated voltage will result in longer—or shorter—lamp life. The following formula predicts the approximate expected lamp life at different voltages. For example, assume that a 120-volt incandescent lamp has a published lamp life of 1000 hours. The calculations show the expected lamp life of this lamp when operated at 130 volts, and the expected lamp life when operated at 110 volts.

- At 130 volts:

$$\text{Expected Life} = \left(\frac{\text{Rated Volts}}{\text{Actual Volts}}\right)^{13.1} \times \text{Rated Life}$$

$$= \left(\frac{120}{130}\right)^{13.1} \times 1000$$

$$= 0.350 \times 1000$$

$$= 350 \text{ hours}$$

- At 110 volts:

$$\text{Expected Life} = \left(\frac{\text{Rated Volts}}{\text{Actual Volts}}\right)^{13.1} \times \text{Rated Life}$$

$$= \left(\frac{120}{110}\right)^{13.1} \times 1000$$

$$= 3.126 \times 1000$$

$$= 3126 \text{ hours}$$

Figure 7-12 It is *not* permitted to install lighting in or on residences where the open-circuit voltage is over 1000 volts. This is a violation of *410.80(B)*.

Incandescent Lamp Lumen Output at Different Voltages

Operating an incandescent lamp at a voltage lower than the lamp's voltage rating will result in longer lamp life. A strong case might be made to install 130-volt lamps, particularly where they are hard to reach, such as flood lights located high up. However, the lumen output is reduced. For example, calculate the approximate lumen output of a 100-watt, 130-volt incandescent lamp that has an initial lumen output of 1750 lumens at rated voltage. The lamp is to be operated at 120 volts.

$$\text{Expected Lumens} = \left(\frac{\text{Actual Volts}}{\text{Rated Volts}}\right)^{3.38} \times \text{Rated Lumens}$$

$$= \left(\frac{120}{130}\right)^{3.38} \times 1750$$

$$= 0.763 \times 1750$$

$$= 1335 \text{ Lumens}$$

Formulas such as these are found in lamp manufacturer's catalogs. These formulas are useful in determining the effect of applied voltage on lamp wattage, line current, lumen output, lumens per watt, and lamp life. A calculator that has a y^x power function is needed to solve these equations.

Keep It Simple!

The formulas for lamp life and lumen output are complicated but here is a quick summary:

- An incandescent lamp operating at a voltage *less* than its rated voltage will last longer but will not burn as bright as it should.

- An incandescent lamp operating at a voltage *greater* than its rated voltage will not last as long but will burn brighter.

LAMP EFFICACY

A lumen is a measurement of light output from a lamp.

One lumen on 1 ft^2 of surface produces 1 *foot-candle*.

Another term you will hear in lighting is *efficacy*. Efficacy is the ratio of light output from a lamp to the electric power it consumes and is measured in lumens per watt. In other words, efficacy is a measurement of input to output. For example, one type of 100-watt lamp produces 1750 lumens at rated voltage, which equates to 17.5 lumens per watt. Another 100-watt lamp produces 1200 lumens at rated voltage, which equates to 12 lumens per watt. The more lumens per watt, the greater the efficacy of the lamp, as shown in Table 7-3.

Although both of the above lamps cost the same to operate, one lamp produces more light. Halogen and compact fluorescent lamps produce more lumens per watt than the typical incandescent lamp, making it possible to get more light using a lower wattage lamp.

LAMP COLOR TEMPERATURE

Lamp color temperature is rated in Kelvin degrees, and the term is used to describe the "whiteness" of

	LUMENS/ WATT	DIMMING	COLOR AND APPLICATION	LIFE (Approx. Hours)	TYPICAL SHAPES
Incandescent	18	Yes	Warm and Natural. Great for general lighting.	500, 1000, 1500; Depends on type of lamp.	Standard, spots, floods, decorative, tubes, globes, PAR (similar to standard spots and floods but much stronger). Use rough service where there is vibration, like garage door openers and ceiling fans.
Halogen	22	Yes	Brilliant white. Excellent for accent and task lighting. Are filled with halogen gas and have an inner lamp allowing the filament to run hotter (whiter).	2000 to 4000	PAR spots and floods, mini-reflector spots and floods, double and single end quartz.
Fluorescent	T12 84 T8 92 T5 104	Yes, but only 40-watt rapid start lamps using special dimming ballast. See Unit 13.	Warm and deluxe warm white, cool and deluxe cool white plus many other shades of white. Great for general lighting, like the Recreation Room in this residence.	6000 to 24,000. Compact fluorescent lamps (CFLs) • can be used with dimmer if so marked. • can last 9 to 13 times longer than incandescent lamps. • produce more lumens per watt, e.g., a 27-watt CFL produces 1800 lumens vs. 1750 lumens for a 100-watt incandescent. • have ballast and lamp in a single screw-in unit. • screw-in type that have an integral ballast can replace most incandescent lamps. • use about 80% less energy and generate 90% less heat than equivalent incandescent lamps. • available in 3-way type. Read instructions on lamp and package for specific restrictions, such as *Not for use with dimmers, Not for use with electronic timers, photocells, or lighted switches,* or *Do not use in recessed or totally enclosed fixtures.*	Straight, U-tube, circular. CFLs have folded tubes. The number of "loops" are referred to as twin, double twin, triple twin, quad twin, and spiral. Some are shaped like typical "A" incandescent lamps. CFLs can save $20 to $50 over their lifetime when compared to equivalent incandescent lamps.

Table 7-3   Comparisons of various lamps' characteristics.

the lamp light. In incandescent lamps, color temperature is related to the physical temperature of the filament. In fluorescent lamps where no hot filament is involved, color temperature is related to the light as though the fluorescent discharge is operating at a given color temperature. The lower the Kelvin degrees, the "warmer" the color tone. Conversely, the higher the Kelvin degrees, the "cooler" the color tone.

Incandescent lamps provide pleasant color

tones, bringing out the warm red flesh tones similar to those of natural light. This is particularly true for the "soft" and "natural" white lamps.

Tungsten filament halogen lamps have a gas filling and an inner coating that reflects heat. This keeps the filament hot with less electricity. Their light output is "whiter." They are more expensive than the standard incandescent lamp.

Fluorescent lamps are available in a wide range of "coolness" to "warmth." Warm fluorescent lamps bring out the red tones. Cool fluorescent lamps tend to give a person's skin a pale appearance.

Fluorescent lamps might be marked daylight D (very cool), cool white CW (cool), white W (moderate), warm white WW (warm). These categories break down further into a deluxe X series (i.e., deluxe warm white—deluxe cool white), specification SP series, and specification deluxe SPX series.

Typical color temperature ratings for lamps are 2800K (incandescent), 3000K (halogen), 4100K (cool white fluorescent), and 5000K (fluorescent that simulates daylight). Note that a halogen lamp is "whiter" than a typical incandescent lamp.

Catalogs from lamp manufacturers provide detailed information about lamp characteristics.

Class P Ballasts

In *410.73(E)*, we find that all fluorescent ballasts installed indoors (except simple reactance-type ballasts), both for new and replacement installations, must have thermal protection built into the ballast by the manufacturer of the ballast, as shown in Figure 7-13. Ballasts provided with built-in thermal protection are listed by the UL as Class P ballasts. Under normal conditions, the Class P ballast has a case temperature not exceeding 194°F (90°C). The thermal protector must open within two hours when the case temperature reaches 230°F (110°C).

Some Class P ballasts also have a nonresetting fuse integral with the capacitor to protect against capacitor leakage and violent rupture. The Class P ballast's internal thermal protector will disconnect the ballast from the circuit in the event of excessive temperature. Excessive temperatures can be caused by abnormal voltage and improper installation, such as being covered with insulation.

The reason for thermal protection is to reduce the hazards of possible fire due to an overheated ballast when the ballast becomes shorted, grounded, covered with insulation, lacking in air circulation, and so on. Ballast failure has been a common cause of electrical fires.

See Figure 7-14 for some important requirements for fluorescent and incandescent luminaires (fixtures).

Outdoor Lighting

Before installing outdoor lighting, check with the local electrical inspector and/or building official

Figure 7-13 An electronic Class P ballast, thermally protected as required in *410.73(E)*. Electronic ballasts provide improved performance in fluorescent lighting installations. Electronic ballast lighting systems are 25–40% more energy efficient than conventional magnetic (core & coil) ballast fluorescent systems. *Courtesy of* OSRAM Sylvania.

Figure 7-14 **Some of the most important UL and *NEC*® requirements for fluorescent and incandescent luminaires (fixtures). Always refer to the UL Standards, the *NEC*,® and the label and/or instructions furnished with the luminaire (fixture).**

to find out if there are any restrictions regarding outdoor lighting.

A virtually unlimited array of outdoor luminaires (fixtures) is available that provide uplighting, downlighting, diffused lighting, moonlighting, shadow and texture lighting, accent lighting, silhouette lighting, and bounce lighting. After the luminaires (fixtures) are selected, the type and color of the lamp is then selected. Some luminaires (fixtures) have specific light "cut-off" data that is useful in determining whether the emitting light will spill over onto the neighbor's property. The method of control must also be considered. Switch control, timer control, dusk-to-dawn control, and motion sensors are ways of turning outdoor lighting on and off.

In recent years, more and more complaints are coming from neighbors claiming that they are being bothered by glare, brightness, and light spillover from their neighbor's outdoor luminaires (fixtures). Security lighting, yard flood lights, driveway lighting, "moonlighting" in trees, and shrub lighting are examples of sources of light that might cross over the property line and be a "nuisance" to the next door neighbor.

Outdoor "lightscape" lighting considered by a homeowner to be aesthetically wonderful might be offensive to the neighbor. Nuisance lighting is also referred to as *light pollution*, *trespassing*, *intrusion*, *glare*, *spill-over*, and *brightness*. Some quiet residential neighborhoods are beginning to look like commercial areas, used car lots, and airport runways because no restrictions govern outdoor lighting methods. This is not a safety issue, and is not addressed in the *NEC.* However, the issue might be found in local building codes.

Many communities are being forced to legislate strict outdoor lighting laws, specifying various restrictions on the location, type, size, wattage, and/or footcandles for outdoor luminaires (fixtures). Checking building codes in your area might reveal requirements such as:

Light Source: The source of light (the lamp) must not be seen directly.

Glare: Glare, whether direct or reflected, such as from floodlights, and as differentiated from general illuminations, shall not be visible at any property line.

Exterior Lighting: Any lights used for exterior illuminations shall direct light away from adjoining properties.

The International Dark-Sky Association has a lot of information regarding light pollution issues on their Web site: www.darksky.org.

Replacing a Luminaire (Lighting Fixture) in an Older Home: Will There Be Problems?

Very possible!

Older homes were wired with conductors that were insulated with rubber (Type R) or thermoplastic (Type T). Both of these older style insulations were rated for maximum 60°C. The heat generated by the lamps over the years literally baked the conductor insulation to a hard, brittle material. If you aren't careful, pulling on or moving these conductors can easily break off the conductor insulation, which will result in a hazardous short-circuit or ground-fault situation.

Handle with care! Being very careful in handling the conductors while making up the splices and mounting the new luminaire (fixture)—possibly sliding some readily available tubing-like electrical insulation—or taping over the conductor insulation with plastic electrical tape might be all that is needed.

Worst case. You may need to do some rewiring using the latest in nonmetallic-sheathed cables Type NM-B, Type NMC-B, or Type NMS-B. If the conductors are in a raceway, replacing the conductors with Type THHN might be a good idea. All of these conductors have a maximum temperature rating of 90°C.

More on this subject is found in Unit 4 in the nonmetallic-sheathed cable section.

REVIEW

1. Is it permissible to install a recessed luminaire (fixture) directly against wood ceiling joists when the label on the luminaire (fixture) does not indicate that the luminaire (fixture) is suitable for insulation to be in direct contact with the luminaire (fixture)? This is a Type Non-IC fixture.

   _____

2. If a recessed luminaire (fixture) without an integral junction box is installed, what extra wiring must be provided? _____

   _____

   _____

   _____

   _____

   _____

   _____

   _____

3. Thermal insulation shall not be installed within _____ in. (mm) of the top or _____ in. (mm) of the side of a recessed luminaire (fixture) unless the luminaire (fixture) is identified for use in direct contact with thermal insulation. This is a Type Non-IC fixture.

4. Recessed luminaires (fixtures) are available for installation in direct contact with thermal insulation. These luminaires (fixtures) bear the UL mark "Type _____."

5. Unless specifically designed, all recessed incandescent luminaires (fixtures) must be provided with factory-installed _____ .

6. Plans require the installation of surface-mounted fluorescent luminaires (fixtures) on the ceiling of a recreation room that is finished with low-density ceiling fiberboard. What sort of mark would you look for on the label of the luminaire (fixure)? _____

   _____

   _____

7. A recessed luminaire (fixture) bears no marking indicating that it is "Identified for Through-Wiring." Is it permitted to run branch-circuit conductors *other* than the conductors that supply the luminaire (fixture) through the integral junction box on the luminaire (fixture)? _____

8. Fluorescent ballasts for all indoor applications must be _____ type. These ballasts contain internal _____ protection to protect against overheating.

9. Additional backup protection for ballasts can be provided by connecting a(an) _____ with the proper size fuse as recommended by the ballast manufacturer.

10. You are called upon to install a number of luminaires (fixtures) in a suspended ceiling. The ceiling will be dropped approximately 8 in. (200 mm) from the ceiling joists. Briefly explain how you might go about wiring these luminaires (fixtures). _____

_____

_____

_____

_____

_____

_____

_____

_____

11. The *Code* places a maximum open-circuit voltage on lighting equipment in or on homes. This maximum voltage is (600) (750) (1000). (Circle the correct answer.) Where in the *NEC*® is this voltage maximum referenced?

_____

_____

12. The letter **E** in a circle on a ballast nameplate indicates that the ballast _____

_____

_____

13. A 120-volt lamp fluorescent ballast for two 40-watt lamps is marked 85 volt-amperes. What is the power factor of the ballast? _____

_____

_____

14. Can an incandescent lamp dimmer be used to control a fluorescent lamp load? _____

15. A good "rule-of-thumb" to estimate the expected life of a motor, a ballast, or other electrical equipment is that for every _____°C above rated temperature, the expected life will be cut in _____ .

16. A post light has a 120-volt, 60-watt lamp installed. The lamp has an expected lamp life of 1000 hours. The homeowner installed a dimmer ahead of the post light, and leaves the dimmer set so the output voltage is 100 volts. The lamp burns slightly dimmer when operated at 100 volts, but this is not a problem. What is the expected lamp life when operated at 100 volts? _____

_____

_____

_____

17. Does your community have exterior outdoor lighting restrictions?
    If yes, what are they? _____

_____

_____

18. Define the following terms.

   a. A *tap conductor* is _____

   _____

   b. A *fixture whip* is _____

   _____

19. Circle the correct answer (True) or (False) for the following statements.

   a. Always match a fluorescent lamp's wattage and type designation to the type of ballast in the luminaire (fixture). (True) (False)

   b. It really makes no difference what type of fluorescent lamp is used, just so the wattage is the same as marked on the ballast. (True) (False)

20. Circle the correct answer (True) or (False) for the following statements.

   a. When selecting a trim for a recessed incandescent luminaire (fixture), select any trim that physically fits and can be attached to the luminaire (fixture). (True) (False)

   b. When selecting a trim for a recessed incandescent luminaire (fixture), select a trim that the manufacturer indicates may be used with that luminaire (fixture). (True) (False)

Lighting Branch-Circuit for the Front Bedroom

OBJECTIVES

After studying this unit, the student should be able to

▶ **understand the meaning of general, accent, task, and security lighting.**

▶ **estimate loads for the outlets connected to a branch-circuit.**

▶ **determine how many receptacles to connect to a branch-circuit.**

▶ **determine how many branch-circuits are needed.**

▶ **draw a cable layout and wiring diagram for a branch-circuit.**

▶ **properly size outlet boxes based on the number of conductors and devices.**

▶ **understand the *NEC*® requirements for luminaires (fixtures) in clothes closets.**

In the first seven units of this text, you studied the *National Electrical Code*® (*NEC*®) requirements and the basics of house wiring. You will now apply all that you have learned to an actual house—room by room—circuit by circuit. You will use the text, the plans, the specifications that are in the back of this book, and the *NEC*® to "bring it all together."

The home used in this text is a typical home—nothing prestigious or elegant. Other than the size of the service entrance, more branch-circuits, and more gadgets, the *NEC*® requirements are the same.

This might be a good time for you to refresh your memory about the arc-fault circuit interrupter (AFCI) requirements for bedrooms. All 120-volt, single-phase, 15- and 20-ampere outlets in bedrooms require AFCI protection, covered in Unit 6.

RESIDENTIAL LIGHTING

Residential lighting is a personal thing. The home-owner, builder, and electrical contractor must meet to decide on what types of luminaires (lighting fixtures) are to be installed in the residence. Many variables (cost, personal preference, construction obstacles, etc.) must be taken into consideration.

Residential lighting can be segmented into four groups.

General Lighting

General lighting provides overall illumination for a given area, such as the front hall, the bedroom hall, the workshop, or the garage lighting in this residence. General lighting can be very basic, which means just getting the job done by providing adequate lighting for the area involved. Or, it can take the form of decorative lighting, such as a chandelier over a dining room table or other decorative luminaires (fixtures) that can be attached to ceiling paddle fans.

Accent Lighting

Accent lighting provides focus and attention to an object or area in the home. Examples are the recessed "spots" over the fireplace. Another example is the track lighting in the living room of this residence, which can accent a picture, photo, painting, or sculpture that might be hung on the wall. To be effective, accent lighting should be at least five times that of the surrounding general lighting.

Task Lighting

This is sometimes referred to as "activity" lighting. Task lighting provides proper lighting where "tasks" or "activities" are performed. The fluorescent luminaires (fixtures) above the workbench, the kitchen range hood, and the recessed luminaire (fixture) over the kitchen sink are all examples of "task" lighting. To avoid too much contrast, task lighting should not exceed three times that of the surrounding general lighting.

Wall Washing. The term, *wall washing*, is used when luminaires (fixtures) are installed in a row, parallel to the wall that is to be "washed" with light. For example, if you wish to "wash" a 12-ft (3.7-m)

wall, as illustrated in Figure 8-1, and you would like the luminaires (fixtures) to be 2 ft (600 mm) from the wall:

$$\frac{12}{2} = 6 \text{ luminaires (fixtures) recommended}$$

Security Lighting

Security lighting generally includes outdoor lighting, such as post lamps, wall luminaires (fixtures), walkway lighting, and all other lighting that serves the purpose of providing lighting for security and safety reasons. Security lighting in many instances is provided by the normal types of luminaires (fixtures) found in the typical residence. In the residence discussed in this text, the outdoor bracket luminaires (fixtures) in front of the garage and next to the entry doors, as well as the post light, might be considered to be security lighting even though they add to the beauty of the residence. Security lighting can be controlled manually with regular wall switches, timers, light sensors, or motion detectors.

Figure 8-1 Illustration of how recessed luminaires (fixtures) can be installed in the ceiling close to a wall to provide beautiful "wall washing."

American Lighting Association

The American Lighting Association offers excellent residential lighting recommendations both online and hardcopy brochures. If you intend to become competent in residential lighting, a visit to their Web site at www.americanlightingassoc.com is a must!

This organization is comprised of many manufacturers of luminaires (fixtures) and electrical distributors that maintain extensive lighting showrooms across the country. A visit to one of these showrooms offers the prospective buyer an opportunity to select from thousands of luminaires (fixtures). Most of the better-quality lighting showrooms are staffed with certified lighting consultants (CLCs) and lighting specialists (LSs) who are highly qualified to make recommendations to homeowners so that they will get the most value for their money.

Find copies of lighting manufacturers' catalogs. Visit their Web sites. They offer many lighting technique suggestions.

The lighting provided throughout this residence conforms to the residential lighting recommendations of the American Lighting Association. Certainly many other variations are possible. For the purposes of studying an entire wiring installation for this residence, all of the load calculations, wiring diagrams, and so on, have been accomplished using the luminaire (fixture) selection as indicated on the plans and in the specifications.

LAYING OUT GENERAL-PURPOSE LIGHTING AND RECEPTACLE CIRCUITS

In most residential installations, circuit layout is usually done by the electrician, always keeping in mind the *NEC®* requirements. In larger, more costly residences, the circuit design might be specified by the architect and included on the plans. Many circuit arrangements are possible.

Here are a few thoughts:

- Supply general-purpose receptacle and lighting outlets with more than one circuit. Mix it up. If one circuit has a problem, the second circuit continues to supply power to the other outlets in that room, as shown in Figure 8-2.

- Keep general-purpose receptacles on one circuit and lighting outlets on another branch-circuit.

Figure 8-2 Wiring layout showing one room that is fed by two different circuits. If one of the circuits goes out, the other circuit will still provide electricity to the room.

- Economize on wiring materials by connecting receptacle outlets back to back as shown in Figure 8-3.

- Run the branch-circuit first to the switch then to the controlled lighting outlet. This keeps the outlet box at the lighting outlet or the wiring compartment on recessed cans less crowded.

Figure 8-3 Receptacle outlets connected back to back. This can reduce the cost of the installation because of the short distance between the outlets. Do not do this in fire-rated walls unless special installation procedures are followed as illustrated in Figure 2-29, Figure 2-30, and Figure 2-31.

Having the grounded circuit conductor present at the switch location makes it easy to install switching devices that require a connection to the grounded circuit conductor.

- Install a receptacle below the room wall switch—an area never blocked by furniture.

- Think about where furniture might be placed. Try to locate receptacles to the side of these so as to be easily accessible after the furniture is in place. Don't let the 6 ft (1.8 m) receptacle requirement hinder your thinking.

- Although not a *Code* issue, it is poor practice to include outlets on different floors on the same circuit. Some local building codes limit this type of installation to lights at the head and foot of a stairway.

Wiring bedrooms became a little more complicated when arc-fault circuit-interrupters (AFCIs) appeared on the scene, as required by *210.12*. The dilemma is putting all bedroom receptacles and lighting outlets on one AFCI-protected branch-circuit, or connecting the receptacle and lighting outlets on different branch-circuits, which means that more than one AFCI circuit breaker will be needed.

The number of receptacle and lighting outlets connected to one branch-circuit is discussed a little later.

Cable Runs

A student studying the great number of wiring diagrams throughout this text will find many different ways to run the cables and make up the circuit connections. Sizing of boxes that will conform to the *Code* for the number of wires and devices is covered in great detail in this text. When a circuit is run "through" a recessed luminaire (fixture) like the type shown in Figure 7-8, that recessed luminaire (fixture) must be identified for "through-wiring." This subject is discussed in detail in Unit 7. When in doubt as to whether the luminaire (fixture) is suitable for running wires to and beyond it to another luminaire (fixture) or part of the circuit, it is recommended that the circuitry be designed so as to end up at the recessed luminaire (fixture) with only the two wires that will connect to that luminaire (fixture).

ESTIMATING LOADS FOR OUTLETS

Although this text is not intended to be a basic electrical theory text, it should be mentioned that, when calculating loads,

$$\text{Volts} \times \text{Amperes} = \text{Volt-amperes}$$

Yet, many times we say that

$$\text{Volts} \times \text{Amperes} = \text{Watts}$$

What we really mean to say is that

$$\text{Volts} \times \text{Amperes} \times \text{Power factor} = \text{Watts}$$

In a pure resistive load such as a simple light bulb, a toaster, an iron, or a resistance electric heating element, the power factor is 100 percent. Then

$$\text{Volts} \times \text{Amperes} \times 1 = \text{Watts}$$

With transformers, motors, ballasts, and other "inductive" loads, wattage is not necessarily the same as volt-amperes.

EXAMPLE: Calculate the wattage and volt-amperes of a 120-volt, 10-ampere resistive load.

SOLUTIONS:
    a. $120 \times 10 \times 1 = 1200$ watts
    b. $120 \times 10 \quad\ \ = 1200$ volt-amperes

EXAMPLE: A 120-volt fluorescent ballast has an input current of 0.34 ampere and an input power rating of 22 watts. Calculate the power factor of the ballast.

SOLUTION:

$$\text{Power Facter} = \frac{\text{Watts}}{\text{Volt-amperes}}$$

$$= \frac{22}{120 \times 0.34} = 0.54 \ (54\% \ PF)$$

Therefore, to be sure that adequate ampacity is provided for in branch-circuit wiring, feeder sizing, and service-entrance calculations, the *Code* requires that we use the term *volt-amperes*. This allows us to ignore power factor and address the *true* current draw that will enable us to determine the correct ratings of electrical equipment.

However, in some instances the terms *watts* and *volt-amperes* can be used interchangeably without creating any problems. For instance, *220.55* recognizes that for electric ranges and other cooking equipment, the kVA and kW ratings shall be considered to be equivalent for the purpose of branch-circuit and feeder calculations.

The load calculation examples in *Annex D* of the *NEC®* show kW values in the list of connected loads, but the actual calculations are done using volt-ampere values. Throughout this text, *wattage* and *volt-amperes* are used when calculating and/or estimating loads.

For the residence in the plans, it is shown that six lighting circuits meet the minimum standards set by the *Code*. However, to provide sufficient capacity, thirteen lighting circuits are to be installed in this residence.

How Many Outlets Are Permitted on One General-Purpose Lighting Branch-Circuit?

This is a tough question, but it can be answered reasonably well using a few simple *Rules of Thumb* discussed below.

The *NEC®* does not specify the maximum number of receptacle outlets or lighting outlets that may be connected to one 120-volt lighting or small-appliance branch-circuit in a residence. It may not seem logical that ten or twenty receptacle outlets and lighting outlets could be connected to one branch-circuit and not be in violation of the *Code*. On our side is the fact that there is much load diversity in residential occupancies. We have no idea what will be plugged into the receptacles. We do know that rarely, if ever, would all receptacle outlets and lighting outlets be used at the same time. Having many "convenience" receptacles is safer because more receptacle outlets virtually eliminate or minimize the use of extension cords, one of the leading causes of electrical fires.

Recessed and surface-mounted incandescent luminaires (fixtures) are marked with their maximum lamp wattage. Fluorescent luminaires (fixtures) are marked with lamp wattage, and ballasts are marked in volt-amperes. In new construction, other than for recessed luminaires (fixtures), we probably do not know what type of surface or hanging luminaires (fixtures) will be selected. Most luminaire (fixture) manufacturers' catalogs provide excellent recommendations for lamp wattages for residential applications.

There are many items such as nightlights (1 to 7½ watts), smoke detectors (5 watts), clocks (5 watts), carbon monoxide detector (60 milliamperes), and closet lights (60 to 75 watts) that use only a small portion of the branch-circuit's capability. It comes down to approximating, estimating, experience, and common sense! It is not an exact science.

Circuit Loading Rules of Thumb

Let's look at a few favorite *Rules of Thumb* that electricians use for determining how many outlets they connect to one general-purpose branch-circuit.

Outlets per Circuit Method. This is the most common method used by electricians across the country. It is very simple. No calculations are needed. Just count the outlets! Connect ten outlets on a circuit where you anticipate fairly heavy wattage lighting loads. Connect fifteen outlets per circuit for circuits that have a lot of low-wattage lighting loads. Somewhere in between might be the right number. This method is really estimating a load of 1 to 1½ amperes per outlet.

An example of this might be a bedroom with six or seven receptacles, one or two closet lights, and possibly a ceiling paddle fan with a light kit. This would probably be one 15-ampere lighting branch-circuit with not much leeway for additional outlets.

Common sense tells us there can be more outlets when the circuit consists of low-wattage loads and fewer outlets when the circuit consists of higher wattage loads. At best, it's a "guesstimate."

How Many Circuits Do You Need Using the Outlets per Circuit Method? If we were to wire a new house that has a count of 80 general-purpose lighting and receptacle outlets, how many 15-ampere lighting branch-circuits do we need?

Where you anticipate somewhat high-wattage loads, figure on connecting ten outlets per circuit.

$$\frac{80}{10} = 8 \quad \text{(a minimum of eight 15-ampere general-purpose lighting circuits)}$$

Where you anticipate low-wattage loads, figure on connecting fifteen outlets per circuit.

$$\frac{80}{15} = 5.3 \quad \text{(a minimum of six 15-ampere general-purpose lighting circuits)}$$

Certainly more circuits than the results of the preceding calculations would be better.

How Many Circuits Do You Need Using per Square Foot Method?

This method is more complicated because it involves calculations. In Unit 3, we discussed how typical 15-ampere lighting branch-circuits in homes could be figured at one 15-ampere branch-circuit for every 600 ft² (55.8 m²). Although not often used in residential wiring, if you choose to install 20-ampere lighting branch-circuits, figure one 20-ampere branch-circuit for every 800 ft² (74.4 m²). This results in a good approximation for the minimum number of lighting branch-circuits needed and is based on 3 volt-amperes per ft² (33 volt-amperes per m² of floor area).

A 2400 ft² house would have a minimum of:

$$\frac{2400}{600} = \text{four 15-ampere general-purpose lighting branch-circuits}$$

With so many electronic and electrical appliances in homes today, this method might prove to be inadequate. For this size house, you would probably want to install more than four 15-ampere general-purpose lighting branch-circuits. It is better to be safe than sorry.

The 80% Rule.

This method is also complicated because it involves calculations. Add up the wattages and/or the amperes of the loads you anticipate will be connected to a given branch-circuit. Then limit the load to not more than 80 percent of the branch-circuit's rating. Although this is a *Code* requirement for continuous loads for commercial and industrial applications as found in *NEC® 210.19(A), 210.20(A),* and *215.2(A), 215.3, 230.42(A)*, it makes good sense to follow this guideline for most residential loads.

A 15-ampere, 120-volt lighting branch-circuit would be calculated as:

$15 \times 0.80 = 12$ amperes maximum connected load

or

12 amperes $\times$ 120 volts = 1440 watts (volt-amperes) maximum connected load

A 20-ampere, 120-volt lighting branch-circuit would be calculated as:

$20 \times 0.80 = 16$ amperes (volt-amperes) maximum connected load

or

16 amperes $\times$ 120 volts = 1920 watts (volt-amperes) maximum connected load

Wrap Up.

In this residence, estimated loads for general-purpose outlets are figured at 120 volt-amperes (1 ampere). Receptacles in the garage and workshop are figured at 180 volt-amperes (1½ amperes) because these outlets will probably supply portable tools.

Obviously the ten to fifteen outlets (1 to 1½ amperes per outlet) would not apply to the required small-appliance circuits in the kitchen, the laundry area, workshop, or similar areas that supply a considerable number of plug-in tools and appliances. These circuits are separate issues and are discussed later on in this text.

Divide Loads Evenly

It is mandatory that loads be divided evenly among the various circuits, *210.11(B)*. The obvious reason is to avoid overload conditions on some circuits when other circuits might be lightly loaded. This is common sense. At the Main Service, don't connect the branch-circuits and feeders to result in, for instance, 120 amperes on phase A and 40 amperes on phase B. Look at the probable and/or calculated loads to attain as close a balance as possible, such as 80 amperes on each A and B phase.

SYMBOLS

The symbols used on the cable layouts in this text are the same as those found on the actual electrical plans for the residence. Refer to Figure 2-5, Figure 2-6, Figure 2-7, Figure 2-8, and Figure 2-9.

Pictorial illustrations are used on all wiring diagrams in this text to make it easy for the reader to complete the wiring diagrams, as in Figure 8-4.

DRAWING A CABLE LAYOUT AND WIRING DIAGRAM

Typical house plans show the electrical symbols on the construction floor plans. Rarely are separate electrical plans provided such as those included in the back of this text. Most residential plans do not provide much detail other than showing the location of receptacles, lighting, and appliances.

For the most part, the electrician gets on the job—studies the plans then figures out how to make

Figure 8-4   Pictorial illustrations used on wiring diagrams in this text.

it work, according to *Code*, and how to do it in a cost-effective way. The electrician decides how to lay out the branch-circuits, the number of outlets on a circuit, how to run the cables, and from where the branch-circuit home runs will be taken.

Skilled electricians, after many years of experience, will generally not prepare detailed wiring diagrams because they have ability to "think" the connections through mentally. They can visualize the size and type of boxes and cables needed.

Most electricians sketch a *Cable Layout*, bearing in mind that he is limited to working with two-wire and three-wire cables.

The following steps will guide you through the simple steps for preparing a cable layout and wiring

diagram. In later units, you will be asked to draw cable layouts and wiring diagrams. This is an exercise in how to make a color-coded cable layout and wiring diagram. Certainly there are other ways to lay out the cable runs for this particular "typical" room.

DRAWING THE WIRING DIAGRAM OF A LIGHTING CIRCUIT

1. Refer to plans and make a cable layout of all lighting and receptacle outlets, as in Figure 8-5.

2. Draw a wiring diagram showing the traveler conductors for all three-way switches, if any, as in Figure 8-6.

Figure 8-5   Typical cable layout.

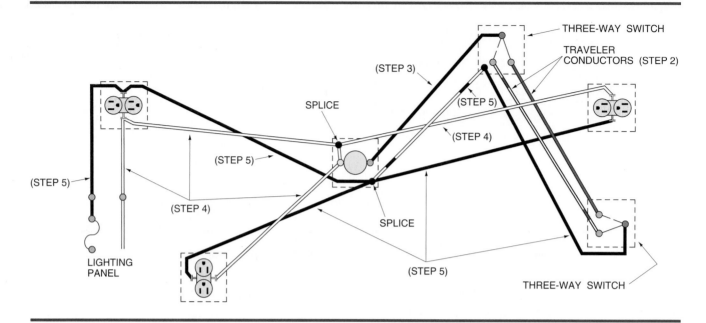

Figure 8-6   Wiring diagram of circuit shown in Figure 8-5.

3. Draw a line between each switch and the outlet or outlets it controls. For three-way switches, do this for one switch only. This is the "switch leg."

4. Draw a line from the grounded terminal in the lighting panel to each current-consuming outlet. This line may pass through switch boxes, but must not be connected to any switches. This is the white grounded circuit conductor, often called the "neutral" conductor.

   Note: An exception to step 4 may be made for double-pole switches. For these switches, all conductors of the circuit are opened simultaneously. They are rarely used in residential wiring.

5. Draw a line from the ungrounded "hot" terminal in the lighting panel to each switch and to each unswitched outlet. For three-way switches, do for one switch only. This is the "feed." The "switch leg" (switch loop, switch return) is covered in step 3.

6. Show splices as small dots where the various wires are to be connected together. In the wiring diagram, the terminal of a switch or outlet may be used for the junction point of wires. In actual wiring practice, however, the *Code* does not permit more than one wire to be connected to a terminal unless the terminal is

a type identified for use with more than one conductor. The standard screw-type terminal is not acceptable for more than one wire, *110.14*.

7. The final step in preparing the wiring diagram is to mark the color of the conductors, as in Figure 8-6. Note that the colors selected—black (B), white (W), and red (R)—are the colors of two- and three-conductor cables (refer to Unit 5). It is suggested that the reader use colored pencils or markers for different conductors when drawing the wiring diagram.

   Use a fine line for the white conductor.

   General comment: All cable layouts in this text are but one way of laying out the circuits. Certainly other layouts are possible. Some circuits were designed to give the student practice in hooking up similar circuits differently.

LIGHTING BRANCH-CIRCUIT A16 FOR THE FRONT BEDROOM

Circuit A16 is a 15-ampere branch-circuit. Overcurrent protection for this circuit is provided by a 15-ampere AFCI circuit breaker in Panel A. AFCIs are discussed in detail in Unit 6. As we look at this circuit, we find five split-circuit receptacle outlets.

These will provide the general lighting through the use of table or other lamps that will be plugged into the switched receptacles, as shown in Figure 8-7.

Televisions, radios, clocks, computers, compact disc players and burners, DVD players, scanners, printers, telephones, and other electrical and electronic items not intended to be controlled by a wall switch will be plugged into the "hot continuously" receptacle.

Next to the wall switch we find the control for the ceiling fan, which also has a luminaire (lighting fixture) as an integral part of the fan, as shown in Figure 8-8. A single-pole switch controls the "ON/OFF" and speed of the fan motor.

The closet light is controlled by a single-pole switch outside and to the right of the closet door.

Note that the outside weatherproof receptacle connected to this circuit is required to have GFCI protection, *210.8(A)(3)*. For this particular installation, there are two ways to accomplish GFCI protection for this receptacle:

1. Install a GFCI receptacle.

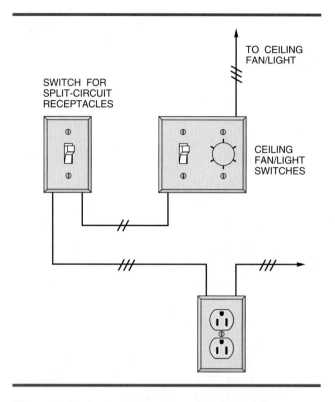

Figure 8-8 Conceptual view of how the front bedroom switch arrangement is to be accomplished.

Figure 8-7 Cable layout for front bedroom circuit, A16.

DESCRIPTION	QUANTITY	WATTS	VOLT-AMPERES
Receptacles @ 120 watts each	5	600	600
Weatherproof receptacle	1	120	120
Closet fixture	1	75	75
Ceiling fan/light	1		
3 - 50-W lamps		150	150
Fan motor (0.75 A @ 120 V)		80	90
TOTALS	8	1025	1035

Table 8-1 Front bedroom: outlet count and estimated load (Circuit A16).

2. Install a combination GFCI/AFCI circuit breaker in the main panel. Be careful when selecting the circuit breaker to be sure the nameplate indicates that it will provide both GFCI and AFCI protection.

If you cannot locate a combination GFCI/AFCI circuit breaker, then install a GFCI outdoor receptacle and an AFCI circuit breaker in the panel.

GFCI and AFCI protection are discussed in Unit 6.

Checking the actual electrical plans, we find one television outlet and one telephone outlet are to be installed in the front bedroom. Television and telephones are discussed in Unit 25. Table 8-1 summarizes the outlets in the front bedroom and the estimated load.

DETERMINING THE SIZE OF OUTLET BOXES, DEVICE BOXES, JUNCTION BOXES, AND CONDUIT BODIES

Box fill was discussed in detail in Unit 2. Here is more on the subject. One of the most common problems found in residential wiring are wires jammed into electrical boxes. Factors that need to be considered are the size and number of conductors, the number of wiring devices, cable clamps, luminaire (fixture) studs, hickeys, and splices. NEC® 314.16 provides the rules for sizing boxes.

Experience has shown that *minimum* does not always make for easy working with the splices, conductors, wiring devices, and carefully putting them all into the box without crowding. Do not skimp on

box sizing. The small additional cost of larger boxes is worth much more than the grief and added time it will take to trim out the box using *minimum* size boxes. You will be much happier with boxes that have more than the *minimum* size. When it comes to box sizing, think big—not small.

The following example shows how to calculate the *minimum* size box for a particular installation, as in Figure 8-9.

1. Add the 14 AWG circuit conductors:
   2 + 3 + 3 =   8
2. Add equipment grounding wires (count one only)   1
3. Add two conductors for the receptacle   2

                    Total   11 Conductors

Note that the pigtails connected to the receptacle in Figure 8-9 need not be counted when determining the correct box size. *NEC® 314.16(B)(1)* states, "*A conductor, no part of which leaves the box, shall not be counted.*"

THE "PIGTAILS" CONNECTED TO THE RECEPTACLE NEED NOT BE COUNTED WHEN DETERMINING THE CORRECT BOX SIZE. THE *CODE* IN *314.16(B)(1)* STATES, "*A CONDUCTOR, NO PART OF WHICH LEAVES THE BOX, SHALL NOT BE COUNTED.*"

Figure 8-9 Determining size of box according to number of conductors.

Once the total number of conductors, plus the volume count required for wiring devices, luminaire (fixture) studs, hickeys, and clamps is known, refer to *Table 314.16(A)* and *(B)* and Figure 2-20 to select a box that has sufficient space. Also refer to Table 2-1 for a summary of box fill requirements. For example, a $4 \times 2\frac{1}{8}$ in. square box with a suitable plaster ring can be used.

The volume of the box plus the space provided by plaster rings, extension rings, and raised covers may be used to determine the total available volume. In addition, it is desirable to install boxes with external cable clamps. Remember that if the box contains one or more devices, such as cable clamps, fixture studs, or hickeys, the number of conductors permitted in the box shall be one less than shown in *Table 314.16(A)* for *each type* of device contained in the box.

GROUNDING OF WALL BOXES

The specifications for the residence and the *NEC*® require that all metal boxes be grounded. Equipment grounding using nonmetallic-sheathed cable is accomplished by properly connecting the bare equipment grounding conductor in the cable to the metal box. An equipment grounding conductor is used *only* to ground the metal box and must *never* be used as a current-carrying conductor.

When using nonmetallic boxes, the equipment grounding conductor in nonmetallic-sheathed cable is connected to the green hexagon-shaped equipment grounding screw found on all switches and receptacles.

Improper connecting of the bare equipment grounding conductor has resulted in electrocutions!

Methods of connecting equipment grounding conductors are shown in many of the illustrations in Units 4 and 5, and in Figure 8-9.

Grounding-type receptacles must be installed on 15-ampere and 20-ampere branch-circuits, *406.3 (A)*.

POSITIONING OF SPLIT-CIRCUIT RECEPTACLES

The receptacle outlets shown in the front bedroom are called two-circuit or split-circuit receptacles. The top portion of such a receptacle is "hot" at all times, and the bottom portion is controlled by the wall switch, as in Figure 8-10. It is recommended

Figure 8-10   Split-circuit wiring for receptacles. The top receptacle is "hot" at all times. The bottom receptacle is controlled by the switch.

that the electrician wire the bottom section of the receptacle as the switched section. As a result, when the attachment plug cap of a lamp is inserted into the bottom switched portion of the receptacle, the cord does not hang in front of the unswitched section.

When split-circuit receptacles are horizontally mounted, which is common when 4-in. square boxes are used because the plaster ring is easily fastened to the box in either a vertical or horizontal position, locate the switched portion to the right.

POSITIONING OF RECEPTACLES NEAR ELECTRIC BASEBOARD HEATING

Electric heating for a home can be accomplished with an electric furnace, electric baseboard heating units, heat pump, or resistance heating cables embedded in plastered ceilings or sandwiched between two layers of drywall sheets on the ceiling.

The important thing to remember at this point is that some baseboard electric heaters may not be permitted to be installed below a receptacle outlet, *210.52 FPN*.

According to *210.52*, receptacles that are part of an electric baseboard heating unit may be counted as one of the required number of receptacles in a given space, but only if the receptacles are not connected to the electric baseboard heating branch-circuit.

See Unit 23 for a detailed discussion of installing electric baseboard heating units and their relative positions below receptacle outlets.

The front bedroom does have a ceiling fan/light on the ceiling. Ceiling fans are discussed in Unit 9.

LUMINAIRES (FIXTURES) IN CLOTHES CLOSETS

The *NEC*® does not require luminaires (fixtures) in clothes closets, *210.70*. However, some local codes and some building codes do. When luminaires (fixtures) are installed in clothes closets, they must be of the correct type and must be installed properly.

Clothing, boxes, and other material normally stored in clothes closets are a potential fire hazard. These items may ignite on contact with the hot surface of exposed lamps. Incandescent lamps have a hotter surface temperature than fluorescent lamps. In *410.8*, there are very specific rules for the location and types of luminaires (lighting fixtures) permitted to be installed in clothes closets. Figure 8-11, Figure 8-12, Figure 8-13, Figure 8-14, and Figure 8-15 illustrate the requirements of *410.8*.

In this residence, the closets in the bedrooms and front hall are 24 in. (600 mm) deep. The storage closet in the recreation room is 30 in. (750 mm) deep. The shelving is approximately 12 in. (300 mm) wide. Because of the required clearances between

Figure 8-11   Typical closet with one shelf and one rod. The shaded area defines "storage space." Dimension A is width of shelf or 12 in. (300 mm) from wall, whichever is greater. Dimension B is below rod, 24 in. (600 mm) from wall. See *410.8(A)*.

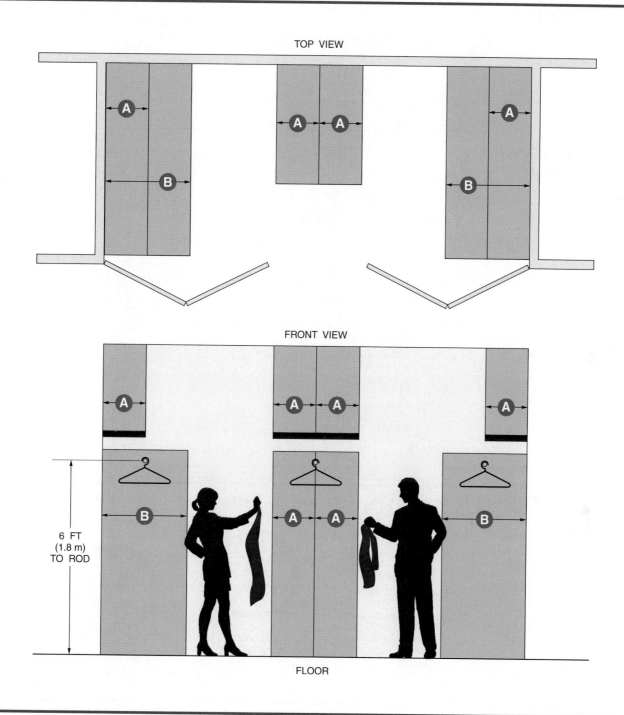

TOP VIEW

FRONT VIEW

6 FT
(1.8 m)
TO ROD

FLOOR

Figure 8-12   Large walk-in closet where there is access to the center rod from both sides. The shaded area defines "storage space." Dimension A is width of shelf or 12 in. (300 mm), whichever is greater. Dimension B is 24 in. (600 mm) from wall. See *410.8(A)*.

luminaires (fixtures) and the storage space, locating luminaires (fixtures) in closets can be difficult. In fact, some clothes closets are so small that luminaires (fixtures) of any kind are not permitted.

Figure 8-14 shows the possibilities: recessed incandescents, recessed fluorescents, surface-mounted incandescents, and surface-mounted fluorescents. These can be on the ceiling, or on the wall space above the closet

LUMINAIRES (FIXTURES) IN CLOTHES CLOSETS

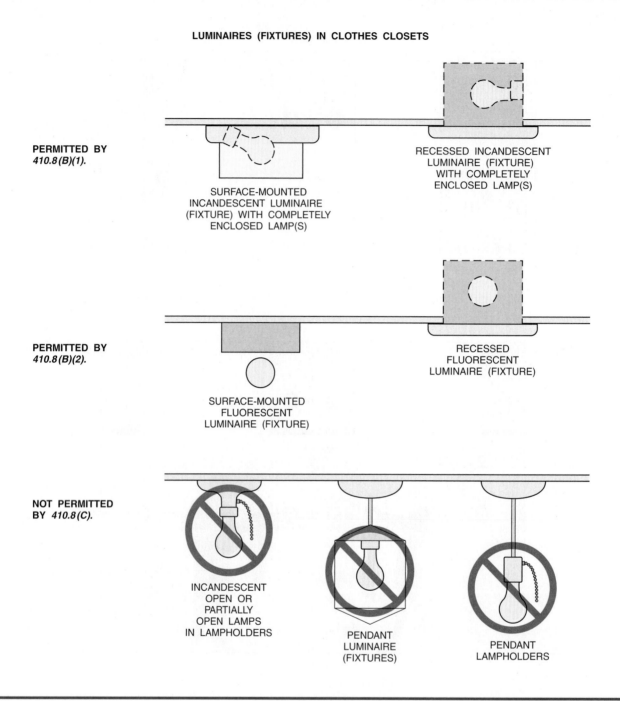

Figure 8-13 Illustrations of the types of luminaires (lighting fixtures) permitted in clothes closets. The *Code* does not permit bare incandescent lamps, lampholders, pendant luminaires (lighting fixtures), or pendant lampholders to be installed in clothes closets.

door. Whichever type is chosen, be sure that the minimum clearances are provided as required by *410.8*.

Figure 8-16 shows typical bedroom-type luminaires (fixtures).

Figure 8-17 shows unique luminaires (fixtures)

referred to as *sconces*. They hug the wall and provide interesting uplighting. They are generally used to complement other lighting. They can be located just about anywhere—hallways, next to a fireplace, foyers, and similar places.

LOCATION OF LUMINAIRES (FIXTURES) IN CLOTHES CLOSETS

MINIMUM CLEARANCES FOR SURFACE-MOUNTED INCANDESCENT LUMINAIRES (FIXTURES), *410.8(D)(1).*

MINIMUM CLEARANCES FOR SURFACE-MOUNTED FLUORESCENT LUMINAIRES (FIXTURES), *410.8(D)(2).*

STORAGE SPACE

MINIMUM CLEARANCES FOR RECESSED COMPLETELY ENCLOSED INCANDESCENT LUMINAIRES (FIXTURES), *410.8(D)(3),* AND RECESSED FLUORESCENT LUMINAIRES (FIXTURES), *410.8(D)(4).*

NOTE:
6 IN. = 150 mm
12 IN. = 300 mm

STORAGE SPACE

Figure 8-14 The above illustrations show the minimum clearances required between luminaires (lighting fixtures) and the storage space in clothes closets. See Figure 8-11 and Figure 8-12 for a definition of storage space.

Figure 8-15 Recessed incandescent closet luminaire (lighting fixture) with pull-chain switch that may be used where a separate wall switch is not installed. *Courtesy of* Progress Lighting.

Figure 8-16 Typical bedroom-type luminaires (lighting fixtures). *Courtesy of* Progress Lighting.

Figure 8-17 Sconces can provide attractive uplighting on a wall. Install approximately 66 in. (1.68 m) to the center from the floor. *Courtesy of* Progress Lighting.

REVIEW

Note: Refer to the *Code* or the plans where necessary.

1. Can the outlets in a circuit be arranged in different groupings to obtain the same result? Why? _____

_____

_____

_____

_____

2. Is it good practice to have outlets on different floors on the same circuit? Why?

   _____

   _____

   _____

   _____

   _____

3. A good rule to follow is to never load a circuit to more than _____ percent of the branch-circuit rating.

4. The *NEC*® does not limit the number of lighting and receptacle outlets permitted on one branch-circuit for residential installations. A guideline that is often used by electricians for house wiring is to allow _____ to _____ amperes per outlet. For a 15-ampere circuit, this results in _____ to _____ outlets on the branch-circuit.

5. For residential wiring, not less than one 15-ampere lighting branch-circuit should be provided for every_____ ft² of floor area. If 20-ampere lighting branch-circuits are used, provide one 20-ampere lighting branch-circuit for every _____ ft² of floor area.

6. For this residence, what are the estimated wattages used in determining the loading of branch-circuit A16?

   Receptacles _____ watts (volt-amperes)
   Closet fixture _____ watts (volt-amperes)

7. a. What is the ampere rating of circuit A16? _____

   b. What is the conductor size of circuit A16? _____

8. The *NEC*® requires what type of unique type of protection for all outlets in residential bedrooms? In which section of the *NEC*® is this requirement found? _____

   _____

   _____

9. How many and what type of receptacles are connected to this circuit? _____

   _____

   _____

   _____

10. What main factors influence the choice of wall boxes?_____

   _____

   _____

   _____

11. How is a wall box grounded? _____

   _____

   _____

   _____

12. What is a split-circuit receptacle? _____

_____

_____

_____

_____

13. Is the switched portion of an outlet mounted toward the top or the bottom? Why?

_____

_____

_____

_____

14. The following questions pertain to luminaires (lighting fixtures) in clothes closets.
    a. Does the *Code* permit bare incandescent lamps to be installed in porcelain keyless or porcelain pull-chain lampholders? _____

    b. Does the *Code* permit bare fluorescent lamp luminaires (fixtures) to be installed?

    _____

    c. Does the *Code* permit pendant lampholders to be installed?_____

    _____

    d. What is the minimum clearance from the storage area to surface-mounted incandescent luminaires (fixtures)? _____

    e. What is the minimum clearance from the storage area to surface-mounted fluorescent luminaires (fixtures)? _____

    f. What is the minimum distance between recessed incandescent or recessed fluorescent luminaires (fixtures) and the storage area? _____

    g. Define the "storage area."

    _____

    _____

    _____

    _____

    _____

    h. If a clothes hanging rod is installed where there is access from both sides, such as might be found in a large walk-in closet, define the storage area under that rod.

    _____

    _____

    _____

    _____

15. How many switches are in the bedroom circuit, what type are they, and what do they control? _____

_____

16. The following is a layout of the lighting circuit for the front bedroom. Using the cable layout shown in Figure 8-8, make a complete wiring diagram of this circuit. Use colored pencils or pens.

17. When planning circuits, what common practice is followed regarding the division of loads? _____
_____

18. The *Code* uses the terms *watts*, *volt-amperes*, *kW*, and *kVA*. Explain their significance in calculating loads. _____
_____
_____
_____
_____

19. How many 14 AWG conductors are permitted in a device box that measures 3 in. × 2 in. × 2¾ in.? _____

20. A 4 in. × 1½ in. octagon box has one cable clamp and one luminaire (fixture) stud. How many 14 AWG conductors are permitted? _____
_____
_____
_____
_____

Lighting Branch-Circuit for the Master Bedroom

OBJECTIVES

After studying this unit, the student should be able to

▶ draw the wiring diagram of the cable layout for the master bedroom.

▶ understand where AFCIs are required for bedrooms.

▶ study *Code* requirements for the installation of ceiling suspended (paddle) fans.

▶ estimate the probable connected load for a room based on the number of luminaires (fixtures) and outlets included in the circuit supplying the room.

▶ gain more practice in determining box sizing based upon the number of conductors, devices, and clamps in the box.

The discussion in Unit 3 and Unit 8 about grouping outlets, estimating loads, selecting wall box sizes, and drawing wiring diagrams can also be applied to the circuit for the master bedroom.

The residence panel schedules show that the master bedroom is supplied by Circuit A19. Circuit A19 is a 15-ampere branch-circuit. Overcurrent protection for this circuit is provided by a 15-ampere AFCI circuit breaker in Panel A. AFCI protection is required for all 120-volt outlets in bedrooms, *210.12*. In Unit 8, we discussed how GFCI/AFCI protection could be attained in the Front Bedroom. The same principle applies to the Master Bedroom. AFCIs and GFCIs are discussed in detail in Unit 6. Because Panel A is located in the basement below the wall switches next to the sliding doors, the home run for Circuit A19 is brought into the outdoor weatherproof receptacle. This results in six conductors in the outdoor receptacle box. The home run could have been brought into the corner receptacle outlet in the bedroom. Again, it is a matter of studying the circuit

to determine the best choice for conservation of cable or conduit in these runs and to economically select the correct size of wall boxes.

LIGHTING BRANCH-CIRCUIT A19 FOR THE MASTER BEDROOM

Figure 9-1, Table 9-1, and the electrical plans show that Circuit A19 has four split-circuit receptacle outlets in this bedroom, one outdoor weatherproof GFCI receptacle outlet, two closet luminaires (fixtures), each on a separate switch, plus a ceiling fan/luminaire (fixture), one telephone outlet, and one television outlet.

In addition, an outdoor bracket luminaire (fixture) is located adjacent to the sliding door and is controlled by a single-pole switch just inside the sliding door.

The split-circuit receptacle outlets are controlled by two three-way switches. One is located next to the sliding door. As in the front bedroom, living room, and study/bedroom, the use of split-circuit

Figure 9-1   Cable layout for master bedroom.

DESCRIPTION	QUANTITY	WATTS	VOLT-AMPERES
Receptacles @ 120 watts each	4	480	480
Weatherproof receptacle	1	120	120
Outdoor bracket luminaire (fixture)	1	150	150
Closet luminaires (fixtures) One 75-W lamp each	2	150	150
Ceiling fan/light	1		
Three 50-W lamps		150	150
Fan motor (0.75 @ 120 V)		80	90
TOTALS	9	1130	1140

Table 9-1   Master bedroom outlet count and estimated load (Circuit A19).

receptacles offers the advantage of having switch control of one of the receptacles at a given outlet, while the other receptacle remains "live" at all times. See Figure 3-4 and Figure 12-14 for definition of a receptacle.

Next to the bedroom door are a three-way switch and the ceiling fan/light control, which is installed in a separate two-gang box, as shown in Figure 9-2.

SLIDING GLASS DOORS AND FIXED GLASS PANELS

Receptacle outlets must be installed so that no point along the floor line in any wall space is more than 6 ft (1.8 m), measured horizontally, from an outlet in that space, including any wall space 24 in. (600 mm) or more in width. Wall space occupied by fixed glass panels in exterior walls are considered to be wall space. Sliding glass panels are not considered to be wall space, the same as any other interior or exterior door. This is covered in *210.52(A)(2)*.

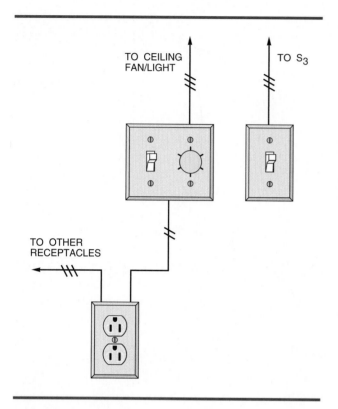

Figure 9-2 Conceptual view of how the switching arrangement is to be accomplished in the master bedroom Circuit A19.

Because fixed glass panels are considered to be wall space, in the master bedroom a receptacle must be located so as to be not more than 6 ft (1.8 m) from the left-hand edge of the fixed glass panel.

SELECTION OF BOXES

As discussed in Unit 2, the selection of outlet boxes and switch boxes is made by the electrician. These decisions are based on *Code* requirements, space allowances, good wiring practices, and common sense.

For example, in the master bedroom, the electrician may decide to install a 4-in. square box with a two-gang raised plaster cover at the box location next to the sliding door. Or two sectional switch boxes ganged together may be installed at that location, as shown in Figure 9-3 and Table 9-2.

The type of box to be installed depends upon the number of conductors entering the box. The suggested cable layout, Figure 9-1, shows that four cables enter this box for a total of fourteen conductors (ten 14 AWG circuit conductors and four equipment grounding conductors).

The requirements for calculating box fill are found in *314.16*. These requirements are summarized in Figure 2-36. Also refer to *Table 314.16(A), 314.16(B)*, and Figure 2-20.

In this example, we have ten circuit conductors, four equipment grounding conductors (counted as

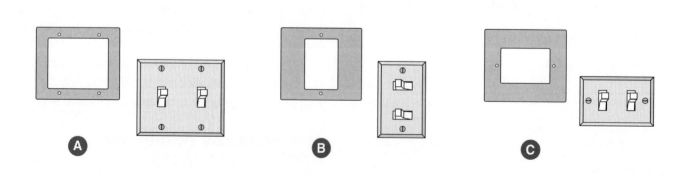

Figure 9-3 (A) shows two switches and wall plate that attach to a two-gang, 4 in.-square raised plaster cover or to two device boxes that have been ganged together. (B) and (C) show two interchangeable-type switches and wall plate that attach to a one-gang, 4 in.-square raised plaster cover or to a single device (switch) box. These types of wiring devices may be mounted vertically or horizontally.

NO. OF GANGS	HEIGHT	WIDTH
1	4½ in. (114.3 mm)	2¾ in. (69.85 mm)
2	4½ in. (114.3 mm)	4⁹⁄₁₆ in. (115.9 mm)
3	4½ in. (114.3 mm)	6⅜ in. (161.9 mm)
4	4½ in. (114.3 mm)	8³⁄₁₆ in. (207.9 mm)
5	4½ in. (114.3 mm)	10 in. (254 mm)
6	4½ in. (114.3 mm)	11¹³⁄₁₆ in. (300 mm)

Table 9-2   Height and width of standard wall plates.

Figure 9-4   Typical home-type ceiling fan. *Courtesy of Broan-NuTone, LLC.*

one conductor), four cable clamps (counted as one conductor), one single-pole switch (counted as two conductors), and one three-way switch (counted as two conductors), for a total of 16 conductors.

1. Count the circuit conductors
   2 + 2 + 3 + 3 =   10
2. Add one for one or more
   equipment grounding conductors   1
3. Add two for each switch (2 + 2)   4
4. Add one for one or more cable clamps 1
                              Total   16

Now look at the Quik-Chek Box Selection Guide (Figure 2-20), and select a box or a combination of gangable device boxes that are permitted to hold 16 conductors.

Possibilities are:

1. Gang two 3 in. × 2 in. × 3½ in. device boxes together (9 × 2 = eighteen 14 AWG conductors allowed).
2. Install a 4 in. × 4 in. × 2⅛ in. square box (fifteen 14 AWG conductors allowed). Note that the raised plaster cover, if marked with its cubic-inch volume, can increase the maximum number of conductors permitted for the combined box and raised cover. See Figure 2-37.

CEILING-SUSPENDED (PADDLE) FANS

For appearance and added comfort, ceiling-suspended (paddle) fans are very popular. Figure 9-4 is a typical ceiling-suspended (paddle) fan. Ceiling-suspended (paddle) fans rotate slowly in the range of approximately 60 r/min to 250 r/min for residential fans. Without forced air movement, warm air rises to the ceiling and cool air drops to the floor. Air movement results in increased evaporation from our skin, and we feel cooler. Fans that hug the ceiling are not quite as efficient in moving air as those that are suspended a few inches or more from the ceiling.

Most ceiling-suspended (paddle) fans have a reversing switch, a "HIGH/MEDIUM/LOW/OFF" speed control pull chain switch for the fan, and an "ON/OFF" pull chain switch for the light accessory kit. Upscale models feature remote "ON/OFF" infinite speed and light control, eliminating the need for "hard wiring" of wall switches. These remote controls are similar to those used with television, VCR, stereo, and cable channel selectors. Wiring for wall switches is discussed later in this unit.

What Direction Should a Ceiling-Suspended (Paddle) Fan Rotate?

Major manufacturers of ceiling-suspended (paddle) fans recommend directing the air downward (counter clockwise) in the summer to benefit from the "wind chill" effect (see Figure 9-5A), and upward (clockwise) in the winter (see Figure 9-5B). Some people like air blowing on them; others do not. Upward or downward, the key is to get the stratified air moving. Try different speeds and direction, and do whatever feels best. Read the manufacturer's suggestions.

SUGGESTED FAN SIZES	
For rooms up to:	**Fan size**
144 ft²	42 in.
225 ft²	44–48 in.
400 ft²	52–54 in.
500 ft²	56 in.

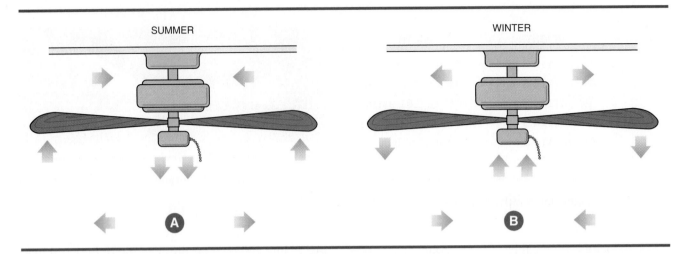

SUMMER   WINTER

Figure 9-5   Suggested rotation of a ceiling suspended paddle fan for winter and summer use.

Supporting Ceiling-Suspended (Paddle) Fans

This is a very important issue. There have been many reports of injuries (contusions, concussions, lacerations, fractures, head trauma, etc.) caused by the falling of a ceiling-suspended (paddle) fan. There is also the possibility of starting an electrical fire when the wires in the outlet box and fan canopy "short-circuit" as the fan falls from the ceiling. These problems are usually traced back to an improper installation.

The safe supporting of a ceiling-suspended (paddle) fan involves three issues:

1. the actual weight of the fan
2. the twisting and turning motion when the fan is started
3. vibration caused by unbalanced fan blades

▶Here's the scoop for installing ceiling-suspended paddle fans. These requirements are found in *NEC® 314.27(D)* and *422.18*, and in the UL White Book.

The outlet box or outlet box system:

- must be listed and marked as "Acceptable for fan support." Always look for this on the product.
- is permitted to be metallic or nonmetallic.
- must not support more than 70 lb. (32 kg).
- if designed to support more than 35 lb. (16 kg) but not more than 70 lb. (32 kg), it will be marked "Acceptable for fan support up to 70 lb. (32 kg)."

- weight limitation includes the fan and light kit.
- must be installed according to the manufacturer's instructions.◀

What Is an Outlet Box System?

This includes such things as the outlet box, hanger, clamps, and other mounting hardware.

New Work.   Install ceiling boxes marked "Acceptable for fan support" in all locations where it is likely that a ceiling-suspended paddle fan might be installed. This generally includes all habitable rooms.

Ceiling boxes installed close to a wall, in hallways, in closets, and similar nonhabitable rooms, and ceiling boxes installed for smoke alarms, fire alarms, and security devices do not require fan support boxes.

Some electricians are tired of trying to outguess what *is likely*, and what the homeowner might do at some later date. They install suitable fan support outlet boxes for all ceiling outlets other than those that are clearly intended for smoke alarms, fire alarms, and security devices.

Old Work.   When replacing an existing luminaire (fixture) with a ceiling-suspended (paddle) fan, make sure that the existing outlet box is listed and marked as suitable for fan support. In older homes, it is unlikely that you will find this marking. You are taking a big chance if you secure a ceiling fan to an outlet box that is not listed. It might fall on someone's head! You may have to support the fan from

Figure 9-6   Typical mounting of a ceiling-suspended (paddle) fan.

the building structure. Carefully read and follow the manufacturer's installation instructions.

See Figure 9-6 for a typical mounting of a ceiling-suspended (paddle) fan.

You always have the choice of hanging a ceiling-suspended (paddle) fan independent of the outlet box as illustrated in Figure 9-7E.

Figure 9-7 shows one type of box/hanger assembly that can be used for new work or for existing installations.

Figure 9-7   (A), (B), (C), and (D) show how a listed ceiling fan hanger and box assembly is installed through a properly sized and carefully cut hole in the ceiling. This is done for existing installations. Similar hanger/box assemblies are used for new work. The hanger adjusts for 16-in. (400-mm) and 24-in. (600 mm) joist spacing but can be cut shorter if necessary. (E) shows a type of box listed and identified for the purpose where the fan is supported from the joist, independent of the box, as required by *422.18* and *314.27(D)* for fans that weigh more than 35 lb (16 kg). *Courtesy of* Pass & Seymour/Legrand.

Figure 9-8 A special outlet box for the support of a luminaire (lighting fixture) or a ceiling-suspended (paddle) fan. There are no ears on the box. Instead, special strong mounting brackets are provided. Note that the mounting holes on the brackets are of two sizes—one set of holes is tapped for No. 8-32 screws [for a luminaire (lighting fixture)], and the other set of holes is tapped for No. 10-32 screws (for a ceiling-suspended [paddle] fan). *Photo courtesy of* **Pass & Seymour/Legrand.**

Figure 9-8 shows an outlet box that is listed for both luminaire (fixture) and fan support. This box has No. 8-32 holes for luminaire (fixture) mounting and No. 10-32 holes for fan support.

Figure 9-9 shows the wiring for a fan/light combination with the supply at the fan/light unit. Figure

9-10 shows the fan/light combination with the supply at the switch. Figure 9-7 shows two types of ceiling fan hanger/box assemblies. Many other types are available.

Figure 9-11 illustrates a typical combination light switch and three-speed fan control. The electrician is required to install a deep 4-in. box with a two-gang raised plaster ring for this light/fan control.

Figure 9-12 illustrates a "slider" switch for speed control of a fan, and a "dual control" that provides four-speed fan control and two-level "hi/lo" dimming for the incandescent lamps. Figure 9-13 shows two other styles of fan/light controls. Dual controls are needed where the ceiling fan also has a luminaire (lighting fixture). These controls fit into a deep single-gang device box, or preferably into a 4-in. square box with single-gang plaster cover. Be careful of "box fill." Make sure the wall box is large enough. Dimmer and speed control devices are quite large when compared to regular wall switches. Do not jam the control into the wall box.

A typical home-type ceiling fan motor draws 50 to 100 watts (50 to 100 volt-amperes). This is approximately 0.4 to slightly over 0.8 ampere.

Ceiling fan/light combinations increase the load requirements somewhat. Some ceiling fan/light units have one lamp socket, whereas others have four or five lamp sockets.

For the ceiling fan/light unit in the master bedroom, 240 volt-amperes were included in the load calculations. The current draw is

$$I = \frac{VA}{V} = \frac{240}{120} = 2 \text{ amperes}$$

Figure 9-9 Fan/light combination with the supply at the fan/light unit.

Figure 9-10 Fan/light combination with the supply at the switch.

For most fans, switching on and off may be accomplished by the pull-switch, an integral part of the fan ("OFF/HIGH/MED/LOW"), or with a solid-state switch having an infinite number of speeds. Some of the more expensive models offer a remote control similar to that used with television, VCRs, stereos, and cable channel selectors.

Most ceiling fan motors have an integral reversing switch to blow air downward or upward.

Do not use a standard incandescent lamp dimmer to control fan motors, fluorescent ballasts, transformers, low-voltage lighting systems (they use transformers to obtain the desired low voltage), motor-operated appliances, or other "inductive" loads. Serious overheating and damage to the motor can result. Fan speed controllers have internal circuitry that is engineered specifically for use on inductive load circuits. Fan speed controllers virtually eliminate any fan motor "hum." Read the label and the instructions furnished with a dimmer to be sure it is suitable for the application.

Figure 9-11 Combination light switch and three-speed fan switch.

Figure 9-12 The "slider" switch is for speed control of a fan. The "dual-control" provides four-speed fan control and a two-level "hi/lo" dimming for an incandescent lamp. *Courtesy of* Pass & Seymour.

Figure 9-13 Two types of fan/light controls. One has a toggle switch for turning a ceiling fan off and on, plus a "slider" that provides three-speed control of the fan. The other has a toggle switch for turning a light off and on, plus two "sliders," one for "ON/OFF" and three-speed control of a ceiling fan, the other for full range dimming of lighting. *Courtesy of* Lutron.

REVIEW

Note: Refer to the *Code* or the plans where necessary.

1. What circuit supplies the master bedroom? _____

2. What unique type of electrical protection is required for all 120-volt outlets in residential bedrooms? Where in the *NEC®* is this requirement found? _____

   _____

3. What type of receptacles are provided in this bedroom? How many receptacles are there? _____

4. How many ceiling outlets are included in this circuit? _____

   _____

5. What wattage for the closet luminaires (fixtures) was used for calculating their contribution to the circuit? _____

6. What is the current draw for the ceiling fan/light? _____

7. What is the estimated load in volt-amperes for the circuit supplying the master bedroom? _____

8. Which is installed first, the switch and outlet boxes or the cable runs? _____

   _____

9. How many conductors enter the ceiling fan/light wall box? _____

10. What type and size of box may be used for the ceiling fan/light wall box? _____

_____

_____

11. What type of covers are used with 4-in. square outlet boxes? _____

_____

_____

12. a. Does the circuit for the master bedroom have a grounded circuit conductor?

    b. Does it have an equipment grounding conductor? _____

    c. Explain the difference between a "grounded" conductor and a "grounding" conductor.

    _____

    _____

    _____

    _____

    _____

    _____

    _____

    _____

13. Approximately how many feet (meters) of two-wire cable and three-wire cable are needed to complete the circuit supplying the master bedroom? Two-wire cable _____ ft (_____ m). Three-wire cable _____ ft (_____ m).

14. If the cable is laid in notches in the corner studs, what protection for the cable must be provided? _____

_____

_____

_____

15. How high are the receptacles mounted above the finished floor in this bedroom?

_____

_____

16. Approximately how far from the bedroom door is the first receptacle mounted? (See the plans.) _____

17. What is the distance from the finished floor to the center of the wall switches in this bedroom? _____

18. The master bedroom features a sliding glass door. For the purpose of providing the proper receptacle outlets, answer the following statements true or false.

    • Sliding glass panels are considered to be wall space.   _____

    • Fixed panels of glass doors are considered to be wall space.   _____

19. What type of receptacle will be installed outdoors, just outside of the master bedroom?

_____

_____

20. Show your calculations of how to select a proper wall box for the closet luminaire (fixture) switch. Keep in mind that the available space between the wood casings is small. See Figure 9-1. _____

_____

_____

_____

_____

_____

_____

_____

21. When an outlet box is to be installed to support a ceiling fan, how must it be marked?

_____

_____

22. The following is a layout of the lighting circuit for the master bedroom. Using the cable layout shown in Figure 9-1, make a complete wiring diagram of this circuit. Use colored pencils or pens.

23. Connect the three-wire cable, fan, light, fan speed control and light switch. Refer to *200.7* to review the permitted use of the white conductor in the cable.

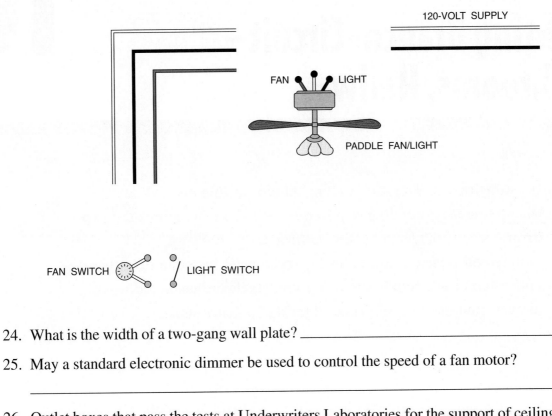

24. What is the width of a two-gang wall plate? _____

25. May a standard electronic dimmer be used to control the speed of a fan motor?

    _____

26. Outlet boxes that pass the tests at Underwriters Laboratories for the support of ceiling-suspended (paddle) fan are listed and marked "Acceptable For Fan Support." When roughing-in the electrical wiring for a new home, this type of outlet box must be installed wherever there is a likelihood of installing a ceiling-suspended (paddle) fan. Name those locations in homes that you consider likely to support a ceiling-suspended (paddle) fan.

    _____

    _____

    _____

Lighting Branch-Circuit—Bathrooms, Hallway

OBJECTIVES

After studying this unit, the student should be able to

▸ list equipment grounding requirements for bathroom installations.

▸ draw a wiring diagram for the bathroom and hallway.

▸ understand *Code* requirements for receptacles installed in bathrooms.

▸ understand *Code* requirements for receptacle outlets in hallways.

▸ discuss fundamentals of proper lighting for bathrooms.

Circuit A14 is a 15-ampere branch-circuit that supplies the lighting outlets and receptacle outlet in the hallway and the vanity lighting in both bathrooms. The receptacle outlets in the bathrooms are supplied by separate circuits A22 and A23. This is a *Code* requirement, discussed later in this unit and in Unit 15.

Table 10-1 summarizes outlets and estimated load for the bathrooms and bedroom hall.

DESCRIPTION	QUANTITY	WATTS	VOLT-AMPERES
Receptacles @ 120 W each	1	120	120
Vanity luminaires (fixtures) @ 200 W each	2	400	400
Hall luminaire (fixture)	1	100	100
TOTALS	4	620	620

Table 10-1 Bathroom and bedroom hall: outlet count and estimated load (Circuit A14). The receptacles in the master bathroom and hall bathroom are not included in this table as they are connected to separate 20-ampere branch-circuits A22 and A23.

Note that each bathroom shows a ceiling heater/light/fan that is connected to separate circuit Ⓐ J and Ⓐ K. These are discussed in detail in Unit 22.

A hydromassage tub is located in the bathroom serving the master bedroom. It is connected to a separate circuit Ⓐ A and is also discussed in Unit 22.

The attic exhaust fan in the hall is supplied by a separate circuit Ⓐ L, also covered in Unit 22.

LIGHTING BRANCH-CIRCUIT A14 FOR THE HALLWAY AND BATHROOMS

Figure 10-1 and the electrical plans for this area of the home show that each bathroom has a luminaire (fixture) above the vanity mirror. Some typical luminaires (fixtures) are shown in Figure 10-2. Of course, the homeowner might decide to purchase a medicine cabinet complete with a self-contained luminaire (lighting fixture). See Figure 10-3, which illustrates how to rough-in the wiring for each type. These luminaires (fixtures) are controlled by single-pole switches at the doors.

Figure 10-1   Cable layout for the master bathroom, hall bathroom, and hall. Layout includes lighting Circuit A14, two circuits (A11 and A12) for the heat/vent/lights, and two circuits (A22 and A23) for the receptacles in the bathrooms. The special-purpose outlets for the hydromassage tub, attic exhaust fan, and smoke detector are covered elsewhere in this text.

Figure 10-2   Typical vanity (bathroom) luminaires (lighting fixtures) of the side bracket and strip types. *Courtesy of* Progress Lighting.

A

MOUNT BOX TO PROVIDE ENOUGH
CLEARANCE BETWEEN LUMINAIRE
(LIGHTING FIXTURE) AND MIRROR

FRAMED OPENING
FOR RECESSED
MEDICINE CABINET

B

BE SURE TO BRING CABLE OR CONDUIT IN
AT PROPER LOCATION. IT IS BEST TO
HAVE THE ACTUAL MEDICINE CABINET OR
INSTALLATION INSTRUCTIONS TO DETERMINE
EXACT LOCATION OF CABLE KNOCKOUT.

CABLE IS
CONNECTED
TO WIRING
COMPARTMENT
KNOCKOUT ON
CABINET.

FRAMED OPENING FOR RECESSED
MEDICINE CABINET THAT COMES COMPLETE
WITH A LUMINAIRE (LIGHTING FIXTURE)

Figure 10-3   Two ways to rough-in wiring for lighting above or on a vanity mirror or a medicine cabinet.

You must remember that a bathroom (powder room) should have proper lighting for shaving, combing hair, grooming, and so on. Mirror lighting can accomplish this, because a mirror will reflect what it "sees." If the face is poorly lit, with shadows on the face, that is precisely what will be reflected in the mirror.

A luminaire (lighting fixture) directly overhead will light the top of one's head, but will cause shadows on the face. Mirror lighting and/or adequate lighting above and forward of the standing position at the vanity can provide excellent lighting in the bathroom. Figure 10-4, Figure 10-5, and Figure 10-6 show pictorial as well as section views of typical soffit lighting above a bathroom vanity.

Bathroom Receptacles

Figure 10-7 shows the definition of a bathroom according to the *NEC.* Review Unit 3 to refresh your memory about the *NEC* requirements for receptacles in bathrooms.

In this text, we have chosen to run a separate 20-ampere branch-circuit A22 to the receptacle in the master bedroom bathroom, a separate 20-ampere branch-circuit A23 to the receptacle in the hall bathroom, and another separate 20-ampere branch-circuit B21 to the receptacle in the powder room located near the laundry. These separate circuits are included in the general lighting load calculations, so no additional load need be added, *220.14(J)*.

WRONG

RIGHT

Figure 10-4   Positioning of bathroom luminaires (lighting fixtures). Note the wrong way and the right way to achieve proper lighting.

TYPICAL INCANDESCENT RECESSED SOFFIT LIGHTING
OVER BATHROOM VANITY

FIXTURE   JUNCTION BOX

END CUTAWAY OF SOFFIT SHOWING RECESSED
INCANDESCENT LUMINAIRES (FIXTURES) IN TYPICAL
SOFFIT ABOVE BATHROOM VANITY. TWO OR
THREE LUMINAIRES (FIXTURES) GENERALLY
INSTALLED TO PROVIDE PROPER LIGHTING.

Figure 10-5   Incandescent soffit lighting. Refer to Unit 7 for minimum clearance *Code* requirements for installation of recessed luminaires (fixtures).

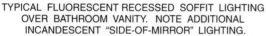

CEMENT BOARD

F40 SP30/RS/WM

DRYWALL

END CUTAWAY VIEW OF SOFFIT ABOVE
BATHROOM VANITY SHOWING RECESSED
FLUORESCENT LUMINAIRES (FIXTURES) CONCEALED
ABOVE TRANSLUCENT ACRYLIC LENS

TYPICAL FLUORESCENT RECESSED SOFFIT LIGHTING
OVER BATHROOM VANITY. NOTE ADDITIONAL
INCANDESCENT "SIDE-OF-MIRROR" LIGHTING.

Figure 10-6   Combination fluorescent and incandescent bathroom lighting.

A THIS IS A BATHROOM.
(BASIN, TOILET)

B THIS IS A BATHROOM.
(BASIN, TOILET, TUB)

C THIS IS NOT A BATHROOM.
(BASIN ONLY)

D THIS IS A BATHROOM EVEN THOUGH A DOOR
SEPARATES THE AREA INTO TWO PARTS.

Figure 10-7   The *Code* in *Article 100* defines a *bathroom* as an area including a basin with one or more of the following: a toilet, a tub, or a shower.

Figure 10-8   In a bathtub or shower area, no parts of cord-connected or cord-suspended luminaires (fixtures) or ceiling-suspended (paddle) fans are permitted in the shaded area. See text for details. Also see Figure 10-9.

Receptacles in Bathtub and Shower Spaces

Because of the obvious hazards associated with water and electricity, receptacles are *not* permitted to be installed in bathtub and shower spaces, *406.8(C)*.

HANGING LUMINAIRES (FIXTURES) IN BATHROOMS

▶*NEC® 410.4(D)* does not permit cord-connected luminaires (fixtures), chain-, cable-, or cord-suspended luminaires (fixtures), lighting track, pendants, or ceiling-suspended (paddle) fans to be located within a zone measured 3 ft (900 mm) horizontally and 8 ft (2.5 m) vertically from the top of a bathtub rim or top of a shower stall sill.◀ Figure 10-8 shows a top view of the 3 ft (900 mm) restriction. Figure 10-9 shows a side view of both permitted and not permitted installations.

Recessed or surface-mounted luminaires (lighting fixtures) and recessed exhaust fans may be located within the restricted zone.

Figure 10-9   In a bathtub or shower area, no parts of cord-connected or cord-suspended luminaires (fixtures) or ceiling-suspended (paddle) fans are permitted in the shaded area. See text for details. Also see Figure 10-8.

HALLWAY LIGHTING

The hallway lighting is provided by one ceiling luminaire (fixture) that is controlled with two three-way switches located at either end of the hall. The home run to Main Panel A has been brought into this ceiling outlet box.

RECEPTACLE OUTLETS IN HALLWAYS

One receptacle outlet has been provided in the hallway as required in *210.52(H)*, which states that "*for hallways of 3 m (10 ft) or more in length, at least one receptacle outlet shall be required.*"

For the purpose of determining the length of a hallway, the measurement is taken down the centerline of the hall, turning corners if necessary, but not passing through a doorway, Figure 10-10.

EQUIPMENT GROUNDING

With few exceptions, fixed and fastened-in-place electrical equipment with exposed non-current-carrying metal parts that could become energized must be grounded. This includes such things as appliances, furnaces, air conditioners, heat pumps, heat exchangers, fans, luminaires (fixtures), electric heaters, and medicine cabinets that have lighting. Equipment that must be grounded is itemized in *250.110*, *250.112*, *250.114* (cord and plug connected) and *Article 410, Part V* (luminaires).

Acceptable equipment grounding conductors are listed in *250.118*, which was previously discussed.

CAUTION: When using nonmetallic-sheathed cable or armored cable, *never* use the bare equipment grounding conductor as a grounded circuit

Figure 10-10   The centerline measurement of the bedroom hallway in this residence is **12 ft 6 in. (3.81 m)**, which requires at least one wall receptacle outlet, *210.52(H)*. This receptacle may be installed anywhere in the hall. The *Code* does not specify a location.

conductor, and *never* use the grounded circuit conductor as an equipment grounding conductor. These two conductors come together only at the main service panel, *never* in branch-circuit wiring. Refer to *250.142*.

Double Insulation in Lieu of Grounding

Appliances that are cord- and plug-connected with a three-wire cord are grounded by the equipment grounding conductor in the cord. Instead of grounding, many small electrical appliances have a two-wire cord, and are "double insulated." Examples of these are some portable electrical tools, razors, tooth brushes, radios, TVs, VCRs, CD players and burners, stereo components, and similar appliances.

Double-insulated appliances provide two levels of protection against electric shock. Accepted levels of protection can be the extra insulation on a conductor, added insulating material, the plastic nonmetallic enclosure of the appliance itself, or even a specified air-gap at an exposed terminal. With double-insulated products, it takes two failures to create a shock hazard to the user. These appliances are tested and clearly marked that they are double insulated in accordance to UL Standard 1097 *Double-Insulated Systems for Use in Electrical Equipment.*

Immersion Detection Circuit Interrupters

Another way to protect people from electrical shock is found on grooming appliances that are equipped with a special attachment plug cap.

Switches in Wet Locations

Do *not* install switches in wet locations, such as in bathtubs or showers, unless they are part of a listed tub or shower assembly, in which case the manufacturer has taken all of the proper precautions, and submitted the assembly to a recognized testing laboratory to undergo exhaustive testing to establish the safety of the equipment. See *404.4*.

REVIEW

1. List the number and types of switches and receptacles used in Circuit A14.

2. There is a three-way switch in the bedroom hallway leading into the living room. Show your calculation of how to determine the box size for this switch. The box contains cable clamps.

3. What wattage was used for each vanity luminaire (fixture) to calculate the estimated load on Circuit A14? _____

4. What is the current draw for the answer given in question 3?

5. Exposed noncurrent-carrying metallic parts of electrical equipment must be grounded if installed within _____ ft (_____ m) vertically or _____ ft (_____ m) horizontally of bathtubs, plumbing fixtures, pipes, or other grounded metal work or grounded surfaces.

6. What color are the faceplates in the bathrooms? Refer to the specifications.

   _____

7. Most appliances of the type commonly used in bathrooms, such as hair dryers, electric shavers, and curling irons, have two-wire cords. These appliances are _____ insulated or _____ _____ _____ _____ protected.

8. a. The *NEC*® in section _____ requires that all receptacles in bathrooms be _____ protected.

   b. The *NEC*® in section _____ requires that all receptacles in bathrooms be connected to one or more separate 20-ampere branch-circuits that serve no other outlets.

   c. The *NEC*® in section _____ permits the additional required 20-ampere branch-circuit for bathroom receptacles to supply more than one bathroom.

   d. The *NEC*® in section _____ permits other electrical equipment to be connected to the additional required 20-ampere branch-circuit for bathroom receptacles, but only if the branch-circuit supplies a single bathroom and the other equipment is located in that same bathroom.

   e. The *NEC*® in section _____ prohibits mounting receptacles in bathrooms face-up in the countertops and work surfaces near basins.

9. Hanging luminaires (lighting fixtures) must be kept at least _____ ft (_____ m) from the edge of the tub as measured horizontally. In bathrooms with high ceilings, where the hanging luminaire (fixture) is installed directly over the tub, it must be kept at least _____ ft (_____ m) above the edge of the tub.

10. The following is a layout of a lighting circuit for the bathroom and hallway. Using the cable layout shown in Figure 10-1, make a complete wiring diagram of this circuit. Use colored pencils or pens.

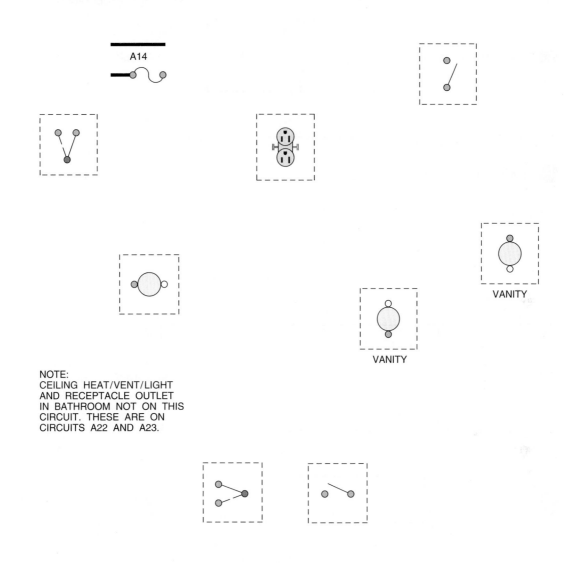

A14

VANITY

VANITY

NOTE:
CEILING HEAT/VENT/LIGHT
AND RECEPTACLE OUTLET
IN BATHROOM NOT ON THIS
CIRCUIT. THESE ARE ON
CIRCUITS A22 AND A23.

BATHROOMS – HALLWAY

11. Circle the correct answer as to whether a receptacle outlet is required in the following hallways.

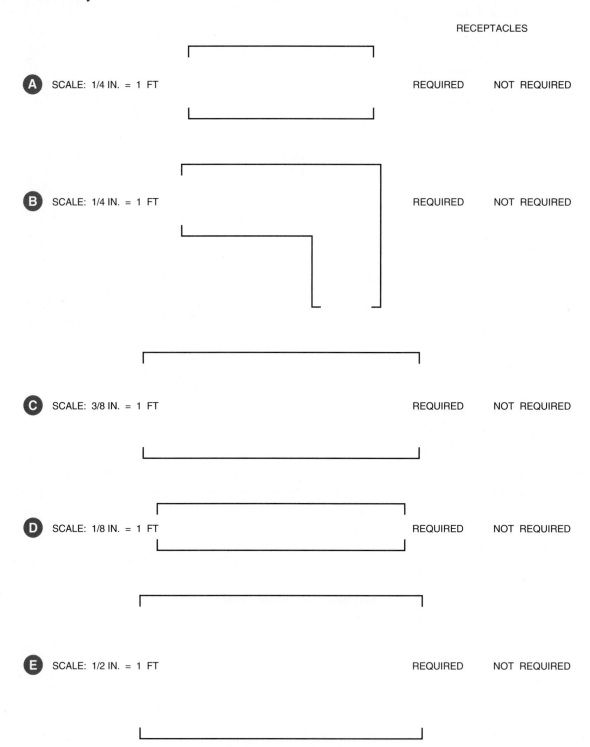

RECEPTACLES

A   SCALE: 1/4 IN. = 1 FT   REQUIRED   NOT REQUIRED

B   SCALE: 1/4 IN. = 1 FT   REQUIRED   NOT REQUIRED

C   SCALE: 3/8 IN. = 1 FT   REQUIRED   NOT REQUIRED

D   SCALE: 1/8 IN. = 1 FT   REQUIRED   NOT REQUIRED

E   SCALE: 1/2 IN. = 1 FT   REQUIRED   NOT REQUIRED

Lighting Branch-Circuit— Front Entry, Porch

OBJECTIVES

After studying this unit, the student should be able to

▶ understand how to install a switch in a doorjamb for automatic "ON/OFF" when the door is opened or closed.

▶ discuss types of luminaires (fixtures) recommended for porches and entries.

▶ complete the wiring diagram for the entry-porch circuit.

▶ discuss the advantages of switching outdoor receptacles from indoors.

▶ define wet and damp locations.

▶ understand box fill for sectional ganged device boxes.

The front entry-porch is connected to Circuit A15. The home run enters the ceiling box in the front entry. From this box, the circuit spreads out, feeding the closet light, the porch bracket luminaire (fixture), the two bracket luminaires (fixtures) on the front of the garage, one receptacle outlet in the entry, and one outdoor weatherproof GFCI receptacle on the porch, as shown in Figure 11-1. Table 11-1 summarizes the outlets and estimated load for the entry and porch.

Note that the receptacle on the porch is controlled by a single-pole switch just inside the front door. This allows the homeowner to plug in such things as strings of ornamental holiday lighting or decorative lighting and have convenient switch control of that receptacle from inside the house, a nice feature.

Although it adds cost, consider an outdoor weatherproof receptacle on both sides of the front entry door. This makes for easy plugging in of outdoor decorations. Some electricians install a receptacle in the soffit. Switched or nonswitched? The decision is up to you.

Typical ceiling luminaires (fixtures) commonly installed in front entryways, where it is desirable to make a good first impression on guests, are shown in Figure 11-2.

DESCRIPTION	QUANTITY	WATTS	VOLT-AMPERES
Receptacles @ 120 W	1	120	120
Weatherproof receptacles @ 120 W	1	120	120
Outdoor porch bracket luminaire (fixture)	1	100	100
Outdoor garage bracket luminaires (fixtures) @ 100 W each	2	200	200
Ceiling luminaire (fixture)	1	150	150
Closet recessed luminaire (fixture)	1	75	75
TOTALS	7	765	765

Table 11-1 Entry and porch: outlet count and estimated load (Circuit A15).

TO OTHER S₃ IN HALL. NOT CONNECTED TO ENTRY-PORCH CIRCUIT.

A15

* SEE NOTE BELOW

S₃S₃

FRONT HALL SWITCH ARRANGEMENT LEFT TO RIGHT FACING SWITCHES
• 3-WAY—LIGHTING ABOVE BASEMENT STAIR LANDING AND AT BOTTOM OF STAIRS
• SINGLE POLE—RECEPTACLE ON PORCH
• 3-WAY—PORCH AND OUTSIDE GARAGE LIGHTING
• 3-WAY—FRONT HALL CEILING LIGHTING

S₃S₃SS₃

GFCI WP

TO S₄ IN GARAGE. NOT CONNECTED TO ENTRY-PORCH CIRCUIT.

* NOTE: TO SWITCHES AT BOTTOM OF STAIRS. NOT CONNECTED TO ENTRY-PORCH CIRCUIT.

S₃S₃

FRONT ENTRY–PORCH

Figure 11-1 Cable layout of the front entry-porch circuit A15. Note that cables from some of the other circuits are shown so that you can get a better idea of exactly how many wires will be found at the various locations.

Typical outdoor wall-bracket-style porch and entrance luminaires (lighting fixtures) are shown in Figure 11-3.

A location exposed to the weather is a wet location. A partially protected area such as an open porch, or under a roof or canopy, is a damp location. For more precise definitions of dry, wet, and damp locations, see *Article 100* of the *NEC*.®

Luminaires (fixtures) installed in wet or damp locations must be installed so that water cannot enter

Figure 11-2   Hall luminaires (lighting fixtures): ceiling mount and chain mount. *Courtesy of* Progress Lighting.

Figure 11-3   Outdoor porch and entrance luminaires (lighting fixtures): wall bracket styles. *Courtesy of* Progress Lighting.

Figure 11-4 Outdoor areas considered to be a *damp* or *wet* location.

or accumulate in wiring compartments, lampholders, or other electrical parts. Luminaires (fixtures) installed in wet locations shall be marked, "Suitable for Wet Locations." Luminaires (fixtures) installed in damp locations shall be marked, "Suitable for Wet Locations" or "Suitable for Damp Locations." This information is found in *410.4(A)*.

Some inspectors consider the area within a 45° angle from a roof line or overhang to be a damp location, and outside the 45° angle to be a wet location. See Figure 11-4.

CIRCUIT A15

Circuit A15 is a rather simple circuit that has two sets of three-way switches: one set controlling the front entry ceiling luminaire (fixture), the other set controlling the porch bracket luminaire (fixture) and the two bracket luminaires (fixtures) on the front of the garage. Also, one single-pole switch controls the weatherproof porch receptacle outlet, and another single-pole switch controls the closet light. In the review questions you will be asked to follow the suggested cable layout for making up all of the circuit connections.

Probably the most difficult part of this circuit is planning and making up the connections at the switch location just inside the front door. Refer to Figure 11-1 for details. Figure 11-5 shows a four-gang nonmetallic box of the type that can be used

Figure 11-5 A 4-gang, nonmetallic box that has a marked volume of 55 cu in.³ (901.3 cm³).

for the front door switch location. Note that the three-way switch for the luminaire (fixture) over the stairway leading to the basement is part of the Recreation Room lighting branch-circuit B12.

There are a lot of conductors, wiring devices, and cable clamps in this box. Let's figure out the proper wall size wall box.

1. Add the circuit conductors
   $2 + 2 + 3 + 3 + 3 + 3 =$ 16

2. Add for equipment grounding wires
   (count 1 only) 1

3. Add eight for the four switches 8

4. Add for cable clamps (count 1 only) $\underline{\hspace{1cm}1}$

   Total 26

To select a box that has adequate cubic-inch capacity for this example, refer to *Table 314.16(A)* in the *NEC*® and Figure 2-20 and Table 2-1 in this text. One possibility would be to use four 3 in. × 2 in. × 3½ in. device boxes ganged together, providing $4 \times 9 = 36$ conductor count based on 14 AWG conductors.

When using nonmetallic boxes, refer to *Table 314.16(B)* and base the volume allowance on 2 in.³ for each 14 AWG conductor and for the volume allowance for the wiring devices, cable clamps, and equipment grounding conductors. In the previous example, we are looking for a cubic-in. volume of $26 \times 2 = 52$ in.³

DOORJAMB SWITCH

An interesting possibility presents itself for the front entry closet. Although the plans show that the closet luminaire (fixture) is turned on and off by a standard single-pole switch to the left as you face the closet, a doorjamb switch could have been installed. A doorjamb switch is usually mounted about 6 feet (1.8 m) above the floor, on the inside of the 2 × 4 framing on the hinge side of the closet door. The electrician will run the cable to this point, and let it hang out until the carpenter can cut the proper size opening into the doorjamb for the box. After the finished woodwork is completed, the electrician then finishes installing the switch. Figure 11-6 illustrates a door switch. Note that the plunger on the switch is pushed inward when the edge of the door pushes on it as the door is closed, shutting off the light. The plunger can be adjusted in or out to make the switch work properly. These doorjamb switches come complete with their own special wall box.

The integral box furnished with door switches is generally suitable for one cable only. Wiring space is very limited. Do not plan on using this box for splices other than those necessary to make up the connections to the switch.

Figure 11-6 Doorjamb switch. Feed at light. Switch loop run from ceiling box to doorjamb box.

REVIEW

1. a. How many circuit wires enter the entry ceiling box? _____
   b. How many equipment grounding conductors enter the entry ceiling box? _____

2. Assuming that an outlet box for the entry ceiling has a fixture stud but has external cable clamps, what size box could be installed? Show calculations for both metal and nonmetallic boxes.

   _____

   _____

   _____

   _____

   _____

3. How many receptacle outlets and lighting outlets are supplied by Circuit A15?

   _____

4. Outdoor luminaires (fixtures) directly exposed to the weather must be marked as:

   a. suitable for damp locations

   b. suitable for dry locations only

   c. suitable for wet locations

   Circle the correct answer.

5. Make a material list of all types of switches and receptacles connected to Circuit A15.

   _____

   _____

   _____

   _____

6. From left to right, facing the switches, what do the switches next to the front door control?

   _____

   _____

   _____

   _____

7. Who is to select the entry ceiling luminaire (fixture)? _____

8. The following layout is for Lighting Circuit A15, the entry, porch, and front garage lights. Using the cable layout shown in Figure 11-1, make a complete wiring diagram of this circuit. Use colored pencils to indicate the color of the conductors' insulation.

ENTRY–PORCH

Lighting Branch-Circuit and Small-Appliance Circuits for the Kitchen

OBJECTIVES

After studying this unit, the student should be able to

▶ understand lighting for a typical kitchen and dining room.

▶ understand the installation and operation of kitchen exhaust fans.

▶ know the *NEC®* requirements for small-appliance branch-circuits in kitchens.

▶ know the *NEC®* requirements for GFCI protection for all receptacles that serve countertops in kitchens.

▶ decide if a separate branch-circuit should be installed for a refrigerator.

▶ decide where it might be desirable to install multiwire branch-circuits and split circuit receptacles in areas where high concentration of plug-in appliances might be used.

LIGHTING CIRCUIT B7

The lighting circuit supplying the kitchen originates at Panel B. The circuit number is B7. The cable layout for this circuit is shown in Figure 12-1. The home run is very short, leading from Panel B in the corner of the recreation room to the switch box to the right of the kitchen sink. From here, the circuit continues upward to the outlet box above one of the square surface-mounted luminaires (fixtures) on the ceiling, then across the ceiling to the outlet box above the eating area where a lighting track will be installed. From this outlet box, the circuit picks up the outdoor bracket luminaire (fixture). The range hood exhaust fan/light is fed from the outlet box above the other square luminaire (fixture) on the ceiling.

As you study the cable layouts in this text, be assured that there are other ways to run the conductors. The thought process should be to sketch a few different cable run possibilities. Then consider how many conductors enter the boxes, the ease or difficulty in making up the connections, and problems you might encounter because of obstructions (pipes and ducts) in the walls or ceilings. When installing recessed luminaires (fixtures), be sure they are listed for "through-wiring" if you intend to use the wiring compartment on the luminaire (fixture) to continue on with the branch-circuit wiring. These luminaires are marked with the maximum number and size of conductors permitted in the wiring compartment.

Table 12-1 shows the outlets and estimated loads for the kitchen lighting Circuit B7.

KITCHEN LIGHTING

Figure 12-1 Cable layout for the kitchen lighting. This is Circuit B7. The appliance circuit receptacles are connected to 20-ampere circuits B13, B15, and B16 and are not shown on this lighting circuit layout.

DESCRIPTION	QUANTITY	WATTS	VOLT-AMPERES
Ceiling luminaires (fixtures) (eight 20-W FL lamps)	2	160	180
Recessed luminaire (fixture) over sink	1	100	100
Lighting track	1	100	100
Outdoor rear bracket luminaire (fixture)	1	100	100
Range hood fan/light Two 60-W lamps Fan motor (1⅓ A @ 120 V)		120 160	120 180
TOTALS	7	740	780

Table 12-1 Kitchen: outlet count and estimated load (Circuit B7).

KITCHEN LIGHTING

The general lighting for the kitchen is provided by two square surface-mounted luminaires (fixtures) as shown in Figure 12-2. These luminaires might have four straight or two U-bent fluorescent lamps. These luminaires (fixtures) are controlled by a single-pole switch located next to the doorway leading to the living room.

Above the breakfast area table is a short section of track lighting controlled by two three-way switches: one located adjacent to the sliding door, the other at the doorway leading to the living room. Figure 12-3 illustrates just a few of the many types of luminaires (lighting fixtures) that might be secured to the track.

Figure 12-2 Typical ceiling luminaire (fixture) of the type installed in the kitchen of this residence. This type of luminaire (fixture) might have four 20-watt fluorescent lamps, two 40-watt U-shaped fluorescent lamps, or four 60-watt incandescent lamps. *Courtesy of* Progress Lighting.

Figure 12-4 shows a variety of chain-suspended luminaires (lighting fixtures) of the type commonly used in dining rooms.

One goal of good lighting is to reduce shadows in work areas. This is particularly important in the kitchen. The two ceiling luminaires (fixtures), the luminaire (fixture) above the sink, plus the luminaire (fixture) within the range hood will provide excellent lighting in the kitchen area. Of course, the track lighting in the eating area offers good lighting for that area.

Undercabinet Lighting

In some instances, particularly in homes with little outdoor exposure, the architect might specify strip fluorescent luminaires (lighting fixtures) under the kitchen cabinets. Several methods may be used to install undercabinet luminaires (lighting fixtures). They can be fastened to the wall just under the upper cabinets (Figure 12-5[A]), installed in a recess that is part of the upper cabinets so that they are hidden from view (Figure 12-5[B]), or fastened under and to the front of the upper cabinets (Figure 12-5[C]). All three possibilities require close coordination with the cabinet installer to be sure that the wiring is brought out of the wall at the proper location to connect to the undercabinet luminaires (lighting fixtures).

Flat fluorescent and incandescent luminaires (lighting fixtures) are available that mount on the underside of upper cabinets. Some of these luminaires (lighting fixtures) are direct connected—others are cord- and plug-connected. These are pretty much out of sight, but provide excellent lighting for countertop work surfaces.

Luminaire (lighting fixture) manufacturers also offer shallow undercabinet track lighting systems in both 120 volts and 12 volts. These systems include track, track lighting heads for small halogen lamps, end fittings, and transformer with cord and plug. Some have two-level, high-low switching to vary the light intensity.

Visiting a lighting showroom, an electrical supply house, a home center, or a hardware store is an exciting experience. The choice of luminaires (lighting fixtures) is almost endless.

Lamp Types

Good color rendition in the kitchen area is achieved by using incandescent lamps, or by using

Figure 12-3 Fixtures that can be hung either single or on a track over dinette tables, such as in the eating area of this kitchen. *Courtesy of* Progress Lighting.

Figure 12-4 Types of luminaires (fixtures) used over dining room tables. *Courtesy of* Progress Lighting.

Figure 12-5   Methods of installing undercabinet fluorescent strip lighting.

warm white deluxe (WWX), warm white (WW), or deluxe cool white (CWX) 3000 K lamps wherever fluorescent lamps are used. See Unit 7 for details regarding energy-saving ballasts and lamps.

FAN OUTLET

Ducted fans exhaust air to the outside. They remove cooking odors, steam, and smoke. Ductless fans remove cooking odors but do not remove steam or smoke.

In this residence, the range hood exhaust fan above the electric range is connected to lighting branch-circuit B7. The speed control, light, and light switch are integral parts of the fan. The electric supply is brought down from the ceiling to the wiring compartment on the fan. For exhaust fans mounted under cabinets along a wall, the electrical supply can be brought into the top or into the back of the fan housing wiring compartment. In either case, a sufficient length of cable must be left during the roughing-in stage to be able to make the final connections in the wiring compartment. Figure 12-6 is a typical range hood exhaust fan.

Fan Noise

The fan motor and air movement through an exhaust fan produce a certain amount of noise. Exhaust fan manufacturers specify a sound-level rating in their descriptive literature. This provides some idea of the noise level one might expect from the fan after it is installed. The sound-level rating unit used to define fan noise is a *sone* (rhymes with tone).

For simplicity, one sone is the noise level of an average refrigerator. The lower the sone rating, the quieter the fan. All manufacturers of exhaust fans provide this information in their descriptive literature.

Figure 12-6   Typical range hood exhaust fan. Speed control and light are integral parts of the unit. *Courtesy of* **Broan-NuTone.**

CLOCK OUTLETS

Battery-operated wall and mantle clocks for the most part have replaced electric clocks. They can be put just about anywhere.

Clock hanger receptacles are still available. In a clock hanger receptacle, the receptacle is set back from the front edge of the faceplate, providing room for the typical small plug cap on the end of a cord. This type of receptacle can be hidden behind a painting or photograph that you want to light up.

SMALL-APPLIANCE BRANCH-CIRCUITS FOR RECEPTACLES IN THE KITCHEN

The *Code* requirements for small-appliance circuits and the spacing of receptacles in kitchens are covered in *210.11(C)(1), 210.52(B),* and *220.52(A).* This was discussed in detail in Unit 3. Important points are:

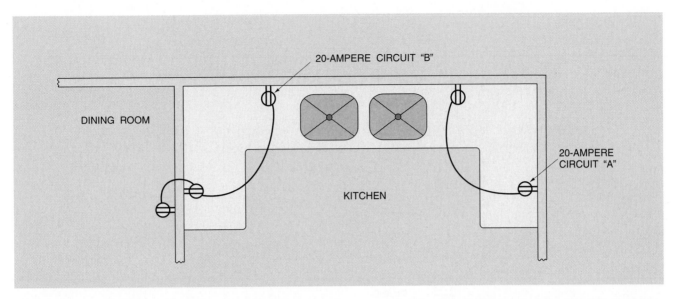

Figure 12-7 A receptacle outlet in a dining room may be connected to one of the two required small-appliance circuits in the kitchen, *210.52(B)(1)*. Countertop receptacles must be supplied by at least two 20-ampere small-appliance circuits, *210.52(B)(3)*.

- At least two 20-ampere small-appliance circuits shall be installed, as in Figure 12-7.

- Either (or both) of the two circuits required in the kitchen is permitted to supply receptacle outlets in other rooms, such as a dining room, breakfast room, or pantry, Figure 12-7.

- These small-appliance circuits shall be assigned a load of 1500 volt-amperes each when calculating feeders and service-entrance requirements.

- The receptacle for a refrigerator in a kitchen is *permitted* to be supplied by an individual branch-circuit.

- A clock outlet may be connected to an appliance circuit when this outlet is to supply and support the clock only (see Figure 3-3).

- Countertop receptacles must be supplied by at least two 20-ampere small-appliance circuits, as in Figure 12-7.

- All 125-volt, single-phase, 15- and 20-ampere receptacles installed in kitchens to serve the countertops shall be GFCI protected, *210.8(A)(6)*, Figure 12-8. A person can receive a deadly electrical shock by accidentally touching a "hot" wire and a grounded surface at the same time. A grounded surface could be a stainless steel sink, a faucet, frames of a range, an oven, a cook-top, a refrigerator, a microwave oven, a countertop food processor, a luminaire (fixture), a metal water pipe, or metal hot/cold air registers.

- A receptacle may be installed for plugging in a gas appliance to serve the ignition and/or clock devices on the appliance. See Figure 3-3.

- The 20-ampere small-appliance circuits are provided to serve plug-in portable appliances. They are not permitted by the *Code* to serve lighting or appliances such as dishwashers, garbage disposals, or exhaust fans.

- A receptacle installed below the sink for easy plug-in connection of a food waste disposer is *not* to be connected to the required small-appliance circuits, *210.52(B)(2)*, and it is *not* required to be GFCI protected, *210.8(A)(6)*. Refer to Unit 3 for spacing requirements for receptacles located in kitchens.

The 20-ampere small-appliance circuits prevent circuit overloading in the kitchen because of the heavy concentration and use of electrical appliances. For example, one type of cord-connected microwave oven is rated 1500 watts at 120 volts. The current required by this appliance alone is:

$$I = \frac{W}{E} = \frac{1500}{120} = 12.5 \text{ amperes}$$

Figure 12-8 In kitchens, *all* 125-volt, single-phase, 15- and 20-ampere receptacles that serve countertop surfaces (walls, islands, or peninsulas) must be GFCI protected. The key words are "that serve countertop surfaces." GFCI personnel protection is *not* required for the refrigerator receptacle, a receptacle under the sink for plugging in a food waste disposer, or a receptacle installed inside cabinets above the countertops for plugging in a microwave oven attached to the underside of the cabinets. ▶All 125-volt, single-phase, 15- and 20-ampere receptacles within 6 ft (1.8 m) of a wet-bar, laundry, or utility room sink must be GFCI protected. See *210.8(A)(6)* and *210.8(A)(7).*◀

In 90.1(B) we find a statement that *Code* requirements are considered necessary for safety. According to the *Code*, compliance with these rules will not necessarily result in an efficient, convenient, or adequate installation for good service or future expansion of electrical use.

Figure 12-9 shows four 20-ampere branch-circuits feeding the receptacles in the kitchen area. These circuits are B13, B15, B16, and B22.

Figure 12-9 shows one way to arrange the 20-ampere branch-circuits in the kitchen area. Obvi-

ously, there are other ways to arrange these branch-circuits. The intent is to arrange the circuits so as to provide the best availability of electrical power for appliances that will be used in the heavily concentrated work areas in the kitchen.

*A*ll receptacles serving countertops in kitchens must be GFCI protected, *210.8(A)(6)*. See Unit 6.

GFCI receptacles or GFCI circuit breakers can provide this protection.

In Figure 12-9, Circuits B13 and B16 are similar in that the circuit is run first to the line side terminals

+ GFCI

KITCHEN RECEPTACLES

Figure 12-9 Cable layout for kitchen receptacles.

of a feedthrough GFCI receptacle. From the load-side terminals of the feedthrough GFCI receptacle, conductors are then run to the other grounding-type receptacle(s) on the circuit. GFCI circuit breakers could have been installed on these circuits, in which case the receptacles would have been grounding-type receptacles. The choice is yours!

Circuit B15 does not require GFCI protection because none of these receptacles serve countertops.

Refrigerator Receptacle

Many nuisance power outages in homes are caused by using too many high wattage small appliances in the kitchen at the same time—and then the refrigerator "kicks in."

The *NEC®* does not help us in this respect. It *permits* the receptacle for the refrigerator to be connected to any one of the two required 20-ampere small-appliance branch-circuits, *210.52(B)(1)*. Some home-type combination refrigerator/freezers with many features draw 11 amperes or more. That doesn't leave much room on a 20-ampere branch-circuit for other small cord- and plug-connected appliances in the kitchen.

The *NEC®* also *permits* this receptacle to be connected to an individual 15- or 20-ampere branch-circuit, *210.52 (B)(1), Exception No. 2*. This is a much better way to do it. Many instructions furnished with these larger home-type refrigerator/freezers require that an individual branch-circuit be provided. In this residence, Circuit B22 is an individual (dedicated) 20-ampere branch-circuit that supplies only the refrigerator receptacle. This is a single receptacle rated 20 amperes.

Following are a few more *NEC®* requirements:

110.3(B) states that *"Listed or labeled equipment shall be installed and used in accordance with any instructions included in the listing or labeling."**

This receptacle located behind the refrigerator does not require GFCI protection since it does not serve a countertop, *210.8(A)(6)*.

Branch-circuit, Individual: A branch-circuit that supplies only one utilization equipment. When an individual branch-circuit is run for the refrigerator, a single receptacle would be required to "Meet Code." A single receptacle ensures that the receptacle will not supply other appliances.

A single receptacle must be rated not less than the branch-circuit rating, *210.21(B)(1)*. That's a 15-ampere receptacle on a 15-ampere branch-circuit—a 20-ampere receptacle on a 20-ampere branch-circuit.

Cord- and plug-connected appliances shall not exceed 80% of the branch-circuit rating, *210.23(A)(1)*. That's 12 amperes on a 15-ampere circuit, and 16 amperes on a 20-ampere circuit.

This individual-branch circuit is not considered to be additional load for service-entrance calculations, *220.14(J)*. The refrigerator load would have been there no matter what circuit it was connected to. The advantage of doing this is that it diverts the refrigerator load from the small-appliance branch-circuits in the kitchen.

Sliding Glass Doors

See Units 3 and 9 for discussions on how to space receptacle outlets on walls where sliding glass doors are installed.

Small-Appliance Branch-Circuit Load Calculations

The 20-ampere small-appliance branch-circuits are figured at 1500 volt-amperes for each two-wire

circuit when calculating feeders and services. The separate 20-ampere branch-circuit for the refrigerator is excluded from this requirement, *220.14(J)* and *220.53(A)*, *Exception*. See Unit 29 for complete coverage of load calculations.

SPLIT-CIRCUIT RECEPTACLES AND MULTIWIRE CIRCUITS

Split-circuit receptacles may be installed where a heavy concentration of plug-in load is anticipated, in which case each receptacle is connected to a separate circuit, as in Figure 12-10 and Figure 12-11.

A popular use of split-circuit receptacles is to control one receptacle with a switch and leave one receptacle "hot" at all times, as is done in the bedrooms and living room of this residence.

Split-circuit GFCI receptacles are not available. Refer to Unit 6, where GFCIs are discussed in great detail.

Therefore, the use of split-circuit receptacles is rather difficult to incorporate where GFCI receptacles are required. A multiwire branch-circuit could be run to a box where proper connections are made, with one circuit feeding one feedthrough GFCI receptacle and its downstream receptacles.

This would be used when it becomes more economical to run one three-wire circuit rather than installing two two-wire circuits. Again, it is a matter of knowing what to do, and when it makes sense to do it, as in Figure 12-12.

When multiwire circuits are installed, do not use the terminals of a receptacle or other wiring device to serve as the splice for the grounded neutral conductor. The *NEC®* in *300.13(B)* states that: *"In multiwire branch circuits, the continuity of a grounded conductor shall not be dependent upon device connections . . . "* Any splicing of the grounded conductor in a multiwire circuit must be made independently of the receptacle or lampholder, or other wiring device. This requirement is illustrated in Figure 12-10. Note that the screw terminals of the receptacle *must* not be used to splice the neutral conductors. The hazards of an open neutral are discussed in Unit 17. Figure 12-13 shows two types of grounding receptacles commonly used for installations of this type.

When multiwire branch-circuits supply more

(A) PROPER WAY TO CONNECT GROUNDED NEUTRAL CONDUCTORS IN A MULTIWIRE BRANCH-CIRCUIT.

(B) IMPROPER WAY TO CONNECT GROUNDED NEUTRAL CONDUCTORS IN A MULTIWIRE BRANCH-CIRCUIT. NOT PERMITTED.

Figure 12-10 Connecting the grounded neutral conductors in a three-wire (multiwire) branch-circuit, *NEC® 300.13(B)*.

120/240 VOLT THREE-WIRE CIRCUIT (MULTIWIRE CIRCUIT)

SPLIT-CIRCUIT RECEPTACLE

* *NEC® 210.4(B)* AND *210.7(B)* REQUIRES THAT A MEANS MUST BE PROVIDED TO SIMULTANEOUSLY DISCONNECT BOTH UNGROUNDED CONDUCTORS AT THE PANELBOARD WHERE THE BRANCH-CIRCUIT ORIGINATES. THIS COULD BE A 2-POLE SWITCH WITH FUSES, A 2-POLE CIRCUIT BREAKER, OR TWO 1-POLE CIRCUIT BREAKERS WITH A LISTED HANDLE TIE.

Figure 12-11 Split-circuit receptacle connected to multiwire circuit.

than one receptacle on the same yoke, a means must be provided to disconnect simultaneously all of the "hot" conductors at the panelboard where the branch-circuit originates. Figure 12-11 shows that 240 volts are present on the wiring device; only 120 volts are connected to each receptacle on the wiring device. ►See *210.4(B)* and *210.7(B)*.◄

The "simultaneous disconnect" rule applies to any receptacles, lampholders, or switches mounted on one yoke.

RECEPTACLES AND OUTLETS

Article 100 defines a receptacle and a receptacle outlet, as illustrated in Figure 3-4, Figure 12-14, and Figure 12-15.

Receptacles are selected for circuits according to the following guidelines. A single receptacle connected to a circuit must have a rating not less than the rating of the circuit, *210.21(B)*. In a residence, typical examples for this requirement for single receptacles are the clothes dryer outlet

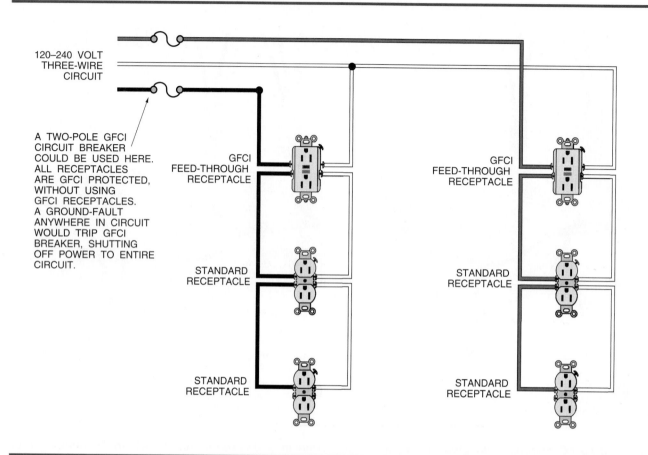

Figure 12-12 This diagram shows how a multiwire circuit can be used to carry two circuits to a box, then split the three-wire circuit into two two-wire circuits. This circuitry does not require simultaneous disconnect of the ungrounded conductors as shown in Figure 12-11.

Figure 12-13 Grounding-type duplex receptacles, 15A–125V rating. Most duplex receptacles can be changed into "split-circuit" receptacles by breaking off the small metal tab on the terminals. The above photo clearly shows this feature.

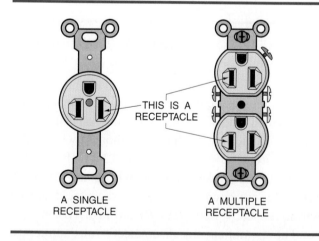

Figure 12-14 Receptacles: See *NEC®* definitions, *Article 100.*

THE BRANCH-CIRCUIT CONDUCTORS IN THE BOX ARE REFERRED TO AS AN *OUTLET.*

WHEN A RECEPTACLE IS CONNECTED TO THE CONDUCTORS, THE OUTLET IS REFERRED TO AS A *RECEPTACLE OUTLET.*

WHEN A LUMINAIRE (LIGHTING FIXTURE) IS CONNECTED TO THE CONDUCTORS, THE OUTLET IS REFERRED TO AS A *LIGHTING OUTLET.*

Figure 12-15 An *outlet* is defined in the *NEC®* as "*A point on the wiring system at which current is taken to supply utilization equipment.*"

(30 amperes), the range outlet (50 amperes), or the freezer outlet (15 amperes).

Circuits rated 15 amperes supplying two or more receptacles shall not contain receptacles rated over 15 amperes. For circuits rated 20 amperes supplying two or more receptacles, the receptacles connected to the circuit may be rated 15 or 20 amperes. See *Table 210.21(B)(3).*

REVIEW

Note: Refer to the *Code* or plans where necessary.

1. If everything on Circuit B7 were turned on, what would be the total current draw?

   _____

   _____

2. From what panel does the kitchen lighting circuit originate? What size conductors are used? _____

   _____

3. How many luminaires (lighting fixtures) are connected to the kitchen lighting circuit?

   _____

4. What color fluorescent lamps are recommended for residential installations?

   _____

   _____

5. a. What is the minimum number of 20-ampere small-appliance circuits required for a kitchen according to the *Code*? _____

  b. How many are there in this kitchen? _____

6. How many receptacle outlets are provided in the kitchen? _____

7. What is meant by the term *two-circuit (split-circuit) receptacles*? _____

   _____

   _____

   _____

   _____

   _____

8. Circle the correct response. Duplex receptacles connected to the 20-ampere small-appliance branch-circuits in kitchens and dining rooms:

   a. may be rated 15 amperes.   (True) (False)

   b. must be rated 15 amperes.   (True) (False)

   c. may be rated 20 amperes.   (True) (False)

   d. must be rated 20 amperes.   (True) (False)

9. In kitchens, a receptacle must be installed at each counter space _____ in. or wider.

10. A fundamental rule regarding the grounding of metal boxes, luminaires (fixtures), and so on, is that they must be grounded when "in reach of _____ ."

11. How many circuit conductors enter the box

   a. where the range hood will be installed? _____

   b. in the ceiling box over which the track will be installed? _____

   c. at the switch location to the right of the sliding door? _____

12. How much space is there between the countertop and upper cabinets? _____

13. Where is the speed control for the range hood fan located? _____

14. Who is to furnish the range hood? _____

15. List the appliances in the kitchen that must be electrically connected. _____

   _____

   _____

   _____

   _____

   _____

   _____

16. Complete the wiring diagram, connecting feedthrough GFCI 2 to also protect receptacle 1, both to be supplied by Circuit A1. Connect feedthrough GFCI 3 to also protect receptacle 4, both to be supplied by Circuit A2. Use colored pencils or markers to show proper color. Assume that the wiring method is EMT, where more freedom in the choice of insulation colors is possible.

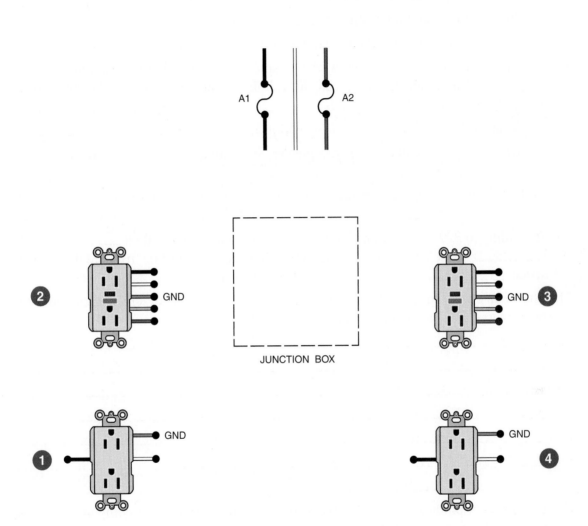

17. Each 20-ampere small-appliance branch-circuit load demand shall be determined at _____ . Choose one.

    a. 2400 volt-amperes   b. 1500 volt-amperes   c. 1920 volt-amperes

18. a. Does the *NEC*® permit the receptacle for the refrigerator to be supplied by one of the required 20-ampere small-appliance branch-circuits?   Yes _____ No _____ *Section* _____.

  b. Does the *NEC*® permit the receptacle for the refrigerator to be supplied by an individual branch-circuit?   Yes _____ No _____ *Section* _____.

  c. What is the practice in your area?

  Run an individual branch circuit? _____

  Connect it to one of the 20-ampere small-appliance branch circuits? _____

19. a. The *Code* requires a minimum of two small-appliance circuits in a kitchen. Is it permitted to connect a receptacle in a dining room to one of the kitchen small-appliance circuits? _____

  What *Code* section applies to this situation? _____

  b. May outdoor weatherproof receptacles be connected to a 20-ampere small-appliance circuit? _____

  c. Receptacles located above the countertops in the kitchen must be supplied by at least _____ 20-ampere small-appliance circuits.

20. According to *210.52*, no point along the floor line shall be more than _____ ft (_____ m) from a receptacle outlet. A receptacle must be installed in any wall space _____ ft (_____ m) wide or greater.

21. The *Code* states that in multiwire circuits, the screw terminals of a receptacle must *not* be used to splice the neutral conductors. Why?

  _____

  _____

  _____

  _____

  _____

  _____

  _____

22. Electric fans produce a certain amount of noise. It is possible to compare the noise levels of different fans prior to installation by comparing their _____ ratings.

23. Is it permitted to connect the white grounded circuit conductor to the equipment grounding terminal of a receptacle? _____

24. What is unique about 120-volt appliances that are two-wire cord- and plug-connected instead of being three-wire cord- and plug-connected? _____

  _____

  _____

  _____

25. a. All 125-volt, single-phase 15- and 20-ampere receptacles installed in a kitchen that serve countertop surfaces must be GFCI protected. (True) (False) Circle the correct answer.

b. All 125-volt, single-phase 15- and 20-ampere receptacles installed within 6 ft (1.8 m) of a wet-bar, laundry, or utility room sink must be GFCI protected. (True) (False) Circle the correct answer.

26. The kitchen features a sliding glass door. For the purpose of providing the proper receptacle outlets, answer the following statements *true* or *false*.

a. Sliding glass panels are considered to be wall space._____

b. Fixed panels of glass doors are considered to be wall space. _____

27. The following is a layout for the lighting circuit for the kitchen. Complete the wiring diagram using colored pens or pencils to show the conductors' insulation color.

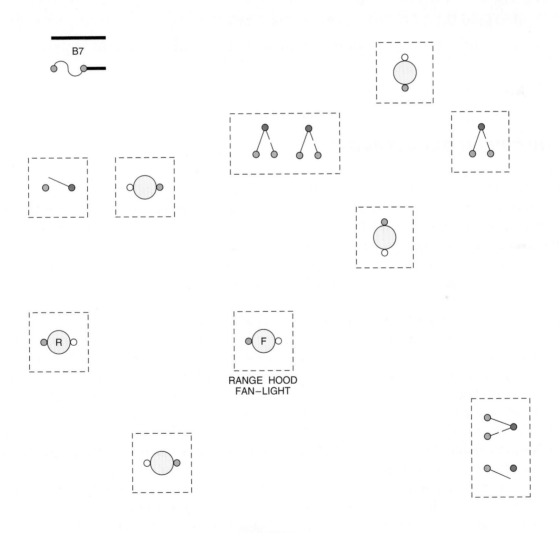

KITCHEN

Lighting Branch-Circuit for the Living Room

OBJECTIVES

After studying this unit, the student should be able to

▶ **lay out the wiring for a typical living room.**

▶ **understand the *NEC*® requirements for track lighting.**

▶ **understand the basics for dimming incandescent and fluorescent lamps.**

LIGHTING CIRCUIT B17 OVERVIEW

The feed for the living room lighting branch-circuit is connected to Circuit B17, a 15-ampere circuit. The home run is brought from Panel B to the weatherproof receptacle outside of the living room next to the sliding door. The circuit is then run to the split-circuit receptacle just inside the sliding door, almost back-to-back with the outdoor receptacle, Figure 13-1.

A three-wire cable is then carried around the living room feeding in and out of the eight receptacles. This three-wire cable carries the black and white circuit conductors plus the red wire, which is the switched conductor. Because these receptacles will provide most of the general lighting for the living room through the use of floor lamps and table lamps, it is advantageous to have some of these lamps controlled by the three switches—two three-way switches and one four-way switch.

The receptacle below the four-way switch in the corner of the living room is required because the *Code* requires that a receptacle be installed in any wall space 24 in. (600 mm) or more in width, *210.52(A)(2)(1)*. It is highly unlikely that a split-circuit receptacle connected for switch control, as are the other receptacles in the living room, will ever be needed. Plus, the short distance to the study/

bedroom receptacle makes it economically sensible to make up the connections as shown on the cable layout.

Accent lighting above the fireplace is in the form of two recessed luminaires (fixtures) controlled by a single-pole dimmer switch. See Unit 7 for installation data and *Code* requirements for recessed luminaires (fixtures).

Table 13-1 summarizes the outlets and estimated load for the living room circuit.

The spacing requirements for receptacle outlets and location requirements for lighting outlets are covered in Unit 3, as is the discussion on the spacing of receptacle outlets on exterior walls where sliding glass doors are involved, *210.52(A)(2)(2)*.

You might want to consider running more than one branch-circuit for the receptacles in the living room. You might also consider using 12 AWG for the "home run" or even for the entire branch-circuit. There very likely might be a concentration of stereo surround sound, TV, CD burners and players, video recorders and players, DVD recorders and players, personal computers, printers, fax machines, copy machines, adding machines, paper shredders, all sorts of other audio/video equipment, and desk lamps. It's a judgment call. Remember, the *Code* is a minimum!

Figure 13-1 Cable layout for the living room. Note that the receptacle on the short wall where the four-way switch is located is fed from the study/bedroom circuit A21.

DESCRIPTION	QUANTITY	WATTS	VOLT-AMPERES
Receptacles @ 120 W Note: Receptacle under S$_4$ is connected to Study/Bedroom circuit.	8	960	960
Weatherproof receptacle	1	120	120
Track light five lamps @ 40 W each	1	200	200
Fireplace recessed luminaires (fixtures) @ 75 W each	2	150	150
TOTALS	12	1430	1430

Table 13-1 Living room outlet count and estimated load (Circuit B17).

TRACK LIGHTING (ARTICLE 410, PART XV)

The plans show that track lighting is mounted on the ceiling of the living room on the wall opposite the fireplace. The lighting track is controlled by a single-pole dimmer switch.

Track lighting provides accent lighting for a fireplace, a painting on the wall, or some item (sculpture or collection) that the homeowner wishes to focus attention on, or it may be used to light work areas such as counters, game tables, or tables in such areas as the kitchen eating area.

Differing from recessed luminaires (fixtures) and individual ceiling luminaires (fixtures) that occupy a very definite space, track lighting offers flexibility

Figure 13-2   End-feed track light.

Figure 13-3   Center-feed track light.

because the actual lampholders can be moved and relocated on the track as desired.

Lampholders from track lighting selected from hundreds of styles are inserted into the extruded aluminum or PVC track at any point (the circuit conductors are in the track), and the plug-in connector on the lampholder completes the connection. In addition to being fastened to the outlet box, the track is generally fastened to the ceiling with toggle bolts or screws. Various track light installations are shown in Figure 13-2, Figure 13-3, and Figure 13-4.

Note on the plans for the residence that the living room ceiling has wood beams. This presents

a challenge when installing track lighting that is longer than the space between the wood beams.

There are a number of things the electrician can do:

1. Install pendant kit assemblies that will allow the track to hang below the beam, as in Figure 13-5.

2. Install conduit conductor fittings on the ends of sections of the track, then drill a hole in the beam through which ½-in. EMT can be installed and connected to the conduit connector fittings, as in Figure 13-6.

Figure 13-4   Plug-in track light.

Figure 13-5   Track lighting mounted on the bottom of a wood beam. The supply wires to the track assembly are fed through the pendant.

Figure 13-6   Track lighting mounted on the ceiling between the wood beams. The two track assembly sections are connected together by means of a conduit that has been installed through a hole drilled in the beam.

3. Run the branch-circuit wiring concealed above the ceiling to outlet boxes located at the points where the track is to be fed by the branch-circuit wiring, as in Figure 13-7.

Where to Mount Lighting Track

Lighting track manufacturers' catalogs and Web sites provide excellent recommendations for achieving the desired lighting results.

Figure 13-7 Track lighting mounted on the ceiling between the wood beams. Each of the two track assembly sections are connected to outlet boxes in the ceiling in the same manner that a regular luminaire (lighting fixture) is hung.

Lighting Track and the *Code*

Lighting track is listed under UL Standard 1574. The *Code* rules for lighting track are found in *410.100* through *410.105.*

Lighting track in residential installations:

- shall be listed by an NRTL.

- shall be installed according to the manufacturers' instructions, per *110.3(B).*

- shall not be installed:
  - where subject to physical damage.
  - in damp or wet locations.
  - concealed.
  - through walls or partitions.
  - less than 5 ft (1.5 m) above finished floor unless protected against physical damage, or if the track is of the low-voltage type operating at less than 30 volts rms open-circuit voltage. See *Article 411* for low-voltage system requirements.
  - In the zone 3 ft (900 mm) horizontally and 8 ft (2.5 m) vertically from the top of a bathtub rim ▶or shower stall threshold. ◀ This is discussed in Unit 10.

- shall be permanently installed and permanently connected to a branch-circuit.

- shall be supported twice if the track is not more than 4 ft (1.2 m), and shall have one additional support for each additional 4 ft (1.2 m) of track.

- shall be grounded. Grounding of equipment is covered in previous units.

- shall not be cut to length in the field unless permitted by the manufacturer. Instructions will specify where and how to make the cut, and how to close off the cut with a proper end cap.

- shall have lampholders designed specifically for the particular track.

- shall not have so many lampholders that the load would exceed the rating of the track.

Figure 13-8 shows cross-sectional views of a typical lighting track. Figure 13-9 shows typical lighting track lampholders.

The track lighting in the living room of this residence is controlled by a single-pole dimmer switch.

Load Calculations for Track Lighting

For track lighting installed in homes, it is not necessary to add additional loads for the purpose of branch-circuit, feeder, or service-entrance calculations. Track lighting is just another form of general lighting. Refer to *220.43(B).*

Figure 13-8 Tracks in cross-sectional views.

Figure 13-9   **Typical lampholders that attach to track lighting systems.** *Courtesy of* Progress Lighting.

In a commercial building, a loading factor of 150 VA for each 2 ft (600 mm) of lighting track must be added to the calculations for branch-circuit, feeder, and service-entrance calculations, *220.43(B)*.

For further information regarding track lighting, refer to *Part XV* of *Article 410*.

Portable Lighting Tracks

Portable lighting tracks, per UL Standard 153 shall:

- not be longer than 8 ft (2.4 m).

- have a cord not less than 5 ft (1.5 m) permanently attached to the track by the manufacturer.

- have overcurrent protection (fuse or circuit breaker) built into the plug cap. Overcurrent

protection in the plug cap is not required if the conductors in the cord are 12 AWG, rated 20 amperes.

- not have its length altered in the field. To do so would void the UL listing.

DIMMER CONTROLS FOR HOMES

Dimmers are used in homes to lower the level of light. There are two basic types: electronic and autotransformer.

Electronic Dimmers

Figure 13-10 shows two types (slide and rotate) of electronic dimmers. These are very popular. They "fit" in a standard switch (device) box. They can

Figure 13-10   **Some typical electronic dimmers. Both perform the same function of varying the voltage to the connected incandescent lamp load; (A) rotating knob; (B) slider knob.** *Courtesy of* Pass & Seymour/Legrand.

replace standard single-pole or three-way toggle switches. See Figure 13-11 for the typical hookup of a single-pole and a three-way dimmer.

The most common type of electronic dimmer for residential use is rated 600 watts, 125 VAC, 60 hertz. They are also available in 1000-watt ratings. For larger wattage requirements, consult manufacturers' literatures and Web sites. The connected load must not exceed the marked wattage rating of the dimmer.

Electronic dimmers offer many features such as LED nightlight, soft start when changing from "off" to the desired setting, and return to the previous setting after a power outage.

Dimmer wiring devices and fan speed controls are physically larger than typical wiring devices (receptacles and switches). Think ahead to make sure you install a large enough wall box to accommodate the dimmer, conductors, and splices that are in the box. To "jam" the dimmer into the box may cause trouble, such as short circuits and ground faults.

Dimmers that install in flush device boxes are intended to control only permanently installed incandescent lighting unless listed for other applications, *404.14(E)*. The instructions will state something like "Do not use to control receptacle outlets." The reasons for this are:

1. Serious overloading could result, because it may not be possible to determine or limit what other loads might be plugged in the receptacle.

2. Dimmers reduce voltage. Reduced voltage supplying a television, radio, stereo components, personal computer, vacuum cleaner, and so on, can result in costly damage to the appliance.

Electronic dimmers for use *only* to control incandescent lamps are marked with a *wattage* rating. Serious overheating and damage to the transformer and to the electronic dimmer will result if an "incandescent only" dimmer is used to control inductive loads.

For example, a dimmer marked 600 watts @ 120 volts is capable of safely carrying 5 amperes.

$$\text{Amperes} = \frac{\text{Watts}}{\text{Volts}} = \frac{600}{120} = 5 \text{ amperes}$$

Dimmers that are suitable for control of "inductive" loads, such as lighting systems that incorporate a transformer (as in low-voltage outdoor lighting and low-voltage track lighting) will be marked in *volt-amperes*. The instructions will clearly indicate that the dimmer can be used on inductive loads.

See Unit 7 for a discussion of watts versus volt-amperes.

Do *not* hook up a dimmer with the circuit "hot." Not only is it dangerous because of the possibility of receiving an electrical shock, but the electronic solid-state dimmer can be destroyed when the splices are being made as a result of a number of "make and breaks" before the actual final connection is completed. The momentary inrush of current each time there is a "make and break" can cause the dimmer's internal electronic circuitry to heat up beyond its thermal capability. As one lead is being connected, the other lead could come in contact with "ground." Unless the dimmer has some sort of "built-in" short-circuit protection, the dimmer is destroyed. The manufacturer's warranty will probably be null and void if the dimmer is worked "hot." It is well worth the time to take a few minutes to de-energize the circuit.

CAUTION: THE POWER MUST BE OFF BEFORE THE DIMMER IS CONNECTED TO THE CIRCUIT.

DIMMER SWITCH

THREE-WAY DIMMER SWITCH

THREE-WAY SWITCH

THREE-WIRE CABLE

Ⓐ LIGHT CONTROLLED BY A SINGLE-POLE DIMMER SWITCH.

Ⓑ LIGHT CONTROLLED BY A THREE-WAY DIMMER SWITCH AND A STANDARD THREE-WAY SWITCH.

Figure 13-11 Use of single-pole and three-way electronic solid-state dimmer controls in circuits.

Electronic dimmers use triacs to "chop" the ac sine wave, so some incandescent lamps might "hum" because the lamp filament vibrates. This hum can usually be eliminated or greatly reduced by turning the dimmer setting to a different brightness level, changing the lamp, installing a "rough service lamp," or as a last resort, changing the dimmer.

Autotransformer Dimmers

Figure 13-12 shows an autotransformer (one winding)-type dimmer. These are physically larger than the electronic dimmer, and require a special wall box. They are not ordinarily used for residential applications unless the load to be controlled is very large. They are generally used in commercial applications for the control of large loads. The lamp intensity is controlled by varying the voltage to the lamp load. When the knob is rotated, a brush contact moves over a bared portion of the transformer winding. As with electronic dimmers, they will not work on direct current, but that is not a problem because all homes are supplied with alternating current. The maximum wattage to be controlled is marked on the nameplate of the dimmer. Overcurrent protection is built into the dimmer control to prevent burnout due to an overload.

Figure 13-13 illustrates the connections for an autotransformer dimmer that controls incandescent lamps.

Figure 13-12   Dimmer control, autotransformer type.

Figure 13-13   Dimmer control (autotransformer) wiring diagram for incandescent lamps.

Dimming Fluorescent Lamps

To dim incandescent lamps, the voltage to the lamp filament is varied (raised or lowered) by the dimmer control. The lower the voltage, the less intense is the light. The dimmer is simply connected "in series" with the incandescent lamp load.

Fluorescent lamps are a different story. The connection is not a simple "series" circuit. To dim fluorescent lamps, special dimming ballasts and special dimmer switches are required. The dimmer control allows the dimming ballast to maintain a voltage to the cathodes that will maintain the cathodes' proper operating temperature, and also allows the dimming ballast to vary the current flowing in the arc. This in turn varies the intensity of light coming from the fluorescent lamp.

There is an exception to the need to use special fluorescent dimming ballasts and special dimmer controls. Some compact fluorescent lamps are available that have an integrated circuit built right into the lamp, allowing the lamp to be controlled by a standard incandescent wall dimmer. These dimmable compact fluorescent lamps can be used to replace standard incandescent lamps of the medium screw shell base type. No additional wiring is necessary. They are designed for use with dimmers, photocells, occupancy sensors, and electronic timers that are marked "Incandescent Only."

Figure 13-14 illustrates the connections for an autotransformer-type dimmer for the control of fluo-

rescent lamps using a special dimming ballast. The fluorescent lamps are F40T12 rapid start lamps. With the newer electronic dimming ballasts, 32-watt, rapid start T8 lamps are generally used.

Figure 13-15 shows an electronic dimmer for the control of fluorescent lamps using a special dimming ballast.

Fluorescent dimming ballasts and controls are available in many types such as: simple "ON-OFF," 100% to 1% variable light output, 100% to 10% variable light output, and two-level 100%/50% light output. Wireless remote control is also an option.

Here are some hints to ensure proper dimming performance of fluorescent lamps:

- Make sure that the lamp reflector and the ballast case are solidly grounded.

- Do not mix ballasts or lamps from different manufacturers on the same dimmer.

- Do not mix single- and two-lamp ballasts.

- "Age" the lamps for approximately 100 hours before dimming. This allows the gas cathodes in the lamps to stabilize. Without "aging" (sometimes referred to as "seasoning"), it is possible to have unstable dimming, lamp striation (stripes), and/or unbalanced dimming characteristics between lamps.

- Because incandescent lamps and fluorescent lamps have different characteristics, they can-

Figure 13-14 Dimmer control wiring diagram for fluorescent lamps.

Figure 13-15   This wiring diagram shows the connections for an electronic dimmer designed for use with a special dimming ballast for rapid start lamps. Always check the dimmer and ballast manufacturers' wiring diagrams, as the color coding of the leads and the electrical connections may be different than shown in this diagram. The dimmer shown is a "slider" type, and offers wide-range (20% to 100%) adjustment of light output for the fluorescent lamp.

not be controlled simultaneously with a single dimmer control.

• Do not control electronic ballasts and magnetic ballasts simultaneously with a single dimmer control.

Always read and follow the manufacturer's instructions furnished with dimmers and ballasts.

Without special dimming fluorescent ballasts and controls, a simple way to have two light levels is to use four-lamp fluorescent luminaires (fixtures), such as those installed in the Recreation Room. Then, control the ballast that supplies the center two lamps with one wall switch, and control the ballast that supplies the outer two lamps with a second wall switch. This is sometimes referred to as an "inboard/outboard" connection. Lamps and ballasts were discussed in Unit 7.

REVIEW

Note: Refer to the *Code* or the plans where necessary.

1. To what circuit is the living room connected? _____

2. How many receptacles are connected to the living room circuit? _____

3. a. How many wires enter the switch box at the four-way switch location? _____

   b. What type and size of box may be installed at this location? _____

   _____

   _____

   _____

4. How many wires must be run between an incandescent lamp and its dimmer control?

   _____

5. Complete the wiring diagram for the dimmer and lamp.

120-VOLT
SUPPLY

6. Is it possible to dim standard fluorescent ballasts? _____

7. a. How many wires must be run between a dimming-type fluorescent ballast and the dimmer control? _____

   b. Is a switch needed in addition to the dimmer control? _____

   _____

   _____

   _____

   _____

8. Explain why fluorescent lamps having the same wattage draw different current values.

   _____

   _____

   _____

   _____

   _____

9. What is the total current consumption of the track lighting and recessed luminaires (fixtures) above the fireplace? Show your calculations.

10. How many television outlets are provided in the living room? _____

11. Where is the telephone outlet located in the living room? _____

_____

12. Solid-state electronic dimmers of the type sold for residential use (shall) (shall not) be used to control fluorescent luminaires (fixtures). These dimmers (are) (are not) intended for speed control of small motors. Circle the correct answers.

13. When calculating a dwelling lighting load on the "volt-ampere per ft^2" basis, the receptacle load used for floor lamps, table lamps, and so forth,
    a. must be added to the general lighting load at 1½ amperes per receptacle.
    b. must be added to the general lighting load at 1500 volt-amperes for every 10 receptacles.
    c. is already included in the calculations as part of the volt-amperes per ft^2.
    Circle the correct letter.

14. A layout of the outlets, switches, dimmers, track lighting, and recessed luminaires (fixtures) is shown in the following diagram. Using the cable layout of Figure 13-1, make a complete wiring diagram of this circuit. Use colored pencils or marking pens to indicate conductors.

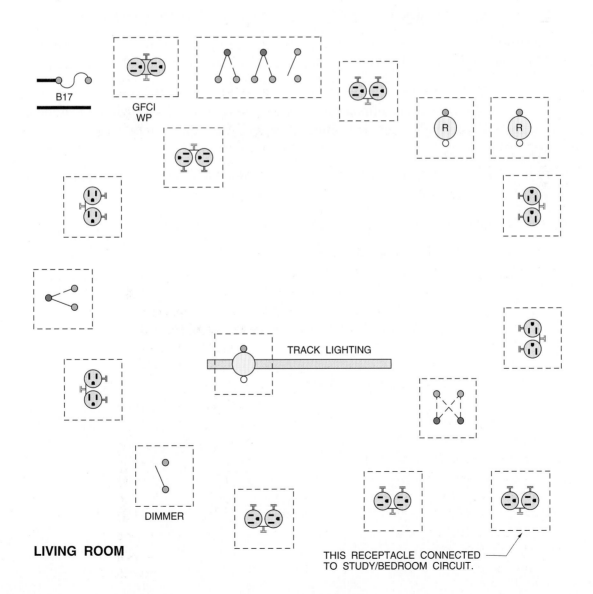

B17

GFCI
WP

TRACK LIGHTING

DIMMER

LIVING ROOM

THIS RECEPTACLE CONNECTED
TO STUDY/BEDROOM CIRCUIT.

15. Prepare a list of television outlets, telephone outlets, and wiring devices shown on the plans and cable layout for the living room area. Include the number of each type present.

_____

_____

_____

_____

_____

_____

16. The living room features a sliding glass door. For the purpose of providing the proper receptacle outlets, answer the following statements *true* or *false*.

    a. Sliding glass panels are considered wall space. _____

    b. Fixed panels of glass doors are considered wall space. _____

17. a. May lengths of track lighting be added when the track is permanently connected?

    _____

    b. May lengths of track lighting be added when the track is cord-connected?

    _____

18. Must track lighting always be fed (connected) at one end of the track? _____

    _____

19. May a standard electronic dimmer be used to control low-voltage lighting that incorporates a transformer, or for controlling the speed of a ceiling fan motor?

    Yes _____   No _____ .

20. Does the *Code* permit cutting lengths of portable track lighting to change their length?

    Yes _____   No _____ .

21. Why should the power be turned off when hooking up a dimmer? _____

    _____

    _____

Lighting Branch-Circuit for the Study/Bedroom

CIRCUIT A21 OVERVIEW

The study/bedroom circuit is so named because it can be used as a study for the present, providing excellent space for home office, personal computer, or den. Should it become necessary to have a third bedroom, the change is easily made.

Circuit A21 originates at Main Panel A. Circuit A21 is a 15-ampere branch-circuit. Overcurrent protection for this circuit is provided by a 15-ampere AFCI circuit breaker in Panel A. AFCIs are discussed in detail in Unit 6. This circuit feeds the study/bedroom at the receptacle in the living room just outside of the study. From this point, the circuit feeds the split-circuit receptacle just inside the door of the study. A three-wire cable runs around the room to feed the other four split-circuit receptacles. Figure 14-1 shows the cable layout for Circuit A21.

The black and white conductors carry the "live" circuit, and the red conductor is the "switch return" for the control of one portion of the split-circuit receptacles.

There are four switched split-circuit receptacles in the study/bedroom controlled by two three-way switches. A fifth receptacle on the short wall leading to the bedroom hallway is not switched. The ready access to this receptacle makes it ideal for plugging in vacuum cleaners or similar appliances.

A ceiling fan/light is mounted on the ceiling. This installation is covered in Unit 9. Table 14-1 summarizes the outlets and estimated load for the study/bedroom circuit.

DESCRIPTION	QUANTITY	WATTS	VOLT-AMPERES
Receptacles @ 120 W each	6	720	720
Closet luminaire (fixture)	1	75	75
Paddle fan/light	1		
Three 50-W lamps		150	150
Fan motor (0.75 A 120 V)		80	90
Valance lighting	1		
Two 40-W fluorescent lamps		80	90
TOTALS	9	1105	1125

Table 14-1 Study/bedroom outlet count and estimated load (Circuit A21).

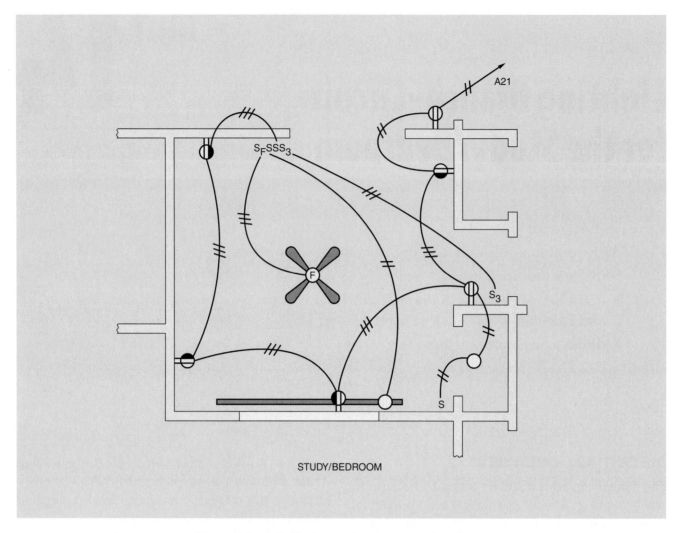

STUDY/BEDROOM

Figure 14-1   Cable layout for the study/bedroom.

As we all know, when someone decides to use any room, such as the study/bedroom, as a home office, a lot of electronic equipment will need to be plugged in. The current draw is very low, but the list for electronic equipment never ends! You will never have enough wall receptacles. Plug-in strips—probably of the surge protection type—are almost always needed.

You could consider:

- running more than one branch-circuit for the receptacles and lighting in the study/bedroom.

- using 12 AWG for the home run or even for the entire branch-circuit(s).

- that all 120-volt, single-phase, 15- and 20-ampere branch-circuits that supply outlets in bedrooms require AFCI protection.

- wiring in quadplex receptacles instead of duplex receptacles.

- roughing in more wall receptacles than the *NEC®* minimum.

- using only plug-in strips that are UL listed.

As previously mentioned, the arrival of AFCI requirements might require new thinking about how to do the wiring for bedrooms.

VALANCE LIGHTING

Figure 14-2 illustrates an interesting indirect fluorescent lighting treatment above the windows and behind the valance board.

If the fluorescent valance lighting is to have a dimmer control, then special dimming ballasts would be required. See the discussion on fluorescent lamp dimming in Unit 13.

The study/bedroom wiring is rather simple. Most of the concepts are covered in previous units in this text.

Figure 14-3 suggests one way to arrange the wiring for the ceiling fan/light control, the valance lighting control, and the receptacle outlet control.

SURGE SUPPRESSORS

Transient voltage surge suppressors, covered in Unit 6, minimize the possibility of voltage surges damaging sensitive electronic equipment.

Figure 14-2   Fluorescent lighting behind valance.

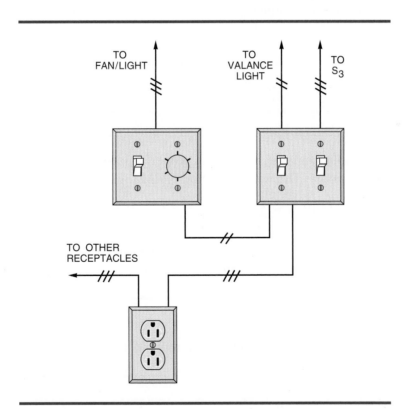

Figure 14-3   Conceptual view of how the study/bedroom switch arrangement is to be accomplished (Circuit A21).

REVIEW

Note: Refer to the *Code* or plans where necessary.

1. Based upon the total estimated load calculations, what is the current draw on the study/bedroom circuit? _____

2. a. The study/bedroom is connected to circuit _____ .

   b. The conductor size for this circuit is _____ .

3. What unique type of electrical protection is required for all outlets in residential bedrooms?

   _____

4. Why is it necessary to install a receptacle in the wall space leading to the bedroom hallway? _____

   _____

   _____

5. Show the calculations needed to select a properly sized box for the receptacle outlet mentioned in question 4. Nonmetallic-sheathed cable is the wiring method.

   _____

   _____

   _____

   _____

   _____

6. Show the calculations necessary to select a properly sized box for the receptacle outlet mentioned in question 4. The wiring method is EMT, usually referred to as "thin-wall conduit."

   _____

   _____

   _____

   _____

   _____

7. Prepare a list of *all* wiring devices used in Circuit A21.

   _____

   _____

   _____

   _____

   _____

   _____

8. Using the cable layout shown in Figure 14-1, make a complete wiring diagram of Circuit A21. Use colored pencils or pens.

FRONT BEDROOM

VALANCE
LIGHT

Dryer Outlet, Lighting and Receptacle Circuits for the Laundry, Powder Room, Rear Entry Hall, and Attic

OBJECTIVES

After studying this unit, the student should be able to

▶ understand the *Code* requirements for bathrooms receptacles.

▶ understand the *Code* requirements for making load calculations and electrical connections for electric dryers.

▶ understand the *Code* requirements for receptacle outlets in laundry areas.

▶ understand the principles of exhaust fans.

▶ understand the *Code* requirements for attic wiring.

LIGHTING CIRCUIT B10

The estimated loads for the various receptacles and luminaires (fixtures) in the laundry, rear entry, powder room, and attic are shown in Table 15-1.

RECEPTACLE CIRCUIT B21

Receptacles in a residential bathroom require special treatment. This was discussed in detail in Unit 3. Now would be a good time for you to review Unit 3.

Circuit B21 is a separate 20-ampere branch-circuit that supplies the receptacle in the powder room.

A bathroom is defined as an area that has a basin, plus one or more of the following: a toilet, a tub, or a shower. Therefore, by definition, the *NEC*® considers the powder room to be a bathroom.

DESCRIPTION	QUANTITY	WATTS	VOLT-AMPERES
Rear entry hall receptacle	1	120	120
Powder room vanity luminaire (fixture)	1	200	200
Rear entrance hall ceiling luminaires (fixtures) @ 100 W each	2	200	200
Laundry room fluorescent luminaire (fixture) Four 40-W lamps	1	160	180
Attic luminaires (fixtures) @ 75 W each	4	300	300
Laundry exhaust fan	1	80	90
Powder room exhaust fan	1	80	90
TOTALS	11	1140	1180

Table 15-1 Laundry-rear entrance-powder room-attic (Circuit B10).

CLOTHES DRYER CIRCUIT ⒶD

The electric clothes dryer requires a separate 120/240-volt, single-phase, three-wire branch-circuit. In this residence, the electric clothes dryer is indicated on the Electrical Plans in the laundry by the symbol ⒶD. It is supplied by Circuit B(1-3) located in Panel B. Figure 15-1 shows the internal components and wiring of a typical electric clothes dryer.

Dryer Connection Methods

Electric clothes dryers can be connected in several ways.

One method is to run a three-wire armored cable directly to a junction box on the dryer provided by the manufacturer, Figure 15-2. This flexible connection allows the dryer to be moved for servicing without disconnecting the wiring.

Another method is to run a conduit (EMT) to a point just behind the dryer. A combination coupling makes the transition from the EMT to a length of flexible metal conduit 2 ft (600 mm) to 3 ft (900 mm) long, as in Figure 15-3. This flexible connection also allows the dryer to be moved for servicing without disconnecting the wiring. A separate equipment grounding conductor must be installed in the conduit to meet the grounding requirements found in *250.118*.

Figure 15-2 Dryer connected by armored cable.

Figure 15-3 Dryer connected by EMT and flexible metal conduit.

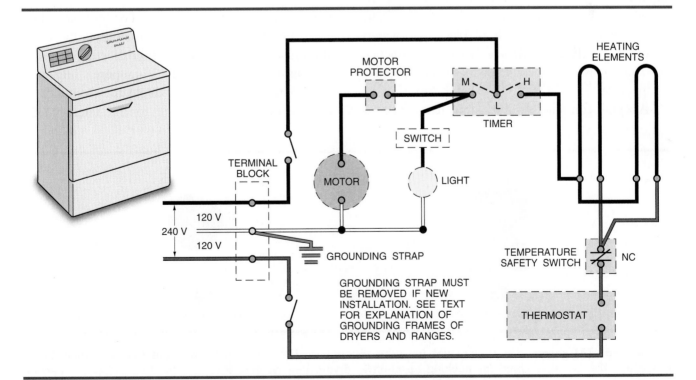

Figure 15-1 Clothes dryer: typical wiring and components.

Figure 15-4 Dryer connection using a cord set.

Probably the most common method is to install a four-wire, 30-ampere receptacle on the wall behind the dryer. A four-wire, 30-ampere cord is then connected to the dryer. This arrangement is referred to as a "cord- and plug-connection" and is shown in Figure 15-4. In conformance to *210.23*, the connected load shall not exceed the branch-circuit rating.

Appliances

All new homes have appliances—some that operate entirely on electricity and some that require electricity as well as another energy source, such as a gas furnace.

Article 422 in the *NEC®* is where to look for the majority of *Code* requirements for appliances. In *Article 422*, you will find quite a few references to other parts of the *NEC,®* such as *Article 430* where motors are involved, and *Article 440* where hermetic-motor compressors (air conditioners and heat pumps) are involved.

As you continue studying this text, specific *Code* rules for appliances are discussed at the time it becomes necessary to mention the rule.

Disconnecting Means

All electrical appliances are required to have a disconnecting means, *422.30*.

For permanently connected motor-driven appliances, *422.32* states that a disconnect switch or a circuit breaker in a panelboard may serve as the disconnecting means, provided the switch or circuit breaker is within sight of the appliance, or is capable of being locked in the "OFF" position. Breaker manufacturers supply "lock-off" devices that fit over the circuit breaker handle. Disconnect switches have provisions for locking the switch "OFF."

The "unit switch" on an appliance is acceptable as a disconnect for an appliance, provided the unit switch has a marked "OFF" position, *422.34*. This certainly turns off the power within the appliance on the load side of the unit switch, but does not shut off the power to the appliance. To consider the unit switch as the appliance's disconnect, another disconnecting means must be provided, and this is permitted to be the service disconnect, *422.34(C)*. If this arrangement is used, the service disconnect serving as the "other" disconnect does not have to be within sight of the appliance, *422.32, Exception*.

Cord- and Plug-Connection. For cord- and plug-connected appliances, *422.33(A)* accepts the cord-and-plug arrangement as the disconnecting means.

Figure 15-4 shows a 30-ampere, four-wire receptacle configuration for an electric clothes dryer installation in conformance to the latest *NEC®*

A

SURFACE OUTLET

B

FLUSH OUTLET

Figure 15-5 (A) shows a surface-mount four-wire receptacle. (B) shows a flush-mount four-wire receptacle. These are the types required for hooking up electric dryers and ranges. Prior to the 1996 *NEC,®* three-wire receptacles and cords were permitted. This is no longer permitted. It is mandatory that a separate equipment grounding conductor be installed to ground the frames of electric clothes dryers and electric ranges. Also see Figure 20-2.

Figure 15-5 shows a surface-mount four-wire receptacle (A) and a flush mount four-wire receptacle (B).

For most residential electric clothes dryers, receptacles and cord sets are rated 30 amperes, 125/250 volts. Some large electric clothes dryers might require a 50-ampere, 125/250-volt receptacle and cord set.

Receptacles for dryers can be surface mounted or flush mounted. Flush-mounted receptacles are quite large, and require a large wall box. A 4-in. square, 1½-in. or 2⅛-in. deep outlet box with a suitable single-gang or two-gang plaster ring generally works fine. Correct box sizing is determined by referring to *Table 314.16(A)* or *Table 314.16(B)*, or to Figure 2-20 and Table 2-1.

Dryer and range receptacles and cord sets are identical. To avoid duplication of text, refer to Unit 20 for a complete description of ratings and blade and slot configurations for three-wire and four-wire dryer and range receptacles, and their associated cord sets.

Load Calculations

In this residence, the dryer is installed in the laundry room. The Schedule of Special Purpose Outlets in the specifications shows that the dryer is rated 5700 watts, 120/240 volts. The schematic wiring diagram in Figure 15-1 shows that the heating element is connected across 240 volts. The motor and lamp are connected to the 120-volt terminals.

It is impossible for the electrician to know the ampere rating of the motor, the heating elements, timers, lights, relays, and so on, and in what sequence they actually operate. This dilemma is answered in the *NEC®* and UL listing requirements for all appliances.

The nameplate must show volts and amperes, or volts and watts, *422.60(A)*. The nameplate must also show the minimum supply circuit conductor's ampacity and the maximum overcurrent protection, *422.62(B)(1)*.

A minimum load of 5000 watts (volt-amperes) or the nameplate rating of the dryer, whichever is larger, is used when calculating branch-circuit, feeder or service-entrance requirements, *220.54*.

The rating of an appliance branch-circuit must not be less than the marked rating on the appliance, *422.10(A)*.

The electric clothes dryer in this residence has a nameplate rating of 5700 watts. This calculates out to be:

$$I = \frac{W}{E} = \frac{5700}{240} = 23.75 \text{ amperes}$$

Rounded up, we have a minimum 30-ampere branch-circuit rating requirement.

Conductor Sizing

The minimum branch-circuit rating for the electric clothes dryer was found to be 30 amperes, and that is the minimum rating of the overcurrent protective device—fuses or circuit breaker. We need to find a conductor that has an ampacity of 30 amperes that will serve as the branch-circuit for the dryer. The conductors will be protected by the 30-ampere overcurrent device.

According to *Table 310.16*, the allowable ampacity for a 10 AWG copper conductor is 35 amperes. The maximum overcurrent protection for these conductors must not exceed 30 amperes, *240.4(D)*.

In this residence, Circuit B(1-3) is a three-wire, 30-ampere, 240-volt branch-circuit.

Nonmetallic-sheathed cable is run concealed in the walls from Panel B to a flush-mounted 4-in. square box behind the dryer location in the laundry room. The nonmetallic-sheathed cable contains three 10 AWG conductors plus a 10 AWG equipment grounding conductor.

Where EMT is used as the wiring method, *Table C1, Annex C* of the *NEC®* shows that three 10 AWG THHN conductors require trade size ½ EMT. The EMT serves as the equipment ground per *250.118*.

Overcurrent Protection for the Dryer

The motor of the dryer has integral thermal overload protection as required by *422.11(G)*. Note the motor protector in Figure 15-1. This protection is required by UL standards and prevents the motor from reaching dangerous temperatures as the result of an overload, bearing failure, or failure to start. Also note the temperature safety switch. This is a high-temperature cut-off that shuts off the heating element should the appliance's thermostat fail to operate. This high temperature cut-off is also a UL requirement.

Overcurrent Protection for Branch-Circuit

NEC® 422.62 requires the appliance to be marked with the minimum supply circuit conductor ampacity and the maximum size overcurrent device.

The rating of the branch-circuit must not be less than the rating marked on the appliance, *422.10(A)*.

In *422.11(A)* we find that the overcurrent protection for the branch-circuit conductors must be sized according to *240.4*, which refers us right back to *Article 422*—Appliances.

The maximum overcurrent protection for 10 AWG copper conductors is 30 amperes, *240.4(D)*.

Grounding Frames of Electric Dryers and Electric Ranges

For protection against electrical shock hazard, frames of electric clothes dryers and electric ranges must be grounded. Some key *Code* references are *250.134, 250.138, 250.140,* and *250.142(B)*.

There are now two methods for grounding the frames of electric dryers and electric ranges, namely *New Installations* and *Existing Installations*.

New Installations. One accepted grounding means is to connect the appliance with a metal raceway such as EMT, flexible metal conduit (Greenfield), armored cable (BX), or other means listed in *250.118*. There are limitations on the use of flexible metal conduit as an equipment grounding conductor. See *250.118* and *348.60*. This topic is covered in Unit 4.

The separate equipment grounding conductor found in nonmetallic-sheathed cable is also an acceptable grounding means. The equipment grounding conductor in nonmetallic-sheathed cable is sized according to *Table 250.122*. See Figure 15-6.

Service-entrance cable is also permitted as a branch-circuit to an electric dryer or electric range. The *Code* rules governing service-entrance cable are covered in Unit 4.

When electric clothes dryers and electric ranges are cord- and plug-connected, the receptacle and the cord set must be four-wire, the fourth wire being the separate equipment grounding conductor.

The small copper bonding strap (link) that is furnished with an electric clothes dryer or electric range *must not* be connected between the neutral terminal and the metal frame of the appliance.

Existing Installations. Prior to the 1996 edition of the *NEC*, *250.60* permitted the frames of electric ranges, wall-mounted ovens, surface cooking units, and clothes dryers to be grounded to the neutral conductor, as in Figure 15-7. By way of *Tentative*

Figure 15-6 New installations of branch-circuit wiring for electric ranges, wall-mounted ovens, surface cooking units, and clothes dryers require that all equipment grounding be done according to *250.134, 250.138, 250.140,* and *250.142(B)*. Receptacles and cord sets must be four-wire (three circuit conductors *plus* equipment grounding conductor).

Interim Amendment No. 53, this special permission was put into effect in July of 1942, and was supposedly an effort to conserve raw materials during World War II. In effect, this special permission allowed the neutral conductor to serve a dual purpose: (1) the neutral conductor and (2) the equipment grounding conductor. This special permission remained in effect until the 1996 *NEC*.

Since then, grounding equipment to a neutral conductor is not permitted!

Existing installations need not be changed, because they were in conformance to the *NEC* at the time.

If an electric clothes dryer or electric range is to be installed in a residence where the dryer branch-circuit wiring had been installed according to pre-1996 *NEC* rules, the wiring *does not* have to be changed. Merely use the grounding strap furnished by the dryer manufacturer to make a connection between the neutral conductor and the frame of the dryer.

RECEPTACLE OUTLETS—LAUNDRY

The *Code* in *210.11(C)(2)* requires that at least one additional 20-ampere branch-circuit be provided for

THREE 10 AWG
NONMETALLIC-SHEATHED
CABLE—NO EQ. GROUND

OUTLET
BOX

COVER FOR
OUTLET BOX

DRYER
RECEPTACLE
OUTLET

Figure 15-7 Prior to the 1996 *National Electrical Code,* it was permitted to ground the junction box to the neutral only if the box was part of the circuit for electric ranges, wall-mounted ovens, surface cooking units, and clothes dryers. The receptacle was permitted to be a three-wire type because the appliance grounding was accomplished by making a connection between the neutral conductor and the metal frame of the appliance. *This is no longer permitted.*

the laundry equipment and that this circuit "shall have no other outlets."

A receptacle outlet must be installed within 6 ft (1.8 m) of the intended location of the washer, *210.50(C)*.

▶All 125-volt, single-phase, 15- and 20-ampere receptacles within 6 ft (1.8 m) of a wet-bar, laundry, or utility room sink must be GFCI protected. See *210.8(A)(6)* and *210.8(A)(7)*. ◀

At least one receptacle outlet must be installed for the laundry equipment, *210.52(F)*.

A load value of 1500 volt-amperes must be included for each two-wire branch-circuit serving laundry equipment, *220.52(B)*. This is for the purpose of computing the size of service-entrance and/or feeders. See the complete calculations in Unit 29.

In the laundry room in this residence, the 20-ampere Circuit B18 supplies the receptacle for the washer. Circuit B18 serves no other outlets. The receptacle is not required to be GFCI protected. Figure 15-8 shows the cable layout for the laundry room.

Figure 15-8 Cable layout for laundry room receptacles.

Circuit B20 is a 20-ampere branch-circuit. It supplies two receptacles in the laundry room and will serve an iron, sewing machine, or other small portable appliance. This circuit also serves the weatherproof receptacle on the outside wall of the laundry area. This branch-circuit is *in addition* to the required separate 20-ampere circuit (B18) for the laundry equipment.

Circuits B18 and B20 are included in the calculations for service-entrance and feeder conductor sizing at a calculated load of 1500 volt-amperes per circuit, *220.52(B)*. This load is included with the general lighting load for the purpose of calculating the service-entrance and feeder conductor sizing, and is subject to the demand factors that are applicable to these calculations. See the complete calculations in Unit 29.

COMBINATION WASHER/DRYERS

Combination washer/dryers take up about half the floor space of a traditional washer and dryer pair.

Some models are stacked and others use the same drum for both wash and dry cycles. Electric washer/dryer combinations generally require a 30-ampere branch-circuit similar to that of a typical electric clothes dryer. Gas washer/ dryer combinations draw 10 or 12 amperes and plug into a standard grounding-type receptacle.

If you know that you will be required to provide a 30-ampere branch-circuit and receptacle for a combination washer/dryer, a 120-volt receptacle and separate 20-ampere branch-circuit must still be provided in conformance to *210.52(F)* and *210.11(C)(2)*.

LIGHTING CIRCUIT

The general lighting circuit for the laundry, powder room, and hall area is supplied by Circuit B10. Note on the cable layout, Figure 15-9, that in order to help balance loads evenly, the attic lights are also connected to Circuit B10.

The types of vanity and ceiling luminaires (fixtures) and wall receptacle outlets, as well as the

Figure 15-9   Cable layout for laundry, rear entrance, powder room, and attic (Circuit B10).

circuitry and switching arrangements, are similar to types discussed in other units of this text and will not be repeated.

The receptacle in the powder room is supplied by Circuit B21, a separate 20-ampere branch-circuit as required by *210.52(D)*, discussed in Unit 10.

Exhaust Fans

Ceiling exhaust fans are installed in the laundry and in the powder room. These exhaust fans are connected to Circuit B10.

The exhaust fan in the laundry will remove excess moisture resulting from the use of the clothes washer, dryer, and ironing.

Exhaust fans may be installed in walls or ceilings, as in Figure 15-10. The wall-mounted fan can be adjusted to fit the thickness of the wall. If a ceiling-mounted fan is used, sheet-metal duct must be installed between the fan unit and the outside of the house. The fan unit terminates in a metal hood or grille on the exterior of the house. The fan has a shutter that opens as the fan starts up and closes as the fan stops. The fan may have an integral pull-chain switch for starting and stopping, or it may be used with a separate wall switch. In either case, single-speed or multispeed control is available. The fan in use has a very small power demand, 90 volt-amperes.

To provide better humidity control, both ceiling-mounted and wall-mounted fans may be controlled with a humidistat. This device starts the fan when the humidity reaches a certain value. When the humidity drops to a preset level, the humidistat turns the fan off.

Some of the more expensive exhaust fans have a built-in sensor that will automatically turn on the exhaust fan at a predetermined humidity level. The fan will automatically turn off when the humidity has been lowered to a preset level. This eliminates the need for a separate wall-mounted humidity control (humidistat).

ATTIC LIGHTING AND PILOT LIGHT SWITCHES

NEC® 210.70(A)(3) requires that at least one lighting outlet must be installed in an attic, and that this lighting outlet contain a switch such as a pull-chain lampholder, or be controlled by a switch located near the entry to the attic. This section of the *Code* addresses attics that are used for storage and attics that contain equipment that might need servicing, such as air-conditioning equipment. *NEC® 210.70(A)(3)* requires that the lighting outlet(s) be installed at or near this equipment.

A 125-volt, single-phase, 15- or 20-ampere receptacle outlet shall be installed in an accessible location within 25 ft (7.5 m) of air-conditioning or heating equipment in attics, in crawl spaces, or on the roof, *210.63*. Connecting this receptacle outlet to the load side of the equipment's disconnecting means is not permitted because this would mean that if you turned off the equipment, the power to the receptacle would also be off and thus would be useless for servicing the equipment.

The residence discussed in this text does not have air-conditioning, heating, or refrigeration equipment in the attic or on the roof.

WALL TYPE

CEILING TYPE

Figure 15-10 Exhaust fans. *Courtesy of* **Broan-NuTone.**

The porcelain lampholders are available with a receptacle outlet for convenience in plugging in an extension cord. However, most electrical inspectors (authority having jurisdiction) would not accept the porcelain lampholder's receptacle in lieu of the required receptacle outlet as stated in *210.63*.

The four porcelain lampholders in the attic are turned on and off by a single-pole switch on the garage wall close to the attic storable ladder. Associated with this single-pole switch is a pilot light. The pilot light may be located in the handle of the switch or it may be separately mounted. Figure 15-11 shows how pilot lamps are connected in circuits containing either single-pole or three-way switches.

Because the attic in this residence is served by a folding storable ladder, switch control in the attic is not required. Where a permanent stairway of six or more risers is installed, *210.70(A)(2)(c)* requires switch control at both levels.

If a neon pilot lamp in the handle (toggle) of a switch does not have a separate grounded conductor connection, then it will glow only when the switch is in the "OFF" position, as the neon lamp will then be in series with the lamp load, Figure 15-12.

The voltage across the load lamp is virtually zero, so it does not burn, and the voltage across the neon lamp is 120 volts, allowing it to glow. When the switch is turned on, the neon lamp is bypassed (shunted), causing it to turn off and the lamp load to turn on.

Use this type of switch when it is desirable to

Figure 15-12   Example of neon pilot lamp in handle of toggle switch.

have a switch glow in the dark to make it easy to locate.

Installation of Cable in Attics

When nonmetallic-sheathed cable is installed in accessible attics, the installation must conform to the requirements of *334.23*. This section refers the reader directly to *320.23*, which describes how the cable is to be protected. See Figure 15-13 and Figure 15-14.

In accessible attics, cables must be protected by guard strips if:

- they are run across the top of floor joists ①.

- they are run across the face of studs ② or rafters ③ within 7 ft (2.1 m) of the floor or floor joists.

Guard strips are *not* required if the cable is run along the sides of rafters, studs, or floor joists ④.

Figure 15-11   Pilot lamp connections.

Figure 15-13   Protection of nonmetallic-sheathed cable or armored cable in an attic. Refer to *334.23* and *320.23*.

In attics not accessible by permanent stairs or ladders, as in Figure 15-13(B), guard strips are required only within 6 ft (1.8 m) of the nearest edge of the scuttle hole or entrance.

Figure 15-13(C) illustrates a cable installation that most electrical inspectors consider to be safe. Because the cables are installed close to the point where the ceiling joists and the roof rafters meet, they are protected from physical damage. It would be very difficult for a person to crawl into this space, or store cartons in an area with such a low clearance. Although the plans for this residence show a 2-ft-wide catwalk in the attic, the owner may decide to install additional flooring in the attic to obtain more

GUARD STRIPS REQUIRED WHEN CABLES ARE RUN ACROSS THE TOP OF JOISTS.

GUARD STRIPS

CABLES RUN
THROUGH BORED HOLES
IN JOISTS ARE CONSIDERED
PROTECTED.

Figure 15-14   Methods of protecting cable installations in accessible attics. See *320.23* and *334.23*.

storage space. Because of the large number of cables required to complete the circuits, it would interfere with flooring to install guard strips wherever the cables run across the tops of the joists, as in Figure 15-14. However, the cables can be run through holes bored in the joists and along the sides of the joists and rafters. In this way, the cables do not interfere with the flooring.

When running cables parallel to framing mem-

bers, be careful to maintain at least 1¼ in. (32 mm) between the cable and the edge of the framing member. This is a *Code* requirement referenced in *300.4(D)* to minimize the possibility of driving nails into the cable. All of *300.4* is devoted to the subject of "Protection Against Physical Damage."

See Unit 4 for more detailed discussion and illustrations relating to the mechanical protection of cables.

REVIEW

Note: Refer to the *Code* or plans where necessary.

1. List the switches, receptacles, and other wiring devices that are connected to Circuit B10. _____

    _____

    _____

    _____

    _____

    _____

    _____

    _____

2. a. Fill in the blank spaces. At least _____ 120-volt receptacle must be installed in a bathroom within _____ in. of the basin, and it must be on the wall _____ to the basin. The receptacle must not be installed _____ on the countertop. The receptacle must be _____ protected.

    b. Circle the correct answers. A separate (15) (20)-ampere branch-circuit must supply the receptacle(s) in bathrooms and (shall) (shall not) serve other loads. If a separate 20-ampere branch-circuit serves a single bathroom, then that circuit (is permitted) (is not permitted) to supply other loads in that bathroom.

3. What special type of switch is controlling the attic lights? _____

    _____

4. When installing cables in an accessible attic along the top of the floor joists _____ _____ must be installed to protect the cables.

5. The total estimated volt-amperes for Circuit B10 has been calculated to be 1300 volt-amperes. How many amperes is this at 120 volts? _____

6. If an attic is accessible through a scuttle hole, guard strips are installed to protect cables run across the top of the joists only within (6 ft [1.8 m]) (12 ft [3.7 m]) of the scuttle hole. Circle correct answer.

7. What section(s) of the *Code* refers (refer) to the receptacle required for the laundry equipment? State briefly the requirements.

    _____

    _____

    _____

    _____

    _____

    _____

    _____

    _____

8. What is the current draw of the exhaust fan in the laundry? _____

9. The following is a layout of the lighting circuit for the laundry, powder room, rear entry hall, and attic. Complete the wiring diagram using colored pens or pencils to indicate conductor insulation color. The GFCI receptacle in the powder room is not shown in the diagram because it is connected to a separate circuit, B21.

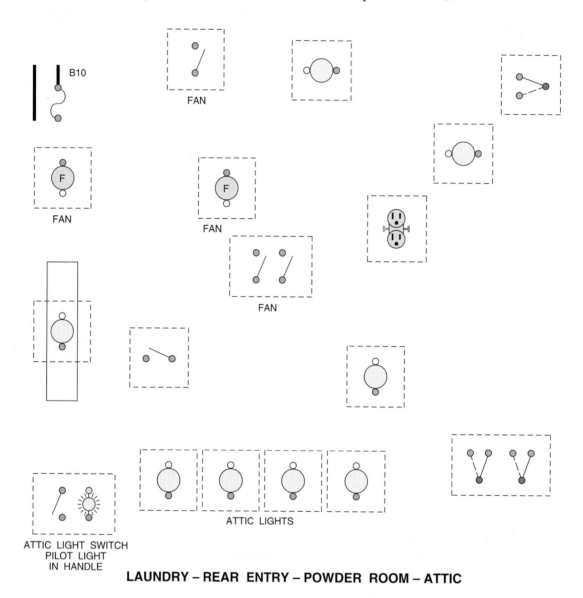

LAUNDRY – REAR ENTRY – POWDER ROOM – ATTIC

10. Laundry receptacle outlets are included in the residential load calculations at a value of _____ volt-amperes per circuit. Choose one: 1000, 1500, 2000 volt-amperes.

11. List the various methods of connecting an electric clothes dryer. _____

_____

12. a. What is the minimum power demand allowed by the Code for an electric dryer if no actual rating is available for the purpose of calculating feeder and service-entrance conductor sizing? _____

   b. What is the current draw? _____

13. What is the maximum permitted current rating of a portable appliance on a 30-ampere branch-circuit? _____

14. What provides motor running overcurrent protection for the dryer? _____

_____

15. a. Must the metal frame of an electric clothes dryer be grounded? _____

  b. May the metal frame of an electric clothes dryer be grounded to the neutral conductor? _____

16. An electric dryer is rated at 7.5 kW and 120/240 volts, three-wire, single-phase. The terminals on the dryer and panelboard are marked 75°C.

  a. What is the wattage rating? _____

  b. What is the current rating? _____

  c. What minimum size type THHN copper conductors are required? _____

  _____

  _____

  _____

  _____

  _____

  d. If EMT is used, what minimum size is required? _____

  _____

  _____

  _____

  _____

  _____

  _____

17. When a metal junction box is installed as part of the cable wiring to a clothes dryer or electric range, may this box be grounded to the circuit neutral? _____

18. A residential air-conditioning unit is installed in the attic.

  a. Is a lighting outlet required? _____

  b. What *Code* section? _____

  c. If a lighting outlet is required, how shall it be controlled? _____

  _____

  d. What *Code* section? _____

  e. Is a receptacle outlet required? _____

  f. What *Code* section? _____

19. The *Code* states that a means must be provided to disconnect an appliance. In the case of an electric clothes dryer, the disconnecting means can be:
  a. a separate disconnect switch located within sight of the appliance.
  b. a separate disconnect switch located out of sight from the appliance. The switch is capable of being locked in the "OFF" position.
  c. a cord- and plug-connected arrangement.
  d. a circuit breaker in a panel that is within sight of the dryer or capable of being locked in the "OFF" position.
  Circle the correct statement(s).

Lighting Branch-Circuit for the Garage

OBJECTIVES

After studying this unit, the student should be able to

▶ understand the basics for lighting a residential garage.

▶ understand the *NEC*® requirements for receptacle outlets in a residential garage.

▶ become familiar with simple landscape lighting.

▶ understand the *NEC*® requirements for underground cable and conduit wiring.

▶ understand the operation and electrical connections for overhead door operators.

LIGHTING BRANCH-CIRCUIT B14

Circuit B14 is a 15-ampere circuit that originates at Panel B in the recreation room and is brought into the wall receptacle box on the front wall of the garage. Figure 16-1 shows the cable layout for Circuit B14. From this point, the circuit is then carried to the switch box adjacent to the side door of the garage. From there the circuit conductors and the switch leg are carried upward to the ceiling lampholders, and then to the receptacle outlets on the right-hand wall of the garage.

The garage ceiling's porcelain lampholders are controlled from switches located at all three entrances to the garage.

The post light in the front of the house is fed from the receptacle located just inside the overhead door. The post light turns on and off by a photocell, an integral part of the post light.

Note that the luminaires (fixtures) on the outside next to the overhead garage door are controlled by

two three-way switches, one just inside the overhead door and the other in the front entry. These luminaires (fixtures) are not connected to the garage lighting circuit.

All of the wiring in the garage will be concealed. Building codes for one- and two-family dwellings require that the walls separating habitable areas and a garage have a fire-resistance rating. In this residence, the walls and ceilings have a fire-resistance rating of 1 hour. Fire resistance rating requirements are discussed in Unit 2.

Table 16-1 summarizes the outlets and the estimated load for the garage circuit.

LIGHTING A TYPICAL RESIDENTIAL GARAGE

The recommended approach to provide adequate lighting in a residential garage is to install one 100-watt lamp on the ceiling above each side of an

Figure 16-1   Cable layout for the garage circuit, B14.

automobile, Figure 16-2. Figure 16-3 shows typical lampholders commonly mounted in garages.

- For a one-car garage, a minimum of two lampholders is recommended.

- For a two-car garage, a minimum of three lampholders is recommended.

- For a three-car garage, a minimum of four lampholders is recommended.

DESCRIPTION	QUANTITY	WATTS	VOLT-AMPERES
Receptacles @ 180 W each	3	540	540
Ceiling lampholders @ 100 W each	3	300	300
Outdoor garage bracket luminaire (fixture) (rear)	1	100	100
Post light	1	100	100
Overhead door opener Motor:	1		
4.8 A × 120 V			576
4.8 A × 120 V × 0.9 PF		518	
two 60-W lamps		120	120
TOTALS	9	1678	1736

Table 16-1 Garage outlet count and estimated load (Circuit B14).

Lampholders arranged in this manner eliminate shadows between automobiles. Shadows are a hazard because they hide objects that can cause a person to trip or fall.

It is highly recommended that these ceiling lampholders be mounted toward the front end of the

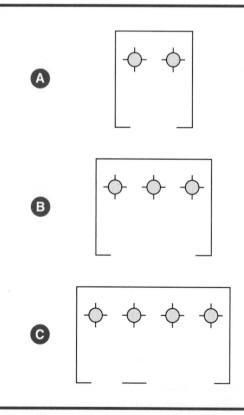

Figure 16-2 Positioning of lights in (A) a one-car garage, (B) a two-car garage, and (C) a three-car garage.

Figure 16-3 Typical porcelain or nonmetallic lampholders that would be installed in garages, attics, basements, crawl spaces, and similar locations. Note that one has a pull chain and one does not.

automobile as it is normally parked in the garage. This arrangement will provide better lighting where it is most needed when the owner is working under the hood. Do not place lights where they will be covered by the open overhead door.

Fluorescent luminaires (fixtures) are quite often used in the warmer parts (south and southwest) of the country. For example, instead of the three porcelain lampholders for conventional incandescent lamps indicated on the electrical plans on the ceiling of the garage, fluorescent strip lights or fluorescent "shop" lights could be installed.

Standard fluorescent lamps installed in a cold garage (northern parts of the country) would be difficult if not impossible to start. If they do start, the lamps will probably "flutter," and the light output will be significantly less than when operated at normal room temperatures. Special lamps and ballasts are required at temperatures below 50–60°F (10–16°C). It is generally cost effective to use incandescent lamps in an unheated garage.

RECEPTACLE OUTLETS IN A GARAGE

At least one receptacle must be installed in an attached residential garage, *210.52(G)*. The garage in this residence has three receptacles. If a detached garage has no electricity provided, then obviously no rules are applicable, but just as soon as a detached garage is wired, then all pertinent *Code* rules become effective and must be followed. These would include rules relating to grounding, lighting outlets, receptacle outlets, GFCI protection, and so on.

NEC® 210.8(A)(2) requires that all 125-volt, single-phase, 15- or 20-ampere receptacles installed in garages shall have ground-fault circuit-interrupter protection. GFCIs are discussed in detail in Unit 6.

There are a few exceptions to *210.8(A)(2)* requiring GFCI protection of receptacles in residential garages. They are:

1. GFCI protection is *not* required if the receptacle is not readily accessible, such as the receptacle installed for the overhead door opener.

2. GFCI protection is *not* required for single or duplex receptacles intended for cord- and plug-connected appliances that are not easily moved and are located in dedicated spaces. Examples include a freezer, refrigerator, or central vacuum equipment. The intent is that these dedicated receptacles will have the appliance cord plugged into it, and therefore would not be used to plug in portable electric tools and extension cords. This exception is difficult to enforce, because the electrician has no idea during the rough-in that a certain receptacle is for the sole purpose of plugging in a specific appliance, say a large freezer. Such appliances are not at the construction site during rough-in.

Note that one GFCI receptacle is installed on the front wall of the garage, in the same box as the single-pole switch for the attic lighting. A second feedthrough GFCI receptacle is installed on the right-hand garage wall near the entry to the house. This GFCI receptacle also provides the required GFCI protection for the other receptacle on the same wall and for the protection of the underground wiring to the post light.

If we were to install one GFCI feedthrough receptacle at the point where Circuit B14 feeds the garage, then the entire circuit would be GFCI protected as required. But nuisance tripping of the GFCI might occur because of:

- the overall length of the circuit wiring.

- possible problems in the overhead door opener.

- possible problems in the underground run to the post light.

- problems in the post lamp itself.

All of these would result in a complete outage on Circuit B14. The electrician must consider the cost of GFCI receptacles and compare it with the cost of additional cable or conduit and wire needed. The possibility of a power outage must also be kept in mind when trying to get by with fewer GFCI receptacles.

All receptacles and switches in the garage are to be mounted 46 in. (1.15 m) to center according to a notation on the First Floor Electrical Plan.

LANDSCAPE LIGHTING

One of the nice things about residential wiring is the availability of a virtually unlimited variety of sizes, types, and designs of outdoor luminaires (lighting fixtures) for gardens, decks, patios, and for accent lighting. Luminaire (fixture) showrooms at electrical distributors and the electrical departments of home centers offer a wide range of outdoor luminaires (lighting fixtures), both 120-volt and low-voltage types. Figure 16-4 and Figure 16-5 illustrate landscape lighting mounted on the ground. See also the following section on outdoor wiring.

UL Standard 1838 covers *Low-Voltage Landscape Lighting Systems* generally used outdoors in wet locations.

UL Standard 2108 covers *Low-Voltage Lighting Systems* for use in dry locations.

Both of these UL standards require the manufacturer to include installation instructions with the product. It is extremely important to read these instructions to be sure the luminaires (lighting fixtures) and associated wiring are installed and used in a safe manner.

Low-voltage lighting products will bear markings such as *Outdoor Use Only*, *Indoor Use Only*, or *Indoor/Outdoor Use*. If marked for use in damp or wet locations, the maximum voltage rating is 15 volts unless live parts are made inaccessible to contact during normal use. Low-voltage luminaires (lighting fixtures) are intended for specific applications, such as surface mounting, suspended installations, or recessed installations, and are so marked. If of the recessed type, they must be installed according to the provisions of *410.66*, as discussed in Unit 7.

Low-voltage lighting systems are covered in *Article 411*. The basic requirements for these systems are that they shall:

- operate at 30 volts rms (42.4 volts peak) or less.

- have one or more secondary circuits, each limited to 25 amperes maximum.

LOW-VOLTAGE DECORATIVE LIGHTING

120- TO 12-VOLT TRANSFORMER

WEATHERPROOF OUTLETS STUBBED OUT OF GROUND

(COVERS SHOWN CLOSED)

COVERS MUST BE WEATHERPROOF WITH PLUG INSERTED OR REMOVED, *406.8(B)(1).*

Figure 16-4 Weatherproof receptacle outlets stubbed out of the ground; either low-voltage decorative landscape lighting or 120-volt PAR luminaires (lighting fixtures) may be plugged into these outlets.

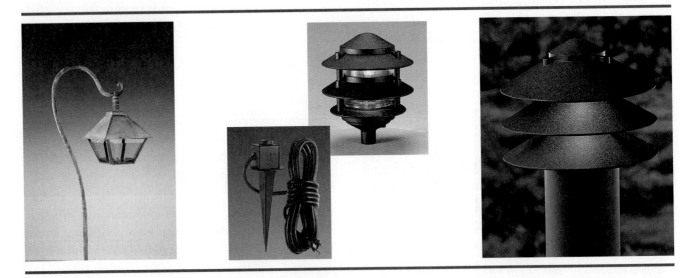

Figure 16-5 Luminaires (fixtures) used for decorative purposes outdoors, under shrubs and trees, in gardens, and for lighting paths and driveways. *Courtesy of* Progress Lighting.

• be supplied through an isolating transformer. A standard step-down transformer does not meet this criteria. Look carefully at the nameplate on the transformer.

• be listed for the purpose.

• not have the wiring concealed or extended through a building unless the wiring meets the requirements of *Chapter 3, NEC.*® An example of this is running Type NMC nonmetallic-sheathed cable for the low-voltage portion of the wiring that is concealed or extended through the structure of the building.

- not be installed within 10 ft (3.0 m) of pools, spas, or water fountains unless specifically permitted in *Article 680*.

- not have the secondary of the isolating transformer grounded.

- not be supplied by a branch-circuit that exceeds 20 amperes.

OUTDOOR WIRING

If the homeowner requests that 120-volt receptacles be installed away from the building structure, the

electrician can provide weatherproof receptacle outlets as illustrated in Figure 16-4.

Outdoor receptacles must have GFCI protection. See Unit 6 for the complete discussion of GFCI protection.

These types of boxes have trade size ½ female threaded openings or hubs in which conduit fittings secure the conduit "stub-ups" to the box. Any unused openings are closed with trade size ½ plugs that are screwed tightly into the threads. Refer to Figure 16-6.

▶All receptacles installed in a wet location must have an enclosure that is weatherproof, whether or not an attachment plug cap is inserted, *406.8(B)(1)*.◀ This requires a self-closing cover that is deep enough to shelter the attachment plug cap of the cord. See Figure 16-4 and Figure 16-7.

CONDUIT BODIES THAT *DO NOT* CONTAIN WIRING DEVICES OR SUPPORT LUMINAIRES (FIXTURES):
- SHALL HAVE THREADED ENTRIES OR HUBS.
- SHALL BE SUPPORTED BY AT LEAST TWO CONDUITS.
- MAY BE SUPPORTED BY ONE RMC, IMC, RNC, OR EMT, BUT ONLY IF THE ENTRY OR HUB ON THE CONDUIT BODY IS NOT LARGER THAN THE TRADE SIZE OF THE CONDUIT.

CONDUIT BODIES THAT *DO* CONTAIN WIRING DEVICES OR SUPPORT LUMINAIRES (FIXTURES):
- SHALL HAVE THREADED ENTRIES OR HUBS.
- SHALL BE SUPPORTED BY AT LEAST TWO CONDUITS.
- MUST BE SUPPORTED WITHIN 18 IN. (450 mm) OF THE ENCLOSURE.
- MAY BE SUPPORTED BY ONE RMC OR IMC, BUT ONLY IF THE ENTRY OR HUB ON THE CONDUIT BODY IS NOT LARGER THAN THE TRADE SIZE OF THE CONDUIT.

NOT A GOOD IDEA TO SUPPORT A CONDUIT BODY WITH ONE CONDUIT. THE CONDUIT BODY COULD EASILY TWIST, RESULTING IN CONDUCTOR INSULATION DAMAGE, AND A POOR GROUND CONNECTION BETWEEN THE CONDUIT AND THE CONDUIT BODY. SUGGEST ADDITIONAL SUPPORT WITH AN ANGLE IRON OR FLAT 1/4 IN. x 1 IN. STEEL ATTACHED TO THE ENCLOSURE AND DRIVEN INTO THE GROUND.

CONDUIT BODIES MAY CONTAIN SPLICES, TAPS, OR DEVICES. BUT ONLY IF MARKED WITH THEIR SPACE VOLUME. USE *TABLE 314.16(B)* FOR DETERMINING THE MAXIMUM NUMBER OF CONDUCTORS.

RMC = RIGID METAL CONDUIT (*ARTICLE 344*)
IMC = INTERMEDIATE METAL CONDUIT (*ARTICLE 342*)
RNC = RIGID NONMETALLIC CONDUIT (*ARTICLE 352*)
EMT = ELECTRICAL METALLIC TUBING (*ARTICLE 358*)

Figure 16-6   Supporting threaded conduit bodies, *314.23(E)* and *(F)*.

Figure 16-7 A recessed type of weatherproof receptacle that meets the requirements of *406.8(B)(1)* and *(B)(2)*, which requires that the enclosure be weatherproof while in use if unattended, where the cord will be plugged in for long periods of time. The "exploded" view illustrates one method of "roughing-in" this type of weatherproof receptacle. *Photos courtesy of* TayMac Corporation.

Where flush-mounted weatherproof receptacles are desired for appearance purposes, a recessed weatherproof receptacle, as illustrated in Figure 16-7, may be used. This type of installation takes up more depth than a standard receptacle and requires special mounting, as indicated in the "exploded" view. Careful thought must be given during the "roughing-in" stages to ensure that the proper depth "roughing-in" box is installed.

Wiring with Type UF Cable
(Article 340)

The plans show that Type UF underground cable, Figure 16-8, is used to connect the post light. The current-carrying capacity (ampacity) of Type

UF cable is found in *Table 310.16*, using the 60°C column. According to *Article 340*, Type UF cable:

- is marked underground feeder cable.

- is available in sizes from 14 AWG through 4/0 AWG (for copper conductors) and from 12 AWG through 4/0 AWG (for aluminum conductors).

- may be used in direct exposure to the sun if the cable is of a type listed for sunlight resistance, or is listed and marked for sunlight resistance.

- may be used with nonmetallic-sheathed cable fittings.

- is flame-retardant.

- is moisture-, fungus-, and corrosion-resistant.

Figure 16-8 Type UF underground cable. *Courtesy of* Southwire Company.

Figure 16-9   Illustration showing how the bare equipment grounding conductor of a Type UF cable is used to ground the metal box.

Figure 16-10   This installation conforms to the *Code*. *NEC® 225.26* does not permit spans of conductors to be run between live or dead trees, but it does not prohibit installing decorative lighting as shown in this figure. *NEC® 410.16(H)* permits outdoor luminaires (lighting fixtures) and associated equipment to be supported by trees. Recognized wiring methods to carry the conductors up the tree must be used.

• may be buried directly in the earth.

• may be used for branch-circuit and feeder wiring.

• may be used in interior wiring for wet, dry, or corrosive installations.

• is installed by the same methods as non-metallic-sheathed cable (*Article 334*).

• must not be used as service-entrance cable.

• must not be embedded in concrete, cement, or aggregate.

• must be buried in the same trench where single-conductor cables are installed.

• shall be installed according to *300.5*.

Equipment fed by Type UF cable is grounded by properly connecting the bare equipment grounding conductor found in the UF cable to the equipment to be grounded, Figure 16-9.

For underground wiring, it is significant to note that whereas the *Code* does permit running up the side of a tree with conduit or cable (when protected from physical damage), as in Figure 16-10, the *Code* does *not* permit supporting conductors between

trees, as in Figure 16-11. Figure 16-12 shows one type of luminaire (lighting fixture) that is permitted for the installation shown in Figure 16-10. Splices are permitted to be made directly in the ground, but only if the connectors are listed by a Nationally Recognized Testing Laboratory for such use, *110.14(B)* and *300.5(E)*.

UNDERGROUND WIRING

Underground wiring is common in residential applications. Examples include wiring for decorative landscape lighting, post lamps, detached buildings such as a garage or tool shed, and low-voltage wiring for lawn sprinkling (irrigation) systems.

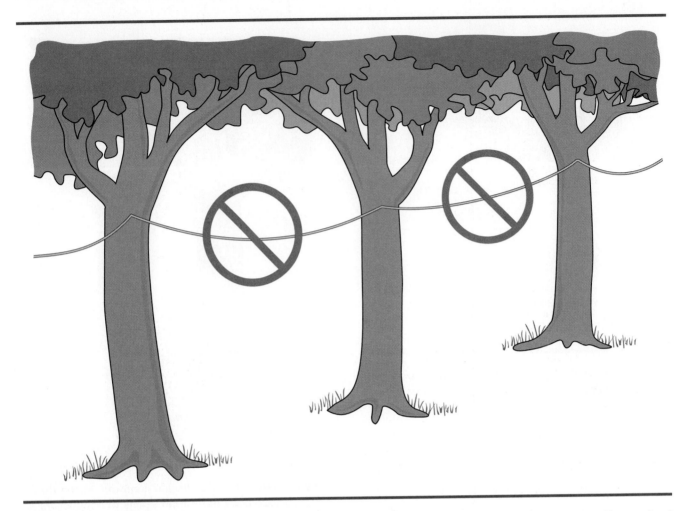

Figure 16-11 *Violation:* Permanent overhead conductor spans are not permitted to be supported by live or dead vegetation, such as trees. ▶Temporary (not over 90 days) holiday decorative lighting branch-circuit conductors or cables are permitted to be supported by trees.◀ See *225.26, 230.10, 590.3(B),* and *590.4(J).*

Figure 16-12 Outdoor luminaire (fixture) with lamps. *Courtesy of* Progress Lighting.

Grounding Equipment Supplied by Underground Wiring

For life safety, proper grounding is important! Figure 16-13 illustrates a lethal *Code* violation. The post lantern is grounded to the white grounded circuit conductor. This violation results in a parallel return path for the current—one path through the ground and the second path through the person. That person might be a child playing around the post. This could result in a possible electrocution!

▶*NEC® 250.4(A)(5)* is very clear about this scenario. It states that *"The earth shall not be considered as an effective ground-fault current path."*◀

Proper grounding of the metal post is achieved using the equipment grounding conductor in the UF cable. Read the instructions that come with the post lantern.

TRANSFORMER

TYPICAL POST LANTERN

"SOME" CURRENT
FLOWS THROUGH
THE PERSON.

IMPROPER CONNECTION
OF POST LANTERN
AND METAL POST TO
GROUNDED CIRCUIT
CONDUCTOR. THE POST
LANTERN AND METAL POST
ARE "LIVE."

THE CIRCUIT IS COMPLETED
THROUGH THE LAMP...
TO THE POST LANTERN...
THEN DOWNWARD ON METAL
POST INTO THE GROUND.

BUT NOTE AT POINTS
Ⓐ & Ⓑ THE CURRENT ALSO
FLOWS THROUGH THE
PERSON... TO THE EARTH...
WHERE THE CURRENTS
JOIN TOGETHER.

MAIN
BONDING
JUMPER

UF CABLE

GROUNDING
ELECTRODE

GROUNDED CIRCUIT
CONDUCTOR "OPEN"

IF THE GROUNDED CIRCUIT CONDUCTOR "OPENS" FOR ANY REASON, THE EARTH PROVIDES AN INEFFECTIVE RETURN PATH
FOR THE CURRENT. THE CURRENT TRAVELS THROUGH THE GROUND BACK TO THE GROUNDING ELECTRODE... TO THE PANEL
GROUNDED NEUTRAL BUS... THROUGH THE SERVICE NEUTRAL CONDUCTOR... THROUGH THE TRANSFORMER... BACK TO THE
SERVICE PANEL... THEN THROUGH THE "HOT" CONDUCTOR TO POST LANTERN AND METAL POST.

**Figure 16-13   VIOLATION! This diagram clearly illustrates the electric shock hazard associated with the _Code_
violation practice of grounding metal objects to the grounded circuit conductor. In this illustration, the post
lantern and metal post have been improperly grounded (connected) to the grounded circuit conductor at (A). The
grounded circuit conductor is "open." One path for the return current is through the metal post into the ground.
The second path is through the person. This is a violation of _250.24(A)(5)_ and _250.142(B)._**

NEC® _250.24(A)(5)_ and _250.142(B)_ prohibit making any connections of equipment to the grounded
circuit conductor anywhere on the load side of the
service disconnect.

How Deep Should Underground Wiring Be?

Digging, aerating, rototilling, and similar yard
work tasks can easily damage underground wiring.

This is extremely hazardous from the electric shock
standpoint. It can also result in costly repairs. The
NEC® addresses minimum depths in great detail.

Table 300.5 shows the minimum depths required
for the various types of wiring methods. Refer to
Figure 16-14 and Figure 16-15.

Across the top of _Table 300.5_ are listed wiring
methods such as direct burial cables or conductors,
rigid metal conduit, intermediate metal conduit,

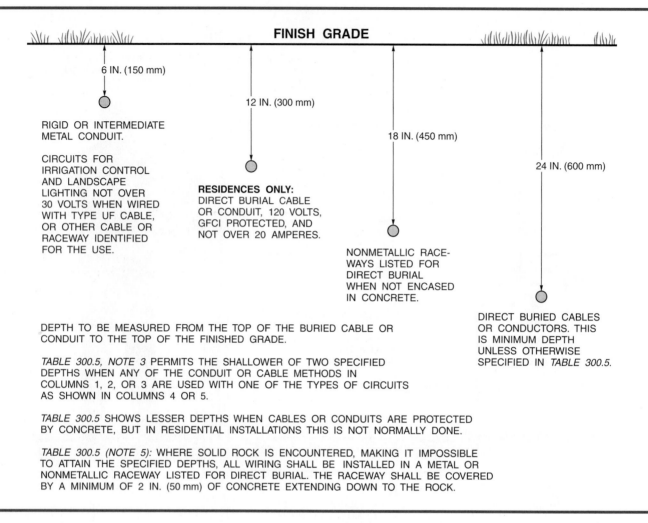

FINISH GRADE

6 IN. (150 mm)

RIGID OR INTERMEDIATE
METAL CONDUIT.

CIRCUITS FOR
IRRIGATION CONTROL
AND LANDSCAPE
LIGHTING NOT OVER
30 VOLTS WHEN WIRED
WITH TYPE UF CABLE,
OR OTHER CABLE OR
RACEWAY IDENTIFIED
FOR THE USE.

12 IN. (300 mm)

RESIDENCES ONLY:
DIRECT BURIAL CABLE
OR CONDUIT, 120 VOLTS,
GFCI PROTECTED, AND
NOT OVER 20 AMPERES.

18 IN. (450 mm)

NONMETALLIC RACE-
WAYS LISTED FOR
DIRECT BURIAL
WHEN NOT ENCASED
IN CONCRETE.

24 IN. (600 mm)

DIRECT BURIED CABLES
OR CONDUCTORS. THIS
IS MINIMUM DEPTH
UNLESS OTHERWISE
SPECIFIED IN *TABLE 300.5.*

DEPTH TO BE MEASURED FROM THE TOP OF THE BURIED CABLE OR
CONDUIT TO THE TOP OF THE FINISHED GRADE.

TABLE 300.5, NOTE 3 PERMITS THE SHALLOWER OF TWO SPECIFIED
DEPTHS WHEN ANY OF THE CONDUIT OR CABLE METHODS IN
COLUMNS 1, 2, OR 3 ARE USED WITH ONE OF THE TYPES OF CIRCUITS
AS SHOWN IN COLUMNS 4 OR 5.

TABLE 300.5 SHOWS LESSER DEPTHS WHEN CABLES OR CONDUITS ARE PROTECTED
BY CONCRETE, BUT IN RESIDENTIAL INSTALLATIONS THIS IS NOT NORMALLY DONE.

TABLE 300.5 (NOTE 5): WHERE SOLID ROCK IS ENCOUNTERED, MAKING IT IMPOSSIBLE
TO ATTAIN THE SPECIFIED DEPTHS, ALL WIRING SHALL BE INSTALLED IN A METAL OR
NONMETALLIC RACEWAY LISTED FOR DIRECT BURIAL. THE RACEWAY SHALL BE COVERED
BY A MINIMUM OF 2 IN. (50 mm) OF CONCRETE EXTENDING DOWN TO THE ROCK.

Figure 16-14   Minimum depths for cables and conduits installed underground. See *Table 300.5* for depths for other conditions.

rigid nonmetallic conduit that is approved for direct burial, special considerations for residential branch-circuits rated 120 volts or less having GFCI protection and overcurrent protection not over 20 amperes, and low-voltage landscape lighting supplied with Type UF cable.

Down the left-hand side of the table we find the location of the wiring methods that are listed across the top of the table.

The notes to *Table 300.5* are very important. Note 4 allows us to select the shallower of two depths when the wiring methods in Columns 1, 2, or 3 are combined with Columns 4 or 5. For example, the basic rule for direct buried cable is that it be covered by a minimum of 24 in. (600 mm). However, this minimum depth is reduced to 12 in. (300 mm) for residential installations when the branch-circuit is not over 120 volts, is GFCI protected, and the

overcurrent protection is not over 20 amperes. This is for residential installations only.

The measurement for the depth requirement is from the top of the raceway or cable to the top of the finished grade, concrete, or other similar cover.

Do not back-fill a trench with rocks, debris, or similar coarse material, *300.5(F)*.

Protection from Damage

Underground direct buried conductors and cables must be protected against physical damage according to the rules set forth in *300.5(D)*.

Installation of Conduit Underground

By definition, an underground installation is a wet location. Cables and insulated conductors installed in underground raceways shall be listed for

SURFACE OF DRIVEWAY OR PARKING AREA

18 IN. (450 mm)

CIRCUITS FOR
IRRIGATION CONTROL
AND LANDSCAPE
LIGHTING NOT OVER
30 VOLTS WHEN WIRED
WITH TYPE UF CABLE,
OR OTHER CABLE OR
RACEWAY IDENTIFIED
FOR THE USE.

12 IN. (300 mm)

18 IN. (450 mm)   18 IN. (450 mm)   18 IN. (450 mm)

DIRECT BURIAL
UF CABLE

NONMETALLIC RACE-
WAYS LISTED FOR
DIRECT BURIAL
WHEN NOT ENCASED
IN CONCRETE.

RIGID OR
INTERMEDIATE
METAL CONDUIT

RESIDENCES ONLY:
RIGID OR INTERMEDIATE
METAL CONDUIT,
DIRECT BURIAL CABLES,
AND NONMETALLIC RIGID
CONDUIT. CIRCUIT
MUST NOT BE OVER
120 VOLTS, MUST BE
GFCI PROTECTED, AND
MUST NOT BE RATED
OVER 20 AMPERES.

WHERE SOLID ROCK IS ENCOUNTERED, MAKING IT IMPOSSIBLE TO ATTAIN THE
SPECIFIED DEPTHS, ALL WIRING SHALL BE INSTALLED IN A METAL OR NONMETALLIC
RACEWAY LISTED FOR DIRECT BURIAL. THE RACEWAY SHALL BE COVERED BY A
MINIMUM OF 2 INCHES (50 mm) OF CONCRETE EXTENDING DOWN TO THE ROCK.

DEPTH TO BE MEASURED FROM THE TOP OF THE BURIED CABLE OR
CONDUIT TO THE TOP OF THE FINISHED DRIVEWAY.

Figure 16-15 Depth requirements under residential driveways and parking areas. These depths are permitted for one- and two-family residences. See *Table 300.5* for depths for other conditions.

use in wet locations. Conductors suitable for use in wet locations will have the letter "W" in their type designation. For example RHW, TW, THW, THHW, THWN, and XHHW. See *300.5(B)* and *310.8(C)*.

All conduit installed underground must be protected against corrosion, *300.6(A)(3)*. The manufacturer of the conduit and UL Standards will furnish information as to whether the conduit is suitable for direct burial. Additional supplemental protection of metal conduit can be accomplished by "painting" the conduit with a nonmetallic coating. See Table 16-2.

When metal conduit is used as the wiring method, the metal boxes, post lantern, and so forth that are properly connected to the metal raceways are considered to be grounded because the conduit serves as the equipment ground, *250.118* (Figure 16-16). See *Article 410, Part V* for details on grounding requirements for luminaires (fixtures).

When nonmetallic raceways are used, then a separate equipment grounding conductor, either green or bare, must be installed to accomplish the adequate grounding of the equipment served. See

THIS METAL BOX IS CONSIDERED
TO BE PROPERLY GROUNDED
WHEN FASTENED PROPERLY TO
THE GROUNDED METAL RACEWAY.

GROUNDED METAL
CONDUIT

PROPER LOCKNUTS
AND BUSHINGS
TIGHTLY FASTENED

Figure 16-16 This illustration shows that a box is grounded when properly fastened to the grounded metal raceway. This is acceptable in *250.118*. More and more *Code*-enforcing authorities and consulting engineers are requiring that a separate equipment grounding conductor be installed in all raceways. This ensures effective grounding of the installation.

Table 300.5 Minimum Cover Requirements, 0 to 600 Volts, Nominal, Burial in Millimeters (Inches)

	Type of Wiring Method or Circuit									
Location of Wiring Method or Circuit	Column 1 Direct Burial Cables or Conductors		Column 2 Rigid Metal Conduit or Intermediate Metal Conduit		Column 3 Nonmetallic Raceways Listed for Direct Burial Without Concrete Encasement or Other Approved Raceways		Column 4 Residential Branch Circuits Rated 120 Volts or Less with GFCI Protection and Maximum Overcurrent Protection of 20 Amperes		Column 5 Circuits for Control of Irrigation and Landscape Lighting Limited to Not More Than 30 Volts and Installed with Type UF or in Other Identified Cable or Raceway	
	mm	in.	mm	in.	mm	in.	mm	in.	mm	in.
All locations not specified below	600	24	150	6	450	18	300	12	150	6
In trench below 50-mm (2-in.) thick concrete or equivalent	450	18	150	6	300	12	150	6	150	6
Under a building	0 (in raceway only)	0	0	0	0	0	0 (in raceway only)	0	0 (in raceway only)	0
Under minimum of 102-mm (4-in.) thick concrete exterior slab with no vehicular traffic and the slab extending not less than 152 mm (6 in.) beyond the underground installation	450	18	100	4	100	4	150 (direct burial) 100 (in raceway)	6 4	150	6
Under streets, highways, roads, alleys, driveways, and parking lots	600	24	600	24	600	24	600	24	600	24
One- and two-family dwelling driveways and outdoor parking areas, and used only for dwelling-related purposes	450	18	450	18	450	18	300	12	450	18
In or under airport runways, including adjacent areas where trespassing prohibited	450	18	450	18	450	18	450	18	450	18

Notes:
1. Cover is defined as the shortest distance in millimeters (inches) measured between a point on the top surface of any direct-buried conductor, cable, conduit, or other raceway and the top surface of finished grade, concrete, or similar cover.
2. Raceways approved for burial only where concrete encased shall require concrete envelope not less than 50 mm (2 in.) thick.
3. Lesser depths shall be permitted where cables and conductors rise for terminations or splices or where access is otherwise required.
4. Where one of the wiring method types listed in Columns 1–3 is used for one of the circuit types in Columns 4 and 5, the shallowest depth of burial shall be permitted.
5. Where solid rock prevents compliance with the cover depths specified in this table, the wiring shall be installed in metal or nonmetallic raceway permitted for direct burial. The raceways shall be covered by a minimum of 50 mm (2 in.) of concrete extending down to rock.

Reprinted with permission from NFPA 70-2005.

IS SUPPLEMENTAL CORROSION PROTECTION REQUIRED?

	In Concrete Above Grade?	In Concrete Below Grade?	In Direct Contact with Soil?
Rigid Conduit[1]	No	No	No[2]
Intermediate Conduit[1]	No	No	No[2]
Electrical Metallic Tubing[1]	No	Yes	Yes[3]

[1] Severe corrosion can be expected where ferrous metal conduits come out of concrete and enter the soil. Here again, some electrical inspectors and consulting engineers might specify the application of some sort of supplemental nonmetallic corrosion protection.

[2] Unless subject to severe corrosive effects. Different soils have different corrosive characteristics.

[3] In most instances, electrical metallic tubing is not permitted to be installed underground in direct contact with the soil because of corrosion problems.

Table 16-2 Corrosion protection required under certain conditions. See *300.6(A)(3)*.

Figure 16-17   Grounding a metal box with a separate equipment grounding conductor.

Figure 16-17. This equipment grounding conductor must be sized according to *Table 250.122, NEC*.®

▶The last sentence of *250.4(A)(5)* states that *"The earth shall not be considered as an effective ground-fault current path."*◀ See Figure 16-18.

Figure 16-19 shows two methods of bringing the conduit into the basement: (1) it can be run below ground level and then can be brought through the basement wall, or (2) it can be run up the side of the building and through the basement wall at the

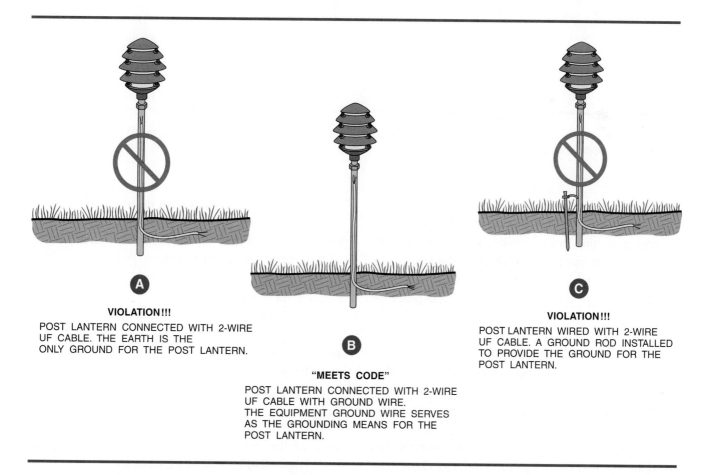

Figure 16-18   ▶These drawings illustrate the intent of the last sentence of *250.4(A)(5)*, which states, "*The earth shall not be considered as an effective ground-fault current path.*"◀

Figure 16-19 Methods of bringing cable and/or conduit through a concrete wall and/or upward from a concrete wall into the hollow space within a framed wall.

ceiling joist level. When the conduit is run through the basement wall, the opening must be sealed to prevent moisture from seeping into the basement. The electrician must decide which of the two methods is more suitable for each installation.

According to the *Code*, any location exposed to the weather is considered a wet location.

Luminaires (fixtures) installed in wet locations shall be marked "Suitable for Wet Locations." These luminaires (fixtures) are constructed so that water cannot enter or accumulate in lampholders, wiring compartments, or other electrical parts, *410.4(A)*.

A damp location is defined as being protected from the weather but subject to moderate degrees of moisture. Partially protected areas such as roofed open porches or areas under canopies are damp loca-

tions. Luminaires (fixtures) in these locations must be marked "*Suitable for Damp Locations.*" A luminaire (fixture) that is "*Suitable for Wet Locations*" is also suitable for damp locations. A luminaire (fixture) that is "*Suitable for Damp Locations*" is not suitable for wet locations.

Check the UL label on a luminaire (fixture) to determine the suitability of the (luminaire) fixture for a wet or damp location.

The post in Figure 16-19 may or may not be embedded in concrete, depending on the consistency of the soil, the height of the post, and the size of the post lantern. Most electricians prefer to embed the base of the post in concrete to prevent rotting of wood posts and rusting of metal posts.

Figure 16-20 illustrates some typical post lights.

Figure 16-20   Post-type lanterns. *Courtesy of* **Progress Lighting.**

OVERHEAD GARAGE DOOR OPERATOR (OPENER) ⒶE

Garage door operators (openers) are listed under UL Standard 325. This is the *Standard for Door, Drapery, Gate, Louver, and Window Operators and Systems.* Residential overhead garage door operators are required to have an inherent "entrapment protection system" that reverses the door in 2 seconds maximum when sensing an obstruction on its way down.

Standard 325 for residential overhead garage door operators also requires that under standard test conditions:

- The *maximum* current draw of the operator shall not exceed 5 amperes, excluding lamps or other devices.*

- The *maximum* current of the motor in locked rotor condition plus the lamps and other devices* shall not exceed 15 amperes.

- The maximum length of the power supply cord is 6 ft (1.8 m).

- The cord must have an equipment grounding conductor and have a grounding-type attachment plug cap.

- The photoelectric sensor shall be mounted not higher than 6 in. (150 mm) above the garage floor.

---

*Such as a capacitor, microprocessor, or photoelectric sensors

- Always follow the manufacturer's installation instructions.

Typical residential operators are advertised as ⅓ and ½ horsepower at 120 volts. The larger the horsepower rating, the more "lift" power. The manufacturer is permitted to show horsepower on the carton, but *not* on the nameplate. Because the horsepower rating is a mechanical rating, the manufacturers' advertised horsepower rating cannot simply be converted to full-load amperes by referring to *NEC® Table 430.248* as we do for conventional electric motors. The ampere rating marked on the nameplate is what counts. This nameplate ampere marking is based on standard test conditions and is the sum of the motor running current, the current draw of the light bulbs, and any other small current draw of additional devices.*

The overhead door operator in this residence is marked 5.8 amperes, 120 volts, and is cord- and plug-connected as shown in Figure 16-21. The carton the unit came in was marked ⅓ horsepower. Overload protection for the motor is an integral part of the motor.

A receptacle is wired in on the ceiling near the operator. This receptacle does not have to be GFCI protected for the simple reason that it is not *readily accessible* as defined in the *NEC.®* See *210.8(A)(2), Exception No. 1.*

Figure 16-21 Overhead door operator with three-wire cord connector.

Figure 16-23 Push-button wiring for overhead door operator.

Although not commonly done, most residential overhead door operators can be "hard-wired" as shown in Figure 16-22.

The receptacle for the overhead door opener is connected to the 15-ampere Garage Branch-Circuit B 14. Although not required by the *NEC,*® many electricians prefer to run a separate branch-circuit. In some localities, a separate circuit is required.

Push Buttons

Push buttons are generally installed next to each door leading into the garage. Key-operated weatherproof switches can also be installed outside, next to the overhead door. The wiring between the overhead

door operator and the push buttons is Class 2 wiring, which is similar to that used for chime and thermostat wiring. See Figure 16-23.

Push buttons should be installed at least 5 ft (1.5 m) or higher above the garage floor so as to be out of reach of small children.

Quite often, electricians will install the low-voltage, Class 2 wiring during the "rough-in" stage so as to conceal this wiring between the push-button locations and the overhead door operator. A neat way to do this is to nail a single-gang, 4-in. square plaster ring to the stud at the desired push-button locations. When the house is "trimmed out," a single-gang faceplate with the push button is installed. See Figure 16-24.

Box-Cover Unit

The overhead door operator unit has overload protection built into it according to UL Standard 325. Additional short-circuit "back-up" protection can be installed in the field. Here's how.

Instead of a standard receptacle on the ceiling near the operator, a box-cover unit can be installed. Box-cover units are available with *switch/fuse* or *receptacle/fuse* combinations, as in Figure 16-21, Figure 16-22, Figure 16-25, and Figure 16-26. Box-cover units provide a convenient disconnecting means in addition to having a fuse holder in which to install a dual-element, time-delay Type S fuse.

Figure 16-22 Connections for overhead door operator.

Figure 16-24 A sketch of how a 4-in. square cover can be used to provide an opening for low-voltage wiring.

Figure 16-25 Box-cover unit that provides disconnecting means plus individual overcurrent protection. *Courtesy of* Cooper Bussmann, Inc.

Size the time-delay fuse in the range of *approximately* 125–150 percent of the overhead door opener's marked ampere rating.

For example, the overhead door operator in this residence is marked 5.8 amperes. Then,

$$5.8 \times 1.25 = 7.25 \text{ amperes}$$

$$5.8 \times 1.5 = 8.7 \text{ amperes}$$

Therefore, installing an 8-ampere, time-delay,

dual-element Type S fuse in the box-cover unit would be appropriate.

The objective is to have a lower ampere rating fuse than the 15- or 20-ampere branch-circuit breaker. Should a short-circuit occur within the overhead door operator, the lower ampere rated fuse will open faster than the 15- or 20-ampere branch-circuit breaker. This takes the overhead door operator off the branch-circuit. The power to the rest of the branch-circuit stays on.

Figure 16-26 This diagram shows the benefits of installing a small-ampere-rated dual element time-delay fuse that is sized to provide backup overload protection for the overhead door operator. If an electrical problem occurs at the overhead door operator, only the 8-ampere fuse will open. The 15-ampere branch-circuit is not affected. All other loads connected to the 15-ampere circuit remain energized.

LOW-VOLTAGE LIGHTING

Low voltage lighting systems that operate at 30 volts or less are covered in *Article 411*. These systems are supplied through an isolating (two-winding) transformer that is connected to a 120-volt, 20-ampere-or-less branch-circuit.

Careful consideration must be given to the length of the low-voltage wiring. When the low-voltage wiring is long, larger conductors are necessary to compensate for voltage drop. If voltage drop is ignored, the lamps will not get enough voltage to burn properly; the lamps will burn dim. Voltage drop calculations are discussed in Unit 4.

You must also be concerned with the current draw of a low-voltage lighting system.

For the same wattage, the current draw of a low-voltage lamp will be much greater than that of a 120-volt lamp. If not taken into consideration, the higher current can cause the conductors to run at dangerous temperatures; that could cause a fire!

EXAMPLE:

A 480-watt load at 120 volts draws:
$$480 \div 120 = 4 \text{ amperes}$$
This is easily carried by a 14 AWG conductor.

A 480-watt load at 12 volts draws:
$$480 \div 12 = 40 \text{ amperes}$$
This requires an 8 AWG conductor.

You can readily see that for this 480-watt example, the conductors must be considerably larger for the 12-volt application to safely carry the current.

Prewired low-voltage lighting assemblies that bear the UL label have the correct size and length of conductors. Adding conductors in the field must be done according to the manufacturers' instructions.

Voltage drop calculations are covered in Unit 4.

REVIEW

Note: Refer to the *Code* or plans where necessary.

1. What circuit supplies the garage? _____

2. What is the circuit rating? _____

3. How many GFCI-protected receptacle outlets are connected to the garage circuit?

   _____

4. a. How many cables enter the wall switch box located next to the side garage door?

   _____

   b. How many circuit conductors enter this box? _____

   c. How many equipment grounding conductors enter this box? _____

5. Show calculations on how to select a proper size box for question 4. What kind of box would you use? _____

   _____

   _____

   _____

   _____

   _____

6. a. How many lights are recommended for a one-car garage? _____ for a two-car garage? _____ for a three-car garage? _____

   b. Where are these lights to be located? _____

   _____

7. From how many points in the garage of this residence are the ceiling lights controlled? _____

8. GFCI breakers or GFCI receptacles are relatively expensive. How would *you* arrange the wiring of the garage circuit to make as economical an installation as possible that complies with the *Code*? _____

_____

_____

_____

_____

9. The total estimated volt-ampere load of the garage lighting branch-circuit B14 draws how many amperes? Show your calculations. _____

_____

_____

_____

_____

_____

10. How high from the floor are the switches and receptacles to be mounted? _____

11. What type of cable feeds the post light? _____

12. All raceways, cables, and direct burial-type conductors require a "cover." Explain the term "cover." _____

_____

_____

_____

_____

_____

13. In the spaces provided, fill in the cover (depth) for the following residential underground installations. See *NEC® Table 300.5*.
    a. Type UF. No other protection. _____ in.
    b. Type UF below driveway. _____ in.
    c. Rigid metal conduit under lawn. _____ in.
    d. Rigid metal conduit under driveway. _____ in.
    e. Electrical metallic tubing between house _____ in.
       and detached garage.
    f. UF cable under lawn. Circuit is 120 volts, _____ in.
       20 amperes, GFCI protected.
    g. Rigid nonmetallic conduit approved for direct _____ in.
       burial. No other protection.
    h. Rigid nonmetallic conduit. Circuit is 120 volts, _____ in.
       20 amperes, GFCI protected.
    i. Rigid conduit passing over solid rock and covered by _____ in.
       2 in. (50 mm) of concrete extending down to rock.

14. What section of the Code prohibits embedding Type UF cable in concrete? _____

    _____

15. The following is a layout of the garage circuit. Using the suggested cable layout, make a complete wiring diagram of this circuit, using colored pens or pencils to indicate conductor insulation colors.

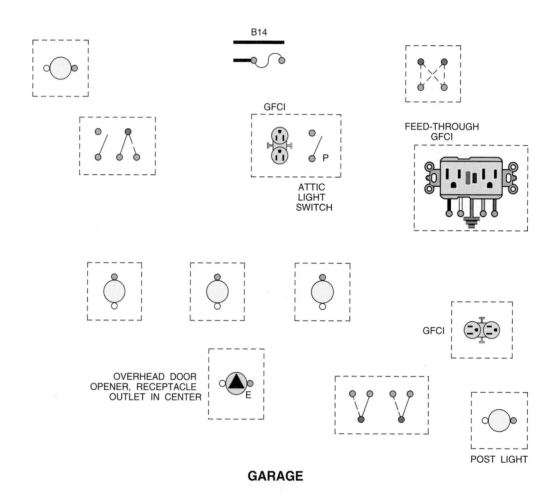

GARAGE

16. When a box-cover unit is installed for an overhead door operator, a time-delay dual-element fuse ampere rating would generally be sized at _____% of the unit's name-plate current rating.

17. When outdoor receptacles are installed where a cord will be plugged in permanently, such as in the case of decorative lighting, the cover must provide weatherproof integrity when the cord is plugged in. True or false? _____ *NEC*® _____ .

18. A homeowner purchased some single-conductor cable that was marked as suitable for direct burial in the ground to wire a post light. A separate equipment grounding conductor to the post light was not carried. Instead, the bottom of the metal post buried two ft into the ground was used as a ground for the post light. The inspector "red tagged" (turned down) the installation. Who is right? The homeowner or the electrical inspector? Explain.

_____

_____

_____

_____

19. What concerns are there when installing a low-voltage lighting system?

_____

_____

_____

UNIT 17

Recreation Room

OBJECTIVES

After studying this unit, the student should be able to

▶ understand three-wire (multiwire) branch-circuits.

▶ understand how to install lay-in luminaires (fixtures).

▶ calculate watts loss and voltage drop in two-wire and three-wire circuits.

▶ understand the term *fixture whips*.

▶ understand the advantages of installing multiwire branch-circuits.

▶ understand problems that can be encountered on multiwire branch-circuits as a result of open neutrals.

RECREATION ROOM LIGHTING (B9, 11, 12)

The recreation room is well-lighted through the use of six lay-in 2 ft × 4 ft fluorescent luminaires (fixtures). These fixtures are exactly the same size as two 2 ft × 2 ft ceiling tiles. The luminaires (fixtures) rest on top of the ceiling tee bars, as in Figure 17-1.

Junction boxes are mounted above a dropped (suspended) ceiling usually within 12 to 24 in. (300 to 600 mm) of the intended luminaire (fixture) location. A flexible conduit referred to as a *fixture whip*, 4 to 6 ft (1.2 to 1.8 m) long, is installed between the junction box and the luminaire (fixture). These fixture whips contain the correct type and size of conductor suitable for the temperature ratings and load required by the *Code* for recessed luminaires (fixtures). See Unit 7 for a complete explanation of *Code* regulations for the installation of recessed luminaires (fixtures), as in Figure 17-2.

Fluorescent luminaires (fixtures) of the type shown in Figure 17-1 might bear a label stating "Recessed Fluorescent Luminaire (Fixture)." The label might also state "Suspended Ceiling Fluorescent Luminaire (Fixture)." Although they might look the same, there is a difference.

Recessed fluorescent luminaires (fixtures) are intended for installation in cavities in ceilings and walls and are to be wired according to *410.64*. These luminaires (fixtures) may also be installed in suspended ceilings if they have the necessary mounting hardware.

Suspended fluorescent luminaires (fixtures) are intended only for installation in suspended ceilings where the acoustical tiles, lay-in panels, and suspended grid are not part of the actual building structure.

Underwriters Laboratories (UL) Standard 1570 covers recessed and suspended ceiling luminaires (fixtures) in detail.

The six recessed "lay-in" fluorescent luminaires (fixtures) in the recreation room are four-lamp luminaires (fixtures) with warm white energy-saving F40SPX30/RS lamps installed in them. The ballasts in these luminaires (fixtures) are high power factor,

338

IMPORTANT: TO PREVENT THE LUMINAIRE (FIXTURE) FROM INADVERTENTLY FALLING, *410.16(C)* OF THE CODE REQUIRES THAT (1) SUSPENDED CEILING FRAMING MEMBERS THAT SUPPORT RECESSED LUMINAIRES (FIXTURES) MUST BE SECURELY FASTENED TO EACH OTHER, AND MUST BE SECURELY ATTACHED TO THE BUILDING STRUCTURE AT APPPROPRIATE INTERVALS, AND (2) RECESSED LUMINAIRES (FIXTURES) MUST BE SECURELY FASTENED TO THE SUSPENDED CEILING FRAMING MEMBERS BY BOLTS, SCREWS, RIVETS, OR SPECIAL LISTED CLIPS PROVIDED BY THE MANUFACTURER OF THE LUMINAIRE (FIXTURE) FOR THE PURPOSE OF ATTACHING THE LUMINAIRE (FIXTURE) TO THE FRAMING MEMBER.

LISTED
STEEL CLIP

Figure 17-1   Typical lay-in fluorescent luminaire (fixture) commonly used in conjunction with dropped suspended ceilings. See *410.16(C)*.

JUNCTION BOX

▶FIXTURE WHIP (NOT OVER 6 FT (1.8 m) LONG, *NEC®* *350.30(A)*, *EXCEPTION NO. 4)*◀

RECESSED LUMINAIRE (FIXTURE)

Figure 17-2   Typical suspended ceiling "lay-in" fluorescent luminaire (fixture) showing flexible fixture whip connection. Conductors in whip must have temperature rating as required on label in luminaire (fixture); for example, "For Supply Use 194°F (90°C) conductors."

energy-saving ballasts. Five of the luminaires (fixtures) are controlled by a single-pole switch located at the bottom of the stairs. One luminaire (fixture) is controlled by the three-way switches that also control the stairwell luminaire (light fixture) hung from the ceiling above the stair landing midway up the stairs.

This was done so that the entire recreation room lighting would not be on continually whenever someone came down the stairs to go into the workshop. This is a practical energy-saving feature.

Circuit B12 feeds the fluorescent luminaires (fixtures) in the recreation room. See the cable layout in Figure 17-3. Table 17-1 summarizes the outlets and estimated load for the recreation room.

How to Connect Recessed and Lay-In Luminaires (Fixtures)

Figure 17-1 and Figure 17-2 show typical lay-in fluorescent luminaires (lighting fixtures). The installation and wiring connections for recessed luminaires (lighting fixtures) and lay-in suspended-ceiling luminaires (fixtures) are covered in Unit 7.

Recessed Luminaires (Fixtures) between Ceiling Joists

Roughing-in recessed luminaires (lighting fixtures) that will be installed between ceiling joists can be quite a challenge, particularly if you are trying to design spacing that makes sense. The conflict is between the 16 in. (400 mm) on center joists and the 24 in. (600 mm) square acoustic ceiling panels and their metal "T" bars and supports. Multiples of 16 in. (400 mm) and 24 in. (600 mm) don't agree except at 4 ft (1.2 m) intervals.

RECREATION ROOM

Figure 17-3 Cable layout for recreation room. The six fluorescent luminaires (fixtures) are connected to Circuit B12. The receptacles located around the room are connected to Circuit B11. The four recessed luminaires (fixtures) over the bar, the three receptacles located on the bar wall, and the receptacles located to the right as you come down the steps are connected to Circuit B9. Circuit B9 and Circuit B11 are three-wire circuits. Junction boxes are located above the dropped ceiling, and to one side of each fluorescent luminaire (fixture). The electrical connections are made in these junction boxes. A flexible fixture "whip" connects between each junction box and each luminaire (fixture). See Figure 17-2.

DESCRIPTION	QUANTITY	WATTS	VOLT-AMPERES
Circuit Number B11			
Receptacles @ 120 W each	7	840	840
Closet luminaires (fixtures) @ 75 W each	2	150	150
Exhaust fan (2.5 A × 120 V)	1	270	300
TOTALS	10	1260	1290
Circuit Number B12			
Luminaire (fixture) on stairway	1	100	100
Recessed fluorescent luminaires (fixtures) with four 40-W lamps each	6	960	1008
TOTALS	7	1060	1108
Circuit Number B9			
Receptacles @ 120 W each	4	480	480
Wet-bar recessed luminaires (fixtures) @ 75 W each	4	300	300
TOTALS	8	780	780

Table 17-1   Recreation room outlets and estimated load (three circuits).

Dropped Ceiling. If there is a "dropped ceiling" that provides sufficient space between the finished ceiling and the bottom edge of the ceiling joists, you probably can space recessed incandescent luminaires (lighting fixtures) easily to be in the exact centers of typical 2 × 2 (600 mm × 600 mm) acoustic panels.

In the case of recessed fluorescent luminaires, their 2 × 4 (600 mm × 1.2 m) size nicely coincides with the ceiling panels.

No Dropped Ceiling. If there is no drop, or if the drop is not as much as the height of the recessed luminaires (fixtures), you have to give a lot of thought to laying out the lighting. You will find that the center-to-center spacing of recessed luminaires (fixtures) will have to be in increments of 4 ft (1.2 m). This spacing will find space between the ceiling joists that are 16 inches on center. With care, the recessed luminaires (fixtures) can be located to be exactly in the center of 2 × 2 (600 mm × 600 mm) acoustic ceiling panels.

RECEPTACLES AND WET BAR (B9–11)

The circuitry for the recreation room wall receptacles and the wet-bar lighting area introduces a new type of circuit. This is termed a *multiwire branch-circuit* or a *three-wire branch-circuit*.

In many cases the use of a multiwire branch-circuit can save money in that one three-wire branch-circuit will do the job of two two-wire branch-circuits.

Also, if the loads are nearly balanced in a three-wire branch-circuit, the neutral conductor carries only the unbalanced current. This results in less voltage drop and watts loss for a three-wire branch-circuit as compared to similar loads connected to separate two-wire branch-circuits.

Figure 17-4 and Figure 17-5 illustrate the benefits of a multiwire branch-circuit relative to watts loss and voltage drop. The example, for simplicity, is a purely resistive circuit, and loads are exactly equal.

The distance from the load to the source is 50 ft. The conductor size is 14 AWG solid, uncoated copper. From *Table 8* in *Chapter 9* in the *NEC*® we find that the resistance of 1000 ft (300 m) of a 14 AWG uncoated copper conductor is 3.07 ohms (10.1 ohms/km). The resistance of 50 ft (15 m) is:

$$\frac{3.07}{1000} \times 50 = 0.1535 \text{ ohms (round off to 0.154)}$$

EXAMPLE 1:
TWO TWO-WIRE BRANCH-CIRCUITS

Watts loss in each current-carrying conductor is

$$\text{Watts} = I^2R = 10 \times 10 \times 0.154 = 15.4 \text{ watts}$$

Watts loss in all four current-carrying conductors is

$$15.4 \times 4 = 61.6 \text{ watts}$$

Voltage drop in each current-carrying conductor is

$$E_d = IR$$
$$= 10 \times 0.154$$
$$= 1.54 \text{ volts}$$

E_d for both current-carrying conductors in each circuit:

$$2 \times 1.54 = 3.08 \text{ volts}$$

Voltage available at load:

$$120 - 3.08 = 116.92 \text{ volts}$$

EXAMPLE 2:
ONE THREE-WIRE BRANCH-CIRCUIT

Watts loss in each current-carrying conductor is

$$\text{Watts} = I^2R = 10 \times 10 \times 0.154 = 15.4 \text{ watts}$$

Watts loss in both current-carrying conductors is

$$15.4 \times 2 = 30.8 \text{ watts}$$

Voltage drop in each current-carrying conductor is

$$E_d = IR$$
$$= 10 \times 0.154$$
$$= 1.54 \text{ volts}$$

E_d for both current-carrying conductors is:

$$2 \times 1.54 = 3.08 \text{ volts}$$

Voltage available at loads:

$$240 - 3.08 = 236.92 \text{ volts}$$

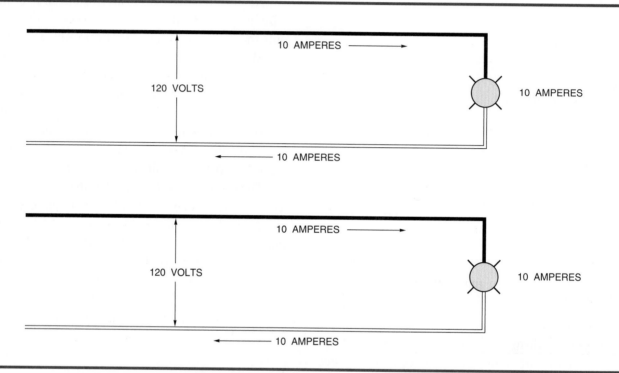

Figure 17-4   Two two-wire branch-circuits.

Figure 17-5   A three-wire branch-circuit.

Voltage available at each load:

$$\frac{236.92}{2} = 118.46 \text{ volts}$$

Advantages of a Three-Wire Branch-Circuit

1. Use less wire for two circuits; there are three conductors instead of four

2. Less voltage drop in the conductors

3. Less watts loss in the conductors

These advantages are why there are three-wire services and feeders, as shown in Panel B in this residence.

Disadvantages of a Three-Wire Branch-Circuit

1. Possible burnout of equipment if the neutral opens

2. Having to work with 240 volts present in outlet and device boxes

3. Easy to make a mistake in connecting the neutral in the panel.

The Recreation Room Wiring

In the recreation room, a three-wire cable carrying Circuit B9 and Circuit B11 feeds from Panel B to the receptacle closest to the panel. This three-wire cable continues to each receptacle wall box along the same wall. At the third receptacle along the wall, the multiwire circuit splits, where the wet-bar Circuit B9 continues on to feed into the receptacle outlet to the left of the wet bar.

The white conductor of the three-wire cable is common to both the receptacle circuit and the wet-bar circuit. The black conductor feeds the receptacles. The red conductor is spliced straight through the boxes and feeds the wet bar area.

Care must be used when connecting a three-wire circuit to the panel. The black and red conductors *must* be connected to the opposite phases in the panel to prevent heavy overloading of the neutral grounded (white) conductor.

The neutral (grounded) conductor of the three-wire cable carries the unbalanced current. This current is the difference between the current in the black wire and the current in the red wire. For example, if one load is 12 amperes and the other load is 10 amperes, the neutral current is the difference between these loads (2 amperes), Figure 17-6.

If the black and red conductors of the three-wire cable are connected to the same phase in Panel A, Figure 17-7, the neutral conductor must carry the total current of both the red and black conductors rather than the unbalanced current. As a result, the neutral conductor will be overloaded. All single-phase, 120/240-volt panels are clearly marked to

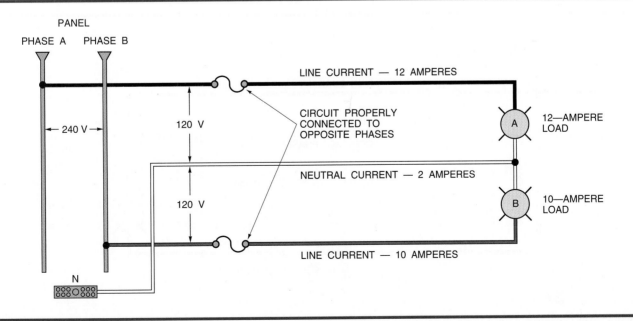

Figure 17-6   Correct wiring connections for a three-wire (multiwire) branch-circuit.

Figure 17-7   Improperly connected three-wire (multiwire) branch-circuit.

help prevent an error in phase wiring. The electrician must check all panels for the proper wiring diagrams.

Figure 17-7 shows how an improperly connected three-wire branch-circuit results in an overloaded neutral conductor.

If an open neutral occurs on a three-wire branch-circuit, some of the electrical appliances in operation may experience voltages higher than the rated voltage at the instant the neutral opens.

For example, Figure 17-8 shows that for an open

neutral condition, the voltage across load A decreases and the voltage across load B increases. If the load on each circuit changes, the voltage on each circuit also changes. According to Ohm's law, the voltage drop across any device in a series circuit is directly proportional to the resistance of that device. In other words, if load B has twice the resistance of load A, then load B will be subjected to twice the voltage of load A for an open neutral condition. To ensure the proper connection, care must be used when splicing the conductors.

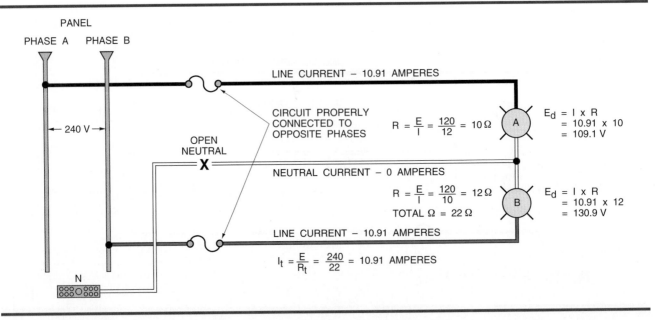

Figure 17-8   Example of an open neutral conductor.

An example of what can occur should the neutral of a three-wire multiwire branch-circuit open is shown in Figure 17-9. Trace the flow of current from phase A through the television set, then through the toaster, then back to phase B, thus completing the circuit. The following simple calculations show why the television set (or stereo or home computer) can be expected to burn up.

$$R_t = 8.45 + 80 = 88.45 \text{ ohms}$$

$$I = \frac{E}{R} = \frac{240}{88.45} = 2.71 \text{ amperes}$$

Voltage appearing across the toaster:

$$IR = 2.71 \times 8.45 = 22.9 \text{ volts}$$

Voltage appearing across the television:

$$IR = 2.71 \times 80 = 216.8 \text{ volts}$$

This example illustrates the problems that can arise with an open neutral on a three-wire, 120/240-volt multiwire branch-circuit.

The same problem can arise when the neutral of the utility company's incoming service-entrance conductors (underground or overhead) opens. The problem is minimized because the neutral of the service is solidly grounded to the metal water piping system within the building. However, there are cases on record where poor service grounding has resulted in serious and expensive damage to appliances within the home because of an open neutral in the incoming service-entrance conductors.

For example, poor service grounding results when relying on a driven iron pipe (that rusts in a short period of time) as the only means of obtaining the service equipment ground. This was quite often done years ago when the electrical code was not as stringent as today's *NEC*.® When working on older homes, check the service ground. It could very well be inadequate, unsafe, and in need of updating to the present *NEC*.®

The integrity of using only ground rods is highly questionable, whether the rod is an iron pipe, a galvanized pipe, or a copper-clad rod.

See Unit 27 for details on the grounding and bonding of service-entrance equipment.

In summary, three-wire branch-circuits are used occasionally in residential installations. The two "hot" conductors of three-wire cable are connected to opposite phases. One white grounded (neutral) conductor is common to both "hot" conductors.

There are two receptacle outlets above the wet bar. ▶All 125-volt, single-phase, 15- and 20-ampere receptacles within 6 ft (1.8 m) of a wet-bar, laundry, or utility room sink must be GFCI protected. See *210.8(A)(6)* and *210.8(A)(7)*.◀ Receptacles must not be installed face-up on the countertop surfaces or work surfaces, *406.4(E)*.

PHASE A   PHASE B

X — OPEN NEUTRAL (POOR CONNECTION)

N

TOASTER
14.2 AMPERES
1704 WATTS
8.45 OHMS
120 VOLTS

TELEVISION SET
1.5 AMPERES
180 WATTS
80 OHMS
120 VOLTS

Figure 17-9   Problems that can occur with an open neutral on a three-wire (multiwire) branch-circuit.

A single-pole switch to the right of the wet bar controls the four recessed luminaires (fixtures) above it. See Unit 7 for details pertaining to recessed luminaires (fixtures).

An exhaust fan of a type similar to the one installed in the laundry is installed in the ceiling of the recreation room to exhaust stale, stagnant, and smoky air from the room.

The basic switch, receptacle, and luminaire (fixture) connections are repetitive of most wiring situations discussed in previous units.

The wall boxes for the receptacles are selected based upon the measurements of the furring strips on the walls. Two-by-two furring strips will require the use of 4-in. square boxes with suitable raised covers. If the walls are furred with 2 × 4s, then possibly sectional device boxes could be used. Select a box that can contain the number of conductors, devices, and clamps to "meet Code." After all, the installation must be safe.

Probably the biggest factor in deciding which wiring method (cable or EMT) will be used to wire the recreation room is "what size and type of wall furring will be used?" See Unit 4 for the in-depth discussion on the mechanical protection required for nonmetallic sheathed cables where the cables will be less than 1¼ in. (32 mm) from the edge of the framing members.

REVIEW

Note: Refer to the *Code* or plans where necessary.

1. What is the total current draw when all six fluorescent luminaires (fixtures) are turned on?

   _____

   _____

2. The junction box that will be installed above the dropped ceiling near the fluorescent luminaires (fixtures) closest to the stairway will have _____ (number of) 14 AWG conductors.

3. Why is it important that the "hot" conductor in a three-wire branch-circuit be properly connected to opposite phases in a panel? _____

   _____

4. In the diagram, Load A is rated at 10 amperes, 120 volts. Load B is rated at 5 amperes, 120 volts.

   a. When connected to the three-wire branch-circuit as indicated, how much current will flow in the neutral? _____

   b. If the neutral should open, to what voltage would each load be subjected, assuming both loads were operating at the time the neutral opened? Show all calculations.

5. Calculate the watts loss and voltage drop in each conductor in the following circuit. Show calculations. Obtain resistance data from *Table 8, Chapter 9,* of the *NEC.*®

```
┌──────────────────────────────────────────────────────────┐
│  50 FT  12 AWG  THHN  SOLID  COPPER  CONDUCTOR             │
│                                                            │
│                                                  15 AMPERE │
│                                                      ☼     │
│                                                            │
│  50 FT  12 AWG  THHN  SOLID  COPPER  CONDUCTOR             │
└──────────────────────────────────────────────────────────┘
```

6. Unless specifically designed, all recessed incandescent luminaires (fixtures) must be provided with factory-installed protection (see Unit 7). _____

7. Where in the *NEC*® would you look for the installation requirements regarding flush and recessed lighting luminaires (fixtures)?

_____

8. What is the current draw of the recessed luminaires (fixtures) above the bar? _____

_____

_____

_____

9. a. VA load for Circuit B9 is _____ volt-amperes.

    b. VA load for Circuit B11 is _____ volt-amperes.

    c. VA load for Circuit B12 is _____ volt-amperes.

10. Calculate the total current draw for Circuit B9, Circuit B11, and Circuit B12.

11. Complete the wiring diagram for the recreation room. Follow the suggested cable layout. Use colored pencils or pens to identify the various colors of the conductor insulation.

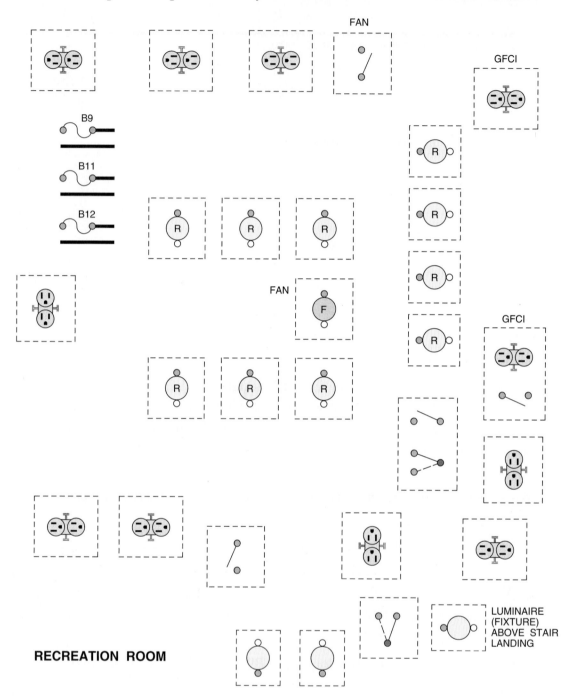

12. a. May a fluorescent luminaire (fixture) that is marked "Recessed Fluorescent Luminaire (Fixture)" be installed in a suspended ceiling? _____

   b. May a fluorescent luminaire (fixture) that is marked "Suspended Ceiling Fluorescent Luminaire (Fixture)" be installed in a recessed cavity of a ceiling? _____

13. The *NEC*® requires GFCI protection for receptacles that serve countertop areas within 6 ft (1.8 m) of the edge of a wet-bar sink, as in the case of the two recreation room receptacles. This requirement is found in *NEC*® _____ .

14. Lay-in fluorescent luminaires (fixtures) that rest on the metal framing members of a suspended ceiling cannot just lay on the framing members. *NEC® 410.16(C)* requires that (1) framing members used to support recessed luminaires (lighting fixtures) must be securely fastened to each other and to the building structure at appropriate intervals, and (2) the luminaires (fixtures) must _____

_____ .

Lighting Branch-Circuit, Receptacle Circuits for Workshop

OBJECTIVES

After studying this unit, the student should be able to

▶ install cable and conduit wiring for a typical nonfinished basement.

▶ understand *NEC®* location requirements for GFCI receptacles in basements.

▶ make conductor fill calculations for raceways.

▶ use derating and correction factors for determining conductor current-carrying capacity.

▶ understand maximum ratings of overcurrent protection for conductors.

▶ install multioutlet assemblies.

The workshop area is supplied by more than one circuit:

A13 Separate branch-circuit for freezer

A17 Lighting

A18 Plug-in strip (multioutlet assembly)

A20 Two receptacles on window wall

The wiring method in the workshop is EMT. Figure 18-1 shows the conduit layout for the workshop.

A single-pole switch at the entry controls all five ceiling porcelain lampholders. However, note on the plans that three of these lampholders have pull-chains, which would allow the homeowner to turn these lampholders on and off as needed (Figure 18-2). They would not have to be on all of the time, which would save energy.

In addition to supplying the lighting, Circuit A17 also feeds the chime transformer, and ceiling exhaust fan. The current draw of the chime transformer is extremely small. Refer to Table 18-1. Refer to Table 18-2 for estimated loads on Circuit A13, Circuit A18, and Circuit A20.

DESCRIPTION	QUANTITY	WATTS	VOLT-AMPERES
Ceiling lights @ 100 W each	5	500	500
Fluorescent luminaires (fixtures) Two 40-W lamps each	2	160	200
Chime transformer	1	8	10
Exhaust fan	1	80	90
TOTALS	9	748	800

Table 18-1 Workshop: outlet count and estimated load. Lighting load (Circuit A17).

Figure 18-1   Conduit layout for workshop. Lighting is Circuit A17. Freezer single receptacle is Circuit A13. The two receptacles on the window wall are connected to Circuit A20. Plug-in strip receptacles and receptacle to the right of Main Panel are connected to Circuit A18. All receptacles are GFCI protected except the single receptacle outlet serving the freezer. A junction box mounted on the ceiling near the fluorescent luminaires (fixtures) above the workbench will provide a convenient place to make up the necessary electrical connections.

WORKBENCH LIGHTING

Two two-lamp, 40-watt fluorescent luminaires (fixtures) are mounted above the workbench to reduce shadows over the work area. The electrical plans show that these luminaires (fixtures) are controlled by a single-pole wall switch. A junction box is mounted immediately above or adjacent to the fluorescent luminaire (fixture) so that the connections can be made readily.

When armored cable or flexible metal conduit is used as the wiring method between the luminaires (fixtures) and the junction box or boxes on the

Figure 18-2 Porcelain and nonmetallic lampholders of both the keyless and pull-chain types. These lampholders are also available with a receptacle outlet, which must be GFCI protected when installed in locations where GFCI protection is required by the *Code*.

DESCRIPTION	QUANTITY	WATTS	VOLT-AMPERES
20-A Circuit, Number A18			
Receptacles next to main panel	1	180	180
A six-receptacle plug-in multioutlet assembly at 1½ A per outlet (180 VA)	1	1080	1080
TOTALS	2	1260	1260
20-A Circuit, Number A20			
Receptacles on window wall @ 180 W each	2	360	360
TOTALS	2	360	360
20-A Circuit, Number A13			
Single receptacle for freezer (*not* GFCI)	1	—	696
TOTALS	1	1620	2316

Table 18-2 Workshop: outlet count and estimated load for three 20-A circuits A13, A18, A20.

ceiling, either wiring method provides the necessary equipment grounding for the fluorescent luminaires (fixtures), provided the metal outlet box is properly grounded. See *250.118* for the many acceptable means to accomplish equipment grounding.

Many workshop fluorescent luminaires (lighting

fixtures) are furnished with a three-wire (black, white, green) flexible cord suitable for the purpose that has the attachment plug cap factory-attached to the cord. These luminaires (fixtures) can be plugged into a grounding-type receptacle mounted on the ceiling immediately above the fluorescent luminaires (fixtures). This provides proper grounding of the luminaire (lighting fixture).

RECEPTACLE OUTLETS

NEC® 210.8(A)(4) and *210.8(A)(5)* require that all 125-volt, single-phase, 15- or 20-ampere receptacles installed in unfinished basements—such as workshop and storage areas that are not considered habitable, and crawl spaces that are at or below ground level—must be GFCI protected.

Exceptions to this requirement are:

- Receptacles that are not readily accessible.

- A single receptacle or a duplex receptacle for two appliances that occupy a dedicated space for each appliance, are cord- and plug-connected, and are not easily moved. Examples include freezers, refrigerators, water softeners, and central vacuum equipment.

- For the receptacle(s) installed for the laundry equipment as required by *210.52(F)* and *210.11(C)(2)*.

See Unit 6 for a full discussion of GFCIs.

NEC® 210.52(G) makes it mandatory to install at least one receptacle outlet in each unfinished basement area. If the laundry area is in the basement, then a receptacle outlet must *also* be installed for the laundry equipment, *210.52(F)*. This additional receptacle must be within 6 ft (1.8 m) of the intended location of the washer, *210.50(C)*, and must be a 20-ampere circuit.

The circuit feeding the required laundry receptacle shall have no other outlets connected to it, *210.11(C)(2)*.

When a single receptacle is installed on an individual branch-circuit, the receptacle rating must not be less than the rating of the branch-circuit, *210.21(B)(1)*.

CABLE INSTALLATION IN BASEMENTS

Unit 4 covers the installation of nonmetallic-sheathed cable (Romex) and armored cable (BX).

Some additional rules apply when nonmetallic-sheathed cable and armored cable are run exposed, because of possible physical damage to the cables.

In unfinished basements, nonmetallic-sheathed cables made up of conductors smaller than two 6 AWG or three 8 AWG must be run through holes bored in joists, or on running boards, *334.15(C)*. In unfinished basements, exposed runs of armored cable must closely follow the building surface or running boards to which it is fastened, *320.15*. *NEC® 320.15* also permits armored cable (BX) to be run on the undersides of joists in basements, attics, or crawl spaces, where supported on each joist and so located as not to be subject to physical damage. This exception is not permitted when using non-metallic-sheathed cable. The local inspection authority usually interprets the meaning of "subject to physical damage."

Review Unit 4 and Unit 15 for additional discussion regarding protection of cables installed in exposed areas of basements and attics.

CONDUIT INSTALLATION IN BASEMENTS

Where metal raceways are used in house wiring, in most cases the raceways will be EMT, *Article 358*, as in Figure 18-3. In some instances, rigid metal conduit (*Article 344*) will be used for a mast service; refer to Unit 27. The following discussion only briefly touches upon metal raceways, because most house wiring across the country is done with non-metallic-sheathed cable, which is covered in detail in Unit 4.

Many local electrical codes do not permit cable wiring in unfinished basements other than where absolutely necessary. For example, a cable may be dropped into a basement from a switch at the head of the basement stairs. Transitions between cable and conduit also must be made. Usually this will be a junction box. The electrician must check the local codes before selecting conduit or cable for the installation, to prevent costly wiring errors.

The electrician should never assume that the *NEC®* is the recognized standard everywhere. Any city or state may pass electrical installation and licensing laws. In many cases these laws are more stringent than the *NEC®*.

RIGID STEEL CONDUIT   ELECTRICAL METALLIC TUBING

Figure 18-3   Types of electrical raceways. *Courtesy of* **Allied Tube & Conduit Corp.**

Outlet boxes may be fastened to masonry walls with lead or plastic anchors, shields, concrete nails, or power-actuated studs.

Conduit may be fastened to masonry or wood surfaces using straps, such as the one-hole strap illustrated in Figure 18-4.

The conduits on the workshop walls and on the ceiling are exposed to view. The electrician must install the exposed wiring in a neat and skillful manner while complying with the following practices:

- The conduit runs must be straight.
- Bends and offsets must be true.
- Vertical runs down the surfaces of the walls must be plumb.

Figure 18-4   A one-hole conduit strap.

- To ensure that the conductor insulation will not be injured, be sure to ream the cut ends of rigid conduit and electrical metallic tubing to remove rough edges and burrs.

All conduits and boxes on the workshop ceiling are fastened to the underside of the wood joists, as in Figure 18-5. Raceways shall be installed as a complete system between pull points before pulling in the wires, as shown in Figure 18-6.

If the electrician can get to the residence before the basement concrete floor is poured, he might find it advantageous to install conduit runs in or under the concrete floor from the main panel to the freezer outlet, the window wall receptacles, and the water pump disconnect switch location. This can be a cost-saving (labor and material) benefit.

Now might be a good time for you to review the *NEC®* rules for electrical metallic tubing.

Metal raceways and armored cables (BX) are acceptable equipment grounding conductors, *250.118*. The runs would have to be continuous between other effectively grounded boxes, panels, and/or equipment. When a nonmetallic wiring method (NM, NMC, UF, etc.) is installed where short sections of metal raceway are provided for mechanical protection—or for appearance—these short sections of raceway need not be grounded, *250.86, Exception No. 2*.

Cable in Ceilings—EMT on Walls

▶In many parts of the country, the wiring method in unfinished basements is nonmetallic-sheathed cable in the ceiling, and EMT on the walls for receptacles and switches. This protects and provides a neat appearance for the cable. Figure 4-20A clearly illustrates the requirements for this type of installation. See *NEC® 334.15(C)*.◀

Outlet Boxes for Use in Exposed Installations

Figure 18-7 illustrates outlet boxes and plaster rings that are used for concealed wiring, such as for the recreation room and first floor wiring in the residence. For surface wiring such as the workshop, various types and sizes of outlet boxes may be used. For example, a handy box, Figure 18-8, could be used for the freezer receptacle. The remaining workshop boxes could be 4-in. square outlet boxes with raised covers, Figure 18-9. The box size and type is determined by the maximum number of conductors contained in the box, *Article 314*.

NEC® 406.4(C) prohibits mounting and supporting a receptacle to a cover (flat or raised) by only one screw unless the device assembly or box cover is listed and identified for securing by a single screw.

Figure 18-5 Raceways, boxes, fittings, and cabinets must be securely fastened in place (points *X*), *300.11(A)*. It is not permissible to hang other raceways, cables, or nonelectrical equipment from an electrical raceway or cable (point *XX*), *300.11(B)* and *(C)*. Refer to Figure 24-10 for an exception that allows low-voltage thermostat cables to be fastened to the conduit that feeds a furnace or other similar equipment. See Unit 17 for discussion of the use of suspended-ceiling support wire for supporting equipment.

INSTALL CONDUIT COMPLETELY BETWEEN BOXES BEFORE PULLING IN CONDUCTORS, *300.18(A)*. NO MORE THAN THE EQUIVALENT OF FOUR QUARTER BENDS (360° TOTAL), IS PERMITTED BETWEEN PULL POINTS. THIS RULE IS FOUND IN *SECTION 26* IN *ARTICLES 342, 344, 348, 352, 358, 350, 356,* AND *362.* BE SURE TO COUNT THE DEGREES OF SMALL OFFSETS AND SADDLE BENDS.

DO NOT PULL CONDUCTORS INTO A PARTIALLY INSTALLED CONDUIT SYSTEM, AND THEN ATTEMPT TO SLIDE THE REMAINING SECTION OF RACEWAY OVER THE CONDUCTORS. THIS IS A VIOLATION OF *300.18(A)*.

Figure 18-6 Do not pull wires into the raceway until the conduit system is completely installed between pull points.

(A) SQUARE BOXES AND SINGLE-GANG RAISED PLASTER COVERS OR RINGS

(B) THREE-GANG BOX WITH RAISED PLASTER COVER OR RING

Figure 18-7 Typical outlet boxes and raised covers. *Courtesy of* Appleton Electric Company.

Figure 18-8   Handy box and covers. *Courtesy of* **Appleton Electric Company.**

Figure 18-9   Four-inch square boxes and raised covers. *Courtesy of* **Appleton Electric Company.**

The listing and identifying for the purpose is a decision that is made by an NRTL, such as Underwriters Laboratories, Inc. In the past, securing a receptacle with a single screw often resulted in a loose receptacle when used over a long period of time. Many short-circuits and ground-faults have been reported because of this.

For the typical raised cover, as illustrated in Figure 18-9, the duplex receptacle would be fastened to the raised cover by two small No. 6-32 screws and nuts, usually furnished with the raised cover. The raised cover has two small "knock-out" holes for these screws. The center No. 6-32 screw could also be used.

If a receptacle is installed on a surface-mounted box where there is direct metal-to-metal contact between the grounded metal box and the metal yoke of the receptacle, the receptacle is considered to be properly grounded without connecting an additional equipment grounding conductor to the equipment grounding terminal on the receptacle, *250.146*. Be sure to remove the small cardboards from the No. 6-32 screws to attain metal-to-metal contact.

Most commonly used outlet, device, and junction boxes have knockouts for trade size ½, ¾, and 1 connectors. The knockout size selected depends upon the size of the conduits entering the box. The conduit size is determined by the maximum percentage of fill of the conduit's total cross-sectional area by the conductors pulled into the conduit.

Supporting equipment and/or raceways to the joists, walls, and so on, is accomplished with proper nails, screws, anchors, and straps. Be careful when trying to support other equipment from raceways as this is not always permitted. Refer to Figure 18-5.

Conduit Fill Calculations

For simplicity, the following conduit fill calculations show in. and in.^{2} To show calculations in in., in.^{2}, mm, and mm^2 would be very confusing.

An electrician must be able to select the proper size raceways based upon the number, size, and type of conductors to be installed in a particular type of raceway. To do this, the electrician must become familiar with the use of certain tables found in *Chapter 9* and *Annex C* of the *NEC.*®

Percent Fill. The *NEC*® has established the following "percent fill" values for conduit and tubing.

- one conductor . . . . 53% fill

- two conductors . . . . 31% fill

- three or more conductors . . . . . 40% fill

- conduit or nipples not over
  24 in. (600 mm). . . . . 60% fill
  Derating of conductor ampacity not required.

These percentages are taken from *Chapter 9, Table 1* of the *NEC.*®

All Conductors the Same Size and Type. When all conductors are the same size and type:

- Determine how many conductors of one size and type are to be installed in the raceway.

- Refer to *Tables C1* through *C12A* in the Annex of the *NEC*® for the type of conduit, tubing, or raceway, and for the specific size and type of conductors.

- Select the proper size raceway directly from the table.

These tables are not difficult to use. Here are a few examples of conductor fill for EMT. Look these up in the tables to confirm the accuracy of the examples.

- Not over five 10 AWG THHN conductors may be pulled into a trade size ½ EMT.

- Not over eight 12 AWG THHN conductors may be pulled into a trade size ¾ EMT.

- Not over three 8 AWG THHN conductors may be pulled into a trade size ½ EMT.

- Not over six 8 AWG THHN conductors may be pulled into a trade size 1 EMT.

Conductors of Different Sizes and/or Types. When conductors are different sizes or types:

- Refer to *Chapter 9, Table 5* of the *NEC.*®

- Find the size and type of conductors used to determine the approximate cross-sectional area in in.² of the conductors.

- Refer to *Chapter 9, Table 8* for the in.² area of bare conductors.

- Total the in.² areas of all conductors.

- Refer to *Chapter 9, Table 4* for the type and size of raceway used.

- Determine the minimum size raceway based on the total in.² area of the conductors and the allowable in.² area fill for the particular type of raceway used.

- The sum must not exceed the allowable percentage fill of the cross-sectional in.² area of the raceway used.

For illustrative purposes, only the data for a few types of wires from *Table 5* are shown in Table 18-3. The dimensions and areas for some EMT from *Table 4* are shown in Table 18-4. You will need to refer to *Chapter 9* and *Annex C* of the *NEC*® for the many other sizes and types of conductors and raceways.

EXAMPLE: If three 6 AWG THW conductors and two 8 AWG THW conductors are to be installed in one EMT, determine the proper size EMT.

SOLUTION:

1. Find the in.² area of the conductors in *NEC*® *Chapter 9, Table 5.*

Three 6 AWG THW conductors	0.0726 in.²
	0.0726 in.²
	0.0726 in.²
Two 8 AWG THW conductors	0.0556 in.²
	0.0556 in.²
Total Area	0.3290 in.²

TYPE	SIZE	APPROX. DIAM. IN.	APPROX. AREA IN.²
THHN	14	0.111	0.0097
THHN	12	0.130	0.0133
THHN	10	0.164	0.0211
THHN	8	0.216	0.0366
THHN	6	0.254	0.0507
THHN	3	0.352	0.0973
THW	6	0.304	0.0726
THW	8	0.266	0.0556

Table 18-3 This sampling of typical conductors and their dimensions is taken from *Table 5, Chapter 9, NEC.*® See *Table 5* for metric values.

ELECTRICAL METALLIC TUBING					
Trade Size (in.)	Internal Diameter (in.)	Total Area 100% (in.^{2})	2 Wires 31% (in.^{2})	Over 2 Wires 40% (in.^{2})	1 Wire 53% (in.^{2})
½	0.622	0.304	0.094	0.122	0.161
¾	0.824	0.533	0.165	0.213	0.283
1	1.049	0.864	0.268	0.346	0.458
1¼	1.380	1.496	0.464	0.598	0.793
1½	1.610	2.036	0.631	0.814	1.079
2	2.067	3.356	1.040	1.342	1.778
2½	2.731	5.858	1.816	2.343	3.105
3	3.356	8.846	2.742	3.538	4.688
3½	3.834	11.545	3.579	4.618	6.119
4	4.334	14.753	4.573	5.901	7.819

Table 18-4   Dimensional and percent fill data for EMT. This is one of the twelve charts found in *Chapter 9, Table 4, NEC,*® for various types of raceways. See *Table 4* for metric values.

2. In *NEC*® *Chapter 9, Table 4*, look at the first chart entitled *Article 358*. Now look in the last column marked "Over 2 Wires—40%." Note that a trade size ¾ EMT holds up to 0.213 in.^{2} of conductor fill, and a trade size 1 EMT holds up to 0.346 in.^{2} of conductor fill. Therefore, a trade size 1 EMT is the minimum size for the combination of conductors in the example.

Many electrical ordinances require that a separate equipment grounding conductor be installed in all raceways, whether rigid or flexible, whether metallic or nonmetallic. This requirement resulted from inspectors finding loose or missing set screws on set-screw type connectors and couplings. Corrosion of fittings also contributes to poor grounding. A separate equipment grounding conductor does take up space, and therefore must be counted when considering conduit fill.

EXAMPLE: A feeder to an electric furnace is fed with two 3 AWG THHN conductors plus one green insulated equipment grounding conductor. This feeder is installed in EMT. The feeder overcurrent protection is 100 amperes. What is the minimum size EMT to be installed?

SOLUTION: According to *Table 250.122*, the minimum size equipment grounding conductor for a 100-ampere feeder is 8 AWG copper. Find the in.^{2} area of the conductors in *Table 5, Chapter 9, NEC*.®

Two 3 AWG THHN	0.0973 in.^{2}
	0.0973 in.^{2}
One 8 AWG THHN	0.0366 in.^{2}
Total Area	0.2312 in.^{2}

From *Table 4, Chapter 9, NEC*®:

40% fill for trade size ¾ EMT = 0.213 in.^{2}
40% fill for trade size 1 EMT = 0.346 in.^{2}

Therefore, install trade size 1 EMT.

Remember, *300.18(A)* states that a conduit system must be completely installed before pulling in the conductors. Refer to Figure 18-6.

Conduit Bodies *[314.16(C)]*

Conduit bodies are used with conduit installations to provide an easy means to turn corners, terminate conduits, and mount switches and receptacles. They are also used to provide access to conductors, to provide space for splicing (when permitted), and to provide a means for pulling conductors (Figure 18-10).

A conduit body:

• must have a cross-sectional area not less than twice the cross-sectional area of the largest conduit to which it is attached, as shown in Figure 18-11(A). This is a requirement only when the conduit body contains 6 AWG conductors or smaller.

Figure 18-10 A variety of common conduit bodies and covers.

- may contain the same maximum number of conductors permitted for the size raceway attached to the conduit body.

- must not contain splices, taps, or devices unless the conduit body is marked with its in.³ capacity, as in Figure 18-11(B), and the conduit body must be supported "in a rigid and secure manner."

- the conductor fill volume must be properly calculated according to *Table 314.16(B)*, which lists the free space that must be provided for each conductor within the conduit body. If the conduit body is to contain splices, taps, or devices, it must be supported rigidly and securely.

CROSS-SECTIONAL AREA OF CONDUIT BODY HERE MUST BE AT LEAST TWICE THE CROSS-SECTIONAL AREA OF CONDUIT HERE. FOR INSTANCE, A TRADE SIZE 1/2 CONDUIT HAS A CROSS-SECTIONAL AREA OF APPROXIMATELY 0.30 IN.² THEREFORE, THE CROSS-SECTIONAL AREA OF THE CONDUIT BODY MUST BE AT LEAST 0.60 SQUARE IN.²

A

8 IN.³

B

Figure 18-11 Requirements for conduit bodies.

DERATING FACTORS FOR MORE THAN THREE CURRENT-CARRYING CONDUCTORS IN CONDUIT OR CABLE

For more than three current-carrying wires in conduit or cable, refer to Table 18-5.

NEC® 310.15(B)(2)(a) states that for the conductors listed in the tables, the maximum allowable ampacity must be reduced when more than three current-carrying conductors are installed in a raceway. The reduction factors based on the number of conductors are noted in Table 18-5. When the conduit nipple does not exceed 24 in. (600 mm), the reduction factor as shown in Figure 18-12 does not apply.

According to *310.15(B)(4)(a)*, a neutral conductor of a multiwire branch-circuit carrying only unbalanced currents shall not be counted in determining current-carrying capacities, as shown in Figure 18-13.

NUMBER OF CURRENT-CARRYING CONDUCTORS	PERCENT OF VALUES IN TABLES. IF HIGH AMBIENT TEMPERATURES ARE PRESENT, ALSO APPLY CORRECTION FACTORS FOUND AT THE BOTTOM OF *Table 310.16.*
4 through 6	80
7 through 9	70
10 through 20	50
21 through 30	45
31 through 40	40
41 and above	35

Table 18-5 Derating factors necessary to determine conductor ampacity when there are more than three current-carrying conductors in a raceway or cable. See *310.15(B)(2)(a)*.

Figure 18-12 The derating factors in Table 18-5 are not required for short lengths of conduit that do not exceed 24 in. (600 mm), *310.15(B)(2)(a), Exception No. 3.* Conductor fill not to exceed 60 percent according to *Chapter 9, Table 1, Note 4.*

EXAMPLE: What is the correct ampacity for size 6 AWG THHN when there are six current-carrying conductors in the conduit?

$$75 \times 0.80 = 60 \text{ amperes}$$

CORRECTION FACTORS DUE TO HIGH TEMPERATURES

When conductors are installed in locations where the temperature is higher than 86°F (30°C), the correction factors noted below *Table 310.16* must apply.

For example, consider that four current-carrying 3 AWG THWN copper conductors are to be installed in one raceway in an ambient temperature of approximately 90°F (32°C). "Ambient" temperature means "surrounding" temperature. This temperature is frequently exceeded in attics where the sun shines directly onto the roof.

The maximum ampacity for these conductors is determined as follows:

- The allowable ampacity of 3 AWG THWN copper conductors from *Table 310.16* = 100 amperes.

- Apply the correction factor for 90°F (32°C) found below *Table 310.16*. The first column shows Celsius; the last column shows Fahrenheit.

  $$100 \times 0.94 = 94 \text{ amperes}$$

- Apply the derating factor for four current-carrying conductors in one raceway.

  $$94 \times 0.80 = 75.2 \text{ amperes}$$

- Therefore, 75.2 amperes is the new corrected ampacity for the example given.

Figure 18-13 A neutral conductor carrying the unbalanced currents from the other conductors need *not* be counted when determining the ampacity of the conductor, according to *310.15(B)(4)(a).* The *Code* would recognize this example as *two current-carrying conductors.* Thus, in the three multiwire circuits shown, the actual total is nine conductors, but only six current-carrying conductors are considered when the derating factor of Table 18-5 is applied.

Some AHJs in extremely hot areas of the country, such as the southwestern desert climates, will require that service-entrance conductors running up the side of a building or above a roof exposed to direct sunlight be "corrected" per the correction factors below *Table 310.16.* Check this out before proceeding with an installation where this situation might be encountered.

OVERCURRENT PROTECTION FOR BRANCH-CIRCUIT CONDUCTORS

The basic rule is that overcurrent protection for conductors shall not be more than the allowable ampacity of the conductor, *240.4.* Allowable ampacities for conductors are found in *Table 310.16.*

If the allowable ampacity of a conductor does not match a standard ampere rating of a fuse or circuit breaker, *240.4(B)* permits the use of the next higher rated fuse or circuit breaker, 800 amperes or less. For example, a conductor having an ampacity of 55 amperes could be protected by a 60-ampere

CONDUCTOR SIZE (Copper)	MAXIMUM AMPERE RATING OF OVERCURRENT DEVICE
14 AWG	15 amperes
12 AWG	20 amperes
10 AWG	30 amperes

Table 18-6 Maximum size overcurrent protection for small size conductors, *240.4(D)*.

overcurrent device. This permission does not apply if the conductors of the branch-circuit supply a number of receptacles for cord- and plug connection of portable loads because it would be impossible to know just how much load might be plugged in.

Small size branch-circuit conductors must be protected according to the values shown in Table 18-6. This table is based on *240.4(D)*.

An exception to this basic maximum size overcurrent protection rule is the protection of motor branch-circuit conductors. For motor branch-circuits, the maximum size overcurrent protection values are found in *Table 430.52*.

If the allowable ampacity of a given conductor is *derated* and/or *corrected*, the overcurrent protection must be based on the new *derated* or *corrected* ampacity.

BASIC *CODE* CONSIDERATIONS FOR CONDUCTOR SIZING AND OVERCURRENT PROTECTION

The following is a quick review of *Code* requirements for conductor sizing and conductor overcurrent protection encountered in house wiring.

- *210.3:* The rating of a branch-circuit is based on the rating of the overcurrent device.

- *210.19(A):* Branch-circuit conductors shall have an ampacity not less than the maximum load to be served.

- *210.20:* An overcurrent device shall not be less than 125% for continuous loads on branch-circuits.

- *210.20(B):* This section refers us to *240.4*, where we find that overcurrent protection for conductors shall not exceed the ampacity of a conductor. See *240.4* for exceptions such as

"next standard size permitted" if the OCD and the conductor ampacity do not match.

- *210.23:* In no case shall the load exceed the branch-circuit ampere rating.

- *215.2(A):* Feeder conductors shall have an ampacity not less than required to supply the load.

- *215.3:* An overcurrent device shall not be less than 125% for continuous loads on feeders. This section refers us to *Part 1* of *240.4*, where we find that overcurrent protection for feeders shall not exceed the ampacity of a conductor. Refer to *240.4* for specific requirements and exceptions to this basic rule.

- *220.4:* The total load shall not exceed the rating of the branch-circuit.

- *240.4:* This section contains all of the overcurrent protection requirements for conductors.

- *230.42:* An overcurrent device shall not be less than 125% for continuous loads on services.

- *424.3(B):* For fixed electric space heating equipment such as an electric furnace, the branch-circuit conductors and the overcurrent protective devices shall not be less than 125 percent of the total load of the motor(s) and heater(s).

- *426.4:* For fixed outdoor electric deicing and snow melting equipment such as heating cables buried in the concrete of a driveway, the branch-circuit conductors and the overcurrent protective devices shall not be less than 125 percent of the total load of the heaters.

- *430.22:* Branch-circuit conductors for motors shall not be less than 125 percent of the motor's full-load current. Motor branch-circuit short-circuit and ground-fault protection is sized according to the percentages listed in *Table 430.52*.

- *Article 440, Part III:* Branch-circuit short-circuit and ground-fault protection for air-conditioning equipment is covered in Unit 23.

- *Article 440, Part IV:* Conductor sizing for air-conditioning equipment is covered in Unit 23.

A few of the above *Code* sections make reference to *continuous loads*. The *Code* defines a continuous load as: "*where the maximum current is expected to continue for three hours or more.*"

Loads in homes are generally not considered to be continuous. However, an electric furnace or snow-melting heating cables buried in the concrete of a driveway can be classified as a continuous load, because they could be on for periods of three hours or more. The 125 percent factor for these has already been included in *424.3(B)* and *426.4*.

Many loads in commercial and industrial installations can be considered to be continuous loads. For example, an electric range in a restaurant might be used continuously 24 hours a day, seven days a week. An electric range in a home would not be used in this manner.

For continuous loads, the overcurrent protective device and the conductors must not be less than 125 percent of the continuous load. For example, a continuous load is 40 amperes. The minimum overcurrent device rating and the minimum conductor ampacity is $40 \times 1.25 = 50$ amperes.

For combination continuous and noncontinuous loads, the overcurrent protective device and the conductors must not be less than 125 percent of the continuous load plus the noncontinuous load. For example, a feeder supplies a continuous load of 40 amperes and a noncontinuous load of 30 amperes. The minimum overcurrent device rating and the minimum conductor ampacity is $(40 \times 1.25) + 30 = 80$ amperes.

EXAMPLE OF DERATING, CORRECTING, ADJUSTING, OVERCURRENT PROTECTION, AND CONDUCTOR SIZING

High ambient temperatures and/or more than three current-carrying conductors in the same raceway or cable call for penalties relative to a conductor's ability to safely carry current.

Suppose that eight THHN copper conductors are installed in one raceway—the conditions are dry and the temperature is expected to reach 100°F (38°C). All conductors are considered to be current-carrying. The expected noncontinuous load on each conductor is 18 amperes. Let's analyze this.

- Consider 14 AWG THHN copper conductors:
  a. The allowable ampacity from *NEC® Table 310.16* is 25 amperes.

b. The maximum overcurrent protection from Table 18-6 in this text or from *NEC® 240.4(D)* is 15 amperes—not adequate to carry the 18-ampere load.
c. To find the corrected ampacity when eight current-carrying conductors are in one raceway, apply the 0.70 adjustment factor (penalty) from Table 18-5 in this text or from *NEC® Table 310.15(B)(2)*.

$$25 \times 0.70 = 17.5 \text{ amperes}$$

d. To find the corrected ampacity for the high ambient temperature, apply the 0.91 correction factor (penalty) from the footnotes to *NEC® Table 310.16*.

$$17.5 \times 0.91 = 15.925 \text{ amperes}$$

You can do this all in one step:

$$25 \times 0.70 \times 0.91 = 15.925 \text{ amperes}$$

Analysis: 14 AWG THHN copper conductors are *too small* for the 18-ampere load. The 15-ampere maximum overcurrent protection *will not carry* the 18-ampere load.

- Consider 12 AWG THHN copper conductors:
  a. The ampacity from *NEC® Table 310.16* is 30 amperes.
  b. The maximum overcurrent protection from Table 18-6 in this text or from *NEC® 240.4(D)* is 20 amperes.
  c. To find the corrected ampacity when eight current-carrying conductors are in one raceway, apply the 0.70 adjustment factor (penalty) from Table 18-5 in this text or from *NEC® Table 310.15(B)(2)*.
  d. To find the corrected ampacity for the high ambient temperature, apply the 0.91 correction factor (penalty) from the footnotes to *NEC® Table 310.16*.

Let's do the calculation in one step:

$$30 \times 0.70 \times 0.91 = 19.11 \text{ amperes}$$

Analysis: 12 AWG THHN copper conductors are all right for the 18-ampere load. The 20-ampere maximum overcurrent protection *will* carry the 18-ampere load. *NEC® 240.4(B)* allows the next higher rated overcurrent device to be used. The next standard rating above 19.11 amperes is 20 amperes.

Another way to approach this example is to divide the load by the penalty factors. This is just another way to find a conductor that will have the proper ampacity for the load to be served.

$$\frac{18}{0.70 \times 0.91} = 28.26 \text{ amperes}$$

We then look in *NEC® Table 310.16* for a THHN conductor that has an ampacity of not less that 28.26 amperes, and we find this to be a 12 AWG.

MULTIOUTLET ASSEMBLY

A multioutlet assembly has been installed above the workbench. For safety reasons, as well as for adequate wiring, it is recommended that any workshop receptacles be connected to a separate circuit. In the event of a malfunction in any power tool commonly used in the home workshop (saw, planer, lathe, or drill), only receptacle Circuit A18 is affected. The lighting in the workshop is not affected by a power outage on the receptacle circuit.

The GFCI feedthrough receptacle to the left of the workbench provides GFCI protection for the entire plug-in strip, *210.8(A)(5)*, as shown in Figure 18-14. GFCI protection could also have been provided by installing a GFCI circuit-breaker in the main panel.

The installation of multioutlet assemblies must conform to the requirements of *Article 380* of the *Code*.

Load Considerations for Multioutlet Assemblies

NEC® Table 220.12 gives the general lighting loads in volt-amperes per ft² (0.093 m²) for types of occupancies. *NEC® 220.14(J)* indicates that *all* general-use receptacle outlets in one, two, and multifamily dwellings are to be considered outlets for general lighting, and as such have been included in the volt-amperes per ft² calculations. Therefore, no additional load need be added.

Because the multioutlet assembly in Figure 18-15 has been provided in the workshop for the purpose of plugging in portable tools, a separate circuit, A18, is provided. This circuit has been shown in the service-entrance calculations at 1500 volt-amperes, similar to the load requirements for small appliance circuits.

Many types of receptacles are available for various sizes of multioutlet assemblies. For example, duplex, single-circuit grounding, split-circuit grounding, and duplex split-circuit receptacles can be used with multioutlet assemblies.

MAIN PANEL

ELECTRICAL METALLIC TUBING

GFCI

PLUG-IN STRIP

Figure 18-14 Detail of how the workbench receptacle plug-in strip is connected by feeding through a GFCI feedthrough receptacle located below the Main Panel. This is Circuit A18.

Figure 18-15 A typical metal multioutlet assembly. Multioutlet assemblies are permitted to be run through (not within) a dry partition when no outlets are located within the partition, and all exposed portions of the assembly can have the cover removed. Nonmetallic mulitoutlet assemblies are also available. *Courtesy of* The Wiremold Company, West Hartford, CT 06110.

Wiring a Multioutlet Assembly

A multioutlet assembly can be connected by bringing the supply into one end as in Figure 18-14, or into the back of the assembly as in Figure 18-16.

Receptacles for multioutlet assemblies can be spaced and wired on the job, in which case the covers are cut to the desired length. Receptacles can also be factory prewired, with six receptacles in a 3-ft (900-mm) long strip, and ten receptacles in a 5-ft (1.5-m) long strip.

For other strip lengths, standard spacing of receptacles for precut assemblies are 6 ft (1.8 m), 9 in. (225 mm), 12 in. (300 mm), or 18 in. (450 mm). These measurements are center to center. Prewired multioutlet assemblies are generally wired with 12 AWG THHN conductors, and have either 15- or 20-ampere receptacles, depending upon how they are specified and ordered.

Multioutlet assemblies are not difficult to install. There is a wide variety of fittings, such as connectors, couplings, ground clamps, blank and end fittings, elbows for turning corners (flat, inside, outside, twist), flush plate adapters, starter boxes, raceway adapters, mounting clips, and so on, to fit just about any application.

Figure 18-16 A multioutlet assembly back plate with the armored cable supply feeding into the back of the assembly.

Figure 18-17 A cord-connected, prewired multioutlet assembly installed above a workbench. *Courtesy of The Wiremold Company, West Hartford, CT 06110.*

Figure 18-18 A multioutlet assembly installed above the backsplash of a kitchen countertop. *Courtesy of The Wiremold Company, West Hartford, CT 06110.*

Figure 18-17 shows a cord-connected multioutlet assembly of the type used in workshops or other locations where there is a need for the convenience of having a number of receptacles in close proximity. These prewired assemblies are available with six receptacles and a circuit breaker, 40 in. long, or ten receptacles and a circuit breaker, 52 in. long. These can also be used under kitchen cabinets. Figure 18-16 shows a multioutlet assembly back plate with the electrical supply feeding into the back of the assembly. Figure 18-18 shows a multioutlet assembly above the backsplash of a kitchen countertop.

EMPTY CONDUITS

Although not truly part of the workshop wiring, *Note 8* to the first-floor electrical plans indicates that two empty trade size 1 EMT raceways are to be installed, running from the workshop to the attic. This installation is unique, but certainly welcomed if at some later date additional wiring for telephone, security, computer, structured wiring, home automation cables, or other similar future needs might be required. The empty raceways will provide the necessary route from the basement to the attic, thus eliminating the need to fish through the wall partitions.

REVIEW

1. a. What circuit supplies the workshop lighting? _____

   b. What circuit supplies the plug-in strip over the workbench? _____

   c. What circuit supplies the freezer receptacle? _____

2. Approximately how much trade size ½ EMT is used for connecting Circuit A13, Circuit A17, Circuit A18, and Circuit A20? _____

3. a. How is EMT fastened to masonry? _____

   _____

   b. The cut ends of rigid and electrical metallic tubing must be _____
      to prevent the insulation on conductors from being damaged.

4. What type of luminaires (lighting fixtures) are installed on the workshop ceiling?

   _____

   _____

5. A check of the plans indicates that two empty trade size 1 EMT conduits are installed
   between the basement and the attic. In your opinion, is this is a good idea? Briefly
   explain your thoughts. _____

   _____

   _____

   _____

   _____

   _____

6. To what circuit are the smoke detectors connected? _____

7. Is it a *Code* requirement to connect smoke detectors to a GFCI-protected circuit?

   _____

8. When a freezer is plugged into a receptacle, be sure that the receptacle is protected by
   a GFCI. Is this statement true or false? Explain.

   _____

   _____

   _____

   _____

   _____

   _____

9. A sump pump is plugged into a single receptacle outlet. This receptacle shall

   a. be GFCI protected

   b. not be GFCI protected

   Circle the correct answer.

10. When a 120-volt receptacle outlet is provided for the laundry equipment in an unfin-
    ished basement, what is the *minimum* number of additional receptacle outlets required
    in that basement? What *Code* reference? _____

    _____

11. Derating factors for conductor ampacities must always be taken into consideration
    when (a) _____ are contained in one raceway.

    Correction factors are applied when (b) _____ .

12. What current is flowing in the neutral conductor at:

    a. _____ amperes?

    b. _____ amperes?

    c. _____ amperes?

    d. _____ amperes?

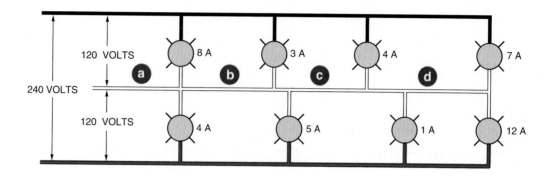

13. When a single receptacle is installed on an individual branch-circuit, the receptacle must have a rating _____ than the rating of the branch-circuit.

14. Calculate the total current draw of Circuit A17 if all luminaires (lighting fixtures) and the exhaust fan were turned on.

15. A conduit body must have a cross-sectional area not less than (two) (three) (four) times the cross-sectional area of the (largest) (smallest) conduit to which it is attached.

    Circle correct answers.

16. A conduit body may contain splices if marked with:

    a. the UL logo.

    b. its in.^{3} area.

    c. the size of conduit entries.

    Circle the correct answer.

17. What size box would you use where Circuit A17 enters the junction box on the ceiling? _____

    _____

18. List the proper trade size of electrical metallic tubing for the following. Assume that it is new work and that THHN insulated conductors are used. See *Chapter 9, Table C1, NEC.*®

    a. Three 14 AWG _____   f. Four 12 AWG _____

    b. Four 14 AWG _____   g. Five 12 AWG _____

    c. Five 14 AWG _____   h. Six 12 AWG _____

    d. Six 14 AWG _____   i. Six 10 AWG _____

    e. Three 12 AWG _____   j. Four 8 AWG _____

19. According to the *Code*, what trade size EMT is required for each of the following combinations of conductors? Assume that these are new installations. Show all calculations. Refer to *Chapter 9, NEC,* *Table 4*, *Table 5*, and *Table 8*.

   a. Three 14 AWG THHN, four 12 AWG THHN

   b. Two 12 AWG THHN, three 8 AWG THHN

   c. Three 3/0 AWG THHN, two 8 AWG THHN

20. a. When more than three current-carrying conductors are installed in one raceway, their allowable ampacities must be reduced according to *NEC®* _____.

   b. The following is a project. We need four 15-ampere branch-circuits. We will use two three-wire multiwire branch-circuits and 14 AWG THHN conductors. We will install the conductors in EMT. The room temperature will not exceed 86°F (30°C). The connected load is nonmotor and is noncontinuous.

   1. What is the minimum trade size EMT? _____

   2. What is the ampacity of the conductors before derating? _____

   _____

3. What is the ampacity of the conductors after derating? _____

_____

4. What is the maximum overcurrent protection permitted for these conductors?

_____

c. Let's increase the conductor size to 12 AWG THHN for the same installation to see what happens.

1. What is the minimum trade size EMT? _____

2. What is the ampacity of the conductors before derating? _____

_____

3. What is the ampacity of the conductors after derating? _____

_____

4. What is the maximum overcurrent protection permitted for these conductors?

_____

d. It (is) (is not) necessary to count equipment grounding conductors, neutral conductors, and/or "travelers" when installing three-way and four-way switches for the purposes of derating the ampacity of these conductors. Circle the correct answer. Explain.

_____

_____

_____

_____

_____

21. What is the minimum number of receptacles required by the *Code* for basements?

_____

22. For laundry equipment in basements, what sort of electrical circuitry is required by the *Code*? _____

_____

_____

_____

_____

_____

23. How many receptacles are required to complete the multioutlet assembly above the workbench? _____

24. A Type NM cable carries a branch-circuit to the outside of a building, where it terminates in a metal weatherproof junction box. From this box, a short length of EMT is installed upward to another metal weatherproof junction box on which is attached a cluster of outdoor flood lights. The first box is located approximately 24 in. (600 mm) above the ground. To "meet Code," do the metal boxes and metal raceway have to be grounded?   Yes _____   No _____   *NEC*® _____ .

25. When raised covers are used for receptacles, does the *Code* permit fastening the receptacle to the cover with the one center No. 6-32 screw? Yes _____ No _____ *NEC*® _____ .

26. The following is a layout of the lighting circuits for the workshop. Using the conduit layout in Figure 18-1, make a complete wiring diagram. Use colored pencils or pens to indicate conductors.

WORKSHOP

27. *NEC® 210.8(A)(5)* requires that GFCI protection be provided for all receptacles in unfinished basements. There are some exceptions. What are these exceptions?

_____

_____

_____

28. Where loads are likely to be on continuously, the *calculated* load for branch-circuits and feeders must be figured at (100 percent) (125 percent) (150 percent) of the continuous load, (plus) (minus) the noncontinuous load. This requirement generally (does) (does not) apply to residential wiring. Circle the correct answers.

Special-Purpose Outlets— Water Pump, Water Heater

WATER PUMP CIRCUIT ▲B

In rural areas where there is no public water supply, dwellings need their own water supply. The Electrical Plans and Schedule of Special Purpose Outlets for this residence show that the well pump is connected to Circuit A 5-7.

After a brief introduction to jet pumps and submersible pumps, we will discuss the key elements that make up a typical motor branch-circuit. For the most part, submersible pumps dominate the residential market.

The manufacturer's installation instructions must always be followed.

JET PUMPS

Figure 19-1 shows the major components of a typical deep-well jet pump.

The pump impeller wheel, ①, forces water down a drive pipe, ②, at a high velocity and pressure to a point just above the water level in the well casing. Just above the water level, the drive pipe curves sharply upward and enters a larger vertical suction pipe, ③. The drive pipe terminates in a small nozzle or "jet," ④. The water emerges from the jet with great force and flows upward though the suction pipe.

Water rises in the suction pipe, drawn up through the tailpipe, ⑤, by the action of the jet. The water rises to the pump inlet and passes through the impeller wheel of the pump. Some of the water is forced down through the drive pipe again. The remaining water passes through a check valve and enters the storage tank, ⑥.

The foot valve, ⑦, prevents water in the pumping equipment from draining back into the well

Figure 19-1   Components of a jet pump.

A 230-VOLT BUS IN PANEL

B 20-AMPERE, TWO-POLE CIRCUIT BREAKER

C 12 AWG THHN CONDUCTORS

D TWO-POLE DISCONNECT SWITCH AND MOTOR OVERLOAD PROTECTION

E TWO-POLE PRESSURE SWITCH

F 1-HP, 230-VOLT, SINGLE-PHASE MOTOR

Figure 19-2   The pump circuit.

when the pump is not operating. The strainer, ⑧, keeps debris out of the system. Pressure switch, ⑨, starts and stops the pump motor.

Figure 19-2 shows the wiring of a typical water pump. Always follow the wiring instructions furnished by the specific pump manufacturer.

SUBMERSIBLE PUMPS

Figure 19-3 shows the major components of a typical submersible pump.

A submersible pump consists of a centrifugal pump, ①, driven by an electric motor, ②. The pump and the motor are contained in one housing, ③, submersed below the permanent water level, ④, within the well casing, ⑤. The pump housing is a cylinder 3–5 in. in diameter and 2–4 ft long. When running, the pump raises the water upward through the piping, ⑥, to the water tank, ⑦. Proper pressure is maintained in the system by a pressure switch, ⑧.

The disconnect switch, ⑨, pressure switch, ⑧, limit switches, ⑩, and controller, ⑪, are installed in a logical and convenient location near the water tank.

Pumps commonly called "two-wire pumps" contain all required starting and protection components and connect directly to the pressure switch—with no aboveground controller. Generally, there are no moving electrical parts within the submersible pump, such as the centrifugal starting switch found in a typical single-phase, split-phase induction motor.

Other pumps, referred to as "three-wire pumps," require an aboveground controller that contains any required components not in the motor, such as a starting relay, overload protection, starting and running capacitors, lightning arrester, and terminals for making the necessary electrical connections.

Most water pressure tanks have a precharged air chamber in an elastic bag (bladder) that separates the air from the water. This ensures that air is not absorbed by the water. Water pressure tanks compress air to maintain delivery pressure in a usable range without cycling the pump at too fast a rate. When in direct contact with the water, air is gradually absorbed into the water, and the cycling rate of

Figure 19-3   Submersible pump.

the pump will become too rapid for good pump life unless the air is periodically replaced. With the elastic bag, the initial air charge is always maintained, so recharging is unnecessary.

Submersible pumps are covered by UL Standard 778.

Submersible Pump Cable

Power to the motor is supplied by a "drop" cable, ⑫, especially designed for use with submersible pumps. This cable is marked "submersible pump cable," and it is generally supplied with the pump. This cable can also be purchased separately. The cable is cut to the proper length to reach between the pump and its controller, as shown in

Figure 19-3, or between the pump and the well-cap, Figure 19-4. When needed, the cable may be spliced according to the manufacturer's specifications.

Submersible water pump cable is "tag-marked" for use within the well casing for wiring deep-well water pumps, where the cable is not subject to repetitive handling caused by frequent servicing of the pump units.

Submersible pump cable is not designed for direct burial in the ground unless it is marked Type USE or Type UF. Should it be required to run the pump's circuit underground for any distance, it is necessary to install Type UF or Type USE cable or a raceway with suitable conductors, and then make up the necessary splices in a listed weatherproof junction box, as in Figure 19-4. Unit 16 covers underground wiring in great detail.

MOTOR CIRCUIT DESIGN

We will now take a look at the *NEC®* requirements for all of the basics for a typical electrical motor branch-circuit.

The pump circuit in this residence originates in Main Service Panel A, Circuit A5-7

Current and Voltage

This information is found on the nameplate and instructions furnished with the pump.

The water pump is this residence is a 1-horsepower (hp), single-phase 115/230-volt pump. Dual-voltage-rated pumps can be supplied by either 115 volts or 230 volts. When connected for the higher voltage, the current draw will be half of that when connected for the lower voltage. For example, in *Table 430.248*, the full-load current rating of a 1-hp, 230-volt, single-phase motor is 8 amperes. The same motor connected to 115 volts will draw 16 amperes. Current is a big factor in determining conductor size, watts loss in the conductors, voltage drop, disconnect switch size, controller size, motor overload protection, and the size and type of branch-circuit overcurrent protection. Some pump controllers have a simple slide switch to change the connections from 115 volts to 230 volts. Pumps sized ½ horsepower are usually prewired for 115 volts. Pumps ¾ hp and larger are usually prewired for 230 volts.

For the purpose of our motor circuit design, we will connect this motor for 230 volts.

SPLICES MADE IN SPACE PROVIDED IN WELL CAP OR JUNCTION BOX.

ALTHOUGH DISCONNECT INSIDE HOUSE IS CAPABLE OF BEING LOCKED "OFF", ANOTHER DISCONNECT * IS REQUIRED ON OUTSIDE OF HOUSE IN SIGHT OF THE WELL HEAD, AND NOT MORE THAN 50 FT (15 M) FROM THE WELL HEAD. ►THE LOCK OFF PROVISION SHALL BE PERMANENT.◄ SEE *430.102(A)* AND *(B)*.

EQUIPMENT GROUNDING CONDUCTOR CONNECTED TO CASING AND TO PUMP

WELL CAP

SEE *NEC® TABLE 300.5* FOR MINIMUM DEPTHS FOR UNDERGROUND WIRING.

PUMP CONTROLLER AND DISCONNECT SWITCH

PUMP CABLE CONTAINS WIRES PLUS EQUIPMENT GROUNDING CONDUCTOR.

- DISCONNECT MUST BE IN SIGHT OF CONTROLLER AND MUST DISCONNECT THE CONTROLLER, *430.102(A)*.

METAL WELL CASING MUST BE BONDED TO PUMP EGC OR TO EQUIPMENT GROUND BUS AT THE PANELBOARD WHERE PUMP CIRCUIT ORIGINATES. *250.112(M).*

TYPE UF OR TYPE USE CABLE, OR RACEWAY WITH PROPER TYPE AND SIZE CONDUCTORS. MUST CARRY EQUIPMENT GROUNDING CONDUCTOR SIZED PER *TABLE 250.122.*

- DISCONNECT MUST BE IN SIGHT OF MOTOR *430.102(B).*

- DISCONNECT MUST OPEN ALL UNGROUNDED (HOT) CONDUCTORS *430.103.*

PUMP AND METAL WELL CASING SHALL BE GROUNDED *250.112(L)* AND *250.112(M).*

- DISCONNECT MAY BE IN SAME ENCLOSURE AS CONTROLLER *430.103.*

PUMP

INLET

MOTOR

- DISCONNECT MUST BE HORSEPOWER RATED *430.109.*

- USE A DISCONNECT WITH A NEMA 3R RATING. THIS RATING IS INTENDED FOR OUTDOOR USE TO PROVIDE A DEGREE OF PROTECTION AGAINST FALLING RAIN AND SLEET, AND IS UNDAMAGED BY ICE WHICH FORMS ON THE ENCLOSURE.

Figure 19-4 Grounding and disconnecting means requirements for submersible water pumps.

Table 430.248 for 115-volt motors covers the voltage range of 110–120 volts. For 230-volt motors, the voltage range is 220–240 volts. You will also find that some motors are marked 115/208-230 volts.

The NEMA Standard MG-1 *Information Guide for General Purpose Industrial AC Small and Medium Squirrel-Cage Induction Motor Standards* is a condensed version of the much large NEMA Standard MG 1-2003 *Motors and Generators.* These standards show that most general-purpose NEMA-rated motors are designed to operate at ±10 percent of the motor's rated voltage.

Important: For determining the conductor size, switch size, and the branch-circuit short-circuit and ground-fault protection, use the current values from the tables in the *NEC.* See *430.6(A)(1).*

For overload protection sizing, use the motor nameplate current value. See *430.6(A)(2).*

The nameplate rating of the pump motor is 8 amperes, 230 volts.

Conductor Size *(430.22)*

From *Table 430.248,* the full-load current rating of a 1-hp, 230-volt, single-phase motor is 8 amperes. To calculate the conductor size, use 125 percent of 8 amperes.

$$1.25 \times 8 = 10 \text{ amperes}$$

Use Type THHN conductors per the Specifications.

Checking *Table 310.16* of the *NEC,* we find that a 14 AWG Type THHN conductor has an ampacity of 25 amperes, more than ample for the pump motor circuit. However, the Schedule of Special Purpose Outlets found in this text, which is part of the Specifications for the residence, indicates that 12 AWG THHN conductors are to be used for the pump circuit. Installing larger size conductors can minimize voltage drop and might be suitable at a later date should the need arise to install a larger capacity pump.

The two 12 AWG THHN conductors are connected to a two-pole, 20-ampere branch-circuit breaker in Panel A, Circuit A(5-7).

Conduit Size

Two 12 AWG Type THHN conductors require trade size ½ EMT, *NEC® Annex C, Table Cl.* This EMT will run across the ceiling from Panel A to the disconnecting means located in the southeast corner of the workshop. The controller and other equipment relating to the water pump are located in this corner of the workshop.

Motor Branch-Circuit Short-Circuit and Ground-Fault Protection
(430.52, Table 430.52)

Fuses and circuit breakers are the most commonly used forms of motor branch-circuit short-circuit and ground-fault protection. Should the motor windings or branch-circuit conductors short-circuit or go to ground, the motor's branch-circuit short-circuit and ground-fault overcurrent device, either fuses or circuit breakers, will open.

Motor branch-circuit short-circuit and ground-fault protection does not protect the motor against overload burnout!

The various choices for motor branch-circuit short-circuit and ground-fault protection are found in *430.52* and *Table 430.52*. In almost all instances, the values found in *Table 430.52* will work just fine. In the event that the sizing calculation results in an unavailable ampere rating fuse or circuit breaker, select the next standard ampere rating, but do not exceed the absolute maximum sizing percentages shown in *430.52*. This will be discussed in more detail after we do the calculations.

From *Table 430.248*, the full-load current rating of a 1-hp, 230-volt, single-phase motor is 8 amperes. Here are the possibilities:

- Non-Time-Delay Fuses (300 percent desired—maximum size: 400 percent)

    $8 \times 3 = 24$ amperes (use 25-ampere fuses)

    The disconnect switch would be a 30-ampere, 2-pole, 250-volt switch.

- Dual-Element, Time-Delay Fuses (175 percent desired—maximum size: 225 percent)

    $8 \times 1.75 = 14$ amperes (use 15-ampere fuses)

    The disconnect switch would be a 30-ampere, 2-pole, 250-volt switch.

- Inverse Time Breakers (250 percent)

    $8 \times 2.5 = 20$-ampere rating

    Maximum setting: 400 percent for FLA of 100 amperes or less.

- Instant Trip Breakers are never used in residential applications.

The motor branch-circuit overcurrent device must be capable of allowing the motor to start, *430.52(B)*. Where the values for branch-circuit protective devices (fuses or breakers) as determined by *Table 430.52* do not correspond to the standard sizes, ratings, or settings as listed in *240.6*, the next higher size, rating, or setting is permitted, *430.52(C)(1), Exception No. 1*.

If, after applying the values found in *Table 430.52* and after selecting the next higher standard size, rating, or setting of the branch-circuit overcurrent device, the motor still will not start, *430.52(C)(1), Exception No. 2* permits using an even larger size, rating, or setting. The maximum values (percentages) are shown in this exception.

A nuisance opening of a fuse, or tripping of a breaker, can easily be predicted by referring to time/current curves for the intended fuse or breaker. This topic is covered in Unit 28.

Motor Overload Protection
(430.31 through 430.44)

Overload protection is needed to keep the motor from dangerous overheating due to overloads or failure of the motor to start. *NEC® 430.31* through *430.44* cover virtually every type and size of motor, starting characteristics, and duty (continuous or noncontinuous) in use today. The overload protection for a motor might be thermal overloads (sometimes called *heaters*) in the controller, electronic sensing overload devices, built-in (inherent) thermal protection, or time-delay fuses.

We find the sizing requirements for motor overload protection in *430.32(A)(1)* and *430.32(B)(1)*. Although there are exceptions, most of the common installations require motor overload protection not to exceed 125 percent of the motor's full-load current draw, as indicated on the nameplate. For the pump motor this would be:

$$1.25 \times 8 = 10 \text{ amperes}$$

Therefore, for backup motor overload protection, we could install 10-ampere, time-delay, dual-element fuses in the disconnect switch. If we are interested only in motor branch-circuit short-circuit and ground-fault protection, we refer to *Table 430.52* and apply the multiplier of 175 percent.

$$8 \times 1.75 = 14 \text{ amperes}$$

We are permitted to go to the next standard size, which are 15-ampere time-delay, dual-element fuses.

Instructions furnished with a particular motor or motor-operated appliance might indicate a maximum size overload protection and a maximum size branch-circuit protection. These values must be followed. Usually the manufacturer provides the necessary motor overload protection in the controller.

Disconnecting Means

The *NEC®* requires that all motors be provided with a means to disconnect the motor from its electrical supply, *Article 430, Part IX*.

The disconnecting means for the water pump is a 30-ampere, two-pole, 250-volt switch mounted on the wall next to the pump controller. See Figure 19-3 and Figure 19-4.

Follow these two simple rules for locating the disconnect switch.

Rule #1. *NEC®* 430.102(A): An individual disconnect must be in sight of the controller and must disconnect the controller. The *NEC®* definition of "In Sight" means that the controller must be visible and not more than 50 ft (15 m) from the disconnect.

Rule #2. *NEC®* 430.102(B): The disconnect must be in sight of the motor and driven machinery. If the disconnect, as required in *430.102(A)*, is in sight of the controller, the

motor, and driven machinery, then that disconnect meets the requirements of both *430.102(A)* and *430.102(B)*.

How to Use the Exception to the "In-Sight" Rule

As with most things, there are exceptions. The exception and the fine print notes to *430.102(B)* tell us that the disconnect need not be in sight of the motor and driven machinery if the disconnect can be individually locked in the open "OFF" position. The locking provision must be of a permanent type installed on the switch or circuit breaker. However, it would be a real stretch of the imagination to apply the exception and FPNs to residential wiring.

Most disconnect switches have this "LOCK-OFF" feature. Listed circuit breaker "LOCK-OFF" devices that fit over the top of the circuit breaker handle also meet this requirement.

Anyone working on the pump needs to be assured that the power is off and stays off until he or she is ready to turn the power back on.

The disconnect switch for the submersible pump for this residence is located next to and in sight of the controller. It is not is sight of the motor. Although the disconnect is capable of being locked in the "OFF" position, this lock-off feature is not always used. This sets the stage for a possible electrocution. For added safety, an additional nonfuse disconnect has been installed on the outside of the house, in view of the pump location, so that anyone working on the pump will be able to clearly see the disconnect switch. And it also is capable of being locked in the "OFF" position. See Figure 19-4.

GROUNDING

Figure 19-4 illustrates how to provide proper grounding of a submersible water pump and the well casing. Proper grounding of the well casing and submersible pump motor will minimize or eliminate stray voltage problems that could occur if the pump motor is not grounded.

Key code sections that relate to grounding water pumping equipment are:

- ▶*NEC®* 250.4(A)(5): The last sentence states that "*The earth shall not be considered as an effective ground-fault current path.*"◀

- *NEC® 250.86:* Metal enclosures for circuit conductors shall be grounded, other than short sections of metal enclosures that provide support or physical protection.
- *NEC® 250.112(L):* Motor-operated water pumps, including submersible pumps, must be grounded.
- *NEC® 250.112(M):* Bond metal well casing to the pump circuit's equipment grounding conductor.
- *NEC® 250.134(B):* Equipment fastened in place or connected by permanent wiring must have the equipment grounding conductor run with the circuit conductors.
- *NEC® 430.241:* Grounding of motors—general.
- *NEC® 430.242:* Grounding of motors—stationary motors.
- *NEC® 430.244:* Grounding of controllers.
- *NEC® 430.245:* Acceptable methods of grounding.

Because so much nonmetallic (PVC) water piping is used today, some electrical codes require that a grounding electrode conductor be installed between the neutral bar of the main service disconnecting means and the metal well casing. This would satisfy the requirements of *250.104* (bonding), *250.50* (grounding electrode system), *250.52* (made and other electrodes), and *250.56* (resistance of electrodes). This then is the service ground and is sized per *Table 250.66* of the *Code*. At the metal casing, the grounding electrode conductor is attached to a lug termination, or by means of exothermic (Cadweld) welding.

WATER HEATER CIRCUIT (Ⓐ)c

Residential electric water heaters are listed under UL Standard 174, *Household Electric Storage Tank Water Heaters*.

Electrical contractors many times are also in the plumbing, heating, and appliance business. They need to know more than how to do the electrical hookup. The following text discusses electrical as well as other data about electric water heaters that will prove useful.

All homes require a supply of hot water. To meet this need, one or more automatic water heaters are generally installed as close as practical to the areas having the greatest need for hot water. Water piping

carries the heated water from the water heater to the various plumbing fixtures and to appliances such as dishwashers and clothes washers.

For safety reasons, in addition to the regular temperature control thermostat that can be adjusted by the homeowner, the *NEC®* and UL Standards require that electric water heaters be equipped with a high temperature cutoff. This high temperature control limits the maximum water temperature to 190°F (88°C). This is 20°F (11°C) below the 210°F (99°C) temperature that causes the "temperature/pressure" relief valve to open. The high temperature cutoff is factory preset and should never be tampered with or modified in the field. In conformance to *422.47* of the *NEC®,* the high temperature limiting control must disconnect all ungrounded conductors.

In most residential electric water heaters, the high temperature cutoff and upper thermostat are combined into one device, as in Figure 19-5.

Combination Pressure/Temperature Safety Relief Valves

Look at Figure 19-5. A pressure/temperature relief valve is an important safety device installed into an opening in the water heater tank within 6 in. (150 mm) of the top. The opening is provided for and clearly marked by the manufacturer. Many water heater pressure/temperature relief valves are factory installed, although some models still require installation of the valve by the installing contractor. Proper discharge piping must be installed from the pressure/temperature relief valve downward to within 6 in. (150 mm) of the floor, preferably near a floor drain. Terminating the discharge pipe to within 6 in. (150 mm) of the floor is important to protect people from injury from scalding water or steam discharge should the relief valve activate for any reason. This also protects electrical equipment in and around the water heater from becoming soaked should the relief valve operate.

Pressure/temperature relief valves are installed for two reasons. Their pressure/temperature sensing features are interrelated.

- *Pressure:* Water expands when heated. Failure of the normal temperature thermostat and the high-temperature limiting device to operate could result in pressures that exceed the rated

Figure 19-5 Typical electric water heater showing location of heating elements and thermostats and electrical connection.

working pressure of the water heater tank and the plumbing system components. The relief valve, properly installed and properly rated to local codes, will open to relieve the excess pressure in the tank to the predetermined setting marked on the relief valve. The rating is stamped on the valve's identification plate.

• *Temperature:* Should runaway water heating conditions occur within the water heater, temperatures far exceeding the atmospheric boiling point of water (212°F [100°C]) can take place. Under high pressures, water will stay liquid at extremely high temperatures. For a typical municipal water supply pressure of 45 psi, water will not boil until it reaches approximately 290°F (143°C). At 290°F (143°C), opening a faucet, a broken pipe, or a ruptured

water heater tank would allow the pressure to instantly drop to normal atmospheric pressure. The superheated water would immediately flash into steam, having about 1600 times more volume than the liquid water occupied. This blast of steam can be catastrophic, and could result in severe burns or death.

Thus, the function of the temperature portion of the pressure/temperature relief valve is to open when a temperature of 210°F (99°C) occurs within the water heater tank. Properly rated and maintained, the pressure/temperature relief valve will allow the water to flow from the water heater through the discharge pipe at a rate faster than the heating process, thereby keeping the water temperature below the boiling point. Should the cold water supply be shut off, the relief valve will allow the steam to escape without undue buildup of temperature and pressure in the tank.

Requirements for relief valves for hot water supply systems are found in the ANSI Standard Z21.22.

Sizes

Residential electric water heaters are available in many sizes, such as 6, 12, 15, 20, 30, 32, 40, 42, 50, 52, 66, 75, 80, 82, 100, and 120 gallons.

Corrosion of Tank

To reduce corrosion of the steel tank, most water heater tanks are glass lined. Glass lining is a combination of silica and other minerals (rasorite, rutile, zircon, cobalt, nickel oxide, and fluxes). This special mix is heated, melted, cooled, crushed, then sprayed onto the interior steel surfaces and baked on at about 1600°F (871°C). To further reduce corrosion, aluminum or magnesium anode rods are installed in special openings, or as part of the hot water outlet fitting, to allow for the flow of a protective current *from* the rod(s) *to* any exposed steel of the tank. This is commonly referred to as *cathodic protection.* Corrosion problems within the water heater tank are minimized as long as the aluminum or magnesium rod(s) remain in an active state. Anodes should be inspected at regular intervals to determine when replacement is necessary, usually when the rod(s) is reduced to about ⅓ of its original diameter, or when the rod's core wire is exposed. Further information can be obtained by contacting the manufacturer, and consulting the installation manual.

Some water heater tanks are fiberglass or plastic lined.

Heating Elements

The wattage ratings of electric water heaters can vary greatly, depending upon the size of the heater in gallons, the speed of recovery desired, local electric utility regulations, and codes. Typical wattage ratings are 1500, 2000, 2500, 3000, 3800, 4500, and 5500 watts.

A resistance heating element might be dual rated. For example, the element might be marked 5500 watts at 250 volts, and 3800 watts at 208 volts.

Most residential-type water heaters are connected to 240 volts except for the smaller 2-, 4-, and 6-gallon

point of use sizes generally rated 1500 watts at 120 volts. Commercial electric water heaters can be rated single-phase or three-phase, 208, 240, 277, or 480 volts.

UL requires that the power (wattage) input must not exceed 105% of the water heater's nameplate rating. All testing is done with a supply voltage equal to the heating element's *rated* voltage. Most heating elements may burn out prematurely if operated at voltages 5% higher than for which they are rated.

To help reduce the premature burnout of heating elements, some manufacturers will supply 250-volt heating elements, yet will mark the nameplate 240 volts with its corresponding wattage at 240 volts. This allows a safety factor if slightly higher than normal voltages are experienced.

Heating Element Construction

Heating elements generally contain a nickel-chrome (nichrome) resistance wire embedded in compacted powdered magnesium oxide to ensure that no grounds occur between the wire and the sheath. The magnesium oxide is an excellent insulator of electricity, yet it effectively conducts the heat from the nichrome wire to the sheath.

The sheath can be made of tin-coated copper, stainless steel, or an iron-nickel-chromium alloy. The later alloy comes under the trade name of Incoloy.® The stainless steel and Incoloy® types can withstand higher operating temperatures than the tin-coated copper elements. They are used in premium water heaters because of their resistance to deterioration, lime scaling, and dry firing burnout.

High watt density elements generally have a U-shaped tube, and give off a great amount of heat per square inch of surface. Figure 19-6(A) illustrates a flange-mounted high-density heating element. Figure 19-6(B) illustrates a high-density heating element that screws into a threaded opening in the tank. High-density elements are very susceptible to lime-scale burnout.

Low watt density elements, as in Figure 19-6(C), generally have a double loop, like a U tube bent in half. They give off less heat per in.^{2}, have a longer life, and are less noisy than high-density elements.

An option offered by at least one manufacturer of electric water heaters is a dual-wattage heating element. A dual-wattage heating element actually

Figure 19-6   Heating elements.

contains two heating elements in the one jacket. One element is rated 3800 watts and one element is rated 1700 watts. Three leads are brought out to the terminal block. The 3800-watt heating element is connected initially, with the option of connecting the second 1700-watt heating element if the wiring to the water heater is capable of handling the higher total combined wattage of 5500 watts. To attain the higher wattage, a jumper (provided by the manufacturer) is connected between two terminals on the terminal block, Figure 19-7.

Any electric heating element coated with lime scale may be noisy during the heating cycle. Limed

up heating elements should be cleaned or replaced, following the manufacturer's instructions.

Do not energize the electrical supply to a water heater unless you are sure it is full of water. Most water heater heating elements are designed to operate only when submersed in water.

Residential electric water heaters are available with one or two heating elements. Figure 19-5 shows two heating elements. Single-element water heaters will have the heating element located near the bottom of the tank. If the water heater has two heating elements, the upper element will be located about two-thirds of the way up the tank.

Figure 19-7 A dual-wattage heating element. The two wires of the incoming 240-volt circuit are connected to L1 and L2. These leads connect to the 3800-watt element. To connect the 1700-watt element, the jumper is connected between L2 and L3. With both heating elements connected, the total wattage is 5500 watts.

Speed of Recovery

Speed of recovery is the time required to bring the water temperature to satisfactory levels in a given length of time. The current accepted industry standard is based on a 90°F (32.22°C) rise. For example, referring to Table 19-1, we find that a 3500-watt element can raise the water temperature 90°F (32.22°C) of a little more than 16 gallons an hour. A 5500-watt element can raise water temperature 90°F (50°C) of about 25 gallons an hour. Speed of recovery is affected by the type and amount of insulation surrounding the tank, the supply voltage, and the temperature of the incoming cold water.

Table 19-2 shows the approximate time it takes to run out of hot water using typical shower heads with various storage capacity residential electric water heaters. Current standards for new shower heads are typically 2.5 gallons per minute. When the flow restrictors are removed, the flow rate can increase to 6–18 gallons per minutes per shower head. There are very low flow rate shower heads available in the range of 1–2 gallons per minute. All of this depends on the water pressure.

An easy way to determine a shower head's flow rate is to turn on the water, hold a large bucket of known capacity under it, and time how long it takes to fill the bucket.

Scalding from Hot Water

The following are Plumbing Code issues but are discussed briefly because of their seriousness.

Temperature settings for water heaters is problematic!

Set the water heater thermostat high enough to kill bacteria on the dishes in the dishwasher and possibly get scalded in the shower. Or set the thermostat to a lower setting and not get scalded in the shower, but live with the possibility that bacteria will still be present on washed dishes.

The Consumer Product Safety Commission (CPSC) reports that 3800 injuries and 34 deaths occur each year due to scalding from excessively hot tap water. Unfortunately, the victims are the elderly and children under the age of 5. They recommend that the setting should not be higher than 120°F (49°C). Some state laws require that the thermostat be preset at no higher than 120°F (49°C). Residential water heaters are all preset by the manufacturer at 120°F (49°C).

Manufacturers post caution labels on their water heaters that read something like this:

> **SCALD HAZARD:** WATER TEMPERATURE OVER 120°F (49°C) CAN CAUSE SEVERE BURNS INSTANTLY OR DEATH FROM SCALDS. SEE INSTRUCTION MANUAL BEFORE CHANGING TEMPERATURE SETTINGS.

To reduce the possibility of scalding, water heaters can be equipped with a thermostatic non-scald mixing valve or pressure-balancing valve that holds a selected water temperature to within one degree regardless of incoming water pressure changes. This prevents sudden unanticipated changes in water temperatures.

Table 19-3 shows time/temperature relationships relating to scalding.

HEATING ELEMENT WATTAGE	GALLONS PER HOUR RECOVERY FOR INDICATED TEMPERATURE RISE					
	60°F	70°F	80°F	90°F	100°F	Equivalent NET Btu output
750	5.2	4.4	3.9	3.5	3.1	2559
1000	6.9	5.9	5.2	4.6	4.1	3412
1250	8.6	7.4	6.5	5.7	5.2	4266
1500	10.3	8.9	7.8	6.9	6.2	5119
2000	13.8	11.8	10.3	9.2	8.3	6825
2250	15.5	13.3	11.6	10.3	9.3	7678
2500	17.2	14.8	12.9	11.5	10.3	8531
3000	20.7	17.7	15.5	13.8	12.4	10,238
3500	24.1	20.7	18.1	16.1	14.5	11,944
4000	27.6	23.6	20.7	18.4	16.6	13,650
4500	31.0	26.6	23.3	20.7	18.6	15,356
5000	34.5	29.6	25.9	23.0	20.7	17,063
5500	37.9	32.5	28.4	25.3	22.5	18,769
6000	41.4	35.5	31.0	27.6	24.6	20,475

NOTES: If the incoming water temperature is 40°F (Northern states), use the 80° column to raise the water temperature to 120°F.

If the incoming water temperature is 60°F (Southern states), use the 60° column to raise the water temperature to 120°F.

For example, approximately how long would it initially take to raise the water temperature in a 42-gallon electric water heater to 120°F where the incoming water temperature is 40°F? The water heater has a 2500-watt heating element.

Answer: 42 ÷ 12.9 = 3.26 hours (approximately 3 hours and 16 minutes)

To keep this table simple, non-cluttered, and easy to read, only Fahrenheit temperatures are shown.

Table 19-1 Electric water heater recovery rate per gallon of water for various wattages.

LENGTH OF CONTINUOUS SHOWER TIME—IN MINUTES—FOR TYPICAL SIZES OF RESIDENTIAL ELECTRIC WATER HEATERS. CHART BASED ON 70% DRAW EFFICIENCY. RECOVERY RATE IS NOT FIGURED IN BECAUSE IT IS NORMALLY NOT A FACTOR IN CONTINUOUS DRAW OF WATER SITUATIONS.

Shower head(s) rate in gallons per minute	Hot water flow rate in gallons per minute based on 120°F. Stored hot water to provide 105°F. mixed water out of shower head.	Hot water flow rate in gallons per minute based on 120°F. Stored hot water to provide 105°F. mixed water out of shower head.	100 GALLON		80 GALLON		66 GALLON		50 GALLON		40 GALLON		30 GALLON	
	40°F. INCOMING WATER	60°F. INCOMING WATER	40°F. INCOMING WATER	60°F. INCOMING WATER	40°F. INCOMING WATER	60°F. INCOMING WATER	40°F. INCOMING WATER	60°F. INCOMING WATER	40°F. INCOMING WATER	60°F. INCOMING WATER	40°F. INCOMING WATER	60°F. INCOMING WATER	40°F. INCOMING WATER	60°F. INCOMING WATER
2.0	1.6	1.5	43	47	34	37	28	31	22	23	17	19	13	14
2.5	2.0	1.9	34	38	28	30	23	25	17	19	14	15	10	11
3.0	2.4	2.3	29	31	23	25	19	21	14	16	11	12	9	9
4.0	3.3	3.0	22	23	17	19	14	15	11	12	9	9	6	7
5.0	4.0	3.8	17	19	14	15	11	12	9	9	7	7	5	6
6.0	4.9	4.5	14	16	11	12	9	10	7	8	6	6	4	5
8.0	6.5	6.0	11	12	9	9	7	8	5	6	4	5	3	4
12.0	9.8	9.0	7	8	6	6	5	5	4	4	3	3	2	2

Example #1: A 40-gallon electric water heater has a 40°F incoming water supply. How many minutes can a 3 GPM shower head be used to obtain 105°F hot water?

Answer: 11 minutes

Example #2: A residence has an 80-gallon electric water heater. The incoming water supply is approximately 60°F. How many minutes can two 2 GPM shower heads be used at the same time to obtain 105°F hot water? Hint: Two 2 GPM shower heads are the same as one 4 GPM shower head.

Answer: 19 minutes

Table 19-2 This chart shows the approximate time in minutes it takes to run out of hot water when using various shower heads, different incoming water temperatures, and different capacity water heaters.

TEMPERATURE	TIME TO PRODUCE SERIOUS BURN
120°F (48.89°C)	Approx. 9½ minutes
125°F (51.67°C)	Approx. 2 minutes
130°F (54.44°C)	Approx. 30 seconds
135°F (57.22°C)	Approx. 15 seconds
140°F (60°C)	Approx. 5 seconds
145°F (62.78°C)	Approx. 2½ seconds
150°F (65.76°C)	Approx. 1⁸∕10 seconds
155°F (68.33°C)	Approx. 1 second
160°F (71.11°C)	Approx. ½ second

Table 19-3 Time/temperature relationships related to scalding of an adult subjected to moving water. For children, the time to produce a serious burn is less than for adults. *Courtesy of* **Shriners Burn Institute, Cincinnati Unit.**

What about Washing Dishes?

Older automatic dishwashers required incoming water temperature of 140°F (60°C) or greater to get the dishes clean. A water temperature of 120°F (49°C) is not hot enough to dissolve grease, activate power detergents, and kill bacteria. That is why commercial and restaurant dishwashers boost temperatures to as high as 180°F (82°C).

Newer residential dishwashers have their own built-in water heaters to boost the incoming water temperature in the dishwasher from 120°F (49°C) to 140°F–145°F (60°C–63°C). This heating element also provides the heat for the drying cycle. Today, a thermostat setting of 120°F (49°C) on the water heater should be satisfactory.

Thermostats/High Temperature Limit Controls

All electric water heaters have a thermostat(s) and a high temperature limit control(s).

Thermostats and high temperature limit controls are available in many configurations. They are available as separate controls and as combination controls.

Figure 19-8 illustrates a combination thermostat/high temperature limit control. This particular control is an interlocking type sometimes referred to as a snap-over type. It is used on two-element water heaters and controls the upper heating element as well as providing the safety feature of limiting the water temperature to a factory preset value. Another thermostat is used to control the lower heating element. When all of the water in the water heater is cold, the upper element heats the upper third of the tank. The lower thermostats are closed, calling for heat, but the interlocking characteristic of the upper thermostat/high limit will not allow the lower heating element to become energized. With this connection, both heating elements cannot be on at the same time. This is referred to as *nonsimultaneous* or *limited demand* operation.

When the water in the upper third of the tank reaches the temperature setting of the upper thermostat, the upper thermostat "snaps over" completing the circuit to the lower element. The lower element then begins to heat the lower two-thirds of the tank. Most of the time, it is the lower heating element that keeps the water at the desired temperature. Should a large amount of water be used in a short time, the upper element comes on. This provides fast recovery of the water in the upper portion of the tank, which is where the hot water is drawn from the tank.

Safety Note: Most residential water heater thermostats are of the single-pole type. They do not open both ungrounded conductors of the 240-volt supply. They open one conductor only, which is all that is needed to open the 240-volt circuit to the heating element. What this means is that a reading of 120 volts to ground will always be present at the terminals of the heating element on a 240-volt, single-phase system such as found in residential wiring. If the heating element becomes grounded to the metal sheath, the element can continue to heat even when the thermostat is in the "OFF" position because of the presence of the 120 volts.

The high temperature limit control opens both ungrounded conductors.

VARIOUS TYPES OF ELECTRICAL CONNECTIONS

There are many variations for electric water heater hookups. Most electric utilities have their own unique time-of-use programs and specific electrical connection requirements for electric water heaters. It is absolutely essential that you contact your local electric utility for this information.

Because of the cumulative large power consumption of electric water heater loads, electric utilities across the country have been innovative in creating special lower rate structures (programs) for electric water heaters, air conditioners, and heat pumps. Electric utilities want to be able to shed some of this load during their peak periods. They will offer the homeowner choices of how the water heater, air conditioner, or heat pump is to be connected, and in some cases, will provide financial incentives (lower rates) to those who are willing to use these restricted "time-of-day" connections. "Time-of-day," "time-of use," and "off-peak" are terms used by different utilities. If homeowners have a large storage capacity electric water heater, they

Figure 19-8 Typical electric water heater controls. The seven terminal control combines water temperature control plus high temperature limit control. This is the type commonly used to control the upper heating element as well as providing a high temperature limiting feature. The two terminal control is the type used to control water temperature for the lower heating element. *Courtesy of* **THERM-O-DISC, Inc., subsidiary of Emerson Electric.**

will have an adequate supply of hot water during those periods of time the power to the water heater is off. If they cannot live with this, they have other options. Check with the local power company for details on their particular programs.

The following text discusses some of the methods in use around the country for connecting electric water heaters, and in some cases, air-conditioning and heat pump equipment.

Utility Controlled

The world of electronics has opened up many new ways for an electric utility to meter and control residential electric water heater, air-conditioning, and heat pump loads. Electronic watt-hour meters can be read remotely, programmed remotely, and can detect errors remotely. The options are endless.

The utility can install watt-hour meters that combine a watt-hour meter and a set of switching contacts. The contacts are switched on and off by the utility during peak hours using a power line carrier signal. Older style versions of this type of watt-hour meter incorporated a watt-hour meter, time clock, and switching contacts. The time clock would be set by the utility for the desired times of the on-off cycle. Today, the metering and the programming for timing and control of the on-off cycling is accomplished through electronic solid-state devices. Electronic meters contain a microprocessor that records, and an optical assembly that scans the rotating disc to record the number of turns and speed of the disc. These electronic digital meters can record the energy consumption during normal hours and during peak hours, allowing the electric utility to bill the homeowner at different rates for different periods of time. One type of residential single-phase watt-hour meter can register four different "time-of-use" rates. This type of meter is shown in Figure 29-2.

One major utility programs their residential meters for "peak periods" from 9 AM to 10 PM, Monday through Friday—except for holidays. Their "off-peak" periods are weekend days and all other hours. Programming of electronic meters can extend out for more than five years. The possibilities for programming electronic meters are endless. Older style mechanical meters used gears to record this data, and did not have the programming capabilities

that electronic meters do as in Figure 19-9, Figure 19-10, and Figure 19-11.

The switching contacts in the meter are connected in *series* with the water heater, air conditioner, or heat pump load that is to be controlled for the special "time-of-day" rates, Figure 19-9, Figure 19-10, and Figure 19-11. The utility determines when they want to have the electric water heater on or off.

Figure 19-9 shows a separate combination meter/time clock for the electric water heater load. The utility sets the clock to be "OFF" during their peak hours of high power consumption. Thus the term *off-peak metering*. Off-peak metering is oftentimes called "time-of-use" or "time-of-day" metering. In exchange for lower electric rates, the homeowner runs the risk of being without hot water. Should homeowners run out of hot water during these off periods, they have no recourse but to wait until the power to the water heater comes on again. For this type of installation, large storage capacity water heaters are needed so as to be able to have hot water during these periods.

Figure 19-10 illustrates a type of meter that contains an integral set of contacts, controlled by the utility via signals sent over the power lines. These meters are available with either 30-ampere or 50-ampere contacts, the contact rating selected to switch the load to be controlled. The electric utility

---

A 240/120-VOLT SERVICE

B MAIN WATT-HOUR METER

C WATER HEATER WATT-HOUR METER

D SERVICE CONDUCTORS TO MAIN PANEL

E TWO-POLE DISCONNECT SWITCH OR CIRCUIT BREAKER

F COMBINATION HIGH TEMPERATURE LIMIT AND UPPER INTERLOCKING THERMOSTAT

G UPPER HEATING ELEMENT

H LOWER THERMOSTAT

I LOWER HEATING ELEMENT

Notations used in Figure 19-9, Figure 19-10, Figure 19-12, and Figure 19-13.

Figure 19-9 Wiring for a typical "off-peak" electric water heater circuit. A separate watt-hour meter/time clock controls the circuit to the water heater. The utility sets the time clock to turn the power off during certain peak periods of the day. This type of circuitry is called "time-of-day" programming.

Figure 19-10 The meter contains a set of contacts that can be controlled by the utility. The water heater circuit is fed from the main panel, through the contacts, then to the water heater.

determines when to shed the load. The homeowner must be willing to live with the fact that the power to the water heater, air conditioner, or heat pump might be turned off during the peak periods.

In Figure 19-11, the utility mounts a radio-controlled relay to the side of the meter socket. One major utility using this method makes use of ten different radio frequencies. When the utility needs to shed load, it can signal one or all of the relays to turn the water heater, air conditioner, or heat pump load off and on. This might only be a few times each year, as opposed to time-clock programs that operate daily. This method is a simple one. One watt-hour meter registers the total power consumption. The homeowner signs up for this program. The program offers a financial incentive, and the homeowner knows up front that the utility might occasionally turn off the load during crucial peak periods.

Customer Controlled

There are some rate schedules where customers can somewhat control their energy costs by doing

Figure 19-11 In this diagram, the utility sends a radio signal to the relay, which in turn can "shed" the load of a residential water heater, air conditioner, or heat pump. If the relay has two sets of contacts, four conductors would be required. Low-voltage relays are also used, in which case low-voltage conductors are run between the radio control unit and a relay in the appliance. For the control of an air conditioner or heat pump, the radio control unit is connected "in series" with the room thermostat. This arrangement can be used with one watt-hour meter using a sliding rate schedule, with a "time-of-day" control. Contact the electric utility for technical data regarding how they want the electrical connections to be made for their particular programs.

their utmost to use electrical energy during off-peak hours. For example, they would not wash clothes, use the electric clothes dryer, take showers, or use the dishwasher during peak (premium) hours. The choice is up to homeowners. To accomplish this, watt-hour meters can be installed that have two sets of dials: one set showing the *total kilowatt hours*, and the second set showing *premium time kilowatt hours*. See Figure 29-2. With this type of meter, the utility will have the total kilowatt hours used as well as the premium time kilowatt hours used. With this information, the utility bills customers at one rate for the energy used during peak hours, and at another rate for the difference between

total kilowatt hours and premium time kilowatt hours. The electric utility will determine and define *premium time*.

Some homeowners have installed time-clocks on their water-heater circuit to shut off during peak premium time hours to be sure they are not using energy during the high rate per kilowatt-hour periods.

Uncontrolled

The simplest and most common uncontrolled method is shown in Figure 19-12. There are no extra meters, switching contacts in the meter, extra disconnect switches, or controllable relays. The water heater circuit is connected to a two-pole branch-circuit breaker in the main distribution panel. The electrical power to the water heater is on 24 hours a day. This is the connection for the water heater in the residence in this text. Instead of separate "low-rate" metering that requires an additional watt-hour meter, the utility offers a sliding scale of rate steps.

Figure 19-12 Water heater connected to two-pole circuit in Main Panel.

EXAMPLE:

METER READING 06-01-2005	00532	
METER READING 05-01-2005	00007	
TOTAL KILOWATT-HOURS USED	00525	
BASIC MONTHLY ENERGY CHARGE		$11.24
1ST. 400 kWh × 0.10206		40.82
NEXT 125 kWh × 0.06690		8.36
AMOUNT DUE		$60.42

State, local, and applicable regulatory taxes would be added to the amount due.

Certain utilities will use the higher energy charge during the June, July, August, and September summer months.

Another uncontrolled method is to use two watt-hour meters, Figure 19-13. The water heater has power 24 hours per day. One meter will register the normal lighting load at the regular residential energy charge. The second meter will register the energy consumed by the water heater, air conditioner, or heat pump at some lower energy rate.

Are Water Heater Control Conductors Permitted in the Same Raceway as Service-Entrance Conductors?

Conductors other than service-entrance conductors are not permitted in the same raceway as the service-entrance conductors, *230.7*. However, *Exception No. 2* to this section does permit load management control conductors that have overcurrent protection to be installed in the service raceway.

ELECTRIC WATER HEATER BRANCH-CIRCUITS
Branch-Circuit Rating

NEC® 422.13 applies to all fixed storage electric water heaters having a capacity of 120 gallons (454.2 L) or less. This would include most typical residential electric water heaters. ▶An electric water heater is a continuous load per *422.13*, which would require that the branch-circuit rating must not be less than 125 percent of the nameplate rating. ◀

Figure 19-13 Twenty-four-hour water heater installation. The total energy demand for the water heater is fed through a separate watt-hour meter at a lower kWh rate than the regular meter.

Disconnecting Means

For obvious safety reasons, *NEC® 422.30* requires that an appliance have a means of disconnecting it from the power source. In most instances, this is a separate disconnect switch.

NEC® 422.31(B) tells us that for permanently connected appliances rated greater than 300 volt-amperes or ⅛ hp, the disconnecting means is permitted to be the branch-circuit switch or circuit breaker for the appliance if the switch or breaker is within sight of the appliance *or* be capable of being locked in the "OFF" position. ▶The locking provision must be permanently installed on or at the switch or circuit breaker used as the disconnecting means.◀

Workmen have been seriously shocked when working on appliances that they thought were deenergized because after turning off the power, someone came along wondering why the switch was off and they turned the power back on.

Listed lock-off devices are available from the various manufacturers of circuit breakers. Disconnect switches have brackets with holes through which a padlock is installed. OSHA lock-off requirements were discussed in Unit 1.

NEC® 422.35 requires that switches and circuit breakers used as the disconnecting means be of the indicating type, meaning they must clearly show that they are in the "ON" or "OFF" position.

Overcurrent Protection and Conductor Size

There are two separate issues that must be considered.

- The overcurrent protective device ampere rating must be calculated.
- The conductor size must be determined.

Do both calculations. Then compare the results of both calculations to make sure that both overcurrent device and conductors are *Code* compliant.

Branch-Circuit Overcurrent Protection

NEC® 422.11(E) states that for single non-motor-operated electrical appliances, the overcurrent

protective device shall not exceed the protective device rating marked on the appliance. If the appliance has no such marking, then the overcurrent device is to be sized as follows:

- If appliance does not exceed 13.3 amperes—20 amperes
- If appliance draws more than 13.3 amperes—150 percent of the appliance rating

For a single non-motor-operated appliance, if the 150 percent sizing does not result in a standard size overcurrent device rating as listed in *240.6*, then it is permitted to go to the next standard size.

EXAMPLE: A water heater nameplate indicates 4500 watts, 240 volts. What is the maximum size fuse permitted by the *Code*?

SOLUTION:

$$I = \frac{W}{E} = \frac{4500}{240} = 18.75 \text{ amperes}$$

▶An electric water heater is a continuous load per *422.13*.◀ The branch-circuit rating must not be less than 125 percent of the water heater's nameplate rating. The minimum branch-circuit rating is:

$$18.75 \times 1.25 = 23.44 \text{ amperes}$$

This would require a minimum 25-ampere fuse or circuit breaker, which is the next standard size larger than the calculated 23.44 amperes. See *240.6(A)* for standard ampere ratings for fuses and circuit breakers.

The maximum overcurrent device for the water heater is:

$$18.75 \times 1.5 = 28.125 \text{ amperes}$$

This would be a 30-ampere fuse or circuit breaker, which is the next standard size larger than the calculated 28.125 amperes. But there is no reason to use a 30-ampere branch circuit fuse or circuit breaker because we have already determined that a 25-ampere fuse or circuit breaker is suitable for the 4500-watt load.

Conductor Size

The conductors supplying the 4500-watt water heater will be protected by the 25-ampere or breaker as determined above.

CONDUCTOR SIZE (THHN Copper)	ALLOWABLE AMPACITY FROM 60°C COLUMN OF *TABLE 310.16*	MAXIMUM AMPERE RATING OF OVERCURRENT DEVICE
14 AWG	20 amperes	15 amperes
12 AWG	25 amperes	20 amperes
10 AWG	30 amperes	30 amperes

Table 19-4 Table showing typical small conductors, their allowable ampacities and the maximum rating overcurrent devices permitted for these conductors.

In *Table 310.16*, we find the allowable ampacity of conductors in the 60°C column as required by *110.14(C)*. In *240.4(D)*, we find the maximum overcurrent protection for small conductors. Putting this all together, we have the following shown in Table 19-4.

For the example, we selected a 25-ampere overcurrent device. A 30-ampere OCD would have been acceptable. Next, we need to find a conductor that is properly protected by a 25- or 30-ampere OCD. The conductor would have to be a minimum 10 AWG Type THHN, which has an allowable ampacity of 30 amperes, more than adequate for the minimum branch-circuit rating of 23.44 amperes. A 10 AWG conductor is properly protected by a maximum 30-ampere overcurrent device.

If a 12 AWG Type THHN had been selected, it would have been suitable for the load but would not have been properly protected by the 25-ampere OCD. The maximum OCD for a 12 AWG is 20 amperes.

You might want to review Unit 4 for a refresher as to why we use the 60°C column of *Table 310.16* for selecting the conductor size.

The Water Heater for This Residence

The water heater for the residence is connected to Circuit A6-8. This is a 20-ampere, two-pole circuit breaker located in the main panel in the workshop. This circuit is straight 240 volts and does not require a neutral conductor.

Checking the Schedule of Special-Purpose Outlets, we find that the electric water heater has two heating elements: a 2000-watt element and a 3000-watt element. Because of the thermostats on the water heater, both elements cannot be energized at the same time. Therefore, the maximum load demand is

$$\text{amperes} = \frac{\text{watts}}{\text{volts}} = \frac{3000}{240} = 12.5 \text{ amperes}$$

For water heaters having two heating elements connected for nonsimultaneous operation, the nameplate on the water heater will be marked with the largest element's wattage at rated voltage. For water heaters having two heating elements connected for simultaneous operation, the nameplate on the water heater will be marked with the total wattage of both elements at rated voltage.

Conductor Size and Raceway Size for the Water Heater in This Residence

The specifications for this residence call for two 12 AWG Type THHN conductors. The branch-circuit overcurrent device is rated 20 amperes.

Table C1 in *Annex C* of the *NEC®* indicates that trade size ½ EMT will be okay for two 12 THHN conductors. For ease of installation, a short length of trade size ½ flexible metal conduit 18 to 24 in. (450 to 600 mm) long is attached to the EMT, and is connected into the knockout provided for the purpose on the water heater.

Cord/Plug Connections Not Permitted for Water Heaters

NEC® *400.7(A)(8)* and *422.16(C)* list the uses where flexible cords are permitted. A key code requirement, often violated, is that flexible cords shall be used only "where the fastening means and mechanical connections are specifically designed to permit ready removal for maintenance and repair, and the appliance is intended or identified for flexible cord connection." Certainly the plumbing does not allow for the "ready removal" of the water heater. The conductors in flexible cords cannot handle the high temperatures encountered on the water heater terminals. Check the instruction manual for proper installation methods.

Equipment Grounding

The electric water heater is grounded through the trade size ½ EMT and the flexible metal conduit (FMC). *Code* references are *250.110*, *250.118*, *250.134*, and *348.60*. The subject of equipment grounding is covered in Unit 4.

EFFECT OF VOLTAGE VARIATION ON RESISTIVE HEATING ELEMENTS

The heating elements in the water heater in this residence are rated 240 volts. A resistive heating element will only produce rated wattage at rated voltage. They will operate at a lower voltage with a reduction in wattage. If connected to voltages above their rating, they will have a very short life.

Ohm's law and the wattage formula show how the wattage and current depend on the applied voltage. Always use *rated* voltage to calculate the resistance, current draw, and wattage of a resistive heating element. Check manufacturers' specifications.

Nichrome wire, commonly used to make heating elements, has a "hot" resistance approximately 10 percent higher than its "cold" resistance.

CAUTION: When actually testing a heating element, never take an ohms reading on an energized circuit. Remove any other connected wires from the heating element terminal block on a water heater to prevent erratic ohms readings. Always check from the heating element leads to ground to confirm that there is no ground fault in the heating element.

Let's take a look at a 3000-watt, 240-volt heating element and calculate its resistance and current draw.

$$R = \frac{E^2}{W} = \frac{240 \times 240}{3000} = 19.2 \text{ ohms}$$

$$I = \frac{E}{R} = \frac{240}{19.2} = 12.5 \text{ amperes}$$

If a different voltage is substituted, the wattage and current values change accordingly.

At 220 volts:

$$W = \frac{E^2}{R} = \frac{220 \times 220}{19.2} = 2521 \text{ watts}$$

$$I = \frac{W}{E} = \frac{2521}{220} = 11.46 \text{ amperes}$$

Another way to calculate the effect of voltage variance is:

$$\text{Correction Factor} = \frac{\text{Applied Voltage Squared}}{\text{Rated Voltage Squared}}$$

Using the previous example:

$$\text{Correction Factor} = \frac{220 \times 220}{240 \times 240} = \frac{48,400}{57,600} = 0.84$$

then $3000 \times 0.84 = 2521$ watts

In resistive circuits, current is directly proportional to voltage and can be simply calculated using a ratio and proportion formula. For example, if a 240-volt heating element draws 12.7 amperes at rated voltage, the current draw can be determined at any other applied voltage, say 208 volts.

$$\frac{208}{240} = \frac{X}{12.7}$$

$$\frac{208 \times 12.7}{240} = 11.006 \text{ amperes}$$

To calculate this example if the applied voltage is 120 volts:

$$\frac{120}{240} = \frac{X}{12.7}$$

$$\frac{120 \times 12.7}{240} = 6.35 \text{ amperes}$$

Also, in a resistive circuit, wattage varies as the square of the current. Therefore, when the voltage on a heating element is doubled, the current also doubles and the wattage increases four times. When the voltage is reduced to one-half, the current is halved and the wattage is reduced to one-fourth.

A 240-volt heating element connected to 208 volts will have approximately three-quarters of the wattage rating than at rated voltage.

EFFECT OF VOLTAGE VARIATION ON MOTORS

The above formulas work only for resistive circuits. Formulas for inductive circuits such as an electric motor are much more complex than this book is intended to cover. For typical electric motors, the following information will suffice (see Table 19-5).

As previously mentioned, NEMA rated electric motors are designed to operate at ±10% of their nameplate voltage.

VOLTAGE VARIATION	FULL-LOAD CURRENT	STARTING CURRENT
110%	7% decrease	10–12% increase
90%	11% increase	10–12% decrease

Table 19-5 Table showing the approximate change in full-load current and starting-current for typical electric motors when operated at under voltage (90 percent) and over voltage (110 percent) conditions.

HEAT PUMP WATER HEATERS

With so much attention given to energy savings, heat pump water heaters have entered the scene.

Residential heat pump water heaters can save 50 to 60 percent over the energy consumption of resistance-type electric water heaters, depending upon energy rates. The heat energy transferred to the water is three to four times greater than the electrical energy used to operate the compressor, fan, and pump.

Heat pump water heaters work in conjunction with the normal resistance-type electric water heater. Located indoors near the electric water heater and properly interconnected, the heat pump water heater becomes the primary source for hot water. The resistance electric heating elements in the electric water heater are the backup source should it be needed.

Heat pump water heaters remove low-grade heat from the surrounding air, but instead of transferring the heat outdoors as does a regular air-conditioning unit, it transfers the heat to the water. The unit shuts off when the water reaches 130°F (54.44°C). Moisture is removed from the air at approximately one pint per hour, reducing the need to run a dehumidifier. The exhausted cool air from the heat pump water heater is about 10°–15°F (−22.22°C–9.44°C) below room temperature. The cool air can be used for limited cooling purposes, or it can be ducted outdoors.

A typical residential heat pump water heater rating is 240 volts, 60 Hz, single-phase, 500 watts.

Heat pump water heaters operate with a hermetic refrigerant motor compressor. The electric hookup is similar to a typical 240-volt, single-phase, residential air-conditioning unit. *Article 440* of the *NEC*® applies. This is covered in Unit 23.

Electric utilities and numerous Web sites are good sources to learn more about the economics of installing heat pump water heaters to save energy.

REVIEW

Note: Refer to the *Code* or the plans where necessary.

WATER PUMP CIRCUIT ⒶB

1. Does a jet pump have any electrical moving parts below the ground level? _____

2. Which is larger, the drive pipe or the suction pipe? _____

3. Where is the jet of the pump located? _____

4. What does the impeller wheel move? _____

5. Where does the water flow after leaving the impeller wheel?

   a. _____

   b. _____

6. What prevents water from draining back into the pump from the tank? _____

7. What prevents water from draining back into the well from the equipment? _____

   _____

8. What is compressed in the water storage tank? _____

9. Explain the difference between a "two-wire" submersible pump and a "three-wire" submersible pump. _____

   _____

   _____

   _____

10. What is a common speed for jet pump motors? _____

11. Why is a 240-volt motor preferable to a 120-volt motor for use in this residence?

    _____

    _____

12. How many amperes does a 1 hp, 240-volt, single-phase motor draw? (See *Table 430.248.*) _____

13. What size are the conductors used for this circuit? _____

14. What is the branch-circuit protective device? _____

15. What provides the running overload protection for the pump motor? _____

    _____

    _____

16. What is the maximum ampere setting permitted for running overload protection of the 1 hp, 240-volt pump motor? _____

17. Submersible water pumps operate with the electrical motor and actual pump located (Circle the correct answer.)
    a. above permanent water level
    b. below permanent water level
    c. half above and half below permanent water level

18. Because the controller contains the motor starting relay and the running and starting capacitors, the motor itself contains _____

19. What type of pump moves the water upward inside of the deep-well pipe? _____

    _____

20. Proper pressure of the submersible pump system is maintained by a _____

    _____

21. Fill in the data for a 16-ampere electric motor, single-phase, no *Code* letters.

    a. Branch-circuit protection, non-time-delay fuses.
        Normal size   _____ A   Switch size _____ A
        Maximum size _____ A   Switch size _____ A

    b. Branch-circuit protection dual-element, time-delay fuses.
        Normal size   _____ A   Switch size _____ A
        Maximum size _____ A   Switch size _____ A

    c. Branch-circuit protection instant-trip breaker.
        Normal setting   _____ A
        Maximum setting _____ A

    d. Branch-circuit protection—inverse time breaker.
        Normal rating   _____ A
        Maximum rating _____ A

    e. Branch-circuit conductor size, Type THHN _____

        Ampacity _____ A

    f. Motor overload protection using dual-element time-delay fuses.

        Maximum size _____ A

22. The *NEC®* is very specific in its requirement that submersible electric water pump motors be grounded. Where is this specific requirement found in the Code?

_____

_____

23. Does the *NEC®* allow submersible pump cable to be buried directly in the ground?

_____

_____

24. Must the disconnect switch for a submersible pump be located next to the well?

_____

_____

25. A metal well casing (shall) (shall not) be bonded to the pump's equipment grounding conductor or by grounding it with a separate equipment grounding conductor run all the way back to the same ground bus in the panel that supplies the pump circuit, *NEC®* _____ .

WATER HEATER CIRCUIT Ⓐc

1. According to *422.47* the high temperature limiting control must disconnect _____ of the ungrounded conductors. The high temperature control limits the maximum water temperature to _____°F (_____°C).

2. A major hazard involved with water heaters is that they operate under whatever pressure the serving water utility supplies. Water stays liquid at temperatures higher than the normal boiling point of 212°F (100°C). If the high temperature limit control failed to operate for whatever reason, should a pipe burst or a faucet be opened, the pressure would instantly drop to normal atmospheric pressure, causing the superheated water to turn into _____ that could result in _____ . To prevent this from happening, water heaters are equipped with pressure/temperature relief valves.

3. Magnesium rods are installed inside the water tank to reduce _____ .

4. The heating elements in electric water heaters are generally classified into two categories. These are _____ density and _____ density elements.

5. An 80-gallon electric water heater is energized for the first time. Approximately how many hours would it take to raise the water temperature 80°F (26.67°C) (from 40°F to 120°F [4.44°C to 48.89°C])? The water heater has a 3000-watt heating element.

_____

_____

6. What term is used by utilities when the water heater power consumption is measured using different rates during different periods of the day? _____

7. Explain how the electric utility in your area meters residential electric water heater loads.

   _____

   _____

   _____

8. For residential water heaters, the Consumer Product Safety Commission suggests a maximum temperature setting of _____ °F (_____°C). Some states have laws stating a maximum temperature setting of _____ °F (_____°C).

9. An 80-gallon electric water heater has 60°F (15.56°C) incoming water. How many minutes can a 3 gallon/minute showerhead be used to draw 105°F (40.56°C) water? The water heater thermostat had been set at 120°F (48.89°C). _____

10. Approximately how long would it take to produce serious burns to an adult with 140°F (60°C) water? _____

11. Two thermostats are generally used in an electric water heater.

    a. What is the location of each thermostat? _____

       _____

    b. What type of thermostat is used at each location? _____

       _____

12. a. How many heating elements are provided in the heater in the residence discussed in this text? _____

    b. Are these heating elements allowed to operate at the same time? _____

13. When does the lower heating element operate? _____

    _____

14. The *Code* states that water heaters having a capacity of 120 gallons (450 L) or less shall be considered _____ duty and, as such, the circuit must have a rating of not less than _____ percent of the rating of the water heater.

15. Why does the storage tank hold the heat so long? _____

16. The electric water heater in this residence is connected for "limited demand" so that only one heating element can be on at one time.

    a. What size wire is used to connect the water heater? _____

    b. What size overcurrent device is used? _____

17. a. If both elements of the water heater in this residence are energized at the same time, how much current will they draw? (Assume the elements are rated at 240 volts.)

b. What size and Type THHN wire is required for the load of both elements? Show calculations.

18. a. How much power in watts would the two elements in question 17 use if connected to 220 volts? Show calculations.

b. What is the current draw at 220 volts? Show calculations.

19. A condominium owner complains of not getting enough hot water. The serviceman checks the voltage at the water heater and finds the voltage to be 208 volts. The serviceman checks further and finds the electrical service on the building to be a 120/208-volt, three-phase, four-wire system. The main electrical panel for each condominium unit is fed with a three-wire supply. The nameplate on the electric water heater is marked 4500 watts, 240 volts. What is the wattage output of the heating element in this water heater when connected to a 208-volt supply?

20. For a single, non-motor-operated electrical appliance rated greater than 13.3 amperes that has no marking that would indicate the size of branch-circuit overcurrent device, what percentage of the nameplate rating is used to determine the maximum size branch-circuit overcurrent device? _____

21. A 7000-watt resistance-type heating appliance is rated 240 volts. What is the maximum size fuse permitted to protect the branch-circuit supplying this appliance?

   _____

22. The *Code* requirements for standard resistance electric water heaters are found in *Article 422* of the *Code*. Because heat pump water heaters contain a hermetic refrigerant motor compressor, the *Code* requirements are found in *Article* _____ of the *Code*.

23. In your area, is the branch-circuit breaker or switch for the electric water heater branch-circuit acceptable as the disconnecting means, or are you required to install a separate disconnect within sight of the water heater? Provide a brief explanation.

   _____

   _____

Special-Purpose Outlets for Ranges, Counter-Mounted Cooking Unit ⬗_G, and Wall-Mounted Oven ⬗_F

OBJECTIVES

After studying this unit, the student should be able to

▸ understand the *NEC*® requirements for installing and connecting freestanding ranges, counter-mounted cooking units, wall-mounted ovens, microwave ovens, and light energy ovens.

▸ make load calculations to determine proper size conductors, overcurrent protection, disconnecting means, and how to achieve proper grounding.

▸ be familiar with the different configurations for 30- and 50-ampere NEMA receptacles and cords.

▸ be aware of the many types of temperature controls found on cooking equipment.

▸ understand significant differences of color coding and terminal identification on foreign-made appliances.

BASIC CIRCUIT REQUIREMENTS FOR ELECTRIC RANGES, COUNTER-MOUNTED COOKING UNITS, AND WALL-MOUNTED OVENS

Electric ranges are covered in *Article 422* of the *NEC*® and by UL Standard 858, *Household Electric Ranges*.

UL Standard 1082 covers *Electric Household Cooking and Food Serving Appliances*.

Connection Methods

Most recognized wiring methods can be used to hook up appliances, such as nonmetallic-sheathed cable, armored cable, EMT, flexible metal or non-metallic conduit, or service-entrance cable. Check local electrical codes to see if there are any restrictions on the use of any of the wiring methods listed above.

Here are some of the more common methods for connecting electric ranges, ovens, and cooktops.

Direct Connection. One method is to run the branch-circuit wiring directly to the wiring compartment on the appliance. When "roughing-in" the branch-circuit wiring, sufficient length must be provided so as to allow flexibility to make up the connections, to place the appliance into position, and for servicing.

Junction Box. Another method is to run the branch-circuit wiring to a junction box (an outlet box) behind, under, or adjacent to the range, oven, or cooktop location. By definition, this junction box is accessible because should the need arise, the junction box can be reached by removing the appliance. The "whip" on the appliance is then connected to the junction box where the splicing of the tap conductors in the whip and the branch-circuit conductors are made. According to UL Standard No. 858, the flexible metal whip must not be less than 3 ft (900 mm) and no longer than 6 ft (1.8 mm). The conductors in the whip are rated for the high temperatures encountered in the appliance.

Depending on the location of the appliances, it might be possible to connect more than one cooking appliance to a single branch-circuit from a single junction box or from two junction boxes. This is illustrated in Figure 20-1.

Cord- and Plug-Connection. Still another very common method is to use a cord- and plug-connection, *422.33(A)*. This is particularly true for free-standing ranges. A cord- and plug connection also serves as the disconnecting means for household cooking equipment. The range receptacle is permitted to be at the rear base of an electric range. Removing the bottom drawer makes this connection accessible, *422.33(B)*. The receptacle must have an ampere rating not less than the rating of the appliance, *422.33(C)*. For household cooking equipment, the demand factors of *Table 220.55* are also applicable to the receptacle, *422.33(C), Exception*. Range and dryer receptacles are identical, and are discussed later in this unit.

Cord sets rated 30 amperes are referred to as *dryer cords*. Cord sets rated 40, 45, or 50 amperes are referred to as *range cords*.

Receptacles rated 30 amperes are referred to as *dryer receptacles*. Receptacles rated 50 amperes are referred to as *range receptacles*.

Nameplate on Appliance

NEC® 422.60(A) requires that the nameplate must show the appliance's rating in volts and amperes, or in volts and watts.

Figure 20-1 A counter-mounted cooktop unit and a wall-mounted oven connected to one branch-circuit. Depending on the location of the appliances, it might be possible to make the installation with one junction box.

Load Calculations

In *210.19(A)(3)* and *Exceptions*, we find that:

- the branch-circuit conductors must have an ampacity not less than the branch-circuit rating.

- the branch-circuit conductors' ampacity must not be less than the maximum load being served.

- the branch-circuit rating must not be less than 40 amperes for ranges of 8¾ kW or more.

- for ranges 8¾ kW or more, the neutral of a three-wire branch-circuit may be 70 percent of the branch-circuit rating, but not smaller than 10 AWG.

- tap conductors connecting ranges, ovens, and cooking unit to a 50-ampere branch-circuit must be rated at least 20 amperes, be adequate for the load, and must not be any longer than necessary for servicing the appliance.

- *NEC® 220.55* permits the use of the values found in *Table 220.55* and its footnotes for the load calculation.

- *Table 220.55* shows the kW-demand for different kW rated household cooking appliances that are over 1¾ kW. Column A is used in most situations. Note that the maximum demand is considerably less than the actual appliance's nameplate rating. This is because rarely, if ever, will all of the heating elements be used at the same time. *Note 3* to the table tells us when we are permitted to use Columns A and B. This is discussed later in this unit.

- *NEC® 220.61* states that for calculating a feeder or service, the neutral load can be figured at 70 percent of the load on the ungrounded conductors for ranges, ovens, and counter-mounted cooktops using *Table 220.55* as the basis for the load calculation.

In *422.10(A)* we find that the branch-circuit rating shall not be less than the marked rating of the appliance. However, there is an exception for household cooking appliances. The last paragraph of this section states that *"Branch circuits for household cooking appliances shall be permitted to be in accordance with Table 220.55."*

Separate Branch-Circuits

When a separate branch-circuit is run to a wall-mounted oven, or to a counter-mounted cooking unit, the branch-circuit size is based on the appliance's full nameplate rating, *Table 220.55, Note 4*.

Wire Size

After making the load calculations to determine the appliance's calculated load in amperes, we then refer to *Table 310.16* to select a conductor that has an allowable ampacity equal to or greater than the calculated load. As mentioned many times in this text, *110.14(C)* as well as the UL Standards require that we use the 60°C column, regardless of the conductor's insulation rating.

An exception to this basic rule is when the equipment is marked for 75°C. This was discussed in detail in Unit 4 in the section entitled "The Weakest Link of the Chain."

Overcurrent Protection

NEC® 422.11(A) states that the branch-circuit overcurrent protection shall not exceed the rating marked on the appliance.

For appliances that have no marked maximum overcurrent protection, then the branch-circuit overcurrent protection is sized at the ampacity of the branch-circuit conductors supplying the appliance. See *240.4(B), (D)*, and *(E)*.

Disconnecting Means

All appliances require a means of disconnecting it from the power source, *422.30*.

NEC® 422.33(A) accepts a cord-and-plug arrangement as the required disconnecting means.

A branch-circuit switch or circuit breaker is acceptable as the disconnecting means if it is within sight of the appliance or is capable of being locked in the "OFF" position, *422.31(B)*. Breaker manufacturers supply "lock-off" devices that fit over the circuit breaker. Individual disconnect switches also have a "lock-off" provision. ▶The lock-off provision must be permanently installed on the breaker or switch.◀

NEC® 422.34 recognizes "unit switches" that have a marked "OFF" position to be an acceptable disconnect for the appliance. A marked-off switch disconnects all ungrounded conductors. This enables a person to work on a specific component of the appliance, knowing that the power is off to that particular component. Obviously, the line side terminals of the appliance are still "hot."

Using the branch-circuit switch or circuit breaker as the disconnecting means shuts off total power to the appliance and is much safer than relying on the "unit switches." Be sure to use the required lock-off requirement.

GROUNDING FRAMES OF ELECTRIC RANGES, WALL-MOUNTED OVENS, AND COUNTER-MOUNTED COOKING UNITS

The frames of electric ranges, wall-mounted ovens, and counter-mounted cooking units must be grounded, *250.140*. Grounding is accomplished through the equipment grounding conductor in non-metallic-sheathed cable, the metal armor of BX, or with metal raceways such as EMT and flexible metal conduit. See *250.118* and *250.134* for acceptable equipment grounding means.

The frames of electric ranges, wall-mounted ovens, counter-mounted cooking units, and clothes dryers *are not* permitted to be grounded to the neutral branch-circuit conductor, *250.140*.

Because of the ever-present confusion over using a separate equipment grounding conductor vs. the old fashioned way of grounding the frame to the neutral conductor of the circuit, you had better review Unit 15, where this subject is covered in detail. You might also want to review Unit 4 where the grounding qualifications for various wiring methods is discussed.

Receptacles for Electric Ranges and Dryers

Because of their similarities, the following text applies to both range and dryer receptacles.

Residential electric ranges, electric dryers, wall-mounted ovens, and surface-mounted cooking units may be connected with cords designed specifically for this use. Depending upon the appliance's wattage rating, the receptacles are usually rated 30 amperes or 50 amperes, 125/250 volts. Cord sets rated 40 amperes or 45 amperes contain 40- or 45-ampere conductors, yet the plug cap is rated 50 amperes. It becomes a cost issue to use a full 50-ampere rated cord set in all cases, or to match the cord ampere rating as close as possible to that of the appliance being connected.

Range and dryer receptacles are available in both surface mount and flush mount.

Blade and Slot Configuration for Plug Caps and Receptacles

Figure 20-2 shows the different configurations for receptacles commonly used for electric range and dryer hookups. NEMA uses the letter "R"

50-AMPERE
3-POLE, 3-WIRE
125/250-VOLT
NEMA 10-50R
PERMITTED
PRIOR TO 1996 *NEC*

50-AMPERE
4-POLE, 4-WIRE
125/250-VOLT
NEMA 14-50R
REQUIRED BY 1996 *NEC*

30-AMPERE
3-POLE, 3-WIRE
125/250-VOLT
NEMA 10-30R
PERMITTED
PRIOR TO 1996 *NEC*

30-AMPERE
4-POLE, 4-WIRE
125/250-VOLT
NEMA 14-30R
REQUIRED BY 1996 *NEC*

Figure 20-2   Illustrations of 30-ampere and 50-ampere receptacles used for cord- and plug-connection of electric ranges, ovens, counter-mounted cooking units, and electric clothes dryers. Four-wire receptacles and four-wire cord sets are required for installations according to the *NEC.* Three-wire receptacles and three-wire cord sets were permitted prior to the 1996 *NEC.* See *250.140.*

to indicate a receptacle, and the letter "P" for plug cap.

- **30-ampere, Three-wire:** The L-shaped slot on the receptacle (NEMA 10-30R) and the L-shaped blade on the plug cap (NEMA 10-30P) are for the white neutral wire.

- **30-ampere, Four-wire:** The L-shaped slot on the receptacle (NEMA 14-30R) and the L-shaped blade on the plug cap (NEMA 14-30P) are for the white neutral wire. The horseshoe-shaped slot on the receptacle and the round or horseshoe-shaped blade on the plug cap are for the equipment ground.

- **50-ampere, Three-wire:** The wide flat slot on the receptacle (NEMA 10-50R) and the matching wide blade on the plug cap (NEMA 10-50P) are for the white neutral wire.

- **50-ampere, Four-wire:** The wide flat slot on the receptacle (NEMA 14-50R) and the matching wide blade on the plug cap (NEMA 14-50P) are for the white neutral wire. The horseshoe-shaped slot on the receptacle and the round or horseshoe-shaped blade on the plug cap are for the equipment ground.

Cord Sets for Electric Ranges and Dryers

A 30-ampere cord set contains:

- **Three-wire:** three 10 AWG conductors. The attachment plug cap (NEMA 10-30P) is rated 30 amperes.

- **Four-wire:** four 10 AWG conductors. The attachment plug cap (NEMA 14-30P) is rated 30 amperes.

A 40-ampere cord set contains:

- **Three-wire:** three 8 AWG conductors. The attachment plug cap (NEMA 10-50P) is rated 50 amperes.

- **Four-wire:** three 8 AWG conductors and one 10 AWG conductor. The attachment plug cap (NEMA 14-50P) is rated 50 amperes.

A 45-ampere cord set contains:

- **Three-wire:** two 6 AWG conductors and one 8 AWG conductor. The attachment plug cap (NEMA 10-50P) is rated 50 amperes.

- **Four-wire:** three 6 AWG conductors and one 8 AWG conductor. The attachment plug cap (NEMA 14-50P) is rated 50 amperes.

A 50-ampere cord set contains:

- **Three-wire:** two 6 AWG conductors and one 8 AWG conductor. The attachment plug cap (NEMA 10-50P) is rated 50 amperes.

- **Four-wire:** three 6 AWG conductors and one 8 AWG conductor. The attachment plug cap (NEMA 14-50P) is rated 50 amperes.

Terminal Identification for Receptacles and Cords

Receptacle and cord terminals are marked:

- "X" and "Y" for the ungrounded conductors.

- "W" for the white grounded conductor. The "W" terminals will generally be whitish or silver (tinned) in color, in accordance with *200.9* and *200.10* of the *NEC*.®

- "G" for the equipment grounding conductor. This terminal is green colored, hexagon shaped, and is marked "G," "GR," "GRN," or "GRND" in accordance with *250.126* and *406.9(B)* of the *NEC*.® The grounding blade on the plug cap must be longer than the other blades so that its connection is made before the ungrounded conductors make connection in accordance with *406.9(D)* of the *NEC*.®

WALL-MOUNTED OVEN CIRCUIT Ⓐ_F

The wall-mounted oven in this residence is located to the right of the kitchen sink.

The branch-circuit wiring, using nonmetallic-sheathed cable, is run from Panel B to the oven location, coming into the cabinet space where the oven will be installed from behind or from the bottom. Earlier in this unit we discussed the various methods for hooking up appliances. The choices are direct connection, junction box, or cord- and plug connection. In most cases, a built-in wall-mounted oven will have a flexible metal conduit attached to it, which is run to a junction box where the electrical connections are completed.

Circuit Number. The circuit number is B (6-8). This information is found in the Schedule of Special

Purpose Outlets contained in the Specifications at the rear of this text, and on the Directory for Panel B found in Unit 27.

Nameplate Data. This appliance is rated 6.6 kW (6600 watts) @ 120/240 volts.

Load Calculations. (Use nameplate rating per *220.55, Note 4.*)

$$I = \frac{W}{E} = \frac{6600}{240} = 27.5 \text{ amperes}$$

Wire Size. The conductors in nonmetallic-sheathed cable are rated for 194°F (90°C). However, in conformance to *334.80*, the allowable ampacity is determined using the 140°F (60°C) column of *Table 310.16*. Here we find that a 10 AWG conductor has an allowable ampacity of 30 amperes—more than adequate for the appliance's calculated load of 27.5 amperes. This will be a 10 AWG three-conductor plus equipment grounding conductor nonmetallic-sheathed cable.

Overcurrent Protection. A two-pole, 30-ampere circuit breaker located in Panel B.

Disconnecting Means. The disconnecting means is the 30-ampere, two-pole circuit breaker in Panel B. If a cord- and plug connection were to be made, that also would serve as a disconnecting means.

Grounding. The equipment grounding conductor contained in the nonmetallic-sheathed cable provides the means for grounding the frame of the oven. According to *Table 250.122*, the equipment grounding conductor for a circuit protected by a 30-ampere overcurrent device is a 10 AWG copper conductor. That is the size EGC contained in 10 AWG nonmetallic-sheathed cable.

COUNTER-MOUNTED COOKING UNIT CIRCUIT Ⓐ G

The cooktop in this residence is located in the island.

The branch-circuit wiring is run from Panel B to the island. The nonmetallic-sheathed cable is concealed above the recreation room ceiling, upward into the island cabinetry, ending in a junction box below the cooking unit. Earlier in this unit we dis-

cussed the various methods for hooking up appliances. The choices are direct connection, junction box, or cord-and-plug connection. In most cases, the cooktop will have a flexible metal conduit attached to it, which also runs to the junction box where the electrical connections are completed.

Note on the plans that a receptacle outlet is to be installed on the side of the island. This receptacle outlet is supplied by small appliance branch-circuit B16. This receptacle is required by *210.52(C)(2)*.

Circuit Number. The circuit number is B (2-4). This information is found in the Schedule of Special Purpose Outlets contained in the Specifications at the rear of this text, and on the Directory for Panel B found in Unit 27.

Nameplate Data. This appliance is rated 7450 watts @ 120/240 volts.

Load Calculations. (Use nameplate rating per *Table 220.55, Note 4.*)

$$I = \frac{W}{E} = \frac{7,450}{240} = 31.04 \text{ amperes}$$

Wire Size. The conductors in nonmetallic-sheathed cable are rated for 90°C. However, according to *334.80*, the allowable ampacity is determined using the 60°C column of *Table 310.16*. Here we find that an 8 AWG conductor has an allowable ampacity of 40 amperes—more than adequate for the appliance's calculated load of 31.04 amperes. This will be an 8 AWG three-conductor plus equipment grounding conductor nonmetallic-sheathed cable.

Overcurrent Protection. A two-pole, 40-ampere circuit breaker located in Panel B.

Disconnecting Means. The disconnecting means is the 40-ampere, two-pole circuit breaker in Panel B. A cord- and plug connection would also be a disconnecting means.

Grounding. The equipment grounding conductor contained in the nonmetallic-sheathed cable provides the means for grounding the frame of the appliance. According to *Table 250.122*, the equipment grounding conductor for a circuit protected by a 40-ampere overcurrent device is a 10 AWG copper conductor. That is the size EGC contained in an 8 AWG nonmetallic-sheathed cable.

FREESTANDING RANGE

Load Calculations

The load calculations for a freestanding electric range are simple.

According to *Table 220.55*, the demand is permitted to be figured at 8 kW for an electric range that is rated 12 kW or less. For example, if an electric range is marked 11.4 kW, the maximum demand can be figured at 8 kW because under normal situations, the surface heating elements, broiler element, baking element, and lights would not all be on at the same time.

$$I = \frac{W}{E} = \frac{8,000}{230} = 34.78 \text{ amperes}$$

Checking *Table 310.16*, we find that an 8 AWG three-conductor plus equipment grounding conductor nonmetallic-sheathed cable with an ampacity of 40 amperes is more than adequate.

Larger Ranges

Load calculations for double-oven ranges are a little more complex than for typical one-oven electric ranges.

Larger double-oven freestanding ranges are probably rated more than 12 kW. Let us consider a range that has a rating of 14,050 watts. This is the same wattage as the previously discussed wall-mounted oven and counter-mounted cooking unit.

The total connected load of this freestanding electric range is:

$$I = \frac{W}{E} = \frac{14,050}{240} = 58.5 \text{ amperes}$$

However, *Table 220.55* permits the calculated load to be considerably less than the connected load.

Table 220.55 and the footnotes provide the information we need to make the load calculation.

1. 14,050 − 12,000 watts = 2050 watts (2 kW)

2. According to *Note 1, Table 220.55*, a 5 percent increase for each kW and major fraction in excess of 12 kW must be added to the load demand in Column C. 2 kW (the kW over 12 kW) × 5% per kW = 10%.

3. 8 kW (from Column C) × 0.10 = 0.8 kW

4. The calculated load is: 8 kW + 0.8 kW = 8.8 kW

5. In amperes, the calculated load is:

$$I = \frac{W}{E} = \frac{8,800}{240} = 36.7 \text{ amperes}$$

Wire Size. The conductors in nonmetallic-sheathed cable are rated for 90°C. However, in conformance to *334.80*, the allowable ampacity is determined using the 60°C column of *Table 310.16*. Here we find that an 8 AWG conductor has an allowable ampacity of 40 amperes—more than adequate for the appliance's calculated load of 36.7 amperes. This will be an 8 AWG three-conductor plus equipment grounding conductor nonmetallic-sheathed cable.

NEC® 210.19(A)(3), Exception No. 2 states that for ranges of 8¾ kW or more, where the maximum demand has been calculated using Column C of *Table 220.55*, the neutral may be reduced to not less than 70 percent of the branch-circuit rating, but never smaller than 10 AWG.

$$40 \times 0.70 = 28 \text{ amperes}$$

Checking *Table 310.16*, we find that a 10 AWG conductor has an allowable ampacity of 30 amperes—more than adequate to carry the calculated neutral load. If the wiring method were a raceway method, this electric range could be served by two 8 AWG phase conductors and one 10 AWG neutral conductor.

Using the 70 percent multiplier is a moot point when installing nonmetallic-sheathed cable or armored cable, because the insulated phase conductors in these cables are all the same size.

Why the Reduced Neutral?

The heating elements are connected across the 240-volt source. Therefore, no neutral current from these elements is flowing in the 240/120-volt supply to the range.

Some loads in a range are 120-volt loads. These might be controls, timers, convention fans, rotisseries, and similar items. These loads do result in a current flow in the neutral conductor of the 240/120-volt supply to the range.

The permission to reduce the neutral conductor supplying an electric range to 70 percent is more than adequate to carry the above mentioned "line-to-neutral" connected loads within the range.

Some manufacturers supply a small 240/120-volt transformer inside the range that is connected across the two "hot" 240-volt conductors. The 120-volt secondary supplies the 120-volt loads within the range. With this arrangement, there is no neutral current flowing in the 240/120-volt supply to the range. In fact, the branch-circuit to this type of electric range really only needs the two "hot" ungrounded conductors. Because you have no way of knowing what type of electric range will be installed, you must install a four-wire branch-circuit consisting of two "hot" ungrounded conductors, one grounded neutral conductor, and one equipment grounding conductor.

When using cable, reducing the neutral conductor is a moot point because all conductors in the cable are the same size, other than the equipment grounding conductor. When using a raceway wiring method, it is possible to take advantage of reducing the size of the neutral conductor.

Overcurrent Protection. A two-pole 40-ampere circuit breaker.

Disconnecting Means. The disconnecting means would be the 40-ampere, two-pole circuit breaker. A cord-and-plug connection behind the base of a freestanding electric range would also serve as a disconnecting means, *422.33(B)*.

Grounding. The equipment grounding conductor contained in the nonmetallic-sheathed cable provides the means for grounding the frame of the appliance. According to *Table 250.122*, the equipment grounding conductor for a circuit protected by a 40-ampere overcurrent device is a 10 AWG copper conductor. That is the size EGC contained in an 8 AWG nonmetallic-sheathed cable.

CALCULATIONS WHEN MORE THAN ONE WALL-MOUNTED OVEN AND COUNTER-MOUNTED COOKING UNIT ARE SUPPLIED BY ONE BRANCH-CIRCUIT

The following discusses how to make the load calculations when more than one wall-mounted oven and counter-mounted cooking unit that are supplied by one branch-circuit.

Load Calculations

Note 4 to *Table 220.55* states that when a single branch-circuit supplies a counter-mounted cooking unit, and not more than two wall-mounted ovens, all located in the same room, the calculation is made by adding up the nameplate ratings of the individual appliances, then treating this total as if it were one electric range.

Figure 20-1 shows a wall-mounted oven and a counter-mounted cooking unit supplied by one 40-ampere branch-circuit.

The total connected load of these two appliances is:

$$I = \frac{W}{E} = \frac{14,050}{240} = 58.5 \text{ amperes}$$

However, since we are permitted to treat the two individual cooking appliances as one, we use the same calculations as we did for the freestanding range previously discussed.

The resulting conductor size, overcurrent protection, disconnecting means, and grounding are the same as for the equivalent freestanding range.

The cost of installing one 40-ampere branch-circuit might be higher than the cost of installing separate branch-circuits to each appliance. There will be additional junction boxes, different size cables, cable connectors, conduit fittings, and splices. In addition, there are restrictions in the *NEC*® on how to size the smaller "tap" conductors.

Factory-installed flexible metal conduit furnished and connected to these built-in cooking appliances will have the proper size and temperature-rated conductors for the particular appliance.

If you provide the flexible connections (whips) in the field, the minimum conductor sizing would be calculated as discussed above for each appliance. After the minimum conductor size is determined, we then must make sure that the conductors meet the "tap" rule. This is important because the "taps" will have overcurrent protection rated greater than the ampacity of the "tap" conductors.

NEC® *210.23(C)* states that fixed cooking appliances that are fastened in place may be connected to 40- or 50-ampere branch-circuits. If "taps" are to be made to a 40- or 50-ampere branch-circuit, the taps must be able to carry the load, and in no case may the taps be less than 20 amperes, *210.19(A)(3)*.

For example, the 40-ampere circuit shown in Figure 20-1 has 10 AWG THHN tap conductors for the oven, and 8 AWG THHN tap conductors for the counter-mounted cooking unit. Each of these taps has an ampacity of greater than 20 amperes.

According to *Table 310.16*, an 8 AWG THHN conductor has an allowable ampacity of 40 amperes, using the 60°C column of the table.

USING A LOAD CENTER

The terms **load center** and **panelboard** are used interchangeably in the electrical trade. They are really one and the same thing. Load center is pretty much a manufacturers' term for panelboard. Panelboards might contain additional features that a typical residential "load center" might not have. Refer to Key Terms in the Appendix in this text for a more complete definition of "load center" and "panelboard."

In this residence, the built-in oven and range are very close to Panel B, in which case it is economical to install a separate branch-circuit to each appliance. However, there are situations where the panel and the built-in cooking appliances are far apart.

The decision becomes one of installation cost (labor and material) as to whether to run separate branch-circuits, or to run one larger feeder and use the "tap" rules.

The tap rules can be ignored if a smaller panelboard is installed as illustrated in Figure 20-3. There will be no "taps"—just branch-circuits to each appliance sized for that particular appliance.

The calculations for the feeder from the main panel to another smaller panelboard are based on the two cooking appliances being treated as one. The conductors from the smaller panelboard are calculated for each appliance as previously discussed.

In homes with a basement, the panelboard could be installed below the kitchen where the appliances are installed.

CALCULATIONS WHEN MORE THAN ONE ELECTRIC RANGE, WALL-MOUNTED OVEN, OR COUNTER-MOUNTED COOKING UNIT ARE SUPPLIED BY A FEEDER OR SERVICE

Although generally not an issue for wiring one- and two-family dwellings, the calculations for services and feeders of larger multifamily dwellings is more complex where a number of household cooking units are installed.

We need to refer to *Table 220.55*. This table has three columns. Column C is almost always used, with one exception. *Exception No. 3* shows that for household cooking units having a nameplate rating of more than 1¾ kW through 8¾ kW, Columns A and B are permitted to be used.

Column A is for household cooking units rated more than 1¾ kW but less than 3½ kW. Column C is for household cooking units rated 3½ kW to 8¾ kW.

When using Columns A and B, apply the demand factors for the number of household cooking units whose kW ratings fall within that column. Then add the results of each column together.

EXAMPLE: Calculate the demand for a feeder that has four 3-kW wall-mounted ovens and four 6-kW counter-mounted cooktops connected to it.

Figure 20-3 Appliances connected to a panelboard.

Step 1: 4 ovens $\times$ 3 kW = 12 kW

12 kW $\times$ 66% = 7.92 kW

Step 2: 4 cooktops $\times$ 6 kW = 24 kW

24 kW $\times$ 50% = 12 kW

Step 3: 7.92 kW + 12 kW = 19.92 kW

Therefore, 19.92 kW is the demand load for the four ovens and four cooktops, even though the actual connected load is 36 kW. Diversity plays a large factor in arriving at the calculated demand load.

Review of Choices for Hooking Up Electric Cooking Appliances

- Run a separate branch-circuit from the main panel to each appliance,

- Run one large branch-circuit to a junction box(es), then make "taps" to serve each appliance. See Figure 20-1.

- Run one large feeder from the main panel to a panelboard located near the appliances, then run a separate branch-circuit to each appliance. See Figure 20-3.

When one considers the complicated calculations, additional splices, taps, working with different size conductors, extra time and material, the best and simplest choice is probably to run a separate branch-circuit from the main panel to each appliance.

MICROWAVE OVENS

Microwave ovens are available in countertop, over-the-range, and over-the-counter models and as part of a freestanding electric range.

The electrician's primary concern is to install the proper size and type of branch-circuit and to provide proper equipment grounding for the appliance.

A freestanding electric range that includes a microwave oven is connected according to the methods previously discussed in this unit.

Countertop microwave ovens are plugged into the most convenient receptacle. If the receptacles that serve the countertops have been wired according to the *NEC,* they would be supplied by at least two 20-ampere, small-appliance branch-circuits, *210.11(C)(1)* and *210.52(B)(1).*

Operating a microwave oven and another appliance at the same time might be too much load for a 20-ampere, small-appliance branch-circuit.

The branch-circuit breaker might trip, or the branch-circuit fuse might open. That is why microwave cooking appliance manufacturers suggest a dedicated 15- or 20-ampere branch-circuit for the appliance.

Microwave ovens fastened to the underside of a cabinet above a countertop or range generally are cord- and plug-connected. They come with a rather short power supply cord. A receptacle outlet must be provided inside the upper cabinet. Instructions furnished with these microwave ovens describe where to position this receptacle. A receptacle installed inside the cabinet is *in addition* to the receptacles required to serve countertop surfaces, *210.52.* This receptacle must not be connected to any of the 20-ampere, small-appliance branch-circuits that serve the countertop areas, *210.52(B)(2).*

If a receptacle is installed inside the upper cabinet, the best choice is to connect the receptacle to a separate 120-volt, 20-ampere branch-circuit for the microwave oven. When a single receptacle is installed on a separate 20-ampere branch-circuit, the receptacle must be rated 20 amperes, *210.21(B)(1).*

NEC® 210.52(C)(5) states that receptacles that serve countertops shall not be located more than 20 in. (500 mm) above the countertop. A receptacle located inside the upper cabinets does not serve the countertop and therefore is not subject to *210.52(C)(5).*

A cord- and plug-connected appliance must not exceed 80 percent of the branch-circuit rating, *210.23(A)(1).* A 20-ampere branch-circuit may supply a maximum cord connected load of 20 $\times$ 0.80 = 16 amperes.

LIGHTWAVE ENERGY OVENS

A recent innovation is an oven that combines and cycles on and off lightwave energy from high-wattage halogen lamps and microwave energy. Cooking time is about four eight times faster than a conventional oven. These ovens are mounted to the underside of an upper cabinet or in a wall cabinet. Some require a dedicated three-wire, 30-ampere, 240/120-volt circuit. A four-wire, 30-ampere cord and plug is furnished with these ovens. A 30-ampere, four-wire, NEMA 14-30R receptacle must be installed in the cabinet above the oven. Lower wattage units require a dedicated conventional 20-ampere, 120-volt branch-circuit.

UL Standard 923 covers microwave and lightwave energy cooking appliances.

SURFACE HEATING ELEMENTS

Figure 20-4 and Figure 20-5 illustrate typical electric range surface units.

Electric heating elements generate a large amount of radiant heat. Heat is measured in British thermal units (Btu). A *Btu* is defined as the amount of heat required to raise the temperature of one pound of water by one degree Fahrenheit. A 1000-watt heating element generates approximately 3412 Btu of heat per hour.

Heating elements in surface-type cooking units typically consist of spiral-wound nichrome resistance wire carefully imbedded in magnesium oxide—a white, chalklike powder. The wire/magnesium oxide is then encased in a nickel-steel alloy sheath, flattened under very high pressure, then formed into coils. In the world of electric ranges, we use the term *surface heating elements*. In the world of gas ranges, the term *burners* is used.

Surface heating elements with one-coil (one heating element) are used with infinite heat controls. A one-coil element is shown in Figure 20-6. Two-coil (two heating elements) surface unit heating elements are used with fixed-heat three-, five-, and seven-position controls. A two-coil element is shown in Figure 20-7. Three-coil (three heating elements) surface unit heating elements are used with three-position switches.

TEMPERATURE CONTROLS

Different types of controls are used to adjust the heating elements on electric ranges.

Infinite-Position Controls

Most electric ranges are equipped with infinite-position temperature controls. These controls contain electrical contacts that open and close constantly, "pulsing" the power supply to the heating element, thus maintaining the desired temperature of the surface heating unit.

Figure 20-6 shows the inner workings of this type of control.

Fixed-Heat Controls

Still in use are rotary controls that have specific "indents" that lock the switch contacts into place as the knob is rotated to the various heat positions. Similar switching arrangements are found in push-button-type controls. These switches vary the connections to the heating elements individually, in series, in parallel, and in series/parallel to attain the various heat levels (high-low-medium, medium-high, etc.) These switches are commonly available in three, five, and seven positions.

Coil Size Selector Controls

Coil size selector controls are used with three-coil heating elements. In position one, the center

Figure 20-4 Typical 240-volt electric range surface heating element used with infinite heat controls.

Figure 20-5 Typical electric range surface unit of the type used with seven-position controls.

Figure 20-6 Typical internal wiring of an infinite heat surface element control.

Figure 20-7 Typical surface unit wiring.

heating element comes on. In position two, the center and middle heating elements come on. In the third position, the center, middle, and outer heating elements come on. The fourth position is "OFF."

Automatic Sensor Controls

This type of surface heating element has a sensor in the center of the unit. A special 12-volt control circuit is provided through a small transformer. When the control knob is set at a specific temperature, a "responder" pulses the heating element off and on to maintain the set temperature.

Caution When Selecting Conductor Size and Type

Recall from previous coverage in this unit and in Unit 4 that, according to UL Standards and *NEC®* *110.14(C)*, when conductor sizes 14 AWG through 1 AWG are installed, their allowable ampacity

(current-carrying capacity) is found in the 60°C column of *Table 310.16*. If you are using nonmetallic-sheathed cable, the ampacity shall be in accordance with the 60°C (140°F) conductor temperature rating, *334.80*.

That's why it is so important to read the appliance manufacturer's installation data carefully for recommended branch-circuit rating, conductor size, and conductor type.

Connecting Foreign-Made Appliances

Be careful when connecting foreign-made appliances. It is quite possible that the color coding of the terminals and conductors are different from what you are accustomed to. Throughout this text, we have discussed color coding as white for the grounded circuit conductor, black and red for the ungrounded "hot" conductor, and green or bare for the equipment grounding conductor. In this country, we use the term *ground*. In Europe, "ground" is referred to as "earth." Instructions might indicate something like this:

Green, green/yellow, or bare = earth

Blue = neutral

Brown = live

The equipment grounding terminal might be marked with the letter "E." Just be aware that there could be differences in the identification of the terminals and conductors.

REVIEW

Note: Refer to the Code or the plans where necessary.

COUNTER-MOUNTED COOKING UNIT CIRCUIT ⓐ𝒢

1. a. What circuit supplies the counter-mounted cooking unit in this residence? _____

   b. What is the rating of this circuit? _____

2. What three methods may be used to connect counter-mounted cooking units? _____

   _____

   _____

   _____

   _____

   _____

   _____

3. Is it permissible to use standard 60°C insulated conductors to connect all counter-mounted cooking units? Why? _____

   _____

   _____

   _____

   _____

4. What is the maximum operating temperature (in degrees Celsius) for a

   a. Type TW conductor? _____

   b. Type THW conductor? _____

   c. Type THHN conductor? _____

5. The *NEC*® permits the following wiring methods to supply household built-in ovens and ranges. True or False (Circle the correct answer.)

   a. Nonmetallic-sheathed cable   (True) (False)

   b. Armored cable   (True) (False)

   c. Flexible metal conduit   (True) (False)

6. For electric ranges that have a calculated demand of 8¾ kW or more, the neutral conductor can be reduced to _____ percent of the branch-circuit rating, but shall not be smaller than _____ AWG.

7. One kilowatt equals _____ Btu per hour.

8. Older style electric ranges had heat control knobs that had "indents" that could be felt when rotating the knob to adjust temperature to three, five, or seven different heat positions. Most new electric ranges have control knobs that do not have "indents." This type of heat control is known as an _____ heat control.

9. When a separate circuit supplies a counter-mounted cooking unit, what *Code* reference tells us that it is "against Code" to apply the demand factors of *Table 220.55* and requires us to compute the load based on the appliance's actual nameplate rating?

_____

_____

WALL-MOUNTED OVEN CIRCUIT ⒶF AND FREESTANDING RANGE

1. To what circuit is the wall-mounted oven connected? _____

2. An oven is rated at 7.5 kW. This is equal to

    a. _____ watts.

    b. _____ amperes at 240 volts.

3. a. What section of the *Code* governs the grounding of a wall-mounted oven? _____

    _____

    b. By what methods may wall-mounted ovens be grounded? _____

    _____

    _____

    _____

    _____

    _____

    _____

4. What is the type and ampere rating of the overcurrent device protecting the wall-mounted oven in this residence? Hint: Refer to the Schedule of Special Purpose Outlets in the Specifications. _____

    _____

    _____

5. Approximately how many ft (m) of cable are required to connect the oven in the residence? _____

    _____

6. When connecting a wall-mounted oven and a counter-mounted cooking unit to one feeder, how long are the taps to the individual appliances? _____

    _____

    _____

7. The branch-circuit load for a single wall-mounted self-cleaning oven or counter-mounted cooking unit shall be the _____ rating of the appliance.

8. A 6-kW counter-mounted cooking unit and a 4-kW wall-mounted oven are to be installed in a residence. Calculate the maximum demand according to *Column A, Table 220.55*. Show all calculations. Both appliances will be connected to the same branch-circuit.

9. The size of the neutral conductor supplying an electric range may be based on _____ percent of the ampacity of the ungrounded conductors.

10. A freestanding electric range is rated 11.8 kW, 120/240 volts. The wiring method is nonmetallic-sheathed cable. Use the 60°C column of *Table 310.16* in conformance to *110.14(C)*. Answer the following questions.

   a. According to Column C, *Table 220.55*, what is the maximum demand? _____ kW

   b. What is the minimum size for the ungrounded copper conductors?   _____ AWG

   c. What is the minimum size neutral copper conductor?   _____ AWG

   d. What is the correct rating for the branch-circuit overcurrent protection?
      _____ amperes

11. A double-oven electric range is rated at 18 kW, 120/240 volts. Calculate the maximum demand according to *Table 220.55*. Show all calculations.

12. For the range discussed in question 11:

    a. What size 140°F (60°C) ungrounded conductors are required? _____

    b. What size 140°F (60°C) neutral conductor is required? _____

    c. What size 167°F (75°C) ungrounded conductors are required? The terminal block and lugs on the range, the branch-circuit breaker, and the panelboard are marked as suitable for 167°F (75°C) wire. _____

    d. What size 167°F (75°C) neutral conductor is required? Terminations are the same as in c. _____

13. For ranges of 8¾ kW or higher rating, the minimum branch-circuit rating is _____ amperes.

14. A nonmetallic-sheathed cable is used to connect a wall-mounted oven. The insulated conductors are 10 AWG. What is the size of the equipment grounding conductor in this cable? _____

15. Match the following statements with the correct letter. Some letters may be used more than once.

_____ A 30-ampere, three-wire receptacle	a. for the equipment grounding conductor
_____ A 50-ampere, four-wire receptacle	b. for the white neutral conductor
_____ The L-shaped blade on a 30-ampere plug cap	c. for the "hot" ungrounded conductors
_____ The horseshoe-shaped slot on a 50-ampere receptacle	d. NEMA 10-30R
_____ A 50-ampere, four-wire plug cap	e. NEMA 14-50R
_____ Terminals "X" and "Y"	f. NEMA 14-50P
_____ Terminal "W"	
_____ Terminal "G"	

16. A receptacle is mounted inside of a kitchen cabinet for the purpose of plugging in a microwave oven. Circle the correct answer.

a. The receptacle may be connected to the same circuit that supplies the other receptacles that serve the countertop areas.

b. The receptacle shall not be connected to the same circuit that supplies the other receptacles that serve the countertop areas.

Special-Purpose Outlets— Food Waste Disposer ⬣H, Dishwasher ⬣I

OBJECTIVES

After studying this unit, the student should be able to

▶ **install circuits for a typical food waste disposer and dishwasher.**

▶ **know the meaning of "continuous feed" and "batch feed" food waste disposers.**

▶ **understand direct connections (hard wired) and cord- and plug connections.**

▶ **understand the meaning of branch-circuit protection and overload protection.**

▶ **be aware of the requirements for providing a means to disconnect appliances.**

▶ **understand the various acceptable methods for grounding appliances.**

▶ **understand the significance of water temperature for dishwashers.**

This unit discusses the circuit requirements for food waste disposers (also called kitchen waste disposer, garbage disposer, garbage disposal, waste disposer, In-Sink-Erator,® Disposall,®) and dishwashers of the type normally installed in homes. *Code* requirements for appliances are found in *Article 422* of the *NEC.*®

These 120-volt major appliances are *not* permitted to be connected to the 20-ampere small-appliance branch-circuits that serve the receptacles above the countertops in the kitchen.

FOOD WASTE DISPOSER ⬣H

Types of electrical connections are shown in Figure 21-1 and Figure 21-2. The food waste disposer out-

let is shown on the plans by the symbol ⬣H. It is rated 120 volts, 7.2 amperes, and it is connected to Circuit B19, a 20-ampere, 120-volt branch-circuit that originates in Panel B.

Controls

Continuous feed food waste disposers are usually controlled by a single-pole wall switch conveniently located above the countertop, as shown in Figure 21-2. The branch-circuit wiring is run first to the switch box. A second cable is then run from the switch box to the junction box on the disposer.

"Batch-feed" food waste disposers are equipped with an integral switch that starts and stops the disposer when the user twists the drain cover into place

FLUSH MOUNTING BOX-COVER UNIT, WITH THREE-WIRE GROUNDING RECEPTACLE AND DUAL-ELEMENT FUSE SIZED AT APPROXIMATELY 125% OF THE FULL-LOAD CURRENT RATING OF THE MOTOR. FUSE ADDS EXTRA OVERCURRENT PROTECTION TO THE OVERLOAD PROTECTION BUILT INTO THE DISPOSER. RECEPTACLE MUST BE ACCESSIBLE AND LOCATED TO AVOID DAMAGE TO CORD.

EGC

BRANCH-CIRCUIT WIRING CONCEALED IN WALL BEHIND CABINETS. WIRING FEEDS INTO FLUSH-MOUNTED SWITCH BOX UNDER COUNTERTOP IN SPACE NEAR FOOD WASTE DISPOSER.

RECEPTACLE MUST BE ACCESSIBLE.

FOOD WASTE DISPOSER

THREE-WIRE GROUNDING-TYPE ATTACHMENT PLUG CAP. PERMITTED TO SERVE AS THE DISCONNECTING MEANS *NEC® 422.33(A)* APPLIANCE MUST COME WITH CORD OR BE MARKED AS SUITABLE FOR FLEXIBLE CORD CONNECTION.

FLEXIBLE CORD MUST BE IDENTIFIED FOR THE PURPOSE, SHALL HAVE A GROUNDING-TYPE ATTACHMENT PLUG CAP, AND MUST NOT BE LESS THAN 18 IN. (450 mm) LONG—NOR LONGER THAN 3 FT (900 mm). EXAMPLES OF SUITABLE THREE-CONDUCTOR CORDS: SP-3, SPE-3, OR SPT-3. SEE *TABLE 400.4* FOR FLEXIBLE CORD TYPES AND PERMITTED USES.

Figure 21-1   A cord- and plug-connected food waste disposer, *422.16(A)* and *(B)(1)*. The same rules apply to built-in dishwashers and trash compactors, except the cord length must be 3 to 4 ft (900 mm to 1.2 m). The receptacle/fuse box-cover unit or a conventional receptacle could be used.

after filling the disposer with waste food. A wall switch is not required. See Figure 21-3.

Although rarely used in residential installations, a plumber could install a flow switch in the cold water line under the sink to ensure that the food waste disposer is not run without water, Figure 21-4. Generally, water should be left running for at least 15 seconds after shutting off the disposer to flush the drain clear of food waste particles.

DISHWASHER ⊕ᵢ

The dishwasher outlet is shown on the plans by the symbol ⊕ᵢ. The dishwasher is connected to Circuit B5, a 20-ampere, 120-volt branch-circuit that origi-

SINGLE-POLE ON-OFF SWITCH LOCATED ABOVE COUNTERTOP NEAR DISPOSER UNIT

FOOD WASTE DISPOSER

120-VOLT SUPPLY

TWO-WIRE AC; IF NM, THEN EQUIPMENT GROUNDING CONDUCTOR REQUIRED

OVERLOAD PROTECTION BUILT INTO DISPOSER UNIT

Figure 21-2   Wiring for a food waste disposer operated by a separate switch located above the countertop near the sink. Many local codes require this wall switch for all installations to be sure that there is a safe means of easily disconnecting the appliance.

SINGLE-POLE, "ON-OFF" CONTROL SWITCH ACTUATED BY TWISTING DRAIN LID

FOOD WASTE DISPOSER

RUNNING OVERCURRENT PROTECTION BUILT INTO THE DISPOSER

120-VOLT SUPPLY

TWO-WIRE AC; IF NM, THEN EQUIPMENT GROUNDING CONDUCTOR REQUIRED.

Figure 21-3   Wiring for a "batch-feed" food waste disposer with an integral "ON-OFF" switch.

Figure 21-4   "Batch-feed" disposer with a flow switch in the cold water line.

nates in Panel B. A typical dishwasher contains a motor, an electric heating element, and, in some models, a small fan to assist in circulating the hot air during the drying cycle. For example, one manufacturer's dishwasher total connected load as marked on the nameplate is 9.2 amperes. This includes the motor, heating element, controls, relays, and an 875-watt heating element.

In most residential dishwashers, the motor and the booster heater operate at the same time. The heating element boosts the incoming hot water from a typical water heater setting of 120° to approximately 140° or higher so as to better clean, dissolve detergents, and somewhat sanitize the dishes. During the dry cycle, only the heating element is on, plus a circulating fan if so equipped.

CODE RULES COMMON TO THE FOOD WASTE DISPOSER AND THE DISHWASHER

Branch-Circuit Rating

The manufacturer will specify the recommended rating of the branch-circuit. Although it is possible to connect both appliances to one branch-circuit, much depends on the current draw of the appliances. The preferred way to connect these appliances is to supply each appliance with a separate 20-ampere branch-circuit. This provides a little spare capacity if the homeowner should decide to add a booster water heater at the sink location later on.

Do not connect the food waste disposer or dishwasher to any of the required 20-ampere small-

appliance branch-circuits in the kitchen. This is a violation of *210.52(B)(2)*. Small-appliance circuits are intended to serve countertops for cord-connected *portable* appliances only. See Unit 12.

Conductors

Per the Specifications in the back of this text, the branch-circuit conductors are 12 AWG. The wiring method might be nonmetallic-sheathed cable, armored cable, or EMT, depending on the wiring method required by the code in your locality.

Direct Connection

See Figure 21-2, Figure 21-3, and Figure 21-4. This is sometimes referred to as "hard wired." These appliances have a junction box (wiring compartment) for the electrical hookup. The branch-circuit supply (cable or flex conduit) is run directly to this box. During the rough-in stage of new wiring, the electrician must make sure that the supply cable or flex conduit is brought in at the proper location. This might be under the sink, brought up from underneath, or brought in behind the appliance. Sufficient length of wiring is necessary to be able to make the electrical connections, and to position the appliance into place. The manufacturer's installation instructions provide all of the necessary information as to where to bring in the power supply.

Cord- and Plug-Connection

See Figure 21-1. Cord- and plug-connection is permitted by *422.16(B)(2)*.

If three-wire cord- and plug-connection is desired for easy disconnecting and servicing, the appliance manufacturers make available a power supply cord kit for this purpose. The appliances would have to be listed as being intended or identified for flexible cord connection, *NEC® 400.7* and *422.16(A)* and *(B)*.

UL lists these appliances for connection with and without a cord and plug assembly.

Cord lengths are restricted by the *NEC®* and UL Standards:

- Food waster disposers: no shorter than 18 in. (450 mm) and no longer than 3 ft (900 mm).

- Dishwashers: no shorter than 3 ft (0.9 m) and no longer than 4 ft (1.2 m)

A duplex grounding-type receptacle is usually installed in an accessible location under the sink for plugging in the dishwasher and food waste disposer. This receptacle does not have to be GFCI protected since it does not serve the countertop.

Remember that *210.23(A)(2)* limits cord-connected equipment to 80 percent of the branch-circuit ampere rating.

Overload Protection

Overload protection is an integral part of the appliance. Overload protection prevents the motor from burning out if it stalls or overheats for any reason. This protection is required in the UL Standard.

Branch-Circuit Protection

Branch-circuit short-circuit and ground-fault protection is provided by the branch-circuit fuse or circuit breaker, sized according to the manufacturers' recommendations. In our case, each appliance is fed by a separate 20-ampere branch-circuit.

Disconnecting Means

All appliances must be provided with some means of disconnecting the appliance, *Article 422, Part III*. The branch-circuit breaker can serve as the disconnecting means if it is within sight of the appliance or be capable of being locked in the "OFF" position. Rarely, if ever, are these appliances within sight of the breaker panel, and rarely are breaker "lock-off" accessories used. ▶Lock-off provisions

must be permanently installed on the breaker or switch, *422.31(B)*.◀ Some local codes actually require that these appliances be cord- and plug-connected so the appliances can easily be disconnected for servicing by pulling out the plug. This requires that a receptacle be installed—usually under the sink in an accessible location.

Figure 21-1 shows a cord- and plug-connection that serves as the disconnecting means.

Figure 21-2 shows a wall switch that serves as the disconnecting means.

If you do install a receptacle(s) below the sink or behind the appliance for cord- and plug connection of a dishwasher or food waste disposer, the receptacle(s) installed do not have to be GFCI protected because they are not there to serve countertops.

Grounding

- *Article 250, Part VI* covers equipment grounding and equipment grounding conductors. In *Part VI*, we find:
  - *250.114:* Grounding requirements for cord- and plug-connected appliances.
  - *250.118:* Lists all of the acceptable methods of grounding equipment, such as the equipment grounding conductor in nonmetallic-sheathed cable, armored cable, or flexible metal conduit.
  - *250.119:* Requires equipment grounding conductor to be bare or have green insulation, or have green insulation with one or more yellow stripes.
  - *250.122:* Shows the minimum size equipment grounding conductors.
  - *250.126:* Requires the equipment grounding conductor terminal to be to be bare or green and hexagon shaped.

- *Article 250, Part VII* covers methods of equipment grounding. In *Part VI*, we find:
  - *250.134:* Discusses grounding of equipment fastened in place, or "hard-wired"; refers us back to *250.118*.
  - *250.138:* Accepts the equipment grounding conductor in flexible cords for cord- and plug-connected appliances.
  - *250.142(B):* Grounding appliances to the grounded circuit conductor is not permitted.

Supplemental Overcurrent Protection and Disconnecting Means

Instead of installing a duplex grounding-type receptacle under the sink, electricians will sometimes install a box-cover "switch/fuse unit" under the sink of a type similar to that shown in Figure 16-25. For a cord- and plug-connection, the box-cover unit would be a "receptacle/fuse" type, as shown in Figure 21-1. This receptacle/fuse unit does not have to be GFCI protected because it does not serve the countertop. Box-cover units provide both supplemental overload and short-circuit protection. They also serve as a convenient accessible disconnecting means, as required by *422.30*, making it easy for the technician to be assured that the circuit is off when working on the dishwasher. Time-delay fuses in this box-cover unit would be sized at approximately 125 percent of the nameplate rating of the dishwasher.

Read the Installation Instructions

Always read the installation instructions, the owners manual, and the nameplate for specific electrical requirements.

PORTABLE DISHWASHERS

In addition to built-in dishwashers, portable models are available. Most portable models are convertible to become built-in units. Portable dishwashers have one hose that connects to the water faucet and a water drainage hose that hangs in the sink. The dishwasher probably will be plugged into the receptacle nearest the sink.

Portable dishwashers are supplied with a three-wire cord and plug. If the three-wire plug cap is plugged into a properly connected three-wire grounding-type receptacle, the dishwasher is adequately grounded.

In older homes that do not have grounding-type receptacles, the use of a portable cord- and plug-connected dishwasher will require the installation of a grounding-type receptacle. Assuming the appliance will be used in the kitchen, the receptacle will also have to be GFCI protected, *210.8(A)(6)*. This topic is covered in Unit 6.

Cord- and plug-connected portable dishwashers are generally rated 10 amperes or less at 120 volts.

WATER TEMPERATURE

Electric water heaters are shipped with the thermostat set as low as possible. When the water heater is installed, the thermostat should be reset to recommended settings such as 120°F (49°C). In fact, because of the problem of people being scalded, many municipalities and states have implemented code requirements that water heater thermostats be set at not over 120°F (49°C). Although this temperature is acceptable for baths and showers, 120°F (49°C) is not hot enough to properly clean dishes and glasses. Some manufacturers of dishwashers provide a feature that turns on the heating element for the final rinse cycle. This raises the water temperature from 120°F (49°C) to the desired 140°F (60°C) to 145°F (63°C).

See Unit 19 for a discussion on water heaters and the dangers of scalding with hot water of varying temperatures.

REVIEW

Note: Refer to the *Code* or the plans where necessary.

FOOD WASTE DISPOSER CIRCUIT ⒶH

1. How many amperes does the food waste disposer draw? _____

2. a. To what circuit is the food waste disposer connected? _____

    b. What size wire is used to connect the food waste disposer? _____

3. Means must be provided to disconnect the food waste disposer. The homeowner need not be involved in electrical connections when servicing the disposer if the disconnecting means is _____

_____

4. How is motor overload protection provided in most food waste disposer units?

_____

5. When running overcurrent protection is not provided by the manufacturer or if additional backup overcurrent protection is wanted, dual-element, time-delay fuses may be installed in a separate box-cover unit. These fuses are sized at not over _____ percent of the full-load rating of the motor.

6. Why are flow switches sometimes installed on food waste disposers? _____

_____

7. Where in the *NEC*® would you look for basic *Code* rules relating to grounding appliances?

_____

8. Do the plans show a wall switch for controlling the food waste disposer? _____

9. A separate circuit supplies the food waste disposer in this residence. Approximately how many ft (m) of cable will be required to connect the disposer?

_____

10. If a receptacle is installed underneath the sink for the purpose of plugging in a cord-connected food waste disposer, must the receptacle be GFCI protected? _____

DISHWASHER CIRCUIT Ⓐ₁

1. a. To what circuit is the dishwasher in this residence connected? _____

   b. What size wire is used to connect the dishwasher? _____

2. The dishwasher in this residence is rated (circle the correct answer)

   a. 5 A   b. 9.2 A   c. 15 A

3. The heating element is rated at (circle one)

   a. 875 watts   b. 1000 watts   c. 1250 watts

4. How many amperes at 120 volts do the following heating elements draw?

   a. 750 watts _____

   b. 1000 watts _____

   c. 1250 watts _____

5. How is the dishwasher in this residence grounded? _____

_____

_____

_____

6. What type of cord is used on most portable dishwashers? _____

7. How is a portable dishwasher grounded? _____

_____

_____

8. Who is to furnish the dishwasher? _____

_____

9. What article of the *Code* specifically addresses electrical appliances? _____

Special-Purpose Outlets for the Bathroom Ceiling Heat/Vent/Lights ⊕K ⊕J, the Attic Fan ⊕L, and the Hydromassage Tub ⊕A

OBJECTIVES

After studying this unit, the student should be able to

▶ explain the operation and control of heat/vent/lights and make electrical connections in conformance to the *NEC.*

▶ understand the operation and control of attic exhaust fans and make electrical connections in conformance to the *NEC.*

▶ understand humidity problems, solutions, and humidistats.

▶ understand *NEC*® requirements for hydromassage bathtub branch-circuits.

BATHROOM CEILING HEATER CIRCUITS ⊕K ⊕J

Both bathrooms contain a combination heater, light, and exhaust fan installed in the ceiling. The heat/vent/light is shown in Figure 22-1. The symbols ⊕K and ⊕J represent the outlets for these units.

Each heat/vent/light contains a heating element similar to the surface burners of electric ranges. The appliances also have a single-shaft motor with a blower wheel, a lamp with a diffusing lens, and a means of discharging air to the outside of the dwelling.

The heat/vent/lights specified for this residence are rated 1500 watts at 120 volts. The unit in the master bedroom bathroom ⊕J is connected to Circuit A12, a 20-ampere, 120-volt circuit. In the front bedroom bathroom, the unit is connected to Circuit A11, also a 20-ampere, 120-volt circuit.

The current rating of the 1500-watt unit is

$$I = \frac{W}{E} = \frac{1500}{120} = 12.5 \text{ amperes}$$

Article 424 of the *NEC*® covers fixed electric space heating. ▶*NEC*® *424.3(B)* considers fixed electric space heating to be a continuous load.◀ Simply stated, this means that the 125 percent multiplier must be applied for the branch-circuit conductors, *210.19(A)*, and the branch-circuit overcurrent device, *210.20(A)*.

Therefore,

12.5 amperes × 1.25 = 15.625 amperes.

We have chosen to supply each of the bathroom ceiling heaters with a 20-ampere, 120-volt branch-circuit. The 12 AWG conductors and the 20-ampere circuit breakers "Meet Code."

Figure 22-1 Heat/vent/light. *Photo courtesy of Broan-NuTone.*

Wiring

At first, the connections look rather complicated. Let's walk through it!

In Figure 22-2, we see the wiring diagram for the heat/vent/light. The 120-volt line is run from Panel A to the wall box. A raceway (flexible or elec-

trical metallic tubing) is run between the wall box where the switches are located to the heat/vent/light unit. In this raceway, five wires are needed—one common white grounded circuit conductor, and one conductor each for the heating element, fan motor, 100-watt lamp, and the 7-watt night light. The four ungrounded switch leg wires may be any color except white, gray, or green. Color coding of conductors is discussed in detail in Unit 5. Because the circuit rating is 20 amperes, 12 AWG conductors are installed. See the Schedule of Special Purpose Outlets contained in the Specifications for this text.

Nontrained electricians often violate the code by running one three-wire and one two-wire cable between the wall box and heat/vent/light. *NEC® 300.3(B)* states that "all conductors of the same circuit and, where used, the neutral and all equipment grounding conductors shall be contained within the same raceway, cable tray, trench, cable, or cord."

The switch location requires a two-gang opening, with the wall box being large enough to accommodate seven 12 AWG conductors, four switches, and cable clamps. For this installation, a 4-in. square, 2⅛ in. deep box with a two-gang raised plaster ring is used. This provides ample room for the conductors and switches. Box sizing was covered in Unit 2 and Unit 8.

Operation of the Heat/Vent/Light

The heat/vent/light is controlled by four switches, as shown in Figure 22-2:

- one switch for the heater
- one switch for the light
- one switch for the exhaust fan
- one switch for the night light

When the heater switch is turned on, the heating element begins to give off heat. The heat activates a bimetallic coil attached to a damper section in the housing of the unit. The heat-sensitive coil expands until the damper closes the discharge opening of the exhaust fan. Air is taken in through the outer grille of the unit. The air is blown downward over the heating element. The blower wheel circulates the heated air back into the room area.

If the heater is turned off so that only the exhaust fan is on, the air is pulled into the unit and is exhausted to the outside of the house by the blower wheel.

NOTE: HEATER CANNOT OPERATE WITHOUT FAN, BUT FAN CAN OPERATE WITHOUT HEATER.

120-VOLT LINE

FAN

HEATER

LAMP

NIGHT LIGHT

CONNECTIONS AS PER MANUFACTURER'S RECOMMENDATIONS

Figure 22-2 Wiring for heat/vent/light.

Exhaust fans and luminaires (fixtures) are available with integral automatic moisture sensors and motion detectors. Step into a dark bathroom, and the light goes on. Step into and run the shower, and the exhaust fan turns on at a predetermined moisture level.

Grounding the Heat/Vent/Light

The heat/vent/light must be properly grounded!

Proper grounding is accomplished by using any of the accepted equipment grounding methods found in *250.118*. For residential wiring, the equipment grounding conductor in nonmetallic-sheathed cable, the armor and bonding wire in armored cable, or flexible metal conduit provides the necessary equipment ground. As stated many times before, NEVER ground an appliance to the white grounded circuit conductor, *250.142(B)*.

ATTIC EXHAUST FAN CIRCUIT Ⓐ_L

Read the Installation Instructions!

Before beginning any installation of electrical equipment, always refer to manufacturers' literature for specific installation instructions.

The attic exhaust fan in this residence has a two-speed, ¼-hp direct-drive motor. "Direct-drive" means that the fan blade is attached directly to the shaft of the motor without the use of pulleys and belts.

The exhaust fan is mounted in the hall ceiling between the master bedroom and the front bedroom, as shown in Figure 22-3(A). Exhaust fans can also be installed in a gable of a house, as shown in Figure 22-3(B), or in the roof.

When the fan is not running, the louvers on the fan and house gable remain in the "CLOSED" position.

When running, the louvers on the fan and house gable "OPEN," removing hot, stagnant, humid, or smoky air from the house by drawing in fresh air through open windows and doors, in turn exhausting the air through the louvers.

An exhaust fan can lower the indoor temperature of the house as much as 10°F to 20°F (6°C to 11°C). On hot days, the air in an attic can reach superheated temperatures of 150°F (66°C) or more. The heat from the attic will radiate and conduct through the ceiling into the living areas. This results in increased loading of the air-conditioning system. These problems are minimized by properly vented exhaust fans. Humidity issues are discussed later on.

Bathroom exhaust fans, vents for clothes dryers, and similar high humidity sources are *never* to be vented directly into the attic or other dead spaces. They are to be vented to the outside.

Sizing a Home-Type Exhaust Fan

Exhaust fan ventilating capability is rated in cubic feet of air per minute (CFM).

To determine the minimum amount of CFM needed, multiply the square footage of the house by 3. For example, an 1800-ft² house would need an exhaust fan rated $1800 \times 3 = 5400$ CFM.

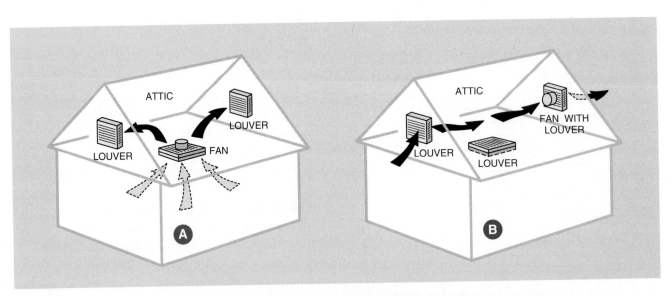

Figure 22-3   Exhaust fan installation in an attic.

To determine the minimum amount of attic exhaust louver square footage area, divide the CFM by 750. In the above example, $5400 \div 750 = 7.2 \ ft^2$.

Never exhaust air into spaces within walls, ceilings, attics, crawl spaces, or garages. The humidity may damage the structure and insulation.

Figure 22-4 shows a typical ceiling-mounted whole-house ventilator exhaust fan.

Branch-Circuit

The exhaust fan in this residence is rated 115 volts nominal, 5.8 amperes (696 VA). It is connected to Circuit A10, a 120-volt, 15-ampere circuit breaker in the main panel. This branch-circuit supplies the attic exhaust fan only. It supplies no other loads.

Conductors (*430.22*)

NEC® requires that the minimum size conductor is to be not less than 125 percent of the motor's full-load current rating.

Therefore,

$$5.8 \times 1.25 = 7.25 \text{ amperes}$$

Table 310.5 shows a minimum size conductor for general wiring to be 14 AWG copper.

Table 310.16 shows that a 14 AWG THHN copper has an allowable ampacity of 25 amperes.

In conformance to these requirements, the Schedule of Special Purpose Outlets tells us that the branch-circuit conductors are 14 AWG conductors. The branch-circuit cable is run to a 4-in. square deep outlet box installed by the electrician near the exhaust fan. A switch/fuse box-cover unit is to be mounted on this outlet box. From this box, a flexible connection like armored cable, flexible metal conduit, or nonmetallic-sheathed cable is made to the factory-installed junction box on the frame of the fan. Depending on the size of junction box on the fan assembly, it might be possible to run the branch-circuit cable directly into that junction box.

Figure 22-4 Ceiling exhaust fan as viewed from the attic. *Courtesy of* Broan-NuTone.

Overload Protection
(430.31 through 430.44)

The exhaust fan motor is provided with integral (built into the motor) overload protection. Overload protection prevents the motor from burning out if it stalls or overheats for any reason. This protection is required in the UL Standard.

NEC® *430.32(C)* states that for automatically started motors, the overload protection must not exceed 125 percent of the full-load running current of the motor. *Table 430.248* shows that a ¼-hp motor has a full-load current draw of 5.8 amperes. Thus, the overload device rating is $5.8 \times 1.25 = 7.25$ amperes. The motors' built-in overload protection meets this requirement.

Motor Branch-Circuit Short-Circuit and Ground Fault Protection
(430.52, Table 430.52)

Motor branch-circuit short-circuit and ground-fault protection is covered in Unit 19. Here is a brief review.

For a circuit breaker, *Table 430.52* indicates a maximum rating of 250 percent of the motor's full-load current rating. Therefore, $5.8 \times 2.5 = 14.5$ amperes. The next standard rating is permitted by *430.52(C)(1)*. This is a 15-ampere circuit breaker according to *240.6(A)*.

For a dual-element, time-delay fuse, *Table 430.52* indicates a maximum rating of 175 percent of the motor's full-load current rating. Therefore, $5.8 \times 1.75 = 11+$. The next standard rating is permitted by *430.52(C)(1)*. This is a 15-ampere, dual-element, time-delay fuse according to *240.6(A)*.

There are exceptions to the above rules that permit even higher rated fuses or circuit breakers under certain conditions as stated in *430.52(C)*.

Disconnecting Means
(Article 430, Part IX)

A switch/fuse box-cover unit is mounted on the 4-in. square outlet box. This meets the requirement that the disconnecting means shall be in sight of the motor, *430.102(B)*. Anyone working on the exhaust fan can easily turn off the power. The 15-ampere circuit breaker in the main panel also serves as a disconnecting means since it can be locked in the "OFF" position.

Grounding *(Article 250, Part VI)*

The exhaust fan must be properly grounded!

Proper grounding is accomplished by using any of the accepted equipment grounding methods found in *250.118*. For residential wiring, the equipment grounding conductor in nonmetallic-sheathed cable, the armor and bonding wire in armored cable, or flexible metal conduit provides the necessary equipment ground. DO NOT ground the exhaust fan metal housing to the white grounded circuit conductor, *250.142(B)*.

Supplemental Overcurrent Protection

Integral overload protection in the motor is the first line of defense against motor burnout and potential fire. A dual-element, time-delay fuse installed in the switch/fuse box-cover unit adjacent to the exhaust fan per the specifications provides backup overload protection as the second line of defense against motor burnout. Dual-element, time fuses are available in many ampere ratings such as 7½, 8, 9, 10, and 12 amperes.

Should the motor fail to start, short-circuit, or go to ground, the fuse in the box-cover switch/fuse unit will open without tripping the branch-circuit breaker back in Panel A. This makes it easy for servicing the unit.

Refer to Table 28-1 in this text for much more information relating to fuses and circuit breakers.

Fan Control

Typical residential fan controls are illustrated in Figure 22-5 and Figure 22-6. These may be mounted adjacent to or above the other wall-mounted lighting switches.

Figure 22-7 shows many additional options available for controlling exhaust fans.

A. A simple "ON-OFF" switch.

B. A speed control switch that allows multiple and/or an infinite number of speeds. (The exhaust fan in this residence is controlled by this type of control.) See also Figure 22-5.

C. A timer switch (S_T) that allows the user to select how long the exhaust fan is to run, up to 12 hours. See also Figure 22-6.

D. A humidity control (H) switch that senses moisture buildups. See "Humidity Control"

Figure 22-5   A typical two-speed fan control. Also available in three-speed, four-speed, and totally variable speed solid state types. *Courtesy of* Broan-NuTone.

Figure 22-6   A timer fan control. Some timer controls have a continuous "on" position. *Courtesy of* Broan-NuTone.

Figure 22-7   Different options for the control of exhaust fans.

later in this unit for further details on this type of switch.

E.   Exhaust fans mounted into end gables or roofs of residences are available with an adjustable temperature control on the frame of the fan. This thermostat generally has a start range of 70°F to 130°F (21°C to 54°C),

and will automatically stop at a temperature 10°F (6°C) below the start setting, thus providing totally automatic "ON-OFF" control of the exhaust fan.

F.   A high temperature automatic heat sensor that will shut off the fan motor when the temperature reaches 200°F (93°C). This is a safety feature so that the fan will not spread a fire. Connect in series with other control switches.

G. A combination of controls. This circuit shows an exhaust fan controlled by an infinite-speed switch, a humidity control (H), and a high temperature heat sensor. Note that the speed control and the humidity control are connected in parallel so that either can start the fan. The high temperature heat sensor is in series so that it will shut off the power to the fan when it senses 200°F (93°C), even if the speed control or humidity control is in the "ON" position.

H. This circuitry combines a humistat (D) and a thermostat (E) in such a way that the homeowner can select between controlling humidity or temperature in the attic. A selector switch is installed in a readily accessible place for ease of use by the homeowner. The switch has a "center off" position. The humidity and temperature sensors are mounted in the attic.

Any of the above variations for controlling an exhaust fan could also include a pilot light to give clear visual indication that the fan is running. This pilot light could be mounted in the same switch box as the speed control or timer control, or it could be mounted next to these switches.

HUMIDITY

Humidity can be a big problem in homes.

Well-insulated homes can experience problems with excess humidity due to the "tightness" of the house. Thermal insulation and vapor barriers that keep moisture out of the insulation must be properly installed. Insulation must be kept dry or its efficiency decreases. Whole house vinyl wraps (Tyvek®) stop airflow through wall cavities, as well as holding out wind-driven rain from getting behind wood, vinyl, brick, and stucco siding.

High humidity is uncomfortable, but a bigger problem is that high humidity promotes the growth of mold and the deterioration of fabrics and floor coverings. Framing members, wall panels, plaster, or drywall may also deteriorate because of the humidity.

Instead of manual "ON-OFF" control of an exhaust fan, a desired humidity level can be maintained automatically with a device called a humidistat. Humidistats have a humidity sensitive element such as human hair (moisture causes hair to contract, which is why hair frizzes when exposed to high humidity), nylon strips as in Figure 22-8 (nylon expands and contracts with a change in humidity), or solid-state components (that change resistance as the humidity changes) for the more sophisticated applications. When the relative humidity reaches a certain set level, the humidity-sensing component reacts, in turn activating the actual switching mechanism such as a small microswitch, which turns the motor on and off. A bimetallic element cannot be used because it reacts to temperature changes only.

Adjustable settings are from 5 percent to 95 percent relative humidity. A comfort level of about 50 percent relative humidity is considered acceptable.

Humidistats having proper voltage rating (120 or 240 volts) and ampere rating are used to automatically switch the motor directly. See Figure 22-9A. They are available in appearance similar to typical speed and timer controls with a knob and faceplate to match the other switch and receptacle faceplates in a home. The wiring is as simple as wiring in a switch leg for switching lighting outlets.

Where the 120-volt power wiring becomes very lengthy because of the distance between the con-

Figure 22-8 Details of a humidity control used with an exhaust fan.

Figure 22-9   Wiring for the humidistat control.

troller and the fan, or for large motors having a high current draw, a relay can be used. See Figure 22-9B. This relay is mounted near the whole house exhaust fan. The branch-circuit wiring is run from the Main Panel to a 4-in. square, 2⅛-in. deep outlet box conveniently located near the fan. A box-cover switch/fuse unit on this outlet box serves as the disconnecting means within sight of the motor as required by *422.30* and *430.10.2* and has a dual-element, time-delay fuse sized at approximately 125 percent of the motor's full-load current rating for motor overload backup protection. Typical home-type whole house exhaust fans do not require relays.

Dehumidifiers used in basements and other damp locations and humidifiers on furnaces use a humidistat to turn the power to the motor on and off.

APPLIANCE DISCONNECTING MEANS

To clean, adjust, maintain, or repair an appliance, it must be disconnected to prevent personal injury. *NEC® 422.30* through *422.35* outline the basic methods for disconnecting fixed, portable, and stationary electrical appliances. Note that each appliance in the dwelling conforms to one or more of the disconnecting methods listed. The more important *Code* rules for appliance disconnects are as follows:

- each appliance must have a disconnecting means.

- the disconnecting means may be a separate disconnect switch if the appliance is permanently connected.

- the disconnecting means may be the branch-circuit switch or circuit breaker if the appliance is permanently connected and not over 1/8 hp or 300 volt-amperes.

- the disconnecting means may be an attachment plug cap if the appliance is cord connected.

- the disconnecting means may be the unit switch on the appliance only if other means for disconnection are also provided. In a single-family residence, the service disconnect serves as the other means.

- the disconnect must have a positive "ON-OFF" position.

- the disconnect must be capable of being locked in the "OFF" position or be within sight of a motor-driven appliance where the motor is more than 1/8 hp. ►The lock-off provision must be permanent.◄

Read *422.30* through *422.35* for the details of appliance disconnect requirements.

HYDROMASSAGE TUB CIRCUIT ▲A

The master bedroom is equipped with a hydromassage tub. A hydromassage bathtub is sometimes referred to as a whirlpool bath.

NEC® 680.2 defines a hydromassage bathtub as "*a permanently installed bathtub equipped with a recirculating piping system, pump and associated equipment. It is designed so that it can accept, circulate, and discharge water upon each use.*"*

In other words, fill—use—drain.

The significant difference between a hydromassage bathtub and a regular bathtub is the recirculating piping system and the electric pump that circulates the water. Both types of tubs are drained completely after each use.

Spas and hot tubs are intended to be filled, then used. They are *not* drained after each use because they have a filtering and heating system.

Electrical Connections

The *NEC®* requirements for hydromassage tubs are found in Unit 30.

The hydromassage tub in this residence is sup-

---

*Reprinted with permission from NFPA 70-2005.

plied by a dedicated branch-circuit A9, a 15-ampere, 120-volt circuit. If you are not sure what the actual current draw is, run a 20-ampere branch-circuit.

The Schedule of Special-Purpose Outlets indicates that the hydromassage bathtub in this residence has a 1/2-hp motor that draws 10 amperes.

The hydromassage bathtub electrical control is prewired by the manufacturer. All the electrician needs to do is run the dedicated 15-ampere, 120-volt, GFCI-protected circuit to the end of the tub where the pump and control are located.

Some manufacturers supply a short flexible conduit that contains one black, one white, and one green equipment grounding conductor, as in Figure 22-10. The electrician installs a junction box near the pump. A dedicated branch-circuit is run to this box, where the branch-circuit wiring is connected to the hydromassage conductors. This junction box must be accessible. The pump and power panel may also need servicing. Access may be from underneath or the end, whichever is convenient for the installation, as in Figure 22-11.

Make sure that the equipment grounding conductor of the circuit is properly connected to the green equipment grounding conductor of the hydromassage tub. *Never connect the green equipment grounding*

Figure 22-10 Typical wiring diagram of a hydromassage tub showing the motor, power panel, and electrical supply leads.

Figure 22-11 The basic roughing-in of a hydromassage bathtub is similar to that of a regular bathtub. The electrician runs a separate 15-ampere, 120-volt GFCI-protected circuit to the area where the pump and control are located. Check the manufacturer's specifications for these data. An access panel from the end or from below is necessary to service the wiring, the pump, and the power panel.

conductor to the grounded (white) branch-circuit conductor.

Some hydromassage tubs come equipped with a cord and plug, in which case the electrician installs a receptacle within 24 in. (600 mm) of the pump. Usually the branch-circuit is protected by a GFCI circuit breaker, although a GFCI receptacle could be used.

Read and follow the manufacturer's installation instructions.

Figure 22-12 is a photograph of a typical hydromassage tub.

Figure 22-12 A typical hydromassage bathtub, sometimes referred to as a whirlpool. *Courtesy of* Trajet.

REVIEW

Note: Refer to the *Code* or the plans where necessary.

BATHROOM CEILING HEATER CIRCUITS Ⓐₖ Ⓐⱼ

1. What is the wattage rating of the heat/vent/light? _____

2. To what circuits are the heat/vent/lights connected? _____

3. Why is a 4-in. square, 2⅛-in. deep box with a two-gang raised plaster ring or similar deep box used for the switch assembly for the heat/vent/light? _____

_____

_____

_____

_____

4. a. How many wires are required to connect the wall switches and the heat/vent/light?

_____

   b. What size wires are used? _____

5. Can the heating element be energized when the fan is not operating? _____

6. Can the fan be turned on without the heating element? _____

7. What device can be used to provide automatic control of the heating element and the fan of the heat/vent/light? _____

8. Where does the air enter the heat/vent/light? _____

9. Where does the air leave this unit? _____

10. Who is to furnish the heat/vent/light? _____

_____

11. For a ceiling heater rated 1200 watts at 120 volts, what is the current draw? _____

_____

12. An electrician wired up a heat/vent/light as described in this unit. The wiring diagram showed that the electrician needed to run five conductors between the switches and the heat/vent/light. The electrician used one 12/3 nonmetallic-sheathed cable and one 12/2 nonmetallic-sheathed cable. Both cables had equipment grounding conductors. During the rough-in inspection, the electrical inspector cited "Noncompliance with *NEC®* *300.3(B)*." What was the problem? _____

_____

_____

ATTIC EXHAUST FAN CIRCUIT Ⓐ∟

1. What is the purpose of the attic exhaust fan? _____

_____

_____

2. At what voltage does the fan operate? _____

3. What is the horsepower rating of the fan motor? _____

4. Is the fan direct- or belt-driven? _____

5. How is the fan controlled? _____

6. What is the rating of the dual-element time-delay fuse in the switch/fuse box-cover unit? _____

7. What is the rating of the running overcurrent protection if the motor is rated at 10 amperes? _____

8. What is the basic difference between a thermostat and a humidistat? _____

_____

_____

9. What size conductors are to be used for this circuit? _____

10. How many ft (m) of cable are required to complete the wiring for the attic exhaust fan circuit? _____

11. May the metal frame of the fan be grounded to the grounded circuit conductor?

_____

12. Which section of the *Code* prohibits grounding equipment to a grounding circuit conductor? _____

HYDROMASSAGE BATHTUB CIRCUIT Ⓐ_A

1. Which circuit supplies the hydromassage bathtub? _____

2. Which of the following statements "meets Code"?

   a. Hydromassage tubs and all of their electrical components shall be GFCI protected.

   b. Hydromassage tubs and all of their electrical components shall not be GFCI protected.

   Circle the correct statement.

3. Which conductor size feeds the hydromassage tub in this residence? _____

4. What is the fundamental difference between a hydromassage bathtub and a spa?

_____

_____

5. Which sections of the *Code* reference hydromassage bathtubs? _____

6. Must the metal parts of the pump and power panel of the hydromassage tub be grounded? _____

_____

_____

_____

7. Is it permissible to connect the hydromassage tub's green equipment grounding conductor to the branch-circuit's grounded (white) conductor? _____

8. All 120-volt, single-phase receptacles within (5 ft [1.5 m]) (10 ft [3.0 m]) (15 ft [4.5 m]) of a hydromassage tub must be GFCI protected. Circle the correct answer.

Special-Purpose Outlets— Electric Heating ⏁M, Air Conditioning ⏁N

There are many types of electric heat available for heating homes (e.g., heating cable, unit heaters, boilers, electric furnaces, duct heaters, baseboard heaters, heat pumps, and radiant heating panels). This unit discusses many of these types. The residence discussed in this text is heated by an electric furnace located in the workshop.

Detailed *Code* requirements are found in *Article 424.*

This text cannot cover in detail the methods used to calculate heat loss and the wattage required to provide a comfortable level of heat in the building. For this residence, the total estimated wattage is 13,000 watts. Depending upon the location of the residence (in the Northeast, Midwest, or South, for example), the heating load will vary.

Electric heating has gained wide acceptance when compared with other types of heating systems. It has a number of advantages. Electric heating is flexible when baseboard heating is used because each room can have its own thermostat. Thus, one room can be kept cool while an adjoining room is warm. This type of zone control for an electric, gas- or oil-fired central heating system is more complex and more expensive.

Electric heating is safer than heating with fuels. The system does not require storage space, tanks, or chimneys. Electric heating is quiet. Electric heat does not add or remove anything from the air. As a result, electric heat is cleaner. This type of heating is considered to be healthier than fuel heating systems that remove oxygen from the air. The only moving part of an electric baseboard heating system is the thermostat. This means that there is a minimum of maintenance.

As with all heating systems, adequate insulation must be provided. Proper insulation can keep electric bills to a minimum. Insulation also helps to keep the residence cool during the hot summer months. The cost of extra insulation is offset through the years by the decreased burden on the air-conditioning equipment.

RESISTANCE HEATING CABLES

Resistance heating cables can be embedded in the plaster or between two layers of drywall. Installation requirements are found in *424.34* through *424.45* of the *NEC*. Because premise wiring (nonmetallic-sheathed cable, etc.) will be located above these ceilings and will be subjected to the heat created by the heating cables, the *Code* requires the following:

- keep the wiring not less than 2 in. (50 mm) above the heated ceiling, and *also* reduce the ampacity of the wiring to the 50°C correction factors found at the bottom of *Table 310.16*, or

- keep the wiring above the thermal insulation that is at least 2 in. (50 mm) thick—in which case the correction factors do not have to be applied.

For ease in identification, *424.35* requires that the leads on heating cables be color coded as follows:

- 120 volts   yellow
- 208 volts   blue
- 240 volts   red
- 277 volts   brown
- 480 volts   orange

ELECTRIC FURNACES

In an electric furnace, the source of heat are the resistance heating elements. The blower motor assembly, filter section, condensate coil, refrigerant line connections, fan motor speed control, high temperature limit controls, relays, and similar components are much the same as those found in gas furnaces.

As with all appliances, many of the *NEC* requirements have already been met by virtue of the product being listed by a recognized testing laboratory. Your job is to make sure the premise wiring is installed according to the *NEC*. The recognized testing laboratory takes care of the safety requirements of the appliance.

Summary of Important *NEC* Rules

Many of the following *NEC* requirements are emphasized in Figure 23-1.

- *NEC 220.51*: When making a calculation for sizing a service or feeder, include 100 percent of the rating of the total connected fixed space heating load. This is shown in Unit 29.

- *Article 250, Part VII*: Contains most of the requirements for equipment and appliances grounding. Grounding is covered in Units 4, 15, 20, 21, and 22.

- *NEC 250.118*: Lists the acceptable methods for equipment and appliance grounding. This is generally the equipment grounding conductor in nonmetallic-sheathed cable, the metal armor and bonding strip in armored cable, or a metal raceway such as EMT or FMC.

- *NEC 422.10(A)*: The branch-circuit rating for an appliance must not be less than the marked rating.

- *NEC 422.11*: The branch-circuit overcurrent protection must be sized according to the value indicated on the nameplate. The electrician need do nothing more than size the conductors and overcurrent protection according to the nameplate data and instructions furnished with the furnace. Overcurrent protection for internal components is provided by the manufacturer.

- *NEC 422.12*: Central heating equipment must be supplied by an individual branch-circuit. Associated equipment such as humidifiers and electrostatic air cleaners are permitted to be connected to the same branch-circuit. The logic behind this "individual circuit" requirement is that if the heating system were to be connected to some other branch-circuit, such as a lighting branch-circuit, a fault on that circuit would shut off the power to the heating system. We find this same requirement in *5.6.4* of *NFPA 54*, the *National Fuel Gas Code*.

- *NEC 422.30*: A disconnecting means must be provided.

- *NEC 422.31(B)*: The disconnecting means must be within sight of the furnace. For electric furnaces, the disconnect might be mounted on the side of the furnace, on an adjacent wall, or the

THERMOSTAT

FIELD WIRING OF LOW-VOLTAGE CLASS 2 CONTROL CIRCUIT CONDUCTORS SHALL NOT BE PLACED IN THE SAME RACEWAY, BOX, OR ENCLOSURE WITH POWER CONDUCTORS EXCEPT WHERE INTRODUCED SOLELY TO CONNECT THE EQUIPMENT, *725.55(A)* AND *EXCEPTION NO. 2.*

TYPICAL ELECTRIC FURNACE

DISCONNECTING MEANS:
● SHALL HAVE AMPERE RATING *NOT LESS* THAN 125% OF THE TOTAL LOAD OF THE MOTOR AND HEATER. SEE NAMEPLATE ON FURNACE.
● MUST DISCONNECT THE EQUIPMENT FROM ALL UNGROUNDED CONDUCTORS.
● MUST BE WITHIN SIGHT OF FURNACE OR MUST BE CAPABLE OF BEING LOCKED IN "OFF" POSITION.
● SHALL INDICATE "ON-OFF."
● FUSES SHALL BE SIZED AT *NOT LESS* THAN 125% OF THE TOTAL LOAD OF THE MOTOR AND HEATERS. SEE NAMEPLATE ON FURNACE.

MUST BE INDIVIDUAL CIRCUIT, *422.12.*

BRANCH-CIRCUIT CONDUCTORS:
● SHALL BE SIZED AT *NOT LESS* THAN 125% OF THE TOTAL LOAD OF THE MOTORS AND HEATERS. SEE NAMEPLATE ON FURNACE.

NAMEPLATE WILL SHOW MANUFACTURER'S NAME, VOLTS AND AMPERES, VOLTS AND WATTS, OR VOLTS AND KILOWATTS.

FURNACE MAY ALSO CONTAIN SUPPLEMENTAL OVERCURRENT PROTECTION SO THAT THE RESISTANCE-TYPE HEATING ELEMENTS ARE FUSED AT NOT MORE THAN 60 AMPERES. THE FURNACE'S RESISTANCE-TYPE HEATING ELEMENTS MUST BE SUBDIVIDED INTO LOADS NOT TO EXCEED 48 AMPERES, *422.11(F).* LISTED FURNACES CONFORM TO THIS REQUIREMENT.

EXAMPLE: WHAT SIZE COPPER CONDUCTORS (THHN), FUSES, AND DISCONNECT SWITCH ARE REQUIRED FOR A FURNACE MARKED 79 AMPERES, 240 VOLT, SINGLE PHASE, 60 CYCLES? TERMINALS ON FURNACE AND SWITCH MARKED 75°C. TO SELECT THE PROPER AMPACITY OF THE CONDUCTORS IN ACCORDANCE WITH *110.14(C),* BE SURE TO USE THE 75°C AMPACITY COLUMN IN *TABLE 310.16.*

ANSWER:
CONDUCTOR SIZE: 79 x 1.25 = 98.75 AMPERES
FROM *TABLE 310.16,* SELECT 3 AWG THHN (100 AMPERES AT 75°C.)

FUSE SIZE: 79 x 1.25 = 98.75 AMPERES
INSTALL 100-AMPERE FUSES.

SWITCH: 100-AMPERE SWITCH

Figure 23-1 ▶An electric furnace is considered to be a *continuous load*, subject to sizing the conductors, disconnecting means, and overcurrent devices at 125 percent of the load, *NEC® 424.3(B).*◀ Central electric heating equipment must be supplied by an individual branch-circuit, *NEC® 422.12.*

circuit breaker in the panel, if the panel is in sight of the furnace. For gas furnaces, the disconnect could simply be a snap switch, switch/fuse mounted in a handy box on the side of the furnace, or the circuit breaker in the panel, if the panel is in sight of the furnace. See *430.102(A)* and *(B).*

● *NEC® 422.32:* Another reminder that the disconnect for motor-driven appliances having a motor greater than ⅛ hp must be within sight of the appliance. See *430.102(A)* and *(B).*

● *NEC® 422.62(B)(1):* The nameplate shall specify the minimum size supply circuit conductor's ampacity and the maximum branch-circuit overcurrent protection.

● ▶*NEC® 424.3(B):* Fixed electric space heating is considered to be a continuous load. The branch-circuit conductors and the rating or setting of overcurrent protective devices supplying fixed electric space-heating equipment consisting of resistance elements with or without a motor shall not be less than 125 percent of the total load of the motors and the heaters. ◄ If applying the 125 percent rule results in a nonstandard rating of overcurrent protective device, then *240.4(B)* permits the use of the next higher rating. See Figure 23-1 for the 125 percent calculations.

- *NEC® 430.60(B)*: The nameplate must show volts and watts or volts and amperes.

- *NEC® 430.102(A)*: The disconnect shall be within sight of the controller and shall disconnect the controller. With a residential furnace, the controls are an integral part of the furnace.

- *NEC® 430.102(B)*: The disconnect shall be located within sight of the motor and driven machinery. There is an exception to the "in sight" requirement, but the exception does not pertain to residential installations.

Figure 23-1 illustrates the key *NEC®* requirements for hooking up an electric furnace.

Supply Voltage

Because proper supply voltage is critical to ensure full output of the furnace's heating elements, voltage must be maintained at not less than 98 percent of the furnace's rated voltage. The effect of voltage drop has already been discussed in Units 4 and 19. Mathematically, for every 1 percent drop in voltage, there will be a 2 percent drop in wattage output of a resistance heating element. Manufacturers' installation instructions usually have tables that show the wattage or kilowatt output at different voltages. Table 23-1 is a typical table.

A simple way to calculate the effect of voltage variance on a resistive heating element is:

$$Correction\ Factor = \frac{Applied\ Voltage\ Squared}{Rated\ Voltage\ Squared}$$

KILOWATT RATING	
@ 240 volts	**@ 208 volts**
12.0	9.0
15.0	11.3
20.0	15.0
25.0	18.8
30.0	22.5

Table 23-1

EXAMPLE: The rating of the heating elements in an electric furnace is 30 kW at 240 volts. What is the kW output if the supply voltage is 208 volts?

Step 1: $Correction\ Factor = \dfrac{208 \times 208}{240 \times 240}$

$$= \frac{43,264}{57,600} = 0.75$$

Step 2: 30 kW × 0.75 = 22.5 kW

Therefore, the actual voltage at the supply terminals of an electric furnace or electric baseboard heating unit will determine the true wattage output of the heating elements.

Cord- and Plug-Connection Not Permitted

Flexible cords are not a recognized wiring method.

NEC® 400.7(A)(8) and *422.16* describe the permitted uses for flexible cords. A key requirement, oftentimes violated, is that flexible cords shall be used only ► *"where the fastening means and mechanical connections are specifically designed to permit ready removal for maintenance and repair, and the appliance is intended or identified for flexible cord connection."* ◄ * Certainly the gas piping and the size of a gas or electric furnace do not allow for "ready removal" of the furnace. Furthermore, the conductors in flexible cords cannot withstand the high temperature encountered on the terminals of a furnace.

ANSI Standard Z21.47 does not allow cord- and plug-connection for gas furnaces.

CONTROL OF ELECTRIC BASEBOARD HEATING UNITS

Line-voltage and low-voltage thermostats can be used to control electric baseboard heating units, as in Figure 23-2. When the current rating or ampere rating of the load exceeds the rating of a line-voltage thermostat, then it is necessary to utilize a relay in conjunction with the low-voltage thermostat. This is illustrated in Figure 23-3. Color-coding requirements are explained later in this unit, and are covered in detail in Unit 5.

Most line-voltage thermostats are ampere rated for noninductive resistive loads. You may need to

*Reprinted with permission from NFPA 70-2005.

Figure 23-2 Thermostats for electric heating systems. A thermostat or other switching device that is marked with an "OFF" position must open all ungrounded conductors of the circuit when turned to the "OFF" position. Typical mounting height is 52 in. (1.3 m) to center. *Courtesy of* Honeywell, Inc.

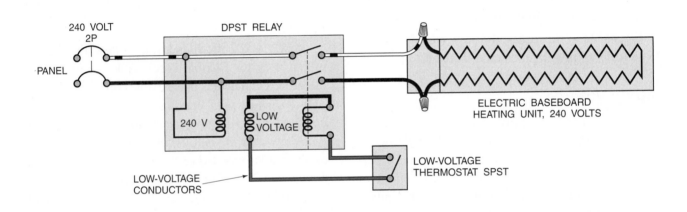

Figure 23-3 Wiring for an electric baseboard heating unit having a low-voltage thermostat and relay. *NEC® 200.7* requires that a white wire used as an ungrounded conductor must be permanently reidentified at points where connections are made or where it is visible and accessible. This becomes necessary when nonmetallic-sheathed cable or armored cable is the wiring method. Reidentification can be accomplished with black paint, black tape, or heat-shrink tubing. Color coding and reidentification of conductors is discussed in Unit 5 and later in this unit.

convert watts to amperes in order to select the right size thermostat. For example, if you are controlling a 2500 watt, 120-volt resistive heating element, or a 5000 watt, 240-volt resistive heating element:

$$I = \frac{W}{E} = \frac{2500}{120} = 20.8 \text{ amperes}$$

—or—

$$I = \frac{W}{E} = \frac{5000}{240} = 20.8 \text{ amperes}$$

The total connected load for this thermostat must not exceed 20.8 amperes.

MARKING THE CONDUCTORS OF CABLES

Two-wire cable contains one white and one black conductor. Because 240-volt circuits are permitted to supply electric baseboard heaters, it would appear that the use of two-wire cable is in violation of the *Code*.

According to *200.7(C)*, two-wire cable is permitted for the 240-volt heaters provided the white conductor is permanently reidentified by paint, colored tape, or other effective means. This is necessary because it might be assumed that an unmarked white conductor is a grounded conductor having no voltage to ground. Actually, the white wire is connected to a "hot" phase in the panel and has 120 volts to ground on it. A person could be subjected to a lethal shock by touching this conductor and the grounded baseboard heater (or any other grounded object) at the same time. The white conductor must be permanently reidentified at the electric heater terminals and at the panels where these cables originate. Figure 23-3, Figure 23-5, and Figure 23-6 show this re-identification. Review the section on "Color Coding" in Unit 5.

CIRCUIT REQUIREMENTS FOR ELECTRIC BASEBOARD HEATING UNITS

Figure 23-3 shows an electric baseboard heater controlled by a low-voltage thermostat. The low-voltage contacts of the relay are connected to the thermostat. The line-voltage contacts of the relay are used to switch the actual heater load. Relays are used whenever the load exceeds the ampere or wattage limitations of a line-voltage thermostat. Color coding for conductors is explained in Unit 5 and is also discussed later in this unit.

Figure 23-4 shows typical electric baseboard heating units. The heater to the right has an integral

thermostat on it. Electric baseboard heating units are available in both 240-volt and 120-volt ratings and in a variety of wattage ratings.

The wiring for an individual baseboard heating unit or group of units is shown in Figure 23-5 and Figure 23-6. A two-wire cable (armored cable or nonmetallic-sheathed cable with ground) would be run from a 240-volt, two-pole circuit in the main panel to the outlet box or switch box installed at the thermostat location. A second two-wire cable runs from the thermostat to the junction box on the heater unit. The proper connections are made in this junction box. Most heating unit manufacturers provide knockouts at the rear and on the bottom of the junction box. The supply conductors can be run through these knockouts. Most baseboard units also have a channel or wiring space running the full length of the unit, usually at the bottom. When two or more heating units are joined together, the conductors are run in this wiring channel. Most manufacturers indicate the type of wire required for these units because of conductor temperature limitations.

Supply conductors generally must be rated 194°F (90°C). This would be Type THHN conductors, or nonmetallic-sheathed cable or armored cable

Figure 23-4   Electric baseboard heating systems.

Figure 23-5   Wiring for a single electric baseboard heating unit. The thermostat is a line-voltage thermostat. See Figure 23-3 caption for color-coding requirements.

Figure 23-6 Wiring for two electric baseboard heating units. Note the circuit connections at the thermostat location. The thermostat is a line-voltage thermostat. See Figure 23-3 caption for color-coding requirements.

that has 194°F (90°C) conductors. The nameplate and/or instructions furnished with the electric baseboard heating unit provides this information.

Most wall and baseboard heating units are available with built-in thermostats, as in Figure 23-4. The supply cable for such a unit runs from the main panel to the junction box on the unit.

Some heaters have receptacle outlets. UL states that when receptacle sections are included with the other components of baseboard heating systems, they must be supplied separately using conventional wiring methods. See Figure 23-7.

A variety of fittings such as internal and external elbows (for turning corners) and blank sections are available from the manufacturer.

Manufacturers of electric baseboard heaters list "blank" sections ranging in length from 2 to 10 ft

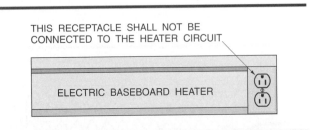

Figure 23-7 Factory-installed receptacle outlets or receptacle outlet assemblies provided by the manufacturer for use with its electric baseboard heaters may be counted as the required receptacle outlet for the space occupied by a permanently installed heater. See the second paragraph of *210.52*.

Figure 23-8 Use of blank baseboard heating sections.

(600 mm to 3.0 m). These blank sections give the installer the flexibility needed to spread out the heater sections, and to install these blanks where wall receptacle outlets are encountered, as in Figure 23-8.

LOCATION OF ELECTRIC BASEBOARD HEATERS IN RELATION TO RECEPTACLE OUTLETS

Listed electric baseboard heaters shall not be installed below wall receptacle outlets unless the instructions furnished with the baseboard heaters indicate that they may be installed below receptacle outlets. See the *Fine Print Note* to *210.52*. The reasoning for generally not allowing electric baseboard heaters to be installed below receptacle outlets is the possible fire and shock hazard resulting when a cord hangs over and touches the heated electric baseboard unit. The insulation on the cord can melt from the heat. See Figure 23-9.

A POSITION ELECTRIC BASEBOARD HEATING UNITS SO THEY WILL *NOT* BE DIRECTLY BELOW A WALL RECEPTACLE OUTLET.

B IF INSTALLED AS SHOWN, ELECTRICAL CORDS COULD COME IN CONTACT WITH THE BASEBOARD UNIT, SUBJECTING THIS CORD TO RUBBING (ABRASION) AND HEAT, WHICH MIGHT RESULT IN FAILURE OF THE INSULATION OF THE CORD, A POTENTIAL FIRE AND SHOCK HAZARD.

BASEBOARD HEATER

BASEBOARD HEATER

RECEPTACLE OUTLET

C AN EXAMPLE OF HOW THE RECEPTACLE OUTLET IN THE WINDOW CORNER OF BEDROOM #1 MIGHT BE INSTALLED.

Figure 23-9   Position of electric baseboard heaters.

Electricians must pay close attention to the location of receptacle outlets and the electric baseboard heaters. They must study the plans and specifications carefully.

They may have to fine-tune the location of the receptacle outlets and the electric baseboard heaters, and/or install blank spacer sections, keeping in mind the *Code* requirements for spacing receptacle outlets and just how much (wattage and length) baseboard heating is required (Figure 23-7 and Figure 23-10).

The *Code* does permit factory-installed or factory-furnished receptacle outlet assemblies as part of permanently installed electric baseboard heaters. These receptacle outlets shall not be connected to the heater circuit (see second paragraph of *210.52*, Figure 23-7 and Figure 23-9).

ELECTRIC BASEBOARD HEATER

Figure 23-10 *Code violation.* Unless the instructions furnished with the electric baseboard heater specifically state that the unit is listed for installation below a receptacle outlet, then this installation is a violation of *110.3(B)*. In this type of installation, cords attached to the receptacle outlet would hang over the heater, creating a fire hazard and possible shock hazard should the insulation on the cord melt. See *Fine Print Note* to *210.52*.

HEAT PUMPS

A heat pump is simply an air conditioner operating in reverse. A "reversing valve" inside the unit changes the direction of the flow of the system's refrigerant.

A "split" heat pump system consists of an outdoor unit with a hermetic motor compressor, a fan, and a coil. The indoor unit consists of a coil inside of the furnace. The movement of air inside the home is accomplished by the blower fan on the furnace. The "split system" is the most common type for residential use.

When all of the components are housed in one outdoor unit, the system is referred to as a "packaged system."

The hermetic motor compressor is the heart of the system. The refrigerant circulates between the indoor and outdoor units, through the coils, tubing, and compressor, absorbing and releasing heat as it travels through the system.

In the winter, the heat pump (the outdoor coil) extracts heat from the outdoor air and distributes it through the warm air ducts in the home. In extremely cold climates, a heat pump is supplemented by electric heating elements in the air handler.

In the summer, the process reverses. The heat pump (indoor coil) absorbs heat and condenses humidity from the indoor air. The heat is transferred to the outside by the refrigerant, and the humidity (condensation) is removed by piping it to a suitable drain.

Heat pump water heaters are discussed in Unit 19.

Because a heat pump contains a hermetic motor compressor, the *Code* requirements found in *Article 440* apply. These are discussed later in this unit.

GROUNDING

Grounding of electrical equipment has already been discussed and will not be repeated here. Grounding is covered in *Article 250* of the *NEC*.® To repeat a caution, *never* ground electrical equipment to the grounded (white) circuit conductor of the circuit.

CIRCUIT REQUIREMENTS FOR ROOM AIR CONDITIONERS

For homes that do not have central air conditioning, window or through-the-wall air conditioners may be installed. These types of room air conditioners are available in both 120-volt and 240-volt ratings. Because room air conditioners are plug-and-cord connected, the receptacle outlet and the circuit capacity must be selected and installed according to applicable *Code* regulations. The *Code* rules for air conditioning are found in *Article 440*. The *Code* requirements for room air-conditioning units are found in *440.60* through *440.65* of the *NEC*.®

The basic *Code* rules for installing these cord-and plug-connected units and their receptacle outlets are as follows:

- the air conditioners must be grounded.

- the air-conditioner rating may not exceed 40 amperes at 250 volts, single phase.

- the rating of the branch-circuit overcurrent device must not exceed the branch-circuit conductor rating or the receptacle rating, whichever is less.

- the air-conditioner load shall not exceed 80 percent of the branch-circuit ampacity if no other loads are served.

- the air-conditioner load shall not exceed 50 percent of the branch-circuit ampacity if other loads are served.

- the attachment plug cap may serve as the disconnecting means.

- The maximum cord length is 10 ft (3.0 m) for 120-volt units, and 6 ft (1.8 m) for 208- or 240-volt units. This is also a UL requirement.

• There is an ever-present potential arcing-fault hazard related to the power supply cords on room air conditioners that could cause a fire or create an electrical shock hazard. In addition to requiring that the air-conditioner unit be properly grounded, *NEC® 440.65* requires that factory-installed LDCI or AFCI devices be provided on single-phase cord-and-plug connected room air conditioners. These devices are permitted to be an integral part of the attachment plug cap, or may be located in the power supply cord within 12 in. (300 mm) of the attachment plug cap.

LCDIs are *Leakage Current Detection and Interruption Protection* devices. They can sense leakage current flowing between or from the cord conductors and interrupt the circuit at a predetermined level of leakage current.

AFCIs are discussed in Unit 6.

Receptacles for Room Air Conditioners

Figure 23-11A is a straight 240-volt receptacle. Figure 23-11B is a combination receptacle. The lower portion is for 120-volt use, and the upper portion is for 240-volt use. Note the different slot configuration that meets the non-interchangeable requirements of *406.3(F)*.

Figure 23-11   Types of receptacles.

CENTRAL HEATING AND AIR CONDITIONING

The residence discussed throughout this text has central electric heating and air conditioning consisting of an electric furnace and a central air-conditioning unit.

The wiring for central heating and cooling systems is shown in Figure 23-12. Basic circuit requirements are shown in Figure 23-13. Note that one branch-circuit runs to the electric furnace and another branch-circuit runs to the air conditioner or heat pump outside the dwelling. Low-voltage wiring is used between the inside and outside units to

Figure 23-12   Connection diagram showing typical electric furnace and air-conditioner/heat pump installation.

DISCONNECTING MEANS:
- SELECT SIZE BASED ON NAMEPLATE RATED-LOAD CURRENT OR BRANCH-CIRCUIT SELECTION CURRENT – WHICHEVER IS GREATER – AND LOCKED ROTOR CURRENT.
- AMPERE RATING OF SWITCH MUST BE AT LEAST 115% OF NAMEPLATE RATED-LOAD CURRENT OR BRANCH-CIRCUIT SELECTION CURRENT – WHICHEVER IS GREATER.
- MUST ALSO BE HORSEPOWER RATED. CHECK *TABLE 430.148*, *TABLE 430.149*, *TABLE 430.150*, AND *TABLE 430.151* TO COMPARE THE RATED-LOAD CURRENT, BRANCH-CIRCUIT SELECTION CURRENT, AND HORSEPOWER RATING.
- MUST BE WITHIN SIGHT OF THE EQUIPMENT AND MUST BE READILY ACCESSIBLE.

BRANCH-CIRCUIT OVERCURRENT PROTECTION:
- MUST BE ABLE TO CARRY STARTING CURRENT.
- FUSES OR CIRCUIT BREAKER MUST BE OF SIZE AND TYPE AS MARKED ON NAMEPLATE.

BRANCH-CIRCUIT CONDUCTORS:
- THE MINIMUM CIRCUIT AMPACITY REQUIRED FOR THE AIR-CONDITIONING UNIT IS FOUND ON THE LABEL. THIS HAS BEEN DETERMINED BY THE MANUFACTURER TAKING INTO CONSIDERATION THE MOTOR COMPRESSOR CURRENT, FAN MOTOR CURRENT, AND HEATER CURRENT. THIS IS GENERALLY 125% OF THE LARGEST MOTOR PLUS THE FULL-LOAD RATING OF THE REST OF THE EQUIPMENT LOADS, SUCH AS FANS AND HEATERS.

TYPICAL AIR-CONDITIONING UNIT

OVERLOAD PROTECTION:
- THIS IS USUALLY AN INTEGRAL PART OF THE EQUIPMENT, SUPPLIED BY THE MANUFACTURER.

LABEL:
- MANUFACTURER'S NAME
- VOLTAGE
- FREQUENCY
- PHASES
- MINIMUM CIRCUIT AMPACITY
- MAXIMUM RATING OF BRANCH-CIRCUITS, SHORT-CIRCUIT, AND GROUND-FAULT PROTECTIVE DEVICE
- WILL STATE "MAXIMUM SIZE FUSE" OR "MAXIMUM SIZE FUSE OR CIRCUIT BREAKER"

FAN MOTOR

HERMETIC REFRIGERANT MOTOR COMPRESSOR

Figure 23-13 Basic circuit requirements for a typical residential-type air conditioner or heat pump. Reading the label is important in that the manufacturer has determined the minimum circuit ampacity and the maximum size and type of fuse or circuit breakers.

provide control of the systems. The low-voltage Class 2 circuit wiring must not be run in the same raceway as the power conductors, *725.55(A)*.

Refer to Unit 24 for more information about low-voltage circuitry.

"Time-of-Use" Lower Energy Rates

Many utilities offer lower energy rates for air-conditioner and/or heat pump loads through special programs, such as "time-of-use" usage and sliding scale energy rates. This is covered in Unit 19.

UNDERSTANDING THE DATA FOUND ON AN HVAC NAMEPLATE

The letters HVAC stand for heating, ventilating, and air conditioning.

Article 440 applies when HVAC equipment employs a hermetic refrigerant motor-compressor(s). *Article 440* is supplemental to the other articles of the *NEC*, and is needed because of the unique characteristics of hermetic refrigerant motor-compressors. The most common examples are air conditioners, heat pumps, and refrigeration equipment.

Hermetic refrigerant motor-compressors:

- combine the motor and compressor into one unit

- have no external shaft

- operate in the refrigerant

- do not have horsepower and full-load current (often referred to as full-load amperes or FLA) ratings like standard electric motors. This is because as the compressor builds up pressure, the current increases, causing the windings to get hot. At the same time, the refrigerant gets colder, passes over the windings, and cools them. Because of this, a hermetically sealed motor can be "worked" much harder than a conventional electric motor of the same size.

Special terminology is needed to understand how to properly make the electrical installation. These special terms are found in *440.2* and in UL Standard 1995. An air-conditioning nameplate from Shiver Manufacturing Company help explain these special terms.

- **Rated-Load Current (RLC):** Rated-load current is established by the hermetic motor-compressor manufacturer under actual operation at rated refrigerant pressure and temperature, rated voltage, and rated frequency. RLC is marked on the nameplate. In most instances, the marked RLC is at least equal to 64.1 percent of the hermetic refrigerant motor-compressor's maximum continuous current (MCC). In our example, the RLC is 17.8 amperes. See *440.2*.

- **Maximum Continuous Current (MCC):** This is the maximum current that the hermetic refrigerant motor-compressor can draw before damage occurs. Manufacturers know the damage capabilities of their equipment, and generally provide the necessary internal overcurrent protection that keeps the hermetic motor-compressor from burning out. This value is 156 percent of the marked RLC or branch-circuit selection current (BCSC). The overload protective system for the hermetic refrigerant motor-compressor must operate for currents in excess of the 156 percent value. Do not look for the MCC value on the nameplate of the end-use equipment. The MCC value might be marked on the motor-compressor. It is nice but not necessary for the electrician to know the MCC value of the motor-compressor. See *440.52*.

- **Branch-Circuit Selection Current (BCSC):** The manufacturer of the HVAC end-use equipment might design better cooling and better heat dissipation into the equipment, in which case the hermetic refrigerant motor-compressor might be capable of being continuously "worked" harder than other equipment not so designed. "Working" the hermetic refrigerant motor-compressor harder will result in a higher current draw. This is safe insofar as the motor-compressor is concerned, but the conductors, disconnect switch, and branch-circuit overcurrent devices must also be capable of safely carrying this higher current draw. This higher value of current is marked on the nameplate BCSC.

  Because the BCSC ampere value is always equal to or greater than the RLC ampere value, this BCSC value must be used instead of the RLC value when selecting conductors, disconnects, overcurrent devices, and other associated electrical equipment. Table 23-2 shows that the BCSC is 19.9 amperes. See *440.2* and *440.4(C)*.

- **Minimum Circuit Ampacity (MCA):** MCA is the minimum circuit ampacity requirement to determine the conductor size and switch rating. MCA data are always marked on the nameplate of the end-use equipment, and are determined by the manufacturer of the end-use equipment as follows:

  MCA = (RLC or BCSC × 1.25) + other loads

  Other loads would include condensing fans, electric heaters, coils, and so forth, that operate concurrently (at the same time).

In our example, the marked MCA is 26.4 amperes, calculated by the manufacturer of the end-use equipment as follows:

$$(19.9 \times 1.25) + 1.5 = 26.375 \text{ amperes}$$
(round off to 26.4)

Referring to *Table 310.16*, using the 60°C column, 10 AWG copper conductors would be adequate for this air-conditioning unit.

SHIVER MANUFACTURING COMPANY						
MODEL NO. XYZ – ELECTRICAL RATINGS –						
	VAC	HZ	PH	RLC	LRC	FLA
Compressor	230	60	1	17.8	107	—
Outdoor fan motor ¼ hp	230	60	1	—	—	1.5
Branch-circuit selection current	19.9 amperes					
Minimum circuit ampacity	26.4 amperes					
Maximum fuse or HACR type brkr.	45.0 amperes					
Operating voltage range:	197 min. 253 max.					

Table 23-2

See *Article 440, Part IV*.

A mistake that electricians often make is to multiply the MCA by 1.25 again. Certainly there is no hazard in doing this, but it might result in an unnecessary higher installation cost. For long runs, larger conductors would keep voltage drop to a minimum.

Outdoor installations of air-conditioning equipment might be in very hot parts of the country. You might want to review Unit 18 for the *NEC*® correction factor requirements where high ambient temperatures are encountered.

- **Maximum Overcurrent Protection (MOP):** The MOP value is the maximum size overcurrent protection, fuse, or circuit breaker. The MOP value is marked on the nameplate and is determined by the manufacturer of the end-use equipment as follows:

$$(\text{RLC or BCSC}) \times 2.25 + \text{other loads}$$

Other loads include condensing fans, electric heaters, coils, and so on, that operate at the same time. In our example, the marked MOP is

Figure 23-14 In a typical air-conditioner unit, the branch-circuit overcurrent device must protect the large components (large wires, large relay, hermetic motor compressor) as well as the small components (small wires, fan motor, crankcase heater) under short-circuit and/or ground-fault situations. It is extremely important to install the proper *size* and *type* of overcurrent device. Read the nameplate on the equipment and the instructions furnished with the equipment!

45 amperes, determined by the manufacturer as follows:

$$(19.9 \times 2.25) + 1.5 = 46.275 \text{ amperes}$$

Rounding down to the next lower standard size, the MOP is 45 amperes. See *Article 440, Part III*.

- **Type of Overcurrent Protection:** Check the nameplate carefully and do what it says! HACR (Heating, Air Conditioning, and Refrigeration) equipment nameplate might indicate "Maximum Size Fuse," "Maximum Size Fuse or Circuit Breaker," or "Maximum Size Fuse or HACR Circuit Breaker."

In the past, circuit breakers were subjected to specific tests unique to HVAC equipment and were marked with the letters HACR. Today, no additional tests are made. All currently listed circuit breakers are now suitable for HVAC application. These circuit breakers and HVAC equipment may or may not show the letters HACR. Because of existing inventories, it will take many years for the marking HACR to disappear from the scene. In the meantime, read and follow the information found on the nameplate of the equipment.

See Figure 23-14, Figure 23-15, Figure 23-16, and Figure 23-17.

Figure 23-16   This installation does not "Meet Code" because the overcurrent device inside the building is not of the type specified on the air-conditioner nameplate. Refer to *NEC® 110.3(B)*.

Figure 23-17   ▶ In the past, air-conditioning equipment and circuit breakers were marked with the letters HACR. As time passes and current inventory is used up, this marking will disappear from the nameplates of equipment and on the circuit-breaker label. In the meantime, follow the manufacturer's installation instructions and the data found on the nameplate. Refer to *NEC® 110.3(B)*. ◀

Figure 23-15   This installation "Meets Code." The disconnect is within sight of the unit. The disconnect contains fuses as specified on the air-conditioner nameplate. Refer to *NEC® 110.3(B)*.

- **Disconnecting Means to Be in Sight:** *NEC®* *440.14* of the *Code* requires that the disconnecting means must be within sight of the unit, and it must be *readily accessible*. See Figure 23-15, Figure 23-16, and Figure 23-17. Mounting the disconnect on the side of the house instead of on the air conditioner itself allows for easy replacement of the air-conditioning unit, should that become necessary. Mounting the disconnect behind the air conditioner or heat pump would not be considered "readily accessible" by most electrical inspectors. Generally, there is just not enough space to work safely on the disconnect. Squeezing behind or leaning across the top of the air conditioner to gain access to the disconnect is certainly not a safe practice. Mounting the disconnect on the outside wall of the house, to one side of the AC unit, is the accepted practice. See *Article 440, Part II*.

- **Disconnecting Means Rating:** Figure 23-18 shows a fusible and a non-fusible "pull-out" disconnect commonly used for residential air-conditioning and/or heat pump installations. The horsepower rating of the disconnecting means must be at least equal to the sum of all of the individual loads within the end-use equip-

ment, at rated load conditions, and at locked rotor conditions, *440.12(B)(1)*.

The ampere rating of the disconnecting means must be at least 115 percent of the sum of all of the individual loads within the end-use equipment, at rated load conditions, *440.12(B)(2)*. Installing a disconnect switch that equals or exceeds the MOP value will in almost all cases be the correct choice. The MOP in our example is 40 amperes. Thus, a 60-ampere disconnect switch is the correct size. In cases where the MOP value is close to the ampere rating of the disconnect switch (i.e., 30, 60, 100, 200, etc.), *locked-rotor current* values must be considered.

- **Locked-Rotor Current (LRC):** This is the maximum current draw when the motor is in a "LOCKED" position. When the rotor is locked, it is not turning. The disconnect switch and controller (if used) must be capable of safely interrupting locked-rotor current. See *440.12* and *440.41*. In our example, the nameplate indicates that the hermetic refrigerant motor-compressor has an LRC of 107 amperes. The fan motor has an FLA of 1.5 amperes, and would have an LRC approximately six times higher. According

A   **B**

Figure 23-18 (A) is a pull-out fusible disconnect. Fuses are inserted into fuse clips on the pull-out device. Insert the pull-out device for "ON," and remove the pull-out device for "OFF." (B) is a non-fusible disconnect, available in 30- and 60-ampere ratings at 240 volts. Both have padlock provisions to prevent unauthorized tampering. *Courtesy of* **Midwest Electric Products.**

to *440.12(B)(1)(b)*, all locked-rotor currents and other loads are added together when combined loads are involved. Therefore, the total locked-rotor current that the disconnect switch in our example would be called upon to interrupt is:

$(1.5 \times 6) + 107 = 116$ locked-rotor amperes

Checking *Table 430.151(A)* for the conversion of locked-rotor current to horsepower, we find that the locked-rotor current for a 230-volt, single-phase, 3-hp motor is 102 amperes, and the locked-rotor current for a 230-volt, single-phase, 5-hp motor is 168 amperes. Selecting a disconnect switch on the basis of the locked-rotor current in our example, the air-conditioning unit is considered to be a 5-hp unit, because 116 falls between 102 and 168. Manufacturers' technical literature indicates that a 30-ampere, 240-volt heavy-duty, single-phase disconnect switch has a 3-hp rating, and a 60-ampere, 240-volt heavy-duty, single-phase disconnect switch has a 10-hp rating. Thus, a 60-ampere disconnect is selected. Note that this is the same size disconnect selected previously on the basis of MOP.

ENERGY RATINGS

Energy ratings indicate the efficiency of heating and cooling equipment. Basically, these ratings are a comparison of output (heating or cooling) to input (electricity, gas, or oil).

Energy Efficiency Rating (EER): Cooling efficiency rating for room air conditioners. The ratio of the rated cooling capacity in Btu per hour divided by the amount of electrical power used in kilowatt hours. The higher the EER number, the greater the efficiency.

Seasonal Energy Efficiency Rating (SEER): Cooling efficiency rating for central air conditioners and heat pumps. SEER is determined by the total cooling of an air conditioner or heat pump in Btu during its normal usage period for cooling divided by the total electrical energy input in kilowatt-hours during the same period. The higher the SEER number, the greater the efficiency.

Annual Fuel Utilization Efficiency (AFUE): Tells you how efficiently a furnace converts

fuel (gas or oil) to heat. For example, an AFUE of 85 percent means that 85 percent of the fuel is used to heat your home, and the other 15 percent goes up the chimney. The higher the efficiency, the lower the operating cost. Old furnaces might have an AFUE rating as low as 60 percent. Mid-efficiency ratings are approximately 80 percent. High-efficiency ratings are 90 percent or higher. Maximum furnace efficiency available is approximately 96.6 percent. The higher the AFUE number, the greater the efficiency.

Heating Seasonal Performance Factor (HSPF): Heating efficiency of a heat pump. HSPF is determined by the total heating of a heat pump in Btu during its normal usage period for heating divided by the total electrical energy input in kilowatt-hours during the same period. The higher the HSPF rating, the greater the efficiency.

NONCOINCIDENT LOADS

Loads such as heating and air-conditioning are not likely to operate at the same time. The *NEC®* recognizes this diversity in *220.60*. Therefore, when calculating a feeder that supplies both types of loads, or when sizing a service-entrance, only the larger of the two loads need be considered. This is discussed in Unit 29 where service-entrance calculations are presented. Of course, the branch-circuit supplying the heating load is sized for that particular load, and the branch-circuit supplying the air-conditioning load is sized for that particular load.

RECEPTACLE NEEDED FOR SERVICING HVAC EQUIPMENT

For servicing HVAC equipment, *NEC® 210.63* requires that a 125-volt, single-phase, 15- or 20-ampere-rated receptacle be installed:

at an accessible location

on the same level as the HVAC equipment

within 25 ft (7.5 m) of the equipment, and

the receptacle shall not be connected to the load side of the equipment disconnecting means

The outdoor receptacles required by *210.52(E)* (one in front and one in back) *might* or *might not* meet the intent of *210.63*. If the HVAC equipment is located on a roof, in an attic, or a similar location, a receptacle must be installed in that location so as to be on the same level as the equipment.

▶A receptacle is not required for one- and two-family homes if the equipment is an evaporative cooler. These require minimal servicing.◀

GAS EXPLOSION HAZARD

Although it might seem a bit out of place in an electrical book to talk about gas explosions, we must talk about it.

Often overlooked is a requirement in *2.7.2(c)* of *NFPA 54*, *National Fuel Gas Code*, that gas meters be located at least 3 ft (900 mm) from sources of ignition. An electric meter or a disconnect switch are possible sources of ignition. Some utility regulations require a minimum of 3 ft (900 mm) clearance between electric metering equipment and gas meters and gas-regulating equipment. It is better to be safe than sorry! Check this issue out with the local electrical inspector and/or the local electric utility before installing the air conditioner or heat pump disconnect on the outside of the house. Also consider the location of the service watt-hour meter.

REVIEW

Note: Refer to the *Code* or the plans where necessary.

ELECTRIC HEAT

1. a. What is the allowance in watts made for electric heat in this residence? _____

   b. What is the value in amperes of this load? _____
   _____

2. What are some of the advantages of electric heating? _____
   _____
   _____
   _____

3. List the different types of electric heating system installations. _____
   _____
   _____
   _____

4. There are two basic voltage classifications for thermostats. What are they? _____
   _____

5. What device is required when the total connected load exceeds the maximum rating of a thermostat? _____

6. The electric heat in this residence is provided by what type of equipment? _____
   _____

7. At what voltage does the electric furnace operate? _____

8. A certain type of control connects electric heating units to a 120-volt supply or a 240-volt supply, depending upon the amount of the temperature drop in a room. These controls are supplied from a 120/240-volt, three-wire, single-phase source. Assuming that this type of device controls a 240-volt, 2000-watt heating unit, what is the wattage produced when the control supplies 120 volts to the heating unit? Show all calculations.

9. What advantages does a 240-volt heating unit have over a 120-volt heating unit?

   _____

   _____

   _____

10. The white wire of a cable may be used to connect to a "hot" ungrounded circuit conductor only if _____

    _____

11. Receptacle outlets furnished as part of a permanently installed electric baseboard heater, when not connected to the heater's branch-circuit, (may) (may not) be counted as the required receptacle outlet for the space occupied by the baseboard heater. Electric baseboard heaters (shall) (shall not) be installed beneath wall receptacle outlets unless the instructions furnished with the heaters indicate that it is acceptable to install the heater below a receptacle outlet. Circle the correct answers.

12. The branch-circuit supplying a fixed electric space heater must be sized to at least _____ percent of the heater's rating according to *NEC®* _____ .

13. Compute the current draw of the following electric furnaces. The furnaces are all rated 240 volts.
    a. 7.5 kW
    b. 15 kW
    c. 20 kW

14. For "ballpark" calculations, the wattage output of a 240-volt electric furnace connected to a 208-volt supply will be approximately 75 percent of the wattage output had the furnace been connected to a 240-volt supply. In question 13, calculate the wattage output of (a), (b), and (c) if connected to 208 volts. _____

    _____

    _____

    _____

15. A central electric furnace heating system is installed in a home. The circuit supplying this furnace:
    a. may be connected to a circuit that supplies other loads.
    b. shall be connected to a circuit that supplies other loads.
    c. shall be connected to a separate circuit.

16. What section of the *Code* provides the correct answer to question 15? _____

17. Electric heating cable embedded in plaster, or "sandwiched" between layers of dry wall, creates heat. Therefore, any nonmetallic-sheathed cable run above these ceilings and buried in insulation must have its current-carrying capacity derated according to the (40°C) (50°C) (60°C) correction factors found at the bottom of *Table 310.16* in the *NEC.* Circle the correct answer.

AIR CONDITIONING

1. a. When calculating air-conditioner load requirements and electric heating load requirements, is it necessary to add the two loads together to determine the combined load on the system? _____

   b. Explain the answer to part (a). _____

   _____

   _____

2. The total load of an air conditioner shall not exceed what percentage of a separate branch-circuit? (circle one)

   a. 75 percent   b. 80 percent   c. 125 percent

3. The total load of an air conditioner shall not exceed what percentage of a branch-circuit that also supplies lighting? (circle one)

   a. 50 percent   b. 75 percent   c. 80 percent

4. a. Must an air conditioner installed in a window opening be grounded if a person on the ground outside the building can touch the air conditioner? _____

   b. What *Code* section governs the answer to part (a)? _____

5. A 120-volt air conditioner draws 13 amperes. What size is the circuit to which the air conditioner will be connected? _____

6. What is the Code requirement for receptacles connected to circuits of different voltages and installed in one building? _____

7. When a central air-conditioning unit is installed and the label states "Maximum size fuse 50-amperes," is it permissible to connect the unit to a 50-ampere circuit breaker?

   _____

8. When the nameplate on an air-conditioning unit states "Maximum Size Fuse or HACR Circuit Breaker," what type of circuit breaker must be used? _____

   _____

9. What section of the *NEC* prohibits the installing of Class 2 control circuit conductors in the same raceway as the power conductors? _____

   _____

10. Match the following terms with the statement that most closely defines the term. Enter the letters in the blank space.

RLC _____   a) The maximum ampere rating of the overcurrent device.

MCA _____   b) The current value to be used instead of the rated-load ampere.

BCSC _____   c) The value used to determine the minimum ampere rating for the branch-circuit.

MCC _____   d) A current value for the hermetic motor compressor that was determined by operating it at rated temperature, voltage, and frequency.

MOP _____   e) The maximum current that the compressor can draw continuously without damage.

11. The disconnect for an air-conditioner or heat pump must be installed _____ _____ of the unit.

Gas and Oil Central Heating Systems

OBJECTIVES

After studying this unit, the student should be able to

▶ understand the basics of typical home warm air and hot water (hydronic) heating systems.

▶ understand and apply the *NEC*® requirements for branch-circuit wiring for central heating systems.

▶ define all of the major components of typical heating systems.

▶ understand and apply the *NEC*® requirements for Class 2 control circuit wiring.

In Unit 23, we discussed various types of electric heating systems. In this unit, we will take a look at gas- and oil-fired systems.

FORCED-WARM AIR FURNACES

Gas- and oil-fired heating systems provide the heat source for forced-warm air furnaces and hot water systems. Gas-fired, forced-warm air furnaces are the most common. Forced-warm air systems move hot air from the furnace through hot air ducts that in turn connect to hot air registers strategically located throughout the home. The room air is then pulled into cold air registers, returning to the furnace through cold air return ducts, where the air is again filtered, heated, and forced out through the hot air ducts.

Air-conditioning, humidity control, and fresh air intake are but a few of the things that can be added to a typical forced-warm air heating system.

HOT WATER SYSTEMS

Hot water systems move hot water through pipes to radiators (i.e., baseboards) and, in some instances, through tubing embedded in the ceiling or in the floor or on the underside of the floor to heat the floors from below. The water is returned to the boiler through a return pipe. Hot water systems are very adaptable to zone control by using a combination of one or more circulating pumps and/or valves for the different zones. Hot water heating systems are referred to as *hydronic* (wet) systems.

PRINCIPLE OF OPERATION

Most residential heating systems operate as follows. A room thermostat is connected to the proper electrical terminals on the furnace or boiler. Inside the equipment, the various controls and valves are interconnected in a manner that will provide safe and adequate operation of the furnace.

Most residential heating systems are "packaged units" in which the burner, hot surface igniter or hot spark igniter, safety controls, valves, fan controls, sensors, high temperature limit switches, fan blower motor, blower fan speed control, draft inducer motor, primary and secondary heat exchanger, and so on, are preassembled, prewired parts of the furnace or boiler. Many of these items are included in a solid-state integrated printed circuit board module, which is the "brain" of the system. This module provides proper sequencing and safety features.

Older models have standing gas pilot lights and thermocouples.

Similar electronic circuitry is used in oil burners where the ignition, the oil burner motor, blower fan motor, high temperature limit controls, hot water circulating pump in the case of hot water systems, and the other devices previously listed are all interconnected to provide a packaged oil burner unit.

Evaporator coils for air-conditioning purposes might be an integral part of the unit, or they might be mounted in the ductwork above the furnace.

Wiring a Residential Central Heating System

Gas and oil-fired heating systems are usually "packaged units," which means that all of the internal wiring of the components has been done by the manufacturer of the unit.

In Unit 23, we discussed the major *NEC®* requirements for wiring an electric furnace.

For the electrician, the "field wiring" for a typical residential furnace or boiler consists of:

1. Installing and connecting the low-voltage wiring between the furnace and the thermostat. This might be a cable consisting of two, three, four, or five small 18 AWG or 20 AWG conductors. Small knockouts are provided on the furnace through which the low-voltage Class 2 wires are brought into the furnace's wiring compartment (see Figure 24-1). Class 2 wiring must not be run through the same raceway or cable as power wiring, *725.55(A).*

2. Installing and connecting the branch-circuit power supply to the furnace. Knockouts are provided on the furnace through which the line voltage power supply is brought into the

THERMOSTAT WIRING INSTALLED BY ELECTRICIAN

ALL INTERNAL COMPONENTS PREWIRED BY MANUFACTURER OF FURNACE

POWER SUPPLY INSTALLED BY ELECTRICIAN

DISCONNECT SWITCH MUST BE WITHIN SIGHT OF FURNACE, *422.31, 424.19,* AND *430.102.*

Figure 24-1   Diagram showing thermostat wiring, power supply wiring, and disconnect switch. ANSI Standard Z21.47 does not allow cord-and-plug connections for a gas furnace.

wiring compartment of the furnace. Depending on local codes, the line voltage power supply wiring might be in EMT, armored cable, or other accepted wiring methods (Figure 24-1).

The disconnecting means must be within sight of the furnace, *422.31, 422.32, 424.19,* and *430.102.* Some furnaces come with a disconnect switch. If not, a toggle switch can be mounted on a box on the side of the furnace. Some electric furnaces come complete with circuit breakers that provide the overcurrent protection for the heating elements as required by UL Standard, and to serve as the disconnecting means required by the *NEC®.*

Separate Branch-Circuit Required

NEC® 422.12 requires that central heating equipment must be supplied by an individual branch-circuit.

The size of the branch-circuit depends on the requirements of the particular furnace. A typical gas-forced warm air furnace might require only a 120-volt, 15-ampere branch-circuit, whereas an electric furnace requires a 240-volt branch-circuit and much larger conductors. Instructions furnished with a given furnace will indicate the branch-circuit requirements.

Some typical wiring diagrams are shown in Figure 24-2, Figure 24-3, Figure 24-4A, and Figure 24-4B. These wiring diagrams show the individual electrical components of a typical residential heating system. In older systems, many of these components were wired by the electrician. In newer

Figure 24-2   Typical wiring diagram for a gas burner, forced hot water system. Always consult the manufacturer's wiring diagrams relating to a particular model heating unit.

Figure 24-3   Typical wiring diagram for a gas burner, forced-warm air system. Always consult the manufacturer's wiring diagrams relating to a particular model heating unit.

Figure 24-4A   Typical wiring diagram of an oil burner heating installation. Always consult the manufacturer's wiring diagrams relating to a particular model heating unit.

Figure 24-4B   An oil burner featuring an electronic microprocessor that controls all facets of the burner operation. Most of the components are an integral part of the unit, thus keeping field wiring to a minimum. All manufacturers provide detailed wiring diagrams for their products. Always consult the manufacturer's wiring diagrams relating to a particular model heating unit.

systems, most components other than the thermostat are prewired as part of a packaged furnace or boiler. The manufacturer's literature includes detailed wiring diagrams for a given model.

MAJOR COMPONENTS

Gas- and oil-fired warm air systems and hot water boilers contain many individual components. Here is a brief description of these components.

Aquastat: An aquastat is a direct immersion water temperature thermostat that regulates boiler or tank temperature in hydronic heating systems, ensuring that the circulating water maintains proper, satisfactory temperature.

Cad Cell: A cad cell is a cadmium sulfide sensor for sensing flame. Cad cells have pretty much replaced the older style stack-mounted bimetal switches.

Circulating Pump: A circulating pump circulates hot water in a hydronic central heating system.

Control Circuit: A control circuit is any electric circuit that controls any other circuit through a relay or an equivalent device; also referred to as a remote-control circuit.

Combustion Chamber: The combustion chamber surrounds the flame and radiates heat back into the flame to aid in combustion.

Combustion Head: A combustion head creates a specific pattern of air at the end of the air tube. The air is directed in such a way as to force oxygen into the oil spray so the oil can burn. A combustion head might also be referred to as the turbulator, fire ring, retention ring, or end cone.

Draft Regulator: A draft regulator is a counterweighted swinging door that opens and closes to help maintain a constant level of draft over the fire.

Fan Control: A fan control is used to control the blower fan on forced-warm air systems. On newer furnaces, the fan control is an electronic timer that starts the fan motor a given time in seconds after the main burner has come on. This timing ensures that the blower will blow warm air. Older type adjustable temperature fan controls are still around.

Fan Motor: The fan motor is the electric motor that forces warm air through the heating ducts. In typical residential heating systems, the motor is provided with integral overload protection. The motor may be single, multiple, or variable speed.

Flue: A flue is a channel in a chimney or a pipe for conveying flame and smoke to the outer air.

Heat Exchanger: A heat exchanger transfers the heat energy from the combustion gases to the air in the furnace or to the water in a boiler.

High-Temperature Limit Control: A high-temperature limit control is a safety device that limits the temperature in the plenum of the furnace to some predetermined safe value as determined by the manufacturer. When this predetermined temperature is reached, the high-temperature limit control shuts off the power to the burner. In newer furnaces, the high-temperature limit control is part of an integrated circuit board and is communicated to by a sensor strategically located in the plenum. Older furnaces might have individual high-temperature limit controls and a separate fan control, or they might have combination high-temperature limit/fan controls.

Hot Surface Igniter: In furnaces that do not have a standing constant pilot light, the hot surface igniter heats to a cherry red/orange color. When it reaches the proper temperature, the main gas valve is allowed to open. If the hot surface igniter does not come on, the main gas valve will not open. Most energy-efficient furnaces use this concept for ignition.

Hydronic System: The hydronic system is a system of heating or cooling that involves transfer of heat by circulating fluid such as water in a closed system of pipes.

Ignition:
Constant: The igniter is designed to stay on continuously.

Intermittent Duty: Defined by UL 296 as "ignition by an energy source that is continuously maintained throughout the time the burner is firing." In other words, the igniter is on the entire time the burner is firing.

Interrupted Duty: Defined by UL 296 as "an ignition system that is energized each time the main burner is to be fired and de-energized at the end of a timed trial for ignition period or after the main flame is proven to be established." In other words, the igniter comes on to light the flame; then, after the flame is established, the igniter is turned off and the main flame keeps burning.

Induced Draft Blower: When the thermostat calls for heat, the induced draft blower starts first to expel any gases remaining in the combustion chamber from a previous burning cycle. It continues to run, pulling hot combustion gases through the heat exchangers, then vents the gases to the outdoors.

Integrated Control: An integrated control is a printed circuit board containing many electronic components. When the thermostat calls for heat, the integrated circuit board takes over to manage the sequence of events that allow the burner to operate safely.

Liquid Immersion Control: This is also referred to as an aquastat. A liquid immersion control controls high water and circulating water temperature in a hot water system. A high-temperature limit control shuts off the burner when a predetermined dangerous high temperature is reached. When acting as a circulating water temperature control, this control makes sure that the water being circulated through the piping system is at the desired temperature—not cold.

Low-Water Control: Also referred to as a low-water cutoff, this device senses low-water levels in a boiler. When a low-water situation occurs, the low-water control shuts off the electrical power to the burner.

Main Gas Valve: The main gas valve is the valve that allows the gas to flow to the main burner of a gas-fired furnace. It may also function as a pressure regulator, a safety shut-off, and as the pilot and main gas valve. Should there be a "flame-out," if the pilot light fails to operate, or in the event of a power failure, this valve will shut off the flow of gas to the main burner.

Nozzle: A nozzle produces the desired spray pattern for the particular appliance in which the burner is used.

Oil Burner: The function of an oil burner is to break fuel oil into small droplets, mix the droplets with air, and ignite the resulting spray to form a flame.

Primary Control: A primary control controls the oil burner motor, ignition, and oil valve in response to commands from a thermostat. If the oil fails to ignite, the controller shuts down the oil burner.

Pump and Zone Controls: Pump and zone controls regulate the flow of water or steam in boiler systems to specific "zones" in the building.

Safety Controls: Safety controls such as pressure relief valves, high-temperature limit controls, low-water cutoffs, and burner primary controls protect against appliance malfunction.

Solenoid: A solenoid is an electrically operated device, which, when the valve coil is energized, a magnetic field is developed, causing a spring-loaded steel valve piston to overcome the resistance of the spring and immediately pull the piston into the stem. The valve is now in the open position. When de-energized, the solenoid coil magnetic field instantly dissipates, and the spring-loaded valve piston snaps closed, stopping oil flow or gas flow to the nozzle. The flame is extinguished, allowing fuel to flow. Usually a spring returns the valve back to the closed position.

Spark Ignition: In spark ignition systems, the spark comes on while pilot gas flows to the pilot orifice. Once the pilot flame is "proven" through an electronic sensing circuit, the main gas valve opens. Once ignition takes place, the sensor will monitor and prove existence of a main flame. If for some reason the main flame goes off, the furnace shuts down. Energy-efficient furnaces may use this concept for ignition.

Switching Relay: A switching relay is used between the low-voltage wiring and line-voltage wiring. When a low-voltage thermostat calls for heat, the relay "pulls in," closing the line-voltage contacts on the relay, which turns on the main power to the furnace or boiler. Usually they have a built-in transformer that provides the low-voltage Class 2 control circuit.

Thermocouple: A thermocouple is used in systems having a standing pilot to hold open a gas valve pilot solenoid magnet. A thermocouple consists of two dissimilar metals connected together to form a circuit. The metals might be iron and copper, copper and iron constantan, copper-nickel alloy and chrome-iron alloy, platinum and platinum-rhodium alloy, or chromel and alumel. The types of metals used depend on the temperatures involved. When one of the junctions is heated, an electrical current flows (Figure 24-5). To operate, there must be a temperature difference between the metal junctions. In a gas burner, the source of heat for the thermocouple is the pilot light. The cold junction of the thermocouple remains open and is connected to the pilot safety shut-off gas valve circuit. A single thermocouple develops a voltage of about 25 to 30 dc millivolts. Because of this extremely low voltage, circuit resistance is kept to a very low value.

Thermopile: More than one thermocouple connected in series is a thermopile. See Figure 24-6. The power output of a thermopile is greater than that of a single thermocouple—typically either 250 or 750 dc millivolts. For example, ten thermocouples (25 dc millivolts each) connected in series result in a 250 dc millivolt thermopile. Twenty-six thermocouples connected in series results in a 750 dc millivolt thermopile.

Thermostat: A thermostat turns the heating/cooling system on and off. Programmable thermostats might have day/night setback settings, a clock, a digital thermometer, humidity control, ventilation, and filtration control. Thermostats are usually mounted approximately 52 in. (1.3 m) above the finished floor. Do not mount thermostats where influenced by drafts, air currents from hot or cold air registers, near fireplaces, near concealed hot or cold water pipes or ducts, in the direct rays of the sun, or on outside walls. See Figure 24-7. Some thermostats contain a small vial of mercury that "tips," so the mercury closes the circuit by bridging the gap between the electrical contacts inside the vial. These thermostats must be kept perfectly level to ensure accuracy. **WARNING: Do not throw mercury thermostats into the trash. The mercury is considered a hazardous waste material. They must be returned to a waste management organization for proper disposal.**

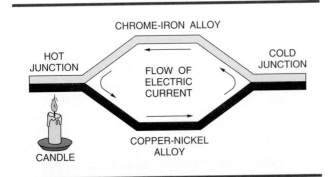

Figure 24-5 Principle of a thermocouple.

Figure 24-6 Principle of a thermopile.

Transformer: The transformer converts (transforms) the branch-circuit voltage to low voltage. For residential applications these are Class 2 transformers, 120 volts to 24 volts. The low-voltage wiring on the secondary of this transformer is Class 2 wiring.

Water Circulating Pump: In hydronic (wet) systems, a circulating pump circulates hot water through the piping system. Multizone systems have more than one pump, each controlled by a thermostat for a particular zone. In multizone systems, a high-temperature control (aquastat) might be set to maintain the boiler water temperature to a specific temperature. Some multizone systems have one circulating pump and use electrically operated valves that open and close on command from the thermostat(s) located in a particular zone(s).

Figure 24-7 A digital electronic programmable thermostat that provides proper cycling of the heating system, resulting in comfort as well as energy savings. This type of thermostat, depending on the model, can be programmed to control temperature, time of operation, humidity, ventilation, filtration, circulation of air, and zone control. Installation and operational instructions are furnished with the thermostat. *Courtesy of Honeywell, Inc.*

CLASS 2 CIRCUITS

Low-voltage circuits fall into three categories—Class 1, Class 2, and Class 3.

Class 1, Class 2, and 3 circuits offer alternative wiring methods regarding voltage, power limitations, wire size, derating factors, overcurrent protection, physical protection, insulation, and materials that differentiate these types of circuits from conventional wiring methods as covered in *Chapter 1* through *Chapter 4* in the *NEC.* Figure 24-8 defines remote control, signaling, and power-limited terminology.

Low-voltage wiring in homes is generally Class 2 wiring. Class 2 wiring is very tolerant. We will limit our discussion to Class 2 wiring. Class 1 and Class 3 circuits are more restrictive than Class 2 circuits and are more commonly found in commercial and industrial applications. You can learn about Class 1 and Class 3 circuits in *Article 725* in the *NEC.*

Wiring for Class 2 Circuits

Electricians describe wire and cable used for low-voltage wiring *bell wire* or *thermostat wire*. What you are really looking for are listed Class 2 conductors and cables, referred to as CL2 or CL2X cables. Examples of Class 2 wiring are the low-voltage conductors and cables used for heating and air conditioning, thermostats, security systems, remote control and signal wiring (chimes), intercom wiring, and other low-voltage applications.

Class 2 conductors are generally made of copper, have a thermoplastic insulation, and have a voltage rating of not less than 150 volts as required by *725.82(G)*. Because the current required for Class 2 circuits is rather small, 18 AWG and 20 AWG conductors are most commonly used. Conductors as small as 24 AWG also are available; however, small conductors and long runs can result in voltage drop problems.

Multiconductor Class 2 cables consist of two to as many as twelve single conductors. These cables are available with or without a protective PVC outer jacket. The type with the outer jacket is preferred because there is less chance of damage to individual conductors, and it gives a neat appearance.

The conductors within these cables are color coded to make circuit identification easy. Table 24-1

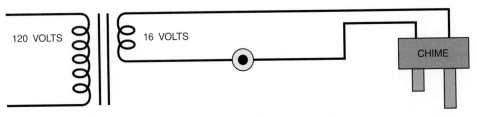

Figure 24-8 The above diagrams explain a remote-control circuit and a signaling circuit.

NUMBER OF CONDUCTORS	COLOR	NUMBER OF CONDUCTORS	COLOR
1	Red	7	Orange
2	White	8	Black
3	Green	9	Pink
4	Blue	10	Gray
5	Yellow	11	Tan
6	Brown	12	Purple

Table 24-1 Table showing the color coding of Class 2 conductors and cables

is a table showing the color coding for the conductors in Class 2 cables.

Here is a summary of the *NEC*® requirements for installing of Class 2 wiring:

- Class 2 circuits are "power-limited" and are considered to be safe from fire hazard and electric shock hazard.

- Class 2 wiring does not have to be in a raceway. If a failure in the wiring should occur, such as a short circuit or an open circuit, there would not be a direct fire or shock hazard.

- Class 2 wiring is not permitted in the same raceway, cable, compartment, outlet box, or fitting with light and power conductors, *725.55(A)*. Even if the Class 2 conductors have the same 600-volt insulation as the power conductors, this is not permitted. See Figure 24-9 and Figure 24-10. Manufacturers of heating and cooling equipment provide separate openings for bringing in the Class 2 wiring and separate openings for bringing in the power supply. *NEC® 725.55(D)* addresses this issue.

- Keep Class 2 conductors at least 2 in. (50 mm) away from light or power wiring. This really pertains to old, open-knob and tube wiring. Today, however, where light and power wiring is installed in Type NMC, Type AC, or in conduit, there is no problem, *725.55(J)*.

- Class 2 wiring shall not enter the same enclosure, unless there is a barrier in the enclosure that separates the Class 2 wiring from the light and power conductors, *725.55(B)*.

- Class 2 circuits are inherently current limiting and do not require separate overcurrent protection. A Class 2 transformer is an example of this.

- Transformers that are intended to supply Class 2 circuits are listed and marked "Class 2 Transformer." These transformers have built-in overcurrent protection to prevent overheating should a short-circuit occur somewhere in the secondary winding or in the secondary circuit wiring. These transformers may be connected to branch-circuits having overcurrent protection not more than 20 amperes, *725.51*.

- Class 2 cables shall have a voltage rating of not less than 150 volts, *725.82(G)*.

Figure 24-9   Most residential air-conditioning units, furnaces, and heat pumps have their low-voltage circuitry classified as Class 2. NEC® 725.55(A) prohibits installing low-voltage Class 2 conductors in the same raceway as the power conductors, even if the Class 2 conductors have 600-volt insulation. Therefore, the conduit running out of the basement wall in the above diagram would not be permitted to contain both the 240-volt power conductors and the 24-volt low-voltage Class 2 conductors. Always read the instructions furnished with these types of appliances to be sure your wiring is in compliance with these instructions, per 110.3(B) and the NEC® Review Unit 4 for grounding methods.

Figure 24-10   The Code in 300.11(B)(2) and 725.58, allows Class 2 control circuit conductors (cables) to be supported by the raceway that contains the power conductors supplying electrical equipment such as the furnace shown.

Securing Class 2 Cables

- Class 2 cables must be installed in a neat and workmanlike manner and shall be adequately supported by the building structure in such a manner that they will not be damaged by normal building use, *725.8.*

- Class 2 cables shall be attached to structural components by straps, staples, hangers, or similar fittings designed and installed so as not to damage the cable, *725.8.*

- Class 2 cables may be secured directly to surfaces with insulated staples or may be installed in raceways.

- The insulation and jacket of Class 2 cables and conductors are rather thin. Be careful during installation not to pierce or crush the wires. Secure these cables and conductors with care, using the proper type of staples. Staple "guns" that use rounded staples are available. They straddle the cable nicely instead of flattening it, which might result in shorted-out wires.

- Do not support Class 2 cables from other raceways, cables, or nonelectric equipment. Taping, strapping, tie wrapping, hanging, or securing Class 2 cables to electrical raceways, piping, or ducts is generally prohibited, *725.58.*

- *NEC®* 725.58 refers back to *300.11(B)(2)*, which allows Class 2 conductors to be supported by the raceway that contains the power supply conductors for the same equipment that the Class 2 conductors are connected to. See Figure 24-10.

- Do not run Class 2 conductors or cables through the same holes in studs and joists, holes that contain nonmetallic-sheathed cable, armored cable, raceways, conduits, or other pipes. There is too much chance of physical damage to the Class 2 cables.

Class 2 Power Source Requirements

Class 2 power source voltage and current limitations, and overcurrent protection requirements, are found in *Chapter 9, Table 11(A)* and *Table 11(B)* of the *NEC®* You should refer to the nameplate data on Class 2 transformers, power supplies, and other equipment to verify that they are UL listed for a particular class.

REVIEW

1. The residence in this text is heated with:

   a. gas

   b. electricity

   c. oil

   Circle the correct answer.

2. The *NEC®* in _____ , _____ , _____ , and _____ requires that a disconnecting means be _____ _____ of the furnace.

3. The *NEC®* in _____ requires that a central heating system be supplied by a _____ branch-circuit.

4. In a home, the low-voltage wiring between the thermostat and furnace is:

   a. Class 1

   b. Class 2

   c. Class 3

   Circle the correct answer.

5. Class 2 circuits are _____-limited and are considered to be _____ from _____ hazard and from _____ _____ .

6. The nameplate on a transformer will indicate whether or not it is a Class 2 transformer. (True) (False) Circle the correct answer.

7. If Class 2 conductors are insulated for 150 volts, and the power conductors are insulated for 600 volts, does the *Code* permit pulling these conductors through the same raceway? (Yes) (No) Circle the correct answer. Give the *NEC*® section number. _____ .

8. If you use 600-volt rated conductors for Class 2 wiring, does the *Code* permit pulling these conductors through the same raceway as the power conductors? (Yes) (No) Circle the correct answer. Give the *NEC*® section number. _____ .

9. The *NEC*® is very strict about securing anything to electrical raceways. Name the *Code* sections that prohibit this practice. *NEC*® _____

   _____

10. Does the *Code* permit attaching the low-voltage thermostat cable to the EMT that brings power to a furnace? (Yes) (No) Circle the correct answer. Give the *NEC*® section numbers. _____

11. Explain what a thermocouple is and how it operates. _____

   _____

   _____

   _____

   _____

Television, Telephone, and Low-Voltage Signal Systems

OBJECTIVES

After studying this unit, the student should be able to

▶ **install residential telephone and television wiring, antennas, and CATV cables, in conformance to *NEC*® requirements.**

▶ **describe the basic operation of satellite antennas.**

▶ **install typical low-voltage wiring for chimes.**

INSTALLING THE WIRING FOR HOME TELEVISION

This unit discusses the basics of installing outlets, cables, receivers, antennas, amplifiers, multiset couplers, and some of the *NEC*® rules for home television. Television is a highly technical and complex field. To ensure a trouble-free installation of a home system, it should be done by a competent television technician. These individuals receive training and certification through the Electronic Technicians Association in much the same way that electricians receive their training through apprenticeship and journeyman training programs.

Some cities and states require that, when more than three television outlets are to be installed in a residence, the television technician making the installation be licensed and certified. Just as electricians can experience problems because of poor connections, terminations, and splices, poor reception problems can arise as a result of poor terminations. In most cases, poor crimping is the culprit. Electrical shock hazards are present when components are improperly grounded and bonded.

According to the plans and for this residence, television outlets are installed in the following rooms:

Front Bedroom	2
Kitchen	1
Laundry	1
Living Room	3
Master Bedroom	2
Recreation Room	3
Study/Bedroom	2
Workshop	1
Total	15

Because TV is a low-voltage system, metallic or nonmetallic standard single-gang device boxes, 4-in. square outlet boxes with a plaster ring, plaster rings only, or special mounting brackets can be installed during the rough-in stage at each likely location of a television set. (More on this later.) See Figure 25-1. For new construction, shielded coaxial cables are installed concealed in the walls. For remodel work, cables can be concealed in the walls by fishing the cables through the walls and installing mounting brackets that are inserted and snapped into place through a hole cut into the wall.

Shielded 75-ohm RG-6 coaxial cable is most often used to hook up television sets to minimize interference and keep the color signal strong. There are different kinds of shielded coaxial cable. Double-shielded cable that has a 100 percent foil

Figure 25-1   Nonmetallic boxes and a nonmetallic raised cover.

shield covered with a 40 percent or greater woven braid is recommended. This coaxial cable has a PVC outer jacket. The older style, flat 300-ohm twin-lead cable, can still be found on existing installations, but the reception might be poor. To improve reception, 300-ohm cables can be replaced with shielded 75-ohm coaxial cables.

For this residence, 15 television outlets are to be installed. A shielded 75-ohm RG-6 coaxial cable will be run from each outlet location back to one central point, as in Figure 25-2.

This central point could be where the incoming Community Antenna Television (CATV) cable comes in, where the cable from an antenna comes in, or where the TV output of a satellite receiver is carried. For rooftop or similar antenna mounting, the cable is run down the outside of the house, then into the basement, garage, or other convenient location where the proper connections are made. CATV companies generally run their incoming coaxial cables underground to some convenient point just inside of the house. Here, the technician hooks up all of the coaxial cables coming from the TV outlets to an amplifier that boosts the signal and improves reception, Figure 25-2, or to a multiset splitter, Figure 25-3. An amplifier may not be necessary if *only* those coaxial cables that will be used are hooked up.

Cable television does not require an antenna on the house because generally the cable company has the proper antenna to receive signals from satellites. The cable company then distributes these signals throughout the community they have contractually agreed to serve. These contracts usually require the cable company to run their coaxial cable to a point

Figure 25-2   A television master amplifier distribution system may be needed where many TV outlets are to be installed. This will minimize the signal loss. In simple installations, a multiset coupler can be used, as shown in Figure 25-3.

Figure 25-3　A three-way CATV cable splitter: one in, three out. Two-way and four-way splitters are also commonly available.

just inside of the house. Inside the house, they will complete the installation by furnishing cable boxes, controls, and the necessary wiring. In geographical areas where cable television is not available, antennas as shown in Figure 25-4 and Figure 25-6 are needed.

Faceplates for the TV outlets are available in styles to match most types of electrical faceplates in the home.

Hazards of Mixing Different Voltages

For safety reasons, video/voice/data wiring must be separated from the 120-volt wiring, *810.18(C)*.

If both 120-volt and low-voltage circuits are run into the same box, then a permanent barrier must be provided in the box to separate the two systems. A better choice is to keep the two systems totally separate using two wall boxes.

Another popular choice is where video/voice/data cables are run to a location right next to a 120-volt receptacle. Mounting brackets that are commercially available can be installed around the electrical device box during the rough-in stage. This lets you trim out a 120-volt receptacle and video/voice/data jacks with one faceplate, as shown in Figure 25-5. These mounting brackets keep the center to center measurements of the device mounting holes for the electrical box and the video/voice/data wiring precisely in alignment. With this method, a wall box is not provided for the video/voice/data wiring. The video/voice/data cables merely come out of the drywall next to the electrical box, and the faceplate takes care of trimming out the receptacle and the video/voice/data wiring jacks.

When wiring a new house, it certainly makes sense to run at least one coaxial cable and one Category 5 cable to a single wall box wherever TV, telephone, and/or computers might be used. This could be a single box located close to a 120-volt receptacle outlet, or it could be a mounting bracket as shown in

Figure 25-4　Older style antennas. Although still available, they have given way to the more popular digital satellite "dish," Figure 25-6(A).

Figure 25-5 A special mounting bracket fastened to the wood stud. The bracket fits nicely over the electrical wall box to accommodate a 120-volt receptacle, two coaxial outlets, and one telephone outlet. A single two-gang faceplate is used for this installation. Many other combinations are possible.

Figure 25-5. Run a separate coaxial cable and a separate Category 5 cable from each location to a common point in the house. You will use a lot of cable, but this will enable you to interconnect the cables as required, such as for a local area network (LAN) to serve more than one computer located in different rooms, or other home automation systems. This is the beginning of "structured wiring" as discussed in Unit 31.

For adding video/voice/data wiring to an existing home, many types of nonmetallic brackets are available that snap into a hole cut in the drywall and lock in place.

More *NEC®* rules are discussed later in this unit. Unit 31 discusses structured wiring using Category 5 unshielded twisted pair (UTP) cable and coaxial shielded cable.

Although shielded coaxial cable minimizes interference, it is good practice to keep the coaxial cables on one side of the stud space, and keep the light and power cables on the other side of the stud space.

Code Rules for Cable Television (CATV) *(Article 820)*

CATV systems are installed both overhead and underground in a community. Then, to supply an individual customer, the CATV company runs coaxial cables through the wall of a residence at some convenient point. Up to this point of entry (*Article 820, Part II*) and inside the building (*Article 820, Part V*),

the cable company must conform to the requirements of *Article 820,* plus local codes if applicable.

Coaxial cable generally used for residential installations is Type CATV or CATVX. Other types are listed in *Table 820.50* of the *NEC.®*

Here are some of the key rules to follow when making a coaxial cable installation.

1. The outer conductive shield of the coaxial cables must be grounded as close to the point of entry as possible, *820.33.*

2. Coaxial cables shall not be run in the same conduits or box with electric light and power conductors, *820.50(A)(1)(2).*

3. Do not support coaxial cables from raceways that contain electrical light and power conductors, *820.55(C).*

4. Keep the coaxial cable at least 2 in. (50 mm) from light and power conductors unless the conductors are in a raceway, nonmetallic-sheathed cable, armored cable, or UF cable, *820.55(A)(2).* This clearance requirement really pertains to old knob-and-tube wiring. The 2-in. (50-mm) clearance is not required if the coaxial cable is in a raceway.

5. Where underground coaxial cables are run, they must be separated by at least 12 in. (300 mm) from underground light and power conductors, unless the underground conductors are in a raceway, Type UF cable, or Type USE cable, *820.12(B).* The 12-in. (300-mm) clearance is not required if the coaxial cable has a metal cable armor.

6. The grounding conductor (*Article 820, Part III* and *Part IV*):

   a. outer conductive shield shall be grounded as close as possible to the coaxial cable entrance or attachment to the building, *820.33.*

   b. shall not be smaller than a 14 AWG copper or other corrosion-resistant conductive material. It need not be larger than 6 AWG copper. The grounding conductor must be insulated. It shall have a current-carrying capacity not less than that of the coaxial cable's outer metal shield. See *NEC® 810.40(A).*

   c. length shall be as short as possible, but not longer than 20 ft (6.0 m). If impossible to keep the grounding conductor to this maximum permitted length, then the *Exception to 820.40(A)(4)* permits installing a separate grounding electrode. These two electrodes shall be bonded together with a bonding jumper not smaller than 6 AWG copper.

d. may be solid or stranded, *820.40(A)(2)*.

e. shall be guarded from physical damage, *820.40(A)(6)*.

f. shall be run in a line as straight as practical, *820.40(A)(5)*.

g. shall be connected to the nearest accessible location on one of the following, *820.40(B)*:

- the building grounding electrode.
- the grounded interior metal water piping pipe, within 5 ft (1.5 m) of where the pipe enters the building. Refer to *250.52*.
- the power service accessible means external to the enclosure.
- the metallic power service raceway.
- the service equipment enclosure.
- the grounding electrode conductor or its metal enclosure.
- the grounding conductor or grounding electrode of a disconnecting means that

is grounded to an electrode according to *250.32*. This pertains to a second building on the same property served by a feeder or branch-circuit from the main electrical service in the first building.

h. If *none* of the options in *g* is available, then ground to any of the electrodes per *250.52*, such as metal underground water pipes, the metal frame of the building, a concrete encased electrode, or a ground ring. Watch out for the maximum length permitted for the grounding conductor, as mentioned above.

CAUTION: DO NOT simply drive a separate ground rod to ground the metal shielding tape on the coax cable. Difference of potential between the coax cable shield and the electrical system ground during lightning strikes could result in a shock hazard as well as damage to the electronic equipment. If for any reason one grounding electrode is installed for the CATV shield grounding and another grounding electrode for the electrical system, bonding these two electrodes together with no smaller than a 6 AWG copper conductor will minimize the possibility that a difference of potential voltage might exist between the two electrodes.

(A)

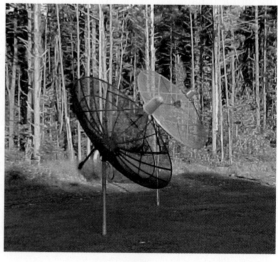

(B)

Figure 25-6 **(A)** A digital satellite system 18-in. (450-mm) antenna, a receiver, and a remote control. *Courtesy of* Thomson Consumer Electronics. **(B)** A large satellite antenna securely mounted to a post that is anchored in the ground. These large antennas are rarely used today. *Courtesy of* Hi-Tech Advisors, www.advisors.com.

SATELLITE ANTENNAS

A satellite antenna is often referred to as a "dish." See Figure 25-6. A satellite dish has a parabolic shape that concentrates and reflects the signal beamed down from one of the many stationary satellites orbiting 22,245 miles (35,800 kilometers) above the equator. Stationary means that the satellite is traveling at the same speed that the earth rotates. The orbit in which satellites travel is called the "Clarke Belt," as shown in Figure 25-7.

The latest satellite technology is the digital satellite system (DSS). A digital system allows the use of small antennas, approximately 18 in. (450 mm) in diameter, that can easily be mounted on a roof, on a chimney, on the side of the house, on a pipe, or on a pedestal, using the proper mounting hardware.

Although rarely used today for residential TV reception, large antennas are still in use. Because of their huge size and weight, proper installation is important.

The basic operation of a satellite system is for a transmitter on earth to beam an "uplink" signal to a satellite in space. Electronic devices on the satellite reamplify and convert this signal to a "downlink" signal, then retransmit the downlink signal back to earth, as shown in Figure 25-8.

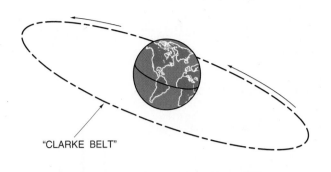

"CLARKE BELT"

Figure 25-7 All satellites travel around the earth in the same orbit called the "Clarke Belt." They appear to be stationary in space because they are rotating at the same speed at which the earth rotates. This is called "geosynchronous orbit (geostationary orbit)." The satellite receives the uplink signal from earth, amplifies the signal, and transmits it back to earth. The downlink signal is picked up by the satellite antenna.

Regardless of the type of antenna used, the line-of-sight from the antenna to the satellite must not be obstructed by trees, buildings, utility poles, or other structures. Instructions furnished with an antenna provide the necessary data for direction and up angle based on ZIP codes.

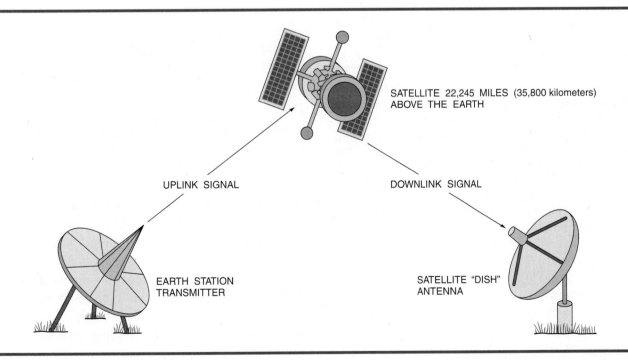

UPLINK SIGNAL

DOWNLINK SIGNAL

SATELLITE 22,245 MILES (35,800 kilometers) ABOVE THE EARTH

EARTH STATION TRANSMITTER

SATELLITE "DISH" ANTENNA

Figure 25-8 The earth station transmitter beams the signal up to the satellite where the signal is amplified, converted, and beamed back to earth.

Figure 25-9   Typical satellite antenna/cable/standard antenna wiring. Connection to a standard antenna cable, CATV cable, or to a satellite cable enables the homeowner to watch one channel while recording another channel. The installation manual for the receiver usually has a number of different hookup diagrams. In this diagram, the A-B switch allows a choice of connecting to a standard antenna or to the CATV cable. This hookup can be desirable should the CATV or satellite reception fail.

Before installing an antenna, check with your local inspection department. Local codes might have restrictions and requirements in addition to the *NEC*.

Figure 25-9 and Figure 25-10 show typical connections between a television set, receiver, VCR,

Figure 25-10   The terminals on the back of a digital satellite receiver. The connections are similar to those in Figure 25-9. Note the difference in that the digital satellite receiver has a modular telephone plug, a feature that uses a toll-free number to update the access card inside the receiver to ensure continuous program service. The telephone can also handle program billing.

and antenna. The manufacturer's installation instructions are always to be followed.

Figure 25-11 illustrates one method of installing a large antenna post in the ground in accordance to a manufacturer's instruction.

CODE RULES FOR THE INSTALLATION OF ANTENNAS AND LEAD-IN WIRES
(ARTICLE 810)

Home television and AM and FM radios generally come complete with built-in antennas. For those locations in outlying areas on the fringe or out of reach of strong signals, it is quite common to install a separate antenna system or a satellite antenna system.

Indoor or outdoor antennas may be used with televisions. The front of an outdoor antenna is aimed at the television transmitting station. When there is more than one transmitting station and they are located in different directions, a rotor is installed. A rotor turns the antenna on its mast so that it can face in the direction of each transmitter. The rotor is controlled from inside the building. The rotor controller's cord is plugged into a regular 120-volt receptacle to obtain power. A three-wire or four-

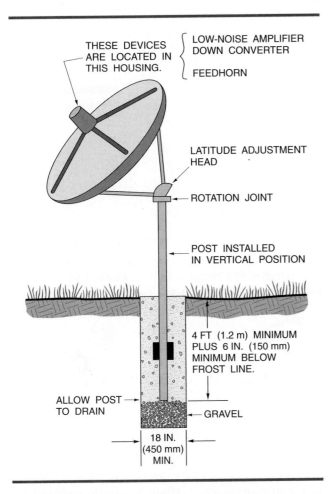

THESE DEVICES
ARE LOCATED IN
THIS HOUSING. { LOW-NOISE AMPLIFIER
DOWN CONVERTER

FEEDHORN

LATITUDE ADJUSTMENT
HEAD

ROTATION JOINT

POST INSTALLED
IN VERTICAL POSITION

4 FT (1.2 m) MINIMUM
PLUS 6 IN. (150 mm)
MINIMUM BELOW
FROST LINE.

ALLOW POST
TO DRAIN

GRAVEL

18 IN.
(450 mm)
MIN.

Figure 25-11 Satellite antenna solidly installed in the ground according to the manufacturer's instructions. The popularity of these large antennas has given way to the small digital satellite system antenna, as shown in Figure 25-6.

conductor cable is usually installed between the rotor motor and the control unit. The wiring for a rotor may be installed during the roughing-in stage of construction, running the rotor's cable into a device wall box, allowing 5 to 6 ft (1.5 to 1.8 m) of extra cable, then installing a regular single-gang switchplate when finishing.

Article 810 covers radio and television equipment. Although instructions are supplied with antennas, the following key points of the *Code* regarding the installation of antennas and lead-in wires should be followed.

1. Antennas and lead-in conductors shall be securely supported, *810.12.*

2. Antennas and lead-in conductors shall not be attached to the electric service mast, *810.12.*

3. Antennas and lead-in conductors shall be kept away from all light and power conductors to avoid accidental contact with the light and power conductors, *810.13.*

4. Antennas and lead-in conductors shall not be attached to any poles that carry light and power wires over 250 volts between conductors, *810.12.*

5. Lead-in conductors shall be securely attached to the antenna, *810.12.*

6. Outdoor antennas and lead-in conductors shall not cross over light and power conductors, *810.13.*

7. Outdoor antennas and lead-in conductors shall be kept at least 24 in. (600 mm) away from open light and power conductors, *810.13.*

8. Where practicable, antenna conductors shall not be run under open light and power conductors, *810.13.*

9. On the outside of a building:

   a. Position and fasten lead-in conductors so they cannot swing closer than 24 in. (600 mm) to light and power conductors having *not* over 250 volts between conductors; 10 ft (3.0 m) if *over* 250 volts between conductors, *810.18(A).*

   b. Keep lead-in conductors at least 6 ft (1.8 m) away from a lightning rod system, *810.18(A),* or bonded together according to *250.60.*

   Note: The clearances in (a) and (b) are not required if the light and power conductors or the lead-in conductors are in a metal raceway or metal cable armor.

10. On the inside of a building:

   a. Keep the antenna and lead-in conductors at least 2 in. (50 mm) from other open wiring (as in old houses) unless the other wiring is in a metal raceway or cable, *810.18(B).*

   b. Keep lead-in conductors out of electric boxes unless there is an effective, permanently installed barrier to separate the light and power wires from the lead-in wire, *810.18(C).*

11. Grounding:

a. All metal masts and metal structures that support antennas shall be grounded, *810.21*.

b. The grounding conductor must be copper, aluminum, copper-clad steel, bronze, or a similar corrosion-resistant material, *810.21(A)*.

c. The grounding conductor need not be insulated. It must be securely fastened in place, may be attached directly to a surface without the need for insulating supports, shall be protected from physical damage or be large enough to compensate for lack of protection, and shall be run in as straight a line as is practicable, *810.21(B)*, *(C)*, *(D)*, and *(E)*.

d. The grounding conductor shall be connected to the nearest accessible location on one of the following:

- the building or structure grounding electrode. Refer to *250.50* for more details.

- the grounded interior metal water pipe (as shown in Figure 25-12), within 5 ft (1.5 m) of where the water pipe enters the building. Refer to *250.52* for more details.

- the metallic power service raceway.

- the service equipment enclosure.

- the grounding electrode conductor or its metal enclosure.

e. If none of *d* is available, then ground to any one of the electrodes per *250.52*, such as metal underground water pipe, metal frame of building, concrete-encased electrode, or ground ring, *810.21(F)(2)*.

f. If neither *d* nor *e* is available, then ground to an effectively grounded metal structure or to any one of the electrodes per *250.52*, a driven rod or pipe, or a metal plate, *810.21(F)(3)*.

g. The grounding conductor may be run inside or outside of the building, *810.21(G)*.

h. The grounding conductor shall not be smaller than 10 AWG copper or 8 AWG aluminum, *810.21(H)*.

Figure 25-12 **Typical connection for the required grounding of the metal shield on coaxial cable to the interior metal water piping. The connection must be made within 5 ft of where the water pipe enters the building.**

CAUTION: **DO NOT simply drive a separate ground rod to ground the metal mast, structure, or antenna. Difference of potential between the metal mast, structure, or antenna and the electrical system ground during a lightning strike could result in a shock hazard as well as damage to the electronic equipment. If any reason permitted by the *Code* results in one grounding electrode for the antenna and another grounding electrode for the electrical system, bond the two electrodes together with a bonding jumper not smaller than 6 AWG copper or equivalent, *810.21(J)*.**

The objective of grounding and bonding to the same grounding electrode as the main service ground is to reduce the possibility of having a difference of voltage potential between the two systems.

TELEPHONE WIRING (SYMBOL ▶|)
(ARTICLE 800)

Since the deregulation of telephone companies, residential "do-it-yourself" telephone wiring has become quite common. Following is an overview of residential telephone wiring.

The telephone company will install the service line to a residence and terminate at a protector device, as in Figure 25-13. The protector protects the system from hazardous voltages. The protector may be mounted either outside or inside the home. Different telephone companies have different rules.

Figure 25-13   Typical residential telephone installation. The telephone company installs and connects the underground cable to the protector. In this illustration, the protector is mounted to the metal service-entrance raceway. This establishes the ground that offers protection against hazardous voltages such as a lightning surge on the incoming line. A cable is then run to a Standard Network Interface, referred to as SNI or just NI. These are shown in (A) and (B) in the diagram. The interior telephone wiring is then connected to the SNI, either with a modular plug as shown in (A), or "hard-wired" through a neoprene grommet on the bottom of (B). The SNI of the type depicted by (B) are preferred because they are weatherproof and can be mounted on the outside of the house.

Always check with the phone company before starting your installation.

The point where the telephone company ends and the homeowner's interior wiring begins is called the *demarcation point*, as shown in Figure 25-13. The preferred demarcation point is outside of the home, generally near the electric meter where proper grounding and bonding can be done. A *network interface* might be of the types depicted in Figure 25-13, at (A) and (B). In the past, interior telephone wiring in homes was installed by the telephone company. Today, the responsibility for roughing-in the interior wiring is usually left up to the electrician. Telephone wiring is *not* to be run in the same raceway with electrical wiring. In addition, it cannot be installed in the same outlet and junction boxes as electrical wiring. Whatever the case, the rules found

in *Article 800* apply. Refer to Figure 25-13, Figure 25-14, and Figure 25-15.

For a typical residence, the electrician will rough-in wall boxes wherever a telephone outlet is wanted. The boxes can be of the "single-gang" types illustrated in Figure 25-1, Figure 2-11, Figure 2-17, Figure 2-19, and Figure 2-32. Concealed telephone wiring in new homes is generally accomplished using multipair cables that are run from each telephone outlet to a common point, as in Figure 25-14(A). The cable can also take the form of a "loop system," as in Figure 25-15. A three-pair, six-conductor twisted cable is shown in Figure 25-16. There are some localities that require that these cables be run in a metal raceway, such as EMT.

The outer jacket is usually of a thermoplastic material. This outer jacket most often is a neutral

STANDARD
NETWORK
INTERFACE (SNI)

JUNCTION BOX

A

JUNCTION BOX FEATURES A SHORT, PREWIRED
CORD THAT PLUGS INTO A STANDARD NETWORK
INTERFACE (SNI). IT ALLOWS EASY CONNECTION
OF ADDITIONAL TELEPHONE CABLES.

B

Figure 25-14 **(A)** shows individual telephone cables run to each telephone outlet from the common connection point. This is sometimes referred to as a star-wired system. **(B)** shows a junction box for easy connection of multiple telephone cables.

STANDARD
NETWORK
INTERFACE (SNI)

JUNCTION BOX

Figure 25-15 A complete "loop system" (daisy-chain) of the telephone cable. If something happens to one section of the cable, the circuit can be fed from the other direction.

color such as white, light gray, or beige that blends in with decorator colors in the home. This is particularly important in existing homes where the telephone cable may have to be exposed. Cables, mounting boxes, junction boxes, terminal blocks, jacks, adaptors, faceplates, cords, hardware, plugs, and so on, are all available through electrical distributors, builders' supply outlets, telephone stores, electronic stores, hardware stores, and similar wholesale and retail outlets. The selection of types of the preceding components is endless.

Telephone Conductors

The color coding of telephone cables is shown in Figure 25-16. Figure 25-17 shows some of the many types of telephone cords available.

Cables are available with various numbers of conductors. Here are a few.

2 pair	4 conductors
3 pair	6 conductors
4 pair	8 conductors
6 pair	12 conductors
12 pair	24 conductors

Telephone circuits require a separate pair of conductors from the telephone all the way back to the phone company's central switching center. To keep to a minimum interference that could come from other electrical equipment, such as electric motors and fluorescent fixtures, each pair of telephone wires (two-wires) is twisted.

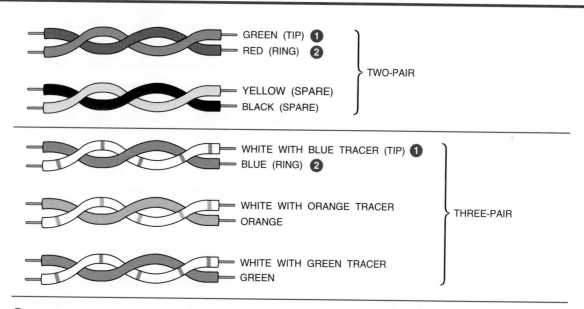

❶ *TIP* IS THE CONDUCTOR THAT IS CONNECTED TO THE TELEPHONE COMPANY'S "POSITIVE" TERMINAL. IT IS SIMILAR TO THE NEUTRAL CONDUCTOR OF A RESIDENTIAL WIRING CIRCUIT.

❷ *RING* IS THE CONDUCTOR THAT IS CONNECTED TO THE TELEPHONE COMPANY'S "NEGATIVE" TERMINAL. IT IS SIMILAR TO THE "HOT" UNGROUNDED CONDUCTOR OF A RESIDENTIAL WIRING CIRCUIT.

Figure 25-16 Color coding for a two-pair and a three-pair, six-conductor twisted telephone cable. The color coding for two-pair telephone cable stands alone. For three or more pair cables, the color coding becomes WHITE/BLUE, WHITE/ORANGE, WHITE/GREEN, WHITE/BROWN, WHITE/SLATE, RED/BLUE, RED/ORANGE, RED/GREEN, RED/BROWN, RED/SLATE, BLACK/BLUE, BLACK/ORANGE, BLACK/GREEN, BLACK/ BROWN, BLACK/SLATE, YELLOW/BLUE, YELLOW/ORANGE, YELLOW/GREEN, YELLOW/BROWN, YELLOW/SLATE, VIOLET/BLUE, VIOLET/ORANGE, VIOLET/GREEN, VIOLET/BROWN, VIOLET/SLATE. In multipair cables, the additional pairs can be used for more telephones, fax machines, security reporting, speakerphones, dialers, background music, etc.

Figure 25-17 Some of the many types of telephone cable available are illustrated. Note that for ease of installation, the modular plugs and terminals have been attached by the manufacturer of the cables. These cords are available in round and flat configurations, depending upon the number of conductors in the cable.

Cross-Talk Problems

A common telephone interference problem in homes is "cross-talk." This is apparent if you hear someone faintly talking in the background when you are using the telephone. Another problem is losing an on-line connection to your computer. The signals traveling in one pair of conductors are picked up by the adjacent pair of conductors in the cord or cable. Quite often, these problems can be traced to older style flat (untwisted) two-line telephone lines. Older style cables could carry audio signals quite well, but they are unsatisfactory for data transmission. These problems are virtually eliminated by using cables and cords that have conductors twisted at proper intervals, such as Category 5 cables. In Figure 25-16, each pair of conductors is twisted, then all of the pairs are twisted again. Generally, you should avoid flat two-line cords.

The recommended circuit lengths are:

24 AWG gauge – – – – not over 200 ft (60 m)

22 AWG gauge – – – – not over 250 ft (75 m)

Ringer Equivalence Number (REN)

Telephones, fax machines, answering machines, ringers, flashers, and so on, that have a sounder (ringer) are assigned an REN number. Generally, the maximum number of RENs on any single telephone line is 5. This maximum number of RENs ensures that all telephones will ring. When too many telephones are connected, the ring signal can become unreliable, resulting in devices failing to respond.

Older style electromechanical ringers were considered to be 1-REN. Electronic ringers have a minimal REN value. Mixing electromechanical and electronic ringers could present a problem as the electromechanical ringers tend to hog the current. It all depends upon your telephone line. Try it and see what happens.

Pay close attention to the total RENs connected in your home. REN values are found on the label of the device.

EXAMPLE:

Four electronic telephones:	$4 \times 0.3 = 1.2$ RENs
One fax machine:	1.5 RENs
One desk top telephone/ Answering machine:	0.7 REN
One desk top telephone:	0.4 RENs
Total	3.8 RENs

Call your telephone company if you think you have a problem because of too many telephones connected to one telephone line.

Plug-in or Permanent Connection?

Most residential telephones plug into a jack of the type illustrated in Figure 25-18 and Figure 25-19. It might be a good idea to have at least one permanently connected telephone, such as a wall phone in the kitchen. If all of the phones were of the plug-in type, it is possible (but highly unlikely) that all of them could be unplugged. As a result, there would be no audible signal.

The electrical plans show that nine telephone outlets are provided as follows:

Front Bedroom	1
Kitchen	1
Laundry	1
Living Room	1
Master Bedroom	1
Rear Outdoor Patio	1
Recreation Room	1
Study/Bedroom	1
Workshop	1

Installation of Telephone Conductors
(*800.51* and *800.52*)

Telephone conductors:

1. are generally Type CM or Type CMX (for dwellings only), listed for telephone installa-

Figure 25-18   Telephone modular jack.

Figure 25-19   Three styles of wall plates for modular telephone jacks: rectangular stainless steel, weatherproof for outdoor use, and circular.

tion as being resistant to the spread of fire. Refer to *Table 800.50* for other types.

2. shall be separated by at least 2 in. (50 mm) from light and power conductors unless the light and power conductors are in a raceway, or in nonmetallic-sheathed cable, Type AC cable, or Type UF cable, *800.55(A)(2)*.

3. shall not be placed in any conduit or boxes with electric light and power conductors unless the conductors are separated by a partition, *800.55(A)(1)(c)*.

4. do not support telephone cables from raceways that contain light and power conductors, *800.55(C)*.

5. should not share the same bored holes as electrical wiring, plumbing, gas pipes, and so on.

6. may be terminated in either metallic or nonmetallic boxes or plaster rings.

7. should be secured using rounded or depthstop plastic staples. Do not use metal staples.

8. do not use nail guns. Do not crush the cable.

9. should be kept away from hot water pipes, hot air ducts, and other heat sources that might harm the insulation.

10. do not run telephone cables in the same stud space as electrical branch-circuit wiring, and keep cables at least 12 in. (300 mm) from power wiring where the cables are run parallel to the power wiring. These cautions were written before twisted pairs and category-rated cables entered the scene. With the advent of *properly* installed twisted and shielded cables, these "old wives' tale" recommendations from the past are probably not an issue.

Follow the specifications and instructions for the cable you are installing.

Conduit

In some communities, electricians prefer to install regular device boxes at a telephone outlet location, then "stub" a trade size ½ EMT to the basement or attic from this box. Then, at a later date, the telephone cable can be fished through the conduit.

Grounding (Article 800, Part IV)

The telephone company will provide the proper grounding of their incoming cable sheath and primary protector, generally using an insulated conductor not smaller than 14 AWG copper or other corrosion resistant material, and not longer than 20 ft (6.0 m).

For one- and two-family dwellings, where it might be impractical to keep the grounding conductor 20 ft (6.0 m) or less, an additional communications grounding electrode must be installed. This additional ground rod must be bonded to the power grounding electrode system with a bonding conductor not smaller than a 6 AWG copper. In just about all instances, all grounding and bonding of telephone equipment is done by the telephone company personnel. Many times, they clamp the primary protec-

tor to the grounded metal service raceway conduit to establish "ground."

Safety

Open-circuit voltage between conductors of an idle pair of telephone conductors is approximately 48 volts dc. The superimposed ringing voltage can reach 90 volts ac. Therefore, always work carefully with insulated tools and stay clear of bare terminals and grounded surfaces. Disconnect the interior telephone wiring if work must be done on the circuit, or take the phone off the hook, in which case the dc voltage level will drop to approximately 7 to 9 volts dc, and there should be no ac ringing voltage delivered.

Wiring for Computers and Internet Access

Wiring for telephones, high-speed Internet access, computers, television, printers, modems, security systems, intercoms, and similar home automation equipment could be thought of as one big project. Planning ahead is critical. Deciding on where this equipment is likely to be located will help design an entire video/voice/data system.

You will want to install Category 5 or Category 5e cables between each telephone outlet and a central distribution point so as to have the flexibility to connect the video/voice/data equipment in any configuration.

In addition to the previously mentioned equipment, there very well might be a scanner, an answering machine, desk lamps, floor lamps, an adding machine, a calculator, a radio, a CD player, a CD "burner," audio/videocassette tape players, VCR and DVD players, and ZIP drives. They draw little current, but all need to be plugged into a 120-volt receptacle! Managing and plugging in that many power cords are a major problem. Give consideration to installing two or three duplex or quadplex 120-volt receptacles at these locations. Multioutlet plug-in strips with surge protection will most likely be needed.

Unit 31 contains much information about home automation and structured wiring.

Bringing Technology Into Your Home

In some areas, the telephone company provides the entire "package" for homes. This might include digital local and long distance telephone service, Internet access, cable modems for personal computers, cable television, digital subscriber lines (DSL) (through existing copper telephone lines), and whatever other great new things may come next.

SIGNAL SYSTEM (CHIMES)

A signaling circuit is described in the *NEC®* as *"any electric circuit that energizes signaling equipment."* Signaling equipment includes such devices as chimes, doorbells, buzzers, code-calling systems, and signal lights.

Door Chimes (Symbol CH)

Present-day dwellings use chimes rather than bells or buzzers to announce that someone is at a door. A musical tone is sounded rather than a harsh ringing or buzzing sound. Chimes are available in single-note, two-note, eight-note (four-tube), and repeater tone styles. In a repeater tone chime, both notes sound as long as the push button is depressed. In an eight-note chime, contacts on a motor-driven cam are arranged in sequence to sound the notes of a simple melody when the chime button is pushed. This type of chime is usually installed in dwellings having three entrances. The chime can be connected so that the eight-note melody sounds for the front door, two notes sound for the side door, and a single note sounds for the rear door. Chimes are also available with clocks and lights.

Electronic chimes may relay their chime tones through the various speakers of an intercom system. When any chime is installed, the manufacturer's instructions must be followed.

The plans show that two chimes are installed in the residence. Two-note chimes are used. Each chime has two solenoids and two iron plungers. When one solenoid is energized, the iron plunger is drawn into the opening of the solenoid. A plastic peg in the end of the plunger strikes one chime tone bar. When the solenoid is de-energized, spring action returns the plunger where it comes to rest against a soft felt pad so that it does not strike the other chime tone bar. Thus, a single chime tone sounds. As the second solenoid is energized, one chime tone bar is struck. When the second solenoid is de-energized, the plunger returns and strikes the second tone bar. A two-tone signal is produced. The plunger then comes to rest between the two tone bars. Generally, two notes indicate front door signaling, and one note indicates rear or side door signaling.

Figure 25-20 shows various push-button styles used for chimes. Many other styles are available. Figure 25-21 shows several types of typical residential-type wall-mounted chimes. The symbols used to indicate push buttons and audible signals on the plans are shown in Figure 25-22.

Figure 25-23 shows how to provide proper backing for chimes. This is done during the rough-in stages of the electrical installation.

Chimes for the Hearing Impaired

Chimes for the hearing impaired are available with accessory devices that can turn on a dedicated lamp at the same time the chime is sounded. Figure 25-24 shows the devices needed to accomplish this. Figure 25-25 shows the wiring details.

Chime Transformers

Chime transformers are covered under UL Standard 1585. Because of their low-voltage rating

Figure 25-20   Push buttons for door chimes. *Courtesy of* **Broan-NuTone, Inc.**

Figure 25-21 Typical residential door chimes. *Courtesy of* Broan-NuTone, Inc.

Figure 25-22 Symbols for chimes and push buttons.

BE SURE TO PROVIDE
ADEQUATE BACKING FOR
SUPPORTING THE CHIME.

"ROUGH-IN" BOX FOR
FLUSH-MOUNTED CHIME.

FOR SMALL CHIMES THE
RECOMMENDED HEIGHT IS
6 ½ FT (2.0 m) TO CENTER.

FOR CHIMES WITH LONG
TUBES, CONSULT
MANUFACTURER INSTALLATION
INSTRUCTIONS.

Figure 25-23   Roughing-in for a flush-mounted chime.

Figure 25-24   Accessory devices can be attached to new or existing chimes to provide visual signals when the chime is sounded. The first illustration shows the module that is connected and mounted inside the chime. The second illustration shows a transmitter that is plugged into a nearby receptacle. The third illustration shows a receiver into which a lamp is plugged.

CHIME

"A" IS A MODULE THAT MOUNTS INSIDE THE CHIME. IT IS CONNECTED TO THE SAME TERMINALS ON THE CHIME AS THE REGULAR CONNECTIONS.

A

TWO WIRES ARE RUN BETWEEN MODULE "A" AND RECEIVER "B." THESE WIRES ARE LOW VOLTAGE, SAME AS REGULAR CHIME WIRES. RUN CONDUCTORS CONCEALED IN WALL, OR ALONG DOOR FRAMES AND BASEBOARDS.

B

TRANSMITTER "B" PLUGS INTO A NEARBY 120-VOLT WALL OUTLET.

C

RECEIVER "C" PLUGS INTO A 120-VOLT WALL OUTLET. LAMP PLUGS INTO RECEIVER. MORE THAN ONE RECEIVER CAN BE PLUGGED IN. THE SIGNAL IS TRANSMITTED THROUGH THE 120-VOLT HOUSE WIRING.

Figure 25-25 Installation diagram for auxiliary devices for use with chimes that can help the hearing impaired. When the chime sounds, the signal is transmitted to the receiver, turning on the lamp that is plugged into the receiver. The "ON" time is adjustable at "C". The lamp should be a "dedicated" lamp that lights up only when the chime sounds.

and power limitation, they are listed as Class 2 transformers. The low-voltage wiring on the secondary side of the transformer is Class 2 wiring. Class 2 circuits are discussed in Unit 24 and in *Article 725.*

Figure 25-26 shows a chime transformer. The top view shows the 120-volt leads. Note the small set screw for ease of mounting into a conduit knockout in an outlet box. The bottom view shows the low-voltage terminals.

Chime transformers used in dwellings are generally rated 16 volts. They have built-in thermal overload protection. Should a short circuit occur in the low-voltage wiring, the overload device opens and closes repeatedly until the short circuit is cleared.

The UL standard requires that all metal parts of a transformer be properly grounded. To accomplish this, some chime transformers have a bare copper equipment grounding conductor in addition to the black and white supply conductors. Internal to the

transformer, this equipment grounding conductor is connected to the metal parts (laminations) of the transformer. This bare equipment grounding conductor must be connected to the equipment grounding conductor or grounding screw in the outlet box. Do not connect this equipment grounding

Figure 25-26 Chime transformer. *Courtesy of* Broan-NuTone.

conductor to the branch-circuit grounded (white) conductor.

Some chime transformers are marked "Install in Metal Box Only."

Additional Chimes

To extend a chime system to cover a larger area, a second or third chime may be added. In the residence, two chimes are used: one chime is mounted in the front hall and a second (extension) chime is mounted in the recreation room. The extension chime is wired in parallel to the first chime. The wires are run from one chime terminal board to the terminal board of the other chime. The terminals are connected as follows: transformer to transformer, front to front, and rear to rear, Figure 25-27.

When more than one chime is installed, it will probably be necessary to install a transformer with a higher volt-ampere (wattage) rating. Do not install a transformer with a higher voltage rating. Check the manufacturers' instructions furnished with the chime transformer.

Do not add another transformer to an existing transformer to solve the problem of multiple chimes not responding properly. The *NEC®* in *725.41(B)* prohibits connecting transformers in parallel unless

specifically listed for interconnection. Residential chime transformers are not listed for interconnection.

If a buzzer (or bell) and a chime are connected to a single transformer and are used at the same time, the transformer will put out a fluctuating voltage. This condition does not allow either the buzzer or the chime to operate properly. The use of a transformer with a larger rating may solve this problem.

The wattage consumption of chimes varies with the manufacturer. Typical ratings are shown in Table 25-1.

Transformers with ratings of 5, 10, 15, 20, and 30 watts (VA) are available. For a multiple chime installation, wattage ratings for the individual chimes are added. The total value is the minimum transformer rating needed to do the job properly.

Low-Voltage Wiring

The low-voltage wiring for a chime(s) is classified as Class 2 wiring as discussed in Unit 24.

Wiring the Chime

The circuit shown in Figure 25-27 is recommended for this chime installation because it provides a hot low-voltage circuit at the front hall location. However, it is not the only way in which these chimes may be connected.

Figure 25-27 shows that a two-wire cable runs from the transformer in the utility room to the front hall chime. A two-wire cable then runs from the chime to both the front and rear door push buttons. A three-wire cable also runs between the front hall chime and the recreation room chime. Because of the hot, low-voltage circuit at the front hall location,

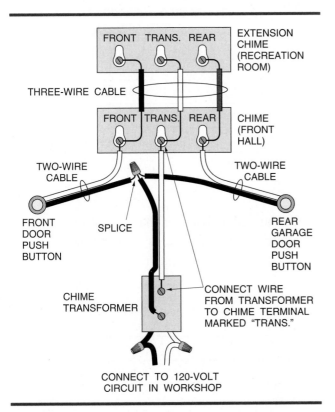

Figure 25-27 Circuit for chime installation.

TYPE OF CHIME	POWER CONSUMPTION
Standard two-note	10 watts
Repeating chime	10 watts
Internally lighted, two lamps	10 watts
Internally lighted, four lamps	15 watts
Combination chime and clock	15 watts
Motor-driven chime	15 watts
Electronic chime	15 watts
Programmable musical	15 watts

Table 25-1 Typical chimes with their power consumption

a chime with a built-in clock can be used. A four-conductor cable may be run to the extension chime so that the owner may install a clock-chime at this location also.

The plans show that the chime transformer is mounted into the knockout on one of the ceiling boxes in the utility room. The line-voltage connections are easy to make here. Some electricians prefer to mount the chime transformer on the top or side of the distribution panelboards. The electrician decides where to mount the transformer after considering the factors of convenience, economy, and good wiring practice.

Don't Conceal Boxes!

It may seem obvious, but the issue of accessibility bears repeating. Electrical equipment junction boxes and conduit bodies must be accessible, *314.29*. It is permitted to have certain electrical equipment such as a chime transformer, junction boxes and conduit bodies above a dropped lay-in ceiling because these are accessible by dropping out a ceiling panel. Never install electrical equipment, junction boxes, or conduit boxes above a permanently closed-in ceiling or within a wall. There may come a time when the electrical equipment and junction boxes need to be accessed. This includes chime transformers.

REVIEW

Note: Refer to the *Code* or the plans where necessary.

TELEVISION CIRCUIT

1. How many television outlets are installed in this residence? _____

2. Which type of television cable is commonly used and recommended? _____
   _____

3. What determines the design of the faceplates used? _____
   _____

4. What must be provided when installing a television outlet and receptacle outlet in one wall box? _____
   _____
   _____
   _____
   _____
   _____

5. From a cost standpoint, which system is more economical to install: a master amplifier distribution system or a multiset coupler? Explain the basic differences between these two systems. _____
   _____
   _____
   _____
   _____
   _____

6. How many wires are in the cable used between a rotor and its controller? _____

7. Digital satellite systems use an antenna that is approximately (18 in. [450 mm]) (36 in. [900 mm]) (72 in. [1.8 mm]) in diameter. Circle the correct answer.

8. List the requirements for cable television inside the house. _____

   _____

   _____

   _____

   _____

   _____

   _____

9. Which article of the *Code* references the requirements for cable community television installation? _____

10. It is generally understood that grounding and bonding together all metal parts of an electrical system and the metal shield of the cable television cable to the same grounding reference point in a residence will keep both systems at the same voltage level should a surge, such as lightning, occur. Therefore, if the incoming cable that has been installed by the CATV cable company installer has the metal shield grounded to a driven ground rod, does this installation conform to the *NEC®*? _____

    _____

11. All television satellites rotate above the earth in (the same orbit) (different orbits). Circle the correct answer.

12. Television satellites are set in orbit (10,000) (18,000) (22,245) miles above the earth, which results in their rotating around the earth at (precisely the same) (different) rotational speed as the earth rotates. This is done so that the satellite "dish" can be focused on a specific satellite (once) (one time each month) (whenever the television set is used). Circle the correct answer.

13. Which section of the *Code* prohibits supporting coaxial cables from raceways that contain light or power conductors? *NEC®* _____

14. When hooking up one CATV cable and another cable from an outdoor antenna to a receiver that has only one antenna input terminal, an _____ switch is usually installed.

TELEPHONE SYSTEM

1. How many locations are provided for telephones in the residence? _____

2. At what height are the telephone outlets in this residence mounted? Give measurement to center.

   _____

   _____

   _____

   _____

3. Sketch the symbol for a telephone outlet.

4. Is the telephone system regulated by the *NEC*®? _____

5. a. Who is to furnish the outlet boxes required at each telephone outlet? _____

   b. Who is to furnish the faceplates? _____

6. Who is to furnish the telephones? _____

7. Who does the actual installation of the telephone equipment? _____

8. How are the telephone cables concealed in this residence? _____

   _____

   _____

   _____

9. The point where the telephone company's cable ends and the interior telephone wiring meets is called the _____ point. The device installed at this point is called a _____ .

10. What are the colors contained in a four-conductor telephone cable assembly and what are they used for? _____

    _____

    _____

    _____

    _____

    _____

    _____

11. Itemize the *Code* rules for the installation of telephone cables in a residence. _____

    _____

    _____

    _____

12. If finger contact were made between the red conductor and green conductor at the instant a "ring" occurs, what shock voltage would be felt? _____

13. What section of the *Code* prohibits supporting telephone wires from raceways that contain light or power conductors? *NEC*® _____

14. The term *cross-talk* is used to define the hearing of the faint sound of voices in the background when you are using the telephone. Cross-talk can be reduced significantly by running (flat cables) (twisted pair cables) to the telephones. Circle the correct answer.

15. Which section of the *Code* prohibits telephone cables from being installed in the same box or enclosure or from being pulled into the same raceway as light and power circuits? _____

SIGNAL SYSTEM

1. What is a signal circuit? _____

   _____

2. What style of chime is used in this residence? _____

3. a. How many solenoids are contained in a two-tone chime? _____

   b. What closes the circuit to the solenoid of a chime? _____

4. Explain briefly how two notes are sounded by depressing one push button (when two solenoids are provided). _____

   _____

   _____

   _____

   _____

   _____

   _____

   _____

5. a. Sketch the symbol for a push button. _____

   b. Sketch the symbol for a chime. _____

6. a. At what voltage do residence chimes generally operate? _____

   b. How is this voltage obtained? _____

7. What is the maximum volt-ampere rating of transformers supplying Class 2 systems?

   _____

8. What two types of chime transformers for Class 2 systems are listed by UL?

   _____

   _____

9. Is the extension chime connected with the front hall chime in series or in parallel?

   _____

10. How many bell wires terminate at

    a. the transformer? _____

    b. the front hall chime? _____

    c. the extension chime? _____

    d. each push button? _____

11. a. What change in equipment may be necessary when more than one chime is connected to sound at the same time on one circuit? _____

    _____

    _____

    _____

    _____

    b. Why? _____

    _____

    _____

    _____

    _____

12. What type of insulation is usually found on low-voltage wires? _____

13. What size wire is installed for signal systems of the type in this residence? _____

14. a. How many wires are run between the front hall chime and the extension chime in the recreation room? _____

    b. How many wires are required to provide a "hot" low-voltage circuit at the extension chime? _____

15. Why is it recommended that the low-voltage secondary of the transformer be run to the front hall chime location and separate two-wire cables be installed to each push button? _____

    _____

16. a. Is it permissible to install low-voltage Class 2 systems in the same raceway or enclosure with light and proper wiring? _____

    b. Which *Code* section covers this? _____

17. a. Where is the transformer in the residence mounted? _____

    b. To which circuit is the transformer connected? _____

18. a. How many feet of two-conductor bell wire cable are required? _____

    b. How many feet of three-conductor bell wire cable are required? _____

19. How many insulated staples are needed for the bell wire if it is stapled every 24 in. (600 mm)? _____

20. Should low-voltage bell wire be pulled through the same holes in studs and joists that also contain nonmetallic-sheathed cable? _____

21. What section of the *Code* prohibits supporting fire alarm conductors from raceways that contain light or power conductors? *NEC®* _____ .

22. Complete this wiring diagram using colored pencils. The two-wire, low-voltage cables contain one red and one white conductor. Use yellow to represent the white conductors.

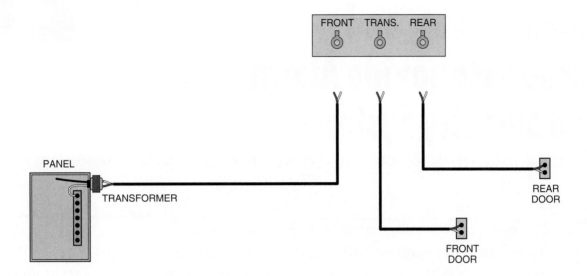

Smoke, Heat, and Carbon Monoxide Alarms, and Security Systems

OBJECTIVES

After studying this unit, the student should be able to

▶ understand the basics of *NFPA 72 (2002)*, the *National Fire Alarm Code*, *NFPA 720 (2003)*, the *Recommended Practice for the Installation of Household Carbon Monoxide Warning Equipment*, and *NFPA 70 (2005)*, the *National Electrical Code.*®

▶ understand the basics of smoke, heat, and carbon monoxide alarms.

▶ understand the location requirements for the installation of smoke, heat, and carbon monoxide alarms for *minimum* acceptable levels of protection.

▶ understand the location requirements for the installation of smoke and heat alarms that *exceed* the minimum acceptable levels of protection.

▶ discuss general requirements for the installation of security systems.

▶ be aware of important UL Standards covering fire warning equipment.

NATIONAL FIRE ALARM CODE

NFPA 72 and *NFPA 720* tells us "where to install" smoke and heat alarms.

NFPA 70 in *Article 760* of the *NEC*® tells us "how to" install the wiring for smoke and heat alarms.

NFPA 72 is the *National Fire Alarm Code.*® It presents the minimum requirements for the proper selection, installation, operation, and maintenance of fire warning equipment that will provide *reasonable* fire safety. The primary function of fire warning equipment is to provide a reliable means to notify the occupants of the presence of a threatening fire and the need to escape to a place of safety before such escape might be obstructed by smoke or other

delaying conditions in the normal path of the way out. *Chapter 11* specifically covers household fire warning equipment. Smoke alarms are required by *NFPA 72*. Heat detectors are not required but are recommended by *NFPA 72*.

Fire warning devices commonly used in a residence are heat detectors, Figure 26-1, and combination smoke alarms/detectors, Figure 26-2.

The more elaborate systems connect to a central monitoring customer service center through a telephone line. These systems offer instant contact with police or fire departments, a panic button for emergency medical problems, low-temperature detection, flood (high-water) detection, perimeter protection, interior motion detection, and other features.

Figure 26-1 Heat detector.

Figure 26-2 Combination smoke alarm/detector.

All detections are transmitted first to the company's customer service center, where personnel on duty monitor the system 24 hours a day, 7 days a week. They will verify that the signal is valid. After verification, they will contact the police department, the fire department, and individuals whose names appear on a previously agreed-upon list that the homeowner prepared and submitted to the company's customer service center.

NFPA Standards are not law until adopted by a city, state, or other governmental body. Some NFPA Standards are adopted totally, and others are adopted partly. Installers must be aware of all requirements in the locality in which they are doing work. Most communities require a special permit for the installation of fire warning equipment and security systems in homes and require registration of the system with the police and/or fire departments. In most instances, fire protection requirements are found in the building codes of a community and are not necessarily spelled out in the electrical code.

Pay particular attention to *11.5.1.1* of *NFPA 72*, which clearly states in part that smoke alarms shall be installed "*Where required by applicable laws, codes, or standards for the specified occupancy.*" This puts the responsibility on the local building department. It might seem confusing to have so many codes. Without question, you must become familiar with your local applicable laws, codes, or standards.

Most communities adopt building codes published by the International Code Council (ICC). These codes contain requirements for the installation of smoke alarms. Your local building department officials can explain what is required for new work and remodel work (alterations, repairs, and additions) where a permit is required. Whereas new work requires interconnected, hard-wired, battery backup smoke alarms, it might be acceptable on remodel jobs to install battery-powered-only smoke alarms for existing areas where interior walls or ceiling finishes are not removed to expose structural framing members. To install, interconnect, and hard-wire smoke alarms in these existing areas could result in damage to walls and ceilings, requiring patching, repainting, or other repairs—an uncalled for and tremendous expense. If there is access from an attic, crawl space, or basement, the inspector might require interconnected, hard-wired, battery backup smoke alarms if you can do the job without having to remove interior finishes. Check this out with your local inspector!

DEFINITIONS

Here are some very important definitions from *NFPA 72* and *NFPA 720*:

An *alarm* is "*A warning of fire danger.*"

A *detector* is "*A device suitable for connection to a circuit that has a sensor that responds to a physical stimulus such as heat or smoke.*"

A **smoke alarm** is "*A single or multiple station alarm responsive to smoke.*"

A **smoke detector** is "*A device that detects visible or invisible particles of combustion.*"

A **heat alarm** is "*A single or multiple station alarm responsive to heat.*"

A *heat detector* is "*A fire detector that detects either abnormally high temperature or rate of temperature rise, or both.*"

A *combination detector* is *"A device that either responds to more than one of the fire phenomenon or employs more than one operating principle to sense one of these phenomenon. Typical examples are a combination of a heat detector with a smoke detector or a combination rate-of-rise and fixed-temperature heat detector."**

A *single station alarm* is *"A detector comprising an assembly that incorporates a sensor, control components, and an alarm notification appliance in one unit operated from a power source either located in the unit or obtained at the point of installation."**

A *multiple station alarm* is *"A single station alarm capable of being interconnected to one or more additional alarms so that the actuation of one causes the appropriate alarm signal to operate in all interconnected alarms."**

A *dwelling unit* is *"A single unit, providing complete, independent living facilities for one or more persons, including permanent provisions for living, sleeping, eating, cooking, and sanitation. A single dwelling unit is a building consisting solely of one dwelling unit. A multiple dwelling is a building containing three or more dwelling units."**

A *living area* is *"Any normally occupiable space in a residential occupancy, other than sleeping rooms or rooms that are intended for combination sleeping/ living, bathrooms, toilet compartments, kitchens, closets, halls, storage or utility spaces, and similar areas."**

A *separate sleeping area* is *"An area of the family living unit in which the bedrooms (or sleeping rooms) are located. Bedrooms (or sleeping rooms) separated by other use areas, such as kitchens or living rooms (but not bathrooms), are considered separate sleeping areas."**

In this chapter, rather than going back and forth between the word "alarm" and "detector," we refer to fire warning devices as alarms, knowing full well that before an alarm can sound, the cause must be detected. Most home-type fire warning devices combine the detector and alarm in one device. The location requirements for smoke and heat alarms are pretty much the same.

---

*Reprinted with permission from NFPA 72-2002 and NFPA 720-2004.

SMOKE, HEAT, AND CARBON MONOXIDE ALARMS

This unit covers the basic requirements for protection in homes against the hazards of fire, heat, smoke, and carbon monoxide. Fire alarm systems for commercial installations are much more complicated than the typical household system.

Household fire alarm requirements are found in *Chapter 11* of *NFPA 72*, the *National Fire Alarm Code*. Specifics for one- and two-family dwellings are found in *11.5.1*.

Fire is the third leading cause of accidental death. Home fires account for the biggest share of these fatalities, most of which occur at night during sleeping hours. Rapidly developing high-heat fires and slow smoldering fires are the culprits. Both produce smoke and deadly gases.

Smoke, heat, and carbon monoxide alarms are installed in a residence to give the occupants *early warning* of the presence of fire or toxic fumes. Fires produce smoke and toxic gases that can overcome the occupants while they sleep. Most fatalities result from the inhalation of smoke and toxic gases, rather than from burns. Heavy smoke reduces visibility.

In nearly all home fires, detectable smoke precedes detectable levels of heat. People sleeping are less likely to smell smoke than people who are awake. The smell of smoke probably will not awaken a sleeping person. Therefore, smoke alarms are considered to be the primary devices for protecting lives.

Heat alarms DO NOT take the place of smoke alarms. Heat alarms are installed *in addition* to smoke detectors.

Another lifesaving device is the carbon monoxide (CO) alarm, which senses dangerous carbon monoxide emitted from a malfunctioning furnace or other source. *NFPA 720* is the *Recommended Practice for the Installation of Household Carbon Monoxide Warning Equipment*.

Home-type smoke, heat, and carbon monoxide devices as a rule have both the detector and the alarm in one device. There also are alarms that combine more than one type of sensor into one unit, such as smoke/heat, smoke/carbon monoxide, or carbon monoxide/explosive gases.

In larger, more complex commercial installations, detectors and alarms are generally separate devices that, when triggered, send a signal to a central control panel, which in turn sounds the general alarm.

As with all electrical installations, always install fire warning equipment that has been listed by a nationally recognized testing laboratory (NRTL).

DETECTOR TYPES

AC/Battery: Required in new construction. These conform to the *NFPA 72* requirement that alarms have two power supplies—regular household 120-volt ac for the primary source of power and 9-volt battery for the secondary source. The alarm "chirps" when the battery gets low. See Figure 26-3.

Interconnected: Required in new construction. Should any smoke, heat, or carbon monoxide alarm trigger, that alarm will sound and simultaneously send a signal to set off all other interconnected alarms. Check the manufacturers' instructions to determine how many units may be interconnected. Never connect more units than the number specified by the manufacturer. Never "mix" different manufacturer's devices. Testing laboratories do not perform tests using more than one manufacturer's devices in the test. The alarm "chirps" when the battery gets low. These units usually

Figure 26-3 An ac–dc smoke alarm that operates on 120 volts ac as the primary source of power and a 9-volt battery as the secondary source of power. In the event of a power outage, the alarm continues to operate on the battery. *Courtesy of* **BRK Electronics.**

have a small three-wire power connector—the wires connect to the field wiring and the other end has a connector that plugs into the back of the unit. See Figure 26-4. The pigtail

Figure 26-4 Interconnected smoke alarms. The two-wire branch-circuit is run to the first alarm, then three-wire nonmetallic-sheathed cable interconnects the other alarms. If one alarm in the series is triggered, all other alarms will sound off.

may have a red or orange-colored "interconnect" wire. The field wiring between alarms is three-wire nonmetallic-sheathed cable that contains a red insulated conductor that is used as the "interconnect" wire.

Plug-in: Not permitted in new construction. These alarms plug into a wall outlet that is not controlled by a wall switch. They operate on 120 volts as the primary source of power and on a 9-volt battery as the secondary source. The alarm "chirps" when the battery gets low.

Battery-Operated Only: Not permitted in new construction. These alarms operate on a 9-volt battery only. The alarm "chirps" when the battery gets low. These alarms are commonly used in existing homes.

For the Hearing Impaired: These alarms have a bright strobe light that provides a visual alarm for the hearing impaired. This complies with the requirements found in the Americans with Disabilities Act (ADA).

TYPES OF SMOKE ALARMS

Two common types of smoke alarms are the **photoelectric type** (sometimes called photoelectronic) and the **ionization type**. They usually contain an indication light to show that the unit is functioning properly. They also may have a test button that simulates smoke, so that when the button is pushed, the detector's smoke-detecting ability as well as its circuitry and alarm are tested.

Some alarms are tested with a magnet. Others are tested with a listed spray from an aerosol can.

Smoke alarms generally do not sense heat, flame, or gas. However, some smoke alarms can be set off by acetylene and propane gas.

Some alarms have a "hush button" that can temporarily silence a nuisance alarm for about 15 minutes.

Photoelectric Type

The photoelectric type of smoke alarm has a light sensor that measures the amount of light in a chamber. When smoke is present, an alarm sounds, indicating a reduction in light due to the obstruction of the smoke. This type of sensor detects smoke from burning materials that produce large quantities of smoke, such as furniture, mattresses, and rags. The photoelectric type of alarm is less effective for

gasoline and alcohol fires, which do not produce heavy smoke. This type of alarm can become more sensitive to smoke as it gets older.

Photoelectric alarms are less likely to sound an alarm from normal cooking unlike the ionization type.

There are two types of photoelectric devices: light obscuration and light scattering.

Ionization Type

The ionization type of alarm contains a low-level radioactive source (less than used in luminescent watch and clock dials), which supplies particles that ionize the air in the detector's smoke chamber. Plates in this chamber are oppositely charged. Because the air is ionized, an extremely small amount of current (millionths of an ampere) flows between the plates. Smoke entering the chamber impedes the movement of the ions, reducing the current flow, which triggers the alarm. This type often sets off a nuisance alarm because cooking routinely gives off small "invisible" smoke particles. This type of alarm can become more sensitive to smoke as it get older.

The ionization type of alarm is effective for detecting small amounts of smoke, as in gasoline and alcohol fires, which are fast-flaming with little or no smoke.

Some smoke alarms are available with both ionization and photoelectric sensors combined. Other smoke alarms are available with smoke and carbon monoxide detection capabilities or smoke and heat detection capabilities in one unit.

TYPES OF HEAT ALARMS

Heat alarms respond to heat—not smoke!

Three types of heat alarms are described as follows.

- Fixed temperature heat detectors sense a specific fixed temperature, such as 135°F (57°C) or 200°F (93°C). They shall have a temperature rating of at least 25°F (14°C) above the normal temperature expected, but not to exceed 50°F (28°C) above the expected temperature. Fixed temperature detectors are sometimes combined with smoke alarms and carbon monoxide alarms.

- Rate-of-rise heat detectors sense rapid changes in temperature (12°F to 15°F per minute) such as those caused by flash fires.

- Fixed/rate-of-rise temperature detectors are available as a combination unit.

- Fixed/rate-of-rise/smoke combination detectors are available in one unit.

INSTALLATION REQUIREMENTS

The following information includes specific recommendations for installing smoke alarms and heat alarms in homes. Remember that smoke alarms are the primary fire warning devices, and heat detectors are installed in addition to the required smoke alarms. As you continue reading, study Figure 26-5, Figure 26-6, Figure 26-7, Figure 26-8 and Figure 26-9. Complete data are found in *NFPA Standard No. 72* and in the instructions furnished by the manufacturer of the equipment.

The Absolute Minimum Level of Protection

Here is the absolute minimum level of protection for smoke alarms in conformance to *NFPA 72* for new residential construction.

- Install multiple station smoke alarms:
  1. in new construction in all sleeping rooms. This is not a requirement for existing one- and two family dwellings.
  2. outside each separate sleeping area, in the immediate vicinity of the sleeping rooms.

3. on each additional story of the dwelling unit, including basements but excluding crawl spaces and unfinished attics.
4. For existing dwellings, listed battery-powered smoke alarms that are not interconnected are generally permitted. This provides a reasonable level of protection as opposed to having no smoke alarms at all. However, your local code might require ac/dc dual-powered interconnected alarms.

- Smoke alarms must have at least two independent sources of power. The primary source is the 120-volt ac circuit, and the secondary (standby) source is the integral battery power supply.

Figure 26-5 Smoke alarms are required in all sleeping areas, outside all sleeping areas, and on every story. The logical location is near the center of the bedroom. However, this may be impossible because of a ceiling luminaire (fixture) and/or ceiling paddle fan. An alternate location is on the wall space above the door, not closer than 4 in. (102 mm) nor more than 12 in. (305 mm) from the ceiling.

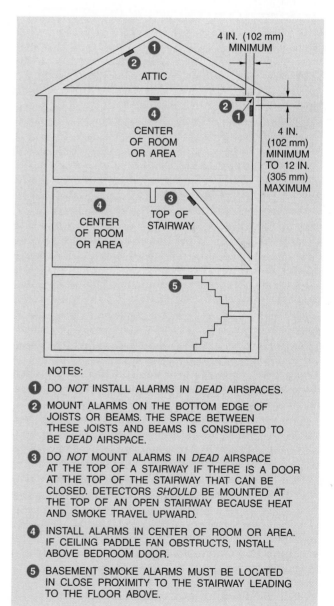

NOTES:

1. DO *NOT* INSTALL ALARMS IN *DEAD* AIRSPACES.

2. MOUNT ALARMS ON THE BOTTOM EDGE OF JOISTS OR BEAMS. THE SPACE BETWEEN THESE JOISTS AND BEAMS IS CONSIDERED TO BE *DEAD* AIRSPACE.

3. DO *NOT* MOUNT ALARMS IN *DEAD* AIRSPACE AT THE TOP OF A STAIRWAY IF THERE IS A DOOR AT THE TOP OF THE STAIRWAY THAT CAN BE CLOSED. DETECTORS *SHOULD* BE MOUNTED AT THE TOP OF AN OPEN STAIRWAY BECAUSE HEAT AND SMOKE TRAVEL UPWARD.

4. INSTALL ALARMS IN CENTER OF ROOM OR AREA. IF CEILING PADDLE FAN OBSTRUCTS, INSTALL ABOVE BEDROOM DOOR.

5. BASEMENT SMOKE ALARMS MUST BE LOCATED IN CLOSE PROXIMITY TO THE STAIRWAY LEADING TO THE FLOOR ABOVE.

Figure 26-6 Recommendations for the installation of heat and smoke alarms.

Figure 26-7 Do not mount smoke or heat alarms in the *dead* airspace where the ceiling meets the wall.

Figure 26-8 Installing smoke and heat detectors/alarms in peaked (cathedral) ceilings.

- Smoke alarms must be interconnected using multiple station devices so that when one alarm in the series triggers, all other alarms in the house will sound. Do not exceed the maximum number of alarms permitted by the manufacturer to be interconnected.

- The alarms shall provide a sound that is audible in all dwelling areas that can be occupied or used.

Figure 26-9 Installing smoke and heat detectors/alarms in sloped ceilings.

Where to Install

In *bedrooms* and *halls outside of sleeping areas*. See Figure 26-5.

In *other rooms and areas*. See Figure 26-6.

On *walls*—not closer than 4 in. (102 mm) but not farther than 12 in. (305 mm) from the adjoining ceiling. Do not install in dead airspace. See Figure 26-6 and Figure 26-7.

On *flat ceilings*—not closer than 4 in. (102 mm) from the adjoining wall. See Figure 26-6 and Figure 26-7.

On *peaked ceilings*—locate with 36 in. (914 mm) horizontally from the peak, but not closer than 4 in. (102 mm) vertically from the peak. A peaked ceiling might also be called a cathedral ceiling. Do not install in dead airspace. See Figure 26-8.

On *sloped ceilings*—a sloped ceiling is defined as having a rise greater than 1 ft in 8 ft (1 m in 8 m). Locate within 36 in. (914 mm) of the high side but not closer than 4 in. (102 mm) from the adjoining wall. Do not install in dead airspace. See Figure 26-9.

How to Wire Smoke and Heat Alarms

In new construction, smoke and heat alarms require both primary (120 volts) and secondary (battery) sources of power. The *National Fire Alarm Code* does not specify where to pick up the 120-volt supply.

NEC® 210.12 requires that *all* 125-volt, single-phase, 15- and 20-ampere electrical outlets in bedrooms be AFCI protected. This includes smoke alarms. Probably the simplest way to hard wire and interconnect smoke alarms and heat alarms when wiring new homes is to pick up one smoke alarm from one of the bedroom branch-circuits, then run the interconnecting wiring to all of the other required and optional smoke alarms. The electrical plans indicate the location of the required and optional smoke and heat alarms.

Some inspectors require a "lock-off" device over the handle of the circuit breaker that serves the alarms so that it will not unintentionally be turned off. This lock-off device is really not necessary. Should the breaker supplying the alarms get turned off intentionally or unintentionally, or if it trips due to trouble in the circuit, the battery backup would power the alarms. When the batteries get "low," the alarms start "chirping."

The usual way to connect alarms is to run the two-wire branch-circuit to the first alarm, then run three-wire nonmetallic-sheathed cable from the first alarm to all other alarms. The third conductor provides the interconnect feature that will sound all alarms should any one of the alarms activate. See Figure 26-4.

Summary of Do's and Don'ts

When installing detectors and alarms, ask yourself, "Where will the smoke and greatly heated air travel?" Because smoke and heated air rise, detectors and alarms must be in the path of the smoke and greatly heated air.

Figure 26-5, Figure 26-6, Figure 26-7, Figure 26-8, and Figure 26-9 illustrate some of the more important issues regarding smoke and heat alarms.

Here is a list of the *Do's* and *Don'ts* in no particular order of importance.

The Do's

1. *Do* install listed smoke alarms, and install them according to the manufacturers' instructions.

2. *Do* install smoke alarms on the ceiling, as close as possible to the center of the room or hallway. Note: Because ceiling paddle fans are often installed in bedrooms, install the alarms on the wall space above the bedroom door, not closer than 4 in (102 mm) nor more than 12 in. (305 mm) from the ceiling.

3. *Do* make sure that the path of rising smoke will reach the alarm when the alarm is installed in a stairwell. This is usually at the top of the stairway. The alarm must not be located in a dead airspace created by a closed door at the head of the stairway.

4. *Do* space according to the instructions furnished with the devices.

5. *Do* install the type that has a "hush" button if located within 20 ft (6.1 m) of a cooking appliance, or install a photoelectric type.

6. *Do* install smoke alarms at the high end of a room that has a sloped, gabled, or peaked ceiling where the rise is greater than 1 ft (305 mm) per 8 ft (2.6 m). Do not install in the dead airspace.

7. *Do* mount on the bottom of open joists or beams.

8. *Do* install a smoke alarm on the basement ceiling close to the stairway to the first floor.

9. *Do* install smoke alarms in split-level homes. A smoke alarm installed on the ceiling of an upper level can suffice for the protection of an adjacent lower level, if the two levels are not separated by a door. Better protection is to install alarms for each level.

10. *Do* install in new construction dual-powered smoke alarms that are hard wired directly to a 120-volt ac source and also have a battery. In existing homes, battery-powered alarms are the most common, but 120-volt ac alarms or dual-powered alarms can be installed. Remember, battery-powered alarms will not operate with dead batteries. Ac-powered alarms will not operate when the power supply is off.

11. *Do* install interconnected smoke alarms in new construction so that the operation of any alarm will cause all other alarms that are interconnected to sound.

12. *Do* install smoke alarms so they are not the only load on the branch-circuit. Connect smoke alarms to a branch-circuit that supplies other lighting outlets in habitable spaces. Because smoke alarms draw such a miniscule amount of current, they can easily be connected to a general lighting branch-circuit, such as a bedroom branch-circuit.

13. *Do* always consider the fact that doors, beams, joists, walls, partitions, and similar obstructions will interfere with the flow of smoke and heat, and, in most cases, create new areas needing additional smoke alarms and heat alarms.

14. *Do* make sure that the gap around the ceiling outlet box is sealed to prevent dust from entering the smoke chamber. This is a big problem for ceiling-mounted alarms.

15. *Do* clean smoke alarms according to the manufacturers' instructions.

16. *Do* consider that the maximum distance between heat alarms mounted on flat ceilings is 50 ft (15 m) and 25 ft (7.5 m) from the detector to the wall. This information is explained in detail by *NFPA 72*. Where obstructions such as beams and joists will interfere with the flow of heat, the 50-ft distance is reduced to 25 ft (7.5 m), and the 25-ft distance is reduced to 12½ ft (3.75 m).

17. *Do* install alarms after construction clean-up of all trades is complete and final.

18. *Do* install interconnected units of the same manufacturer. Different manufacturers' units may or may not be compatible.

19. *Do* connect all interconnected units to the same branch-circuit. Different circuits cannot be shared.

The Do Nots

1. *Do not* install smoke alarms in the dead airspace at the top of a stairway that can be closed off by a door.

2. *Do not* install within 36 in. (914 mm) of a door to a kitchen or to a bathroom containing a shower or tub.

3. *Do not* locate where the smoke alarm will be subject to temperature and/or humidity that exceeds the limitations stated by the manufacturer.

4. *Do not* install within a 36-in. (914-mm) horizontal path from a supply hot air register. Install outside the direct flow of air from these registers.

5. *Do not* install within a horizontal path 36 in. (914 mm) from the tip of the blade of a ceiling-suspended paddle fan.

6. *Do not* install where smoke rising in a stairway could be blocked by a closed door or an obstruction.

7. *Do not* place the edge of a ceiling-mounted smoke alarm or heat alarm closer than 4 in. (102 mm) from the wall.

8. *Do not* place the top edge of a wall-mounted smoke alarm or heat alarm closer than 4 in. (102 mm) from the ceiling and farther than 12 in. (305 mm) down from the ceiling. Some manufacturers recommend placement not farther down than 6 in. (153 mm).

9. *Do not* install smoke alarms or heat detectors on an outside wall that is not insulated, or one that is poorly insulated. Instead mount the alarms on an inside wall.

10. *Do not* install smoke alarms or heat alarms on a ceiling where the ceiling will be excessively cold or hot. Smoke and heat will have difficulty reaching the alarms. This could be the case in older homes that are not insulated or are poorly insulated. Instead mount the alarms on an inside wall.

11. *Do not* install smoke alarms or heat alarms where the ceiling meets the wall because this is considered dead airspace where smoke and heat may not reach the detector.

12. *Do not* connect smoke alarms or heat alarms to wiring that is controlled by a wall switch.

13. *Do not* install smoke alarms or heat alarms where the relative humidity exceeds 85 percent, such as in bathrooms with showers, laundry areas, or other areas where large amounts of visible water vapor collect. Check the manufacturer's instructions.

14. *Do not* install smoke alarms or heat alarms in front of air ducts, air conditioners, or any high-draft areas where the moving air will keep the smoke or heat from entering the detector.

15. *Do not* install smoke alarms or heat alarms in kitchens where the accumulation of house-

hold smoke can result in setting off an alarm, even though there is no real hazard. The person in the kitchen will know why the alarm sounded, but other people in the house may panic. This problem exists in multifamily dwelling units, where unwanted triggering of the alarm in one dwelling unit might cause people in the other units to panic. The photoelectric type may be installed in kitchens but must not be installed directly over the range or cooking appliance. A better choice in a kitchen is to install a heat alarm.

16. *Do not* install smoke alarms where the temperature can fall below 32°F (0°C) or rise above 120°F (49°C) unless the detector is specifically identified for this application.

17. *Do not* install smoke alarms in garages where vehicle exhaust might set off the detector. Instead of a smoke alarm, install a heat detector.

18. *Do not* install smoke alarms in airstreams that will pass air originating at the kitchen cooking appliances across the alarm. False alarms will result.

19. *Do not* install smoke alarms or heat detectors on ceilings that employ radiant heating.

20. *Do not* install smoke alarms or heat alarms in a recessed location.

21. *Do not* connect smoke alarms or heat alarms to a switched circuit or a circuit controlled by a dimmer.

MAINTENANCE AND TESTING

Once installed, smoke, heat, and carbon monoxide alarms must be tested periodically and maintained (blow out accumulated dust) to make sure they are operating properly. This is a requirement of *NFPA 72*, which states, "*Homeowners shall inspect and test smoke alarms and all connected appliances in accordance with the manufacturer's instructions at least monthly*."

When to replace: In conformance to *10.4.6* and *11.8.1.4.(5)* in *NFPA 72*, smoke alarms "*shall be replaced when they fail to respond to tests*" and

"*shall not remain in service longer than 10 years from the date of installation*."

Note: The ten-year mandatory replacement requirement is often overlooked. This requirement is based on studies of tens of thousands of smoke alarms in operation to determine the acceptable number of failures over many years. Code committees and manufacturers analyzed the data and agreed that ten years in service provided a reasonable lifetime. The ten-year replacement program, along with regular testing per the manufacturer's instruction results in very few homes going unprotected for any extended period of time. Hard-wired and battery-operated alarms are equally affected by age.

The expected life of lithium batteries used in some alarms is said to be ten years.

After You Install the System, Then What?

After installing the system, there is still more to do. *NFPA 72* requires that the installer of the fire warning system provide the homeowner with the following information:

1. An instruction booklet illustrating typical installation layouts.

2. Instruction charts describing the operation, method, and frequency of testing and maintenance of fire warning equipment.

3. Printed information for establishing an emergency evacuation plan.

4. Printed information to inform owners where they can obtain repair or replacement service, and where and how parts requiring regular replacement, such as batteries or bulbs, can be obtained within 2 weeks.

5. Information that, unless otherwise recommended by the manufacturer, smoke alarms shall be replaced when they fail to respond to tests.

6. Instruction that smoke alarms shall not remain in service longer than ten years from the date of installation.

This information is usually part of the instructions furnished by the manufacturer of the alarms.

Exceeding Minimum Levels of Protection

The following are recommendations for attaining levels of fire warning protection that exceed the minimum level stated previously and include guidelines for installing smoke and heat detectors in a home.

- Install smoke alarms in *all* rooms, basements, hallways, heated attached garages, and storage areas. Installing alarms in these locations will increase escape time, particularly if the room or area is separated by a door(s). In some instances, smoke alarms are installed in attics and crawl spaces.

- Install heat alarms in kitchens because conventional smoke alarms can nuisance sound an alarm.

- Install heat alarms in garages because gasoline fires give off little smoke.

- Consider special alarms that light up or vibrate for occupants who are hard of hearing.

- Consider smoke alarms that have an "escape" light.

- Consider low temperature alarms (i.e., 45°F or 7°C) that can detect low temperatures should the heating system fail. The damage caused by frozen water pipes bursting can be extremely costly.

Building Codes

In most instances the various building codes published by the International Code Council (ICC) (formerly BOCA, ICBO, and SBCCI) adopt *NFPA 72* and *NFPA 720* by reference, but it would be wise to check with your local AHJ to find out if there are any differences between these codes that might affect your installation. See Unit 1 for information about these building code organizations. Also refer to the Web site list following the Appendix of this text.

Many companies specialize in installing complete fire and security systems. They also offer system monitoring at a central office and can notify the police or fire department when the system gives the alarm. How elaborate the system should be is up to the homeowner.

The ICC Code in 1202.5 requires that smoke alarms be connected to a circuit that also supplies lighting outlets in habitable spaces.

Manufacturers' Requirements

Here are a few of the responsibilities of the manufacturers of smoke alarms and heat alarms. Complete data are found in *NFPA 72*.

- The power supply must be capable of operating the signal for at least 4 minutes continuously.

- Battery-powered units must be capable of a low-battery warning "chirp" of at least one chirp per minute for 7 consecutive days.

- Direct-connected 120-volt ac alarms must have a visible indicator that shows "power-on."

- Alarms must not signal when a power loss occurs or when power is restored.

Note: In *4.3.1* in *NFPA 72*, we find that we must:

- always install fire alarm equipment that is "Listed for the purpose."

- always install fire alarm equipment in accordance with the manufacturers' installation and maintenance instructions.

In this residence, the smoke alarms and heat detectors are connected to Circuit A17 located in Panel "A" in the Workshop. Run two conductors (the circuit) to the first alarm, then interconnect the other alarms by running three conductors between all other detectors.

According to the plans for this residence, smoke alarms are located in:

Bedrooms	3
Hall between bedrooms	1
Top of basement stairs	1
Recreation room	1
Workshop	1

Heat alarms are located in:

Kitchen	1
Garage	1
Total	9

CARBON MONOXIDE ALARMS

Carbon monoxide alarms are installed in addition to smoke and heat alarms!

Carbon monoxide (CO) is referred to as the "silent killer"! It is responsible for more deaths that any other single poison.

When a heat or smoke alarm is triggered, it is generally not difficult to determine the source of the problem. But when a carbon monoxide alarm goes off, it is hard to determine the source.

Carbon monoxide is odorless, colorless, and tasteless—undetectable by any of a person's five senses—taste, smell, sight, touch, hearing, but is highly poisonous. Carbon monoxide replaces oxygen in the bloodstream, resulting in brain damage or total suffocation.

Carbon monoxide is produced by the incomplete burning of fuels. Common sources of carbon monoxide in a home can come from a malfunctioning furnace, gas appliances, kerosene heaters, automobile or other gas engines, charcoal grills, fireplaces, or a clogged chimney.

Symptoms of carbon monoxide poisoning are similar to flulike illnesses such as dizziness, fatigue, headaches, nausea, diarrhea, stomach pains, irregular breathing, and erratic behavior.

Carbon monoxide has about the same characteristics as air. It moves about just like air.

Carbon monoxide alarms detect carbon monoxide from any source of combustion. They are not designed to detect smoke, heat, or gas. The Consumer Product Safety Commission (CPSC) recommends that at least one carbon monoxide alarm be installed in each home, and preferably one on each floor of a multilevel house. Some cities require this. If only one carbon monoxide alarm is installed, it should be in the area just outside individual bedrooms.

Carbon monoxide alarms monitor the air in the house and sounds a loud alarm when carbon monoxide above a predetermined level is detected. They provide early warning before deadly gases build up to dangerous levels.

For residential applications, carbon monoxide alarms are available for hard wiring (direct connection to the branch-circuit wiring), plug-in (the male attachment plug is built into the back of the detector for plugging directly into a 120-volt wall receptacle outlet), and combination 120-volt and battery-operated units. They are available with the interconnect feature for interconnecting with smoke and heat alarms. Some carbon monoxide alarms also have an explosive gas sensor.

NFPA 720 is the *Recommended Practice for the Installation of Household Carbon Monoxide (CO) Warning Equipment.* This standard contains the installation recommendations for carbon monoxide alarms. Here is a summary of these recommendations.

Installation "Do's" for Carbon Monoxide Alarms

1. *Do* install carbon monoxide alarms that are tested and listed in conformance to UL Standard 2034.

2. *Do* install carbon monoxide alarms according to the manufacturer's instructions.

3. *Do* use household electricity as the primary source of power. In existing homes, monitored battery units are permitted.

4. *Do* install carbon monoxide alarms in or near bedrooms and living areas.

5. *Do* install carbon monoxide alarms in locations where smoke alarms are installed.

6. *Do* test carbon monoxide alarms once each month or as recommended by the manufacturer.

7. *Do* remove carbon monoxide alarms before painting, stripping, wallpapering, or using aerosol sprays. Store in a plastic bag until the project is completed.

8. *Do* carefully vacuum carbon monoxide alarms once each month or as recommended by the manufacturer.

9. *Do* interconnect alarms where multiple alarms are installed.

Installation "Do Nots" for Carbon Monoxide Alarms

1. *Do not* connect carbon monoxide alarms to a switched circuit or a circuit controlled by a dimmer.

2. *Do not* install carbon monoxide alarms closer than 6 in. (150 mm) from a ceiling. This is considered dead airspace.

3. *Do not* install carbon monoxide alarms in garages, kitchens, or furnace rooms. This could lead to nuisance alarms, may subject the detector to substances that can damage or contaminate the alarm, or the alarm might not be heard.

4. *Do not* install carbon monoxide alarms within 15 ft (4.5 m) of a cooking or heating gas appliance.

5. *Do not* install carbon monoxide alarms in dusty, dirty, or greasy areas. The sensor inside of the alarm can become coated or contaminated.

6. *Do not* install carbon monoxide alarms where the air will be blocked from reaching the alarm.

7. *Do not* install carbon monoxide alarms in dead airspaces such as in the peak of a vaulted ceiling or where a ceiling and wall meet.

8. *Do not* install carbon monoxide alarms where the detector will be in the direct airstream of a fan.

9. *Do not* install carbon monoxide alarms where temperatures are expected to drop below 40°F (4°C) or get hotter than 100°F (38°C). Their sensitivity will be affected.

10. *Do not* install carbon monoxide alarms in damp or wet locations such as showers and steamy bathrooms. Their sensitivity will be affected.

11. *Do not* mount carbon monoxide alarms directly above or near a diaper pail. The high methane gas will cause the detector to register, and this is not carbon monoxide.

12. *Do not* clean carbon monoxide alarms with detergents or solvents.

13. *Do not* spray carbon monoxide alarms with hair spray, paint, air fresheners, or other aerosol sprays.

FIRE ALARM SYSTEMS

These are the more comprehensive installations involving a central control panel, alarms, sensors, detectors as covered in the *National Fire Alarm Code, NFPA 72, Chapters 1* through *10*.

Fire alarm systems are also covered in *Article 760 of the NEC.*

NEC 760.2 defines a fire alarm circuit as: "*The portion of the wiring system between the load side of the overcurrent device or the power-limited supply and the connected equipment of all circuits powered and controlled by the fire alarm system. Fire alarm circuits are classified as either non power-limited or power-limited.*"*

---

*Reprinted with permission from NFPA 70-2005.

Fire-protective signaling systems installed in homes have power-limited fire alarm (PLFA) circuits where the power output is limited by the listed power supply for the system. Here are the key requirements in *Article 760, Part III*:

- Must have power output capabilities limited to the values specified in *Chapter 9, Table 12(A)* and *Table 12(B)*. Refer to *760.41*. Residential fire-protective signaling systems are generally "Power-limited."

- Do not connect the equipment to a GFCI-protected branch-circuit, *760.41*.

- Wiring on the supply side of the equipment is installed according to the conventional wiring methods found in *Chapter 1* through *Chapter 4* of the *NEC.* Refer to *760.51*.

- The branch-circuit overcurrent protection supplying the fire alarm system shall not exceed 20 amperes, *760.51*.

- Cables may be run on the surface or be concealed and shall be adequately supported and protected against physical damage, *760.52(B)(1)*.

- When run exposed within 7 ft (2.1 m) of the floor, the cable shall be securely fastened at intervals not over 18 in. (450 mm), *760.52(B)(1)*.

- Cables shall not be run in the same raceway, cable, outlet box, device box, or be in the same enclosure as light and power conductors unless separated by a barrier, or introduced solely to connect that particular piece of equipment. Refer to *760.55*. This is also a ruling in *725.55*.

- Cables shall not be supported by taping, strapping, or attaching by any means to any electrical conduits, or other raceways, *760.58*.

- Cables shall be supported by structural components, *760.8*.

- Wiring on the load side shall be insulated solid or stranded copper conductors of the types listed in *Table 760.82(A)*.

- *Table 760.61* shows different types of cable that can be substituted for one another when necessary.

- If installed in a duct or plenum, the cable must be plenum rated, *760.61(A)*.

- Type FLP cable is permitted in one- and two-family dwellings, *760.61(B)(3)*.
  - Conductors in cable shall not be smaller than 26 AWG. Single conductors shall not be smaller than 18 AWG. Refer to *760.57* and *760.82(B)*.
  - Cable minimum voltage rating is 300 volts, *760.82(C)*.
- Remove all abandoned cables. This is extremely important since abandoned cables provide fuel for a fire. Removing abandoned cables is required by *725.2, 725.3(B), 725.61(E), 760.2, 760.3(A), 800.2, 800.3(C), 820.2,* and *820.3*.

It makes a lot of sense to contact your local inspection authority before installing a fire warning system to be sure the system you intend to install meets the requirements of the local codes! Contractors who specialize in this kind of work are well aware of the *Code* requirements and will install the proper types of cables.

SECURITY SYSTEMS

It is beyond the scope of this text to cover all types of residential security systems. What follows is a typical system.

Professionally installed and homeowner installed security systems can range from a simple system to a very complex one. Features and options include master control panels, remote controls, perimeter sensors for doors and windows, motion sensors, passive infrared sensors, wireless devices, interior and exterior sirens, bells, electronic buzzers, strobe lights to provide audio (sound) as well as visual detection, heat sensors, carbon monoxide alarms, smoke alarms, glass break sensors, flood sensors, low temperature sensors, and lighting modules. Figure 26-10 shows some of these devices. Figure 26-11 shows a schematic one-line diagram for a typical security system.

Monitored security systems are connected by telephone lines to a central station, a 24-hour monitoring facility. If the alarm sounds, security professionals will contact the end user to verify the alarm before sending the appropriate authorities (fire, police, or ambulance) to the home.

The decision about how detailed an individual residential security system should be generally begins with a meeting between the homeowner and the security consultant or licensed electrical contractor before the actual installation. Most security system installers, consultants, and licensed electrical contractors are familiar with a particular manufacturer's system.

The systems are usually explained and even demonstrated during an in-home security presentation. They are then professionally installed by the security system company or licensed electrical contractor.

Do-it-yourself kits that contain many of the features provided by the professionally installed systems can be found at home centers, hardware stores, electrical distributors, and electronics stores. Depending on the system, it may or may not be possible to connect to a central monitoring facility.

The wiring of a security system consists of small, easy-to-install low-voltage, multiconductor cables made up of 18 AWG conductors. These conductors should be installed after the regular house wiring is completed to prevent damage to these smaller cables. Usually the wiring can be done at the same time as the chime wiring is being installed.

Security system wiring comes under the scope of *Article 725*.

When wiring detectors such as door entry, glass break, floor mat, and window foil detectors, circuits are electrically connected in series so that if any part of the circuit is opened, the security system will detect the open circuit. These circuits are generally referred to as "closed" or "closed loop." Alarms, horns, and other signaling devices are connected in parallel, because they will all signal at the same time when the security system is set off. Heat detectors and smoke detectors are generally connected in parallel because any of these devices will "close" the circuit to the security master control unit, setting off the alarms.

The instructions furnished with all security systems cover the installation requirements in detail, alerting the installer to *Code* regulations, clearances, suggested locations, and mounting heights of the systems' components.

Always check with the local electrical inspector to determine if there are any special requirements in your locality relative to the installation of security systems.

The wireless key allows the end user to disarm and arm the system with the press of a button. There is no need to remember a security code and it can also be used to operate lights and appliances.

Monitored smoke alarms are on 24 hours a day, even if the burglary protection is turned off.

Magnetic contacts are often located at vulnerable entry points such as doors.

Sirens, such as these placed inside the home, alert families to emergencies.

Motion detectors are interior security devices that sense motion inside the home.

A wireless panic button allows the user to press a button to signal for help.

The wireless keypad is portable so you can arm or disarm the system from the yard or driveway.

Glassbreak detectors, available in a variety of sizes, will sound the alarm when they detect the sound of breaking glass.

Figure 26-10 *Photos courtesy of* First Alert Professional Security Systems. First Alert is a Registered Trademark of the First Alert Trust.

Sprinkling Systems Code

Home sprinkling systems are covered by *NFPA 13D*, which is the *Standard for the Installation of Sprinkler Systems in One- and Two-Family Dwellings and Manufactured Homes*.

Standards Relating to Fire, Smoke, Carbon Monoxide, and Security Devices

UL 217—Standard for Single and Multiple Station Alarms

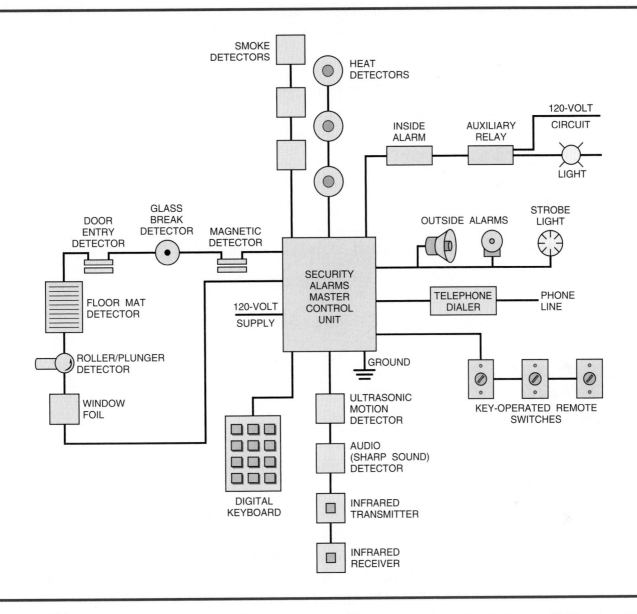

Figure 26-11 Diagram of typical residential security system showing some of the devices available. Complete wiring and installation instructions are included with these systems. Check local code requirements in addition to following the detailed instructions furnished with the system. Most of the interconnecting conductors are 18 AWG.

UL 268—Standard for Smoke Detectors for Fire Protective Signaling Systems

UL 365—Standard for Police Station Connected Burglar Alarm Units

UL 521—Standards for Heat Detectors for Fire Protection Signaling Systems

UL 539—Standard for Single and Multiple Station Heat Detectors

UL 609—Standard for Local Burglar Alarm Units and Systems

UL 827—Standard for Central Station Alarm Services

UL1023—Standard for Household Burglar Alarm System Units

UL 1610—Standard for Central Station Burglar Alarm Units

UL 1641—Standard for Installation and Classification of Residential Burglar Alarm Systems

UL 2034—Standard for Single and Multiple Station Carbon Monoxide Alarms

REVIEW

1. In your own words, explain the terms *alarm* and *detector*.

   _____

   _____

   _____

2. The requirements for household fire alarm systems is found in what *NFPA* Code and in what chapter of this Code?

   _____

3. Do heat alarms and carbon monoxide alarms take the place of smoke alarms? Explain.

   _____

   _____

4. Name two basic ways that smoke alarms are powered.

   _____

   _____

5. Name the two types of smoke alarms.

   _____

   _____

6. List the absolute minimum level of smoke alarm protection in a new one-family dwelling.

   _____

   _____

7. Circle the correct answer from the following statements.
   a. Smoke alarms installed in new one- and two-family homes shall be battery operated only so as not to be affected by a power outage.
   b. Smoke alarms installed in new one- and two-family homes shall be dual powered by a 120-volt circuit and a battery.

8. Circle the correct answer from the following statements.
   a. Smoke alarms installed in new one- and two-family homes shall be interconnected so that if any one of them is triggered, all other alarms will also sound off.
   b. Smoke alarms installed in new one- and two-family homes shall not be interconnected so that each alarm will operate independently of all other alarms in the home.

9. Circle the correct answer from the following statements.
   a. Always install smoke and fire alarms in dead airspaces.
   b. Never install smoke and fire alarms in dead airspaces.

10. Circle the correct answer from the following statements.
    a. Mount wall-mounted smoke alarms so that the top edge is not closer than 4 in. (102 mm) from the ceiling and not more than 12 in. (305 mm) from the ceiling.
    b. Mount wall-mounted smoke alarms anywhere on the wall but not lower than 36 in. (914 mm) from the ceiling.

11. Cooking and baking in the kitchen can produce quite a bit of smoke. The best choice for a smoke alarm in the kitchen is the photoelectric type. The better choice would be to install a   (heat alarm) (carbon monoxide alarm). Circle the correct answer.

12. An important but often overlooked requirement in the *National Fire Alarm Code, NFPA 72* is that alarms must be replaced after being in service for: (Circle the correct answer.)

    a. not more than 5 years

    b. not more than 10 years

    c. not more than 15 years.

13. Carbon monoxide is _____, and is _____, _____, and _____.

14. Circle the correct answer from the following statements.

    a. Carbon monoxide is heavier than air.

    b. Carbon monoxide will always rise to the ceiling.

    c. Carbon monoxide will drop to the floor.

    d. Carbon monoxide is about the same weight as air.

15. The more complex household fire alarm systems are covered in what article of the *National Electrical Code®*? Circle the correct answer.

    a. *Article 72*

    b. *Article 310*

    c. *Article 760*

16. Circle the correct answer from the following statements.

    a. Always connect a fire alarm circuit or system to a power supply that has ground-fault circuit interrupter (GFCI) protection.

    b. Never connect a fire alarm circuit or system to a power supply that has ground-fault circuit interrupter (GFCI) protection.

    c. It makes no difference whether or not the branch-circuit is GFCI or AFCI protected because the alarms are required to have battery backup power. In the event of a loss of power for whatever reason, the alarms would still be operative.

17. Circle the correct answer from the following statements.

    a. Conductors for fire alarm systems shall not be installed in the same raceways, cables, or electrical boxes as the light and power conductors.

    b. It is permissible to install fire alarm conductors in the same raceways, cables, or electrical boxes as the light and power conductors because the conductors for fire alarm systems covered in *Article 760* are small and would easily fit into other electrical raceways and boxes.

18. Fire alarm cable generally used in more complex residential fire alarm and security systems is marked   (FLP) (low-voltage bell wire) (THHN). Circle the correct answer.

19. When installing a fire alarm system or a complete security system package in a home, always follow the installation instructions from:

    a. the manufacturer of the product and applicable codes

    b. your neighbor, because he knows a lot about codes and standards

    c. the man at the home center who sold you the product

Service-Entrance Equipment

OBJECTIVES

After studying this unit, the student should be able to

▶ understand the *NEC®* terminology and requirements for installing all types of residential electrical services.

▶ calculate the proper size of residential service conductors and raceways.

▶ understand and install the required grounding and bonding for residential electrical services.

▶ understand the meaning of the term "UFER" ground.

▶ know the *NEC®* requirements for grounding electrical equipment in a separate building.

▶ calculate the cost of using electricity.

▶ understand the meaning of conductor "withstand rating."

This chapter covers a lot of ground. Mastering the material in this chapter is quite an accomplishment. Be patient, understanding, and thorough as you tackle it. Congratulate yourself when you have completed this chapter.

An electric service is required for all buildings containing an electrical system and receiving electrical energy from a utility company. The *NEC®* defines the term **service** as "*the conductors and equipment for delivering electric energy from the serving utility to the wiring system of the premises served.*" The point where the utility's supply ends and the customer's premise wiring begins is called the *service point*. The utility company generally must be contacted to determine where they want the meter to be located.

OVERHEAD SERVICE

The *Code* defines a **service drop** as the overhead service conductors, including splices, if any, which are connected from an outdoor support to the service-entrance conductors at the structure.

An overhead service includes the service raceways, fittings, meter, meter socket, the main service disconnecting means, and the service conductors between the main service disconnecting means and the point of attachment to the utility's service-drop conductors. Usually, residential-type watt-hour meters are located on the exterior of a building. In some cases, the entire service-entrance equipment may be mounted outside the building. This includes the watt-hour meter and the disconnecting means. Figure 27-1 illustrates the *Code* terms for the various components of a service entrance. Figure 27-2 illustrates the service entrance, main panel, subpanel, and grounding for this service.

Overhead service-drop conductors might be attached to a through-the-roof mast-type service, or to the side of the house. In either case, proper clearance is required above the ground, roof, fences, windows, sidewalks, decks, and so on.

Figure 27-1   *Code* terms for services. See Figure 27-2 and Figure 27-7 for grounding requirements.

MAST-TYPE SERVICE

The mast service, Figure 27-1, is a commonly used method of installing an overhead service entrance. The mast service is often used on buildings with low roofs, such as ranch-style dwellings, to ensure adequate clearance between the ground and the lowest service conductor.

The service raceway is run through the roof, as shown in Figure 27-1, using a roof flashing and neoprene seal fitting, as illustrated in Figure 27-3(A). This fitting keeps water from seeping in where the raceway penetrates the roof. The conduit is securely fastened to comply with code requirements. Many types of fittings are readily available at electrical supply houses for securely fastening raceways and other electrical equipment to wood, brick, masonry, and other surfaces. Figure 27-3(B) shows a typical through-the-wall support fitting.

Clearance Requirements for Mast Installations

Several factors determine the maximum length of conduit that can be installed between the roof support and the point where the service-drop conductors are attached. These factors are the service-drop length, the system voltage, the roof pitch, and

METER PEDESTAL AND COPPER
S.E. CONDUCTORS TWO 2/0 AWG THW,
THWN, XHHW ONE 1 AWG BARE
NEUTRAL FURNISHED AND INSTALLED
BY ELECTRICAL CONTRACTOR

METER

THESE
CONDUCTORS
FURNISHED
AND INSTALLED
BY UTILITY

PAD-MOUNT
TRANSFORMER

COPPER CONDUCTORS–
THREE 3 AWG THHN OR THWN

TRADE SIZE 1 EMT

PANEL B

PANEL A

125 A
M.L.O.

EQUIPMENT
GROUNDING
CONDUCTOR
BUS

200
A
MAIN

MAIN
BONDING
JUMPER

NEUTRAL BUS

NEUTRAL BUS
INSULATED FROM
ENCLOSURE

GROUNDING ELECTRODE CONDUCTOR–
4 AWG COPPER, *TABLE 250.66.*

GROUNDING ELECTRODE CONDUCTOR
TO A SUPPLEMENTAL GROUND ROD
NEED NOT BE LARGER THAN 6 AWG
COPPER, *250.53(E).*

SUPPLEMENTAL GROUNDING ELECTRODE –
MINIMUM 20 FT (6.0 m) OF 4 AWG
BARE COPPER ENCASED IN BOTTOM OF
CONCRETE FOOTING, *250.52(A)(3).*

CONNECT GROUNDING ELECTRODE CONDUCTOR
TO FIRST 5 FT (1.52 m) OF UNDERGROUND
METAL WATER PIPE. THE FIRST 5 FEET MAY
INCLUDE THE WATER METER. BOND AROUND
THE WATER METER, *250.52(A)(1).*

5 FT (1.52 m)

A METAL UNDERGROUND WATER PIPING SYSTEM
USED AS THE GROUNDING ELECTRODE MUST BE
SUPPLEMENTED, *250.53(D)(2).*

DRIVEN GROUND ROD SERVING AS SUPPLEMENTAL
GROUNDING ELECTRODE, *250.52(A)(5), 250.53(A),
250.53(D)(2), 250.53(G), 250.56.*

SUPPLEMENTAL GROUND ROD MAY BE BONDED
TO THE GROUNDING ELECTRODE CONDUCTOR,
THE GROUNDED SERVICE-ENTRANCE CONDUCTOR,
THE NONFLEXIBLE GROUNDED SERVICE-ENTRANCE
RACEWAY, OR ANY GROUNDED SERVICE-ENTRANCE
ENCLOSURE, *250.53(D)(2).*

DO NOT USE THE INTERIOR METAL
WATER PIPING BEYOND THE FIRST
5 FT (1.52 m) (FROM WHERE IT
ENTERS THE BUILDING) AS A
"CONDUCTOR" TO INTERCONNECT
THE SUPPLEMENTAL ELECTRODES
(ROD OR CONCRETE ENCASED)
AND THE UNDERGROUND METAL
WATER PIPE ELECTRODE, *250.50
AND 250.50(A)(1).*

Figure 27-2   The service-entrance, main panel, subpanel, and grounding for the service in this residence. Note that because the integrity of underground water pipes is questionable, the *Code* in *250.53(D)(2)* requires a supplemental grounding electrode. Shown are two common types of supplemental grounding electrodes: a ground rod and a concrete-encased grounding electrode. In many instances, the concrete-encased grounding electrode is better than the water pipe ground, particularly since the advent of nonmetallic (PVC) water piping systems.

the size and type of conduit (aluminum or steel), as shown in Figure 27-4.

Service Mast as Support *(230.28)*

The bending force on the conduit increases with an increase in the distance between the roof support and the point where the service-drop conductors are attached. The pulling force of a service drop on

a mast service conduit increases as the length of the service drop increases. As the length of the service drop decreases, the pulling force on the mast service conduit decreases.

If extra support is not to be provided, the mast service conduit must be at least 2 in. (50 mm) in diameter. This size prevents the conduit from bending due to the strain of the service-drop conductors.

Figure 27-3 **(A)** is a neoprene seal with metal flashing to use where the service raceway comes through the roof. **(B)** is a through-the-wall support for securing the service raceway. **(C)** is a clamp that fastens to the service raceway to which a guy wire can be attached for support of the service riser. **(D)** is an insulator that can be screwed through the roof into a roof truss or rafter. A guy wire is attached to it and to the clamp **(C)**. Commonly referred to as a "sill plate," **(E)** is used to protect service-entrance cable where it enters the building.

Figure 27-4 **Clearance requirements for service-drop conductors passing over residential roofs, *230.24(A)*, where voltage between conductors does not exceed 300 volts.**

If extra support is needed, it is usually in the form of a guy wire attached to a mast fitting, as in Figure 27-3(C), and to a roof fitting, Figure 27-3(D). Again, many support fittings and devices are available at electrical supply houses.

When service-entrance cable is used, a fitting of the type shown in Figure 27-3(E) can be used to protect the cable from physical damage where it penetrates the outside wall of the house. When used as a through-the-roof service mast, rigid metal conduit or intermediate metal conduit is not required to be supported within 3 ft (900 mm) of the service head, *344.30(A)* and *342.30(A)*. The roof mast kit provides adequate support when properly installed. A guy wire might be necessary if the service-drop conductors are long or heavy.

Only power service-drop conductors are permitted to be attached to and supported by the service mast, *230.28*. Refer to Figure 27-5.

All fittings used with raceway-type service masts must be identified for use with service masts, *230.28*.

Consult the utility company and electrical inspection authority for information relating to their specific requirements for clearances, support, and so on, for service masts.

The *NEC®* rules for insulation and clearances apply to the service-drop and service-entrance conductors. For example, the service conductors must be insulated, except where the voltage to ground does not exceed 300 volts. In this case, the grounded neutral conductors are not required to have insulation.

Clearance above Roofs: *NEC® 230.24(A)* gives the required clearances for the service conductors passing over the roof of a dwelling. See Figure 27-4.

Clearance from Ground: *NEC® 230.24(B)* gives the required clearances for service conductors above ground. See Figure 27-6.

Clearance from Windows: *NEC® 230.24(B)* gives the required clearances for service conductors from windows and other building openings. See Figure 27-6.

The installation requirements for a typical service entrance are shown in Figure 27-6 and Figure 27-7.

Conflict of Codes

Electricians must abide by the *NEC*. Utilities must go along with the *National Electrical Safety Code*. These codes have different clearance requirements for service-drop conductors and drip loops. When involved with an overhead service, better check with your electrical inspector to determine which code is enforced in your area.

Figure 27-5 **Only power service-drop conductors are permitted to be attached to and supported by the service mast. Do *not* attach or support television cables, telephone cables, or anything else to the mast, *230.28*.**

OPEN SERVICE CONDUCTORS OR MULTI-CONDUCTOR SERVICE CABLE WITHOUT AN OVERALL OUTER JACKET SHALL HAVE A CLEARANCE OF NOT LESS THAN 3 FT (900 mm) FROM WINDOWS THAT ARE DESIGNED TO BE OPENED, DOORS, PORCHES, BALCONIES, LADDERS, STAIRS, FIRE ESCAPES, OR SIMILAR LOCATIONS.

NOTE: CLEARANCES ARE DESIGNATED BY LETTERS A THROUGH G

CLEARANCES FOR SERVICE DROPS – RESIDENTIAL – 120/240 VOLT SINGLE PHASE:
A = 10 FT (3.0 m) MINIMUM
B = 12 FT (3.7 m) MINIMUM
C = 10 FT (3.0 m) MINIMUM

D = 18 FT (5.5 m) MINIMUM
E = 3 FT (900 mm) MINIMUM
F = CONDUCTORS RUN ABOVE TOP LEVEL OF WINDOW ARE CONSIDERED OUT OF REACH FROM THAT WINDOW. 3 FT (900 mm) CLEARANCE NOT REQUIRED.
G = 10 FT (3.0 m) MIN. FROM DRIP LOOP TO FINISH GRADE

NOTE: ELECTRIC UTILITIES FOLLOW THE NATIONAL ELECTRICAL SAFETY CODE (NESC). THE CLEARANCE REQUIREMENTS IN THE NESC ARE DIFFERENT THAN THOSE IN THE *NATIONAL ELECTRICAL CODE®* (*NEC®*). THE DECIDING FACTOR MIGHT BE: IS THE INSTALLATION CUSTOMER INSTALLED, OWNED, AND MAINTAINED, OR IS THE INSTALLATION UTILITY INSTALLED, OWNED, AND MAINTAINED?

Figure 27-6   Clearances from ground for a typical residential service in accordance with *230.24*. Clearances (A) and (C) of 10 feet (3.0 m) permitted only if the service-entrance drop conductors are supported on and cabled together with a grounded bare messenger wire where the voltage to ground *does not* exceed 150 volts. This 10 ft (3.0 m) clearance must also be maintained from the lowest point of the drip loop. Clearance (B) of 12 ft (3.7 m) is permitted over residential property such as lawns and driveways where the voltage to ground *does not* exceed 300 volts. The circled insert shows clearances from building openings, *230.9*.

UNDERGROUND SERVICE

Underground service means the cable installed underground from the point of connection to the system provided by the utility company.

New residential developments often include underground installations of the high-voltage electrical systems. The conductors in these distribution systems end in the bases of pad-mounted transformers, Figure 27-8. These transformers are placed at the rear lot line or in other inconspicuous locations in the development. The transformer and primary high-voltage conductors are generally installed by, and are the responsibility of, the utility company.

The conductors installed between the pad-mounted transformer and the meter are called **service lateral conductors**. Normally, the electric utility furnishes and installs the service laterals. In many areas, the utility will install both the service lateral conductors and the communication cables in the same trench. In some areas, but not too commonly, service laterals, communication cables, and gas lines are run in the same trench. Figure 27-8 shows a typical underground installation.

The wiring from the external meter to the main service equipment is the same as the wiring for a service connected from overhead lines, as in Figure 27-7. Some local codes may require conduit to be installed underground from the pole to the service-entrance equipment. *NEC®* requirements for underground services are given in *Article 230, Part III*. The underground conductors must be suitable for direct burial in the earth, Type USE single-conductor or Type USE cable containing more than one conductor, *Article 338*.

POINT OF ATTACHMENT TO BUILDINGS, *230.26*, SHALL BE NOT LESS THAN 10 FT (3.0 m) ABOVE FINISHED GRADE

CONDUCTORS EXPOSED TO WEATHER AND SUNLIGHT MUST BE SUITABLE FOR WET LOCATIONS AND MUST BE SUNLIGHT RESISTANT.

ONLY POWER SERVICE-DROP CONDUCTORS ARE PERMITTED TO BE ATTACHED TO A SERVICE MAST, *230.28*.

SERVICE-ENTRANCE CONDUCTORS:
• 100 AMPERE MINIMUM, *230.79(C)*.
• GROUNDED NEUTRAL MAY BE INSULATED OR BARE, SECTIONS *230.30* AND *230.41*.
• SEE TABLE *310.15(B)(6)* FOR SPECIAL AMPACITY RATINGS.

OVERHEAD SERVICE LOCATIONS, *230.54*

CLEARANCES OF SERVICE DROPS, *230.24*

THREADED HUB

CONDUIT SIZE: *CHAPTER 9. TABLES 1, 4, 5,* AND *8.* ALSO *TABLES C2* THROUGH *C12A. ANNEX C TO NEC.®*

MOUNT METER ACCORDING TO UTILITY RULES. NEUTRAL BONDED TO METER ENCLOSURE

CL

MAX. 5 FT (1.5 m)
MIN. 3 FT (900 mm)

IDEAL HEIGHT TO CENTER 50/52 IN. (1.27/1.32 m) VERIFY WITH UTILITY.

MUST BE MARKED AS SUITABLE FOR USE AS SERVICE ENTRANCE EQUIPMENT, *230.66*.

MUST BE IDENTIFIED "MAIN DISCONNECT" (OR SIMILAR WORDING), *230.70(B)*.

BONDING-BUSHING NOT REQUIRED HERE IF THREADED HUB

MAIN SWITCH AND OVERCURRENT PROTECTION, *230.70* THROUGH *230.95* AND ALSO *408.36*. DO NOT INSTALL IN BATHROOMS, *240.24(E)* AND *230.70(A)(2)*.

PRESSURE CONNECTORS, NO SOLDER, *230.81*.

DISCONNECT MEANS LOCATED AT READILY ACCESSIBLE LOCATION NEAREST POINT OF ENTRANCE OF SERVICE CONDUCTORS, *230.70(A)(1)*.

BONDING, *250.90* THROUGH *250.104*, ALSO INSULATING BUSHINGS, *300.5, 312.6(C)*

FUSES OR CIRCUIT BREAKERS MUST HAVE INTERRUPTING RATING SUFFICIENT FOR VOLTAGE EMPLOYED AND CURRENT WHICH MUST BE INTERRUPTED, *110.9* AND *110.10*.

EQUIPMENT GROUNDING CONDUCTOR BUS, *408.3(C)* AND *408.40*.

MAIN BONDING JUMPER, *250.28*, AND *408.3(C)*.

NEUTRAL BAR GROUNDED TO ENCLOSURE, *250.24(A), 250.142, 408.40,* AND *408.3(C)*.

GROUND LEVEL

GROUNDING ELECTRODE CONDUCTOR
   MATERIAL: *250.62*
   HOW TO INSTALL: *250.64*
   SIZE: *250.66*
   CONNECTION: *250.68, 250.70*

GROUNDING ELECTRODE CONDUCTER MUST BE ONE CONTINUOUS LENGTH WITHOUT SPLICE. A SPLICE WITH AN IRREVERSIBLE COMPRESSION TYPE CONNECTOR OR EXOTHERMIC WELDING IS PERMITTED, *250.64(C)*.

BONDING, *ARTICLE 250, PART V*. BONDING JUMPERS MUST BE LONG ENOUGH TO ALLOW REMOVAL OF METER WITHOUT DISCONNECTING BOND, *250.53(D)(1), 250.68(B), 250.104*.

IF GROUNDING ELECTRODE CONDUCTOR IS IN METAL RACEWAY, BOND BOTH ENDS. *250.64(E)*

GROUND CLAMPS

GROUND CLAMPS

COLD WATER PIPE

HOT WATER PIPE

WATER HEATER

THE GROUNDING ELECTRODE CONDUCTOR MUST BE CONNECTED SOMEWHERE ON THE FIRST 5 FT (1.52 m) OF METAL WATER PIPE SERVICE WHERE IT ENTERS THE BUILDING, *250.52(A)(1)*.

BONDING JUMPER, *250.104*.

GROUND ROD

SUPPLEMENTAL GROUND ROD MINIMUM LENGTH 8 FT (2.5 m), *250.52(A)(5)*

CONNECTION TO GROUNDING ELECTRODE, *250.68* AND *250.70*.

GROUNDING ELECTRODE CONDUCTOR THAT CONNECTS TO SUPPLEMENTAL GROUND ROD NEED NOT BE LARGER THAN 6 AWG COPPER, *250.53(E)*.

GROUND ROD
   SIZE: *250.52(A)(5)*
   DEPTH: *250.53(G)*
   SUPPLEMENTAL: *250.54*
   RESISTANCE: *250.56*

CONCRETE ENCASED GROUNDING CONDUCTOR ALSO PERMITTED, *250.52(A)(3)*.

Figure 27-7   The wiring of a typical service-entrance installation and *Code* rules for the system grounding. Refer also to Figure 27-2 and Figure 27-22. Today, most services are installed underground, as shown in Figure 27-8.

SERVICE LATERALS NOT ENCASED IN CONCRETE THAT ARE BURIED 18 IN. (450 mm) OR MORE BELOW GRADE SHALL HAVE THEIR LOCATION IDENTIFIED BY A WARNING RIBBON THAT IS PLACED IN THE TRENCH AT LEAST 12 IN. (300 mm) ABOVE THE UNDERGROUND INSTALLATION, *300.5(D)(3)*.

METER

IDEAL HEIGHT TO CENTER 50/52 IN. (1.27 m/1.32 m) VERIFY WITH UTILITY

METER ENCLOSURE

GRADE

18 IN. (450 mm) MINIMUM

MAIN SERVICE-ENTRANCE PANEL

METER ENCLOSURE MOUNTED ON SIDE OF RESIDENCE

MAIN SERVICE-ENTRANCE PANEL

METER ENCLOSURE TO BELOW GROUND LINE

CAUTION: BURIED ELECTRIC LINE

PAD-MOUNTED TRANSFORMER

UNDERGROUND SERVICE-ENTRANCE CONDUCTORS (SERVICE LATERAL) RUN FROM THE METER TO THE PAD-MOUNTED TRANSFORMER PLACED ON LOT LINE AT REAR OF RESIDENCE: THESE CONDUCTORS USUALLY ARE INSTALLED AND FURNISHED BY THE ELECTRIC UTILITY COMPANY.

CONDUIT RUN FROM METER ENCLOSURE TO MAIN SERVICE-ENTRANCE PANEL

TYPE "USE" SINGLE-CONDUCTOR OR TYPE "USE" CABLE

Figure 27-8 Underground service.

If the electric utility installs the underground service conductors, the work must comply with the rules established by the utility. Utilities follow the National Electrical Safety Code, NESC. These rules may not be the same as those given in *NEC® 90.2*.

The *NEC®* in *90.2(B)(5)* excludes installations under the exclusive control of an electric utility where such installations: (a) consist of service drops or service laterals and associated metering, or (b) are located in legally established easements, rights-of-way, or by other agreements either designated by or recognized by public service commissions, utility commissions, or other regulatory agencies having jurisdiction for such installations, or (c) are on property owned or leased by the electric utility for the purpose of communications, metering, generation, control, transformation, transmission, or distribution of electric energy.

When the underground conductors are installed by the electrician, *230.49* applies. This section deals with the protection of conductors against physical damage. *NEC® 230.49* also refers to *300.5*, which covers all situations involving underground wiring, such as the sealing of raceways where they enter a building; see the "Meter/Meter Base" section in this unit.

MAIN SERVICE DISCONNECT LOCATION

In conformance to *230.70(A)(1)* and *230.70(A)(2)*, the main service disconnecting means shall:

- be located inside the building at a readily accessible location nearest the point of entrance of the service conductors, Figure 27-9, or
- be located outside the building at a readily accessible location, and
- not be installed in a bathroom.

If for some reason it is necessary to locate service-entrance equipment some distance inside the

Figure 27-9 Main service disconnect location, *230.70*. Do *not* install in a bathroom. Refer to *230.70(A)(2)* and *240.24(E)*.

Figure 27-10 A typical 120/240-volt, single-phase, three-wire load center. This load center has a main circuit breaker. Note the neutral terminal strips on which to terminate the white branch-circuit grounded conductors, and the equipment grounding terminal strips on which to terminate the equipment grounding conductors of nonmetallic-sheathed cable. See *230.66. Courtesy of* Schneider Electric.

building, *NEC®* 230.6 considers service-entrance conductors to be outside of a building if the service-entrance conductors are one of the following:

- installed under not less than 2 in. (50 mm) of concrete beneath the building, or

- installed within the building in a raceway encased in at least 2 in. (50 mm) of concrete or brick, or

- installed in conduit, and covered by at least 18 in. (450 mm) of earth beneath the building.

Circuit Directory/ Circuit Identification

▶*NEC®* 408.4 is pretty clear as to its requirements on the subject. It requires that *"Every circuit and circuit modification shall be legibly identified as to its clear, evident, and specific purpose or use. The identification shall include sufficient detail to allow each circuit to be distinguished from all others. The identification shall be included in a circuit directory that is located on the face of the inside of the panel door in the case of a panelboard, and located at each switch on a switchboard."* ◀*

Figure 27-10 shows a typical 240/120-volt, single-phase, three-wire panelboard.

Figure 27-11 shows the circuit identification for Main Service Panel A.

NEC® 110.22 also required that disconnect switches be "legibly marked," describing what the disconnect is for.

Disconnect Means (Panel A)

The requirements for disconnecting electrical services are covered in *230.70* through *230.82. NEC® 230.71(A)* requires that the service disconnecting means consist of not more than six switches or six circuit breakers mounted in a single enclosure, in a group of separate enclosures in the same location, or

---

*Reprinted with permission from NFPA 70-2005.

Figure 27-11 Circuit schedule of Main Service Panel "A."

in or on a switchboard. This permits the disconnection of all electrical equipment in the house with not more than six hand operations. Some local codes take exception to the six-disconnect rule, and require a single main disconnect. Some localities require the main disconnecting means to be located outside. If located outside, *230.70(A)(1)* requires that the service disconnect must be located on the building, and must be readily accessible.

You need to check this out with the local authority that has jurisdiction in your area.

In damp and wet locations, to prevent the accumulation of moisture that could lead to rusting of the enclosure and damage to the equipment inside the enclosure, *312.2(A)* specifies that there be at least a ¼-in. (6-mm) air space between the wall and a surface-mounted enclosure. Most disconnect switches, panelboards, meter sockets, and similar equipment have raised mounting holes that provide the necessary clearance. This is clearly visible in the open-meter socket in Figure 27-15 of the "Meter/Meter Base" section in this unit. The *NEC®* (in Definitions) considers masonry in direct contact with the earth to be a wet location. The *Exception* to *312.2(A)* allows nonmetallic enclosures to be mounted without an air space on concrete, masonry, tile, and similar surfaces.

NEC® 408.34(A) defines a lighting and appliance branch-circuit panelboard as one that has more than 10 percent of the overcurrent devices rated less than 30 amperes with provisions for the connection of neutral conductors. *NEC® 408.36(A)* states that a lighting and appliance branch-circuit panelboard *shall not* have more than two main circuit breakers or two sets of fuses for the panelboards overcurrent protection and disconnecting means.

A lighting and appliance branch-circuit panelboard *shall not* have more than forty-two overcurrent devices in it, *408.35*. A two-pole breaker is counted as two overcurrent devices.

NEC® 230.79 tells us that the service disconnect shall have a rating not less than the load to be carried, determined in accordance with *Article 220*. We show how these calculations are made in Unit 29.

For a one-family residence, the minimum size main service disconnecting means is 100 amperes, three-wire, *NEC® 230.79(C)*.

Service equipment must be marked that it is suitable for use as service equipment, *230.66*.

NEC® 230.70(B) requires that the main service disconnect must be permanently marked to identify it as service disconnect.

When nonmetallic-sheathed cable is used, a panelboard must have a terminal bar for attaching the equipment grounding conductors, *408.20*. See Figure 27-12.

For this residence, Panel A serves as the main disconnecting means for the service. This panelboard consists of a main circuit breaker and many branch-circuit breakers. Figure 27-10 is a photo of this type of load center. Figure 27-11 is the schedule of the circuits connected to Panel A. Panel A is located in the Workshop.

Working Space. To provide for safe working conditions around electrical equipment, *110.26* contains a number of rules that must be followed. For residential installations, there are some guidelines for working space.

- A working space not less than 30 in. (750 mm) wide and 3 ft (900 mm) deep must be provided in front of electrical equipment, such as Panel A and Panel B in this residence.

- This working space must be kept clear and not used for storage.

- Do not install electrical panels inside of cabinets, nor above shelving, washers, dryers, freezers, work benches, and so forth.

- Do not install electrical panels in bathrooms.

- Do not install electrical panels above or close to sump pump holes.

- The hinged cover (door) on the panel shall be able to be opened to a full 90°.

- A "zone" equal to the width and depth of the electrical equipment, from the floor to a height of 6 ft (1.8 m) above the equipment or to the structural ceiling, whichever is lower, shall be dedicated to the electrical installation. No piping, ducts, or equipment foreign to the electrical installation shall be located in this zone. This zone is dedicated space intended for electrical equipment only! See *110.26(F)*.

Mounting Height. Disconnects and panelboards must be mounted so the center of the main discon-

TERMINAL FOR CONNECTION OF EQUIPMENT GROUNDING CONDUCTOR (BARE) OF NONMETALLIC-SHEATHED CABLE (TYPE NM, NMC, UF). NOT MORE THAN THREE EQUIPMENT GROUNDING CONDUCTORS PER SET-SCREW, PER UL PANELBOARD STANDARD NO. 67. TERMINAL BAR MUST BE GROUNDED TO PANEL ENCLOSURE. THE EQUIPMENT GROUND BAR SHALL NOT BE CONNECTED TO THE NEUTRAL BAR EXCEPT AT SERVICE EQUIPMENT, OR IF PANEL IS LOCATED IN A SEPARATE BUILDING.

AT SERVICE-ENTRANCE EQUIPMENT, GROUNDING CONDUCTOR BUS AND NEUTRAL CONDUCTOR BUS MUST BE BONDED TOGETHER; THIS IS FURNISHED BY THE MANUFACTURER OF THE PANEL.

TERMINAL FOR CONNECTION OF NEUTRAL CONDUCTORS. MUST BE WHITE OR GRAY INSULATED EXCEPT NEUTRAL SERVICE-ENTRANCE CONDUCTOR MAY BE BARE. ONLY ONE NEUTRAL CONDUCTOR PER SET-SCREW. MUST BE INSULATED FROM ENCLOSURE EXCEPT AT SERVICE-ENTRANCE EQUIPMENT.

SERVICE-ENTRANCE NEUTRAL CONDUCTOR

DO NOT CONNECT EQUIPMENT GROUNDING CONDUCTORS TO NEUTRAL BAR UNLESS BAR IS IDENTIFIED FOR THE PURPOSE AND IS LOCATED IN MAIN PANEL WHERE NEUTRAL AND GROUNDING ELECTRODE CONDUCTOR ARE PERMITTED OR REQUIRED TO BE INTERCONNECTED.

GROUNDED BRANCH-CIRCUIT CONDUCTORS. ONE CONDUCTOR PER TERMINAL, *408.41*

MAIN BONDING JUMPER TO ENCLOSURE, *250.28*. IF SCREW, MUST BE GREEN, *250.28(B)*.

GROUNDING ELECTRODE CONDUCTOR

Figure 27-12   Connections of service neutral, branch-circuit neutral, and equipment grounding conductors at panelboards, *408.40* and *408.41*.

nect handle in its highest position is not higher than 6 ft 7 in. (2.0 m) above the floor, *404.8*.

Headroom.   To provide for safe adequate working space and easy access to service equipment and electrical panels, the *NEC®* in *110.26(E)* requires that there be at least 6½ ft (2.0 m) of headroom. If you are installing service equipment in an existing dwelling, services of 200 amperes or less are permitted to be located in areas where the headroom is less than 6½ ft (2.0 m).

Lighting Required for Service Equipment and Panels.   Working on electrical equipment with inadequate lighting can result in injury or death. In *110.26(D)*, there is a requirement that illumination be provided for service equipment and panelboards. But the *Code* does not spell out how much illumination is required. It becomes a judgment call on the part of the electrician and/or electrical inspector. Note that the electrical plans for the workshop show a porcelain pull-chain lampholder to the front of and off to one side of the main service Panel A. In many cases, adjacent lighting, such as the fluorescent luminaires (fixtures) in the recreation room, is considered to be the required lighting for equipment such as Panel B.

Panelboard "Set Back" in Walls

▶Panelboards are sometimes installed within a wall for flush mounting. The set-back and repair opening rules for panelboards are found in *312.3* and *312.4*:

- In plaster, drywall, or plasterboard: repair wall so no gap or opening is greater than ⅛ in. (3 mm).
- In combustible walls: flush.
- In wall of concrete, tile, or other noncombustible material: set back not more than ¼ in. (6 mm).◀

Close Openings in Panelboards

Inside a panelboard, there are live parts! There is a real fire and shock hazard when there are openings (knockouts) on the sides, top, or bottom of a panelboard or when circuit breaker openings in the cover are inadvertently broken out.

▶*NEC®* *408.7* requires that "*Unused opening for circuit breakers and switches shall be closed using identified closures, or other approved means that provide protection substantially equivalent to the wall of the enclosure.*"◀*

Identified means "*Recognizable as suitable for the specific purpose, function, use, environment,*

---

*Reprinted with permission from NFPA 70-2005.

application, and so forth, where described in a particular Code requirement."

Approved means *"Acceptable to the authority having jurisdiction."*

Panel B

When a second panelboard is installed, it is oftentimes called a subpanel.

For installation where the main service panel is a long distance from areas that have many circuits and/or heavy load concentration, as in the case of the kitchen and laundry of this residence, it is recommended that at least one additional panel be installed near these loads. This results in the branch-circuit wiring being short. Line losses (voltage and wattage) are considerably less than if the branch-circuits had been run all the way back to the main panel.

The cost of material and labor to install an extra panel should be compared to the cost of running all of the branch-circuit "home runs" back to the main panel.

Panel B in this residence is located in the recreation room. It is fed by three 3 AWG THHN or THWN conductors run in a trade size 1 EMT originating at Panel A. This feeder is protected by a 100-ampere, 2-pole circuit breaker located in Panel A.

Do not install panelboards in closets, *240.24(D)*. There is a fire hazard because of the presence of ignitable materials.

Do not install panelboards in bathrooms, *240.24(E)*. Damaging moisture problems may result from the presence of water, and shock hazard may result from both the presence of water and the close proximity of metal faucets, and other plumbing fixtures.

Panel B is shown in Figure 27-13. The circuit schedule for Panel B is found in Figure 27-14.

Figure 27-13 Typical 240/120-volt, single-phase, three-wire panelboard, sometimes called a load center. This panel does not have a main disconnect, and is referred to as a *main lug only (MLO)* panel. The branch-circuit breakers are not shown. This is the type of panel that could be used as Panel B in the residence discussed in this text. Most panels have a separate terminal bus on which to terminate the equipment grounding conductors of nonmetallic-sheathed cable, as in Figure 27-12. *Courtesy of* **Schneider Electric.**

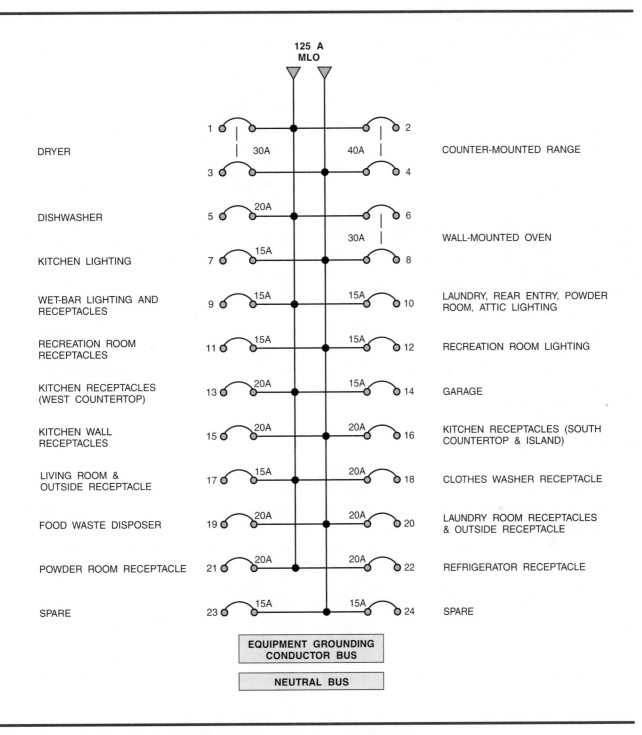

Figure 27-14   Circuit schedule of Panel B.

SERVICE-ENTRANCE CONDUCTOR SIZING

NEC® 230.42(A) states that the minimum size service-entrance conductors must be of sufficient size to carry the load as calculated according to *Article 220*.

NEC® 230.79(C) calls for a minimum service of 100 amperes for a one-family residence.

The standard and optional methods for calculating the minimum size of service-entrance conductors are discussed in detail in Unit 29.

RUNNING CABLES INTO TOP OF SERVICE PANEL

See Figure 4-20 for running nonmetallic-sheathed cables into the top of a panelboard.

SERVICE-ENTRANCE OVERCURRENT PROTECTION

Overcurrent protection (fuses and circuit breakers) is discussed in Unit 28.

SERVICE-ENTRANCE RACEWAY SIZING

This residence is served with an underground service. The conduit running from the meter to Main Panel A must be sized correctly for the conductors it will contain. In the case of a through-the-roof mast service, the conduit might be sized for mechanical strength reasons rather than conductor fill. All utilities have requirements for size and guying of mast services. Table 27-1 shows the calculation for sizing the service-entrance conduit.

METER/METER BASE

The electric utility must be contacted before the installation of the service equipment begins. They will determine when, where, and how they intend to connect to the homeowner's service. Because of the ever-increasing number of wood decks and concrete patios being added to homes, many utilities prefer that the meter be mounted on the side of the house rather than in the rear. If there is a raised deck, the meter could be only a short distance above the deck, making it both subject to physical damage and hard to read. If concrete patios are poured over underground utility cables, it is very difficult to repair or

replace these cables should there be problems. Utilities are happy to furnish manuals and brochures detailing their requirements for services that are not found in the *NEC.*®

The electric utility furnishes the watt-hour meter. The meter base, also referred to as a meter socket, is usually furnished and installed by the electrical contractor, although some utilities furnish the meter base to the electrical contractor for installation.

Figure 27-8 shows a service supplied from an underground service. Figure 27-9 discusses the requirement that the main disconnect must be as close as possible to the point of entrance of the service-entrance conductors. Figure 27-15 shows a typical meter socket for an overhead service, usually mounted at eye level on the outside wall of the house. The service raceway or service-entrance cable is connected to the threaded hub (boss) on top of the meter socket. Figure 27-16 shows how to seal the raceway where it enters the building to keep out moisture.

Combination meter socket/main breaker in one enclosure is also available.

Raintight

Outdoor locations are generally wet locations. However, under an overhang is usually not considered to be a wet location. ▶If you are using EMT outdoors in a wet location, the fittings (couplings and connectors) must be listed as "Raintight." If the fittings and the container they come in are not marked "Raintight," then the fittings have not been listed for raintight applications.◀ Raintight fittings are generally of the compression type that have a special sealing ring.

▶*NEC*® 225.22 states that *"Raceways on exteriors of buildings or other structures shall be arranged to drain and shall be raintight in wet locations."*◀

Ⓐ CONDUCTOR SIZE BASED UPON AMPACITY PER *310.15(B)(6)*		**Ⓑ CONDUCTOR SIZE BASED UPON AMPACITY PER *TABLE 310.16***	
TWO 2/0 AWG THWN COPPER	0.2223 in² 0.2223 in²	TWO 3/0 AWG THWN COPPER	0.2679 in² 0.2679 in²
ONE 1 AWG BARE COPPER	0.0870 in²	ONE 1 AWG BARE COPPER	0.0870 in²
TOTAL	0.5316 in²	TOTAL	0.6228 in²
EMT	TRADE SIZE 1¼	EMT	TRADE SIZE 1½

Table 27-1 Calculations for sizing the EMT between Main Panel A and the meter base. (A) uses the smaller size conductors permitted by *310.15(B)(6)(B)*. (B) uses conductors sized per the standard calculations. Dimensional data for the conductors are found in *Chapter 9, Tables 1, 4, 5,* and *8* of the *NEC.*®

Figure 27-15 Typical residential self-contained meter socket. Combination meter socket/main breaker in one enclosure is also available. *Courtesy of* Milbank Manufacturing Co.

NEC® *300.5(G)* AND *300.7* STATE THAT WHEN RACEWAYS PASS THROUGH AREAS HAVING GREAT TEMPERATURE DIFFERENCES, SOME MEANS MUST BE PROVIDED TO PREVENT PASSAGE OF AIR BACK AND FORTH THROUGH THE RACEWAY. NOTE THAT OUTSIDE AIR IS DRAWN IN THROUGH THE CONDUIT WHENEVER A DOOR OPENS. COLD OUTSIDE AIR MEETING WARM INSIDE AIR CAUSES THE CONDENSATION OF MOISTURE. THIS CAN RESULT IN RUSTING AND CORROSION OF VITAL ELECTRICAL COMPONENTS. EQUIPMENT HAVING MOVING PARTS, SUCH AS CIRCUIT BREAKERS, SWITCHES, AND CONTROLLERS, IS ESPECIALLY AFFECTED BY MOISTURE. THE SLUGGISH ACTION OF THE MOVING PARTS IN THIS EQUIPMENT IS UNDESIRABLE.

SEALANT SHALL BE IDENTIFIED FOR SUCH USE, AND SHALL NOT HAVE AN ADVERSE EFFECT ON THE CONDUCTOR INSULATION, *230.8.*

NEC® *230.8* REQUIRES SEALS WHERE SERVICE RACEWAYS ENTER FROM AN UNDERGROUND DISTRIBUTION SYSTEM.

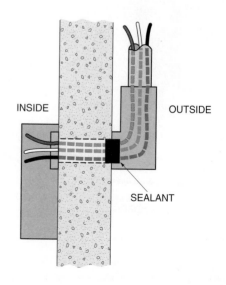

INSIDE

OUTSIDE

SEALANT

Figure 27-16 Installation of conduit through a basement wall.

Pedestals

Probably more common for residential services is the "pedestal," as shown in Figure 27-17. Usually mounted on the outside wall of the house, it could also be installed on the lot line between residential properties. Contact the electric utility on this issue.

For underground services, the utility generally runs the underground lateral service-entrance conductors in a trench from a pad-mount transformer to the line-side terminals of the meter base. The electrician then takes over to complete the service installation by running service-entrance conductors from the load side of the meter to the line side of the main service disconnect.

Quite often, the electric utility and the telephone company have agreements whereby the electric utility will lay both power cables and telephone cables in the same trench. The electric utility carries both types of cables on their underground installation vehicles. This is a much more cost effective way to do this, rather than having each utility dig and then backfill their own trenches.

Often overlooked is a requirement in *NFPA Standard No. 54* (*Fuel Gas Code*®), *Section 2.7.2(c)*, requiring that gas meters be located at least 3 ft (900 mm) from sources of ignition. Examples of sources of ignition might be an air-conditioner disconnect where an arc can be produced when the disconnect is opened under load. Furthermore, some electric utilities require a minimum of 3 ft (900 mm) clearance between electric metering equipment and gas meters and gas regulating equipment. It is better to be safe than sorry. Check this issue with the local electrical inspector if there might be a problem with the installation you are working on.

COST OF USING ELECTRICAL ENERGY

A watt-hour meter is always connected into some part of the service-entrance equipment to record the amount of energy used. Billing by the utility is generally done on a monthly basis. The meter might be mounted on the side of the house, or on a pedestal somewhere on the premise on the lot line. The utility makes this decision. The residence discussed in this text has one meter, mounted on the back of the house near the sliding doors of the master bedroom.

The kilowatt (kW) is a convenient unit of electrical power. One thousand watts (w) is equal to one kilowatt. The watt-hour meter measures and records both wattage and time.

For residential metering, most utilities have rate schedules based on "cents per kilowatt-hour." Stated another way: how much wattage is being used and for how long?

Burning a 100-watt light bulb for 10 hours is the same as using a 1000-watt electric heater for 1 hour. Both equal one kilowatt-hour.

$$\text{kWh} = \frac{\text{watts} \times \text{hours}}{1000} = \frac{100 \times 10}{1000} = 1 \text{ kWh}$$

$$\text{kWh} = \frac{\text{watts} \times \text{hours}}{1000} = \frac{1000 \times 1}{1000} = 1 \text{ kWh}$$

In these examples, if the electric rate is 8 cents per kilowatt-hour, the cost to operate the 100-watt

THE ELECTRIC UTILITY PROVIDES THE UNDERGROUND SERVICE LATERAL

Figure 27-17 Installing a metal "pedestal" allows for ease of installation of the underground service-entrance conductors by the utility. *Photo Courtesy of* Milbank Manufacturing Co.

light bulb for ten hours and the cost to operate the electric heater for one hour are the same—8 cents. Both loads use one kilowatt-hour of electricity.

Simple Formula for Calculating the Cost of Using Electricity

The cost of electrical energy used can be calculated as follows:

$$\text{Cost} = \frac{\text{watts} \times \text{hours used} \times \text{cost per kWh}}{1000}$$

EXAMPLE: Find the cost of operating a color television set for 8 hours. The label on the back of the television set indicates 175 watts. The electric rate is $0.10494 per kilowatt-hour.

$$\text{Cost} = \frac{175 \times 8 \times \$0.10494}{1000} = \frac{\$0.1469}{(\text{approx. } 15¢)}$$

EXAMPLE: Find the approximate cost of operating a central air conditioner per day that on average runs 50 percent of the time during a 24-hour period on a typical hot summer day. The unit's nameplate is marked 240 volts, single-phase, 23 amperes. The electric rate is $.09 per kilowatt-hour. The steps are:

1. The time the air conditioner operates each day is: $24 \times 0.50 = 12$ hours.

2. Convert the nameplate data to use in the calculations:

$$\text{Watts} = \text{volts} \times \text{amperes} = 240 \times 23$$
$$= 5520 \text{ watts}$$

3. $\text{Cost} = \dfrac{\text{watts} \times \text{hours used} \times \text{cost per kWh}}{1000}$

$$= \frac{5520 \times 12 \times \$0.09}{1000} = \$5.96$$

Note: The answers to these examples are approximate because they were based on watts. Power factor and efficiency factors were not included as they generally are unknown when coming up with rough estimates of an electric bill. For all practical purposes, the answers are acceptable.

Table 27-2 is an example of what a typical monthly electric bill might look like.

Some utilities increase their rates during the hot summer months when the air-conditioning load is high. Other utilities provide a second meter for

GENERIC ELECTRIC COMPANY
Anyplace, USA

Days of Service	from 06-01-2005	to 06-28-2005	Due Date 07-25-2005
Present reading			84,980
Previous reading			83,655
Kilowatt-hours used			1,325
Rate/kWh	1st 400 kWh @ $0.10494		$41.98
	remaining 925 kWh @ $0.06168 . .		57.05
Energy charge			99.03
Basic service charge (single-dwelling)			8.91
State tax			4.24
Total current charges			$112.18
Total amount due by 07-25-2005			$112.18
Total amount due after 07-25-2005			$117.79

Table 27-2 A typical monthly electric bill.

specific loads such as electric water heaters, air conditioners, heat pumps, or total electric heat. Other utilities use electronic watt-hour meters that have the capability of registering kilowatt-hour consumption during a specific "time-of-use." These electronic watt-hour meters might have up to four different "time-of-use" periods, each period having a different "cents per kilowatt-hour" rate.

Other charges that might appear on a "light bill" might be a fuel adjustment charge based on a "per kWh" basis. Such charges enable a utility to recover from the consumer extra expenses it might incur for fuel costs used in generating electricity. Fuel charges can vary with each monthly bill without the utility having to apply to the regulatory agency for a rate change.

GROUNDING

Article 250 covers grounding and bonding. To better understand grounding and bonding, it is necessary to understand the definitions found in *250.2*.

Ground-Fault: An unintentional, electrically conducting connection between an ungrounded or grounded conductor of an electrical circuit and the normally non-current carrying conductors, metallic enclosures, metallic raceways, metallic equipment, or earth.

Ground-Fault Current Path: An electrically conductive path from the point of a ground fault on a wiring system through normally non-current-carrying conductors, equipment, or the earth to the electrical supply source.

Effective Ground-Fault Current Path: An intentionally constructed, permanent, low-impedance electrically conductive path designed and intended to carry current under ground-fault conditions from the point of a ground fault on a wiring system to the electrical supply source.

We find in *250.4* the fundamental reasons for grounding electrical systems and electrical equipment. All residential wiring involves a grounded electrical system. In *250.4(A)*:

1. *Electrical System Grounding.* Electrical systems that are grounded shall be connected to earth in a manner that will limit the voltage imposed by lightning, line surges, or unintentional contact with higher voltage lines and that will stabilize the voltage to earth during normal operation.

2. *Grounding of Electrical Equipment.* Non-current-carrying conductive materials enclosing electrical conductors or equipment, or forming part of such equipment, shall be connected to earth so as to limit the voltage to ground on these materials.

3. *Bonding of Electrical Equipment.* Non-current-carrying conductive materials enclosing electrical conductors or equipment, or forming part of such equipment, shall be connected together and to the electrical supply source in a manner that establishes an effective ground-fault current path.

4. *Bonding of Electrically Conductive Materials and Other Equipment.* Electrically conductive materials that are likely to become energized shall be connected together and to the electrical supply source in a manner that establishes an effective ground-fault current path.

5. *Effective Ground-Fault Current Path.* Electrical equipment and wiring and other electrically conductive material likely to become energized shall be installed in a manner that creates a permanent, low-impedance circuit capable of safely carrying the maximum ground-fault cur-

rent likely to be imposed on it from any point on the wiring system where a ground-fault may occur to the electrical supply source. ▶*NEC® 250.4(A)(5)* is very clear about this scenario. It states in part, that "*The earth shall not be considered as an effective ground-fault current path.*" ◀

A few other key *Code* sections relating to system grounding are:

▶*Grounding Electrode:* A grounding electrode is defined in the *NEC®* as "*A device that establishes an electrical connection to the earth.*" ◀

250.20(B)(1): This section requires that an electrical system be grounded where the maximum voltage to ground on the ungrounded "hot" conductors does not exceed 150 volts. The serving electric utility grounds the midpoint of their transformer. This mid-point becomes the system *neutral*.

The electrical system in a typical home is single-phase, three-wire, 120/240 volts. The voltage between the ungrounded "hot" phase conductors is 240 volts, and the voltage between the ungrounded "hot" phase conductors and the neutral (the grounded conductor) is 120 volts. The grounded neutral service conductor is again grounded at the main service.

The electrical system in some large multi-family buildings (condos, apartments, etc.) is derived from a three-phase, four-wire, 120/208-volt system. Here, the voltage between the ungrounded "hot" phase conductors is 208 volts, and the voltage between the ungrounded "hot" phase conductors and the neutral (the grounded conductor) is 120 volts. The supply to each dwelling unit is referred to as a single-phase, three-wire, 120/208-volt system.

250.26(2): This section requires that the *neutral conductor* be grounded for single-phase, three-wire systems. As mentioned earlier, the transformer mid-point (neutral) is grounded at the utility transformer and again at the main service.

Proper grounding means that overcurrent protective devices will operate fast when responding to ground faults. Effective grounding occurs when a low-impedance (ac resistance) ground path is provided. Ohm's law verifies that, in a given circuit, "the lower the impedance, the higher the value of current." As the value of ground-fault current increases, there is an increase in the speed with which a fuse will open or a circuit breaker will trip. This is called an *inverse time* relationship.

Clearing a ground fault or short circuit is important because arcing damage to electrical equipment as well as conductor insulation damage is closely related to a value called *ampere-squared-seconds* (I^2t), where

I = the current in amperes flowing phase to ground, or phase to phase.

t = the time in seconds that the current is flowing.

Thus, there will be less equipment and/or conductor damage when the fault current is kept to a low value, and when the time that the fault current is allowed to flow is kept to a minimum. The impedance of the circuit determines the *amount* of fault current that will flow. The speed of operation of a fuse or circuit breaker determines the amount of *time* the fault current will flow.

Grounding electrode conductors and equipment grounding conductors carry an insignificant amount of current under normal conditions. However, when a ground fault occurs, these grounding conductors as well as the actual circuit conductors must be capable of carrying whatever value of fault current might flow (how much?) for the time (how long?) it takes for the overcurrent protective device to clear the fault and reduce the fault current to zero.

This potential hazard is recognized in the *Note* below *Table 250.122*, which states that "*Where necessary to comply with 250.4(A)(5) or 250.4(B)(4), the equipment grounding conductor shall be sized larger than given this table.*" This note calls attention to the fact that equipment grounding conductors may need to be sized larger than indicated in the table when high-level ground-fault currents are possible.

What is a high-level ground fault? To determine the available short-circuit and ground-fault current, a short-circuit calculation is necessary. It is also necessary to know the time/current characteristic of the overcurrent protective device. This is discussed in *Electrical Wiring—Commercial* (©Thomson Delmar Learning).

This is referred to as the conductor's "withstand rating."

Table 250.66 and *Table 250.122* are based upon the fact that copper conductors and their bolted connections can withstand:

- one ampere
- for 5 seconds
- for every 30 circular mils

Thermoplastic insulation on a copper conductor rated 75°C can safely withstand:

- one ampere
- for 5 seconds
- for every 42.25 circular mils

To exceed these values will result in damage to the conductors, with possible burn-off and loss of the grounding or bonding path.

When bare equipment grounding conductors are in the same raceway or cable as insulated conductors, always apply the insulated conductor withstand rating limitation.

Properly selected fuses and circuit breakers for normal residential installations generally will protect conductors and other electrical equipment against the types of ground faults and short circuits to be expected in homes. Fault currents can be quite high in single-family homes where the pad-mounted transformer is located close to the service equipment, and in multifamily dwellings (apartments, condos, town houses) making it necessary to take equipment short-circuit ratings and conductor withstand rating into consideration.

The subject of conductor and equipment withstand rating is covered in greater depth in *Electrical Wiring—Commercial* (©Thomson Delmar Learning).

Grounding Electrode System

Article 250, Part III covers the requirements for grounding electrode systems, grounding electrode conductors, and ground rods. It is very lengthy. A grounding electrode system "ties" everything together. The service-entrance neutral conductor is grounded, this neutral is bonded to the main panel, and a grounding electrode conductor is connected to a grounding electrode. All metal water piping systems installed in or attached to a building and all metal gas pipes are bonded together. Should any one of these become disconnected, the integrity of the grounding system is maintained through the other paths. Here are a few key points.

- *250.90* states that *"bonding shall be provided where necessary to ensure electrical continuity and the capacity to conduct safely any fault current likely to be imposed."*

- *250.92(A)* explains what parts of a service must be bonded together.

- *250.94* explains what is acceptable as a bonding means.

- *250.96(A)* states in part that bonding of metal raceways, cable armor, enclosures, frames, fittings, and so on, that serve as the grounding path *"shall be effectively bonded where necessary to ensure electrical continuity and the capacity to conduct safely any fault current likely to be imposed on them."**

By having all metal parts bonded together for a grounding electrode system, potential differences between non-current-carrying metal parts is virtually eliminated. In addition, the grounding electrode system serves as a means to bleed off lightning, stabilize the system voltage, and ensure that the overcurrent protective devices will operate.

Figure 27-18 and the following steps illustrate

*Reprinted with permission from NFPA 70-2005.

what can happen if the electrical system is not properly grounded and bonded:

1. A "live" wire contacts the gas pipe. The bonding jumper "A" has not been installed.

2. The gas pipe now has 120 volts on it. The pipe is energized. It is "hot."

3. The insulating joint in the gas pipe results in a poor path to ground. Assume a resistance (impedance) of 8 ohms.

4. The 20-ampere overcurrent device does not open:

$$I = \frac{E}{R} = \frac{120}{8} = 15 \text{ amperes}$$

5. If a person touches the "live" gas pipe and the water pipe at the same time, current flows through the person's body. If the body resistance is 12,000 ohms, the current is:

$$I = \frac{E}{R} = \frac{120}{12,000} = 0.01 \text{ ampere}$$

This amount of current passing through a human body can cause death. See Unit 6 for a discussion regarding electric shock.

6. The overcurrent device is now "seeing" $15 + 0.01 = 15.01$ amperes and does not open.

7. If the grounding electrode system concept had been used, bonding jumper "A" would have kept the voltage difference between the water

Figure 27-18 System grounding.

pipe and the gas pipe at zero. The overcurrent device would have opened. Checking *Table 8, Chapter 9, NEC,* we find the dc resistance of 1000 feet (305 m) of an uncoated 4 AWG copper wire is 0.308 ohms. The resistance of 100 ft (30 m) is 0.0308 ohms. The resistance of 10 ft (3.0 m) is 0.00308 ohms. The current would be:

$$I = \frac{E}{R} = \frac{120}{0.00308} = 38,961 \text{ amperes}$$

In an actual installation, the total impedance of *all* parts of the circuit would be much higher than these simple calculations. A much lower current would result. The value of current would be enough to cause the overcurrent device to open.

Where to Connect the Grounding Electrode Conductor

NEC 250.24(A) tells us that the grounding electrode conductor shall be connected to the grounded (neutral) service conductor. *NEC* 250.24(A)(1) goes on to tell us that the connection is permitted to be made at any accessible point from the load end of the service drop or service lateral to, and including, the terminal or bus to which the grounded (neutral) service conductor is connected at the service disconnecting means.

All residential panelboards have a:

- neutral bus for the white grounded circuit conductors, and

- an equipment grounding bus for the bare equipment grounding conductors when nonmetallic-sheathed cable is used as the wiring method.

For just about all residential services, the connection of the grounding electrode conductor is made to the neutral bus. *NEC* 250.24(A)(4) permits the grounding electrode conductor to be connected to the equipment grounding bus if the main bonding jumper is a wire or busbar. Many residential panelboards have a green hexagon-shaped No.10-32 screw that becomes the main bonding jumper between the neutral bus and the enclosure when properly installed.

Figure 27-7 and Figure 27-12 show the grounding electrode conductor connected to the neutral bus in the main service panelboard.

Grounding and Bonding The Electrical System in a Typical Residence

Figure 27-2, Figure 27-7, Figure 27-19, and Figure 27-20 show the *NEC* rules applicable to residential electrical service grounding and bonding as required by *Article 250, Part III*.

A metal underground water piping system 10 ft (3.0 m) or longer, in direct contact with the earth, is acceptable as the primary grounding electrode, *250.52(A)(1)*. The connection of the grounding electrode conductor must be made on the first 5 ft (1.52 m) of where the metal underground water piping enters the building, *250.52(A)(1)* and *250.54*.

When metal underground water piping is used as the primary grounding electrode, it must be supplemented by at least one additional grounding electrode, *250.53(D)(2)*. In this residence, a driven ground rod is the supplemental grounding electrode, as permitted by *250.52(A)(5)* and *250.54*.

In Figure 27-19, the supplemental grounding electrode is a concrete-encased 4 AWG bare copper conductor at least 20 ft (6.0 in) long laid near the bottom of the concrete footing. This is quite often referred to as a UFER ground. For homes built on a concrete slab, it is common practice to bring one end of a reinforcing bar (called "re-bar") upward out of the slab at an appropriate location, making an easy connection point for the grounding electrode conductor. This is permitted by *250.52(A)(3)*.

Watch out when using ground rods! If the resistance to ground of the first ground rod is more than 25 ohms, at least one additional ground rod must be added, *250.56*. There are a number of manufacturers of testers that measure ground/earth resistance.

More Ground Rod Rules for Residential Application

- The most common ground rods are copper-clad steel. Copper makes for an excellent connection between the rod and the ground clamp. The steel gives it strength to withstand being driven into the ground. Galvanized and stainless steel rods are also available. Aluminum rods are not permitted.

- Must be at least ½ in. (13 mm) in diameter.

- Must be installed below the permanent moisture level if possible.

Figure 27-19 A typical electrical service grounded to an underground metal water piping supply. The required supplemental grounding electrode is the concrete-encased 4 AWG bare copper conductor. See text for details.

- Must not be less than 8 ft (2.5 m) in length.

- Must be driven to a depth so that at least 8 ft (2.5 m) is in contact with the soil.

- If solid rock is encountered, drive rod at an angle not greater than 45 degrees from vertical, or lay it in a trench that is at least 2½ ft (750 mm) deep.

- Drive rod so upper end is flush with or just below ground level. If upper end is exposed, the ground rod, the ground clamp, and the grounding electrode conductor must be protected from physical damage.

- If more than one rod is needed, keep at least 6 ft (1.8 m) apart. Driving the rods close together does not significantly change the resistance values of a single rod. Actually, it is better to space multiple rods the length of the rods. For example, when driving two 8-ft (2.5-m) ground rods, space them 8 ft (2.5 m) apart.

Table 27-3 shows accurate multipliers for multiple ground rods spaced rod length apart. To use this chart, divide the resistance value of one rod by the number of rods used, then apply the multiplier.

STREET SIDE OF
WATER METER

BONDING
JUMPER

GROUNDING ELECTRODE
CONDUCTOR FROM MAIN
SERVICE PANEL NEUTRAL BAR
TO CONNECTION ANYWHERE
ON THE FIRST 5 FT (1.52 m) OF
METAL WATER PIPE AFTER IT
ENTERS THE BUILDING. THE FIRST
FIVE FEET MAY INCLUDE THE
WATER METER, *250.52(A)(1)*

NEC® 250.68(B) REQUIRES THAT
THESE BONDING JUMPERS BE OF
SUFFICIENT LENGTH TO PERMIT
REMOVAL OF THE METER OR WATER
HEATER WITHOUT LOSING THE
INTEGRITY OF THE BONDING PATH.

HOT
WATER
LINE

COLD WATER LINE

BONDING
BUSHING

METER
LOCATED ON
OUTSIDE OF
RESIDENCE

GROUND
CLAMPS

GROUND BUS

NEUTRAL BUS

MAIN
SERVICE
PANEL
LOCATED
INSIDE
RESIDENCE

BOND COLD, HOT,
AND GAS PIPES PER
250.104(A) AND *(B)*. SIZE
BONDING CONDUCTOR
PER *TABLE 250.66*

GAS
LINE

DIELECTRIC
UNIONS*

TEMPERATURE/
PRESSURE
RELIEF VALVE

ANODE

DIP TUBE

MAIN
BONDING
JUMPER

SUPPLEMENTAL
GROUNDING
ELECTRODE
CONDUCTOR

GAS
INLET

WATER
HEATER

SUPPLEMENTAL GROUND ROD

THE SUPPLEMENTAL GROUNDING ELECTRODE
CONDUCTOR MAY BE CONNECTED:
1. TO THE NEUTRAL BUS IN THE MAIN
   SERVICE PANEL, **OR**
2. TO THE NEUTRAL BUS IN THE METER BASE, BUT
   ONLY IF ACCEPTABLE TO THE ELECTRIC UTILITY.
3. **DO ONE OR THE OTHER—NOT BOTH.**

*SOME MFGS. PROVIDE
DIELECTRIC FITTINGS IN
THEIR WATER HEATERS
TO STOP STRAY ELECTRICAL
CURRENTS AND TO MINIMIZE
GALVANIC CORROSION.

Figure 27-20   One method that may be used to provide proper electrical system grounding of service-entrance equipment, and bonding of the cold and hot water piping and gas piping.

NUMBER OF RODS	MULTIPLIER
2	1.16
3	1.29
4	1.36

Table 27-3   Multipliers for multiple ground rods.

EXAMPLE: The ground/earth resistance reading of one 8 ft (2.5 m) ground rod is 30 ohms. This exceeds the *NEC®* maximum of 25 ohms. A second ground rod is driven and connected in parallel to the first ground rod. What is the approximate ground/earth resistance of these two ground rods when spaced 8 ft (2.5 m) apart?

$$\frac{30}{2} \times 1.16 = 17.4 \text{ ohms}$$

- A grounding electrode conductor that is the sole (only) connection to a driven ground rod need not be larger than 6 AWG copper or 4 AWG aluminum, *250.53(E)*.

In some parts of the country, a water pipe ground is considered unreliable. The concern is the prolific use of insulating fittings and nonmetallic water piping services. These parts of the country have found that a concrete encased electrode provides excellent grounding capabilities. Refer to *250.52(A)(3)*.

Other Acceptable Supplemental Grounding Electrodes

If the driven ground rod is not selected as the required supplemental grounding electrode, then any one of the following is acceptable:

- The metal frame of a building (this residence is constructed of wood), *250.50(A)(2)*.

- At least 20 ft (6.0 m) of steel reinforcing bars, ½ in. (12.7 mm) minimum diameter, or at least 20 ft (6.0 m) of bare copper conductor not smaller than 4 AWG. These must be encased in concrete at least 2 in. (50 mm) thick that is in direct contact with the earth, such as near the bottom of a foundation or footing. A bare copper conductor encased in the concrete footing of a building is commonly referred to as a UFER ground, named after the man who developed the concept, *250.52(A)(3)*. Reinforcing bars are permitted to be bonded together by the usual steel tie wires or other effective means.

- At least 20 ft (6.0 m) of bare copper wire, minimum size 2 AWG, buried directly in the earth at least 2½ ft (750 mm) deep. This is called a *ground ring*. See *250.52(A)(4)* and *250.52(G)*.

- Ground plates must be at least 2 ft² (0.186 m²), *250.52(A)(6)*. At least one additional plate must be laid if the resistance of one plate is found to exceed 25 ohms, *250.56*. The plate must be buried at least 2½ ft (750 mm) deep.

- *250.53(D)(2)* tells us that a metal underground water piping system must have a supplemental electrode installed. An underground metal water piping system is the only grounding electrode that needs to be supplemented. Other types of grounding electrodes, such as a concrete-encased electrode, do not need a supplemental electrode.

Some Issues Regarding Bonding of Metal Pipes Inside or Attached to the Outside of the Home

Metal water piping and metal gas piping must be properly bonded, *250.104(B)*.

Over the years, underground natural gas services installed from the street to the home might have been black iron pipe, copper tubing, or plastic piping designed for gas underground installations. You might even find lead water and gas pipes in homes built prior to 1963, but not likely after 1963. Today, metal gas piping installed underground has a factory-applied corrosion protection coating and/or wrapping as required by *NFPA 54, Natural Fuel Gas Code*. Most underground gas pipes today are plastic.

Homes have been blown off their foundations as a result of ignition and explosion of leaking gas from underground piping. The culprit causing the leak usually can be traced to corrosion. Where gas is found to be leaking from an underground copper pipe, the gas company workmen pull the copper pipe out of the ground, pulling in its place an approved plastic pipe. A 14 AWG yellow tracer wire is pulled in (buried) with the plastic pipe so as to be able to locate the underground nonmetallic piping at some later date.

Underground gas line piping is not a major concern for electricians! What is of concern to the electrician is that the metal water and gas piping within or on the home must be bonded together.

Here are the *NEC®* requirements that deal with bonding metal water and gas piping.

- *250.52(B)(1):* Metal underground gas piping is not permitted to be used as a grounding electrode.

  Similarly, the *National Fuel Gas Code, NFPA 54, Section 6.13.2* states that "*Gas piping shall not be used as a grounding conductor or electrode.*"

  To prevent galvanic corrosion of underground metal gas piping, gas utilities install dielectric (insulating) fittings at the line side port of the gas meter. A dielectric fitting "isolates" the underground metal gas piping from the interior metal piping and prevents the corrosion problem. Today, gas companies generally install plastic piping underground. This solves the corrosion problem present when metal piping is installed underground.

- *250.104(A):* Requires that interior metal water piping be bonded, and that the bonding conductor be sized according to *Table 250.66*. Figure 27-20 clearly shows this bond.

  To control galvanic corrosion of the water heater tank and any directly connected steel, galvanized, and/or copper piping, you will usually find dielectric fittings (unions) in the water heater's cold and hot water lines. Always install the bonding jumper *above* the dielectric fittings.

- *250.104(B):* All metal piping, including gas piping, shall be bonded. Figure 27-20 clearly shows this bond.

- *250.104(B):* The bonding conductor for metal gas piping shall be sized according to *Table 250.122* for the ampere rating of the circuit that might energize the metal piping.

    Similarly, the *National Fuel Gas Code, NFPA 54,* in *Section 6.13.1* requires that *"Each above ground portion of gas piping that is likely to become energized shall be electrically continuous and bonded to an effective ground-fault current path. Gas piping shall be considered to be bonded when it is connected to gas utilization equipment that is connected to the equipment grounding conductor of the circuit supplying the equipment."**

    The logic to this requirement is that if the "hot" conductor of the electrical circuit supplying a gas utilization equipment such as a gas furnace comes in contact with the metal frame of the appliance, the ground-fault current path through the equipment grounding conductor

---

*Reprinted with permission from NFPA 70-2005.

will cause the branch-circuit overcurrent device to clear the fault. In the meantime, bonding maintains an equal voltage potential between the metal frame of the gas appliance and its metal gas supply piping. Equal voltage potential is also maintained between the faulted appliance and other nearby equipment.

If there is no likelihood of the gas pipe becoming energized (no gas appliance that is also served by electricity), then no bonding of the gas piping is required.

- *250.104(B), FPN: "Bonding all piping and metal air ducts within the premises will provide additional safety."* Fine print notes are not mandatory but certainly provide excellent recommendations to improve safety.

The ground-fault current return path in Figure 27-21 is the EMT. Where nonmetallic-sheathed cable is used, the equipment grounding conductor in the cable serves as the ground-fault current return path. Where armored cable is used, the armor plus the bonding strip is the ground-fault current return path. See *NEC® 250.118* for other accepted equipment grounding conductors.

In a typical home, there is usually a large number of bonds between the hot and cold water pipes.

BRANCH-CIRCUIT
-20 AMPERES
-120 VOLTS

TRADE SIZE
½ EMT
ACCEPTABLE
AS EQUIPMENT
GROUNDING
CONDUCTOR,
250.118(4).

CLASS 2 LOW-VOLTAGE
CONTROL CIRCUIT CABLE

GAS PIPING IS
PROPERLY
BONDED BY
THE EMT PER
250.104(B) AND
6.13.1 IN THE
NATURAL FUEL
GAS CODE,
NFPA 54.

DISCONNECT

GAS
SHUT-OFF
VALVE

Figure 27-21 Gas piping is considered to be properly bonded by the appliance's branch-circuit equipment grounding conductor. The equipment grounding conductor in this illustration is the EMT, which provides an excellent ground-fault current return path.

There are interconnections at metallic mixer faucets, water heaters, clothes washers, and similar plumbing connections. Many plumbers use short lengths of copper tubing to bridge a stud space or joist space to support and keep water lines in place. They solder (tack) the water lines to these cross pieces. This bonds the hot and cold water pipes in a number of places inside the walls.

Although the above *apparent* bonds might be acceptable in some areas, the plumber and the plumbing are relied on to do a job that is the electrician's responsibility.

Figure 27-20 illustrates the bonding together of the cold and hot water metal pipes and the metal gas piping. This bonding jumper is sized per *Table 250.66*. This installation is easy and eliminates the questionable bonding/grounding of the many interconnections of the metal water and gas piping through various removable gas appliances. This bonding above the water heater *does not* mean that the gas pipe is serving as a grounding electrode.

Don't depend on the plumber to do your job!

Do it right! Follow Figure 27-20 for reliably bonding together all metal water piping and metal gas piping.

NEC® 300.6 addresses corrosion that an electrician needs to be concerned with when working in wet locations, underground, in concrete, and similar installations.

Grounding Conductors

A **grounding conductor** is defined as "*A conductor used to connect equipment or the grounded circuit of a wiring system to a grounding electrode or electrodes.*" There are two types of grounding conductors.

1. An **equipment grounding conductor** is defined as "*The conductor used to connect non-current-carrying metal parts of equipment, raceways, and other enclosures to the system grounded conductor, the grounding electrode conductor, or both, at the service equipment or at the source of a separately derived system.*"*

   NEC® 250.118 recognizes many types of equipment grounding conductors. Copper or aluminum conductors, rigid metal conduit, EMT, the metal armor of armored cable, and flexible metal conduit all qualify as equipment grounding conductors. Figure 5-30, Figure 5-32, Figure 5-33, Figure 5-34, Figure 27-12, and Figure 30-4, illustrate equipment grounding conductors.

   Equipment grounding is discussed throughout this text in the appropriate unit where a specific piece of electrical equipment or appliance requires grounding. Some examples of this are the range, cooktop, dryer, food waste disposer, dishwasher, and water pump. Equipment grounding conductors are sized according to *Table 250.122*.

2. ►A **grounding electrode conductor** is the conductor that connects the grounding electrode to the equipment grounding conductor or to the grounded conductor of the circuit, or to both. This is done at the main service or at a second building supplied by a branch-circuit(s) or feeder from the electrical service in the first building.◄ Acceptable grounding electrodes are listed in *250.52(A)(1)* through *(7)*. Reference to grounding electrode conductors is made in Figure 27-1, Figure 27-2, Figure 27-7, Figure 27-12, Figure 27-19, Figure 27-20, Figure 27-36, and Figure 30-4. See also the text on "Grounding Electrical Equipment at a Second Building" in this unit.

   Grounding electrode conductors:

   - are sized according to *Table 250.66*.
   - shall not be spliced unless done so with an irreversible compression-type connector listed for the purpose, or by exothermic welding, *250.28, 250.52(D)(3)*, and *250.64(C)*.

3. ►A **grounding electrode** is defined as "*a device that establishes an electrical connection to the earth.*"◄

In this residence, a 4 AWG armored copper grounding electrode conductor runs from Main Panel A, across the workshop ceiling to the water pipe, where it is terminated with a listed ground clamp.

Ground Clamps

Ground clamps used for bonding and grounding must be listed for the purpose. The use of solder to

*Reprinted with permission from NFPA 70-2005.

Table 250.66 Grounding Electrode Conductor for Alternating-Current Systems

| Size of Largest Ungrounded Service-Entrance Conductor or Equivalent Area for Parallel Conductors[a] (AWG/kcmil) | | Size of Grounding Electrode Conductor (AWG/kcmil) | |
Copper	Aluminum or Copper-Clad Aluminum	Copper	Aluminum or Copper-Clad Aluminum[b]
2 or smaller	1/0 or smaller	8	6
1 or 1/0	2/0 or 3/0	6	4
2/0 or 3/0	4/0 or 250	4	2
Over 3/0 through 350	Over 250 through 500	2	1/0
Over 350 through 600	Over 500 through 900	1/0	3/0
Over 600 through 1100	Over 900 through 1750	2/0	4/0
Over 1100	Over 1750	3/0	250

Notes:
1. Where multiple sets of service-entrance conductors are used as permitted in 230.40, Exception No. 2, the equivalent size of the largest service-entrance conductor shall be determined by the largest sum of the areas of the corresponding conductors of each set.
2. Where there are no service-entrance conductors, the grounding electrode conductor size shall be determined by the equivalent size of the largest service-entrance conductor required for the load to be served.
[a]This table also applies to the derived conductors of separately derived ac systems.
[b]See installation restrictions in 250.64(A).

(Reprinted with permission from NFPA 70-2005)

Figure 27-23 Armored grounding conductor connected with ground clamp to water pipe. *Courtesy of* Thomas & Betts.

Figure 27-24 Ground clamp of the type used to bond (jumper) around water meter. *Courtesy of* Thomas & Betts.

make up bonding and grounding connections is not acceptable because under high levels of fault current, the solder would probably melt, resulting in loss of the integrity of the bonding and/or grounding path.

Various types of ground clamps are shown in Figure 27-22, Figure 27-23, Figure 27-24, and Figure 27-25. These clamps and their attachment to the grounding electrode must conform to *250.70*.

Figure 27-22 Typical ground clamps used in residential systems.

Figure 27-25 Ground clamp of the type used to attach ground wire to well casings. *Courtesy of* Thomas & Betts.

Connection of Grounding and Grounded Conductors in Main Panel and Subpanel

According to the second paragraph of *408.20*, equipment *grounding* conductors (the bare copper equipment grounding conductors found in non-metallic-sheathed cable) shall not be connected to the *grounded* conductor (the neutral conductor) terminal bar (neutral bar) unless the bar is identified for the purpose, and is located where the *grounded* conductor is connected to the *grounding* electrode conductor. In this residence, this occurs in Main Panel A. The green bonding screw furnished with Panel A is installed in Panel A. This bonds the neutral bar, the ground bus, the grounded neutral conductor, the grounding electrode conductor, and the panel enclosure together, as in Figure 27-12. *Grounding* conductors and *grounded* conductors are not to be connected together anywhere on the load side of the main service disconnect, *250.24(5)* and *250.142(B)*. The green bonding screw furnished with Panel B will *not* be installed in Panel B. More on this a little later in this unit.

Sheet Metal Screws Not Permitted for Connecting Grounding Conductors

▶Sheet metal screws shall not be used to connect grounding conductors or connection devices to enclosures, *250.8*.◀ Sheet metal screws do not have the same fine thread that No. 10-32 machine screws have, which match the tapped No. 10-32 threaded holes in outlet boxes, device boxes, and other enclosures. Sheet metal screws "force" themselves into a hole instead of nicely threading themselves into a pretapped matching hole. Sheet metal screws have *not* been tested for their ability to safely carry ground-fault currents as required by *250.4(A)(5)*.

BONDING

The *NEC®* defines **bonding** as *"the permanent joining of metallic parts to form an electrically conductive path that ensures electrical continuity and the capacity to conduct safely any current likely to be imposed."** *

Bonding jumpers accomplish this in a number of ways:

1. Equipment bonding "ties" equipment together to keep voltage buildup on the equipment the same, so there will be no difference in potential between the various pieces of equipment. For example, to bond two metal raceways together, an **equipment bonding jumper** is installed where there is a break in the metal conduit run, as in Figure 27-26. Bonding equipment together does *not* necessarily mean that the equipment is properly grounded.

2. A **circuit bonding jumper** connects conductors of a circuit together. This jumper must have the same or greater current-carrying ability than the circuit conductors, Figure 27-27.

3. A **main bonding jumper** is defined as "the connection between the grounded circuit conductor and the equipment grounding conduc-

---

*Reprinted with permission from NFPA 70-2005.

Figure 27-26 Bonding jumper (equipment).

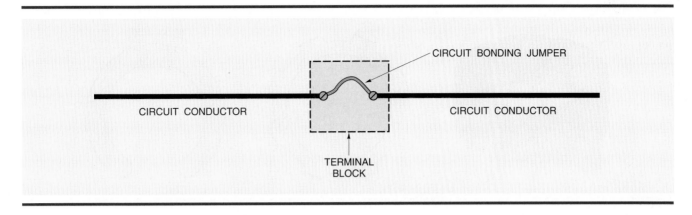

Figure 27-27   Bonding jumper (circuit).

tor at the service." Main bonding jumpers are clearly illustrated in Figure 27-2, Figure 27-7, Figure 27-12, Figure 27-19, and Figure 27-20.

4. Bonding for swimming pools is shown in Figure 30-4.

At the main service-entrance equipment, the grounded neutral conductor must be bonded to the metal enclosure. For most residential panels, this main bonding jumper is a bonding screw that is furnished with the panel. This bonding screw is inserted through the neutral bar into a threaded hole in the back of the panel. This bonding screw must be green and must be clearly visible after it is in place, *250.28(B)*.

The required main bonding jumpers for services must not be smaller than the grounding electrode conductor, *250.28(D)*. The lugs on bonding bushings are based on the ampacity of the conductors that would normally be installed in that particular size of raceway. The lugs become larger as the trade size of the bushing increases.

Bonding at service-entrance equipment is very important because service-entrance conductors do not have overcurrent protection at their line side, other than the electric utility company's primary transformer fuses. Overload protection for service-entrance conductors is at their load end. High available fault currents can result in severe arcing, which is a fire hazard. For all practical purposes, available short-circuit current is limited only by (1) the kVA rating and impedance of the transformer that sup-

plies the service equipment, and (2) the size, type, and length of the service conductors between the transformer and service equipment. Fault currents can easily reach 20,000 amperes or more at the main service in residential installations. Much higher fault current is available in multifamily dwellings such as apartments and condominiums.

Figure 27-7 and Figure 27-20 show the hot and cold metal water pipes bonded together. Some electrical inspectors will consider the hot/cold mixing faucets and the water heater as an acceptable means of bonding the hot and cold water metal pipes together. Most, however, will require a bonding jumper as illustrated in the figures, consistent with the thinking that electrical grounding and bonding shall not be dependent upon the plumbing trade.

Fault current calculations are presented later in this unit. The text *Electrical Wiring—Commercial* (©Thomson Delmar Learning) covers fault current calculations in greater depth.

NEC® *250.92(A)* and *(B)* lists the parts of the service-entrance equipment that must be bonded. These include the meter base, service raceways, service cable armor, and main disconnect. *NEC®* *250.94* lists the methods acceptable for bonding all of the previous equipment.

Grounding/Bonding Bushings

Grounding bushings have a means (lug) for connecting a grounding or bonding conductor to it. A grounding bushing might also have a means (one or more set screws) that ensure reliable bonding of the

Figure 27-28 An insulating grounding/bonding bushing with grounding lug, a bonding locknut, and a bonding wedge.

bushing to a metal raceway, in which case the bushing serves as a grounding and bonding bushing. These bushings are available with an insulated throat. See Figure 27-28. These are the most commonly used for bonding residential electrical services.

Bonding Bushings, Bonding Locknuts, Bonding Wedges

Bonding bushings, bonding locknuts, and bonding wedges have one or more set screws that "dig in" and secure the bushing to the electrical equipment enclosure. These products do not have a means for connecting a grounding or bonding conductor.

Why Are Grounding/Bonding Bushings Installed?

Grounding/bonding bushings, bonding wedges, grounding conductors, and bonding jumpers are installed to ensure a low-impedance path to ground if a fault occurs. Bonding jumpers are required where there are concentric or eccentric knockouts, or where reducing washers are installed. *NEC®* *300.4(F)* and *312.6(C)* state that if 4 AWG or larger conductors are installed in a raceway, an insulating bushing or equivalent must be used, shown in Figure 27-29. Insulating bushings protect the conductors from abrasion where they pass through the bushing. Combination metal/insulating bushings can be used. Some EMT connectors have an insulated throat.

If the bushing is made of insulating material only, as in Figure 27-29, then two locknuts must be used, as shown in Figure 27-30.

Figure 27-29 Insulating bushings.

LOCKNUTS, ONE
INSIDE AND
ONE OUTSIDE

JUNCTION BOX

INSULATING
BUSHING

(USE DOUBLE LOCKNUTS WHEN INSULATING
BUSHING IS NONMETALLIC)

Figure 27-30 The use of double locknuts plus an insulating bushing. See *300.4(F)* and *312.6(C)*.

Installing a Grounding Electrode Conductor from Main Service Disconnect to Metal Underground Water Pipe that Will Serve as the Grounding Electrode

The *Code* specifies in *NEC® 250.64* how a grounding electrode conductor is to be installed, as shown in Table 27-4.

In order to carry high values of ground-fault current, the ground path must have as low an impedance as possible. *NEC® 250.4(A)(5)* requires that the ground path be *effective*. This means that the path must be permanent and continuous, must have the capability of carrying any value of fault current that it might be called upon to carry, and shall have an impedance low enough to limit the voltage to ground and to ensure that the protective overcurrent device will operate. *NEC® 250.90* and *250.96* have similar requirements. For example, the metal raceway enclosing a grounding electrode conductor must be continuous from the main disconnect to the ground clamp, or be made continuous by proper bonding.

NEC® 250.92(A)(3) requires that the metal raceway be bonded at both ends. At first thought, it might appear that simply installing a grounding electrode conductor in a metal raceway makes for a neat and workmanlike installation. However, unless the enclosing metal raceway is properly bonded on both ends, the ground path will be a high impedance path. Bonding the enclosing metal raceway at one end only will result in a ground path impedance approximately twice that of bonding the metal raceway on both ends. Remember—the higher the impedance, the lower the current flow and the longer it takes the overcurrent device to clear the fault.

Past testing (Soares Book on Grounding, IAEI, and IEEE data) has shown that for a 300-ampere ground fault, approximately 295 amperes flowed in the metal raceway, and 5 amperes flowed in the enclosed grounding electrode conductor. Another test revealed that for a 590-ampere ground fault, 585 amperes flowed in the metal raceway, and 5 amperes flowed in the enclosed grounding electrode conductor. The tendency for current to flow along the surface of a conductor is called "skin effect." Proper bonding at both ends of a metal raceway is necessary to ensure a low impedance path that will safely transfer the high magnitudes of ground-fault current flowing in the metal conduit to the ground bus in the main panel, and/or to the ground clamp.

Figure 27-31, Figure 27-32, Figure 27-33, Figure 27-34, and Figure 27-35 illustrate accepted methods of installing a grounding electrode conductor between the main panel and the ground clamp. Additional information concerning grounding electrode conductors can be found in Unit 4.

4 AWG OR LARGER	6 AWG	SMALLER THAN 6 AWG
May be attached to the surface on which it is carried. Does not need additional mechanical protection unless exposed to severe physical damage, which is highly unlikely in the case of residential installations.	If not subject to physical damage, may be run along the surface of the building structure without additional protection. If subject to physical damage, install in rigid metal conduit, electrical metallic tubing, intermediate metal conduit, rigid non-metallic conduit, or cable armor.	Must be installed in rigid metal conduit, electrical metallic tubing, intermediate metal conduit, rigid non-metallic conduit, or cable armor.

A TYPICAL ARMORED GROUNDING ELECTRODE CONDUCTOR.

Table 27-4 Installing a grounding electrode conductor.

GROUNDING ELECTRICAL EQUIPMENT AT A SECOND BUILDING

Detached Garage—Grounding

A detached garage or other second building on the same residential property is almost always served by the same electrical service.

For dwellings, a detached garage is not required to have any electrical supply. However, if you do run

Figure 27-31 This installation "meets Code."

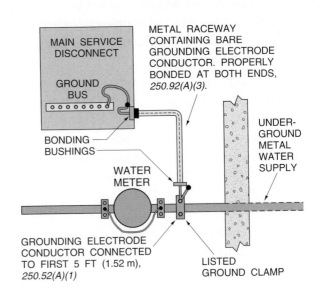

Figure 27-33 This installation "meets Code."

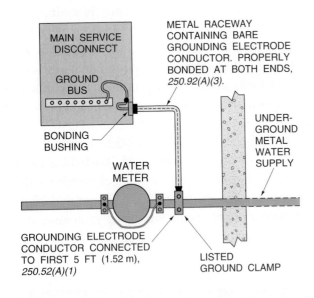

Figure 27-32 This installation "meets Code."

Figure 27-34 This installation *does not* "meet Code."

electric power to a detached garage, there are specific *Code* requirements.

We discussed the receptacle and lighting require-

ments for garages in Unit 3. We will now discuss the grounding requirements for a second building, *250.32*. See Figure 27-36.

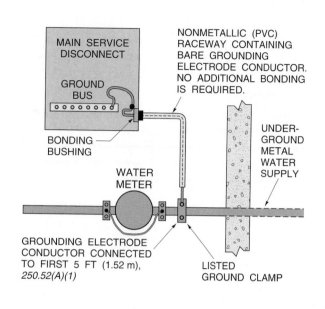

Figure 27-35 This installation "meets Code."

Branch-Circuit to Detached Garage

If only one branch-circuit is run to a detached garage or other second building, the simplest way to provide proper and adequate grounding for the equipment in that detached garage is to install an underground cable Type UF that contains an equipment grounding conductor. A grounded metal raceway between the house and the outbuilding may also serve as the equipment grounding conductor.

With either of these two methods, a separate grounding electrode at the second building is not required, *250.32(A), Exception.*

Feeder to Detached Garage

When a feeder supplies a panel in a detached garage or other second building that has more than one branch-circuit, proper grounding at the second building is accomplished by:

- installing a grounded metal raceway, which is an acceptable equipment grounding conductor, between the main building panel and the panel in the second building, or

- installing a separate equipment grounding conductor (in a cable or in a nonmetallic raceway) between the main building panel and the panel in the second building.

In either case, a grounding electrode is required at the second building. A grounding electrode conductor is run between the grounding electrode and the equipment grounding bus in the panel in the second building.

The grounding electrode conductor is sized according to *Table 250.66,* but does not have to be larger than the largest ungrounded supply conductor, *250.32(E).*

The neutral bus and the equipment grounding bus in the second building are not connected together. If these two buses were tied together in the second building, there would be more than one path (a parallel path) for neutral current to flow back on. Take a look at Figure 27-36. Imagine a tie between the neutral bus and the equipment grounding bus in the second building. Now imagine that there is some value of neutral current. You can readily see that this neutral current would have many ways to return to the source at the main building panel. Some current would return on the neutral conductor, some on the equipment grounding conductor, some

WHEN A SECOND BUILDING IS INVOLVED, EQUIPMENT GROUNDING AT THE SECOND BUILDING CAN BE ACCOMPLISHED IN TWO WAYS:
1. RUN A SEPARATE EQUIPMENT GROUNDING CONDUCTOR, OR
2. INSTALL A GROUNDING ELECTRODE AT THE SECOND BUILDING.

KEEP NEUTRAL ISOLATED

MAIN PANEL IN MAIN BUILDING

RUN SEPARATE EQUIPMENT GROUNDING CONDUCTOR (GREEN OR BARE) SIZED PER *TABLE 250.122*, OR INSTALL ACCEPTABLE METAL RACEWAY. SEE *250.118*.

SUB-PANEL IN SECOND BUILDING

WHITE
BLACK
RED
GREEN OR BARE

MAIN

FEEDER

EQ. GROUND BUS

NEUTRAL BUS

IN MAIN PANEL, THE PANEL ENCLOSURE, NEUTRAL BUS AND GROUND BUS ARE TIED TOGETHER (MAIN BONDING JUMPER).

REGROUND NEUTRAL AT SECOND BUILDING TO PANEL AND TO GROUNDING ELECTRODE *ONLY* IF:
A. THERE IS NO CONTINUOUS PATH (SUCH AS A METAL WATER PIPE) BETWEEN FIRST AND SECOND BUILDING.
B. THERE IS NO EQUIPMENT GROUNDING CONDUCTOR BETWEEN FIRST AND SECOND BUILDING.

MAIN

EQ. GROUND BUS

NEUTRAL BUS

GROUNDING ELECTRODE CONDUCTOR

BONDING JUMPER USUALLY A NO. 10-32 SCREW FURNISHED BY MFG. OF PANEL

GROUNDING ELECTRODE CONDUCTOR SIZED PER *TABLE 250.66*

SUPPLEMENTAL GROUND ROD

UNDERGROUND METAL WATER PIPE

GROUNDING ELECTRODE NOT NEEDED AT SECOND BUILDING IF THERE IS ONLY ONE BRANCH-CIRCUIT THAT INCLUDES AN EQUIPMENT GROUNDING CONDUCTOR.

GROUND ROD

Figure 27-36   Grounding at a second building, *250.32*.

on the grounding electrode conductor—to the grounding electrode—and back through the earth, and some through a metal water pipe if one is present. These multiple paths of neutral return current are not permitted.

If there is no metal path such as a metal water pipe between the two buildings, and no equipment grounding conductor is run between the two buildings, a grounding electrode at the second building is required, and the neutral bus and the equipment grounding bus are connected together. A grounding electrode conductor is run between the grounding bus and the grounding electrode.

If you do not run a separate equipment grounding conductor to a residential garage (or other second building), the grounded conductor of the feeder must not be smaller than the sizes given in *Table 250.122*. This is to ensure a good ground-fault return path. For a feeder to a detached residential garage or other second building, you probably would size the grounded neutral conductor the same size as the two ungrounded conductors. The grounded feeder conductor need not be larger than the largest ungrounded supply conductor.

All of the above requirements are found in *230.32*, and are illustrated in Figure 27-36.

REVIEW

Note: Refer to the *NEC,*® the text, and/or the plans where necessary.

1. Define the "service point." _____

    _____

    _____

    _____

    _____

2. Define "service-drop conductors." _____

    _____

    _____

    _____

    _____

3. Who is responsible for determining the service location? _____

4. a. The service head must be located (above) (below) the point where the service-drop conductors are spliced to the service-entrance conductors. Circle one.

    b. What *Code* section provides the answer to Part (a)? _____

5. What is a mast-type service entrance? _____

    _____

    _____

6. When a conduit is extended through a roof, must it be guyed? _____

7. a. What size and type of conductors are installed for this service? _____

    _____

    b. What size conduit is installed? _____

    c. What size grounding electrode conductor is installed? (not neutral) _____

    d. Is the grounding electrode conductor insulated, armored, or bare? _____

8. How and where is the grounding electrode conductor attached to the water pipe?

    _____

    _____

9. What are the minimum distances or clearances for the following?

    a. Service-drop clearance over private driveway _____

    b. Service-drop clearance over private sidewalks _____

    c. Service-drop clearance over alleys _____

    d. Service-drop clearance over a roof having a roof pitch of not less than 4/12. (Voltage between conductors does not exceed 300 volts.) _____

    e. Service-drop horizontal clearance from a porch _____

    f. Service-drop clearance from a fence that can be climbed upon _____

10. What are the minimum size ungrounded conductors using Type THW copper for the following residential electrical services? The panels are marked 75°C. For the 400-ampere service, install two conductors in parallel per phase. Use two raceways. Install three conductors in each raceway.

AMPERE RATING OF SERVICE	SERVICE-ENTRANCE CONDUCTORS SIZED PER *TABLE 310.16*	SERVICE-ENTRANCE CONDUCTORS SIZED PER *TABLE 310.15(B)(6)*
100		
200		
400		

11. What is the minimum size copper grounding electrode conductor for each of the following residential electrical services? Refer to *Table 250.66*. The ungrounded conductors are Type THW copper.

AMPERE RATING OF SERVICE	GROUNDING ELECTRODE CONDUCTOR WHEN THE SERVICE-ENTRANCE CONDUCTORS ARE SIZED PER *TABLE 310.16*	GROUNDING ELECTRODE CONDUCTOR WHEN THE SERVICE-ENTRANCE CONDUCTORS ARE SIZED PER *TABLE 310.15(B)(6)*
100		
200		
400		

*Table 310.15(B)(6) is only for 120/240-volt, single-phase residential services and feeders. This table does not apply to services and feeders other than residential.

12. What is the recommended height of a meter socket from the ground? _____

13. a. May the bare grounded neutral conductor of a service be buried directly in the ground? _____

    b. What section of the *Code* covers this? _____

14. How far must mechanical protection be provided when underground service conductors are carried up a pole? _____

15. a. A service disconnect may consist of not more than how many switches or circuit breakers? _____

    b. Must these devices be in one enclosure? _____

    c. What type of main disconnect is provided in this residence? _____

    d. Does your city permit more than one service disconnect? _____

16. Complete the following table by filling in the columns with the appropriate information.

	CIRCUIT NUMBER	AMPERE RATING	POLES	VOLTS	CONDUCTOR SIZE
A. Living room receptacle outlets					
B. Workbench receptacle outlets					
C. Water pump					
D. Attic exhaust fan					
E. Kitchen lighting					
F. Hydromassage tub					
G. Attic lighting					
H. Counter-mounted cooking unit					
I. Electric furnace					

17. a. What size conductors supply Panel B? _____

    b. What size raceway? _____

    c. Is this raceway run in the form of electrical metallic tubing or rigid conduit?

    _____

    d. What size overcurrent device protects the feeders to Panel B? _____

18. How many electric meters are provided for this residence? _____

19. a. According to the *NEC*,® is it permissible to ground a residential rural electrical service to driven ground rods only when a metallic water system is available?

    _____

    b. What sections of the *Code* apply? _____

20. What table in the *NEC*® lists the sizes of grounding electrode conductors required for electrical services of various sizes? Table _____

21. *NEC*® 250.53(D)(2) requires that a supplemental ground be provided if the available grounding electrode is: (Circle the correct answer.)

    a. metal underground water pipe
    b. building steel
    c. concrete encases ground

22. Do the following conductors require mechanical protection?

    a. 8 AWG grounding electrode conductor _____

    b. 6 AWG grounding electrode conductor _____

    c. 4 AWG grounding electrode conductor _____

23. Why is bonding of service-entrance equipment necessary? _____

_____

_____

24. What special types of bushings are required on service entrances? _____

_____

25. When conductors 4 AWG or larger are installed in conduit, what additional provision is required on the conduit ends? _____

_____

26. What minimum size copper bonding jumpers must be installed to bond properly the electrical service for the residence discussed in this text? _____

_____

27. a. Where is Panel A located? _____

b. On what type of wall is Panel A fastened? _____

c. Where is Panel B located? _____

d. On what type of wall is Panel B fastened? _____

28. When conduits pass through the wall from outside to inside, the conduit must be _____ to prevent air circulation through the conduit.

29. Briefly explain why electrical systems and equipment are grounded. _____

_____

_____

_____

_____

30. In general, systems are grounded so that the maximum voltage to ground does not exceed: (Circle the correct response.)
    a. 120 volts   b. 150 volts   c. 300 volts

31. To ensure a complete grounding electrode system: (Circle the correct response.)
    a. everything must be bonded together
    b. all metal pipes and conduits must be isolated from one another
    c. the service neutral is grounded to the water pipe only

32. An electric clothes dryer is rated at 5700 watts. The electric rate is 10.091 cents per kilowatt-hour. The dryer is used continuously for 3 hours. Find the cost of operation, assuming the heating element is on continuously.

33. A heating cable rated at 750 watts is used continuously for 72 hours to prevent snow from freezing in the gutters of the house. The electric rate is 8.907 cents per kilowatt-hour. Find the cost of operation.

34. When used as service equipment, a panelboard (load center) must be _____ that is suitable for use as service equipment, *230.66*. If the panelboard (load center) contains the main service disconnect, it must be clearly marked _____ _____ , *230.70(B)*.

35. Here are five commonly used terms in the electrical industry. Enter the letter of the term that corresponds to its definition.
    a. Grounding electrode conductor
    b. Main bonding jumper
    c. Grounded circuit conductor
    d. Equipment grounding conductor
    e. Lateral service conductors

    _____ The neutral conductor.

    _____ The term used to define underground service-entrance conductors that run between the meter and the utilities connection.

    _____ The conductor (sometimes a large threaded screw) that connects the neutral bar in the service equipment to the service-entrance enclosure.

    _____ The conductor that runs between the neutral bar in the main service equipment to the grounding electrode (water pipe, ground rod, etc.).

    _____ The bare copper conductor found in nonmetallic-sheathed cable.

36. The electrician mounted the disconnect switch for a central air-conditioning unit 8 ft (2.5 m) above the ground. This was easy to do because all he had to do was run the conduit across the ceiling joists in the basement, through the outside wall, and directly into the back of the disconnect switch. The electrical inspector turned the job down, citing *NEC*® _____ that requires that the disconnecting operating handle must not be higher than _____ feet above the ground.

37. a. According to the *NEC*® definition of a wet location, a basement cement wall that is in direct contact with the earth is considered to be a wet location. A panelboard mounted on this wall must have at least [¼ in. (6 mm)], [½ in. (13 mm)] [1 in. (25 mm)] space between the wall and the panel. Circle the correct answer.

    b. How is this required spacing accomplished? _____
    _____

38. When using re-bars as the concrete encased electrode, the re-bars must be (a) insulated with plastic material so they will not rust, or (b) bonded together by the usual steel tie wires or other effective means. Underline the correct answer.

39. The circuit directory in a panelboard (may) (shall) be filled out according to *NEC*® _____ . Circle the correct answer, and enter the correct *NEC*® section number.

40. A main service panel is located in a dark corner of a basement, far from the basement light. In your opinion, does this installation meet the requirements of *110.26(D)*? Explain your answer. _____

_____

41. Does the electric utility in your area allow the location of the meter to be on the back side of a residence? _____

42. The term used when the utility charges different rates during different periods during the day is _____ .

43. A ground rod is driven below the meter outside of the house. An armored grounding electrode conductor connects between the meter base and the ground rod. The ground clamp is buried under the surface of the soil. This is permitted if the ground clamp is _____ for direct burial according to *NEC®* _____ .

44. Copper-coated steel ground rods are the most commonly used grounding electrodes. These rods must:

    a. be at least _____ in. (_____ mm) in diameter, *250.52(A)(5)*.

    b. be driven to a depth of at least _____ ft (_____ m) unless solid rock is encountered, in which case the rod may be driven at a _____-degree angle, or it may be laid in a trench that is at least _____ ft (_____ mm) deep, *250.53(G)*.

    c. be separated by at least _____ ft (_____ m) when more than one rod is driven, *250.56*.

    d. have a ground resistance of not over _____ ohms for one rod, *250.56*.

45. What section of the *Code* prohibits the use of sheet metal screws as a means of attaching grounding conductors to enclosures? *NEC®* _____

46. Which section of the *NEC®* prohibits using the space below and in front of an electrical panel as storage space? *NEC®* _____ .

47. What section of the *NEC®* prohibits using an underground metal gas pipe as the grounding electrode for an electrical service? _____

48. When wiring a gas furnace, what additional steps, if any, are necessary in order to make sure that the gas piping supplying the furnace is adequately bonded? Explain.

_____

_____

_____

Overcurrent Protection– Fuses and Circuit Breakers

OBJECTIVES

After studying this unit, the student should be able to

▶ **understand the important** *NEC®* **requirements for fuses and circuit breakers.**

▶ **discuss the four possible circuit conditions.**

▶ **understand the various types and operation of fuses and circuit breakers.**

▶ **know when to use single-pole and two-pole circuit breakers.**

▶ **understand the term** *interrupting rating* **for fuses and circuit breakers.**

▶ **calculate available short-circuit current using a simple formula.**

▶ **understand** *series-rated* **panelboards.**

▶ **understand the meaning of** *selective coordination* **and** *nonselective coordination***.**

THE BASICS

The *NEC®* covers overcurrent protection in *Article 240*. Overcurrent protection for residential main services, branch-circuits, and feeders is provided by circuit breakers or fuses. These are the "safety valves" of an electrical circuit.

Fuses and circuit breakers are sized by matching their ampere ratings to conductor ampacities and connected load currents. They sense overloads, short circuits, and ground faults and protect the wiring and equipment from reaching dangerous temperatures.

A fuse will function only one time. It does its job of protecting the circuit. When a fuse opens, find out what caused it to open, fix the problem, and replace it with the size and type that will provide proper overcurrent protection. Do not keep replacing a fuse without finding out why the fuse blew in the first place.

When a circuit breaker trips, find out what caused it to trip, fix the problem, then reset it. Do not repeatedly reset the breaker again and again into a fault. This is looking for trouble.

KEY *NEC®* REQUIREMENTS FOR OVERCURRENT PROTECTION

- *NEC® Table 210.24:* This table shows the maximum overcurrent protection for branch-circuit conductors.

- *NEC® 240.4:* Overcurrent protection is sized according to the ampacity of a conductor.

- *NEC® 240.4(B):* Where the standard ampere rating of fuses or circuit breaker does not match the ampacity of the conductor, the next higher ampere rating is permitted. This is permitted for 800 amperes or less.

- *NEC® 240.4(D):* The maximum overcurrent protection of small conductors is:

CONDUCTOR SIZE	MAXIMUM OVERCURRENT PROTECTION
14 AWG copper	15 amperes
12 AWG copper	20 amperes
10 AWG copper	30 amperes

- *NEC® 240.6:* Lists the standard ampere ratings for fuses and circuit breakers.

- *NEC® 240.20(A):* Overcurrent protection shall be provided for each ungrounded ("hot") conductor.

- *NEC® 240.21:* Overcurrent protection shall be at the point where a conductor receives its supply. Refer to this section for exceptions. Note that the overcurrent protection in the main service panelboard is located at the load end of the service-entrance conductors. This is permitted because the *NEC®* accepts the fact that short-circuit protection of service-entrance conductors is provided by the utility's primary transformer fuse.

- *NEC® 240.22:* Overcurrent protection devices not permitted in the grounded conductor.

- *NEC® 240.24(A):* Overcurrent protection devices shall be accessible.

- *NEC® 240.24(D):* Overcurrent protective devices shall not be near easily ignitable material, such as in clothes closets.

- *NEC® 240.24(E):* Overcurrent devices shall not be located in bathrooms.

- *NEC® 230.70(A)(2):* Service disconnect shall not be located in bathrooms.

- *NEC® 230.79(C):* 100-ampere minimum size service for a one-family dwelling.

- *NEC® 230.90(A):* Overcurrent protection shall be provided for each ungrounded ("hot") service-entrance conductor.

- *NEC® 230.91:* The main service overcurrent device is usually an integral part of the service disconnecting means. For large services, the service disconnecting means could be a separate disconnect switch.

- *NEC® Table 310.15(B)(6):* This table shows special ampacities permitted for service-entrance conductors for three-wire, single-phase, 120/240-volt services for dwellings. Overcurrent protection for these conductors is not to exceed the special ampacities.

FIVE CIRCUIT CONDITIONS

Normal: Normal loading of a circuit is when the current flowing is within the capability of the circuit and/or the connected equipment. In Figure 28-1 we have a 15-ampere circuit carrying 10 amperes.

An *overload* is a condition where the current flowing is more than the circuit and/or connected equipment is designed to safely carry. Figure 28-2 shows a 15-ampere circuit carrying 20 amperes. A momentary overload is harmless, but a continuous overload condition will cause the conductors and/or equipment to overheat—a potential cause for fire. The current flows through the "intended path"—the conductors and/or equipment.

A *short circuit* is a condition when two or more conductors come in contact with one another, resulting in a current flow that bypasses the connected load. A short circuit might be two "hot" conductors coming together, or it might be a "hot" conductor and a "grounded" conductor coming together. In either case, the current flows outside of the "intended path." The only resistance (impedance) is that of the conductors, the source, and the arc. This low resistance results in high levels of short-circuit current. The heat generated at the point of the arc can result in a fire, as in Figure 28-3.

A *ground fault* is a condition when a "hot" conductor comes in contact with a grounded surface, such as a grounded metal raceway, metal water pipe, sheet metal, etc., as in Figure 28-4. The current flows outside of the "intended path." A ground fault can result in a flow of current greater than normal, in which case the overcurrent device will open. A ground fault can also result in a current flow less than normal, in which case the overcurrent device may not open.

A *GFCI* is an example of a device that protects against very low levels of ground-fault current passing through the human body when a person touches a "live" wire and a grounded surface. See Unit 6.

An *open circuit* is a condition where the circuit is not closed somewhere in the circuit, as in Figure 28-5, and no current can flow.

Ⓐ NORMAL CIRCUIT

THE CONDUCTORS CAN SAFELY CARRY THE CURRENT. THEY DO NOT GET HOT. THE 15-AMPERE FUSES DO NOT OPEN.

$$I \text{ (THROUGH CIRCUIT)} = \frac{E}{R} = \frac{240}{24 + 0.001 + 0.001 + 0.001} = \frac{240}{24.003} = 9.99875 \text{ AMPERES}$$

Figure 28-1   A normally loaded circuit.

Ⓑ OVERLOADED CIRCUIT

THE CONDUCTORS BEGIN TO GET HOT, BUT THE 15-AMPERE FUSES WILL OPEN BEFORE THE CONDUCTORS ARE DAMAGED.

$$I \text{ (THROUGH CIRCUIT)} = \frac{E}{R} = \frac{240}{12 + 0.001 + 0.001 + 0.001} = \frac{240}{12.003} = 19.995 \text{ AMPERES}$$

Figure 28-2   An overloaded circuit.

Ⓒ **SHORT CIRCUIT**

THE CONDUCTORS GET EXTREMELY HOT. THE INSULATION WILL MELT OFF AND THE CONDUCTORS WILL MELT UNLESS THE FUSES OPEN IN A VERY SHORT (FAST) PERIOD OF TIME. CURRENT-LIMITING OVERCURRENT DEVICES WILL LIMIT THE AMOUNT OF "LET-THROUGH" CURRENT BY OPENING SO FAST (FRACTION OF A CYCLE) THAT THE FULL VALUE OF FAULT CURRENT WILL NOT BE REACHED.

CONDUCTORS RATED 15 AMPERES
CONDUCTOR RESISTANCE = 0.001 OHM

15 A FUSE

80,000 AMPERES

240-VOLT SOURCE

SOURCE RESISTANCE = 0.001 OHM

15 A FUSE

$24\ \Omega$

RESISTANCE OF SHORT CIRCUIT = ZERO OHMS

$$\frac{R_1 \times R_2}{R_1 + R_2} = \frac{24 \times 0}{24 + 0} = \frac{0}{24} = 0$$

CONDUCTORS RATED 15 AMPERES
CONDUCTOR RESISTANCE = 0.001 OHM

$$I \text{ (THROUGH CIRCUIT)} = \frac{E}{R} = \frac{240}{0.001 + 0.001 + 0.001} = \frac{240}{0.003} = 80,000 \text{ AMPERES}$$

Figure 28-3   Note that the connected load is short-circuited.

Ⓓ **GROUND FAULT**

"HOT" CONDUCTOR

SOURCE

GROUNDED CIRCUIT CONDUCTOR

"HOT" CONDUCTOR COMES IN CONTACT WITH METAL RACEWAY OR OTHER METAL OBJECT. IF THE RETURN GROUND PATH HAS LOW RESISTANCE (IMPEDANCE), THE OVERCURRENT DEVICE PROTECTING THE CIRCUIT WILL CLEAR THE FAULT. IF THE RETURN GROUND PATH HAS HIGH RESISTANCE (IMPEDANCE), THE OVERCURRENT DEVICE WILL NOT CLEAR THE FAULT. THE METAL OBJECT WILL THEN HAVE A VOLTAGE TO GROUND THE SAME AS THE "HOT" CONDUCTOR HAS TO GROUND. IN HOUSE WIRING, THIS VOLTAGE TO GROUND IS 120 VOLTS. PROPER GROUNDING AND GROUND-FAULT CIRCUIT INTERRUPTER PROTECTION IS DISCUSSED ELSEWHERE IN THIS TEXT. THE CALCULATION PROCEDURE FOR A GROUND FAULT IS THE SAME AS FOR A SHORT CIRCUIT, HOWEVER, THE VALUES OF "R" CAN VARY GREATLY BECAUSE OF THE UNKNOWN IMPEDANCE OF THE GROUND RETURN PATH. LOOSE LOCKNUTS, BUSHINGS, SET SCREWS ON CONNECTORS AND COUPLINGS, POOR TERMINATIONS, RUST, ETC., ALL CONTRIBUTE TO THE RESISTANCE OF THE RETURN GROUND PATH, MAKING IT EXTREMELY DIFFICULT TO DETERMINE THE ACTUAL GROUND-FAULT CURRENT VALUES.

Figure 28-4   The insulation on the "hot" conductor has come in contact with the metal conduit. This is termed a "ground fault."

E OPEN CIRCUIT

THE CIRCUIT HAS AN "OPEN." NO CURRENT WILL FLOW IN THE CIRCUIT.

CONDUCTORS RATED 15 AMPERES
CONDUCTOR RESISTANCE = 0.001 OHM

15 A FUSE

X "OPEN"

240-VOLT SOURCE

24 Ω   LOAD RESISTANCE = 24 OHMS

SOURCE RESISTANCE = 0.001 OHM

15 A FUSE   ZERO AMPERES

$$I \text{ (THROUGH CIRCUIT)} = \frac{E}{R} = \frac{240}{\infty} = \text{ZERO AMPERES*}$$

*THE SYMBOL ∞ REPRESENTS AN IMMEASURABLY HIGH VALUE. AN OPEN CIRCUIT HAS SUCH A HIGH RESISTANCE VALUE THAT ORDINARY OHMMETERS CANNOT READ IT. WE CAN IGNORE THE RESISTANCE VALUES OF THE OTHER COMPONENTS OF THE CIRCUIT, AND USE INFINITY FOR THE "R" VALUE.

CAUTION: LINE VOLTAGE (IN THIS CASE 240 VOLTS) APPEARS ACROSS THE "OPEN."

Figure 28-5   The circuit is "open" where marked "X."

FUSES

Fuses are a reliable and economical form of overcurrent protection. *NEC® 240.50* through *240.54* provides the requirements for plug fuses, fuseholders, and adapters. Here is a brief recap:

Edison-Base Plug Fuses

- Plug fuses may be used (1) in circuits that do not exceed 125 volts between conductors, and (2) in circuits having a grounded neutral where no conductor operates at over 150 volts to ground. In a residential 120/240-volt electrical system, the voltage between the two "hot" conductors is 240 volts, and the voltage from either "hot" conductor to ground is 120 volts.

- The fuses shall have ampere ratings of 0 through 30 amperes.

- Plug fuses shall have a hexagonal configuration somewhere on the fuse when rated at 15 amperes or less.

- The screw shell of the fuseholder must be connected to the load side of the circuit.

- Edison-base plug fuses may be used only to replace fuses in existing installations where there is no sign of overfusing or tampering.

- Typical applications for plug-fuses are using box-cover units where you need to provide close overcurrent protection such as for an attic fan, a gas furnace, an appliance, and similar loads. This already has been discussed elsewhere in this text.

Type S Fuses

All new plug fuse installations shall be Type S fuses. That is because the ampere ratings of conventional Edison-base plug fuses are all interchangeable—any ampere rating up to and including 30 amperes.

- Type S fuses are classified at 0 through 15 amperes, 16 through 20 amperes, and 21 through 30 amperes. The reason for this classification is given in the following paragraph. In the 0–15 ampere range, there are many ampere ratings to choose from, excellent for protecting motors. See Table 28-1. Type S fuses are used with a corresponding size adapter.

TYPE S FUSE INFORMATION

Type S Fuse Ampere Ratings	Type S Adapter Rating	Type S Fuse Ampere Ratings That Fit Into This Adapter
³⁄₁₀, ½, ⁸⁄₁₀, ⁴⁄₁₀, ⁶⁄₁₀, 1	1	1 ampere & smaller
1⅛, 1¼	1¼	All smaller
1⁴⁄₁₀, 1⁶⁄₁₀	1⁶⁄₁₀	All smaller
1⁸⁄₁₀, 2	2	1⁸⁄₁₀, 2
2¼, 3½	2½	1⁸⁄₁₀, 2, 2¼, 2½
2⁸⁄₁₀, 3²⁄₁₀	3²⁄₁₀	1⁸⁄₁₀, 2, 2¼, 2½, 2⁸⁄₁₀, 3²⁄₁₀
3½, 4	4	3½, 4
4½, 5	5	3½, 4, 4½, 5
5⁶⁄₁₀, 6¼	6¼	3½, 4, 4½, 5, 5⁶⁄₁₀, 6¼
7, 8	8	7, 8
9, 10	10	7, 8, 9, 10
12, 14	14	7, 8, 9, 10, 12, 14
15	15	15
20	20	20
25	30	20, 25, 30
30	30	20, 25, 30

Table 28-1   This table shows the many ampere rating combinations for Type S fuses and adapters.

When the electrician installs fusible equipment, the ampere rating of the circuit(s) must be determined. After determining the ampere rating of the Type S fuse to be used, an adapter of the proper size is inserted into the Edison-base fuseholder. The proper Type S fuse is then screwed into the adapter. Because of the adapter, the fuseholder is nontamperable and noninterchangeable. For example, assume that a 15-ampere adapter is inserted for a 15-ampere branch-circuits; it is impossible to substitute a Type S fuse with a higher ampere rating without removing the 15-ampere adapter.

A Type S fuse and adapter is shown in Figure 28-6.

Fuse Characteristics

Fuses have different time-current characteristics. The term "time-current" refers to how long it will take a fuse or circuit breaker to open under different current values.

- Nontime-delay: This type of fuse has one link (fusible element). One part of the link is "necked down" so when it is called upon to open, it will open in the weakest part of the link—the necked down portion. Nontime-delay fuses are not the best choice for motor circuits because of the high starting inrush current of motors. See the "W" fuse, Figure 28-7.

- Time-delay, loaded link: This type of fuse has one link (fusible element) that is "loaded" with a heat sink next to the "necked down" portion of the link. This "load" acts as a heat sink that

Figure 28-6   Type S fuse and adapter. *Courtesy of* Cooper Bussmann, Inc.

Figure 28-7   Three types of plug fuses: single-element, nontime-delay (W); dual-element, time-delay (T); loaded link, time-delay (TL). *Courtesy of* Cooper Bussmann, Inc.

absorbs a considerable amount of heat before the "necked down" portion of the link melts open. This heat sink provides the fuse with time delay. See the "TL" fuse, Figure 28-7.

- Time-delay, dual-element: One fuse element opens quickly when a short circuit, heavy overload, or ground fault occurs. The other element in series with the first element opens slowly on overload conditions. Dual-element, time-delay fuses are an excellent choice for motor circuits because they will not open needlessly on momentary overloads. See the "T" fuse, Figure 28-7.

Cartridge Fuses

Cartridge fuses are covered in *NEC® 240.60* and *240.61*.

Cartridge fuses are available with the same three basic types of time-current characteristics as plug fuses.

The most common type of dual-element cartridge fuse is shown in Figure 28-8. They are available in 250-volt and 600-volt sizes with ratings from 0 through 600 amperes.

Large ampere rating cartridge fuses may have more than one fuse element.

Time-delay, dual-element fuses provide a time-delay of about 10 seconds at a current of 500 percent of the fuse rating before it opens. This provides accurate protection for prolonged overloads.

Dual-element fuses are used on motor and appliance circuits where the long time-delay characteristic is required. Single-element fuses are more suitable for circuits that are not expected to have high inrush currents.

Cartridge fuses are available in Class H and Class R types. Refer to Table 28-2.

General-Purpose Cartridge Fuses

These are single-element cartridge fuses that have no intentional time-delay designed into them as are dual-element, time-delay fuses. These are the type of fuses used in conventional disconnect switches.

They are available in ampere ratings from ⅛ through 600 amperes, in both 250-volt and 600-volt ratings.

Those rated through 60 amperes are classified as K5 and have an interrupting rating of 50,000 amperes. Sizes 0 through 60 have a brass ferrule on each end. See Figure 28-9.

Those rated 65–600 are classified as class H and have an interrupting rating of 10,000 amperes. These sizes have a blade coming out of each end and are therefore referred to as "knife blade" fuses. See Figure 28-9.

Other Classes of Fuses

Class CC, Class G, and Class T fuses are special-purpose fuses that are found in original equipment, or in commercial and industrial panelboards, multi-metering equipment, and motor controllers. They are used in original equipment because of their small physical size. If you need more information about these types of fuses, consult the manufacturer's technical literature and/or web sites. These fuses are rarely used in residential applications but are mentioned here to make you aware of them.

Figure 28-8   Cartridge-type dual-element fuse (A) is a 250-volt, 100-ampere fuse. The cutaway view in (B) shows the internal parts of the fuse. When this type of fuse has a rejection slot in the blades of 70–600 ampere sizes, or a rejection ring in the end ferrules of the 0–60 ampere sizes, it is a UL Class RK1 or RK5 fuse. *Courtesy of Cooper Bussmann, Inc.*

TYPE OF OVERCURRENT DEVICE	INTERRUPTING RATING
Circuit breakers	Are marked with their interrupting rating if other than 5000 amperes.
Plug fuses (Edison base and Type S)	10,000 amperes ac RMS symmetrical.
	Plug fuses 15 amperes and less will have a hexagon shape in the window—or the fuse body of the fuse will have the hexagon shape.
Class H cartridge fuses	10,000 amperes ac RMS symmetrical.
	This is the most common type of low-cost cartridge fuse.
	▶Renewable link Class H cartridge fuses are permitted only on existing installations where there is no evidence of overfusing or tampering.◀
	Rarely, if ever, will these be found in residential electrical systems.
	The interrupting rating is marked when other than 10,000 amperes.
Class R cartridge fuses	The letter "R" stands for "Rejection." Knife blade types have a notch in the blade. Ferrule types have an annular ring in one of the ferrules. This means that if the switch or fuseholder is designed for Class R fuses, then ordinary Class H fuses will not fit.
	Further broken down into RK1 and RK5 categories.
	RK fuses are available with interrupting ratings of 50,000, 100,000, and 200,000 amperes ac RMS symmetrical.
	The interrupting rating is marked on the fuse.

Note: Using a fuse or circuit breaker having a high interrupting rating does not necessarily constitute a safe installation. The entire assembly (panelboard, switch, controller) is marked with its short-circuit rating when used with a specific type of overcurrent device. That is why it is so important to "Read the Label."

Table 28-2   The interrupting ratings of the more common fuses and circuit breakers.

Figure 28-9   A 60-ampere and a 100-ampere general purpose, one-time fuse. *Courtesy of* Cooper Bussmann, Inc.

Sometimes poor connections at the terminals or loose fuse clips can be found because the fuse tubing will appear to be charred. This problem can be dangerous and should be corrected as soon as possible. Quite often, the heat resulting from poor connections in the fuseclips or fuseholder will cause the brass or copper fuseclips or fuseholders to turn from a bright, shiny appearance to that of antique brass or copper.

Fuse Ampere Ratings

The standard ampere ratings for fuses are given in *240.6(A)*. In addition to the standard ratings listed, there are many other "in-between" ampere ratings available to match specific load requirements. Check a fuse manufacturer's catalog for other available ampere ratings.

CIRCUIT BREAKERS

Installations in dwellings normally use thermal-magnetic circuit breakers. On a continuous overload, a bimetallic element in such a breaker moves until it unlatches the inner tripping mechanism of the breaker. Momentary small overloads do not cause the element to trip the breaker. If the overload is heavy or if there is a short circuit, a magnetic coil in the breaker causes it to interrupt the branch-circuit almost instantly. Figure 28-10 illustrates a typical single-pole and a two-pole circuit breaker.

NEC® 240.80 through 240.86 give the requirements for circuit breakers. The following points are taken from these sections.

- Circuit breakers shall be trip-free so that even if the handle is held in the "ON" position, the internal mechanism will trip to the "OFF" position.

- Breakers shall indicate clearly whether they are on or off.

- A breaker shall be nontamperable so that it cannot be readjusted (trip point changed) without dismantling the breaker or breaking the seal.

- The rating shall be durably marked on the breaker. For small breakers rated at 100 amperes or less and 600 volts or less, the rating must be molded, stamped, or etched on the handle or on another part of the breaker that will be visible after the cover of the panel is installed.

- Every breaker with an interrupting rating other than 5000 amperes shall have this rating marked on the breaker.

- Circuit breakers rated at 120 volts and 277 volts and used for fluorescent loads shall not be used as switches unless marked "SWD."

Most circuit breakers are ambient temperature compensated. This means that the tripping point of the breaker is not affected by an increase in the surrounding temperature. An ambient-compensated breaker has two elements. One element heats up due to the current passing through it and the heat in the surrounding area. The other element heats up because of the surrounding air only. The actions of these elements oppose each other. Thus, as the tripping element tends to lower its tripping point because of external heat, the second element opposes the tripping element and stabilizes the tripping point. As a result, the current through the tripping element is the only factor that causes the element to open the circuit. It is a good practice to turn the breaker on and off periodically to "exercise" its moving parts.

NEC® 240.6(A) gives the standard ampere ratings of circuit breakers.

How Much Load Can a Circuit Breaker Safely Carry?

The *NEC*® permits 100 percent loading on an overcurrent device *only* if that overcurrent device is listed for 100 percent loading. See *NEC*® 210.20 and 215.3. At the time of this writing, Underwriters Laboratories (UL) has no listing of residential-type molded-case circuit breakers that are suitable for 100 percent loading. A *maximum* loading of 80 percent is good practice—less is even better!

One Two-Pole or Two Single-Pole Circuit Breakers

Use two-pole circuit breakers on three-wire 120/240-volt branch-circuits (two "hots" and one "neutral") for loads such as electric ranges, electric ovens, and electric clothes dryers.

Use two-pole circuit breakers for straight 240-volt branch-circuits (two "hots") for such loads as electric water heaters, electric furnaces, electric baseboard heaters, air conditioners, and heat pumps.

The use of "handle ties," which connect the handles of two single-pole circuit breakers together so

Figure 28-10   Typical molded-case circuit breakers.
Courtesy of **Schneider Electric.**

both poles trip together, is permitted for a few very specific applications.

Handle ties *do* allow two single-pole circuit breakers to be switched simultaneously to the "OFF" or "ON" positions.

Handle ties generally *do not* simultaneously trip both handle tied single-pole circuit breakers on overloads, short circuits, or ground faults on one of the circuit breakers. This could result in a false sense of security for someone working on an appliance where two single-pole circuit breakers with a handle tie has tripped. He thinks that the branch-circuit is totally in the "OFF" position, when in fact it is quite possible that one of the circuit breakers may still be "ON."

Identified Handle Ties and the *Code*

▶ New in the 2005 *NEC®* is that handle ties for circuit breakers must be *identified. Identified* means they have been tested and listed by a nationally recognized testing laboratory (NRTL) for use on a specific manufacturer's circuit breaker. ◀

- *210.4(B):* Where more than one device or equipment (switches, receptacles, pilot lights, etc.) is on the same yoke, a means must be provided to simultaneously disconnect all ungrounded conductors. Identified handle ties on a circuit breaker meet this requirement. A disconnect switch also meets this requirement.

- *210.4(C):* A multiwire branch-circuit is only permitted to serve line-to-neutral loads. Exceptions to this are if the branch-circuit supplies only one piece of equipment, or if the overcurrent device will simultaneously disconnect all ungrounded conductors of the multiwire branch-circuit.

- *225.33(B):* Identified handle ties permitted to disconnect all conductors of an outside multiwire branch-circuit or feeder with no more than six operations of the hand.

- *230.71(B):* Identified handle ties permitted to disconnect all conductors of a service as long as it takes no more than six operations of the hand to shut everything off.

- *240.15:* Requires that a circuit breaker open all ungrounded conductors of a circuit.

- *240.16(A):* Individual single-pole circuit breakers, with or without handle ties, permitted to protect the ungrounded conductors of a multiwire branch-circuit only when serving single-phase, line-to-neutral loads. See *210.4(B).*

- *240.16(B):* Individual single-pole circuit breakers with identified handle ties permitted to protect the ungrounded conductors for line-to-line loads for single-phase circuits.

With all of the above restrictions on the use of handle ties, they should only be used as "the last resort." If you have to use handle ties, use only those that are *identified.*

INTERRUPTING RATINGS FOR FUSES AND CIRCUIT BREAKERS

Here are some very important *NEC®* sections regarding interrupting ratings.

- *NEC® 110.9:* **Interrupting Rating.** *"Equipment intended to interrupt current at fault levels shall have an interrupting rating sufficient for the nominal circuit voltage and the current that is available at the line terminals of the equipment. Equipment intended to interrupt current at other than fault levels shall have an interrupting rating at nominal circuit voltage sufficient for the current that must be interrupted."**

- *NEC® 110.10:* **Circuit Impedance and Other Characteristics.** *"The overcurrent protective devices, the total impedance, the component short-circuit current ratings, and other characteristics of the circuit to be protected shall be selected and coordinated to permit the circuit protective devices used to clear a fault to do so without extensive damage to the electrical components of the circuit. This fault shall be assumed to be either between two or more of the circuit conductors, or between any circuit conductor and the grounding conductor or enclosing metal raceway. Listed products applied in accordance with their listing shall be considered to meet the requirements of this section."**

- UL Standard No. 67 requires that panelboards be marked with their short-circuit current rating.

Panelboards are available in two types: "fully-rated" and "series-rated." Figure 28-11, Figure 28-12, Figure 28-13, and Figure 28-14 explain the meaning of these terms.

---

*Reprinted with permission from NFPA 70-2005.

The overcurrent protective device must be able to interrupt the current that may flow under any condition (overload or short circuit). Such interruption must be made with complete safety to personnel and without damage to the panel or switch in which the overcurrent device is installed.

CAUTION: Overcurrent devices with inadequate interrupting ratings are, in effect, bombs waiting for a short circuit to trigger them into an explosion. Personal injury may result and serious damage will be done to the electrical equipment.

Interrupting Ratings

Interrupting ratings of overcurrent devices, as shown in Table 28-2, are given in root-mean-square (RMS) values.

The interrupting rating of a fuse or circuit breaker does not indicate the short-circuit rating of a panelboard. These are stand-alone ratings. The short-circuit rating of a panelboard is derived by testing the panel as an assembly with breakers and/or fuses installed in the panel. The short-circuit rating of a panelboard is marked on its label.

Figure 28-11 A *fully-rated* system using main fuses and branch-circuit fuses. The entire assembly (fuses and panelboard) is tested, listed, and marked with a maximum short-circuit rating. Look for the marking on the panelboard. This type of combination can be found in commercial and industrial facilities.

Figure 28-12 A *series-rated* system using main fuses and branch-circuit breakers. The entire assembly (main fuses, branch-circuit breakers, and panelboard) is tested, listed, and marked with a maximum short-circuit rating. Look for the marking on the panelboard. The lower interrupting rating of the branch-circuit breakers is acceptable because the combination has been tested and listed as a series-rated system. This type of combination can be found in light commercial installations such as multi-metering service equipment.

AVAILABLE FAULT CURRENT
20,800 AMPERES

SERVICE EQUIPMENT MUST
HAVE A SHORT-CIRCUIT RATING
OF AT LEAST 20,800 AMPERES

MAIN BREAKER MUST HAVE
AT LEAST A 20,800-AMPERE
INTERRUPTING RATING

BRANCH-CIRCUIT
BREAKERS MUST HAVE AT
LEAST A 20,800-AMPERE
INTERRUPTING RATING

MEETING REQUIREMENTS OF *110.9* AND *110.10*
FOR BREAKER MAIN/BREAKER BRANCH-CIRCUITS.
THIS PANEL IS "FULLY-RATED," NOT "SERIES-RATED."

Figure 28-13 A *fully-rated* system using a main circuit breaker circuit and branch-circuit breakers. The entire assembly (main breaker, branch-circuit breakers, and panelboard) is tested, listed, and marked with a maximum short-circuit rating. Look for the marking on the panelboard. This type of combination can be found in commercial and industrial installations.

AVAILABLE FAULT CURRENT
20,800 AMPERES

PANEL IS UL LISTED
AS "SERIES-CONNECTED"
22/10

SERVICE EQUIPMENT MUST
HAVE A SHORT-CIRCUIT RATING
OF AT LEAST 20,800 AMPERES

MAIN BREAKER
RATED 22,000-AMPERE
INTERRUPTING RATING

BRANCH-CIRCUIT
BREAKERS RATED
10,000-AMPERE
INTERRUPTING RATING

MEETING REQUIREMENTS OF *110.9*, *110.10*, *110.22*, AND *220.86*
FOR SERIES RATED BREAKER MAIN/BREAKER BRANCH-CIRCUITS.
THIS PANEL IS "SERIES-RATED," NOT "FULLY-RATED."

Figure 28-14 A *series-rated* system using a main circuit breaker and branch-circuit breakers. The entire assembly (main breaker, branch-circuit breakers, and panelboard) is tested, listed, and marked with a maximum short-circuit rating. Look for the marking on the panelboard. The lower interrupting rating of the branch-circuit breakers is acceptable because the assembly has been listed as a series-rated system. This type of combination is most commonly found in typical residential and light commercial applications.

SHORT-CIRCUIT CURRENTS

This text does not cover in detail the methods of calculating short-circuit currents. For a detailed discussion, see *Electrical Wiring—Commercial* (©Thomson Delmar Learning). The ratings required to determine the maximum available short-circuit current delivered by a transformer are the kVA and impedance values of the transformer. The size

and length of wire installed between the transformer and the overcurrent device must be considered as well.

The transformers used in modern electrical installations are efficient and have very low impedance values. A low-impedance transformer having a given kVA rating delivers more short-circuit current than a transformer with the same kVA rating and a higher impedance. When an electrical service is connected to a low-impedance transformer, the problem of available short-circuit current is very serious.

The examples given in Figure 28-11, Figure 28-12, Figure 28-13, and Figure 28-14 show that the available short-circuit current at the transformer secondary terminals is 20,800 amperes.

HOW TO CALCULATE SHORT-CIRCUIT CURRENT

Short-circuit current is also referred to as fault current.

Calculating fault currents is just as important as calculating load currents. Overloads will cause conductors and equipment to run hot, shorten their life, and will eventually destroy them. Fault currents that exceed the interrupting rating of the overcurrent protective devices and panel rating can cause violent electrical explosions of equipment with the potential of serious injury to people standing near it. Fire hazard is also present.

The local electric utility and the electrical inspector are good sources of information when short-circuit current is to be determined.

A simplified method is given as follows to determine the approximate available short-circuit current at the terminals of a transformer. This method assumes that there is an infinite amount of primary short-circuit current available (infinite bus), and a bolted fault at the secondary transformer terminals.

1. Determine the normal full-load secondary current delivered by the transformer.

   For single-phase transformers:

   $$I = \frac{kVA \times 1000}{E}$$

   For three-phase transformers:

   $$I = \frac{kVA \times 1000}{E \times 1.73}$$

   Where I = current, in amperes
       kVA = kilovolt-amperes (transformer nameplate rating)
       E = secondary line-to-line voltage (transformer nameplate rating)

2. Using the impedance value given on the transformer nameplate, find the multiplier to determine the short-circuit current.

   $$multiplier = \frac{100}{percent\ impedance}$$

3. The short-circuit current = normal full-load secondary current × multiplier.

EXAMPLE: A transformer is rated at 100 kVA and 120/240 volts. It is a single-phase transformer with an impedance of 1 percent (from the transformer nameplate). Find the short-circuit current.

For a single-phase transformer:

$$I = \frac{kVA \times 1000}{E}$$

$$I = \frac{100 \times 1000}{240}$$

$$= 416\ amperes,\ full\text{-}load\ current$$

The multiplier for a transformer impedance of 1 percent is:

$$multiplier = \frac{100}{percent\ impedance}$$

$$= \frac{100}{1}$$

$$multiplier = 100$$

The short-circuit current = I × multiplier
                         = 416 amperes × 100
                         = 41,600 amperes

Thus, the available short-circuit current at the terminals of the transformer is 41,600 amperes.

This value decreases as the distance from the transformer increases.

If the transformer impedance is 1.5 percent, the multiplier is:

$$multiplier = \frac{100}{1.5} = 66.6$$

The short-circuit current = 416 × 66.6
                         = 27,706 amperes

If the transformer impedance is 2 percent:

$$multiplier = \frac{100}{2} = 50$$

The short-circuit current = 416 × 50
                         = 20,800 amperes

(**Note:** A short-circuit current of 20,800 amperes is used in Figure 28-11, Figure 28-12, Figure 28-13, and Figure 28-14.)

CAUTION: The line-to-neutral short-circuit current at the secondary of a single-phase transformer is approximately 1½ times greater than

the line-to-line short-circuit current. For the previous example,

Line-to-line short-circuit current = 20,800 amperes

Line-to-neutral short-circuit current
= 20,800 × 1.5 = 31,200 amperes

It is significant to note that UL Standard 1561 allows the marked impedance on the nameplate of a transformer to vary plus or minus (±) 10% from the actual impedance value of the transformer. Thus, for a transformer marked "2%Z," the actual impedance could be as low as 1.8%Z to as much as 2.2%Z. Therefore, the available short-circuit current at the secondary of a transformer as calculated previously should be increased by 10 percent if one wishes to be on the safe side when considering the interrupting rating of the fuses and breakers to be installed.

An excellent point-to-point method for calculating fault currents at various distances from the secondary of a transformer that uses different size conductors is covered in detail in *Electrical Wiring—Commercial* (©Thomson Delmar Learning). It is simple and accurate. The point-to-point method can be used to calculate both single-phase and three-phase faults currents.

For an easy and fun to use computer program for making short-circuit calculations, visit the Bussmann Web site at www.Bussmann.com. Look for **Application Info**...then **Software**... then **Short-Circuit Calculations**.

PANELBOARDS . . . WHAT ARE THEY?

The terms *panel*, *panelboard*, and *load center* are all used, and for all practical purposes mean the same thing. The *NEC®* and the UL Standards use the term *panelboard*. Neither the *NEC®* nor the UL Standards contain the term *load center*. A load center is really a panelboard with less gutter (wiring) space and less depth and width, and generally does not have features such as integral relays, remote-controlled circuit breakers, load-shedding devices, and other similar optional features. However, some load centers are available for home automation applications that do have certain remote control features. Circuit breakers used in load centers are usually of the "plug-on" type as opposed to the "bolt-on" type found in commercial and industrial panelboards. Load centers generally are for use in a shallow wall

(2 × 4 studs) such as found in residential and light commercial construction. They are available with a main circuit breaker or main fuses, or with main lugs only (MLO).

Panelboards are tested and listed under UL Standard 67.

The most popular residential panelboard installed nowadays are *series-rated* as shown in Figure 28-14. The most common rating is the 22/10 panelboard. The main circuit breaker has an interrupting rating of 22,000 amperes. The branch-circuit breakers have an interrupting rating of 10,000 amperes. The series rating of 22/10 is adequate for most typical residential main panelboards. If you are installing a large electrical service where a large kVA rated transformer is involved, don't take a chance! You should have real concern for services over 200 amperes. Obtain the available fault current on the line side of the service equipment from the local electric utility, then select panels, fuses, and circuit breakers suitable for the available fault current.

Table 28-3 compares the requirements for different types of panelboards.

Selective Coordination/Nonselective Coordination

These are terms used by the real pros!

Selective Coordination: Under overload, short-circuit, or ground-fault conditions, only the overcurrent device nearest the fault opens. The main fuses or circuit breaker remain closed. Refer to *240.2*. See definition in *NEC® Article 100*.

Nonselective Coordination: Under overload, short-circuit, or ground-fault conditions, the branch breaker or fuses might open, the main breaker or fuses might open, or both might open. For example, a short circuit in a branch-circuit could also cause the main breaker to trip, resulting in a total power outage. Series-rated systems are susceptible to nonselectivity when available fault-currents are high.

The theory behind selective and nonselective systems is covered in detail in *Electrical Wiring— Commercial* (©Thomson Delmar Learning). Time-current curves for fuses and circuit breakers, circuit-breaker "unlatching time" data, and fuse melting/total clearing data are discussed.

	FUSED MAIN/FUSED BRANCHES (Fully-Rated) FIGURE 28-11	FUSED MAIN/BREAKER BRANCHES (Series-Rated) FIGURE 28-12	BREAKER MAIN/BREAKER BRANCHES (Fully-Rated) FIGURE 28-13	BREAKER MAIN/BREAKER BRANCHES (Series-Rated) FIGURE 28-14
Must be listed and marked as suitable for use as service equipment when used as service equipment. Does not need this marking when used as a subpanel.	YES	YES	YES	YES
Must have short-circuit current rating adequate for the available fault current at the line side of the panelboard.	YES	YES	YES	YES
Must be marked with its short-circuit current rating.*	YES	YES	YES	YES
Must be marked "Series-Rated."	NO	YES Marking: Caution—Series Combination System Rated _____ Amperes.	NO	YES Marking: Caution—Series Combination System Rated _____ Amperes.
Main fuses or circuit breaker must have interrupting rating adequate for the available fault current on the line side of the panelboard.	YES	YES	YES	YES
Branch fuses or breakers must have interrupting rating adequate for the available fault current on the line side of the panelboard.	YES	NO. Main fuses might have 200,000-ampere interrupting rating. Branch breakers might have 10,000-ampere interrupting rating. Other combinations available. Check manufacturers' data.	YES	NO. Main breaker might have 22,000-ampere interrupting rating. Branch breakers might have 10,000-ampere interrupting rating. Other combinations available. Check manufacturers' data.
Main fuses are current-limiting. See 240.2.	YES	YES	NOT APPLICABLE	NOT APPLICABLE
Must be sized as determined by service-entrance calculations and/or local codes.	YES	YES	YES	YES
Check manufacturer's data for time-current characteristic curves for fuses and breakers, and for "unlatching times" of breakers.	ALWAYS A GOOD IDEA.	ALWAYS A GOOD IDEA.	ALWAYS A GOOD IDEA.	ALWAYS A GOOD IDEA.

*If a panelboard has no marked short-circuit rating, then its short-circuit rating is based upon the lowest interrupting-rated device installed.

Table 28-3   Comparison of panelboard types.

REVIEW

1. What is the minimum size service for a one-family dwelling? _____

2. a. What is a Type S fuse? _____

   _____

   b. Where must Type S fuses be installed? _____

   _____

3. a. What is the maximum voltage permitted between conductors when using plug fuses?

   _____

   b. May plug fuses (Type S) be installed in a switch that disconnects a 120/240-volt clothes dryer? _____

   c. Give a reason for the answer to (b). _____

   _____

   _____

4. Will a 20-ampere, Type S fuse fit properly into a 15-ampere adapter? _____

5. A time-delay, dual-element fuse will hold 500 percent of its rating for approximately (5 seconds) (10 seconds) (20 seconds). Circle the correct answer.

6. What part of a circuit breaker causes the breaker to trip

   a. on an overload? _____   b. on a short circuit? _____

7. What is meant by an ambient-compensated circuit breaker? _____

   _____

   _____

   _____

8. List the standard sizes of fuses and circuit breakers up to and including 100 amperes.

   _____

   _____

9. When hooking up a 240-volt electric baseboard heater, you should use: (Circle the correct answer)

   a. a two-pole circuit breaker

   b. two single-pole breakers and a handle tie

10. Using the method shown in this unit, what is the approximate short-circuit current available at the terminals of a 50-kVA single-phase transformer rated 120/240 volts? The transformer impedance is 1 percent.

    a. line-to-line?

    b. line-to-neutral?

11. State four possible combinations of service equipment that meet the requirements of *110.9* and *110.10* of the *Code*.

    a. _____   c. _____

    b. _____   d. _____

12. Which *Code* section states that all overcurrent devices must have adequate interrupting ratings for the available fault current to be interrupted? _____

13. All electrical components have some sort of "withstand rating." The *NEC®* and UL Standards refer to this as the components short-circuit current rating. This rating indicates the ability of the component to withstand a specified amount of fault current for the time it takes the overcurrent device upstream of the component to open the circuit. *NEC®* _____ refers to short-circuit current rating.

14. Arcing fault damage is closely related to the value of _____ .

15. The utility company has provided a letter to the contractor stating that the available fault current at the line side of the main service-entrance equipment in a residence is 17,000 amperes RMS symmetrical, line-to-line. In the space following each statement, write in "Meets Code" or "Violation" of *110.9* and *110.10* of the *Code*.

   a. Main breaker has a 10,000-ampere interrupting rating; branch breakers have a 10,000-ampere interrupting rating. _____

   b. Main current-limiting fuse has a 200,000-ampere interrupting rating; branch breakers have a 10,000-ampere interrupting rating. The panel is "Series-Rated."

   _____

   c. Main breaker has a 22,000-ampere interrupting rating; branch breakers have a 10,000-ampere interrupting rating. The panel is marked "Series-Connected."

   _____

16. While working on the main panel with the panel energized, the electrician inadvertently causes a direct short circuit (line-to-ground) on one of the branch-circuit breakers. The available fault current at the main service equipment is rather high. The panel is labeled "Series-Connected." The main breaker is rated 100 amperes. Both branch-circuit breaker and main breaker trip "OFF." This condition is referred to as being (selective) (nonselective). Circle the correct answer.

17. Repeat question 13 on a main service panel that consists of 100-ampere main current-limiting fuses and breakers for the branch-circuits. The branch-circuit trips; the 100-ampere fuses do not open. This condition is referred to as being (selective) (nonselective). Circle the correct answer.

18. *NEC®* _____ prohibits the connecting of fuses and/or circuit breakers into the grounded circuit conductor.

19. In general, overcurrent protective devices are inserted where the conductor receives its

   _____ .

20. a. Does the *NEC®* permit panelboards to be installed in clothes closets? _____

   b. Does the *NEC®* permit panelboards to be installed in bathrooms? _____

21. Overcurrent devices must be accessible. (True) (False). Circle the correct response.

Service-Entrance Calculations

The various load values determined in earlier units of this text are now used to illustrate the proper method of determining the size of the service-entrance conductors for the residence. The calculations are based on *NEC*® requirements. The student must check local and state electrical codes for any variations in requirements that may take precedence over the *NEC*.®

SIZING OF SERVICE-ENTRANCE CONDUCTORS AND SERVICE DISCONNECTING MEANS

NEC® *230.42* tells us that service-entrance conductors must be sufficient to carry the load as calculated according to *Article 220*, and shall have an ampacity not less than the rating of the service disconnect.

For a one-family residence, the minimum size main service disconnecting means is 100 amperes, three-wire, *230.79(C)*.

There are two methods for calculating service size for homes:

1. The conventional method is found in *Article 220, Part III*.

2. The optional method is found in *Article 220, Part IV*.

Method 1 *(Article 220, Part III)*

All of the volt-ampere values in the following calculations are taken from previous units of this text.

Single-Family Dwelling Service-Entrance Calculations

1. General Lighting Load, *220.12*

   3232 ft^2 @ 3 VA per ft^2 = 9696 VA

   Note: Included in this floor area calculation are all lighting outlets and general-use receptacles. Do not include open porches, garages, or unused or unfinished spaces not adaptable for future use. See *NEC*® *220.12*, *Table 220.12*, and *220.14(J)*.

2. Minimum Number of Lighting Branch-Circuits

   9696 VA ÷ 120 volts = 80.8 amperes

   then, $\dfrac{\text{amperes}}{15} = \dfrac{80.8}{15}$ = 5.387 (round up to 6) 15-ampere branch-circuits

3. Small-Appliance Load, *210.11(C)(1)* and *220.52(A)*
   (Minimum of two 20-ampere branch-circuits)

Kitchen	3
Workshop	2**

   5 branch-circuits @ 1500 VA each = 7500 VA

4. Laundry Branch-Circuit, *210.11(C)(2)*, *220.50(F)*, and *220.52(B)*
   (Minimum of one 20-ampere branch-circuit)

   1 branch-circuit(s) @ 1500 VA = 1500 VA

   1* Clothes washer @ 1500 VA = 1500 VA

   Total = 3000 VA

5. Total General Lighting, Small Appliance, and Laundry Load

   Lines 1 + 3 + 4 = 20,196 VA

6. Net calculated General Lighting, Small Appliance, and Laundry Loads (less ranges, ovens, and "fastened-in-place" appliances). Apply demand factors from *Table 220.42*.

   a. First 3000 VA @ 100% = 3000 VA

   b. Line 5 20,196 − 3000 = 17,196 @ 35% = 6019 VA

   Total a + b = 9019 VA

7. Electric Range, Wall-Mounted Ovens, Counter-Mounted Cooking Units *(Table 220.55)*

Wall-mounted oven	7450 VA
Counter-mounted cooking unit	6600 VA
Total	14,050 VA (14 kW)

   14 kW exceeds 12 kW by 2 kW

   2 kW × 5% = 10% increase, therefore: 8 kW + 0.8 kW = 8.8 kW = 8,800 VA

---

* This is in addition to the required minimum of one laundry branch-circuit. Thus, the laundry is served by a 20-ampere branch-circuit B18 for the automatic washer and another 20-ampere branch-circuit B20 supplying two receptacles for plugging in an electric iron, steamer, or clothes press.

** Although not true small-appliance circuits as defined by the *NEC*,® it is better to add these circuits in as 1500-VA small-appliance circuits rather than including them in the general lighting load calculations. This results in a more adequate load calculation.

8. Electric Dryer, *220.54* = 5700 VA

9. Electric Furnace, *220.54*, Air Conditioner, Heat Pump, *Article 440*

   Air conditioner: 26.4×240 = 6336 VA
   Electric Furnace: 13,000 VA
   (Enter largest value, *220.60*) = 13,000 VA

10. Net Calculated General Lighting, Small Appliance, Laundry, Ranges, Ovens, Cooktop Units, HVAC

    Lines 6 + 7 + 8 + 9 = 36,519 VA

11. List "Fastened-in-Place" Appliances *in addition* to
    Electric Ranges, Air Conditioners, Clothes Dryers, Space Heaters.

Appliance	VA Load
Water Heater	3000 VA
Dishwasher 9.2×120 =	1104 VA
Garage Door Opener 5.8×120 =	696 VA
Food Waste Disposer 7.2×120 =	864 VA
Water Pump 8×240 =	1920 VA
Hydromassage Tub 10×120 =	1200 VA
Attic Exhaust Fan 5.8×120 =	696 VA
Heat/Vent/Lights 1500×2 =	3000 VA
Freezer 5.8×120 =	696 VA
Total	13,176 VA

12. Apply 75% Demand Factor, *220.53*, if Four or More "Fastened-in-Place" Appliances. If Less than Four, Figure at 100%. Do not include electric ranges, clothes dryers, space heating, or air-conditioning equipment.

    Line 11 $13,176 \times 0.75$ = 9882 VA

13. Total Calculated Load (Lighting, Small Appliance, Ranges, Dryers, HVAC, "Fastened-in-Place" Appliances)

    Line 10 36,519 + Line 12 9882 = 46,401 VA

14. Add 25% of Largest Motor (*220.50* and *430.24*)
    This is the water pump motor:

    $8 \times 240 \times 0.25$ = 480 VA

15. Grand Total Line 13 + Line 14 = 46,881 VA

16. Minimum Ampacity for Ungrounded Service-Entrance Conductors

    $$\text{Amperes} = \frac{\text{Line 15}}{240} = \frac{46,881}{240} = 195.4 \text{ amperes}$$

17. Ungrounded Conductor Size (copper)   2/0 AWG THWN

Note: This could be a 3/0 AWG THW, THHW, THWN, XHHW, or THHN per *Table 310.16*, or a AWG 2/0 (same types) using *Table 310.15(B)(6)*. *Table 310.15(B)(6)* may only be used for 120/240-volt, three-wire, residential single-phase service-entrance conductors, service-lateral conductors, and feeder conductors that serve as the main power feeder to a dwelling unit. This table *cannot* be applied to the feeder to Panel B because that feeder carries only part of the load in the residence. We have assumed that the equipment is marked 75°C. We read the ampacity from the 75°C column in *Table 310.16* as stated in *110.14(C)*.

18. Minimum Ampacity for Neutral Service-Entrance Conductor, *220.61* and *310.15(B)(6)*. Do not include straight 240-volt loads.

    a. Line 6   9019 VA

    b. Line 7 8800 × 0.70   =   6160 VA

    c. Line 8 5700 × 0.70   =   3990 VA

    d. Line 11 (include only 120-volt loads)

Freezer	696 VA
Food Waste Disposer	864 VA
Garage Door Opener	696 VA
Heat/Vent/Light	3000 VA
Attic Fan	696 VA
Hydromassage tub	1200 VA
Dishwasher	1104 VA
Total	8256 VA

    e. Line d total @ 75% Demand Factor per *220.60*: 8256 × 0.75 = 6192 VA

    f. Add 25% of largest 120-volt motor
       This is the hydromassage pump motor:
       10 × 120 × 0.25   =   300 VA

    g. Total a + b + c + e + f   = 25,492 VA

$$\text{Amperes} = \frac{\text{volt} - \text{amperes}}{\text{volts}} = \frac{25{,}492}{240} = 106.2 \text{ amperes}$$

19. Neutral Conductor Size (copper), *220.61*   1 AWG

Note: The calculations indicate that a 2 AWG copper neutral conductor would be adequate based upon the unbalanced load calculations. *NEC® 310.15(B)(6)* states that the neutral conductor is permitted to be smaller than the ungrounded "hot" conductors if the requirements of *215.2, 220.61,* and *230.42* are met. *NEC® 220.61* states that a feeder or service neutral load shall be the maximum unbalance of the load determined by *Article 220*. The specifications for this residence specify that a 1 AWG neutral conductor be installed. The neutral conductor must not be smaller than the grounding electrode conductor, *250.24(B)(1)*. Where bare conductors are used with insulated conductors, their ampacity is based on the ampacity of the other insulated conductors in the raceway, *310.15(B)(3)*.

20. Grounding Electrode Conductor Size (copper), *Table 250.66*   4 AWG

21. Conduit (See Table 27-1 for calculations) trade size 1¼.

Table 310.15(B)(6) provides special consideration for 120/240-volt, three-wire, single-phase service entrance, service lateral, and feeder conductors that serve as the main power supply to a dwelling unit. Table 29-1 in this text replicates *Table 310.15(B)(6)* from the *NEC.* The current ratings for a given size wire are higher than the allowable ampacities found in *Table 310.16* because of the tremendous diversity of electricity use in homes.

Table 310.15(B)(6) cannot be used to determine the feeder size to Panel B because this feeder does not carry the main power to the home.

Be careful. Locations exposed to the weather or underground are considered to be wet locations, per the definition in the *NEC.* Therefore, most electrical inspectors will require that service-entrance conductors, from the line side terminals of the service disconnect to the service head, be suitable for wet locations. This would mean that the conductors must have a "W" designation in their type lettering. See *310.8(C).*

SPECIAL AMPACITY RATINGS FOR RESIDENTIAL 120/240-VOLT, THREE-WIRE, SINGLE-PHASE SERVICE-ENTRANCE CONDUCTORS, SERVICE-LATERAL CONDUCTORS, FEEDER CONDUCTORS, AND CABLES THAT SERVE AS THE MAIN POWER FEEDER TO A DWELLING UNIT.

Copper Conductor Size (AWG) for RHH, RHW-2, THHN, THHW, THW, THW-2, THWN, THWN-2, XHHW, XHHW-2, SE, USE, USE-2	Aluminum or Copper-Clad Aluminum Conductors (AWG)	Allowable Special Ampacity
4	2	100
3	1	110
2	1/0	125
1	2/0	150
1/0	3/0	175
2/0	4/0	200
3/0	250 kcmil	225
4/0	300 kcmil	250
250 kcmil	350 kcmil	300
350 kcmil	500 kcmil	350
400 kcmil	600 kcmil	400

Table 29-1 This table is similar to the table in *Table 310.15(B)(6)*, and is *only* for residential 120/240-volt, three-wire, single-phase service-entrance conductors, service-lateral conductors, and residential feeder conductors that carry the main power to the dwelling.

High Temperatures

▶In certain parts of the country, such as the southwestern desert climates where extremely hot temperatures are common, the AHJ will probably enforce *310.8(D)* and *310.10* regarding temperature limitations of insulated conductors. This means that conductors in raceways or cables exposed to direct sunlight (on a roof or side of building) be corrected (derated) according to the "Correction Factors" found below *Table 310.16.* See Figure 29-1. ◀

EXAMPLE: If the U.S. weather bureau lists the average summer temperature as 113°F (45°C), then a correction factor of 0.82 must be applied. For instance, a 3/0 XHHW copper conductor per *Table 310.16* is 200 amperes at 86°F (30°C). At 113°F (45°C) the conductor's ampacity is

$$200 \times 0.82 = 164 \text{ amperes}$$

A properly sized conductor capable of carrying 200 amperes safely requires

$$\frac{200}{0.82} = 243.9 \text{ amperes}$$

Therefore, according to *Table 310.16* and the applied correction factor, a 250-kcmil Type XHHW copper conductor is required.

Check this with the electrical inspection department where this situation might be encountered.

Neutral Conductor

Why is the neutral conductor permitted to be smaller than the "hot" ungrounded conductors?

The neutral conductor must be able to carry the calculated maximum neutral current, *NEC* *220.61.* Straight 240-volt loads are not included in the calculation of neutral conductors. The calculations resulted in a calculated load of 195 amperes for the ungrounded conductors, and 106 amperes for the neutral conductor. Referring to *Table 310.16*, this could be two 3/0 AWG THW, THHW, THWN, XHHW, or THHN copper conductors and one 2 AWG neutral conductor. The neutral conductor may be insulated or bare per *230.30* and *230.41.*

NEC *310.15(B)(6)* also permits the neutral conductor to be smaller than the "hot" ungrounded conductors, but only if it can be verified by calculations that the requirements of *215.2, 220.61,* and *230.42* are met. We verified this in our calculations for the residence discussed in this text.

Figure 29-1 Example of high-temperature location.

In general, service-entrance conductors must be insulated, *230.30*, *230.41*, and *310.2*. However, the exceptions to these rules permit the use of bare neutral conductors.

Where bare conductors are used with insulated conductors in raceways or cables, their allowable ampacity is limited to the allowable ampacity for the adjacent insulated conductors, *310.15(B)(3)*.

For example, for a service consisting of two 3/0 AWG Type THWN conductors and one 1/0 AWG bare neutral conductor, the allowable ampacity of the bare neutral conductor is found in *Table 310.16* in the THWN column.

3/0 AWG THWN phase conductors   200 amperes

1/0 AWG bare neutral conductor   150 amperes

The specifications for this residence call for two 2/0 AWG ungrounded "hot" conductors and one 1 AWG neutral conductor.

The service neutral conductor serves another purpose. In addition to carrying the unbalanced current, the neutral conductor must also be capable of safely carrying any fault current that it might be called upon to carry, such as a line-to-ground fault in the service equipment. Sizing the neutral conductor only for the normal neutral unbalance current could result in a neutral conductor too small to safely carry fault currents. The neutral conductor might burn off, causing serious voltage problems,

damage to equipment, and the creation of fire and shock hazards.

In sizing neutral conductors for dwellings, this generally is not a problem. But installing reduced size neutrals on commercial and industrial services can present a major problem. The grounded conductor of *any* service must never be smaller than the required grounding electrode conductor for that particular service. Refer to *230.23*, *230.31*, *230.42(C)*, *250.2*, *250.4*, and *250.24(B)*.

Subpanel B Calculations

General lighting load
  (*220.12, 220.43*)
  1420 sq. ft. @ 3 volt-
    amperes per ft^2 =   4260 volt-amperes
  (kitchen, living room,
    laundry, rear entry
    hall, powder room,
    recreation room)

Small-appliance load
  [*220.52(A)*]
  Kitchen   3
  Laundry   1   [*220.52(B)*]
  Automatic
    washer   1   [*220.52(B)*]
       Total   5 @ 1500 VA
                    per
                    circuit =   7500 volt-amperes

Total general lighting and
   small-appliance load = 11,760 volt-amperes
Application of demand
   factors (*Table 220.42*)
   3000 volt-amperes
      @100% =   3000 volt-amperes
   11,760 − 3000 = 8760
      @ 35% =   3066 volt-amperes

A   Net calculated load (less
    range and "fastened-in-
    place" appliances)   6066 volt-amperes

Wall-mounted oven and
   counter-mounted range:
   Table 220.55, Note 4:
   Wall-mounted oven   7450 volt-amperes
   Counter-mounted range   6600 volt-amperes
            Total   14,050 volt-amperes

14 kW exceeds 12 kW by 2 kW:
   2 kW × 5% = 10%
      increase
   8 kW × 0.10 = 0.8 kW
   (*Column C, Table 220.55*)

B   8 + 0.8 = 8.8 kW =   8800 volt-amperes

C   Net calculated load (with
    range) = A + B = C   14,866 volt-amperes

Dryer (*220.54*)   5700 volt-amperes
Dishwasher
   Motor: 7.2 × 120 =
      864 volt-amperes
   Heater: 1000 watts
      (volt-amperes)
      (maximum demand
      is 1000 VA because
      motor and heater do
      not operate at the
      same time, *220.60*)   1000 volt-amperes
Food waste disposer
   7.2 × 120 =   864 volt-amperes
Garage door opener
   5.8 × 120 =   696 volt-amperes
Total "fastened-in-place"
   appliance load   8260 volt-amperes

Lighting, small-appliance
   circuits, ranges, dryer = 14,866 volt-amperes

D   Add 25% of largest motor.
    7.2 × 120 × 0.25 =   216 volt-amperes

E   Net calculated load
    (lighting, small appli-
    ance circuits, ranges,
    dryer, dishwasher, food
    waste disposer, garage
    door opener)   23,342 volt-amperes

$$\text{Amperes} = \frac{\text{volt-amperes}}{\text{volts}} = \frac{23{,}342}{240} = 97.3 \text{ amperes}$$

The feeder conductors supplying Panel B could be 3 AWG THHW, THW, THHN, THWN, or XHHW per *Table 310.16*. We have not shown the calculations for the neutral to Panel B. The specifications and also Figure 27-2 require that three 3 AWG THHN feeder conductors supply Panel B.

Method 2 (Optional Calculations — Dwellings) *(Article 220, Part IV)*

A second method for determining the load for a one-family dwelling is given in *Article 220, Part IV*. This method simplifies the calculations. But remember, the minimum size service for a one-family dwelling is 100 amperes, *230.42(B)* and *230.79(C)*.

Let's take a look at *Part IV* of *Article 220*. This is an alternative method of calculating service loads and feeder loads. It is referred to as the *optional method*.

NEC® 220.80 tells us that we are permitted to calculate service-entrance conductors and feeder conductors (both phase conductors and the neutral conductor) using an optional method. *NEC® 220.82* addresses single-family dwellings. *NEC® 220.84* addresses multifamily dwellings.

NEC® 220.52(B) tells us to

1. include 1500 volt-amperes for each 20-ampere small-appliance circuit.

2. include 3 volt-amperes per ft² for lighting and general-use receptacles.

3. include the nameplate rating of appliances:
   • that are fastened in place, or

- that are permanently connected, or
- that are located to be connected to a specific circuit, such as: ranges, wall-mounted ovens, counter-mounted cooking units, clothes dryers, and water heaters.

4. include nameplate ampere rating or kVA for motors and all low-power-factor loads. The intent of the reference to low-power factor is to address such loads as low-cost, low-power-factor fluorescent ballasts. It is always recommended that high-power-factor ballasts be installed.

NEC® 220.82(C) tells us that for heating and air-conditioning loads, select the largest load from a list of six types of loads:

1. 100 percent of air conditioning and cooling.

2. 100 percent of heat pump compressors and supplemental heating unless both cannot operate at the same time.

3. 100 percent electric thermal storage heating if expected to be continuous at full nameplate value.

4. 65 percent of central electric space heating that also has supplemental heating in heat pumps where the compressor and supplemental heating will not operate at the same time.

5. 65 percent of electric space heating of less than four separately controlled units.

6. 40 percent of electric space heating if four or more separately controlled units.

So let's begin our optional calculation for the residence discussed in this text. The residence has an air conditioner and an electric furnace.

Air conditioner

$26.4 \times 240 = 6336$ volt-amperes

Electric furnace

$13,000 \times 0.65 = 8450$ volt-amperes

Therefore we will select the electric furnace load for our calculations because it is the largest load. It is also a noncoincidental load, as defined in *220.60*. We can omit the air-conditioner load from our calculations from here on.

We now add up all of the other loads:

General lighting load 3232 ft² @ 3 volt-amperes per ft² =	9696 volt-amperes
Small-appliance circuits (7) @ 1500 volt-amperes each	10,500 volt-amperes
Wall-mounted oven (nameplate rating)	6600 volt-amperes
Counter-mounted cooking unit (nameplate rating)	7450 volt-amperes
Water heater (nameplate rating)	3000 volt-amperes
Clothes dryer (nameplate rating)	5700 volt-amperes
Dishwasher (maximum demand, heater only)	1000 volt-amperes
Food waste disposer 7.2 × 120	864 volt-amperes
Water pump 8 × 240	1920 volt-amperes
Garage door opener 5.8 × 120	696 volt-amperes
Heat-vent-lights (2) 1500 × 2	3000 volt-amperes
Attic exhaust fan 5.8 × 120	696 volt-amperes
Hydromassage tub 10 × 120	1200 volt-amperes
Total other loads	52,322 volt-amperes

We can now complete our optional calculation:

Enter electric furnace load 13,000 × 0.65 =	8450 volt-amperes
Plus first 10 kVA of all other loads at 100%	10,000 volt-amperes
Plus remainder all other loads at 40%:	
52,322 − 10,000 = 42,322	
42,322 × 0.40	16,929 volt-amperes
Total	35,379 volt-amperes

$$\text{Amperes} = \frac{\text{volt-amperes}}{\text{volts}} = \frac{35,379}{240} = 147.4 \text{ amperes}$$

Using the optional calculation method, and referring to *Table 310.15(B)(6)*, we find that for the 147+ amperes calculated load, a 1 AWG copper conductor with any of the insulation types listed has a special ampacity of 150 amperes and could be installed for this service entrance. This seems a bit small, and that is why some local electrical codes do not permit the use of this special ampacity table.

The specifications for the residence discussed in this text call for a full 200-ampere service consisting of two AWG 2/0 THW, THHN, THWN, or XHHW copper phase conductors and one 1 AWG bare copper neutral conductor.

Existing Homes

When adding new loads in an existing home, *220.83* provides an optional calculation method that may be used to confirm that the existing service is large enough to handle the additional load. The procedure is a mirror image of the optional calculation method we discussed above.

Service-Entrance Conductor Size Table

Table 29-2 has been taken from one major city that prefers to show minimum service-entrance conductor size requirements rather than having electrical contractors make calculations each and every time they install a service.

It is within the realm of local authorities to publish code requirements specifically for their communities.

Main Disconnect

The main service disconnecting means in this residence is rated 200 amperes, discussed in Unit 27.

Grounding Electrode Conductor

Grounding electrode conductors are sized according to *Table 250.66* of the *Code*.

As an example, this residence is supplied by 2/0 AWG copper service-entrance conductors. Checking *Table 250.66*, we find that the minimum grounding electrode conductor must be a 4 AWG copper conductor.

The grounding electrode conductor for the service in this residence is a 4 AWG copper armored ground cable, discussed in Unit 27.

TYPES OF WATT-HOUR METERS

Figure 29-2 shows different styles of watt-hour meters. The first, (A), is a typical solid-state residential kilowatt-hour-only meter. The second, (B), is a single-phase watt-hour meter that has two sets of registers: the electromechanical register records the total kilowatt-hours, and the electronic register records and displays kilowatt-hour consumption for up to four different rate schedules. The third, (C), is a solid-state programmable polyphase watt-hour meter that can register kilowatt-hour consumption for up to five different rate schedules, demand, reactive measurements, and load profile is so equipped. Utilities that offer special "time-of-use" rates install meters of the second and third types.

Wiring diagrams for the metering and control of electric water heaters, air conditioners, and heat pumps are discussed in Unit 19. Electric utilities across the country have different requirements for hooking up these loads. Always consult your local utility for their latest information regarding electric service.

SERVICE-ENTRANCE CONDUCTOR SIZE						
	COPPER			ALUMINUM		
Size (amperes)	Phase (Hot) Conductors	Neutral Conductor	Conduit Trade Size	Phase (Hot) Conductors	Neutral Conductor	Conduit Trade Size
100	2 AWG	4 AWG	1½	1 AWG	2 AWG	1½
125	1 AWG	4 AWG	1½	00 AWG	1 AWG	1½
150	0 AWG	4 AWG	1½	000 AWG	0 AWG	2
175	00 AWG	2 AWG	2	0000 AWG	00 AWG	2
200	000 AWG	2 AWG	2	250 kcmil	000 AWG	2

Table 29-2   Service-entrance conductor sizing used in some cities so that service-entrance calculations do not have to be made for each service-entrance installation.

Table 250.66 Grounding Electrode Conductor for Alternating-Current Systems

Size of Largest Ungrounded Service-Entrance Conductor or Equivalent Area for Parallel Conductors[a] (AWG/kcmil)		Size of Grounding Electrode Conductor (AWG/kcmil)	
Copper	Aluminum or Copper-Clad Aluminum	Copper	Aluminum or Copper-Clad Aluminum[b]
2 or smaller	1/0 or smaller	8	6
1 or 1/0	2/0 or 3/0	6	4
2/0 or 3/0	4/0 or 250	4	2
Over 3/0 through 350	Over 250 through 500	2	1/0
Over 350 through 600	Over 500 through 900	1/0	3/0
Over 600 through 1100	Over 900 through 1750	2/0	4/0
Over 1100	Over 1750	3/0	250

Notes:

1. Where multiple sets of service-entrance conductors are used as permitted in 230.40, Exception No. 2, the equivalent size of the largest service-entrance conductor shall be determined by the largest sum of the areas of the corresponding conductors of each set.

2. Where there are no service-entrance conductors, the grounding electrode conductor size shall be determined by the equivalent size of the largest service-entrance conductor required for the load to be served.

[a]This table also applies to the derived conductors of separately derived ac systems.

[b]See installation restrictions in 250.64(A).

(Reprinted with permission from NFPA 70-2005)

Figure 29-2   Photo A shows a solid-state watt-hour meter. Photo B shows a single-phase watt-hour meter that has two sets of registers: one registers the total kilowatt-hours, the second records and displays kilowatt-hour consumption for up to four different rate schedules. Photo C shows a solid-state programmable polyphase watt-hour meter that is capable of registering five different time-of-use rates, demand, reactive measurement, and load profile if so equipped. *Courtesy of* Landis & Gyr, Inc.

READING WATT-HOUR METERS

Reading a digital watt-hour meter is easy. Just read the numbers.

For dial-type meters, starting with the first dial, record the last number the pointer has just passed. Continue doing this with each dial until all dial readings have been recorded. The reading on the five-dial watt-hour meter in Figure 29-3 is 18,672 kilowatt-hours.

If the meter reads 18,975 one month later, as indicated in Figure 29-4, by subtracting the previous reading of 18,672, it is found that 303 kilowatt-hours of electricity were used during the month. The 303 kilowatt-hours of electricity used is multiplied by the rate per kilowatt-hour, and the power company bills the consumer for the energy used. The utility may also add a fuel adjustment charge or other charges.

Figure 29-3 The reading of this five-dial meter is 18,672 kilowatt-hours.

Figure 29-4 One month later, the meter reads 18,975 kilowatt-hours, indicating that 303 kilowatt-hours were used during the month.

REVIEW

Note: Refer to the *Code* or the plans where necessary.

1. According to *NEC®* _____ , the minimum size service required for a one-family dwelling is _____ amperes.

2. a. What is the unit load per ft² for the general lighting load of a residence?

   _____

   b. What are the demand factors for the general lighting load in dwellings? _____

   _____

   _____

3. a. What is the ampere rating of the circuits that are provided for the small-appliance loads? _____

    b. What is the minimum number of small appliance circuits permitted by the *Code*?

        _____

    c. How many small-appliance circuits are included in this residence? _____

4. Why is the air-conditioning load for this residence omitted in the service calculations?

    _____

    _____

    _____

    _____

    _____

    _____

5. What demand factor may be applied when four or more fixed appliances are connected to a service, in addition to an electric range, air conditioner, clothes dryer, or space heating equipment? *(220.53)* _____

    _____

6. What load may be used for an electric range rated at not over 12 kW? *(Table 220.55)*

    _____

7. What is the calculated load for an electric range rated at 16 kW? *(Table 220.55)* Show calculations.

8. What is the calculated load when fixed electric heating is used in a residence? *(220.51)*

    _____

9. On what basis is the neutral conductor of a service entrance determined? _____

    _____

    _____

    _____

    _____

10. Why is it permissible to omit an electric space heater, water heater, and certain other 240-volt equipment when calculating the neutral service-entrance conductor for a residence? _____

    _____

    _____

    _____

11. a. What section of the *Code* contains an optional method for determining residential service-entrance loads? _____

    b. Is this section applicable to a two-family residence? _____

12. Calculate the minimum size of copper service-entrance conductors and grounding electrode conductor required for a residence containing the following: floor area is 24 ft × 38 ft (7.3 m × 11.6 m). The dwelling will have:

    - a 12-kW, 120/240-volt electric range

    - a 5-kW, 120/240-volt electric clothes dryer consisting of a 4-kW, 240-volt heating element, a 120-volt motor, and 120-volt lamp that have a combined load of 1 kW

    - a 2200-watt, 120-volt sauna heater

    - six individually controlled 2-kW, 240-volt baseboard electric heaters

    - a 12-ampere, 240-volt air conditioner

    - a 3-kW, 240-volt electric water heater

    - a 1000-watt, 120-volt dishwasher

    - a 7.2-ampere, 120-volt food waste disposer

    - a 5.8-ampere, 120-volt attic exhaust fan

    Determine the sizes of the ungrounded "hot" conductors and the neutral conductor. Use Type THWN. Use *Table 310.15(B)(6)*. Also determine the correct size grounding electrode conductor.

    Do not forget to include the required two 20-ampere small-appliance circuits and the laundry circuit.

    The panelboard is marked 167°F (75°C).

    You may use the blank *Single-Family Dwelling Service-Entrance Calculation* form in the Appendix of this text, or use the form as a guide to make your calculations in the proper steps.

    Two _____ AWG THWN ungrounded "hot" conductors

    One _____ AWG THWN or bare neutral conductor

    One _____ AWG grounding electrode conductor

STUDENT CALCULATIONS

13. Read the meter shown. Last month's reading was 22,796. How many kilowatt-hours of electricity were used for the current month?

14. After studying this unit, we realize that electric utilities across the country have many ways to meter residential customers. They may have lower rates during certain hours of the day. A common term used to describe these programs is _____

_____

Swimming Pools, Spas, Hot Tubs, and Hydromassage Baths

OBJECTIVES

After studying this unit, the student should be able to

▶ **recognize the importance of proper swimming pool wiring with regard to human safety.**

▶ **discuss the hazards of electrical shock associated with faulty wiring in, on, or near pools.**

▶ **understand and apply the basic *NEC®* requirements for the wiring of swimming pools, spas, hot tubs, and hydromassage bathtubs.**

Water and electricity do not mix!

To protect people against the hazards of electric shock associated with swimming pools, spas, hot tubs, and hydromassage bathtubs, special *Code* requirements are necessary.

A picture is worth a million words. A detailed drawing of the *NEC®* requirements for swimming pools appears on Sheet 10 of 10 in the Plans found in the back of this text. Refer to this drawing often as you study this unit.

ELECTRICAL HAZARDS

A person can suffer an immobilizing or lethal shock in a residential-type pool in either of two ways.

1. *Direct contact:* An electrical shock can be deadly to someone in a pool who touches a "live" wire or touches an object such as a metal ladder, metal fence, metal parts of a luminaire (fixture), metal enclosures of electrical equipment, or appliances that have become "live" by an ungrounded conductor coming in contact with the metal enclosure or exposed metal

parts of an appliance. Figure 30-1 shows a person in the water touching a faulty appliance.

2. *Indirect contact:* An electrical shock can be lethal to someone in a pool if the person is merely in the water when voltage gradients in the water are present. This is illustrated in Figure 30-2.

As shown in Figure 30-2, "rings" of voltage radiate outward from the radio to the pool walls and bottom of the pool. These rings can be likened to the rings that form when a rock is thrown into the water. The voltage rings or gradients range from 120 volts at the radio to zero volts at the pool walls and bottom. The pool walls and bottom are assumed to be at ground or zero potential. The gradients, in varying degrees, are found throughout the body of water. Figure 30-2 shows voltage gradients in the pool of 90 volts and 60 volts. This figure is a simplification of the actual situation in which there are many voltage gradients. In this case, the voltage differential, 30 volts, is an extremely hazardous value. The person in the pool immersed in these voltage

Figure 30-1   Touching a "live" faulty appliance can cause lethal shock.

Figure 30-2   Voltage gradients surrounding a person in the pool can cause severe shock, drowning, or electrocution.

gradients, is subject to severe shock, immobilization (which can result in drowning), or actual electrocution. Tests conducted over the years have shown that a voltage gradient of 1½ volts per foot can cause paralysis.

The shock hazard to a person in a pool is quite different from that of the normal "touch" shock hazard to a person not submersed in water. The water makes contact with the entire skin surface of the body rather than just at one "touch" point. Skin wounds, such as cuts and scratches, reduce the body's resistance to shock to a much lower value

than that of the skin alone. Body openings such as ears, nose, and mouth further reduce the body resistance. As Ohm's law states, for a given voltage, the lower the resistance, the higher the current.

KEY *NEC*® REQUIREMENTS— WIRING FOR SWIMMING POOLS

Article 680 covers swimming pools, wading pools, hydromassage bathtubs, fountains, therapeutic and decorative pools, hot tubs, and spas.

Here are some of the general *Code* rules. Comprehensive details for swimming pool requirements are found on Sheet 10 of 10 of the Plans at the back of this text.

General, Part I: Contains general requirements that apply to all types of swimming pools, spas, hot tubs, hydromassage tubs, and similar water related equipment. *Parts II, III, IV, V, VI,* and *VII* discuss the wiring requirements for specific equipment.

Scope, *680.1:* The scope tells us exactly what is covered by *Article 680.*

Definitions, *680.2:* There are many one-of-a-kind terms listed in *680.2.* These terms might seem like a foreign language. It makes sense to read these definitions carefully to make it easier to understand the rest of *Article 680.*

Other Articles, *680.3:* Many of the basic requirements of Chapter 1 through Chapter 4 in the *NEC*® are modified by special rules in *Article 680* that are unique to swimming pools, wading pools, hydromassage bathtubs, therapeutic pools, decorative pools, hot tubs, and spas. This section of the *NEC*® contains a convenient table that lists other articles and chapters that apply.

Ground-Fault Circuit Interrupters, *680.5:* GFCIs shall be listed circuit breakers or receptacles, or other listed types. GFCIs were discussed in Unit 6. Throughout *Article 680,* you will find the need for GFCIs.

Be careful when connecting underwater luminaires (fixtures). Once you run conductors beyond (on the load side of) a GFCI device protecting that wiring, you are not permitted to have other conductors in the same raceway, box, or enclosure. The exceptions to this rule are: (1) the other conductors are protected by GFCIs; (2) the other conductors are grounding conductors; (3) the other conductors are supply conductors to a feedthrough type GFCI; or (4) the GFCI is installed in a panelboard, where obviously there will be a "mix" of other circuit conductors that are not GFCI protected. This is covered in *680.23(F)(3).*

Grounding, *680.6:* In general, grounding shall be done as required in *Article 250.* But there are modifications that specifically address swimming pools.

Cord- and Plug-Connected Equipment, *680.7:* Here we find permission to use flexible cords to connect certain equipment. Underwater luminaires (fixtures) are not included in this permission.

Overhead Conductors, *680.8:* Overhead open conductors must be kept out of reach. Clearances are shown in Plan 10 of 10 in the back of this text.

Electric Pool Water Heaters, *680.9:* The branch-circuit conductor ampacity and the rating of the branch-circuit overcurrent devices must not be less than 125 percent of the total nameplate rated-load ampere rating of the electric heater. Figure 30-3 shows the requirements for unit heaters, radiant heaters, and embedded heating cables.

Underground Wiring Location, *680.10:* Underground wiring is not permitted under pools or withing 5 ft (1.5 m) horizontally from the inside wall of the pool. There are a few exceptions.

▶**Maintenance Disconnecting Means, *680.12:*** A disconnecting means must be provided for pool-associated equipment. They must be readily accessible and within sight of the equipment they disconnect. Pool lighting is not covered by *680.12.*◀

Permanently Installed Pools, Part II. Here we find the rules governing permanently installed pools such as for motors; clearances for overhead lighting; underwater lighting; existing installations; receptacles; requirements for switching; equipment associated with the pool; bonding, grounding, and audio equipment; and electric heaters. Electric deck heater *Code* requirements are shown in Figure 30-3.

Grounding and bonding together all metal parts in and around pools ensures that these metal parts are at the same voltage potential to ground, greatly reducing the shock hazard. Stray voltage and voltage gradient (see Figure 30-2) problems are kept to a minimum. Proper grounding and bonding also facilitates the opening of the overcurrent protection devices should a fault occur in the circuit(s).

You will see on Sheet 10 of 10 of the Plans that a metal conduit by itself is not considered an adequate grounding means for equipment in and around pools.

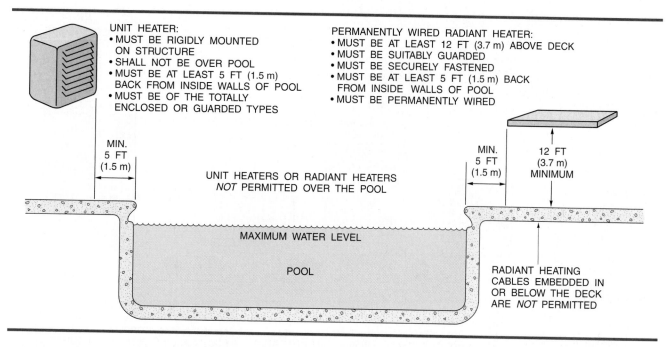

UNIT HEATER:
• MUST BE RIGIDLY MOUNTED ON STRUCTURE
• SHALL NOT BE OVER POOL
• MUST BE AT LEAST 5 FT (1.5 m) BACK FROM INSIDE WALLS OF POOL
• MUST BE OF THE TOTALLY ENCLOSED OR GUARDED TYPES

PERMANENTLY WIRED RADIANT HEATER:
• MUST BE AT LEAST 12 FT (3.7 m) ABOVE DECK
• MUST BE SUITABLY GUARDED
• MUST BE SECURELY FASTENED
• MUST BE AT LEAST 5 FT (1.5 m) BACK FROM INSIDE WALLS OF POOL
• MUST BE PERMANENTLY WIRED

MIN. 5 FT (1.5 m)

UNIT HEATERS OR RADIANT HEATERS *NOT* PERMITTED OVER THE POOL

MIN. 5 FT (1.5 m)

12 FT (3.7 m) MINIMUM

MAXIMUM WATER LEVEL

POOL

RADIANT HEATING CABLES EMBEDDED IN OR BELOW THE DECK ARE *NOT* PERMITTED

Figure 30-3 Deck-area electric heating within 20 ft (6.0 m) from inside edge of swimming pools, *680.27(C)*.

Figure 30-4 is a detailed illustration showing the grounding of important metal parts of a permanently installed pool.

Figure 30-5 shows a common bonding grid of a permanently installed swimming pool. Note that the bonding conductor does not need to be connected to any service equipment or any grounding electrode. It merely ties all metal parts together.

Storable Pools, Part III. ▶Here we find the *Code* rules for storable pools. Storable pools are just that: they are used on or above the ground, and can be dismantled and stored.

• All electrical equipment used with storable pools must be GFCI protected.

• All 125-volt receptacles within 20 ft (6.0 m) of inside walls of a storable pool must be GFCI protected.

• Receptacles are not permitted within 10 ft (3.0 m) from the inside walls of a storable pool.

• Receptacles rated 15 or 20 amperes, 125 through 250 volts, that supply pool pump motors, shall be GFCI protected.◀

Spas and Hot Tubs, Part IV. What's the difference? In reality, there is no difference. Many years ago, hot tubs were constructed of wood such as redwood, teak, cypress, or oak. Along came acrylics,

plastics, fiberglass, and other man-made products. The name "spa" was created. Today, both names are used interchangeably depending upon where you live. They have therapeutic jets that create whirling currents of hot water, a filtering system, and some have mood lighting, sound systems, and waterfalls. Spas and hot tubs are not designed or intended to have the water drained or discharged after each use. Here are the code rules for spas and hot tubs. (See Figure 30-6.)

Outdoor Installations, *680.42*: Shall be wired according to *Parts I* and *II*, except for the following:

(A) Flexible connections using flexible conduit and cord-and-plug connections are permitted.

(B) Different metal components where there is metal-to-metal mounting of the same metal frame are acceptable as the required bond.

(C) For one-family dwellings, a self-contained spa or hot tub or a packaged spa or hot tub assembly may be connected using conventional wiring methods that contain a copper equipment grounding conductor not smaller than 12 AWG. Conventional wiring methods are found in Chapter 3 of the *NEC*.®

Nonmetallic-sheathed cable with 14 AWG conductors has a 14 AWG equipment grounding conductor.

Figure 30-4 A "trail" of the grounding requirements from a wet-niche luminaire back to the main service panelboard.

Nonmetallic-sheathed cable with 12 AWG conductors has a 12 AWG equipment grounding conductor.

Indoor Installations, *680.43*: Shall be wired according to *Parts I* and *II*, except for the following:

(A) Receptacles:

• At least one 125-volt, 15- or 20-ampere receptacle connected to a general-purpose branch-circuit must be installed at least 5 ft (1.5 m) but not more than 10 ft (3.0 m) from the inside wall of the spa or hot tub.

• All other 125-volt, 15- or 20-ampere receptacles must be located at least 5 ft

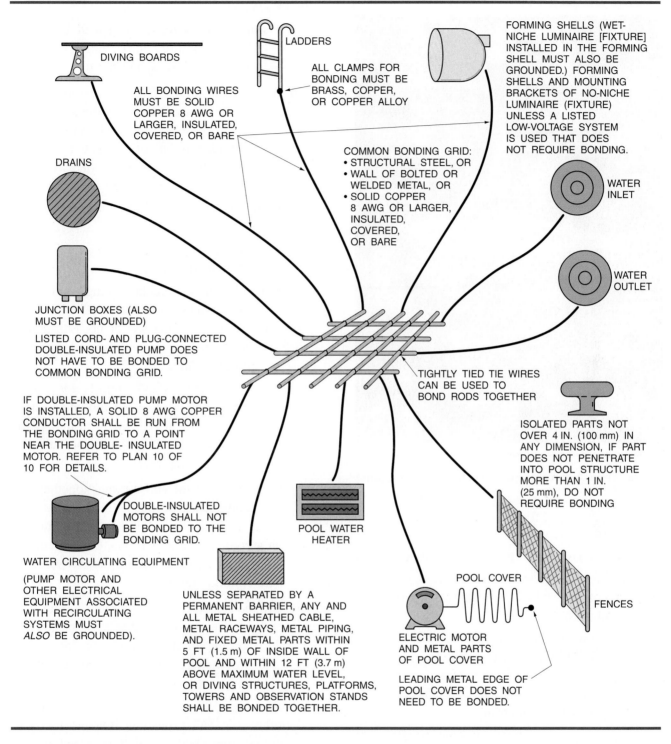

DIVING BOARDS

ALL BONDING WIRES MUST BE SOLID COPPER 8 AWG OR LARGER, INSULATED, COVERED, OR BARE

DRAINS

JUNCTION BOXES (ALSO MUST BE GROUNDED)

LISTED CORD- AND PLUG-CONNECTED DOUBLE-INSULATED PUMP DOES NOT HAVE TO BE BONDED TO COMMON BONDING GRID.

IF DOUBLE-INSULATED PUMP MOTOR IS INSTALLED, A SOLID 8 AWG COPPER CONDUCTOR SHALL BE RUN FROM THE BONDING GRID TO A POINT NEAR THE DOUBLE- INSULATED MOTOR. REFER TO PLAN 10 OF 10 FOR DETAILS.

DOUBLE-INSULATED MOTORS SHALL NOT BE BONDED TO THE BONDING GRID.

WATER CIRCULATING EQUIPMENT

(PUMP MOTOR AND OTHER ELECTRICAL EQUIPMENT ASSOCIATED WITH RECIRCULATING SYSTEMS MUST *ALSO* BE GROUNDED).

LADDERS

ALL CLAMPS FOR BONDING MUST BE BRASS, COPPER, OR COPPER ALLOY

COMMON BONDING GRID:
• STRUCTURAL STEEL, OR
• WALL OF BOLTED OR WELDED METAL, OR
• SOLID COPPER 8 AWG OR LARGER, INSULATED, COVERED, OR BARE

FORMING SHELLS (WET-NICHE LUMINAIRE [FIXTURE] INSTALLED IN THE FORMING SHELL MUST ALSO BE GROUNDED.) FORMING SHELLS AND MOUNTING BRACKETS OF NO-NICHE LUMINAIRE (FIXTURE) UNLESS A LISTED LOW-VOLTAGE SYSTEM IS USED THAT DOES NOT REQUIRE BONDING.

WATER INLET

WATER OUTLET

TIGHTLY TIED TIE WIRES CAN BE USED TO BOND RODS TOGETHER

ISOLATED PARTS NOT OVER 4 IN. (100 mm) IN ANY DIMENSION, IF PART DOES NOT PENETRATE INTO POOL STRUCTURE MORE THAN 1 IN. (25 mm), DO NOT REQUIRE BONDING

POOL WATER HEATER

UNLESS SEPARATED BY A PERMANENT BARRIER, ANY AND ALL METAL SHEATHED CABLE, METAL RACEWAYS, METAL PIPING, AND FIXED METAL PARTS WITHIN 5 FT (1.5 m) OF INSIDE WALL OF POOL AND WITHIN 12 FT (3.7 m) ABOVE MAXIMUM WATER LEVEL, OR DIVING STRUCTURES, PLATFORMS, TOWERS AND OBSERVATION STANDS SHALL BE BONDED TOGETHER.

POOL COVER

ELECTRIC MOTOR AND METAL PARTS OF POOL COVER

LEADING METAL EDGE OF POOL COVER DOES NOT NEED TO BE BONDED.

FENCES

Figure 30-5 **Equipotential bonding the metal parts of a permanently installed swimming pool is intended to eliminate voltage gradients in the pool area. See** *NEC*® *680.26* **and Plan 10 of 10.**

(1.5 m), measured horizontally, from the inside wall of the spa or hot tub.

• All 125-volt receptacles rated 30 amperes or less located within 10 ft (3.0 m) of the inside walls of a spa or hot tub must be GFCI protected.

• A receptacle that supplies power to a spa or hot tub must be GFCI protected.

(B) Luminaires (fixtures), lighting outlets, and ceiling-suspended paddle fans above spa or hot tub within 5 ft (1.5 m) from the inside walls of the spa or hot tub:

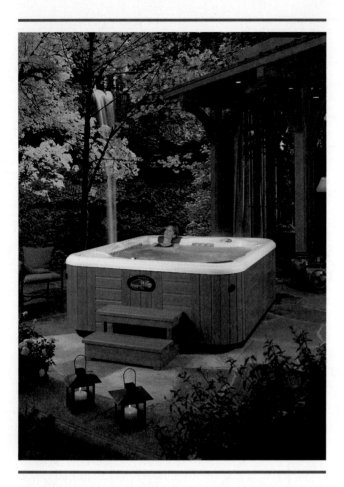

Figure 30-6   A typical spa. *Courtesy of* Hot Spring Spas.

- not less than 12 ft (3.7 m) above if not GFCI protected.
- not less than 7 ft 6 in. (2.3 m) if GFCI protected.
- if less than 7 ft 6 in. (2.3 m), must be suitable for damp locations and be:
  - recessed luminaires (fixtures) with a glass or plastic lens and nonmetallic or isolated metal trim.
  - surface-mounted luminaires (fixtures) with a glass or plastic globe, a nonmetallic body, or a metallic body isolated from contact.
- if underwater lighting is installed, the rules for regular swimming pools apply, in which case refer to Sheet 10 of 10 of the Plans.

(C) Switches:
- wall switches must be kept at least 5 ft (1.5 m) measured horizontally from the inside edge of the spa or hot tub.

(D) Bonding: Bonding requirements are similar to that of conventional swimming pools. Bonding keeps everything at the same potential. Excluded are small conductive surfaces not likely to become energized such as drains and air and water jets. Bond together:
- all metal fittings within or attached to the spa or hot tub structure.
- metal parts of associated electrical equipment including pump motors.
- metal conduits, metal piping, and other metal surfaces within 5 ft (1.5 m) of the inside walls of the spa or hot tub, unless separated by a permanent barrier, such as a wall or building.
- electrical devices and controls *not* associated with the pool or hot tub and located not less than 5 ft (1.5 m) from the spa or hot tub. Bonding to the spa or hot tub system also is acceptable.

(E) Bonding all of the above things together can be accomplished by means of threaded piping and fittings, by metal-to-metal mounting on a common base or frame, or by a copper 8 AWG or larger solid, insulated, covered or bare bonding jumper.

(F) Ground all electrical equipment within 5 ft (1.5 m) of the inside wall of a spa or hot tub. Also ground all electrical equipment associated with the circulating system of the spa or hot tub.

Protection, 680.44: The word "protection" refers to GFCI protection. Outlets for spas and hot tubs shall be GFCI protected, unless the unit has integral GFCI protection. A few other exceptions are mentioned in *680.44*.

Fountains, Part V: The residence discussed in this text does not have a fountain. Permanently installed fountains come under the requirements of *Part V*. For fountains that share water with a regular swimming pool, the wiring must conform to the requirements found in *Part II* of *Article 680*, which covers permanently installed swimming pools.

Small self-contained portable fountains that have no dimension over 5 ft (1.5 m) do not come under the requirements of *Part V*.

Pools and Tubs for Therapeutic Use, Part VI: The residence discussed in this text does not have any therapeutic equipment.

Hydromassage Bathtubs, Part VII: A hydromassage tub is intended to be filled, used, then drained after each use. These are also known as whirlpool bathtubs. Figure 22-12 illustrates one manufacturer's hydromassage bathtub.

- Protection, *680.71*: The word "protection" refers to GFCI protection. Hydromassage bathtubs, together with their associated electrical components, must be GFCI protected.

- Protection, *680.71*: All 125-volt, single-phase receptacles that are located within 5 ft (1.5 m) measured horizontally from the inside walls of the hydromassage bathtub must be GFCI protected.

- Other Electrical Equipment, *680.72*: A hydromassage bathtub does not constitute any more of a shock hazard than a regular bathtub. All wiring (fixtures, switches, receptacles, and other equipment) in the same room but not directly associated with the hydromassage bathtub is installed according to all the normal *Code* requirements covering installation of that equipment in bathrooms.

- Accessibility, *680.73*: Electrical equipment associated with a hydromassage bathtub must be accessible without damaging the building structure or finish.

- ▶Bonding, *680.74*: All metal piping systems and all grounded metal parts in contact with the circulating water shall be bonded together using a copper 8 AWG or larger solid, insulated, covered, or bare.◀

This residence has a hydromassage tub in the master bathroom. The wiring of it is covered in Unit 22.

GETTING TRAPPED UNDER WATER

There is another real hazard, not truly electrical in nature, but solvable electrically.

Drownings have occurred when a person is trapped under water by the tremendous suction at an unprotected drain. The grate over the drain may have been removed, or is broken. One report indicated that three adult males could not pull a basketball from a drain because of the suction!

To address this hazard, *680.41* requires that a clearly labeled emergency shutoff switch for the control of the recirculation system and jet system for spas and hot tubs shall be installed not less than 5 ft (1.5 m) away, adjacent to, and within sight of the spa or hot tub. Although this requirement does not apply to single-family dwellings, some states require this disconnect for single-family dwelling pools.

Another hazard associated with spas and hot tubs is prolonged immersion in hot water. If the water is too hot and/or the immersion too long, lethargy (drowsiness, hyperthermia) can set in. This can increase the risk of drowning. UL Standard 1563 establishes the maximum water temperature at 104°F (40°C). The suggested maximum time of immersion is generally 15 minutes. Instructions furnished with spas and hot tubs specify the maximum temperature and time permitted for using the spa or hot tub. Hot water, in combination with drugs and alcohol, presents a real hazard to life. Caution must be observed at all times.

UNDERWRITERS LABORATORIES STANDARDS

UL standards of interest are:

UL 676	Underwater Lighting Fixtures
UL 943	Ground-Fault Circuit Interrupters
UL 1081	Swimming Pool Pumps, Filters, and Chlorinators
UL 1241	Junction Boxes for Underwater Pool Luminaires (Fixtures)
UL 1563	Electric Spas, Equipment Assemblies, and Associated Equipment
UL 1795	Hydromassage Bathtubs

REVIEW

Note: Refer to the *Code* or the plans where necessary.

1. Most of the requirements for the wiring of swimming pools are covered in *NEC*®
   Article _____ .

2. Name the two ways in which a person may sustain an electrical shock when in a pool.

   a. _____

   b. _____

3. Using the text, the Plans in the back of this text, and the *NEC*,® determine if the
   following items must be grounded. Place an "X" in the correct space.

	True	False
a. Wet-and-dry-niche luminaires (fixtures).	_____	_____
b. Electrical equipment located within 5 ft (1.52 m) of inside walls of pool.	_____	_____
c. Electrical equipment located within 10 ft (3.0 m) of inside walls of pool.	_____	_____
d. Recirculating equipment and pumps.	_____	_____
e. Luminaires (fixtures) and ceiling fans installed more than 15 ft (4.5 m) from inside walls of pool.	_____	_____
f. Junction boxes, transformers, and GFCI enclosures that contain conductors that supply wet-niche luminaires (fixtures).	_____	_____
g. Panelboards that supply the electrical equipment for the pool.	_____	_____
h. Panelboards 20 ft (6.0 m) from the pool that do not supply the electrical equipment for the pool.	_____	_____

4. Is it permitted to run other conductors in the same conduit as the GFCI protected
   conductors that run to underwater luminaires (fixtures)? _____

5. *NEC*® _____ prohibits running electrical conduits under a swimming pool.

6. What is the purpose of grounding and bonding? _____

   _____

   _____

7. Does the *NEC*® require that, after all of the metal parts of a swimming pool are bonded
   together to form a common grounding grid, all of this grid must be connected to a
   grounding electrode? _____

8. May conductors be run above the pool? Explain. _____

   _____

   _____

9. What is the closest distance from the inside wall of a pool that a receptacle may be
   installed ? _____

10. Receptacles located within 20 ft (6.0 m) from the inside wall of a pool must be _____ protected.

11. Luminaires (fixtures) installed above a new outdoor swimming pool must be mounted not less than :

    a. 10 ft. (3.0 m)   b. 12 ft. (3.7 m)

    above the maximum water level. Circle correct answer.

12. Junction boxes and enclosures for tranformers and GFCIs have one thing in common. They all (are made of brass) (have threaded hubs) (must be mounted at least 8 in. [200 mm] above the deck). Circle the correct answer.

13. For indoor spas and hot tubs, the following statements are either true or false. Check one.

	True	False
a. Receptacles may be installed within 5 ft (1.5 m) from the edge of the spa or hot tub.	_____	_____
b. All receptacles within 10 ft (3.0 m) of the spa or hot tub must be GFCI protected.	_____	_____
c. Any receptacles that supply power to pool equipment must be GFCI protected.	_____	_____
d. Wall switches must be located at least 5 ft (1.5 m) from the pool.	_____	_____
e. In general, luminaires (fixtures) above a spa shall not be less than 7 ft 6 in. (2.3 m) above the maximum water level if the circuit is GFCI protected.	_____	_____

14. Bonding and grounding of electrical equipment in and around spas and hot tubs (is required by the *NEC*®) (is decided by the electrician). Circle correct answer.

15. Where a spa or hot tub is installed in an existing bathroom or other suitable location, an existing receptacle outlet within 5 ft (1.5 m) of the tub is permitted to remain, but only if the circuit feeding the receptacle has _____ protection.

16. Underwater pool luminaires (fixtures) shall be installed so that the top of the lens is not less than _____ in. below the normal water line, unless they are listed for a lesser depth.

17. a. Hydromassage bathtubs and their associated electrical components (shall) (shall not) be protected by a GFCI. Circle the correct answer.

    b. All receptacles located within 5 ft (1.5 m) of a hydromassage tub (shall) (shall not) be protected by a GFCI. Circle the correct answer.

Wiring for the Future: Home Automation Systems

OBJECTIVES

After studying this unit, the student should be able to

▶ understand the basics of X10 systems.

▶ understand the basics of a "structured wiring" system.

▶ understand the types of category-rated cables.

▶ understand some of the terminology used in the home automation industry.

▶ understand the basics of "wireless."

Wiring for the future comes under many names: *Home Automation, Advanced Home Automation, Structured Wiring, Voice-Data-Video (VDV), Smart House* to name a few.

An increasing trend in upper-end, upscale residential wiring is the installation of advanced home automation systems. These systems are generally installed in larger, more expensive homes, but are also suitable for the typical home on a limited basis.

Unit 31 is an introduction to advanced home automation wiring. It is beyond the scope of this book to cover the subject in detail because of the many manufacturer's products available today. Some of these products can be interchanged; others cannot. There are texts, videos, and Web sites devoted specifically to advanced home wiring. Web sites (listed after the Appendix of this text) identified with an * are related to home automation. You will want to browse these Web sites and download specific information that fits your needs.

We can hear about, read articles about, and even attend seminars about home automation wiring. Names such as ActiveHome, Avaya's HomeStar, CEBus, Decora Home Controls (Leviton), Echelon, Elan, Enerlogic, Gemini, Greyfox Systems, Home Base, Honeywell Total-Home, IBM Home Director, IES Technologies, LightTouch, LonWorks, Lutrons RadioRA, OnQ, Radio Shack, "Plug 'N Power," PowerMark, Stargate, Smart-House, SmartLinc, Stanley Lightminder, and X-10 Powerhouse are associated with home automation systems. The list is endless!

Everyone is familiar with individual control devices. Examples are wall switches, thermostats, motion detectors, occupancy sensors, timers, TV remote controllers, and similar devices. They control one "thing."

Now we have additional systems for controlling lighting (on, off, and dimming), heating, cooling, ventilation, appliances, entertainment, and security in a home. Control of single or multiple loads from any number of locations is possible. Modules are available that can be controlled through a computer.

Today, there are different types of advanced home systems for residential wiring.

THE X10 SYSTEM

X10 technology was invented in Scotland in 1978. The original X10 patent expired in 1997, so it is now an open standard for any manufacturer to apply to their products.

X10 technology is simple and is the least costly home automation system, especially for retrofit wiring in existing homes. For the most part, special wiring is not required for basic functions. For many applications, all that is necessary is to plug in the various components as needed.

This system transmits carrier wave signals that are superimposed on the regular 120/240-volt, 60-cycle branch-circuit wiring in the home. Components communicate with each other over the existing electrical wiring without a central controller.

Transmitting devices take a small amount of power from the 120-volt power line, modulate or change it to a higher frequency, and then superimpose this high-frequency signal back onto the ac circuit. At the other end, a receiving device responds to the high-frequency burst. Xl0 is basically a one-way communications system. Some devices "talk" and others "listen."

The control of appliances, audio/video equipment, outdoor and indoor lighting, and receptacles, as well as the arming of a security system, is possible from just about any place in the home. You might call it a "plug-n-play" system. Components are easily added as needed.

An X10 system might include programmers; receiving modules (Figure 31-1); transmitters; keypad wall or tabletop controller (Figure 31-2); dimmers; single-pole, three-way, four-way, and double-pole switches; on/off/dim lamp modules; on/off appliance modules; receptacle modules; thermostats; timers; surge protection devices; burglar alarm devices; motion sensors (turn on lights, sound an alarm); wireless controllers; handheld remotes; photocells for light/darkness sensing; drapery controllers; telephone responders; tie-in with a personal computer; voice activation; and so on. A typical system of this type will have the capability of 256 easily adjusted different "addresses," more than adequate for just about any size home. Receivers and switches generally mount in standard wall boxes. X10 wall switches and receptacles can replace conventional switches and receptacles. Make sure that the wall box has ample space for the larger X10 devices so the wires in the wall box will not be jammed.

To ensure trouble-free operation, a phase-coupler (Figure 31-3) connected at the main panel is used to provide proper strength command signals to both sides of the ac wiring.

Figure 31-1   Various types of typical receiver modules used with X10 systems.

Figure 31-2   A typical X10 tabletop keypad controller makes it easy to transmit preprogrammed signals to receiver modules.

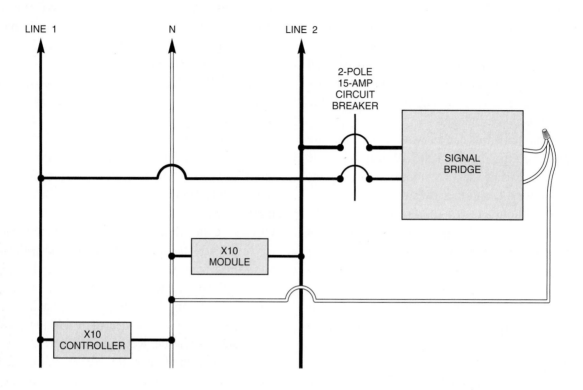

Figure 31-3 A signal bridge phase coupler connected to the main panel to ensure proper signals on both sides of an ac line.

Limitations of X10 Technology

X10 modules are susceptible to interference from an ac power line. Many things can generate "noise" on 120-volt power lines, including fluorescent luminaire (fixture) ballasts, appliances, and wireless intercoms. And because standard home wiring (most often Type NM cable) is unshielded, it can pick up unintended induced signals in the same way that an antenna does, even from devices that are not attached to the same circuit.

In fact, where two or more houses are connected to the secondary of the same utility transformer, signals could be transmitted from one house to another. X10 filters are available that install at the service panel to block interference from outside the house.

X10 signals are relatively slow. The maximum data rate is about 60 bits per second as compared to systems that handle 10,000 bps. This means there can be a noticeable delay between the time you activate a control and the time a controlled action takes place.

STRUCTURED RESIDENTIAL WIRING SYSTEMS

Another type of system is referred to as "structured wiring" or "infrastructure wiring," and is completely separate from the conventional branch-circuit wiring in the home. It is nicely suited for new construction. Some electricians refer to this concept as "future proofing," which means wiring the new home during the original construction (rough-in) stage for all the applications that the homeowner may want now or at some later date.

These systems are wired with special cables such as unshielded twisted pair (UTP), shielded twisted pair (STP), speaker wire, coaxial, fiber optic (FO), and cables that contain multiple types of conductors, as shown in Figure 31-4. The UL Standard refers to FO cables as *optical fiber cables*. A single pair of FO cables can handle the same amount of voice traffic as 1400 pairs of copper conductors.

Figure 31-4 A typical cable consisting of ac power, telephone, coaxial cable, and low-voltage control conductors.

Structured premises wiring systems, sometimes called "home LANs" (for local area networks), are intended to allow distribution of high-speed data, audio, and video signals throughout the home. They offer such advantages as faster Internet access for home computers and convenient networking of data devices (computers, printers, and fax machines) in home offices. Structured premises wiring systems get their name from ANSI/TIA/EIA standards 568 and 570, which describe "structured" or standardized cabling systems for businesses and homes (see box on home wiring standards).

While different systems may not be interchangeable, they all provide the same wiring infrastructure and similar product interfaces such as cables, FO cables, modules, receptacles, switches, outlets, connectors, jacks, keypads, and so forth.

Here are a few of the common devices used in a "structured" wiring installation.

Combination Outlets

Telecommunications, coaxial, and sometimes FO connectors can be combined in the same wall plates, with multiple cables to each outlet location, as shown in Figure 31-5.

Central Distribution

Cables from room outlets are run to a central cabinet, sometimes referred to as a "service-center." Except for outlets, most system components are located at this cabinet, which also functions as a patch panel for making most wiring changes at a central location, as shown in Figure 31-6.

Figure 31-5 A single-gang combination wall outlet with two telecommunications outlets and two coaxial cable outlets. Many other combinations are available.

You will need a 120-volt receptacle near the location where the central distribution panel is located.

Standards

The EIA/TIA 570-A *"Telecommunications Cabling Standard"* is available from Global Engineering, 15 Inverness Way East, Englewood, CO

Figure 31-6 A "service-center" is the hub of a structured wiring system with telecommunications, audio, video, and home automation controllers installed.

CABLE TYPES AND INSTALLATION RECOMMENDATIONS

Communication cable is tested and listed in accordance with UL Standard 444. Optical fiber cable is tested and listed in accordance to UL Standard 1651. These standards are mirror images of the EIA and TIA standards. Cables that have passed the UL testing will bear a "listing" marking.

If the cable has also passed the data transmission performance category program, the cable will be surface-marked and tag-marked, "Verified to UL Performance Category Program."

Some communities are installing underground hybrid FO/coaxial cable systems that combine telephone, CATV, and community intranet capabilities.

The categories are shown in Table 31-1.

Proper selection and installation of any type of audio/video/data cable is extremely important. The

Category 1:	Four-pair non-twisted cable. Referred to as *quad wire*. OK for audio and low-speed data. Not suitable for modern audio, data, and imaging applications. This is the old fashioned Plain Old Telephone Service (POTS) cable. If you have any, get rid of it!
Category 2:	Usually four-pair with slight twist to each pair. OK for audio and low-speed data.
Category 3:	UTP*. Data networks up to 16 MHz.
Category 4:	UTP*. Data networks up to 20 MHz.
Category 5:	UTP*. Data networks up to 100 MHz. By far the most popular, having four pairs with 24 AWG copper conductors, solid (for structured wiring) and stranded (for patch cords). Cable has an outer PVC jacket.
Category 5e:	UTP*. Basically an enhanced Cat 5 manufactured to tighter tolerances for higher performance. Data networks up to 350 MHz. This is detailed in the new TIA/EIA 568A-5 Standard. Will eventually replace Cat 5 for new installations.
Category 6:	UTP*. Four twisted pairs. Data networks up to 250 MHz. Has a spline between the four pairs to minimize interference between the pairs.
Category 7:	Four twisted pairs surrounded by a metal shield. Data networks up to 750 MHz. Specifications not yet standardized at the time of publishing this text.

* UTP means unshielded twisted-pair.

Table 31–1 Types of category-rated cables.

80112-5776 (Phone numbers are 1-800-854-7179 and 1-303-397-7956). This standard was jointly developed by the Electronic Industries Association (EIA) and the Telecommunications Industry Association (TIA). It describes a wiring system that uses twisted four-pair telecommunications cabling, modular jack outlets, and star wiring (home runs) to each room from the distribution panel.

The NEMA Standard WC 63.1-2000 entitled *Performance Standard for Twisted Pair Premise Voice and Data Communications Cable* contains the minimum electrical performance as well as material and mechanical specifications, definitions, and test methods for these types of cables.

final performance will be that of the weakest link in the system. Improper installation, connection hardware, outlets, and other components can reduce the performance to that of an old-fashioned phone system.

Although you may not experience problems with audio transmission, using the proper cable for data transmission is critical. Losing your on-line Internet connection for no apparent reason could very well be pinpointed to older style untwisted-pair connecting cables that are unable to handle high-speed data transmission. Using high-quality cables is extremely important.

For residential applications, a minimum of one Category 5 or 5e (preferred) unshielded twisted-pair (UTP) cable with four twisted pairs of 24 AWG copper conductors or two Category 5 (or 5e) UTP cables with two twisted pairs of 24 AWG copper conductors should be installed for the "home runs" between each outlet location and a central distribution panel where all of the audio/video/data cables are run. Also install one or two RG-6 quad-shielded coax cable to each outlet location. Figure 31-7 illus-

trates a typical bundled cable with two coaxial cables and two Category 5E cables.

There are still widespread problems with existing telephone lines outside of the home causing havoc with on-line Internet connections. There is not much you can do about this, other than complain to the telephone company.

Installation Recommendations

Because it is so easy to damage the cable, thus affecting its transmission properties, here are some cautions for installing any audio/video/data cables:

- Keep the runs as short as possible.

- Run the cables in one continuous length from Point A to Point B without splices.

- Do not to make sharp bends in the cable. Make slow sweeping bends: radius 4D for Category 5 cable, radius 10D for coaxial cable.

- Do not pull so hard as to damage the conductor insulation and the cable jacket (25 lb. is the maximum it can stand).

 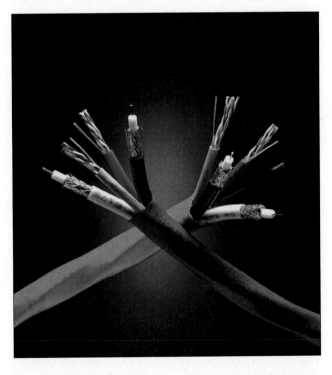

Figure 31-7 Typical bundled cables showing both coaxial and Category 5E cables. *Courtesy of* **Genesis Cable Systems.**

- Do not kink or knot the cable.

- Use rounded or depth-stop plastic staples.

- Where a number of cables are bundled, use tie-wraps loosely around the bundle, then nail the tie-wrap to the framing member.

- Do not use nail guns or metal staples. This can only lead to trouble. Use cable-tie guns.

- Always untwist the least amount, about ½ in. (13 mm) of the conductors at point of termination. The tighter the twist in a cable, the less distortion and interference.

- Do not run these cables through the same holes as power wiring, conduit, pipes, and ducts. There is too much possibility of damage to the cables.

- Keep cables away from hot water pipes, hot air ducts, and other sources of high heat.

- Keep cables at least 12 in. (300 mm) from power wiring where the cables are run parallel to the power wiring. This caution was written before twisted pairs and category-rated cables entered the scene. With the advent of properly installed twisted and shielded Category 5 and 5e cables, this recommendation from the past is probably not an issue. Category 5 and 5e cables are highly immune to interference.

- Run a dedicated 15-ampere 120-volt branch-circuit to the central distribution panel that will be hub of the audio/video/data system.

- Always check the specifications and recommendations of the cable you are installing.

- Consider installing two or three trade size 2 PVC empty conduits from the basement to the attic. Additional cables can be installed at a later date without time-consuming fishing of cables or damaging walls and ceilings.

Many types of bundled cables are available that contain a multitude of cables. Examples are one RG-6 (75-ohm) coaxial cable and one Cat 5 cable; two RG-6 (75-ohm) coaxial cables, two Cat 5 cables, and one ground; two FO cables, two RG-6 (75-ohm) coaxial cables, two Cat 5 cables, and one ground; or two RG-6 (75-ohm) coaxial cables, and one Cat 5 cable. The combinations are endless, and you need to check out manufacturers' catalogs to find the cable that meets your needs.

TERMINOLOGY

The letters RG stand for *Radio Grade*.

The jacks (connectors) for Cat 5 cables are generally the RJ-11 six-pin jack and the RJ-45 eight-pin jack. The letters RJ stand for *Registered Jack*.

Connectors for coaxial cable are referred to as *F* connectors. The threaded type is preferred over the push-on type.

The letter M means *mega*, which is one million (for short: *meg*).

The letter G means *giga*, which is one billion (for short: *gig*).

The letters *bps* mean bits per second.

The letters Hz means *hertz*, which are cycles per second.

The letters MHz means *megahertz*, which are millions of cycles per second.

A bit is the smallest unit of measurement in the binary system. The expression is derived from the term *binary digit*.

It takes eight bits to make one byte. A *byte* is the amount of data required to describe a single character of text.

The letters Mbps means *megabits per second*. For example, 20 Mbps means that twenty million (20,000,000) *bits* of data are being transmitted per second over copper or FO cable.

There is no direct correlation between *megahertz* and *megabits*. Different encoding schemes represent information in different ways.

WIRELESS

Structured wiring involves a separate system of conductors and cables that run between controllers, sensors, and all of the other equipment. X10 technology uses the existing wiring in your home. The devices are electronically coupled to the premise wiring in your house. Both systems use wireless remote control devices using infrared technology. Infrared (IR) technology is the heart of most wireless systems, although some operate on RF. "RF" stands for radio frequency. The main drawback of IR is that you must have a straight unobstructed line of sight between the remote and the equipment being controlled. With RF devices, you do not need a straight, unobstructed line of sight. RF sends out a radio signal that goes around corners and through walls and doors. Both systems can control TVs, radios, sound systems, VCR players, CD players, DVD players, lighting, ceiling paddle fans, garage door openers, small relays that plug into a wall receptacle, remote car locking systems, etc. Sound familiar? You may not have thought about it, but you are already in the world of home automation.

SUMMARY

Wireless or Wired? The choice is yours!

Since home automation systems vary, visit the many Web sites marked with an * listed after the Appendix of this text.

An excellent text on the subject is *Premises Cabling*, available from Delmar. See Data and Voice Communications Cabling and Fiber Optics in the front material of this text for other related books.

REVIEW

1. One type of home automation system is referred to as:
   a. X10   b. X20   c. HAS

2. When using an X10 system, which of the following statements is most correct?
   a. A totally independent wiring system must be installed.
   b. The devices generally are "plug-n-play," requiring no additional wiring.
   c. An X10 system is ideally suited for installations in new construction.

3. A home automation system that requires a totally separate wiring system is referred to as:
   a. an X10 system.
   b. a structured system.
   c. a DC-powered system.

4. The most recommended "category-rated" cable for wiring from various outlets back to the main "service-center" for the interconnection of computers is:
   a. Category 1.
   b. Category 2.
   c. Category 3.
   d. Category 5 or 5e.

5. When installing Category 5 rated cable, the minimum bending radius generally recommended is:
   a. 4 times the diameter of the cable.
   b. 6 times the diameter of the cable.
   c. 10 times the diameter of the cable.

6. When installing coaxial cable, the minimum bending radius generally recommended is:
   a. 4 times the diameter of the cable.
   b. 6 times the diameter of the cable.
   c. 10 times the diameter of the cable.

7. When selecting cables for a structured wiring system or computer hookups, the best choice is to select a cable that has:
   a. untwisted pairs of conductors.
   b. twisted pairs of conductors.

8. Match each of the following terms with its description by drawing a line between them.

   a. RG   * cyles per second

   b. RJ   * the smallest unit of measurement in the binary system.

   c. bps   * registered jack

   d. mbps   * radio grade

   e. Hz   * bits per second

   f. bit   * megabits per second

Standby Power Systems

OBJECTIVES

After studying this unit, the student should be able to

▶ understand some of the safety issues concerning optional standby power systems.

▶ understand the basics of standby power.

▶ understand the types of standby power systems.

▶ understand wiring diagrams for portable and standby power systems.

▶ understand transfer switches, disconnecting means, and sizing recommendations.

▶ understand the *National Electric Code®* requirements for standby power systems.

The text that follows is a general discussion and overview of home-type temporary generator systems. It is not possible in this unit to cover all the variables and details involved with the many types of systems available in the marketplace today. Always follow the manufacturer's installation instructions. Always turn off the power when hooking up standby systems. Do not work on "live" equipment. Do not stand in water or otherwise work on electrical equipment with wet hands. Do not touch bare wires. Operate the generator only outside in open air. Exhaust gases contain deadly carbon monoxide, the same as an automobile engine. Carbon monoxide can cause severe nausea, fainting, or death. It is particularly dangerous because it is an odorless, colorless, tasteless, nonirritating gas. Typical symptoms of carbon monoxide poisoning are dizziness, headache, light-headedness, vomiting, stomachache, blurred vision, inability to speak clearly. If you suspect carbon monoxide poisoning,

remain active. Do not sit down, lie down, or fall asleep. Breathe fresh air—fast!

Store gasoline only in approved red containers clearly marked "GASOLINE."

Always install equipment such as generators, panelboards, power inlets, and transfer switches that are listed by a nationally recognized testing laboratory.

WHY STANDBY (TEMPORARY) POWER?

Everyone has experienced a power outage, sometimes more often than one would like. You are left in the dark. Your furnace does not operate. You are concerned about water pipes freezing. Your refrigerator and freezer are not running. Your sump pump is not operating, and your basement is flooding. Your overhead garage door does not operate.

What can you do? The Y2K crisis made everyone aware of the consequences of a major power outage. Generators were carried out of home centers and other dealers at a furious pace. Now that the Y2K crisis is history, it still makes sense to consider a standby power system.

A few questions need to be answered. In your home, which loads are critical and must continue to operate in the event of a utility power outage? You need to know this so you can select a generator panel that will have enough branch-circuits. Only you can make this decision. Another question is, how large must the generator be to serve the loads that are considered critical in your home? Still another question is, how simple or complicated a standby system do you want?

WHAT TYPES OF STANDBY POWER SYSTEMS ARE AVAILABLE?

The Simplest

Home centers usually carry the simplest and most economical portable generators, as shown in Figure 32-1. These consist of a gasoline-driven motor/generator set that must be started manually. Some models have the recoil "pull-to-start" feature; others have battery electric start. These are the types construction workers use for temporary power on job sites. Depending upon the size, these smaller generators might be capable of running for four to eight hours. This type of portable generator is commonly referred to as "back-up" power.

These small portable generators consist of a gasoline engine driving a small electrical generator, just like a gasoline engine drives the blades of a lawn mower or snow thrower. Some generator sets have one or more standard 15- or 20-ampere single or duplex grounding-type receptacles, some have GFCI receptacles, and others have 20- and 30-ampere twistlock-type receptacles. Some are rated 120 volts, while others are rated 120/240 volts. These generators are available in sizes up to about 5000 watts. Also available are models with more "bells and whistles" such as oil alerts, adjustable output voltage, larger fuel tanks, and so on.

The procedure for this type is to start up the generator, then plug an extension cord into the receptacle and run it to whatever critical cord- and plug-connected load needs to operate.

If you are not home when a power outage occurs, you are out of luck!

The Next Step Up

The next step up involves both permanent wiring and cord- and plug-connected wiring.

This is by far the most popular when it comes to

Figure 32-1 A portable generator serving standby power to a 120-volt sump pump.

PORTABLE GENERATOR

CORD (2-WIRE PLUS EQUIPMENT GROUND)

SUMP PUMP

standby power for homes. In the event of a power outage, you must manually start the generator, flip the transfer switching device over from normal power to standby power, and plug in the power "patch" cord (see Figure 32-2).

The permanent wiring part of the installation involves connecting specific branch-circuits to a separate generator panel (usually located right next to the main panel), installing and properly connecting a transfer switch (discussed later), and installing a power inlet receptacle (discussed later). The generator panel serves those circuits that you have selected as critical to operate in the event of a power outage. This panel is sometimes referred to as a generator

panel, a selected load panel, an emergency panel, and sometimes a critical load panel (see Figure 32-3).

The cord- and plug-connected part of the installation consists of merely a flexible power patch cord that runs between the generator and the power inlet receptacle. This cord and the generator are set up as needed.

These are gasoline-powered generators that use a flexible four-wire rubber cord (12 AWG for 5000-watt generators, 10 AWG for 7500-watt generators) that plugs into a male polarized twistlock receptacle on the generator. The other end of the cord plugs into a polarized twistlock receptacle mounted on the outside of the house in a weatherproof power inlet

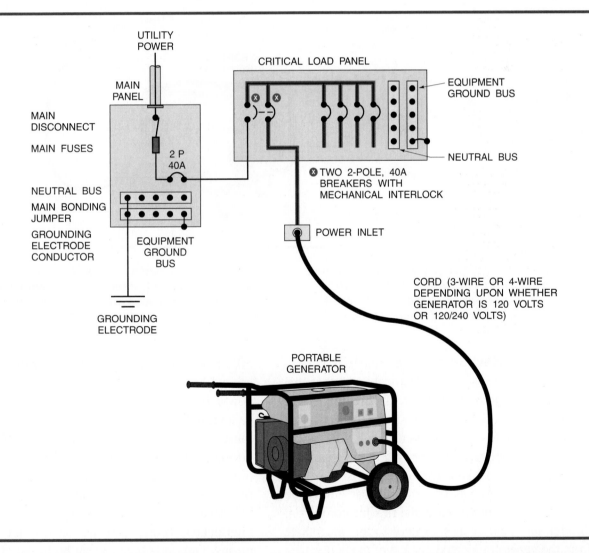

Figure 32-2 A portable generator supplying standby power to a critical load panel. The circuit breakers in the critical load panel must be manually turned to the temporary power position. These breakers are equipped with a mechanical interlock so there will be no feedback of electricity between the generator power and the utility normal power.

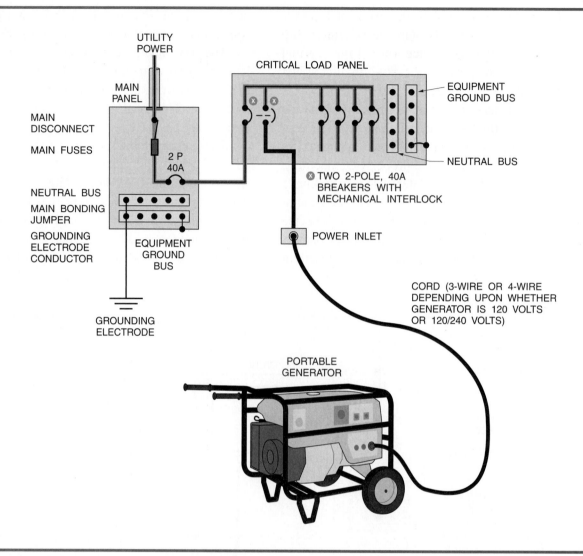

UTILITY
POWER

MAIN
PANEL

CRITICAL LOAD PANEL

EQUIPMENT
GROUND BUS

MAIN
DISCONNECT

MAIN FUSES

2 P
40A

NEUTRAL BUS

NEUTRAL BUS
MAIN BONDING
JUMPER

GROUNDING
ELECTRODE
CONDUCTOR

EQUIPMENT
GROUND
BUS

TWO 2-POLE, 40A
BREAKERS WITH
MECHANICAL INTERLOCK

POWER INLET

CORD (3-WIRE OR 4-WIRE
DEPENDING UPON WHETHER
GENERATOR IS 120 VOLTS
OR 120/240 VOLTS)

GROUNDING
ELECTRODE

PORTABLE
GENERATOR

Figure 32-3   A critical load panel served by normal power from the main service panel. A portable generator is shown plugged into the power inlet. To switch to standby power, the circuit breakers in the critical load panel must be manually turned to the standby position. These breakers are equipped with a mechanical interlock so there will be no feedback of electricity between the generator power and the utility normal power.

box. This polarized twistlock receptacle is permanently connected to a special electrical generator panel in which the critical branch-circuits are connected. The generator panel will have a transfer mechanism plus from four to twenty branch-circuit breakers. It all depends upon the wattage rating of the generator. Some generator panels have the polarized twistlock receptacle mounted as part of and just below the generator panel. In this case, the flexible cord is run through an open window or door to this receptacle from the generator located outside of the house.

During normal operation, this generator panel is fed from a two-pole circuit breaker in the main panel. This two-pole breaker might be rated 30, 40,

50, or 60 amperes, depending upon the number of critical branch-circuits to be supplied. When utility power is lost, this generator panel is fed from the generator. When the electric utility loses power, you must *manually* turn the "transfer switch" from its normal power position to its temporary power position. One manufacturer of electrical equipment furnishes a panel that contains two 30-ampere, two-pole circuit breakers that are mechanically interlocked. These serve as the transfer switching means. Only one breaker can be in the "ON" position at the same time. One of these breakers brings the normal power to the generator panel. The second breaker brings the power from the generator to the generator panel. When one breaker is turned "OFF," the other is turned "ON." This panel will have four or more branch-circuit breakers.

With this system, the procedure generally is to first plug the four-wire cord into the polarized twist-lock receptacle on the generator, plug the other end into the power inlet receptacle, and then start the generator, which is always located outdoors. The transfer switch in the generator panel is then switched to the "GEN" position.

When normal power is restored, turn the transfer switch in the generator panel back to its normal position, turn off the generator according to the manufacturer's instructions, unplug the power cord from the generator, and finally, unplug the other end of the cord from the power inlet receptacle.

Important: If you intend to have a GFCI breaker-protected branch-circuit transferred to the generator panel, be sure to connect this circuit to a GFCI breaker in the generator panel.

Because GFCIs and AFCIs require a neutral conductor, make sure the neutral conductor is properly connected between the main panel neutral bus and the neutral bus in the critical panel.

The neutral bus in the generator panel is insulated from the generator panel enclosure.

The ground bus in the generator panel is bonded to the generator panel enclosure.

Standby generators of this type can have run times as long as 16 hours. Again, it depends upon the loading. The manufacturer's installation and operating instructions will provide this information.

If you are not home when the power outage occurs, you are out of luck!

The Top of The Line

Standby power served by a permanently installed generator is truly standby power. These types of generator systems are covered under UL Standard 2200, *Stationary Engine Generator Assemblies*.

Standby power is a system that allows you to safely provide electrical power to your home in the event of a power outage. Instead of running extension cords from a portable generator located outside of the house to whatever critical loads (lights, sump pumps, refrigerators, etc.), you set up the critical branch-circuit wiring in a generator panel in such

a manner that all that is necessary to do in the event of a power outage is to transfer the critical loads from normal utility power to generator power. The diagrams in this unit show this.

The ultimate and most expensive in standby power systems is one where the homeowner has nothing to do in the event of a power outage. Standby power comes on when the normal utility power goes off. There is complete peace of mind! The generator for this type of system is permanently installed on a concrete pad at a suitable location outside of the home, convenient to making up all of the electrical connections to serve the selected critical loads. These systems generally run on natural gas (Figure 32-4), or propane (liquid petroleum gas—LPG) and in rare cases, on diesel fuel. All the conductors between the generator, the transfer switch, and the generator panel installed for the selected critical loads are permanently wired. The automatic transfer switch monitors the incoming electric utility's voltage. Should a power outage occur, the standby generator automatically starts, the transfer switch "transfers," and the critical loads continue to function. When normal power is restored, the transfer switch automatically "transfers" back to the normal power source, and the generator shuts down.

Since the fuel supply is a permanently installed natural gas line, these systems can run indefinitely in conformance to the manufacturer's specifications.

To be totally automatic, a solid-state controller that is part of the transfer switch immediately senses the loss of normal power and "tells" the standby

Figure 32-4 A natural gas-driven generator.

generator to start, and to "tell" the transfer switch to transfer from normal power to standby power. This controller senses when normal power is restored, allowing the transfer switch to return to normal power and to shut off the standby generator. All of the circuitry is reset, and it is now ready for the next power interruption.

WIRING DIAGRAMS FOR A TYPICAL STANDBY GENERATOR

For simplicity, the diagrams in Figure 32-5 and Figure 32-6 are one-line diagrams. Actual wiring consists of two ungrounded conductors and one grounded "neutral" conductor. Equipment grounding of all of the components is accomplished through the metal raceways that interconnect the components. See *250.118*.

The neutral bus in a panel that serves critical loads must *not* be connected to the metal panel enclosure. Connecting the neutral, the grounding electrode conductor, and the equipment grounding conductors together is permitted only in the main service panel.

If you were to bond the neutral conductor and the metal enclosures of the equipment (panelboard, transfer switch, and generator) together beyond the main service panelboard, you will have created a parallel path. A parallel path means that some of the normal return current and fault current will flow on the grounded neutral conductor and some will flow on the metal raceways and other enclosures. This is not a good situation! See *NEC® 250.6*.

Manufacturers of generators in most cases do not connect the generator neutral conductor lead to the metal frame of the generator. Instead, they con-

Figure 32-5 A natural gas-driven generator connected through a transfer switch. The transfer switch is in the normal power position. Power to the critical load panel is through the main panel, then through the transfer switch.

nect it to an isolated terminal. Then it is up to you to determine how to connect the generator in compliance with the *NEC®* and/or local electrical codes.

Most inspectors will permit the internal bond in a portable generator to remain in place. For permanently installed generators, inspectors will generally not permit a bond between the generator neutral and the metal frame of the generator.

This is a tricky subject. Different electrical inspectors interpret the *NEC®* rules differently. You had better check this out with your local electrical inspector.

A sign must be placed at the service-entrance main panel indicating where the standby generator is located, and what type of standby power it is. See *702.8(A)*.

The total time for complete transfer to standby power is approximately 45–60 seconds.

To further give the homeowner assurance that the standby generator will operate when called upon, some systems provide automatic "exercising" of the system periodically, such as once every 7 or 14 days for a run time of 7 to 15 minutes.

WARNING: **When an automatic type of standby power system is in place and set in the automatic mode, the engine may crank and start at any time without warning. This would occur when the utility power supply is lost. To prevent possible injury, be alert, aware, and very careful when working on the standby generator equipment or on the transfer switch. Always turn the generator disconnect to the "OFF" position, then lock out and tag out the switch, warning others not to turn the switch back "ON." In the main panel, turn it "OFF," then lock out and tag out the circuit breaker feeding the transfer switch.**

Figure 32-6   A natural gas-driven generator connected through a transfer switch. The transfer switch is in the standby position, feeding power to the critical load panel.

TRANSFER SWITCHES

A transfer switch shifts electrical power from the normal utility power source to the standby power source. A transfer switch isolates the utility source from the standby source in such a way that there is no feedback from the generator to the utility's system, or vice versa. When normal power is restored, the transfer switch resets itself, and is ready for the next power outage.

Transfer switches for typical residential applications might have ratings of 40 to 200 amperes. Some have wattmeters for balancing generator loads.

For home installations, transfer switches are double-pole, double-throw (DPDT). This type of transfer switch does not break the "neutral" conductor. Technically speaking, the *NEC®* refers to this type of system as a *nonseparately derived system*. A nonseparately derived system is properly grounded through the grounding electrode system of the normal premise wiring. A system where the neutral is disconnected by the transfer switch is referred to as a *separately derived system* and must be regrounded according to *250.30*. Separately derived systems are common in commercial and industrial installations.

If a transfer switch is connected on the line side of the main service disconnect, it must be listed as being suitable for use as service equipment. In Figure 32-5 and Figure 32-6, the transfer switch is not on the line side of the main service disconnect, and therefore does not have to be listed for use as service equipment.

A transfer switch must "break-before-make." Without the "break-before-make" feature, a dangerous situation is present and could lead to destruction of the generator, personal injury, or death. If the transfer switch does not separate the utility line from the standby power line, utility workers working on the line could be seriously injured. For low-cost, non-automatic transfer systems, the transferring means might consist of two two-pole circuit breakers that are mechanically interlocked so both cannot be "ON" at the same time.

The rating of the transfer switch must be capable of safely handling the load to be served, otherwise the transfer switch could get so hot as to cause a fire. This is nothing new; it is the same hazard as overloading a conductor. The transfer switch must also be capable of safely interrupting the available fault-current that the generator is capable of delivering.

A typical transfer switch might take 10–15 seconds to start the generator. This eliminates nuisance start ups during momentary utility power outages. After start-up, another 10–15-second delay is provided to allow the voltage to stabilize. After start-up and warm-up, full transfer takes place. Similar time delay features are used for the return to normal power. Some systems can do the full transfer in a few seconds.

Only listed transfer switches are permitted to be used. The UL Product Standard is 1008, *Transfer Switch Equipment*.

DISCONNECTING MEANS

A disconnecting means is required for generators, *NEC® 445.18*. This section states that *"Generators shall be equipped with disconnect(s) by means of which the generator and all protective devices and control apparatus are able to be disconnected entirely from the circuits supplied by the generator, except where both of the following conditions apply:*

1. *The driving means for the generator can be readily shut down*

2. *The generator is not arranged to operate in parallel with another generator or other source of voltage."**

   NEC® 225.34 requires that all disconnecting means shall be grouped.

   NEC® 230.72(A) also requires that all disconnecting means shall be grouped. If a disconnect is located outside, it must be on the building or within sight of the building.

Where the normal power main service disconnecting means is located outside of the home and the standby generator is located nearby, the electrical inspector may permit the disconnect at the generator to be in conformance to *NEC® 230.72(B)*.

Where the normal power main service disconnecting means is located inside of the home, some electrical inspectors will require that a disconnect for the standby power from the generator be installed near the main service equipment.

---

*Reprinted with permission from NFPA 70-2005.

NEC® 225.34(B) and *230.72(B)* both state that the required disconnect for the standby generator be located remote from the normal service disconnect. This is to lessen the possibility of accidentally shutting off both disconnects at the same time.

Having one disconnecting means inside the home and another outside of the home present quite a challenge for firemen or others wanting to totally shut off the power to the home.

CONDUCTOR SIZE FROM STANDBY GENERATOR

The conductors that run from the standby generator to the transfer switch and to the generator panel that contains the branch-circuit overcurrent devices shall be sized not less than 115 percent of the generator's nameplate current rating. See *NEC® 445.13*.

EXAMPLE: a 7500-watt, single-phase, 120/240-volt generator.

Step 1. Calculate the generator's current rating.

$$\text{Amperes} = \frac{\text{Watts}}{\text{Volts}} = \frac{7500}{240} = 31.25 \text{ Amperes}$$

Step 2. $31.25 \times 1.15 = 35.94$ amperes.

Step 3. Refer to *NEC® Table 310.16* to determine the minimum size conductors.

In the 140°F (60°C) column of *Table 310.16*, we find that an 8 AWG copper conductor has an allowable ampacity of 40 amperes. The discussion on when to use the 140°F (60°C) and when to use the 167°F (75°C) columns of *NEC® Table 310.16* is covered in Unit 4.

The neutral conductor is permitted to be sized according to *220.61*, which under certain conditions allows the neutral conductor to be smaller than the phase conductors. This information is found in *NEC® 445.13*.

GENERATOR SIZING RECOMMENDATIONS

No matter what type or brand of generator you purchase, you will need to size it properly. There is no better way to do this than to follow the manufacturer's recommendations. Generators for home use are generally rated in watts. The manufacturer of the generator has taken into consideration that watts = volts ×

amperes. Some manufacturers suggest that after adding up all of the loads to be picked up by the generator, add another 20 percent capacity for future loads.

Wattage ratings of appliances vary greatly, as shown in Table 32-1. Resistive loads such as toasters, heaters, and lighting do not have an initial high inrush surge of current. Motors, on the other hand, do have a high inrush surge of starting current, which lasts for only a few seconds until the motor gets up to speed. This inrush must be taken into consideration when selecting a generator. Here are some typical *approximate* values for household loads. For heating-type appliances, the typical operating wattage values are used. For motor-operated appliances, the starting wattage values should be used. Verify actual wattage ratings by checking the nameplate on the appliance. Verify that the connected loads do not exceed the generator's marked capacity.

Most generators are capable of handling a momentary "inrush" of an extra 50 percent of their rating. Here again, check the manufacturer's specifications.

THE *NATIONAL ELECTRICAL CODE®* REQUIREMENTS

The *NEC®* addresses optional standby systems in *Article 702*. This includes those systems that are "permanently installed in their entirety," and those systems that are intended for "connection to a premises wiring system from a portable alternate power supply."

- *NEC® 702.4*: All equipment shall be approved for the intended use.

- *NEC® 702.5*: All of the equipment must have the capacity and rating for the loads that will be supplied by the equipment. The equipment shall be suitable for the maximum available fault-current at its terminals.

- *NEC® 702.6*: Transfer equipment shall be suitable for the intended use and designed to prevent the interconnection of the generator and the normal utility supply.

- *NEC® 702.7*: Some systems, where practical, require audible and visual signal devices to indicate when the normal power system has been transferred to the standby system. The manufacturer of the equipment provides these features.

	TYPICAL OPERATING WATTAGE REQUIREMENTS	STARTING WATTAGE REQUIREMENTS
Blender	600	700
Broiler	1600	1600
Central Air Conditioner		
10,000 BTU	1500	4500
20,000 BTU	2500	7500
24,000 BTU	3800	11,000
32,000 BTU	5000	15,000
40,000 BTU	6000	18,000
Coffee Maker	900–1750	900–1750
CD Player	50–100	50–100
Clothes Dryer		
Electric	5000 @ 240 volts	6500
Gas	700	2200
Computer	300–800	300–800
Curlers	50	50
Dehumidifier	250	350
Dishwasher	1500	2500
Electric		
Drill	250–750	300–900
Fry Pan	1300	1300
Blanket	50–200	50–200
Space Heater	1650	1650
Water Heater	5000–8000 @ 240 volts	5000–8000 @ 240 volts
Electric Range		
6-in. surface unit	1500	1500
8-in. surface unit	2100	2100
Oven	4500	4500
Lights	Add wattage of bulbs	Add wattage of bulbs
Furnace Fan		
1/8 Horsepower Motor	300	900
1/6 Horsepower Motor	500	1500
1/4 Horsepower Motor	600	1800
1/3 Horsepower Motor	700	2100
1/2 Horsepower Motor	875	2650
Garage Door Opener		
1/3 Horsepower	725	1400
1/4 Horsepower	550	1100
Hair Dryers	300–1650	300–1650
Iron	1200	1200
Microwave Oven	700–1500	1000–2300
Radio	15 to 500 (with components)	15 to 500
Refrigerator	700–1000	2200
Security, Home	25–100	25–100
Sump Pump		
1/3 Horsepower	800	2400
1/2 Horsepower	1050	3150
Television	300	300
Toaster		
4-Slice	1500	1500
2-Slice	950	950

Table 32-1 Wattage ratings of appliances *(continues)*

	TYPICAL OPERATING WATTAGE REQUIREMENTS	STARTING WATTAGE REQUIREMENTS
Automatic Washer	1150	3400
Well Pump Motor		
⅓ Horsepower	800	2400
½ Horsepower	1050	3150
1 Horsepower	1920 @ 240 volts	5500
Vacuum Cleaner	1100	1600
VCR	150–250	150–250
Window Fan	200	300

Table 32-1 *(continued)*

▶Signal devices are not required for portable standby generators, *702.7, Exception.*◀

- *NEC® 702.8(A):* A sign is required at the service-entrance equipment to indicate the type and location of the standby generator.

- *NEC® 702.8(B):* Where the grounded circuit conductor connected to the optional standby generator is connected to a grounding electrode conductor at a location remote from the standby source, there shall be a sign at the grounding location that shall identify all standby and normal sources connected at that location. For most residential standby systems, this connection is made on the neutral bus in the generator panel. This neutral bus in turn has a conductor run to the neutral bus in the main panel, where the neutral bus and equipment grounding bus are connected together, and a grounding electrode conductor is connected and run to the grounding electrode of the system. The neutral bus in the generator panel is isolated from the panel enclosure. The equipment grounding bus in the generator panel is bonded to the panel enclosure.

- *NEC® 702.9:* It is permissible to run standby power conductors and normal power conductors in the same raceway or enclosure.

- *NEC® 702.10(B):* Where a portable optional standby source is used as a nonseparately derived system, the equipment grounding conductor shall be bonded to the system grounding electrode. The equipment grounding conductor (green or bare) from the standby generator to the generator panel is connected to the equipment grounding bus in the generator panel. The equipment grounding bus in the generator panel is bonded to the generator panel enclosure. In turn, the equipment grounding bus in the generator panel is bonded back to the main panel either by a metallic wiring method, or by a separate equipment grounding conductor. At the main panel, the grounding electrode conductor is connected to both the equipment ground bus and the neutral bus.

- ▶*NEC® 702.11* tells us that if an optional outdoor housed standby system generator is:
  - equipped with a readily accessible disconnecting means, and
  - is located within sight of the building it serves,

  an additional disconnecting means is not required where the ungrounded conductors from the generator serve or pass through the building it serves. See *225.31* and *225.32.*◀

Permits

Permanently installed generators require a considerable amount of electrical work and gas line piping. The installation will probably require applying for a permit so that proper inspection by the authorities can be made. Check with your local building official.

Sound Level

Because generators produce a certain level of "noise" (decibels) when running, you will want to check with your building authorities to make sure the generator you choose is in compliance with local codes relative to any sound ordinance that might be applicable. Sound level information is provided by the manufacturer in their descriptive literature.

REVIEW

Note: Refer to the *Code* or the plans where necessary.

1. The basic safety rule when working with electricity is to _____
   _____ .

2. Where would the logical location be for running a portable generator?

3. The best advice to follow is to always use _____ equipment.
   Insert the appropriate word in the blank space: listed, cheapest, smallest.

4. How would you define the term "standby power?" _____
   _____

5. Describe in simple terms the three types of standby power systems.
   _____
   _____
   _____
   _____
   _____

6. Is it permitted to ground the neutral conductor of the standby generator to the metal
   enclosure of the generator, transfer switch, or critical (generator) panel? Explain.
   _____
   _____
   _____

7. Briefly explain the function of a transfer switch.
   _____
   _____
   _____

8. When a transfer switch transfers to standby power, the electrical connection inside the
   switch (a) maintains connection to the normal power supply as it makes connection to
   the standby power supply, or (b) breaks the connection to the normal power supply
   before it makes connection to the standby power supply. Circle (a) or (b) as the correct
   answer.

9. A typical transfer switch for residential application is (a) a double-pole, double-throw
   switch (DPDT) (b) a single-pole, double-throw switch (SPDT) or (c) a double-pole,
   single-throw switch (DPST). Circle the correct answer.

10. The technical *NEC*® definition of a system in which the neutral conductor is not switched
    is referred to as (a) a separately derived system (b) a nonseparately derived system.
    Circle the letter indicating the correct definition.

11. The *NEC®* requires that a _____ means be provided for standby power systems. In the case of portable gen sets, this might be as simple as pulling out the plug on the extension cord that is plugged into the receptacle on the gen set. For permanently installed standby power generators, this might be on the gen set and/or separately provided inside or outside of the home.

12. The conductors running from a permanently installed standby power generator to the transfer switch shall not be less than (100%) (115%) (125%) of the generator's output rating.

Specifications for Electrical Work– Single-Family Dwelling

1. **GENERAL:** The "General Clause and Conditions" shall be and are hereby made a part of this division.

2. **SCOPE:** The electrical contractor shall furnish and install a complete electrical system as shown on the drawings and/or in the specifications. Where there is no mention of the responsible party to furnish, install, or wire for a specific item on the electrical drawings, the electrical contractor will be responsible completely for all purchases and labor for a complete operating system for this item.

3. **WORKMANSHIP:** All work shall be executed in a neat and workmanlike manner. All exposed conduits shall be routed parallel or perpendicular to walls and structural members. Junction boxes shall be securely fastened, set true and plumb, and flush with finished surface when wiring method is concealed.

4. **LOCATION OF OUTLETS:** The electrical contractor shall verify location, heights, outlet and switch arrangements, and equipment prior to rough-in. No additions to the contract sum will be permitted for outlets in wrong locations, in conflict with other work, and so on. The owner reserves the right to relocate any device up to 10 ft (3.0 m) prior to rough-in, without any charge by the electrical contractor.

5. **CODES:** The electrical installation is to be in accordance with the latest edition of the *National Electrical Code,*® all local electrical codes, and the utility company's requirements.

6. **MATERIALS:** All materials shall be new and shall be listed and bear the appropriate label of Underwriters Laboratories, Inc., or another nationally recognized testing laboratory for the specific purpose. The material shall be of the size and type specified on the drawings and/or in the specifications.

7. **WIRING METHOD:** Wiring, unless otherwise specified, shall be nonmetallic-sheathed cable, armored cable, or electrical metallic tubing (EMT), adequately sized and installed according to the latest edition of the *National Electrical Code*® and local ordinances.

8. **PERMITS AND INSPECTION FEES:** The electrical contractor shall pay for all permit fees, plan review fees, license fees, inspection fees, and taxes applicable to the electrical installation and shall be included in the base bid as part of this contract.

9. **TEMPORARY WIRING:** The electrical contractor shall furnish and install all temporary wiring for handheld tools and construction lighting per latest OSHA standards and *Article 590, NEC,*® and include all costs in base bid.

10. **WORKSHOP:** Workshop wiring is to be installed in EMT using steel compression gland fittings.

11. **NUMBER OF OUTLETS PER CIRCUIT:** In general, not more than ten (10) lighting and/or receptacle outlets shall be connected to any one lighting branch-circuit. Exceptions may be made in the case of low-current-consuming outlets.

12. **CONDUCTOR SIZE:**   General lighting branch-circuits shall be 14 AWG copper protected by 15-ampere overcurrent devices. Small-appliance circuits shall be 12 AWG copper protected by 20-ampere overcurrent devices. Conductors shall be Type THHN. For service-entrance conductors, see Clause 28.

     All other circuits: conductors and overcurrent devices as required by the *Code*.

13. **LOAD BALANCING:**   The electrical contractor shall connect all loads, branch-circuits, and feeders per the Panel Schedule, but shall verify and modify these connections as required to balance connected and calculated loads to within 10 percent variation.

14. **SPARE CONDUITS:**   Furnish and install two empty trade size 1 EMT conduits between workshop and attic for future use.

15. **GUARANTEE OF INSTALLATION:**   The electrical contractor shall guarantee all work and materials for a period of one full year after final acceptance by the architect/engineer, electrical inspector, and owner.

16. **APPLIANCE CONNECTIONS:**   The electrical contractor shall furnish all wiring materials and make all final electrical connections for all permanently installed appliances such as, but not limited to, the furnace, water heater, water pump, built-in ovens and ranges, food waste disposer, dishwasher, and clothes dryer.

     These appliances are to be furnished by the owner.

17. **CHIMES:**   Furnish and install two (2) two-tone door chimes where indicated on the plans, complete with two (2) push buttons and a suitable chime transformer. Allow $150.00 for above items. Chimes and buttons to be selected by owner.

18. **DIMMERS:**   Furnish and install dimmer switches where indicated.

19. **EXHAUST FANS:**   Furnish, install, and provide connections for all exhaust fans indicated on the plans, including, but not limited to, ducts, louvers, trims, speed controls, and lamps. Included are the recreation room, laundry, rear entry powder room, range hood, and bedroom hall ceiling fans. Allow a sum of $700.00 in the base bid for this. This allowance does not include the two bathroom heat/vent/light units.

20. **LUMINAIRES (FIXTURES):**   A luminaire (fixture) allowance of $1,500.00 shall be included in the electrical contractor's bid for all surface-mounted luminaires (fixtures) and post lanterns, not including the ceiling paddle fans. Luminaires (fixtures) to be selected by owner. This allowance includes the three (3) medicine cabinets.

     The electrical contractor shall:

     a. furnish and install all incandescent and fluorescent recessed luminaires (fixtures).

     b. install all surface, recessed, track, strip, pendant, and hanging luminaires (fixtures), and post lanterns.

     c. furnish and install two 2-lamp fluorescent luminaires (fixtures) above the workbench.

     d. furnish and install all lamps except as noted. Fluorescent lamps and incandescent lamps shall be energy-efficient type. Lamps for ceiling paddle fans to be furnished by owner.

     e. furnish and install all porcelain pull-chain and keyless lampholders.

     This luminaire (fixture) allowance does not include the two bathroom ceiling heat/vent/light units. See Clause 21.

21. **HEAT/VENT/LIGHT CEILING UNITS:** Furnish and install two heat/vent/light units where indicated on the plans, complete with switch assembly, ducts, and louvers required to perform the heating, venting, and lighting operations as recommended by the manufacturer.

22. **PLUG-IN STRIP:** Where noted in the workshop, furnish and install a multioutlet assembly with six outlets.

23. **SWITCHES, RECEPTACLES, AND FACEPLATES:** All flush switches shall be of the quiet ac-rated toggle type. They shall be mounted 46 in. (1.15 m) to center above the finished floor unless otherwise noted.

    Receptacle outlets shall be mounted 12 in. (300 mm) to center above the finished floor unless otherwise noted. All convenience receptacles shall be of the grounding type. Furnish and install, where indicated, ground-fault circuit interrupter (GFCI) receptacles to provide ground-fault circuit protection as required by the *National Electrical Code.®* All wiring devices are to be provided with ivory handles or faces and shall be trimmed with ivory faceplates except in the kitchen, where chrome-plated steel faceplates shall be used.

    Receptacle outlets, where indicated, shall be of the split-circuit design.

24. **TELEVISION OUTLETS:** Furnish and install 4-in. square, 1½-in.-deep outlet boxes with single-gang raised plaster covers at each television outlet where noted on the plans. Mount at the same height as receptacle outlets. Furnish and install 75-ohm coaxial cable to each television outlet from a point in the workshop near the main service-entrance switch. Allow 6 ft (1.8 m) of cable in workshop. Furnish and install television plug-in jacks at each location. Faceplates are to match other faceplates in home. All remaining work done by others.

25. **TELEPHONES:** Furnish and install a 3-in.-deep device box or 4-in. square box, 1½ in. deep, with a suitable single-gang raised plaster cover at each telephone location, as indicated on the plans.

    Furnish and install four-conductor 18 AWG copper telephone cables from the telephone company's point of demarcation near the service-entrance panel to each designated telephone location. Terminate in a proper modular jack, complete with faceplates that match the electrical device faceplates. Allow 6 ft (1.8 m) of cable to hang below ceiling joists. Telephone company to furnish, install, and connect any and all equipment (including grounding connection) up to and including their Standard Network Interface (SNI) device.

    Installation shall be in accordance with any and all applicable *National Electrical Code®* and local code regulations.

26. **MAIN SERVICE PANEL:** Furnish and install in the Workshop where indicated on the plans one (1) 200-ampere, 120/240-volt, single-phase, three-wire panelboard with a 200-ampere main circuit breaker. Panelboard shall be series-rated for 22,000 amperes available fault current.

    Furnish and install all active and spare breakers as indicated on the Panel Schedule.

27. **SERVICE-ENTRANCE UNDERGROUND LATERAL CONDUCTORS:** To be furnished and installed by the utility. Meter enclosure (pedestal type) to be furnished by the utility and installed by the electrical contractor where indicated on plans. Electrical contractor to furnish and install all panels, conduits, fittings, conductors, and other materials required to complete the service-entrance installation from the demarcation point of the utility's equipment to and including the Main Panel.

28. **SERVICE-ENTRANCE CONDUCTORS:** Service-entrance conductors supplied by the electrical contractor shall be two 2/0 AWG THHN/THWN phase conductors and one 1 AWG bare neutral conductor. Install trade size 1½ EMT from Main Panel A to the meter pedestal.

29. **BONDING AND GROUNDING:** Bond and ground service-entrance equipment in accordance with the latest edition of the *National Electrical Code,*® local, and utility code requirements. Install one 4 AWG copper grounding electrode conductor.

30. **SUBPANEL:** Furnish and install in the Recreation Room where indicated on the plans one (1) 20-circuit, 125-ampere, 120/240-volt, single-phase, three-wire MLO load center. Panelboard shall be series-rated for 22,000 amperes available fault current.

    Furnish and install all active and spare breakers as indicated on the Panel Schedule.

    Feed panelboard with three 3 AWG THHN or THWN conductors from a 100-ampere, two-pole breaker in Main Panel A. Install conductors in trade size 1 EMT.

31. **CIRCUIT IDENTIFICATION:** All panelboards shall be furnished with typed-card directories with proper designation of the branch-circuit loads, feeder loads, and equipment served. The directories shall be located in the panel in a holder for clear viewing.

32. **SEALING PENETRATIONS:** The electrical contractor shall seal and weatherproof all penetrations through foundations, exterior walls, and roofs.

33. **FINAL SITE CLEAN-UP:** Upon completion of the installation, the electrical contractor shall review and check the entire installation, clean equipment and devices, and remove surplus materials and rubbish from the owner's property, leaving the work in neat and clean order and in complete working condition. The electrical contractor shall be responsible for the removal of any cartons, debris, and rubbish for equipment installed by the electrical contractor, including equipment furnished by the owner or others and removed from the carton by the electrical contractor.

34. **SMOKE DETECTORS:** Furnish and install all smoke detectors and associated wiring per manufacturer's instructions and all codes. See Clause 5. Detectors to be of the ac/dc type. Interconnect to provide simultaneous signaling.

35. **SPECIAL-PURPOSE OUTLETS:** Install, provide, and connect all wiring for all special-purpose outlets. Upon completion of the job, all luminaires (fixtures) and appliances shall be operating properly. See plans and other sections of the specifications for information as to who is to furnish the luminaires (fixtures) and appliances.

36. **AFCI AND GFCI:** Furnish and install ground-fault circuit interrupter (GFCI) and arc-fault circuit interrupter (AFCI) protection as required by the latest edition of the *National Electrical Code.*®

37. **CEILING PADDLE FANS:** The electrical contractor shall furnish and install outlet boxes identified as suitable for fan support at all locations where a ceiling suspended paddle fan is likely to be installed. The ceiling suspended paddle fans, controls, timers, lamps for same to be furnished by the owner and installed by the electrical contractor.

SCHEDULE OF SPECIAL-PURPOSE OUTLETS

SYMBOL	DESCRIPTION	VOLTS	HORSE-POWER	APPLIANCE AMPERE RATING	TOTAL APPLIANCE WATTAGE RATING (OR VA)	CIRCUIT AMPERE RATING	POLES	WIRE SIZE THHN	CIRCUIT NUMBER	COMMENTS
Ⓐ_A	Hydromassage tub, master bedroom	120	½	10	1200	15	1	14	A9	Connect to separate circuit. Class "A" GFCI circuit breaker.
Ⓐ_B	Water pump	240	1	8	1920	20	2	12	A(5–7)	Run circuit to disconnect switch on wall adjacent to pump. Install 10-ampere dual-element fuses as "back-up" protection. Pump controller has integral overload protection.
Ⓐ_C	Water heater: top element 2000 W. Bottom element 3000 W.	240	–0–	12.5	3000	20	2	12	A(6–8)	Connected for limited demand. Both elements cannot operate at the same time.
Ⓐ_D	Dryer	120/240	120 V ⅙ Motor Only	23.75	5700 Total	30	2	10	B(1–3)	Provide flush-mounted 30-A dryer receptacle.
Ⓐ_E	Overhead garage door opener	120	See unit for explanation of overhead door operator horse-power rating	5.8	696	15	1	14	B14	Unit comes with 3-W cord. Provide box-cover unit (fuse/switch). Install Fustat Type S fuse, 8 amperes. Unit has integral protection. Fustat fuses are additional "back-up" protection. Connect to garage lighting circuit.
Ⓐ_F	Wall-mounted oven	120/240	–0–	27.5	6600	30	2	10	B(6–8)	
Ⓐ_G	Countertop range	120/240	–0–	31	7450	40	2	8	B(2–4)	
Ⓐ_H	Food waste disposer	120	⅓	7.2	864	20	1	12	B19	Controlled by S.P. switch on wall to the right of sink.
Ⓐ_I	Dishwasher	120	⅓	9.2 Includes 875 W heater	1104	20	1	12	B5	
Ⓐ_J	Heat/vent/light master B.R. bath	120	–0–	12.5	1500	20	1	12	A12	
Ⓐ_K	Heat/vent/light front B.R. bath	120	–0–	12.5	1500	20	1	12	A11	
Ⓐ_L	Attic exhaust fan	120	¼	5.8	696	15	1	14	A10	Run circuit to 4-in. square box. Locate near fan in attic. Provide box-cover unit. Install Fustat Type S fuse, 8 amperes. Unit has integral protection. Fustat fuses provide additional back-up protection.
Ⓐ_M	Electric furnace	240	⅓ Motor Only	Motor 3.5 Htr. 50.7 Total 54.2	13000	70	2	4	A(1–3)	The overcurrent device and branch-circuit conductors shall not be less than 125% of the total load of the heaters and motor. 54.2 × 1.25 = 67.75 (424.3(B)).
Ⓐ_N	Air conditioner	240	–0–	26.4	6336	45	2	10	A(2–4)	
Ⓐ_O	Freezer	120	¼	5.8	696	15	1	14	A13	Install single receptacle outlet. Do not provide GFCI protection.

USEFUL FORMULAS

TO FIND	SINGLE PHASE	THREE PHASE	DIRECT CURRENT
AMPERES when kVA is known	$\dfrac{kVA \times 1000}{E}$	$\dfrac{kVA \times 1000}{E \times 1.73}$	not applicable
AMPERES when horsepower is known	$\dfrac{HP \times 746}{E \times \% \text{ eff.} \times pf}$	$\dfrac{HP \times 746}{E \times 1.73 \times \% \text{ eff.} \times pf}$	$\dfrac{HP \times 746}{E \times \% \text{ eff.}}$
AMPERES when kilowatts are known	$\dfrac{kW \times 1000}{E \times pf}$	$\dfrac{kW \times 1000}{E \times 1.73 \times pf}$	$\dfrac{kW \times 1000}{E}$
HORSEPOWER	$\dfrac{I \times E \times \% \text{ eff.} \times pf}{746}$	$\dfrac{I \times E\ 1.73 \times \% \text{ eff.} \times pf}{746}$	$\dfrac{I \times E \times \% \text{ eff.}}{746}$
KILOVOLT AMPERES	$\dfrac{I \times E}{1000}$	$\dfrac{I \times E \times 1.73}{1000}$	not applicable
KILOWATTS	$\dfrac{I \times E \times pf}{1000}$	$\dfrac{I \times E \times 1.73 \times pf}{1000}$	$\dfrac{I \times E}{1000}$
WATTS	$E \times I \times pf$	$E \times I \times 1.73 \times pf$	$E \times I$

$$\text{ENERGY EFFICIENCY} = \frac{\text{Load Horsepower} \times 746}{\text{Load Input kVA} \times 1000}$$

$$\text{POWER FACTOR (pf)} = \frac{\text{Power Consumed}}{\text{Apparent Power}} = \frac{W}{VA} = \frac{kW}{kVA} = \cos\varnothing$$

I = Amperes   E = Volts   kW = Kilowatts   kVA = Kilovolt-amperes
HP = Horsepower   % eff. = Percent Efficiency   pf = Power Factor
e.g., 90% eff. is 0.90   e.g., 95% pf is 0.95

EQUATIONS BASED ON OHM'S LAW:

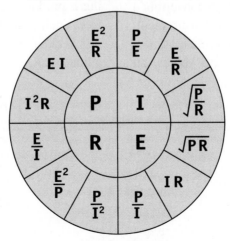

P = POWER, IN WATTS
I = CURRENT, IN AMPERES
R = RESISTANCE, IN OHMS
E = ELECTROMOTIVE FORCE,
      IN VOLTS

METRIC SYSTEM OF MEASUREMENT

The metric system is known as the International System of Units (SI), taken from the French Le Système International d'Unites. The following conversion table was developed from the latest National Institute of Standards and Technology (NIST) publications. In the table, whenever the slant line (/) is found, say it as "per."

The NIST Web site is www.nist.gov.

The *NEC*® shows both inch-pound and metric units. The term *inch-pound* refers to the way dimensions (i.e., inches, feet, yards), weights (i.e., ounces, pounds), and volume (i.e., cubic inches, cubic feet) are given in the United States. Inch-pound units are also called U.S. Customary units. Whenever possible, use the measurements found in the *NEC*® instead of making your own conversions.

Throughout the *NEC*,® inch-pound values were converted to metric units in accordance to specific rules established by the National Fire Protection Association.®

Soft Metric Conversion

A *soft conversion* is a direct mathematical conversion. A soft conversion is the result of taking a product already designed, manufactured to inch-pound values, then converting those values to metric units. The products dimension does not change in any way. Only the measurement units change.

Converting an inch-pound measurement to a metric measurement always results in bizarre, fractional values. The same holds true when converting a metric measurement to an inch-pound measurement.

In the *NEC*,® to obtain a more convenient numerical expression and eliminate awkward values, an "exact metric conversion" is often rounded (up or down) to some value within acceptable tolerances of the original value, making sure that the final measurement does not violate a "maximum" or "minimum" *Code* requirement. This is called a *soft conversion*.

Here is an example of a soft conversion where there is a real safety issue.

EXAMPLE: The minimum clearance from combustible material for a NON-IC recessed luminaire (fixture) is ½ in. (13 mm). One-half inch is exactly 12.70 mm. The *NEC*® rounded this measurement up to 13 mm to obtain a more convenient, easy-to-use numerical expression.

Here is an example of a soft conversion where safety is not an issue.

EXAMPLE: Type NM cable shall be secured at intervals of 4½ ft (1.4 m) and not more than 12 in. (300 mm) from an electrical box. The exact conversion of 4½ ft is 1.3716 m and of ½ in. is 304.8 mm. These values would be cumbersome to use, so the *NEC*® rounded off the measurements to obtain convenient numerical expressions.

Hard Metric Conversion

A *hard metric conversion* involves a change in physical size of a product or measurement. An inch-pound standard size is replaced with an accepted metric standard size for a particular purpose.

A hard metric dimension is attained by designing a product to metric dimensions. No conversion from inch-pound units is involved. For example, a box is designed to measure $50 \times 100 \times 100$ mm. It's a done deal.

Multiply This Unit(s)	By This Factor	To Obtain This Unit(s)
acre	4 046.9	square meters (m²)
acre	43 560	square feet (ft²)
ampere hour	3 600	coulombs (C)
angstrom	0.1	nanometer (nm)
atmosphere	101.325	kilopascals (kPa)
atmosphere	33.9	feet of water (at 4°C)
atmosphere	29.92	inches of mercury (at 0°C)
atmosphere	0.76	meter of mercury (at 0°C)
atmosphere	0.007 348	ton per square inch
atmosphere	1.058	tons per square foot
atmosphere	1.0333	kilograms per square centimeter
atmosphere	1.0333	kilograms per square meter
atmosphere	10 333	pounds per square inch
atmosphere	14.7	kilopascals (kPa)
bar	100	

Multiply This Unit(s)	By This Factor	To Obtain This Unit(s)
barrel (oil, 42 U.S. gallons)	0.158 987 3	cubic meter (m³)
barrel (oil, 42 U.S. gallons)	158.987 3	liters (L)
board foot	0.002 359 737	cubic meter (m³)
bushel	0.035 239 07	cubic meter (m³)
Btu	778.16	foot-pounds
Btu	252	gram-calories
Btu	0.000 393 1	horsepower-hour
Btu	1 054.8	joules (J)
Btu	1.055 056	kilojoules (kJ)
Btu	0.000 293 1	kilowatt-hour (kWh)
Btu per hour	0.000 393 1	horsepower (hp)
Btu per hour	0.293 071 1	watt (W)
Btu per degree Fahrenheit	1.899 108	Kilojoules per kelvin (kJ/K)
Btu per pound	2.326	Kilojoules per kilogram (kJ/kg)
Btu per second	1.055 056	kilowatts (kW)
calorie	4.184	joules (J)
calorie, gram	0.003 968 3	Btu
candela per feet squared (cd/ft²)	10.763 9	candelas per meter squared (cd/m²)

Note: The former term *candlepower* has been replaced with the term candela.

candela per meter squared (cd/m²)	0.092 903	candela per feet squared (cd/ft²)
candela per meter squared (cd/m²)	0.291 864	footlambert*
candela per square inch (cd/in²)	1 550.003	candelas per square meter (cd/m²)

Celsius = (Fahrenheit − 32) × 5/9
Celsius = (Fahrenheit − 32) × 0.555555
Celsius = (0.556 × Fahrenheit) − 17.8

Note: The term *centigrade* was officially discontinued in 1948, and was replaced by the term *Celsius*. The term *centigrade* may still be found in some publications.

centimeter (cm)	0.032 81	foot (ft)
centimeter (cm)	0.393 7	inch (in.)
centimeter (cm)	0.01	meter (m)
centimeter (cm)	10	millimeters (mm)
centimeter (cm)	393.7	mils
centimeter (cm)	0.010 94	yard
circular mil	0.000 005 067	square centimeter (cm²)
circular mil	0.785 4	square mil
circular mil	0.000 000 785 4	square inch (in²)
cubic centimeter (cm³)	0.061 02	cubic inch (in³)
cubic foot per second	0.028 316 85	cubic meter per second (m³/s)
cubic foot per minute	0.000 471 947	cubic meter per second (m³/s)
cubic foot per minute	0.471 947	liter per second (L/s)
cubic inch (in³)	16.39	cubic centimeters (cm³)
cubic inch (in³)	0.000 578 7	cubic foot (ft³)
cubic meter (m³)	35.31	cubic feet (ft³)
cubic yard per minute	12.742 58	liters per second (L/s)
cup (c)	0.236 56	liter (L)
decimeter	0.1	meter (m)
dekameter	10	meters (m)
Fahrenheit = (9/5 Celsius) + 32		
Fahrenheit = (Celsius × 1.8) + 32		
fathom	1.828 804	meters (m)
fathom	6.0	feet
foot	30.48	centimeters (cm)
foot	12	inches
foot	0.000 304 8	kilometer (km)
foot	0.304 8	meter (m)
foot	0.000 189 4	mile (statute)
foot	304.8	millimeters (mm)
foot	12 000	mils
foot	0.333 33	yard
foot, cubic (ft³)	0.028 316 85	cubic meter (m³)
foot, cubic (ft³)	28.316 85	liters (L)
foot, board	0.002 359 737	cubic meter (m³)
cubic feet per second (ft³/s)	0.028 316 85	cubic meter per second (m³/s)
cubic feet per minute (ft³/min)	0.000 471 947	cubic meter per second (m³/s)
cubic feet per minute (ft³/min)	0.471 947	liter per second (L/s)

Multiply This Unit(s)	By This Factor	To Obtain This Unit(s)
foot, square (ft²)	0.092 903	square meter (m²)
footcandle	10.763 91	lux (lx)
footlambert˙	3.426 259	candelas per square meter (cd/m²)
foot of water	2.988 98	kilopascals (kPa)
foot pound	0.001 286	Btu
foot pound-force	1 055.06	joules (J)
foot pound-force per second	1.355 818	joules (J)
foot pound-force per second	1.355 818	watts (W)
foot per second	0.304 8	meter per second (m/s)
foot per second squared	0.304 8	meter per second squared (m/s²)
gallon (U.S. liquid)	3.785 412	liters (L)
gallons per day	3.785 412	liters per day (L/d)
gallons per hour	1.051 50	milliliters per second (mL/s)
gallons per minute	0.063 090 2	liter per second (L/s)
gauss	6.452	lines per square inch
gauss	0.1	millitesla (mT)
gauss	0.000 000 064 52	weber per square inch
grain	64.798 91	milligrams (mg)
gram (g) (a little more than the weight of a paper clip)	0.035 274	ounce (avoirdupois)
gram (g)	0.002 204 6	pound (avoirdupois)
gram per meter (g/m)	3.547 99	pounds per mile (lb/mile)
grams per square meter (g/m²)	0.003 277 06	ounces per square foot (oz/ft²)
grams per square meter (g/m²)	0.029 494	ounces per square yard (oz/yd²)
gravity (standard acceleration)	9.806 65	meters per second squared (m/s²)
quart (U.S. liquid)	0.946 352 9	liter (L)
horsepower (550 ft•lbf/s)	0.745 7	kilowatt (kW)
horsepower	745.7	watts (W)
horsepower hours	2.684 520	megajoules (MJ)
inch per second squared (in./s²)	0.025 4	meter per second squared (m/s²)
inch	2.54	centimeters (cm)
inch	0.254	decimeter (dm)
inch	0.083 3	feet
inch	0.025 4	meter (m)
inch	25.4	millimeters (mm)
inch	1 000	mils
inch	0.027 78	yard
inch, cubic (in.³)	16 387.1	cubic millimeters (mm³)
inch, cubic (in.³)	16.387 06	cubic centimeters (cm³)
inch, cubic (in.³)	645.16	square millimeters (mm²)
inches of mercury	3.386 38	kilopascals (kPa)
inches of mercury	0.033 42	atmosphere
inches of mercury	1.133	feet of water
inches of water	0.248 84	kilopascal (kPa)
inches of water	0.073 55	inch of mercury
joule (J)	0.737 562	foot pound-force (ft•lbf)
kilocandela per meter squared (kcd/m²)	0.314 159	lambert˙
kilogram (kg)	2.204 62	pounds (avoirdupois)
kilogram (kg)	35.274	ounces (avoirdupois)
kilogram per meter (kg/m)	0.671 969	pound per foot (lb/ft)
kilogram per square meter (kg/m²)	0.204 816	pound per square foot (lb/ft²)
kilogram meter squared (kg•m²)	23.730 4	pounds foot squared (lb•ft²)
kilogram meter squared (kg•m²)	3 417.17	pounds inch squared (lb•in²)
kilogram per cubic meter (kg/m³)	0.062 428	pound per cubic foot (lb/ft³)
kilogram per cubic meter (kg/m³)	1.685 56	pound per cubic yard (lb/yd³)
kilogram per second (kg/s)	2.204 62	pounds per second (lb/s)
kilojoule (kJ)	0.947 817	Btu
kilometer (km)	1 000	meters (m)
kilometer (km)	0.621 371	mile (statute)
kilometer (km)	1 000 000	millimeters (mm)
kilometer (km)	1 093.6	yards
kilometer per hour (km/h)	0.621 371	mile per hour (mph)
kilometer squared (km²)	0.386 101	square mile (mile²)
kilopound-force per square inch	6.894 757	megapascals (MPa)
kilowatts (kW)	56.921	Btus per minute
kilowatts (kW)	1.341 02	horsepower (hp)
kilowatts (kW)	1 000	watts (W)

Multiply This Unit(s)	By This Factor	To Obtain This Unit(s)
kilowatt-hour (kWh)	3 413	Btus
kilowatt-hour (kWh)	3.6	megajoules (MJ)
knots	1.852	kilometers per hour (km/h)
lamberts*	3 183.099	candelas per square meter (cd/m²)
lamberts*	3.183 01	kilocandelas per square meter (kcd/m²)
liter (L)	0.035 314 7	cubic foot (ft³)
liter (L)	0.264 172	gallon (U.S. liquid)
liter (L)	2.113	pints (U.S. liquid)
liter (L)	1.056 69	quarts (U.S. liquid)
liter per second (L/s)	2.118 88	cubic feet per minute (ft³/min)
liter per second (L/s)	15.850 3	gallons per minute (gal/min)
liter per second (L/s)	951.022	gallons per hour (gal/hr)
lumen per square foot (lm/ft²)	10.763 9	lux (lx)
lumen per square foot (lm/ft²)	1.0	footcandles
lumen per square foot (lm/ft²)	10.763	lumens per square meter (lm/m²)
lumen per square meter (lm/m²)	1.0	lux (lx)
lux (lx)	0.092 903	lumen per square foot (foot-candle)
maxwell	10	nanowebers (nWb)
megajoule (MJ)	0.277 778	kilowatt hours (kWh)
meter (m)	100	centimeters (cm)
meter (m)	0.546 81	fathom
meter (m)	3.2809	feet
meter (m)	39.37	inches
meter (m)	0.001	kilometer (km)
meter (m)	0.000 621 4	mile (statute)
meter (m)	1 000	millimeters (mm)
meter (m)	1.093 61	yards
meter, cubic (m³)	1.307 95	cubic yards (yd³)
meter, cubic (m³)	35.314 7	cubic feet (ft³)
meter, cubic (m³)	423.776	board feet
meter per second (m/s)	3.280 84	feet per second (ft/s)
meter per second (m/s)	2.236 94	miles per hour (mph)
meter, squared (m²)	1.195 99	square yards (yd²)
meter, squared (m²)	10.763 9	square feet (ft²)
mho per centimeter (mho/cm)	100	siemens per meter (S/m)

Note: The older term *mho* has been replaced with *siemens*. The term *mho* may still be found in some publications.

micro inch	0.025 4	micrometer (μm)
mil	0.002 54	centimeters
mil	0.000 083 33	feet
mil	0.001	inches
mil	0.000 000 025 40	kilometers
mil	0.000 027 78	yards
mil	25.4	micrometers (μm)
mil	0.025 4	millimeter (mm)
miles per hour	1.609 344	kilometers per hour (km/h)
miles per hour	0.447 04	meter per second (m/s)
miles per gallon	0.425 143 7	kilometer per liter (km/L)
miles	1.609 344	kilometers (km)
miles	5 280	feet
miles	1 609	meters (m)
miles	1 760	yards
miles (nautical)	1.852	kilometers (km)
miles squared	2.590 000	kilometers squared (km²)
millibar	0.1	kilopascal (kPa)
milliliter (mL)	0.061 023 7	cubic inch (in³)
milliliter (mL)	0.033 814	fluid ounce (U.S.)
millimeter (about the thickness of a dime)	0.1	centimeter (cm)
millimeter (mm)	0.003 280 8	foot
millimeter (mm)	0.039 370 1	inch
millimeter (mm)	0.001	meter (m)
millimeter (mm)	39.37	mils
millimeter (mm)	0.001 094	yard
millimeter squared (mm²)	0.001 550	square inch (in²)
millimeter cubed (mm³)	0.000 061 023 7	cubic inches (in³)
millimeter of mercury	0.133 322 4	kilopascal (kPa)

Multiply This Unit(s)	By This Factor	To Obtain This Unit(s)
ohm	0.000 001	megohm
ohm	1 000 000	micro ohms
ohm circular mil per foot	1.662 426	nano ohms meter (nΩ•m)
oersted	79.577 47	amperes per meter (A/m)
ounce (avoirdupois)	28.349 52	grams (g)
ounce (avoirdupois)	0.062 5	pound (avoirdupois)
ounce, fluid	29.573 53	milliliters (mL)
ounce (troy)	31.103 48	grams (g)
ounce per foot squared (oz/ft²)	305.152	grams per meter squared (g/m²)
ounces per gallon (U.S. liquid)	7.489 152	grams per liter (g/L)
ounce per yard squared (oz/yd²)	33.905 7	grams per meter squared (g/m²)
pica	4.217 5	millimeters (mm)
pint (U.S. liquid)	0.473 176 5	liter (L)
pint (U.S. liquid)	473.177	milliliters (mL)
pound (avoirdupois)	453.592	grams (g)
pound (avoirdupois)	0.453 592	kilograms (kg)
pound (avoirdupois)	16	ounces (avoirdupois)
poundal	0.138 255	newton (N)
pound foot (lb•ft)	0.138 255	kilogram meter (kg•m)
pound foot per second	0.138 255	kilogram meter per second (kg•m/s)
pound foot squared (lb•ft²)	0.042 140 1	kilogram meter squared (kg•m²)
pound-force	4.448 222	newtons (N)
pound-force foot	1.355 818	newton meters (N•m)
pound-force inch	0.112 984 8	newton meter (N•m)
pound-force per square inch	6.894 757	kilopascals (kPA)
pound-force per square foot	0.047 880 26	kilopascal (kPa)
pound per cubic foot (lb/ft³)	16.018 46	kilograms per cubic meter (kg/m³)
pound per foot (lb/ft)	1.488 16	kilograms per meter (kg/m)
pound per foot squared (lb/ft²)	4.882 43	kilograms per meter squared (kg/m²)
pound per foot cubed (lb/ft³)	16.018 5	kilograms per meter cubed (kg/m³)
pound per gallon (U.S. liquid)	119.826 4	grams per liter (g/L)
pound per second (lb/s)	0.453 592	kilogram per second (kg/s)
pound inch squared (lb•in²)	292.640	kilograms millimeter squared (kg•mm²)
pound per mile	0.281 849	gram per meter (g/m)
pound square foot (lb•ft²)	0.042 140 11	kilogram square meter (kg/m²)
pound per cubic yard (lb/yd³)	0.593 276	kilogram per cubic meter (kg/m³)
quart (U.S. liquid)	946.353	milliliters (mL)
square centimeter (cm²)	197 300	circular mils
square centimeter (cm²)	0.001 076	square foot (ft²)
square centimeter (cm²)	0.155	square inch (in²)
square centimeter (cm²)	0.000 1	square meter (m²)
square centimeter (cm²)	0.000 119 6	square yard (yd²)
square foot (ft²)	144	square inches (in²)
square foot (ft²)	0.092 903 04	square meter (m²)
square foot (ft²)	0.111 1	square yard (yd²)
square inch (in²)	1 273 000	circular mils
square inch (in²)	6.4516	square centimeters (cm²)
square inch (in²)	0.006 944	square foot (ft²)
square inch (in²)	645.16	square millimeters (mm²)
square inch (in²)	1 000 000	square mils (m²)
square meter (m²)	10.764	square feet (ft²)
square meter (m²)	1 550	square inches (in²)
square meter (m²)	0.000 000 386 1	square mile
square meter (m²)	1.196	square yards (yd²)
square mil	1.273	circular mils
square mil	0.000 001	square inch (in²)
square mile	2.589 988	square kilometers (km²)
square millimeter (mm²)	1 973	circular mils
square yard (yd²)	0.836 127 4	square meter (m²)
tablespoon (tbsp)	14.786 75	milliliters (mL)
teaspoon (tsp)	4.928 916 7	milliliters (mL)
therm	105.480 4	megajoules (MJ)
ton (long) (2 240 lb)	1 016.047	kilograms (kg)
ton (long) (2 240 lb)	1.016 047	metric tons (t)
ton, metric	2 204.62	pounds (avoirdupois)
ton, metric	1.102 31	tons, short (2 000 lb)

Multiply This Unit(s)	By This Factor	To Obtain This Unit(s)
ton, refrigeration	12 000	Btus per hour
ton, refrigeration	4.716 095 9	horsepower-hours
ton, refrigeration	3.516 85	kilowatts (kW)
ton (short) (2 000 lb)	907.185	kilograms (kg)
ton (short) (2 000 lb)	0.907 185	metric ton (t)
ton per cubic meter (t/m³)	0.842 778	ton per cubic yard (ton/yd³)
ton per cubic yard (ton/yd³)	1.186 55	tons per cubic meter (ton/m³)
torr	133.322 4	pascals (Pa)
watt (W)	3.412 14	Btus per hour (Btu/hr)
watt (W)	0.001 341	horsepower
watt (W)	0.001	kilowatt (kW)
watt-hour (Wh)	3.413	Btus
watt-hour (Wh)	0.001 341	horsepower-hour
watt-hour (Wh)	0.001	kilowatt-hour (kW/h)
yard	91.44	centimeters (cm)
yard	3	feet
yard	36	inches
yard	0.000 914 4	kilometer (km)
yard	0.914 4	meter (m)
yard	914.4	millimeters
yard, cubic (yd³)	0.764 55	cubic meter (m³)
yard, squared (yd²)	0.836 127	meter squared (m²)

* These terms are no longer used, but may still be found in some publications.

SINGLE-FAMILY DWELLING SERVICE-ENTRANCE CALCULATIONS

1. General Lighting Load (*220.12*).

_____ ft² @ 3 VA per ft²   = _____ VA

Note: Included in this floor area calculation are all lighting outlets and general-use receptacles. Do not include open porches, garages, or unused or unfinished spaces not adaptable for future use. See *NEC® 220.12, Table 220.12*, and *220.14(J)*.

2. Minimum Number of Lighting Branch-Circuits.

$$\frac{\text{Line 1}}{120} = \frac{\text{_____}}{120}$$ = _____ amperes

then, $\dfrac{\text{amperes}}{15} = \dfrac{\text{\_\_\_\_\_}}{15}$ = _____ 15-ampere branch-circuits

3. Small-Appliance Load [*210.11(C)(1)* and *220.52(A)*].
(Minimum of two 20-ampere branch-circuits)

_____ branch-circuits @ 1500 VA each   = _____ VA

4. Laundry Branch-Circuit [*210.11(C)(2)* and *220.52(B)*].
(Minimum of one 20-ampere branch-circuit)

_____ branch-circuit(s) @ 1500 VA each   = _____ VA

5. Total General Lighting, Small-Appliance, and Laundry Load.
Lines 1 + 3 + 4   = _____ VA

6. Net Calculated General Lighting, Small-Appliance, and Laundry Loads (less ranges, ovens, and "fastened-in-place" appliances). Apply demand factors from *Table 220.42*.

a. First 3000 VA @ 100%   = _____3000_____ VA

b. Line 5 _____ − 3000 = _____ @ 35% = _____ VA

Total a + b = _____ VA

7. Electric Range, Wall-Mounted Ovens, Counter-Mounted Cooking Units (*Table 220.55*).   = _____ VA

8. Electric Clothes Dryer (*Table 220.54*).   = _____ VA

9. Electric Furnace (*220.54*).
Air Conditioner, Heat Pump (*Article 440*).
(Enter largest value, *220.60*)   = _____ VA

10. Net Calculated General Lighting, Small-Appliance, Laundry, Ranges, Ovens, Cooktop Units, HVAC.

Lines 6 + 7 + 8 + 9   = _____ VA

11. List "Fastened-in-Place" Appliances *in addition* to Electric Ranges, Electric Clothes Dryers, Electric Space Heating, and Air-Conditioning Equipment.

Appliance	VA Load
Water heater:	= _____ VA
Dishwasher:	= _____ VA
Garage door opener:	= _____ VA
Food waste disposer:	= _____ VA
Water pump:	= _____ VA
Gas-fired furnace:	= _____ VA
Sump pump:	= _____ VA
Other: _____	= _____ VA
_____	= _____ VA
_____	= _____ VA
_____	= _____ VA
Total	= _____ VA

12. **Apply 75% Demand Factor (*220.53*) if Four or More "Fastened-in-Place" Appliances. If Less Than Four, Figure @ 100%.** Do not include electric ranges, electric clothes dryers, electric space heating, or air-conditioning equipment.

    Line 11 Total: _____ × 0.75   = _____ VA

13. **Total Calculated Load (Lighting, Small-Appliance, Ranges, Dryer, HVAC, "Fastened-in-Place" Appliances).**

    Line 10 _ _____ + Line 12 _____   = _____ VA

14. **Add 25% of Largest Motor (*220.50* and *430.24*).**

    _____ × 0.25   = _____ VA

15. **Grand Total Line 13 + Line 14.**   = _____ VA

16. **Minimum Ampacity for Ungrounded Service-Entrance Conductors.**

    $$\text{Amperes} = \frac{\text{Line 15}}{240} = \frac{\text{_____}}{240}$$ = _____ amperes

17. **Ungrounded Conductor Size (copper).**   _____ AWG

    Note: *Table 310.15(B)(6)* may be used only for 120/240-volt, three-wire, residential single-phase service-entrance conductors, service lateral conductors, and feeder conductors that serve as the main power feeder to a dwelling unit.

18. **Minimum Ampacity for Neutral Service-Entrance Conductor, *220.61* and *310.15(B)(6)*. Do Not Include Straight 240-Volt Loads.**

    a. Line 6:   = _____ VA

    b. Line 7: _____ @ 0.70   = _____ VA

    c. Line 8: _____ @ 0.70   = _____ VA

    d. Line 11: (Include only 120-volt loads.)

       _____   _____ VA

       _____   _____ VA

       _____   _____ VA

       _____   _____ VA

       _____   _____ VA

                 Total   _____ VA

    e. Line d total @ 75% demand factor if four or more per *220.60*, otherwise use 100%.

       _____ × 0.75   = _____ VA

    f. Add 25% of largest 120-volt motor.

       _____ × 0.25   = _____ VA

                         Total = _____ VA

    g. Total a + b + c + e + f.   = _____ VA

$$\text{Amperes} = \frac{\text{Line g}}{240} = \frac{\underline{\hspace{3cm}}}{240} \qquad = \underline{\hspace{3cm}} \text{ amperes}$$

19. **Neutral Conductor Size (copper) (*220.22*).**   _____ AWG

    Note: *NEC® 310.15(B)(6)* permits the neutral conductor to be smaller than the ungrounded "hot" conductors if the requirement of *215.2, 220.61*, and *230.42* are met. *NEC® 220.61* states that a feeder or service neutral load shall be the maximum unbalance of the load determined by *Article 220*. When bare conductors are used with insulated conductors, the conductors' ampacity is based on the ampacity of the other insulated conductors in the raceway, *310.15(B)(3)*. The neutral conductor must not be smaller than the grounding electrode conductor, *250.24(B)(1)*.

20. **Grounding Electrode Conductor Size (copper) (*Table 250.66*).**   _____ AWG

21. **Raceway Size.**   _____ Trade Size

    Obtain dimensional data from *Table 1, Table 4, Table 5*, and *Table 8, Chapter 9, NEC®*

MEMBERSHIP APPLICATION
For Inspector Member/Associate Member

INTERNATIONAL ASSOCIATION
ELECTRICAL INSPECTORS
Established 1928
Keystone of the Electrical Industry

Annual Membership Dues $90

Services from the IAEI for you and your organization include all this and more:

- IAEI News magazine subscription
- Educational Opportunities
- Networking
- Career Advancement
- Great Discount Savings

PLEASE PRINT

Name - Last	First	M.I.	Chapter, where you live or work, if known
Title			(Division, where appropriate)
E-mail Address			
Employer			For Office Use Only — Section No. / Ch. No. / Div. No.
Address of Applicant			If previous member, give last membership number
City	State or Province	Zip or Postal Code	and last year of membership
Endorsed by RAY C. MULLIN			Endorser's Membership Number **533800**
Applicant's Signature			(Area Code) Telephone Number - -

☐ Check ☐ Money Order ☐ Diners Club
☐ MasterCard ☐ Visa ☐ Amex ☐ Discover

Name on Card	Charge Card Number	Exp. Date

☐ Inspector $90   ☐ Associate $90

Inspector Members MUST sign below:

☐ I,_____ , meet the qualifications for inspector member type as described below.

Inspector members must regularly make electrical inspections for preventing injury to persons or damage to property on behalf of a governmental agency, insurance agency, rating bureau, recognized testing laboratory or electric light and power company.

Contact the IAEI Customer Service Department for information on our special membership categories - Section, National, and International Member; Sustaining Member (Bronze, Silver, Gold, or Platinum); and Inspection Agency Member.

Email addresses and telephone numbers are not required but are requested to facilitate IAEI's communicating with members and conducting IAEI/member business.

Mail to: IAEI, P. O. Box 830848 Richardson, TX 75083-0848   For information call: (972) 235-1455 (8-5 CT)
www.iaei.org • email: iaei@iaei.org

MA2003

The following is a list of terms used extensively in the electrical trades. Many of these terms are unique to the electrical industry. Words marked with an * are defined in the *National Electrical Code.* Definitions are found in *Article 100, NEC.*

Accent Lighting: Directional lighting to emphasize a particular object or to draw attention to a part of the field in view.

Accessible (equipment)*: Admitting close approach: not guarded by locked doors, elevation, or other effective means.

Accessible (wiring methods)*: Capable of being removed or exposed without damaging the building structure or finish, or not permanently closed in by the structure or finish of the building.

Accessible, Readily*: Capable of being reached quickly for operation, renewal, or inspections, without requiring those to whom ready access is requisite to climb over or remove obstacles or to resort to portable ladders, chairs, and so forth.

Addendum: Modification (change) made to the construction documents (plans and specifications) during the bidding period.

Adjustment Factor: A multiplier (penalty) that is applied to conductors when there are more than three current-carrying conductors in a raceway or cable. These multipliers are found in *Table 310.15(B)(2)*. An *adjustment factor* is also referred to as a *derating factor*.

AL/CU: Terminal marking on switches and receptacles rated 30 amperes and greater suitable for use with aluminum, copper, and copper-clad aluminum conductors. If not marked, suitable for copper conductors only.

Ambient Lighting: Lighting throughout an area that produces general illumination.

Ambient Temperature: The environmental temperature surrounding the object under consideration.

American National Standards Institute (ANSI): An organization that identifies industrial and public requirements for national consensus standards and coordinates and manages their development, resolves national standards problems, and ensures effective participation in international standardization. ANSI does not itself develop standards. Rather, it facilitates development by establishing consensus among qualified groups. ANSI ensures that the guiding principles—consensus, due process, and openness—are followed.

Ampacity*: The current, in amperes, that a conductor can carry continuously under the conditions of use without exceeding its temperature rating.

Ampere: The measurement of intensity of rate of flow of electrons in an electric circuit. An ampere is the amount of current that will flow through a resistance of one ohm under a pressure of one volt.

Appliance*: Utilization equipment, generally other than industrial, normally built in standardized sizes or types, that is installed or connected as a unit to perform one or more functions such as clothes washing, air conditioning, food mixing, deep frying, and so forth.

Approved*: Acceptable to the authority having jurisdiction (AHJ).

Arc-Fault Circuit Interrupter (AFCI)*: A device intended to provide protection from the effects of arc faults by recognizing characteristics unique to arcing and by functioning to de-energize a circuit when an arcing fault is detected.

Architectural Drawing: A line drawing showing plan and/or elevation views of the proposed building for the purpose of showing the overall appearance of the building.

As-Built Plans: When required, a modified set of working drawings that are prepared for a construction project that includes all variances from the original working drawings that occurred during the project construction.

Authority Having Jurisdiction (AHJ):   The organization, office, or individual responsible for approving equipment, materials, an installation, or a procedure. An AHJ is usually a governmental body that has legal jurisdiction over electrical installations. The AHJ has the responsibility for making interpretations of the rules, for deciding upon the approval of equipment and materials, and for granting special permission if required. Where a specific interpretation or deviation from the *NEC®* is given, get it in writing. See *90.4* of the *NEC®*.

Average Rated Life (lamp):   How long it takes to burn out the lamp, for example, a 60-watt lamp is rated 1000 hours. The 1000-hour rating is based on the point in time when 50 percent of a test batch of lamps burn out and 50 percent are still burning.

AWG: The American Wire Gauge. Previously known as the Brown & Sharpe Gauge. The smaller the AWG number, the larger the conductor. There are forty electrical conductor sizes from 36 AWG through 0000 AWG. Starting from the smallest size, 36 AWG, each successive size is approximately 1.26 times larger than the previous AWG size. *Table 8* of *Chapter 9* in the *NEC®* lists conductor sizes commonly found in the electrical industry. Conductor sizes through 0000 AWG are shown using the AWG designation. Conductors larger than 0000 AWG are shown in circular mil area. The 1.26 relationship between sizes is easily confirmed in *Table 8* by checking conductor sizes 4 AWG through 0000 AWG.

Backlight: Illumination from behind a subject directed substantially parallel to a vertical plane through the optical axis of the camera.

Ballast: Used for energizing fluorescent lamps. It is constructed of a laminated core and coil windings or solid-state electronic components.

Ballast Factor: The percentage of rated light output that can be achieved with a specific ballast.

Bare Lamp: A light source with no shielding.

Black Line Print: Another term for a photocopy or computer-aided drawing (CAD).

Blueprints: A term used to refer to a plan or plans. A photographic print in white on a bright blue ground or blue on a white ground used especially for copying maps, mechanical drawings, and architects' plans. See text for further explanation.

Bonding*:   The permanent joining of metallic parts to form an electrically conductive path that will ensure electrical continuity and the capacity to conduct safely any current likely to be imposed.

Bonding Jumper*:   A reliable conductor to ensure the required electrical conductivity between metal parts required to be electrically connected.

Bonding Jumper, Circuit*:   The connection between portions of a conductor in a circuit to maintain the required ampacity of the circuit.

Bonding Jumper, Equipment*:   The connection between two or more portions of the equipment grounding conductor.

Bonding Jumper, Main*:   The connection between the grounded circuit conductor and the equipment grounding conductor at the service.

Branch-Circuit*:   The circuit conductors between the final overcurrent device protecting the circuit and the outlet(s).

Branch-Circuit, Appliance*:   A branch-circuit supplying energy to one or more outlets to which appliances are to be connected; such circuits are to have no permanently connected luminaires (fixtures) not a part of an appliance.

Branch-Circuit, General Purpose*: A branch-circuit that supplies a number of outlets for lighting and appliances.

Branch-Circuit, Individual*: A branch-circuit that supplies only one utilization equipment.

Branch-Circuit—Multiwire*: *"A branch-circuit consisting of two or more ungrounded conductors having a voltage between them, and a grounded conductor having equal voltage between it and each ungrounded conductor of the circuit and that is connected to the neutral or grounded conductor of the system."***

Branch-Circuit Selection Current*: Branch-circuit selection current is the value in amperes to be used instead of the rated-load current in determining the ratings of motor branch-circuit conductors, disconnecting means, controllers, and branch-circuit short-circuit and ground-fault protective devices wherever the running overload protective device permits a sustained current greater than the specified percentage of the rated-load current. The value of branch-circuit selection current will always be equal to or greater than the marked rated-load current.

Brightness: In common usage, the term "brightness" usually refers to the strength of sensation that results from viewing surfaces or spaces from which light comes to the eye.

Candela: The international (SI) unit of luminous intensity. Formerly referred to as a "candle," as in "candlepower."

Candlepower: A measure of intensity mathematically related to lumens.

Carrier: A wave having at least one characteristic that may be varied from a known reference value by modulation.

Carrier Current: The current associated with a carrier wave.

Circuit Breaker*: A device designed to open and close a circuit by nonautomatic means and to open the circuit automatically on a predetermined overcurrent without damage to itself when properly applied within its rating.

CC: A marking on wire connectors and soldering lugs indicating they are suitable for use with copper-clad aluminum conductors only.

CC/CU: A marking on wire connectors and soldering lugs indicating they are suitable for use with copper or copper-clad aluminum conductors only.

CU: A marking on wire connectors and soldering lugs indicating they are suitable for use with copper conductors only.

CO/ALR: Terminal marking on switches and receptacles rated 15 and 20 amperes that are suitable for use with aluminum, copper, and copper-clad aluminum conductors. If not marked, they are suitable for copper or copper-clad conductors only.

Conductor*:

   Bare: A conductor having no covering or electrical insulation whatsoever. (See "Conductor, Covered.")

   Covered: A conductor encased within material of a composition or thickness that is not recognized by this *Code* as electrical insulation. (See "Conductor, Bare.")

   Insulated: A conductor encased within material of a composition and thickness that is recognized by this *Code* as electrical insulation.

Construction Documents: A term used to represent all drawings, specifications, addenda, and other pertinent construction information associated with the construction of a specific project.

Construction Drawing: See Drawings.

Construction Types: (Also see *Annex E, NEC.®*)

   Type I: Fire-resistive construction. A building constructed of noncombustible materials such as reinforced concrete, brick, stone, etc., and having any metal members properly

---

**Reprinted with permission from NFPA 70-2005.

"fireproofed" with major structural members designed to withstand collapse and to prevent the spread of fire. Type I construction is subdivided into Types IA and IB.

Type II: Noncombustible construction. A building having all structural members, including walls, floors, and roofs, of noncumbustible materials such as reinforced concrete, brick, stone, etc., and not qualifying as fire-resistive construction. Type II construction is divided into Types IIA, IIB, and IIC.

Type III: Construction where the exterior walls are of concrete, masonry, or other noncombustible material. The interior structural members are constructed of any approved materials such as wood or other combustible materials. Type III construction is subdivided into Types IIIA and IIIB.

Type IV: Construction where the exterior walls are constructed of approved noncombustible materials. The interior structural members are of solid or laminated wood. Type V construction is divided into Types VA and VB.

Type V: Construction where the exterior walls, loadbearing walls, partitions, floors, and roofs are constructed of any approved material. This is wood construction as found in typical residential construction.

Continuous Load*: A load where the maximum current is expected to continue for three hours or more.

Contract: An agreement between two or more parties, especially one that is written and enforceable by law.

Contract Modifications: After the agreement has been signed, any additions, deletions, or modifications of the work to be done are accomplished by change order, supplemental instruction, and field order. They can be issued at any time during the contract period.

Contractor: A properly licensed individual or company that agrees to furnish labor, materials, equipment, and associated services to perform the work as specified for a specified price.

Correction Factor: A multiplier (penalty) that is applied to conductors when the ambient temperature is greater than 86°F (30°C) . These multipliers are found at the bottom of *Table 310.16* of the *NEC*.®

CSA: This is the Canadian Standards Association that develops safety and performance standards in Canada for electrical products, similar to but not always identical to those of (Underwriters Laboratories) UL in the United States.

Current: The flow of electrons through an electrical circuit, measured in amperes.

Derating Factor: A multiplier (penalty) that is applied to conductors when there are more than three current-carrying conductors in a raceway or cable. These multipliers are found in *Table 310.15(B)(2)*. A *derating factor* is also referred to as an *adjustment factor*.

Details: Plans, elevations, or sections that provide more specific information about a portion of a project component or element than smaller scale drawings.

Device: ▶A unit of an electrical system that is intended to carry or control but not utilize energy.◀

Diagrams: Nonscaled views showing arrangements of special system components and connections not possible to clearly show in scaled views. A *schematic diagram* shows circuit components and their electrical connections without regard to actual physical locations. A *wiring diagram* shows circuit components and the actual electrical connections.

Dimmer: A switch with components that permits variable control of lighting intensity. Some dimmers have electronic components; others have core and coil (transformer) components.

Dimming Ballast: Controls light output of fluorescent lamps.

Direct Glare: Glare resulting from high luminances or insufficiently shielded light sources in the field of view. Usually associated with bright areas, such as luminaires (fixtures), ceilings, and windows that are outside the visual task of the region being viewed.

Disconnecting Means*: A device, or group of devices, or other means by which the conductors of a circuit can be disconnected from their source of supply.

Downlight: A small, direct lighting unit that guides the light downward. It can be recessed, surface-mounted, or suspended.

Drawings: (1) A term used to represent that portion of the contract documents that graphically illustrates the design, location, and dimensions of the components and elements contained in a specific project. (2) A line drawing.

Dry Niche Luminaire (fixture)*: A luminaire (lighting fixture) intended for installation in the wall of a pool or fountain in a niche that is sealed against the entry of pool water.

▶**Dwelling Unit:** One or more rooms for the use of one or more persons as a housekeeping unit with space for eating, living, cooking, and sleeping and permanent provisions for sanitation.◀

Efficacy: The total amount of light energy (lumens) emitted from a light source divided by the total lamp and ballast power (watts) input, expressed in lumens per watt.

Elevations: Views of vertical planes, showing components in their vertical relationship, viewed perpendicularly from a selected vertical plane.

Feeder*: All circuit conductors between the service equipment, the source of a separately derived system, or other power-supply source and the final branch-circuit overcurrent device.

Fill Light: Supplementary illumination to reduce shadow or contract range.

Fine Print Note (FPN)*: Explanatory material is in the form of Fine Print Notes (FPN). Fine Print Notes are informational only and are not enforceable as a requirement by the *Code*.

Fire Rating: The classification indicating in time (hours) the ability of a structure or component to withstand fire conditions.

Flame Detector: A radiant energy-sensing fire detector that detects the radiant energy emitted by a flame.

Fluorescent Lamp: A lamp in which electric discharge of ultraviolet energy excites a fluorescing coating (phosphor) and transforms some of that energy to visible light.

Footcandle: The unit used to measure how much total light is reaching a surface. One lumen falling on 1 ft^2 of surface produces an illumination of one footcandle.

Fully Rated System: Panelboards are marked with their short-circuit rating in RMS symmetrical amperes. In a fully-rated system, the panelboard short-circuit current rating will be equal to the lowest interrupting rating of any branch-circuit breaker or fuse installed. All devices installed shall have an interrupting rating greater than or equal to the specified available fault-current. (See "Series-Rated System.")

Fuse: An overcurrent protective device with a fusible link that operates to open the circuit on an overcurrent condition.

General Conditions: A written portion of the contract documents set forth by the owner, stipulating the contractor's minimum acceptable performance requirements including the rights, responsibilities, and relationships of the parties involved in the performance of the contract. General conditions are usually included in the book of specifications but are sometimes found in the architectural drawings.

General Lighting: Lighting designed to provide a substantially uniform level of illumination throughout an area, exclusive of any provision for special lighting.

Glare: The sensation produced by luminance within the visual field that is sufficiently greater than the luminance to which the eyes are adapted to cause annoyance, discomfort, or loss of visual performance and visibility.

Ground*: A conducting connection, whether intentional or accidental, between an electrical circuit or equipment and the earth, or to some conducting body that serves in place of the earth.

Grounded*: Connected to earth or to some conducting body that serves in place of the earth.

Grounded Conductor*:   A system or circuit conductor that is intentionally grounded.

Ground Fault*:   An unintentional, electrically conducting connection between an ungrounded or grounded conductor of an electrical circuit and the normally non-current-carrying conductors, metallic enclosures, metallic raceways, metallic equipment, or earth.

Ground-Fault Circuit-Interrupter (GFCI)*:   A device intended for the protection of personnel that functions to de-energize a circuit or portion thereof within an established period of time when a current to ground exceeds the values established for a Class A device.

Ground-Fault Current Path*:   An electrically conductive path from the point of a ground fault on a wiring system through normally noncurrent-carrying conductors, equipment, or the earth to the electrical supply source.

Ground-Fault Current Path, Effective*:   An intentionally constructed, permanent, low-impedance electrically conductive path designed and intended to carry current under ground-fault conditions from the point of a ground fault on a wiring system to the electrical supply source and that facilitates the operation of the overcurrent protection device or ground fault detectors on high-impedance grounded systems.

Ground-Fault Protection of Equipment (GFPE)*:   A system intended to provide protection of equipment from damaging line-to-ground fault currents by operating to cause a dis-connecting means to open all ungrounded conductors of the faulted circuit. This protection is provided at current levels less than those required to protect conductors from damage through the operation of a supply circuit overcurrent device.

Grounding Conductor*:   A conductor used to connect equipment or the grounded circuit of a wiring system to a grounding electrode or electrodes.

Grounding Conductor, Equipment*:   The conductor used to connect the non-current-carrying metal parts of equipment, raceways, and other enclosures to the system grounded conductor, the grounding electrode conductor, or both, at the service equipment or at the source of a separately derived system.

▶**Grounding Electrode Conductor*:**   The conductor used to connect the grounding elec-trode(s) to the equipment grounding conductor, to the grounded conductor, or to both, at the service, at each building or structure where supplied by a feeder(s) or branch-circuit(s), or at the source of a separately derived system.◀

HACR: This term has become obsolete. In the past, when a circuit breaker was subjected to additional tests unique to HVAC equipment, it was marked with the letters HACR. The let-ters stood for Heating, Air Conditioning, and Refrigeration. The HVAC equipment would also be marked with the letters HACR if that is the type of overcurrent protection required by the manufacturer of the HVAC equipment.

  Today, all standard molded case circuit breakers tested and listed to UL 489 are suitable for use on HVAC equipment. The letters *HACR* will disappear from the scene as time goes on.

Halogen Lamp: An incandescent lamp containing a halogen gas that recycles tungsten back onto the tungsten filament surface. Without the halogen gas, the tungsten would normally be deposited onto the bulb wall.

Heat Alarm: A single or multiple station alarm responsive to heat.

Hermetic Refrigerant Motor-Compressor: A combination consisting of a compressor and motor, both of which are enclosed in the same housing, with no external shaft or shaft seals; the motor operating in the refrigerant.

High Intensity Discharge Lamp (HID):   A general term for a mercury, metal-halide, or high-pressure sodium lamp.

High-Pressure Sodium Lamp: A high-intensity discharge light source in which the light is primarily produced by the radiation from sodium vapor.

Horsepower (tools): Horsepower is a measurement of motor torque multiplied by speed. Horsepower also refers to the rate of work (power) an electric motor is capable of delivering. See Torque, Rated Horsepower, and Maximum Developed Horsepower.

Horsepower, Rated: Rated horsepower is a motor's running torque at its rated running speed. The motor can be run continuously at its rated horsepower without overheating. If the motor is required to give an extra spurt of effort while running, the motor is overloaded and develops extra horsepower to compensate. The most horsepower that can be drawn from a motor to handle this extra effort is its maximum developed horsepower. See Torque, Horsepower and Maximum Developed Horsepower.

Horsepower, Maximum Developed: If a motor is required to work harder than its idle speed, it will be overloaded and must develop extra horsepower. The most horsepower that can be expected from a motor to handle this extra effort is referred to as its maximum developed horsepower. See Horsepower, Torque, and Rated Horsepower.

Household Fire Alarm System: A system of devices that produces an alarm signal in the household for the purpose of notifying the occupants of the presence of a fire so that they will evacuate the premises.

Identified* (conductor): The identified conductor is the insulated grounded conductor.
- For sizes 6 AWG or smaller, the insulated grounded conductor shall be identified by a continuous white or gray outer finish or by three continuous white stripes on other than green insulation along its entire length.
- For sizes larger than 6 AWG, the insulated grounded conductor shall be identified either by a continuous white or gray outer finish or by three continuous white stripes on other than green insulation along its entire length, or at the time of installation by a distinctive white marking at its terminations.

Identified* (equipment): Recognizable as suitable for the specific purpose, function, use, environment, application, and so forth, where described in a particular *Code* requirement.

Identified* (terminal): The identification of terminals to which a grounded conductor is to be connected shall be substantially white in color. The identification of other terminals shall be of a readily distinguishable different color.

IEC: The International Electrotechnical Commission is a worldwide standards organization. These standards differ from those of Underwriters Laboratories. Some electrical equipment might conform to a specific IEC Standard, but may not conform to the UL Standard for the same item.

Illuminance: The amount of light energy (lumens) distributed over a specific area expressed as footcandles (lumens/ft^2) or lux (lumens/m^2).

Illumination: The act of illuminating or the state of being illuminated.

Immersion Detection Circuit Interrupter (IDCI): A device integral with grooming appliances that will shut off the appliance when the appliance is dropped in water.

Incandescent Filament Lamp: A lamp that provides light when a filament is heated to incandescence by an electric current. Incandescent lamps are the oldest form of electric lighting technology.

Indirect Lighting: Lighting by luminaires (fixtures) that distribute 90 to 100 percent of the emitted light upward.

Inductive Load: A load that is made up of coiled or wound wire that creates a magnetic field when energized. Transformers, core and coil ballasts, motors, and solenoids are examples of inductive loads.

In Sight*: Where this *Code* specifies that one equipment shall be "in sight from," "within sight from," or "within sight," and so forth, of another equipment, the specified equipment is to be visible and not more than 50 ft (15 m) from the other.

Instant Start: A circuit used to start specially designed fluorescent lamps without the aid of a starter. To strike the arc instantly, the circuit utilizes higher open-circuit voltage than is required for the same length preheat lamp.

International Association of Electrical Inspectors (IAEI): A not-for-profit and educational organization cooperating in the formulation and uniform application of standards for the safe installation and use of electricity, and collecting and disseminating information relative thereto. The IAEI is made up of electrical inspectors, electrical contractors, electrical apprentices, manufacturers, and governmental agencies.

Interrupting Rating*: The highest current at rated voltage that a device is intended to interrupt under standard test conditions.

Ionization Detector: This type of detector triggers an alarm when oxygen and nitrogen particles are ionized in the ionization chamber. The internal circuitry measures a minute amount of electrical current between two plates. When smoke enters the ionization chamber the current is reduced, triggering the alarm. This type of detector works well for detecting small amoutns of smoke such as that resulting from gasoline fires. Consult the manufacturer's literature for more information.

Isolated Ground Receptacle: A grounding type device in which the equipment ground contact and terminal are electrically isolated from the receptacle mounting means.

Kilowatt (kW): One thousand watts equals one kilowatt.

Kilowatt-hour (kWh): One thousand watts of power in one hour. One 100-watt lamp burning for ten hours is one kilowatt-hour. Two 500-watt electric heaters operated for one hour is one kilowatt-hour.

Labeled*: Equipment or materials to which has been attached a label, symbol, or other identifying mark of an organization that is acceptable to the authority having jurisdiction (AHJ) and concerned with product evaluation that maintains periodic inspection of production of labeled equipment or materials and by whose labeling the manufacturer indicates compliance with appropriate standards or performance in a specified manner.

Labor and Material Payment Bond: A written form of security from a surety (bonding) company to the owner, on behalf of an acceptable prime or main contractor or subcontractor, guaranteeing payment to the owner in the event the contractor fails to pay for all labor, materials, equipment, or services in accordance with the contract. (see Performance Bond and Surety Bond).

Lamp: A generic term for a man-made source of light.

Light: The term generally applied to the visible energy from a source. Light is usually measured in lumens or candlepower. When light strikes a surface, it is either absorbed, reflected, or transmitted.

Lighting Outlet*: An outlet intended for the direct connection of a lampholder, a luminaire (fixture), or a pendant cord terminating in a lampholder.

Listed*: Equipment, materials, or services included in a list published by an organization that is acceptable to the authority having jurisdiction (AHJ) and concerned with evaluation of products or services, that maintains periodic inspection of production of listed equipment or materials or periodic evaluation of services, and whose listing states that either the equipment, material, or services meets appropriate designated standards or has been tested and found suitable for a specified purpose.

Load: The electric power used by devices connected to an electrical system. Loads can be figured in amperes, volt-amperes, kilovolt-amperes, or kilowatts. Loads can be intermittent, continuous intermittent, periodic, short-time, or varying. See the definition of "Duty" in the *NEC*.

Load Center: A common name for residential panelboards. A load center may not be as deep as a panelboard and generally does not contain relays or other accessories as are available for panelboards. Circuit breakers "plug in" as opposed to the "bolt-in" types used in panelboards. Manufacturers' catalogs will show both load centers and panelboards. The UL standards do not differentiate.

Locked Rotor Current (LRC): The steady-state current taken from the line with the rotor locked and with rated voltage and frequency applied to the motor.

Low-Pressure Sodium Lamp: A discharge lamp in which light is produced from sodium gas operating at a partial pressure.

Lumens: The SI unit of luminous flux. The units of light energy emitted from the light source.

Luminaire (fixture)*: A complete lighting unit consisting of a lamp or lamps together with the parts designed to distribute the light, to position and protect the lamps and ballast (where applicable), and to connect the lamps to the power supply. Prior to the *National Electrical Code* adopting the International System (SI) definition of *luminaire*, the commonly used term in the United States was and in most instances still is "lighting fixture." It will take years for the electrical industry to totally change and feel comfortable with the term *luminaire*.

Luminaire (fixture) Efficiency: The total lumen output of the luminaire (fixture) divided by the rated lumens of the lamps inside the luminaire (fixture).

Lux: The SI (International System) unit of illumination. One lumen uniformly distributed over an area one square meter in size.

Mandatory Rules: Required. The terms *shall* or *shall not* are used when a *Code* statement is mandatory.

Maximum Continuous Current (MCC): A value that is determined by the manufacturer of hermetic refrigerant motor compressors under high load (high refrigerant pressure) conditions. This value is established in the *NEC* as being no greater than 156 percent of the marked rated amperes or the branch-circuit selection current.

Maximum Overcurrent Protection (MOP): A term used with equipment that has a hermetic motor compressor(s). MOP is the maximum ampere rating for the equipment's branch-circuit overcurrent protective device. The ampere rating is determined by the manufacturer, and is marked on the nameplate of the equipment. The nameplate will also indicate if the overcurrent device can be a circuit breaker, a fuse, or either.

Mercury Lamp: A high-intensity discharge light source in which radiation from the mercury vapor produces visible light.

Metal-Halide Lamp: A high-intensity discharge light source in which the light is produced by the radiation from mercury, together with the halides of metals such as sodium and candium.

Minimum Circuit Ampacity (MCA): A term used with equipment that has a hermetic motor compressor(s). MCA is the minimum ampere rating value used to determine the proper size conductors, disconnecting means, and controllers. MCA is determined by the manufacturer, and is marked on the nameplate of the equipment.

National Electrical Code® (NEC®): The electrical code published by the National Fire Protection Association. This *Code* provides for practical safeguarding of persons and property from hazards arising from the use of electricity. It does not become law until adopted by federal, state, and local laws and regulations.

National Electrical Manufacturers Association (NEMA): NEMA is a trade organization made up of many manufacturers of electrical equipment. They develop and promote standards for electrical equipment.

National Fire Protection Association (NFPA): Located in Quincy, MA. The NFPA is an international Standards Making Organization dedicated to the protection of people from the ravages of fire and electric shock. The NFPA is responsible for developing and writing the *National Electrical Code*,® *The Sprinkler Code*,® *The Life Safety Code*,® *The National Fire Alarm Code*,® and over 295 other codes, standards, and recommended practices. The NFPA may be contacted at (800) 344-3555 or on their Web site at www.nfpa.org.

Nationally Recognized Testing Laboratory (NRTL): The term used to define a testing laboratory that has been recognized by OSHA; for example, Underwriters Laboratories (UL).

Neutral: In residential wiring, the neutral conductor is the grounded conductor in a circuit consisting of three or more conductors. There is no neutral conductor in a two-wire circuit, although electricians many times refer to the white grounded conductor as the neutral.

The voltages from the ungrounded conductors to the grounded conductor are of equal magnitude.

See the definition of "Multiwire Branch-Circuit" in the *NEC.*®

No Niche Luminaire (fixture)*: A luminaire (fixture) intended for installation above or below the water without a niche.

Noncoincidental Loads: Loads that are not likely to be on at the same time. Heating and cooling loads would not operate at the same time. See *220.21* of the *NEC.*®

Notations: Words found on plans to describe something.

Occupational Safety and Health Act (OSHA): This is the code of federal regulations developed by the Occupational Safety and Health Administration, U.S. Department of Labor. The electrical regulations are covered in *Part 1910, Subpart S*. In order to shorten and simplify the regulations, *Part 1920, Subpart S* contains only the most common performance requirements of the *NEC.*® The *NEC*® must still be referred to in conjunction with OSHA regulations.

Ohm: A unit of measure for electric resistance. An ohm is the amount of resistance that will allow one ampere to flow under a pressure of one volt.

Outlet*: A point on the wiring system at which current is taken to supply utilization equipment.

Overcurrent*: Any current in excess of the rated current of equipment or the ampacity of a conductor. It may result from overload, short circuit, or ground fault.

Overcurrent Device: Also referred to as an overcurrent protection device. A form of protection that operates when current exceeds a predetermined value. Common forms of overcurrent devices are circuit breakers, fuses, and thermal overload elements found in motor controllers.

Overload*: Operation of equipment in excess of the normal, full-load rating, or of a conductor in excess of rated ampacity that, when it persists for a sufficient length of time, would cause damage or dangerous overheating. The current flow is contained in its normal intended path. A fault, such as a short circuit or a ground fault, is not an overload.

Panelboard*: A single panel or group of panel units designed for assembly in the form of a single panel, including buses and automatic overcurrent devices; equipped with or without switches for the control of light, heat, or power circuits; designed to be placed in a cabinet or cutout box placed in or against a wall or partition and accessible only from the front.

Performance Bond: (1) A written form of security from a surety (bonding) company to the owner, on behalf of an acceptable prime or main contractor or subcontractor, guaranteeing payment to the owner in the event the contractor fails to perform all labor, materials, equipment, or services in accordance with the contract. (2) The surety companies generally reserve the right to have the original prime or main or subcontractor remedy any claims before paying on the bond or hiring other contractors.

Permissive Rules: Allowed but not required. Terms such as *shall be permitted* or *shall not be required* are used when a *Code* statement is permissive.

Photoelectric Detector: This type of detector triggers an alarm when light is blocked between its internal light source and sensor. Typically, heavy smoke will block the light. The action of blocking light is similar to that of the safety feature on a garage door operator. Consult the manufacturer's literature for more information.

Photometry: The pattern and amount of light that is emitted from a luminaire (fixture), normally represented as a cross-section through the luminaire (fixture) distribution pattern.

Plan: (1) A line drawing (by floor) representing the horizontal geometrical section of the walls of a building. The section (a horizontal plane) is taken at an elevation to include the relative positions of the walls, partitions, windows, doors, chimneys, columns, pilasters, etc. (2) A plan can be thought of as cutting a horizontal section through a building at an eye level elevation.

Plans: A term used to represent all drawings including sections and details and any supplemental drawings for complete execution of a specific project.

Power Supply: A source of electrical operating power including the circuits and terminations connecting it to the dependent system components.

Preheat: A type of ballast that is easily identified because it has a starter. One type of starter is automatic with two buttons on one end; the other type is a manual ON/OFF switch that has a momentary "make" postion jsut beyond the "ON" position.

Preheat Fluorescent Lamp Circuit: A circuit used on fluorescent lamps wherein the electrodes are heated or warmed to a glow stage by an auxiliary switch or starter before the lamps are lighted.

Prints: A term used to refer to a plan or plans.

Proposal: A written offer from a bidder to the owner, preferably on a prescribed proposal form, to perform the work and to furnish all labor, materials, equipment and/or service for the prices and terms quoted by the bidder.

Punch List (Inspection List):   A list prepared by the owner or his or her authorized representative of items of work requiring immediate corrective or completion action by the contractor.

Qualified Person*:   One who has skills and knowledge related to the construction and operation of the equipment and has received safety training on the hazards involved.

Raceway*:   An enclosed channel of metallic or nonmetallic materials designed expressly for holding wires, cables, or busbars, with additional functions as permitted in this *Code*. Raceways include, but are not limited to, rigid metal conduit, rigid nonmetallic conduit, intermediate metal conduit, liquidtight flexible conduit, flexible metallic tubing, flexible metal conduit, electrical nonmetallic tubing, electrical metallic tubing, underfloor raceways, cellular concrete floor raceways, cellular metal floor raceways, surface raceways, wireways, and busways.

Rapid Start: The most common type of ballast used today that does not require a starter.

Rapid Start Fluorescent Lamp Circuit: A circuit designed to start lamps by continuously heating or preheating the electrodes. Lamps must be designed for this type of circuit. This is the modern version of the "trigger start" system. In a rapid-start two-lamp circuit, one end of each lamp is connected to a separate starting winding.

Rate of Rise Detector: A device that responds when the temperature rises at a rate exceeding a predetermined value.

Rated-Load Current: The rated-load current for a hermetic refrigerant motor-compressor is the current resulting when the motor-compressor is operated at the rated load, rated voltage, and rated frequency of the equipment it serves.

Receptacle*:   A receptacle is a contact device installed at the outlet for the connection of an attachment plug. A single receptacle is a single contact device with no other contact device on the same yoke. A multiple receptacle is two or more contact devices on the same yoke.

Receptacle Outlet*: An outlet where one or more receptacles are installed.

Resistive Load: An electric load that opposes the flow of current. A resistive load does not contain cores or coils of wire. Some examples of resistive loads are the electric heating elements in an electric range, ceiling heat cables, and electric baseboard heaters.

Root-Mean-Square (RMS): The square root of the average of the square of the instantaneous values of current or voltage. For example, the RMS value of voltage, line to neutral, in a home is 120 volts. During each electrical cycle, the voltage rises from zero to a peak value ($120 \times 1.4142 = 169.7$ volts), back through zero to a negative peak value, then back to zero. The RMS value is 0.707 of the peak value ($169.7 \times 0.707 = 120$ volts). The root-mean-square value is what an electrician reads on an ammeter or voltmeter.

Schedules: Tables or charts that include data about materials, products, and equipment.

Scope: A written range of view or action; outlook; hence, room for the exercise of faculties or function, capacity for achievement; all in connection with a designated project.

Sections: Views of vertical cuts through and perpendicular to components showing their detailed arrangement.

Series-Rated System: Panelboards are marked with their short-circuit rating in RMS symmetrical amperes. A series-rated panelboard will be determined by the main circuit breaker or fuse, and branch-circuit breaker combination tested in accordance to UL Standard 489. The series rating will be less than or equal to the interrupting rating of the main overcurrent device, and greater than the interrupting rating of the branch-circuit overcurrent devices. (See "Fully Rated System.")

Service*: The conductors and equipment for delivering energy from the serving utility to the wiring system of the premises served.

Service Conductors*: The conductors from the service point to the service disconnecting means.

Service Drop*: The overhead service conductors from the last pole or other aerial support to and including the splices, if any, connecting to the service-entrance conductors at the building or other structure.

Service Equipment*: The necessary equipment, usually consisting of a circuit breaker or switch and fuses and their accessories, located near the point of entrance of supply conductors to a building or other structure, or an otherwise defined area, and intended to constitute the main control and means of cutoff of the supply.

Service Lateral Conductor*: The underground service conductors between the street main, including any risers at a pole or other structure or from transformers, and the first point of connection to the service-entrance conductors in a terminal box or meter or other enclosure with adequate space, inside or outside the building wall. Where there is no terminal box, meter, or other enclosure with adequate space, the point of connection shall be considered to be the point of entrance of the service conductors into the building.

Service Point*: The point of connection between the facilities of the serving utility and the premises wiring.

Shall: Indicates a mandatory requirement.

Short Circuit: A connection between any two or more conductors of an electrical system in such a way as to significantly reduce the impedance of the circuit. The current flow is outside of its intended path, thus the term "short circuit." A short circuit is also referred to as a fault.

Should: Indicates a recommendation or that which is advised but not required.

Site: The place where a structure or group of structures was, or is to be located (as in construction site).

Smoke Alarm: A single or multiple station alarm responsive to smoke.

Smoke Detector: A device that detects visible or invisible particles of combustion.

Sone: A unit of loudness equal to the loudness of a sound of one kilohertz at 40 decibels above the threshold of hearing of a given listener.

Specifications: Text setting forth details such as description, size, quality, performance, workmanship, and so forth. Specifications that pertain to all of the construction trades involved might be subdivided into "General Conditions" and "Supplemental General Conditions." Further subdividing the specifications might be specific requirements for the various contractors such as electrical, plumbing, heating, masonry, and so forth. Typically, the electrical specifications are found in Division 16.

Split-Circuit Receptacle: A receptacle that can be connected to two branch-circuits. These receptacles may also be used so that one receptacle is live at all times, and the other receptacle is controlled by a switch. The terminals on these receptacles usually have breakaway tabs so the receptacle can be used either as a split-circuit receptacle or as a standard receptacle.

Standard Network Interface (SNI): A device usually installed by the telephone company at the demarcation point where their service leaves off and the customer's service takes over. This is similar to the service point for electrical systems.

Starter: A device used in conjunction with a ballast for the purpose of starting an electric discharge lamp.

Structure*: That which is built or constructed.

Subcontractor: A qualified subordinate contractor to the prime or main contractor.

Surety Bond: A legal document issued to ensure the completion of an act by another person. Contractors usually are required to purchase surety bonds, if they are working on public projects. A surety bond guarantees to one party that another (the contractor) will perform specified acts, usually within a stated period of time. The surety company typically becomes responsible for fulfillment of a contract if the contractor defaults. In the case of a public works project, such as a road, that means that the surety bond protects taxpayers should a contractor go out of business.

Surface-Mounted Luminaire (fixture): A luminaire (fixture) mounted directly on the ceiling.

Surge Suppressor: A device that limits peak voltage to a predetermined value when voltage spikes or surges appear on the connected line.

Suspended (pendant) Luminaire (fixture): A luminaire (fixture) hung from a ceiling by supports.

Switches*:

General-Use Snap Switch: A form of general-use switch constructed so that it can be installed in device boxes or on box covers, or otherwise used in conjunction with wiring systems recognized by this *Code*.

General-Use Switch: A switch intended for use in general distribution and branch-circuits. It is rated in amperes, and it is capable of interrupting its rated current at its rated voltage.

Motor-Circuit Switch: A switch, rated in horsepower, capable of interrupting the maximum operating overload current of a motor of the same horsepower rating as the switch at the rated voltage.

Symbols: Graphic representations that stand for or represent other things. A symbol is a simple way to show such things as lighting outlets, switches, and receptacles on an electrical plan. The American Institute of Architects has developed a very comprehensive set of symbols that represent just about everything used by all building trades. When an item cannot be shown using a symbol, then a more detailed explanation using a notation or inclusion in the specifications is necessary.

Task Lighting: Lighting directed to a specific surface or area that provides illumination for visual tasks.

Terminal: A screw or a quick-connect device where a conductor(s) is intended to be connected.

Thermocouple: A pair of dissimilar conductors so joined at two points that an electromotive force is developed by the thermoelectric effects when the junctions are at different temperatures.

Thermopile: More than one thermocouple connected together. The connections may be series, parallel, or both.

Torque: A measurement of rotation or turning force. Torque is measured in ounce-inches (oz.-in.), ounce-feet (oz.-ft) and pound-feet (lb.-ft). See Horsepower, Rated Horsepower, and Maximum Developed Horsepower.

Transient Voltage Surge Suppressor (TVSS): A device that clamps transient voltages by absorbing the major portion of the energy (joules) created by the surge, allowing only a small, safe amount of energy to enter the actual connected load.

Troffer: A recessed lighting unit, usually long and installed with the opening flush with the ceiling. The term is derived from "trough" and "coffer."

UBC (Uniform Building Code): One of the family of codes and related publications published by the International Conference of Building Officials (ICBO) and other organizations, such as the International Association of Plumbing and Mechanical Officials (IAPMO) and the National Fire Protection Association (NFPA), which have similar goals as far as code publications are concerned. The Uniform Building Code is designed to be compatible with these other codes, as together they make up the enforcement tools of a jurisdiction.

UL: Underwriters Laboratories (UL) is an independent not-for-profit organization that develops standards and tests electrical equipment to these standards.

UL-Listed: Indicates that an item has been tested and approved to the standards established by UL for that particular item. The UL Listing Mark may appear in various forms, such as the letters UL in a circle. If the product is too small for the marking to be applied to the product, the marking must appear on the smallest unit container in which the product is packaged.

UL-Recognized: Refers to a product that is incomplete in construction features or limited in performance capabilities. A "Recognized" product is intended to be used as a component part of equipment that has been "listed." A "Recognized" product must not be used by itself. A UL product may contain a number of components that have been "Recognized."

Ungrounded Conductor: The conductor of an electrical system that is not intentionally connected to ground. This conductor is referred to as the "hot" or "live" conductor.

Volt: The difference of electric potential between two points of a conductor carrying a constant current of one ampere, when the power dissipated between these points is equal to one watt. A voltage of one volt can push one ampere through a resistance of one ohm.

Voltage (of a circuit)*: The greatest root-mean-square (effective) difference of potential between any two conductors of the circuit concerned.

Voltage (nominal)*: A nominal value assigned to a circuit or system for the purpose of conveniently designating its voltage class (e.g., 120/240 volts, 480Y/277 volts, 600 volts).

The actual voltage at which a circuit operates can vary from the nominal within a range that permits satisfactory operation of equipment.

Voltage to Ground*: For grounded circuits, the voltage between the given conductor and that point or conductor of the circuit that is grounded; for ungrounded circuits, the greatest voltage between the given conductor and any other conductor of the circuit.

Voltage Drop: Also referred to as IR drop. Voltage drop is most commonly associated with conductors. A conductor has resistance; when current is flowing through the conductor, a voltage drop will be experienced across the conductor. Voltage drop across a conductor can be calculated using Ohm's law: $E = IR$.

Volt-ampere: A unit of power determined by multiplying the voltage and current in a circuit. A 120-volt circuit carrying 1 ampere is 120 volt-amperes.

Watt: A measure of true power. A watt is the power required to do work at the rate of 1 joule per second. Wattage is determined by multiplying voltage times amperes times the power factor of the circuit: $W = E \times I \times PF$.

Watertight: Constructed so that moisture will not enter the enclosure under specified test conditions.

Weatherproof*:   Constructed or protected so that exposure to the weather will not interfere with successful operation.

   (FPN): Rainproof, raintight, or watertight equipment can fulfill the requirements for weatherproof where varying weather conditions other than wetness, such as snow, ice, dust, or temperature extremes, are not a factor.

Wet Niche Luminaire (fixture)*:   A luminaire (lighting fixture) intended for installation in a forming shell mounted in a pool.

Working Drawing(s):   A drawing sufficiently complete with plan and section views, dimensions, details, and notes, so that whatever is shown can be constructed and/or replicated without instructions but subject to clarifications, see drawings.

Work Order: A written order, signed by the owner or his or her representative, of a contractual status requiring performance by the contractor without negotiation of any sort.

Zoning: Restrictions of areas or regions of land within specific geographical areas based on permitted building size, character, and uses as established by governing urban authorities.

The following list of Internet World Wide Web sites has been included for your convenience. These are current as of the time of printing. Web sites are a "moving target" with continual changes. We do our utmost to keep these Web sites current. If you are aware of Web sites that should be added, deleted, or that are different than shown, please let us know. Most manufacturers will provide catalogs and technical literature upon request either electronically or by mail. You will be amazed at the amount of electrical and electronic equipment information that is available on these Web sites.

You can search the Web using a search engine such as http://www.google.com, http://www.dogpile.com, http://www.yahoo.com, or http://www.askjeeves.com. Just type in a key word (for example: fuses) and you will be led to many Web sites relating to fuses. It's easy!

Web sites marked with an asterisk * are closely related to the "home automation" market.

ACCUBID	http://www.accubid.com
Acme Electric Corporation	http://www.acmepowerdist.com
ActiveHome*	http://www.X10-beta.com/activehome/
Active Power	http://www.activepower.com
Adalet	http://www.adalet.com
Advance Transformer Co.	http://www.advancetransformer.com
Advanced Cable Ties	http://www.advancedcableties.com
AEMC Instruments	http://www.aemc.com
AFC Cable Systems	http://www.afcweb.com
Air-Conditioning & Refrigeration Association	http://www.ari.org
Alcan Cable	http://www.cable.alcan.com
Alflex	http://www.alflex.com
Allen-Bradley	http://www.automation.rockwell.com
Allen-Bradley	http://www.ab.com
Allied Moulded Products	http://www.alliedmoulded.com
Allied Support Products	http://www.alliedsupport.com
Allied Tube & Conduit	http://www.alliedtube.com
Alpha Wire	http://www.alphawire.com
ALTO Lamps	http://www.altolamp.com
The Aluminum Association, Inc.	http://www.aluminum.org
American Council for an Energy Efficient Economy	http://www.aceee.org
American Gas Association	http://www.aga.org
American Heart Association	http://www.americanheart.org
American Institute of Architects	http://www.aia.org
American Insulated Wire Corp.	http://www.aiwc.com
American Lighting Association	http://www.americanlightingassoc.com
American National Standards Institute	http://www.ansi.org
American Pipe & Plastics, Inc.	http://www.ampipe.com
American Power Conversion	http://www.apcc.com
American Public Power Assoc.	http://www.appanet.org

American Society of Heating, Refrigerating, and Air-Conditioning Engineers	http://www.ashrae.org
American Society for Testing and Materials	http://www.astm.org
American Society of Mechanical Engineers	http://www.asme.org
American Society of Safety Engineers	http://www.asse.org
AMP*	http://www.amp.com
Amprobe	http://www.amprobe.com
Angelo	http://www.angelobrothers.com
Anixter (data & telecommunications cabling)*	http://www.anixter.com
Appleton Electric Co.	http://www.appletonelec.com
ArcWear	http://www.arcwear.com
Arlington Industries, Inc.	http://www.aifittings.com
Arrow Fasteners	http://www.arrowfastener.com
Arrow Hart Wiring Devices	http://www.arrowhart.com
Assisted Living: The World of Assisted Technology	http://www.abledata.com
Associated Builders & Contractors	http://www.abc.org
Association of Cabling Professions	http://www.wireville.com
Association of Home Appliance Manufacturers	http://www.aham.org
Automated Home Technologies*	http://www.autohometech.com
Automatic Switch Co. (ASCO)	http://www.asco.com
AVO Training Institute	http://www.avotraining.com
Baldor Motors & Drives	http://www.baldor.com
R. W. Beckett Corporation (oil burner technical info.)	http://www.beckettcorp.com
Belden Wire & Cable Co.	http://www.belden.com
Bell, Hubbell Inc.	http://www.hubbell-bell.com
Bender	http://www.bender.org
Best Power	http://www.bestpower.com
BICC General	http://www.generalcable.com
BICSI (Building Industry Consulting Services International, a telecommunications assoc.)*	http://www.bicsi.org
BidStreet	http://www.BidStreetUSA.com
Black and Decker	http://www.blackanddecker.com
The Blue Book (construction)	http://www.thebluebook.com
Bodine Company	http://www.bodine.com
BOCA	http://www.bocai.org
boltswitch, inc.	http://www.boltswitch.com
Brady Worldwide, Inc.	http://www.whbrady.com
Bridgeport Fittings Inc.	http://www.bptfittings.com
Broan	http://www.broan.com
Bryant – Hubbell Inc.	http://www.hubbell-bryant.com
Building Industry Consulting Service International, Inc.	http://www.bisci.org
BuildPoint	http://www.buildpoint.com
Burndy Electrical Products	http://www.fciconnect.com
Bussmann, Cooper Industries	http://www.bussmann.com
Cable Telecommunications Engineers*	http://www.scte.org
Caddy	http://www.erico.com
Cadweld	http://www.erico.com

Canadian Standards Association	http://www.csa-international.org
Cantex	http://www.cantexinc.com
Carlon, A Lamson & Sessions Co.	http://www.carlon.com
Carol Cable	http://www.generalcable.com
Casablanca Fan Company	http://www.casablancafanco.com
Caterpillar	http://www.cat.com
Caterpillar	http://www.cat-engines.com
CEBus Industry Council*	http://www.cebus.org
CEE News magazine	http://www.ceenews.com
Center for Disease Control (CDC)	http://www.cdc.gov
Centralite*	http://www.centralite.com
CEPro*	http://www.ce-pro.com
Channellock Inc	http://www.channellock.com
ChannelPlus*	http://www.multiplextechnology.com/ channelplus/
Chromalox	http://www.chromalox.com
Circon Systems Corporation*	http://www.circon.com
Clifford of Vermont*	http://www.cliffordvt.com
Coleman Cable	http://www.colemancable.com
Columbia Lighting	http://www.columbia-ltg.com
CommScope* (coax cable mfg.)	http://www.commscope.com
Communications Systems, Inc.	http://www.commsystems.com
Computer and Telecommunications Page*	http://www.cmpcmm.com/cc/
ConnectHome*	http://www.connecthome.com
Construction Specifications Institute	http://www.csinet.org
Construction Terms/Glossary	http://www.constructionplace.com/ glossary.html
Consumer Electronic Association (CEA)*	http://www.ce.org
Consumer Product Safety Commission (CPSC)	http://www.cpsc.gov
Continental Automated Buildings Association (CABA)*	http://www.caba.org
Contrast Lighting (recessed)	http://www.contrastlighting.com
Copper (technical information)	http://energy.copper.org
Copper Development Assoc.	http://www.copper.org
Cooper Industries	http://www.cooperindustries.com
Here you will find links to all Cooper Divisions for electrical products, lighting, wiring devices, power and hand tools.	
Craftsman Tools	http://www.craftsman.com
Crouse-Hinds	http://www.crouse-hinds.com
Cummins Power Generation*	http://www.cumminspower.com
Custom Electronic Design & Installation Association*	http://www.cedia.net
Custom Electronic Professional*	http://www.ce-pro.com
Cutler-Hammer	http://www.cutler-hammer.com
DABMAR	http://www.dabmar.com
Daman Tools	http://www.damantools.com
Danaher Power Solutions	http://www.danaherpoersolutions.com
Daniel Woodhead Company	http://www.danielwoodhead.com
Day-Brite Lighting	http://www.daybrite.com

Delmar Publishers	http://www.delmarlearning.com
Delmar (electrical)	http://www.delmarelectric.com
Dept. of Energy	http://www.energy.gov
Dimension Express (to obtain roughing-in dimensions from all appliance manufacturers)	http://www.dexpress.com
F. W. Dodge	http://www.fwdodge.com
DOMOSYS Corporation*	http://www.domosys.com
Douglas Lighting Control	http://www.douglaslightingcontrol.com
Dremel	http://www.dremel.com
Dual-Lite	http://www.dual-lite.com
Dynacom Corporation	http://www.dynacomcorp.com
Eaton (links to all divisions)	http://www.eatonelectrical.com
Edison Electric Institute	http://www.eei.org
EC Online	http://www.econline.com
EC & M magazine	http://www.ecmweb.com
Edwards Signaling and Security Products	http://www.edwards-signals.com
ELAN Home Systems*	http://www.elanhomesystems.com
Electric-Find	http://www.electric-find.com
Electric Smarts	http://www.electricsmarts.com
Electri-flex	http://www.electriflex.com
Electrical Apparatus Service Assoc.	http://www.easa.com
Electrical Contracting & Engineering News	http://www.ecenmag.com
Electrical Contractor Magazine	http://www.ecmag.com
The Electrical Contractor Network	http://www.electrical-contractor.net
Electrical Designers Reference	http://www.edreference.com
Electrical Distributor Magazine	http://www.tedmag.com
Electrical Reliability Services, Inc. (formerly Electro-Test)	http://www.electro-test.com
Electrical Safety	http://www.electrical-safety.com
Electrician.com	http://www.electrician.com
Electro-Test Inc.	http://www.electro-test.com
Electronic Design Technology*	http://www.edt.biz
Electronic House*	http://www.electronichouse.com
Electronic Industry Alliance*	http://www.eia.org
Emerson	http://www.emersonfans.com
Encore Wire Limited	http://www.encorewire.com
Energy Efficiency & Renewable Energy Network (U.S. Dept. of Energy)	http://www.eere.energy.gov
Energy Star	http://www.energystar.gov
Environmental Protection Agency	http://www.epa.gov
Erico	http://www.erico.com
Essex Group, Inc.*	http://www.essexgroup.com
Estimation Inc.	http://www.estimation.com
ETL Semko	http://www.etlsemko.com
Exide Technologies	http://www.exide.com
Factory Mutual Insurance Co.	http://www.fmglobal.com
Faraday	http://www.faradayfirealarms.com
FCI/Burndy (telecom connectors)	http://www.fciconnect.com
Federal Communications Commission*	http://www.fcc.gov
Federal Pacific	http://www.federalpacific.com

Home Controls, Inc.*	http://www.homecontrols.com
Home Director, Inc.*	http://www.homedirector.net
HomeStar*	http://www.homestar.net
Homestore*	http://www.homestore.com
Home Touch*	http://www.home-touch.com
Home Toys*	http://www.hometoys.com
Honeywell	http://www.honeywell.com
Honeywell*	http://www.honeywell.com/yourhome
Housing Urban Development	http://www.hud.gov.
Houston Wire & Cable Company	http://www.houwire.com
Hubbell Incorporated (links to all divisions)	http://www.hubbell.com
Hubbell Incorporated	http://www.hubbell-premise.com
Hubbell Lighting Incorporated	http://www.hubbell-ltg.com
Hubbell Wiring Devices	http://www.hubbell-wiring.com
Hunt Control Systems, Inc.	http://www.huntdimming.com
Hunter Fans	http://www.hunterfan.com
Husky	http://www.mphusky.com
ICBO	http://www.icbo.org
Ideal Industries, Inc.	http://www.idealindustries.com
Illuminating Engineering Society of North America	http://www.iesna.org
ILSCO	http://www.ilsco.com
Independent Electrical Contractors	http://www.ieci.org
Institute of Electrical and Electronic Engineers, Inc.	http://www.ieee.org
Instrumentation, Systems, and Automation Society	http://www.isa.org
Insulated Tools (certified insulated products)	http://www.insulatedtools.com
Intel's Home Network	http://www.intel.com/anypoint
Intermatic. Inc.	http://www.intermatic.com
International Association of Electrical Inspectors (IAEI)	http://www.iaei.org
International Association of Plumbing & Mechanical Officials	http://www.iapmo.org
International Brotherhood of Electrical Workers	http://www.ibew.org
International Code Council (ICC)	http://www.iccsafe.org
International Dark-Sky Association	http://www.darksky.org
International Electrotechnical Commission (IEC)	http://www.iec.ch/
International Organization for Standardization (ISO)	http://www.iso.org
Intertec Testing Services (ETL)	http://www.etlsemko.com
Jacuzzi	http://www.jacuzzi.com
Jefferson Electric	http://www.jeffersonelectric.com
Johnson Controls, Inc.	http://www.johnsoncontrols.com
Joslyn	http://www.joslynmfg.com
Juno Lighting, Inc.	http://www.junolighting.com
KatoLight	http://www.katolight.com
Kenall Lighting	http://www.kenall.com
Kichler Lighting	http://www.kichler.com

Kidde (home safety)	http://www.kiddeus.com
King Safety Products	http://www.kingsafety.com
King Wire Inc.	http://www.kingwire.com
Klein Tools, Inc.	http://www.kleintools.com
Kohler Generators	http://www.kohler.com
Leviton Mfg. Co., Inc. (links to all divisions)	http://www.leviton.com
Lew Electric	http://www.lewelectric.com
Liebert	http://www.liebert.com
Lighting	http://www.lightsearch.com
Lighting Design Forum	http://www.qualitylight.com
Lighting Research Center	http://www.lrc.rpi.edu
Lightning Protection Institute	http://www.lightning.org
Lightolier	http://www.lightolier.com
Lightolier Controls*	http://www.lolcontrols.com
Litetouch*	http://www.litetouch.com
Lithonia Lighting	http://www.lithonia.com
Littelfuse, Inc.	http://www.littelfuse.com
LonWorks *	http://www.lonworks.echelon.com
Lucent Technologies*	http://www.lucent.com
Lumisistemas, S.A. de C.V.	http://www.lumisistemas.com
Lutron Electronics Co., Inc.	http://www.lutron.com
Lutron Home Theater*	http://www.ultimatehometheater.com
MagneTek (links to all divisions)	http://www.magnetek.com
Makita Tools	http://www.makita.com
Manhattan/CDT	http://www.manhattancdt.com
McGill	http://www.mcgillelectrical.com
Megger	http://www.megger.com/us/index.asp
MET Laboratories, Inc.	http://www.metlabs.com
MI Cable Company	http://www.micable.com
Midwest Electric Products, Inc.	http://www.midwestelectric.com
Milbank Manufacturing Co.	http://www.milbankmfg.com
Milwaukee Electric Tool	http://www.milwaukeetool.com
Minerallac	http://www.minerallac.com
Molex Inc.	http://www.molex.com
Motor Control	http://www.motorcontrol.com
Motorola Lighting Inc.	http://www.motorola.com
National Association of Electrical Distributors	http://www.naed.org
National Association of Home Builders	http://www.nahb.org
National Association of Mfg.	http://www.nam.org
National Association of Radio and Telecommunications Engineers, Inc.	http://www.narte.com
National Conference of States on Building Codes and Standards	http://www.ncsbcs.org
National Electrical Contractors Association (NECA)	http://www.necanet.org
National Electrical Manufacturers Association (NEMA)	http://www.nema.org
National Electrical Safety Foundation	http://www.nesf.org
National Fire Protection Association	http://www.nfpa.org
National Fire Protection Association *National Electrical Code*® information	http://www.nfpa.org/nec/

National Institute for Occupational Safety & Health	http://www.cdc.gov/niosh
National Institute of Standards and Technology (NIST)	http://www.nist.gov
National Joint Apprenticeship and Training Committee (NJATC)	http://www.njatc.org
National Lighting Bureau	http://www.nlb.org
National Resource for Global Standards	http://www.nssn.org
National Safety Council	http://www.nsc.org
National Spa and Pool Institute	http://www.nspi.org
National Systems Contractors Association*	http://www.nsca.org
NECDirect	http://www.necdirect.org
Newton's Electrician	http://www.electrician.com
Nora Lighting	http://www.noralighting.com
NuTone	http://www.nutone.com
Occupation Health & Safety	http://www.ohsonline.com
Occupational Safety & Health Administration	http://www.osha.gov
Okonite Co.	http://www.okonite.com
Olflex Cable	http://www.olflex.com
Onan Corporation	http://www.onan.com
ONEAC Corporation	http://www.oneac.com
OnQ Technologies*	http://www.onqtech.com
Optical Cable Corporation*	http://www.occfiber.com
Ortronics Inc.	http://www.ortronics.com
Osram Sylvania Products, Inc.	http://www.sylvania.com
Overhead door information: Door & Access System Manufacturers Association	http://www.dasma.com
OZ/Gedney	http://www.o-zgedney.com
Panasonic Lighting	http://www.panasonic.com/lighting
Panasonic Technologies	http://www.panasonic.com
Panduit Corporation	http://www.panduit.com
Pass and Seymour, Legrand	http://www.passandseymour.com
Pelican Rope Works	http://www.pelicanrope.com
Penn-Union Corp.	http://www.penn-union.com
Phase-A-Matic, Inc.	http://www.phase-a-matic.com
Philips	http://www.philips.com
Philips Lamps	http://www.ALTOlamp.com
Philips Lighting Company	http://www.philipslighting.com
Philtek Power Corporation	http://www.philtek.com
Phoenix Contact	http://www.phoenixcon.com
Pigeon Hole, The	http://www.thepigeonhole.com
Plant Engineering On-Line	http://www.plantengineering.com
Popular Home Automation*	http://www.pophome.com
Porter Cable	http://www.porter-cable.com
Power Quality Monitoring	http://www.pqmonitoring.com
Power-Sonic Corporation	http://www.power-sonic.com
Premise Wiring Products	http://www.unicomlink.com
Prescolite	http://www.prescolite.com
Pringle Electrical Mfg. Co.	http://www.pringle-elec.com
Progress Lighting	http://www.progresslighting.com
Radix Wire Company	http://www.radix-wire.com

RCA*	http://www.rca.com
RACO	http://www.racoinc.com
Rayovac Corporation	http://www.rayovac.com
Refrigeration Service Engineering Society	http://www.rses.org
Regal Fittings	http://www.regalfittings.com
Reliable Power Meters	http://www.reliablemeters.com
Remke Industries, Inc.	http://www.remke.com
Rhodes, M. H. (timers)	http://www.mhrhodes.com
Rigid Tool Company	http://www.ridgid.com
Robertson Transformer Co.	http://www.robertsontransformer.com
Robicon	http://www.robicon.com
Robroy Industries – Conduit Div.	http://www.robroy.com
Romex Cable	http://www.biccgeneral.com
RotoZip Tool Corp.	http://www.rotozip.com
Russelectric	http://www.russelectric.com
Safety Technology International, Inc.	http://www.sti-usa.com
S&C Electric Company	http://www.sandc.com
SBCCI	http://www.sbcci.org
Sea Gull Lighting	http://www.seagulllighting.com
Seatek, Inc.	http://www.SeatekCo.com
SECURITY*	
Caddx*	http://www.caddx.com
Digital Monitoring Products*	http://www.dmpnet.com
Home Automation*	http://www.homeauto.com
Honeywell Security*	http://www.security.honeywell.com/sce
NAPCO Security Systems*	http://www.napcosecurity.com
Sensor Switch Inc.	http://www.sensorswitchinc.com
Siemens (main site)	http://www.siemens.com
Siemens Energy & Automation	http://www.sea.siemens.com
The Siemon Company*	http://www.siemon.com
Shat-R-Shield	http://www.shat-r-shield.com
Silent Knight	http://www.silentknight.com
Simplex Grinnell	http://www.simplexgrinnell.com
Simpson Electric Co.	http://www.simpsonelectric.com
Skil Tools	http://www.skil.com
Smart Home*	http://www.smarthome.com
Smart Home Pro*	http://www.smarthomepro.com
Smart House, Inc.*	http://www.smart-house.com
Smoke & Fire Extinguisher Signs	http://www.smokesign.com
Sola/Hevi Duty	http://www.sola-hevi-duty.com
Southwire Co.	http://www.mysouthwire.com
SP Products, Inc.	http://www.spproducts.com
A.W. Sperry Instruments, Inc.	http://www.awsperry.com
Square D, Groupe Schneider	http://www.squared.com
Square D (Homeline)	http://www.squared.com/retail
Square D, Power Logic	http://www.powerlogic.com
Stanley Tools	http://www.stanleyworks.com
Starfield Controls*	http://www.starfieldcontrols.com
Steel Tube Institute	http://www.steeltubeinstitute.org
Straight Wire*	http://www.straightwire.com
Sunbelt Transformers	http://www.sunbeltusa.com

Superior Electric	http://www.superiorelectric.com
Superior Essex Group	http://www.superioressex.com
Suttle	http://www.suttleonline.com
Sylvania	http://www.sylvania.com
Sylvania Lighting Int'l.	http://www.sli-lighting.com
Technical Consumer Products Inc.	http://www.springlamp.com
Telecommunications Industry Association (TIA)*	http://www.tiaonline.org
Terms /Construction /Glossary	http://www.constructionplace.com/ glossary.html
Thermostats, Communicating*	http://www.homeauto.com
This Old House	http://www.thisoldhouse.org
Thomas & Betts Corporation	http://www.tnb.com
Thomas Lighting	http://www.thomaslighting.com
Thomas Register	http://www.thomasregister.com
3M Electrical Products	http://www.3m.com/elpd
Tork	http://www.tork.com
Touch-Plate Lighting Controls	http://www.touchplate.com
Trade Service Corporation	http://www.tradeservice.com
Trade Service Systems	http://www.tradepower.com
Triplett Corporation	http://www.triplett.com
TYCO (links to all divisions)	http://www.tyco.com
Tyton Hellermann	http://www.tytonhellermann.com
UEI Test Equipment	http://www.ueitest.com
Underwriters Laboratories	http://www.ul.com
UL Standards Dept	http://ulstandardsinfonet.ul.com
Unistrut Corp	http://www.unistrut.com
Unity Manufacturing	http://www.unitymfg.com
Universal Lighting Technologies	http://www.universalballast.com
USA Wire and Cable, Inc.	http://www.usawire-cable.com
U.S. Department of Energy (DOE)	http://www.doe.gov
U.S. Department of Labor	http://www.dol.gov
U.S. Department of Labor, Employment & Training Administration	http://www.doleta.gov
U.S. Environment Protection Agency (EPA)	http://www.epa.gov
Ustec*	http://www.ustecnet.com
Vantage Automation & Lighting Control*	http://www.vantagecontrols.com
Watt Stopper	http://www.wattstopper.com
Waukesha Electric Systems	http://www.waukeshaelectric.com
Werner Ladder Company	http://www.wernerladder.com
Westinghouse Lighting	http://www.westinghouselighting.com
Wheatland Tube Company	http://www.wheatland.com
Wireless Industry*	http://www.wireless.about.com
Wireless Thermostats*	http://www.wirelessthermostats.com
Wiremold Company, The	http://www.wiremold.com
Woodhead Industries	http://www.danielwoodhead.com
X-10*	http://www.x10.com
X-10 Pro Automation Products*	http://www.x10pro.com
Zenith Controls, Inc.*	http://www.geindustrialsystems.com
Zircon	http://www.zircon.com

Code Index

Note: Page numbers in **bold** reference Figures

Index

Note: Page numbers in **bold type** reference non-text material, such as tables and figures